Solid State Lighting Technology and Application Series

Volume 4

Series Editors
Guo Qi Zhang, Eindhoven, The Netherlands
Wu Ling, Beijing, China

Driven by societal needs and challenges, fast development of LED and semiconductors technologies, lighting is going through a radical transformation. This transformation is characterized by continuous and aggressive efficacy increase and cost reduction to replace traditional light sources by LED; revolution of forms and fixtures; seamless integration of LED light sources with semiconductors to create digitalized and smart lighting systems and solutions; emerging of "more than Illumination" functions and applications by using LED for healthcare and wellbeing, horticulture and food, communication, safety and security, etc. The success of this lighting transformation depends on not only the industry and market development, but also the creation and exploitation of new fundamental knowledge and talent development. This book series is the concerted action of global experts from industry and academia. It aims to provide the state-of-the-art fundamental theory and knowledge, the latest technology development, best industrial practices and guidance for academic research, technology and industrial development. It will cover all aspects of development of SSL, such as materials, device technologies, packaging and modules, luminaries & systems, thermal management, testing, reliability, total life time management, equipment, power management, sensors, communication and control, design and architecture, and other newly emerged scientific domains.

More information about this series at http://www.springer.com/series/8864

Jinmin Li • G. Q. Zhang
Editors

Light-Emitting Diodes

Materials, Processes, Devices and Applications

Springer

Editors
Jinmin Li
Chinese Academy of Sciences
State Key Laboratory of Solid State Lighting
Solid State Lighting R&D Center
Beijing, China

G. Q. Zhang
Delft University of Technology
DELFT, Noord-Brabant
The Netherlands

ISSN 2196-4203 ISSN 2196-4211 (electronic)
Solid State Lighting Technology and Application Series
ISBN 978-3-319-99210-5 ISBN 978-3-319-99211-2 (eBook)
https://doi.org/10.1007/978-3-319-99211-2

Library of Congress Control Number: 2018962860

This Springer imprint is published by the registered company Springer Nature Switzerland AG
The registered company address is: Gewerbestrasse 11, 6330 Cham, Switzerland

Foreword

From the report of red light-emitting diodes (LEDs) by Nick Holonyak in the 1960s to the Nobel Prize in Physics awarded for the invention of blue LEDs in 2014, we have witnessed many historical milestones of LED research, development, and industrialization. It is no exaggeration to say that solid-state lighting (SSL) based on compound semiconductor materials has revolutionized artificial lighting. Benefits of the progress in lighting technology on civilization have led to a greatly improved standard of living for all of mankind. For instance, SSL has enabled the realization of bright and energy-saving white light sources, full-color display, and projection. Besides, many novel and emerging applications have been possible thanks to the SSL technology, such as visible light communications, LED-enabled medical treatment, and lighting for semiconductor manufacturing, agriculture, fish industry, horticulture, and animal breeding. These achievements have dramatically expanded the application boundary of lighting, marking a brand new horizon of illumination. At the same time, they have also triggered new challenges for future research and technology development.

Contributors to this book are global leading SSL engineers and scientists. They provide an overview of the latest developments in the relevant areas, as well as their personal views about future development directions. It covers wide aspects of the device from deep ultraviolet to the visible spectrum made from compound semiconductor materials. This book is divided into four parts, and the concept covers all key processes of the research, from epitaxial growth of the materials on different substrates, structure design and optimization, packaging, and reliability to the emerging applications. This book will be of interest to scientists and engineers working on LED technology and applications, as well as graduate students in material science, optical engineering, applied physics, and electrical engineering.

Nagoya University, Nagoya, Japan Hiroshi Amano

Contents

Chapter 1
GaN Substrate Material for III–V Semiconductor Epitaxy Growth

Rong Zhang and Xiangqian Xiu

1.1 Introduction

Wide bandgap semiconductor is known as the third-generation semiconductor material, as the engine and key technology of high-temperature, high-frequency, high-power semiconductor devices and IT industry. Wide bandgap semiconductor refers to the wide bandgap in 2.0–6.0 eV semiconductor materials, including SiC, GaN, ZnO, AlGaN, etc. The excellent properties, such as big bandgap, high breakdown electric field intensity, high saturated electron drift speed, big thermal conductivity, small dielectric constant, strong radiation resistance, and good chemical stability, make them well suitable for the radiation-resistant, high-frequency, high-power, and high-density integration of semiconductor devices. Based on the wide bandgap, the blue, green, and ultraviolet light-emitting devices (LEDs) and light detector have been developed and commercialized.

The III-N materials enable new semiconductor devices with previously unobtainable performance capabilities in terms of light output, power handling, and efficiency, and these attributes will make possible the reinvention of existing technologies in ways that benefit many facets of our lives. More application fields accelerate the marketization. While there are many potential applications for these materials, the biggest applications appear to be light generation and the control of electrical power. Driven by optoelectronics, particularly GaN-based laser diodes (LDs) and high-brightness GaN-on-GaN LEDs, the market for bulk GaN substrates is expected to grow at a CAGR (compound annual growth rate) of 10% from 2017 to 2022 and could reach more than 100 M$ in 2022, according to a recent market survey from Yole Développement.

R. Zhang (✉) · X. Xiu
School of Electronic Science and Engineering, Nanjing University, Nanjing, China
e-mail: rzhang@nju.edu.cn

J. Li, G. Q. Zhang (eds.), *Light-Emitting Diodes*, Solid State Lighting Technology and Application Series 4, https://doi.org/10.1007/978-3-319-99211-2_1

In this paper, we summarize the major efforts in GaN substrate development and present the current progress in producing large-scale high-quality GaN substrate materials. We discuss the advantages of using the native GaN substrates with respect to the current heteroepitaxial nitride-based devices.

1.1.1 Importance of GaN Substrates

Bulk crystals of nitrides cannot be obtained by well-known direct synthesis methods such as Czochralski or Bridgman growth from stoichiometric melts, because of the extreme physical conditions, such as high melting temperature and very high decomposition pressure at the melting point (molten gallium nitride under 2225 °C and the pressure of 6 GPa). Growth of semiconductor-quality crystals at these conditions would be practically impossible. Therefore the crystals have to be grown by the methods requiring lower temperatures. In the past, large-scale high-quality GaN films are mostly grown on the heterogeneous substrate for a long time.

Sapphire is the most commonly used substrate for the hetero-epitaxy of GaN but is far as a kind of ideal substrate for GaN-based LEDs except the low cost. Its two biggest disadvantages are that the obvious difference of the lattice constant and the thermal expansion coefficient compared with GaN. This will cause the strain in the GaN epitaxial layer, and strain can produce point defects and dislocations, which reduces the quality of GaN epitaxial layers and InGaN quantum wells. The defects and dislocations can limit the light-output efficiency of LED devices.

As a substrate for sapphire, there are two very important obstacles in the manufacturing of high-performance LED. One is the electrical insulating properties of sapphire. Sapphire substrate with the high resistivity is not ideal for the LED devices, because the top contact electrode restrained its transverse current density for the inversion structure of LED devices. This will lead to the current crowding effect and local hotspots, which will be unfavorable factors for the LED luminous efficiency and the maximum brightness of the device. The second, sapphire, has the characteristics of big thermal resistance and also can seriously hinder the ability of heat dissipation, which in turn will further restrict the luminous efficiency and the life and brightness of the devices.

In order to solve these problems, people invented all kinds of technology to improve the luminous efficiency, life, and power. According to the above two obstacles, high-end technology used by LED chip maker is GaN epitaxial layer transferred to the substrates of good electrical and thermal conductivity properties. Usually by laser stripping and wafer bonding process, the LED chip structure will be laser stripping down from sapphire substrate and then transferring to bond with another kind of thermally conductive substrate. The laser stripping process involves usage of pulsed 248 nm KrF excimer laser. The laser is strongly adsorbed at the the sapphire/GaN interface and results in decomposition of the 100 nm thick GaN thin layer, which lead to the separation of sapphire and GaN. The technology is widely applied in the high-power LED device.

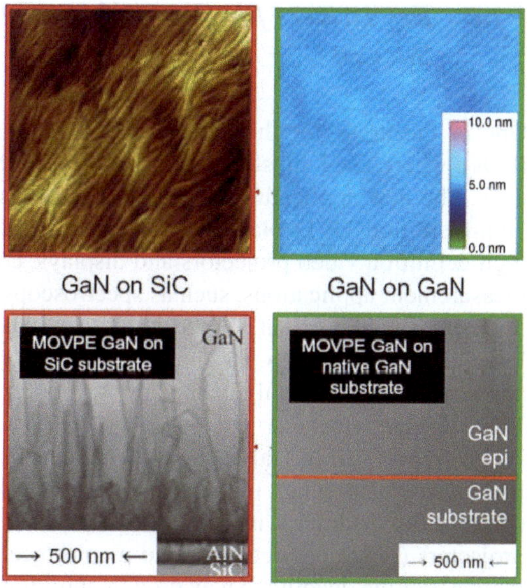

Fig. 1.1 Homogeneous epitaxial GaN films have higher quality and lower dislocation density, even compared with the best foreign substrate SiC

In fact, the most suitable substrate materials for the fabrication of LED should be native GaN substrate. Availability of lattice-matched GaN substrates can bring not only a high quality in GaN-based devices to improving device performance through eliminating the structural defects and providing better vertical electrical and thermal conductivity; moreover, it will lead to cost-efficient device processing through simplification of the epitaxial growth process by eliminating the need for low-temperature buffer layers.

As shown in Fig. 1.1, even if compared with the best foreign substrate, homogeneous epitaxial GaN has a higher quality and lower dislocation density [1]. It can improve the quality of the epitaxial layer and quantum well because of the proper lattice and thermodynamics match. It also has almost all of other advantages as follows. Although GaN substrate is more expensive, the area of one single chip on GaN substrate wafer is greatly reduced, which means that a single chip can use as three traditional chips. LEDs based on the GaN substrate have greater drive current, higher luminous efficiency, and simple process to the very good application prospect. Usually, GaN substrate is n-type conductive, which is beneficial to make vertical structured LED. Compared with traditional vertical structure in production of sapphire substrate to use laser liftoff process, it can bring the simplified processes and can reduce costs, raising the yield to achieve the higher cost performance.

In recent 10 years, researchers around the world have made an important progress in the study of the freestanding GaN substrate. And in the next few years, GaN substrates suitable for the devices with a diameter from 3 in. to 6 in. will be mass production.

1.1.2 *Key Drivers for GaN Substrate Commercialization Success*

One of the areas for the application of GaN substrates is in GaN laser diodes (LDs), where GaN substrates provide a clear and immediate benefit in the earlier days. The new applications for GaN-based LDs are primarily for next-generation, high-definition DVD players and recorders, although other applications include high-definition video projectors and displays, commercial printing, and testing and measurement applications, such as spectroscopy and bio-sensing.

There is no doubt that LED technology will take market share over the traditional lamp and tube business. In either a cautious or a more aggressive scenario, LED applications will certainly be the key drivers for the bulk GaN market, according to Yole Développement.

The report "Bulk GaN substrate market 2017" from Yole notes that while the Blu-ray segment which in the past was the GaN-based laser industry's main driver continues to decline, it is expected to be offset by nascent, growing segments like projectors (office projector, mobile pico projector, head-up display, etc.) and automotive lighting, leading to new growth opportunities for bulk GaN substrates. When new devices are developed and demonstrated and new areas of commercial demand are generated, this can provide further motivation for substrate manufacturers to increase manufacturing capacities. For example, Fig. 1.2 shows the forecast for TIE (2 in. equivalent) GaN wafer volume for device applications, with unit volumes at nearly 350,000 in 2022. This volume forecast represents a tremendous opportunity for GaN substrates but will require significant cost reduction in order to meet expected cost requirements.

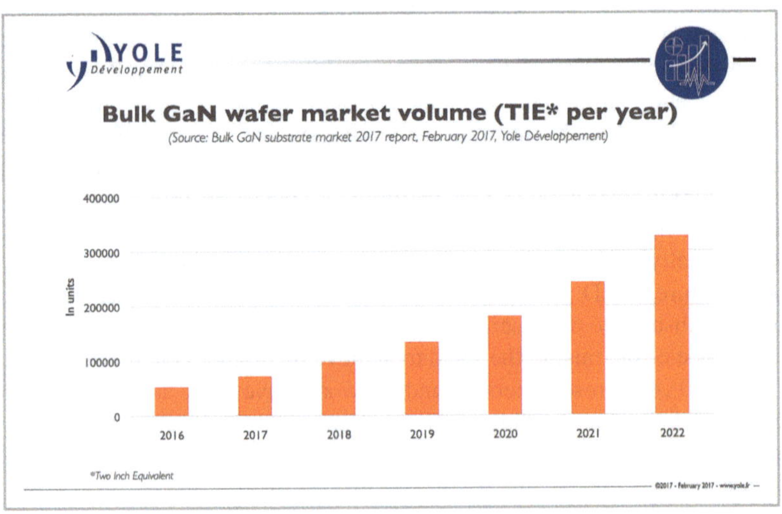

Fig. 1.2 TIE GaN wafer forecast for device applications. Source: Yole Développement

Considerable technology development is needed to enable penetration of GaN devices into the solid-state lighting (SSL) and power electronics markets. The performance of the GaN devices must be improved in many areas: LED luminous efficacy at high lumen output, FET and diode breakdown voltage and on-resistance, and power switching reliability, among others. The latest report reveals more and more GaN-based devices fabricated on advanced GaN substrate. At the same time, with the increase of demand of blue laser diode, UV LED, and high-power, high-frequency devices, the amount of GaN or AlN substrates is increasing. Only the high quality of substrate can provide high-performance device because of lattice matching and excellent thermal management features. Bulk GaN substrates are expected to be used for many devices addressing these market opportunities, but there are numerous challenges facing the adoption of bulk GaN substrate technology.

Above all, substrate cost must also be significantly reduced in order to become economically practical in the device applications, which is particularly challenging due to incumbent GaN device technology based on foreign substrates. GaN substrate materials provide clear and immediate benefits for GaN-based LD applications over sapphire-based approaches, particularly at higher operating current density, and GaN substrate manufacturers are currently focused on addressing this market. Other benefits for LEDs, RF FETs, power devices, and photodetectors have been shown, but commercialization of these devices requires high volumes of low-cost substrates. These improvements will occur through the continued development of bulk growth techniques, such as HVPE, ammonothermal growth, solution growth, and combinations thereof.

As these improvements are made, the potential for III-N devices to do more (more light, more power) with less (higher energy efficiency, less electricity, less waste heat) will be enabled.

1.2 The Technical Routes for GaN Substrate Materials

Because the homo-epitaxy on GaN substrates has better advantages compared with the hetero-epitaxy, then research about GaN substrate is of great practical value. Now the basic ways for GaN substrate materials have generally the following three technical routes (as shown in Fig. 1.3) (Substrates for Nitride Epitaxy, IWN2008, Switzerland, 2008).

1.2.1 *Native GaN Substrates*

One of the most important approaches to obtain GaN substrate wafers is to obtain a larger dimension of columnar bulk crystals or boule. A lot of experimental studies have shown that GaN at the top of the columnar crystal has very high quality. Bulk GaN crystals can be obtained by high-temperature and high-pressure solution method, ammonothermal growth, Na-flux growth, or the combination of a variety

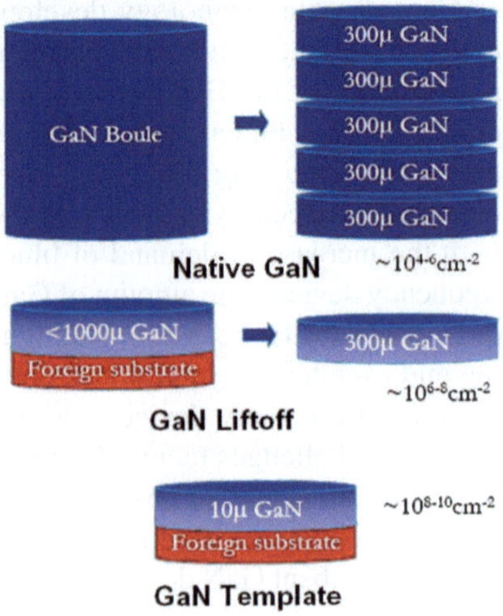

Fig. 1.3 Three technical routes for GaN substrate materials (IWN2008, Switzerland, 2008)

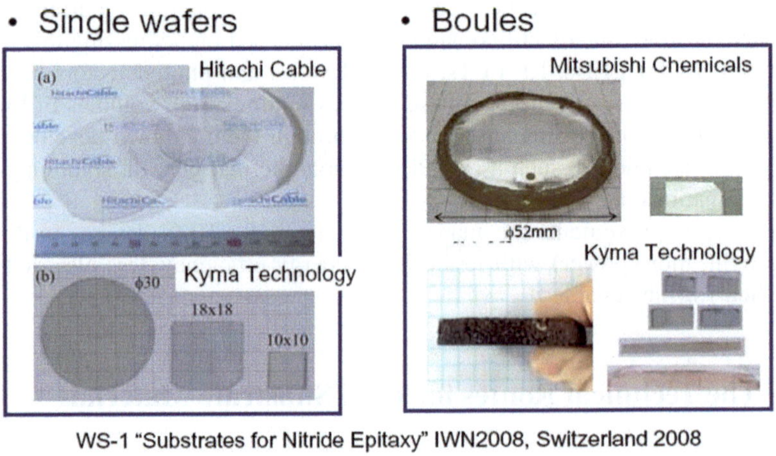

Fig. 1.4 Native GaN substrates reported by some companies (IWN2008, Switzerland, 2008)

of ways. High-quality GaN chips can be got by cutting GaN crystal into flakes and polishing, and the dislocation density for GaN crystal in different ways is usually between 10^4 and 10^6 cm^{-2}. Thicker pear-shaped or columnar crystals can not only produce more chips, but also can further reduce the defect density. So the most ideal solution to solve the problems of GaN substrate is to obtain single crystal GaN substrate chips by the abovementioned methods. But although GaN substrates have been reported (as shown in Fig. 1.4), GaN native substrates with a large size of more than 2 in. have not been commercialized. As shown in Fig. 1.4, the large size of columnar crystal is still mainly grown for a relatively long time by the method of hydride vapor-phase epitaxy (HVPE) or the combination of HVPE and other technologies.

Among all the bulk growth techniques under investigation today for nitrides, HVPE is the most promising technique since it utilizes a process with more favorable conditions, such as the low pressure and relatively low growth temperature. Because of the high growth rate, its growth process is also cost-effective. Other techniques, such as Na-flux or ammonothermal growth, are still under development.

1.2.2 GaN Liftoff Substrate Wafers

Until now, hydride vapor-phase epitaxy is considered as a highly practical method by growing the GaN thick films for quasi-bulk substrates after the separation from the foreign substrates or obtaining GaN boules with a thickness of more than centimeter grades to be sliced to fabricate native bulk substrates.

The biggest advantage of the technique is its ability to produce high-quality material at high growth rates due to a high surface migration of the halide species. The versatility of HVPE as a growth method is motivated for both device applications and substrate application.

HVPE-GaN thick film is usually grown on foreign substrate with the thickness of more than 500 μm and less than 1000 μm at a growth rate of between several microns and 1000 μm. And then the foreign substrate is peeled by laser liftoff (such as sapphire substrate), chemical corrosion (such as Si and GaAs substrate), or mechanical thinning or self-separation by utilizing the change of stress. After the thinning or polishing of self-standing GaN thick film, GaN substrate wafers would be obtained. Due to the lattice mismatch from the heterogeneous substrate, the lattice distortion introduced by the atom cannot completely eliminate the stress, so the dislocation density is roughly between 10^6 and 10^8 cm^{-2}. Now, 2 in. GaN wafers by HVPE method have been commercialized, but the price is quite higher. Due to the technical reasons, the yield is low. The HVPE-GaN wafers are mainly used in industry of GaN LDs; the mass application in the field of LEDs is not yet mature.

1.2.3 GaN Templates

GaN template refers to the compound substrate grown on heterogeneous substrate, which the thickness is more than 10 μm (Fig. 1.5). And compared with MOCVD GaN of the thickness of a few microns, the dislocation density is slightly lower and generally 10^8 cm^{-2}. The increase of the thickness will further reduce the dislocation density, and the dislocation density of free standing GaN substrate can reduce about two orders of magnitude (Fig. 1.6). The results show that with the increase of the thickness of GaN template from 5 to 50 to 150 μm, the dislocation density is reduced from 10^9 cm^{-2} to 5×10^8 cm^{-2} to 1×10^8 cm^{-2}. The thickness of GaN templates is generally less than 50 μm, and GaN template with the thickness of more than 50 μm will be easily cracked due to large lattice and thermal mismatch between

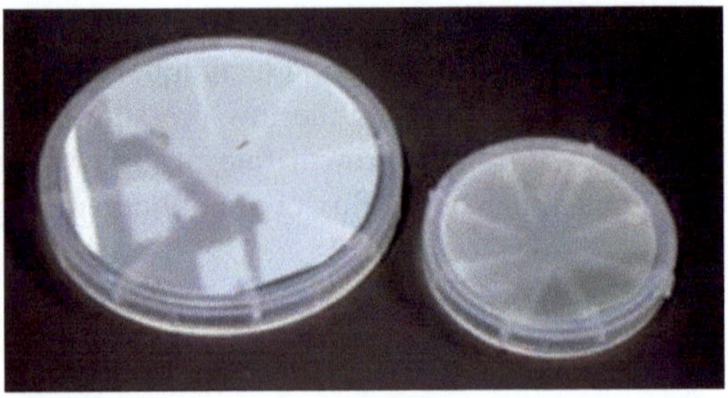

Fig. 1.5 GaN templates on sapphire substrate

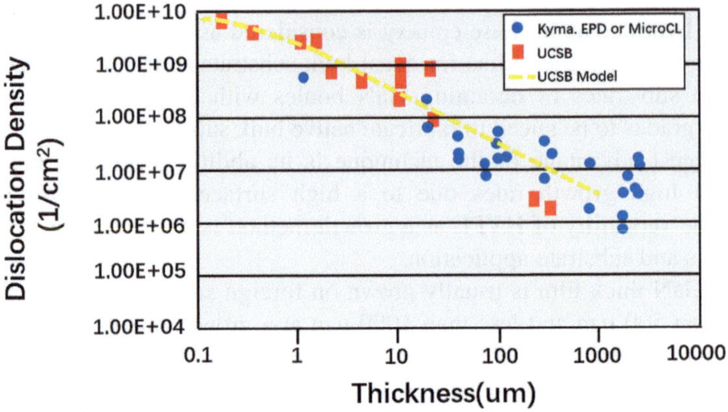

Fig. 1.6 Comparison of the dislocation density of different GaN films (source: Cree, Inc.)

GaN and sapphire. Due to the low cost and high quality, GaN templates by HVPE have the very big potential applications.

1.3 Major Methods for the Growth of GaN Substrate

GaN at high temperatures is decomposed into Ga and N_2, only melting in the physical conditions of more than 2200 °C and 6 GPa above nitrogen pressure. So the traditional Czochralski or Bridgman method cannot be used for the growth of GaN single crystal. At present, GaN is generally grown by hetero-epitaxy of MOCVD and MBE. In order to get high-quality GaN film, almost all the growth of technology and the substrates have been tried. But the high defects in the hetero-epitaxy GaN can limit the improvement of the GaN-based device performance. At the same time, research on the growth technologies for bulk GaN crystals has never stop, and a lot of trial and effort has been carried out, including the liquid-phase and gas-phase growth.

1.3.1 The Liquid-Phase Growth

The common liquid-phase growth methods are mainly high-pressure nitrogen solution growth (HPNSG), Na-flux dissolved agent method, and ammonothermal method, which all these three methods can prepare high-quality bulk GaN crystal.

As the homo-epitaxy substrate, GaN obtained by the liquid-phase growth will overcome the deficiency of the hetero-epitaxy. In addition, the photoelectric properties of GaN crystal by the liquid-phase growth are excellent, which is suitable for the high-performance devices. So the liquid-phase methods have also attracted more attention for the preparation of high-quality GaN crystal.

1.3.1.1 High-Pressure Nitrogen Solution Growth (HPNSG)

In the liquid phase growth technology of gallium nitride, high-pressure nitrogen solution growth is the most commonly used method for single crystal growth. Early in the mid-1970s, GaN single crystal with a diameter of about 1 mm was obtained by this technology. Its growth principle is shown in Fig. 1.7 (left); to put the high-pressure nitrogen gas into the molten state of Ga in the high temperature of crucible, enough N can be dissolved into the high-temperature molten metal Ga; nitrogen molecules on the surface of the molten Ga would be decomposed into N atoms and react with Ga to form GaN. GaN single crystal would be prepared at a certain temperature gradient for the local supersaturation of N atom. At that time, the temperature and pressure were only 1200 °C and 10.8 GPa. Because GaN is very stable at high temperature and high pressure, the dissolvability of N atoms is very

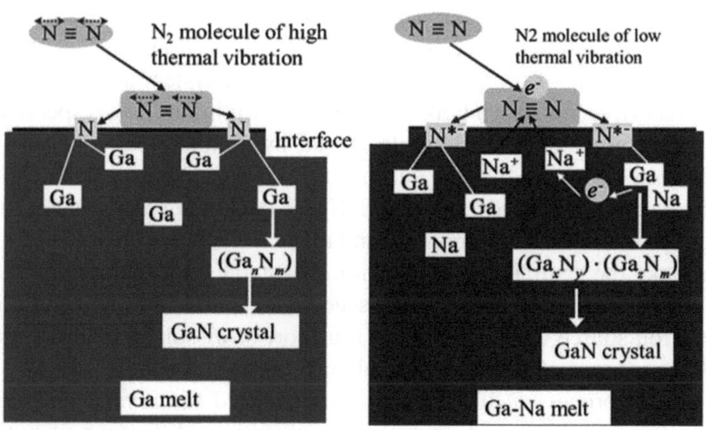

Fig. 1.7 Growth mechanism of GaN by HPNSG (T = 1300–1600 °C, $PN_2 > 1$ GPa) (left) and Na-flux method (T = 600–850 °C,$PN_2 > 0.1$ GPa) (right)

low in the molten metal Ga (atomic number scores less than 1%) [2], which leads to a decline of GaN crystal growth rate (about 1 mm/24 h) [3]. So it must be long enough to grow the large-size and high-quality GaN single crystal. Of course, due to the growth conditions of the high pressure, the crystal defect is extremely low and suitable for the preparation of high-quality crystal [4].

After more than 10 years of development, as the experimental conditions change, the temperature and pressure have been effectively improved to 2300 °C and 4 GPa. GaN single crystal by HPNSG method is generally a hexagonal flake structure, along with the [0001] direction of wurtzite structure. A suitable temperature gradient can improve the growth rate of [0001] direction at a certain extent. But too large temperature gradient will affect the stability of the growth, which will lead to the honeycomb structure in the GaN surface [2]. HPNSG GaN single crystal has good crystallization quality commonly by XRD rocking curve test; the full width at half maximum along the major axis and the vertical straight in the axis direction is only, respectively, 0. 028° and 0.017°, which is similar to the quality of the crystal.

High-pressure nitrogen solution method can be used for the preparation of high-quality GaN crystals, but this growth of GaN requires very high pressure in the range of GPa which sets rigorous standards for the growth facilities, the high production costs is not conducive to large-scale industrial production.

1.3.1.2 Ammonothermal Growth

From the angle of the crystal growth, the basic principle of ammonothermal crystal growth is similar with that of hydrothermal method. The ammonothermal process is a solvothermal process that allows the solubilization of polycrystalline III-nitride nutrient or feedstock in supercritical ammonia under high pressure by utilizing a solubilizing agent or mineralizer. The dissolved nutrient is then transported to the region of crystallization, where it recrystallizes on seed crystals or through self-seeding. A schematic of the internally heated growth system and a photo of a 2 in. diameter crack-free, as-grown bulk crystal are shown in Fig. 1.8, respectively [5, 6].

The growth process of GaN single crystal by ammonothermal method can be summarized as follows: dissolve the needed III-nitride nutrient or feedstock in supercritical ammonia fluid (such as KNH_2) containing mineralizer to form a saturated solution under high pressure, and then take appropriate technical measurements to make the saturated solution into metastable supersaturated solution, and begin to grow GaN single crystal on the seed crystal. Here, the choice of mineralizer agent, reaction source distribution, and transport of dissolved material are the key factors in the process of ammonothermal GaN crystal growth.

The specific process is to seal the nutrient, seed the crystal and mineralizer solution in the autoclave, and control the temperature difference of nutrient area and seed crystal area, respectively, so as to accelerate the convection of the solute and seed for the crystal growth. For various kinds of growth process, the largest difference is caused by the different mineralizer, because the solubility of crystal is different from temperature dependence in different mineralization agent solution.

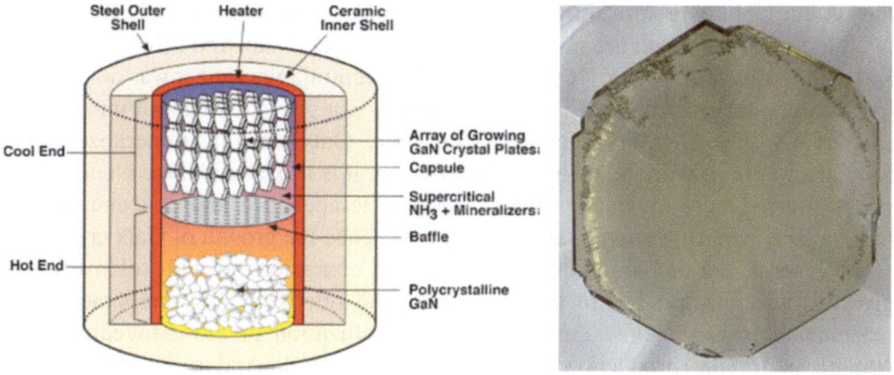

Fig. 1.8 (**a**) Schematic of high-pressure ammonothermal GaN growth apparatus employed at Soraa; (**b**) As-grown nominal 2 in. diameter GaN(0001) crystal produced using the ammonothermal method

The ammonothermal growth is conducted by transporting the mass from high- to low-temperature zones, and this method has been reported to possess several advantages, including lower dissolution density due to solubilizing in the higher-temperature zone, lower contamination probability due to recrystallization in the lower-temperature zone, lower probability for cubic formation, and absence of disastrous effect of the basic solution on the autoclave materials, which results in the improvements of safety and cost-effectiveness. In contrast, the acidic approach requires special, usually costly, Au or Pt liners, and the safety operation still remains problematic.

When the crystal in mineralizer solution has the solubility of positive temperature coefficient (higher solubility while varying with temperature), the bottom of the autoclave used for nutrient is generally designed as the high-temperature zone, and the top of the autoclave as low-temperature area is used to hang the seed crystal. Usually the crystal nucleation can be easily controlled for the growth of the solubility of positive temperature coefficient. When the crystal in mineralizer solution has a negative solubility coefficient (the solubility of crystal drops while varying with temperature), it should be seed crystal hanging in the high-temperature area and put nutrient at low-temperature area.

GaN solubility in pure ammonia is small because of ammonia with weak polarity; thus in the ammonothermal method, pure ammonia solution is not suitable for GaN crystal growth. To refer the hydrothermal method for the selection principle of mineralizer, ammonothermal GaN crystal growth also needs to choose a suitable mineralizer. The type of the mineralizer used determines the main type of the ammonothermal approach, with respective advantages and disadvantages. Selection of mineralizer is conformed to the following two points: (1) the solubility and solubility temperature coefficient of the basic crystal material in the ammonia solution is enough to achieve an effective growth rate; (2) in the ammonothermal system, crystalline materials are the only stable solid phase; thus it will not affect the system's phase relationship. At present in ammonothermal GaN crystal growth, two kinds of mineralizer are mainly adopted, a kind of alkali XNH (X = Li, Na, K) and another for acidic NHX (X = C1, Br, I). The neutral mineralizer was not suitable for GaN crystal growth. GaN crystal is generally not grown by neutral mineralizer, such as KX (X = C1, Br) and NaX (X = C1, Br). Only when the mineralizer is KI, a small amount of wurtzite GaN is generated, but it is mixed with zinc blende GaN.

Of course, there are many factors, such as temperature and concentration of mineralizer, impurity, growth, and so on, to influence the growth of GaN crystal.

In general, the ammonothermal technique is characterized with relatively low growth rate capacity; but in this technique, the relatively low temperature and reasonable pressure allow controllable recrystallization at close to equilibrium conditions for the production of high-quality material. And the technique also allows large-scale wafer growth and the multiple-seeded growth in a single run. Up to now, Poland Ammono Inc. has been able to achieve the commercial mass production of 2 inch GaN wafer with m- and c- direction. But as a result, the ammonothermal growth needs GaN seed crystal, and it is affected by seed crystal size. GaN single crystal with more than 2 in. in diameter is still in the development stage. Nevertheless, the approach is expected to be capable of handling higher growth rates at higher pressures.

The quality of this type of material is still in the beginning of detailed evaluation. Given the fact that a reasonable large crystal is a result of a very long process, optimization of the growth conditions and a systematic series for thorough analysis will take more time.

1.3.1.3 Na-Flux Method

An alternative solution growth approach by using Na flux instead of nitrogen solution was proposed by Yamane et al. [7] several years ago. So far, as a self-nucleation growth, it is quite similar to that of high-pressure solution growth approach.

On the basis of high-pressure nitrogen solution method, the researchers developed a Na-flux method, and the growth mechanism is shown in Fig. 1.7 (right). Na is added in the molten Ga solution. Because of the very high reducibility of Na, N_2 molecules can be ionized and easily dissolved into molten Ga-Na solution under low temperature and low pressure, which thus greatly increase the N solubility in the solution. So GaN crystal will be synthetized at relatively low pressure. The results of GaN crystal by this self-nucleation growth are quite similar to that obtained by high-pressure solution growth approach. But the advantages of Na-flux method are mostly related to the more favorable growth conditions of typically 800 °C and 5 MPa. In the 1990s, Yamane et al. [7] got the GaN crystal by solvent method using Na as cosolvent for the first time.

In the early stages of Na-flux research period, the crystal nucleation position is uncertain; the crystal has often been found in the crucible wall. So the liquid-phase epitaxy (LPE) has been developed in order to limit the nucleation on the seeds.

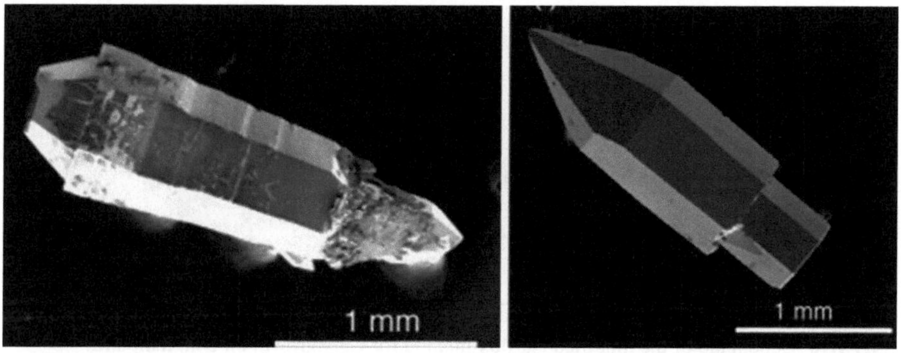

Fig. 1.9 GaN seeded growth with a (left) Na-Ga premixed melt and (right) with the Ga melt and Na vapor

Aoki et al. studied the influence of the growth temperature and N_2 pressure on the crystal morphology and growth rate during the epitaxial growth of seed crystal. However, increasing the solubility of the nitrogen in the molten Ga-Na solution, at the same time, the initial GaN seed and GaN epi-crystal also become much easier to decompose, which make the crystal surface rough before and after the crystal growth. In order to overcome the shortcomings, Yamada et al. [8, 9] proposed the Na-vapor assisted GaN growth to obtain the high-quality surface of GaN crystal, and the obvious difference can be seen from Fig. 1.9.

In the early study, GaN grown by Na-flux method is the colored crystal due to the lack of N in the process of crystal growth, because N solubility is not high in the solutions. In order to improve the N solubility in the Ga-Na melt, joining the alkali metal element such as Li and Ca into the Ga-Na melt in the hybrid cosolvent significantly increased the N solubility in the melt and improved the permeability of GaN crystals [10].

In the absence of the GaN seed, two growth habits were demonstrated—prisms or platelets—with length and lateral sizes of a few millimeters. In order to increase the size of GaN crystal, F. Kawamura and Y. Mori et al. used HVPE-GaN substrate as a seed crystal for the epitaxial growth of 2–4 in. GaN film. Kawamun et al. [11, 12] developed a set of equipment for large-size GaN single crystal substrate. The schematic diagram of the equipment is shown in Fig. 1.10, and one of the most critical improvements is to put the heating system (crucible and electric furnace) in the stainless steel compression cavity with a water cooling system, by shaking the cavity to stir the melt at the same time. After this modification, the growth system can be used to grow large-size GaN chip. In the case of using GaN film grown by MOCVD, growth rates up to 20 μm/h were obtained, and a boule with a thickness of 3 mm was shown.

In order to improve the crystal quality, Kawamura et al. [14] join carbon (C) to inhibit the polycrystalline growth. By optimizing the growth conditions, Mori et al. [13, 15] have successfully carried out the epitaxial growth of 2–4 in. GaN on the large size of seed crystal, GaN crystals are highly uniform and free of cracks, and

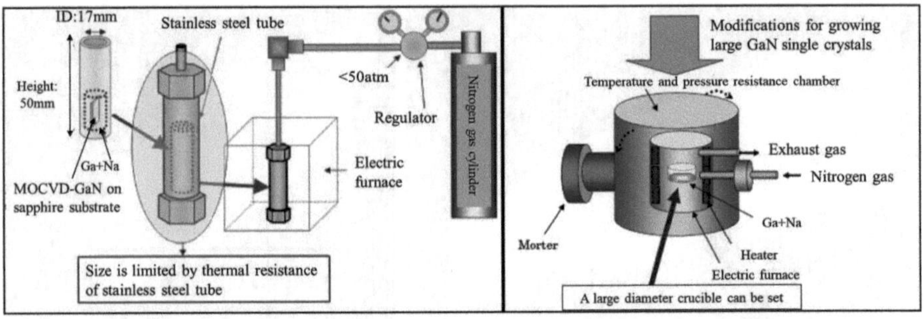

Fig. 1.10 Schematic of the modified Na-flux method for GaN growth on 2 in. templates

the thickness can be up to more than 750 μm, as shown in Fig. 1.29. GaN crystal has a good permeability and lower impurities, although the dislocation density of the initial seed crystal HVPE-GaN is up to 10^8 cm^{-2} orders of magnitude, but the chip dislocation density decreased to 2.3×10^5 cm^{-2} after growth by the liquid-phase epitaxy (LPE). A decrease of dislocation density from 10^8 cm^{-2} in the seeds to 10^4 cm^{-2} in the regrown crystal was reported and explained in terms of a change in the dominating growth mode from vertical in the seed to lateral in the Na-flux solution growth [14]. It is also found that Na and C additives are not doped into the GaN crystals by the test of the secondary ion mass spectrometry.

Although its dislocation density of GaN crystal by Na-flux method has been greatly reduced, but with the demand for higher crystal quality of GaN-based photoelectric device, lower dislocation density or even no dislocation GaN crystal is needed. Similar to the Czochralski method for the growth of Si ingot, a necking technique was adopted to realize the growth of large-size GaN crystal without dislocations [16], as shown in Fig. 1.11. A dislocation generally propagates perpendicularly to the growing surface. Therefore, in the initial growth layer that consists of numerous (10–11) facets, the dislocations propagate perpendicularly to the (10–11) facets, and they are bent. Usually in the necking technique, the dislocation begins to extend like a bent neck from the growth surface and end in the sapphire wall, so that further propagation does not occur. The growth of (10–11) facets in this initial growth layer and the dislocation behavior are unique phenomena of the Na-flux method. Because the area of crystallography is without continuity, and the bent dislocation will end at the interface between the initial growth layer and single crystal, almost all of the dislocations from GaN seed crystal will eliminate inside the hole, and the final GaN crystals show no dislocation. In addition, the dislocations that propagate in the (0001) direction also terminate at the boundary between the initial growth layer and the preferentially grown single grain so that propagation of dislocations to a single grain cannot occur. Almost all dislocations which originated from the GaN seed layer are eliminated in the holes (i.e., the necking region), resulting in a GaN crystal free of threading dislocations. Imanishi et al. [17] have already succeeded in growing dislocation-free bulk GaN crystals with a diameter of more than 2 cm and a height of about 1.2 cm by the GaN small

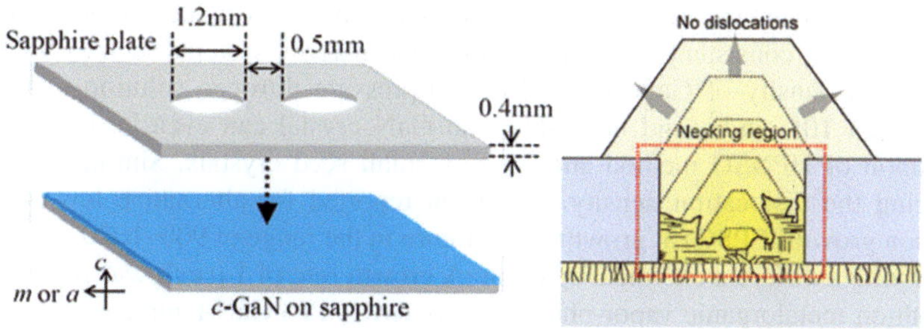

Fig. 1.11 Schematic drawings of the configuration of two GaN point seeds and that of the growth model with a necking technique

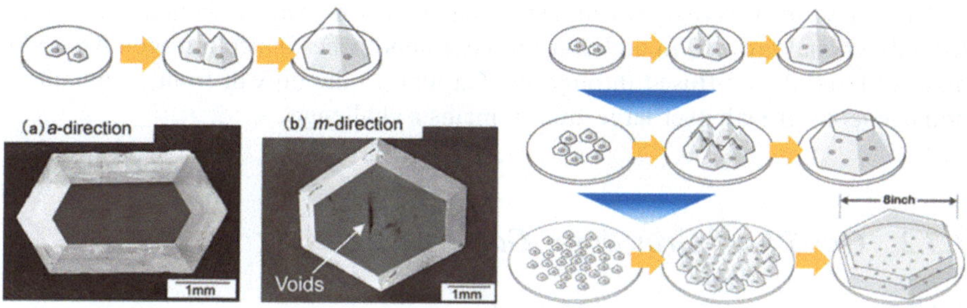

Fig. 1.12 Schematic of coalescence growth from two isolated small seed crystals and SEM images of the sliced crystals (left), and schematic of coalescence growth from many periodically arranged small GaN seed crystals (right) [15]

seed crystal growth. The necking technique in the Na-flux GaN growth may be a major breakthrough for fabricating dislocation-free GaN ingots.

In order to grow large-scale GaN crystal without dislocations, they put forward the joint growth of multiple isolated small seed crystals for the large-scale GaN crystals. Figure 1.12 (left) is the experiment schematic of two seed crystals' joint growth; it can be found that GaN crystal quality is very high along one orientation and poor along the orientation of m-direction with a lot of defects [15]. Recently, they got high-quality GaN crystals from three isolated small seed crystals along a joint growth orientation, as shown for SEM images of GaN crystals in Imanishi et al. [17]. Obviously the joint growth is possible for manufacturing large diameter GaN crystals without dislocations. As shown in Fig. 1.12 (right), it is the schematic for GaN crystal by a periodic array of small seed crystals' growth [15].

One advantage of Na-flux method is simple for the growth equipment and low pressure for crystal growth, which can grow a large-scale, high-quality GaN crystal and even GaN crystals without dislocations using small seed crystal growth. So this method is gradually becoming the key technology of high-quality GaN crystals.

Another advantage of Na-flux method is its ability to reduce some structural defect density, consistent with reports for other material systems such as SiC. The dislocation density of GaN grown by high-pressure nitrogen solution method is lower than 10^2 cm^{-2}, and no dislocation GaN crystal can even be prepared by the union of HPNSG method and Na-flux small seed crystals. Similar results of reducing the dislocation density have been reported for alternative low-pressure solution growth (LPSG) at growth temperature in the range of 900–1020 °C [18] by employing a different chemical solution. A growth rate of 1 μm/h was reported for growth on metalorganic vapor-phase epitaxy (MOVPE) GaN template on sapphire up to 3 in. in diameter. Also, at nearly atmospheric pressure and temperature of about 800 °C, employing another chemical solvent, seeded growth on HVPE-GaN seeds was demonstrated [19].

Up to now, the reported growth rates of a few micrometers per hour are not suited for bulk crystal growth. In addition, in these approaches, since nitrogen has to be dissolved and then diffused through the Ga melt to the growth front, maintaining homogeneous growth over large areas remains a challenge.

1.3.2 Gas-Phase Growth for GaN Substrates

1.3.2.1 Gas-Phase Transport Method

Since Johnson and others synthesized polycrystalline GaN with NH_3 and molten Ga, many people began to study the gas-phase transport method for GaN single crystal growth, which is to transmit the Ga vapor at high temperature to react with NH_3 for GaN single crystal. If the molten Ga metal is replaced with GaN powder as Ga resource, it is known as "sublimation method." The growth rate by the sublimation method is relatively slow but steady. But the sublimation method itself has a technical problem, which must keep a continuous supply of GaN powder during the growth. While molten Ga metal is used as Ga source, it can get the maximum growth rate, but not very much stable. Here, the high growth rate is thought to form the Ga droplets, and the instability of the growth rate is attributed to the formation of GaN shell in the surface of molten Ga metal, causing the growth stopping. Besides, the reaction of Ga and NH_3 at the surface of Ga metal will cause the splash of Ga droplets on the surface of GaN crystal, which will change the growth mechanism.

Now the improved gas-phase transport process is mostly studied, called the sublimation sandwich transport method (SST) [20], as shown in Fig. 1.13. In SST, a powder or polycrystalline material (Ga, GaN, or their mixture) is evaporated, transported across a small gap by a thermal gradient, and recrystallized on a substrate. The evaporation of metallic Ga in the ammonia ambient is a very convenient alternative to grow the thick, high-quality GaN films. This technique has all the characteristics of a vapor-phase epitaxy performed at an unusual high temperature.

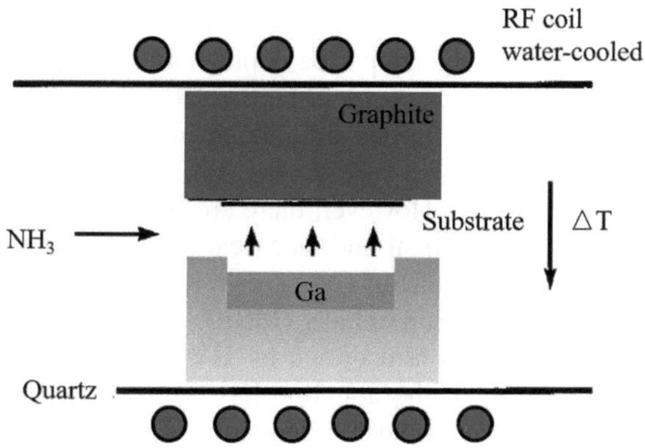

Fig. 1.13 Schematic of the high-temperature vapor-phase epitaxy reactor. The graphite crucible containing the metallic Ga and the substrate is heated inductively via a water-cooled coil. Ga vapor is transported over a short gap of a few millimeters

The distance between the Ga source and the substrate is only several millimeters, the pressure of reaction tube is an atmospheric pressure, and the growth temperature is 1170–1270 °C. The SST growth rate is more than 100 μm/h, and the fastest rate is up to about 1100 μm/h. The SST method is simple, low cost, and faster, but the SST GaN crystal quality is also very poor. There are still some technical problems to solve, not up to the practical level.

1.3.2.2 Hydride Vapor-Phase Epitaxy (HVPE)

Although ammonothermal growth and Na-flux method seem to have more developing potential for the preparation of single crystal GaN substrate, it is not suitable for commercial application because of the small size (≤2 in.) and low growth rate (50 μm per day). Therefore, atmospheric hydride vapor-phase epitaxy is still the main method to obtain large-scale, high-quality self-standing GaN substrate. As a chemical vapor-phase transport, HVPE method has not only high growth rate, compared with other methods of substrate preparation, but also versatility as a method for the fabrication of both substrates and devices, with a larger application direction and development prospects.

In history, HVPE methods have played an important role in the research and development of arsenic and phosphide semiconductor materials. Although the extensive research and full development have been done for HVPE growth for arsenide and phosphide, little is known in terms of nitride growth. The reason is that there is no suitable substrate and the chemical growth is different, leading to HVPE growth of the nitride more complex. In spite of this, HVPE is still the first and the most common method of GaN epitaxial growth until the early 1980s.

The first growth of GaN by HVPE was reported by Maruska et al. in 1969, using a sapphire substrate by modified hydride equipment for gallium arsenide (GaAs) and indium phosphide (InP) [21]. In the 1970s to the early 1980s, the researchers attempted to grow GaN thick film on sapphire and then made the self-supported GaN substrate [22]. However, the HVPE-GaN film has a large number of crystal defects, high background carrier concentration, and poor crystal quality. At the same time, it is difficult to realize p-type doping and effective doping control, which hinders the production and research of GaN-related devices.

Due to the serious difficulties in reducing the intrinsic carrier concentration and realizing the p-type doping, HVPE technology was almost entirely abandoned after the 1980s. However, there are still some reports about HVPE growth of GaN film, and its optical and electrical properties reported can still be the best results compared with that of GaN grown by metal organic vapor-phase epitaxy (MOVPE) or molecular beam epitaxy (MBE). In fact, even by the standards of the 1970s, the quality of GaN film grown by HVPE is still remarkable [23, 24]. From the late 1990s, most of the attention has been focused on the hetero-epitaxy of GaN on foreign substrates such as sapphire, Si, $NdGaO_3$, and GaAs to realize freestanding GaN wafers. The first thick GaN (approximately 400 μm) with a smooth surface was reported by Detchprohm et al. in 1992, using a ZnO buffer layer on sapphire [25]. Usui et al. [26] achieved a freestanding GaN wafer of 2 in. diameter size by the use of a new technique, epitaxial lateral overgrowth (ELO), to reduce the density of dislocations significantly, and by a laser irradiation from behind to remove GaN thick layer from the sapphire substrate.

HVPE system equipment is relatively simple and convenient to maintain, and with the improvement of growth technology, the crystal quality of HVPE-GaN has been improved unceasingly. Its growth rate is high and it is easy to obtain homogeneous and large-scale GaN thick film, which make it very hopeful to solve the problem of self-standing GaN substrate. Meanwhile, the existence of halide in the growth environment of HVPE can significantly affect the equilibrium of GaN reaction and improve the ELO growth rate. Using the HVPE-ELO technology, GaN thick film can also be grown at high growth rate in situ growth, and then low-density GaN substrates can be obtained by a variety of substrate separation technologies (such as corrosion, laser liftoff technology or mechanical polishing technology, etc.). HVPE, therefore, becomes the important method to prepare freestanding GaN substrates for the large-scale application of GaN-based optoelectronic devices.

1.4 HVPE for GaN Substrate Materials

1.4.1 Chemical Reaction in the Growth of GaN by Hydride Vapor-Phase Epitaxy

HVPE for the growth of nitrides is a chemical vapor deposition method, and the reaction is usually carried out within the atmospheric quartz reactor. The reaction process is to introduce the hydrogen chloride under carrier gas into the reactor, reacting with metal gallium in the low-temperature zone, to flow the generated gallium chloride to the high-temperature zone and being mixed with ammonia to grow GaN. Unreacted gas is absorbed by the tail gas treatment system.

Nitride growth is actually using metal chloride disproportionation reaction, by raising the temperature to improve the stability of low compounds. Ga chloride has a variety of valence states (such as GaCl, $GaCl_3$, etc.), and GaCl is stable at high temperatures in HVPE system, decomposing into $GaCl_3$ below a certain temperature. The above features make us adjust the temperature of the reaction chamber, realize the GaCl generation, and transfer and deposit GaN. So, in usual nitride HVPE growth system, they are divided into a high and low temperature reaction chamber area.

Therefore, the novel of HVPE method is the primary reactant (GaCl) generated in the reactor, where the liquid metal Ga react with HCl gas in the range of 800–900 °C. Then GaCl with the carrier gas are carried into the substrate and mixed with NH_3, depositing GaN on the substrate. Substrate temperature is kept at 900–1100 °C, and N2 or H2 is the carrier gas. The chemical reaction is as follows:

$$\mathrm{Ga\,(s) + HCl\,(g) = GaCl\,(g) + \tfrac{1}{2}H_2}$$

(low-temperature zone)

$$\mathrm{GaCl\,(g) + NH_3\,(g) = GaN\,(s) + HCl\,(g) + H_2\,(g)}$$

(high-temperature zone)

The following six species were chosen as the necessary vapor species in analyzing the HVPE growth of GaN: GaCl, $GaCl_3$, NH_3, HCl, H_2, and inert gas (IG) such as nitrogen or helium. Here, group III elements such as chloride (usually a single chloride) were transported to the substrate, so hydride vapor-phase epitaxy is usually known as chlorine gas-phase epitaxy. Due to relatively low saturated vapor pressure of metal chloride at room temperature, chloride molecules can easily condensed on the substrate at room temperature, and HVPE growth usually adopts a hot wall reaction chamber for in situ synthesis of chloride.

Using in situ synthesis chloride and hot wall reaction chamber, the complex gas transportation system and assistant heating systems can be avoided. In addition, if the metal chloride was pre-synthesized out the reactor, the strong absorbent and corrosive metal chloride will make their preservation and transportation very difficult. However, high corrosive HCl gas can also cause some difficulties. For example, if there is air leakage into the gas path or reactor, HCl will soon destroy the reactor. Based on this, some of the research team reported synthetic $GaCl_3$ instead of in situ synthesis of HCl to GaCl [27–29]. Using these precursors, researchers grow single crystal GaN and improve the purity of metal chloride source.

Based on the purity and operability, HCl is usually more common than chlorine. Because the Chloride has higher saturation vapor pressure than bromide and iodide, the chloride is the most commonly used halogen for transportation. Bromide and iodide have a lower decomposition temperature, and it is reported that they are often used in the growth at low temperature [30, 31].

In hydride vapor-phase epitaxy system, the chemical reaction will produce large amounts of NH_4Cl, $GaCl_3$,$GaCl_3$, and NH_3, which will be condensed and even block the exhaust system, unless they are heated to a high enough temperature (>340 °C) or the reaction is under low-pressure conditions.

Although there are a lot of difficulties in the design of HVPE system due to the chemical reaction, it also provides a unique advantage. For example, unlike MOVPE, HVPE growth process is no carbon in nature, which is easier to grow high-purity GaN film. In addition, strong corrosive chlorine and hydrogen chloride can help to remove the excess growth of metal components on the substrates, so as to suppress the formation of rich gallium materials or separated Ga droplet. The self-stabilization effect and high mobility of adsorbed GaCl can be used to explain that HVPE is much higher in the growth rate than other epitaxial methods. Because of the self-stabilization effect in rich gallium conditions, GaN film by HVPE has a higher quality than that of other epitaxial methods (such as MBE or MOVPE). Although HVPE growth rate is at least one order of magnitude higher than other methods, the dosage of NH_3 is far lower. The impurities from NH3 and the cost are both significantly reduced in the growth of the VPE nitrides.

The consciously introduction of additional HCl to the growth zone at high temperature can effectively improve the Cl/Ga ratio, which plays a very important role in the growth of high-quality GaN film. The introduction of HCl may enhance the GaCl local saturation and improve the lateral growth, leading to a larger grain size. The early experiment research from Nanjing University showed that the introduction of additional HCl can improve the surface morphology and the crystal quality. Some patents point out that additional HCl can also reduce the pre-reacted GaN particles in space before depositing on the substrate. In addition, the presence of HCl can also enhance the decomposition of other impurities.

1.4.2 Hydride Gas-Phase Epitaxial Growth System

It should be pointed out that HVPE-GaN growth with groups' III–V semiconductor materials is completely different in nature. For example, the HVPE synthesis of GaAs, As_4, and As_2 molecules with volatile and chemical activity formed by thermal decomposition of arsenic compounds can continue to participate in the growth of thin films. But in the process of HVPE-GaN growth, the N_2 molecules formed by NH_3 decomposition are very stable in the temperature range which we are interested in. In fact, HVPE-GaN growth is based on the feasibility of the NH_3 relatively low decomposition, which may be helpful for transporting the reactive nitrogen effectively to the growth surface of substrate [32], to prevent the generation of black-rich Ga GaN film [33]. Another difficulty of HVPE-GaN growth is that there are unforeseen gas-phase reaction, space particles, and wall deposition problems except the formation of GaN.

HVPE growth system generally consists of four parts: the furnace, reactor, gas configuration system, and exhaust gas system. There are two types of design including the horizontal and vertical HVPE system. As shown in the abovementioned disproportionation reaction, HVPE growth system is designed into a dual

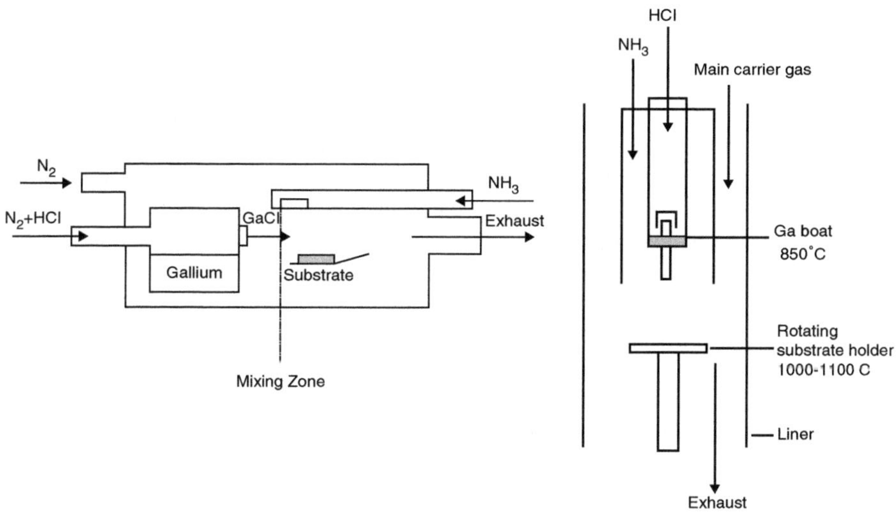

Fig. 1.14 Schematic diagram of (left) horizontal HVPE and (right) vertical HVPE

temperature zone: high- and low-temperature area. Low-temperature zone is used for generating GaCl and high-temperature zone for the deposition for GaN. The growth mechanism in different HVPE system is the same in nature. Figure 1.14 (left) shows the schematic diagram of the horizontal growth system (named as horizontal HVPE) [34], which is widely used in the growth of groups' III–V and II–VI material. At present, most of the reported HVPE for the growth of GaN is based on the horizontal HVPE growth system from RCA laboratory. Figure 1.14 (right) is the schematic diagram of vertical HVPE growth system (named as vertical HVPE) [35], which can improve the uniformity of GaN film by more easily introducing the rotation of the substrate. Substrate can also be dropped down and cooled slowly in ammonia atmosphere, to reduce the decomposition of GaN film.

1.4.3 The Growth and Doping of HVPE Nitrides

One of the difficulties for the growth of high-quality GaN by HVPE is the nucleation on the substrate. A lot of experiments show that the heterogeneous nucleation is very critical for the growth of high-quality GaN. If without any pretreatment, GaN directly on the sapphire is from highly transparent to brown. And the high-temperature growth after depositing the low-temperature buffer layer can often form polycrystalline GaN. Several kinds of different pretreatment, such as GaCl pretreatment of substrate surface and ZnO buffer layer technology [25, 36–38], are used to improve the heterogeneous nucleation density and quality of GaN on sapphire (0001).

In general, HVPE-GaN has the very high intrinsic shallow donor levels, usually in the level of 10^{18}–10^{20} cm^{-3}. Therefore, a great deal of doping research has been done for a suitable acceptor dopant to compensate for the donor and for p-type doping to realize the p-n junction. However, as a hot wall epitaxy, HVPE growth is vulnerable to the quartz reactor and gas-phase dopant in the chemical reaction. Some oxide doping agent might be more stable than SiO_2 in thermodynamics. Possible interactions will occur, such as the reduction of quartz reactor components, deposition of oxidized dopant on the pipe wall, and the corrosion of the quartz tube, leading to n-type Si impurities into the epitaxial layer, etc.

Although p-type conductivity has not been obtained, Mg doping in HVPE-GaN has been reported recently [39, 40]. Some studies by ion implantation in GaN show that zinc may be the third choice of shallow acceptor dopant. As a result, most acceptor doping research is focused on the introduction of Zn acceptor [41]. But the large ionization energy zinc acceptor and typically high intrinsic donor level in GaN made Zn doping for GaN p-type conductivity very difficult.

The HVPE technique was proven to be capable to produce n-type, p-type, and semi-insulating material. Hydride vapor-phase epitaxy is not only used to grow GaN but also to grow other nitride compounds (such as InN and InGaN, AlN and AlGaN, etc.) and the doping growth of nitrides.

Cubic GaN Cubic GaN has a unique advantage in the cleavage and doping. But compared with wurtzite GaN, cubic GaN in thermodynamics is metastable and only observed in the film. Several preparation techniques available have been adopted for the growth of cubic GaN, but the quality is very poor, because there is no lattice match suitable substrate. So far, there have been several research reports, using hydride vapor-phase epitaxy technology to grow high-quality cubic GaN thick film [42, 43]. And the most commonly used substrate is GaAs (100), because it can deposit a homogeneous epitaxial layer first on GaAs to modify the surface polishing damage, forming a nitride buffer layer to prevent further intrusion of ammonia gas. At that time, it was generally supposed that a low-temperature buffer layer of GaN could not be grown on a foreign substrate using HVPE system since only non-equilibrium growth system realized a GaN buffer layer. Here, the researchers think that it is necessary to control the lower growth rate (<4 μm/h) to prevent the formation of wurtzite GaN [44]. Further work is to obtain a device quality GaN material.

AlN-AlGaN Studies show that the growth of AlN-based nitrides is more complex by hydride vapor-phase epitaxy. The schematic diagram of horizontal HVPE for the growth of AlN substrate is shown in Fig. 1.15. And there are very strong interactions between AlCl and hot quartz reactor, which can lead to the excessive impurities introduced into the film and may decrease the quality of quartz reactor. Only a small amount of research has been done about the growth of AlN [45] and AlGaN [46, 47] by hydride vapor-phase epitaxy. Some companies, such as TDI, Crystal IS, The Fox Group, etc., have made progress in the growth of AlN template or AlN substrates by HVPE. Due to patent restrictions of AlN single crystal, TDI turned to the growth of AlN template rather than freestanding AlN substrate, as shown in Fig. 1.16. AlN

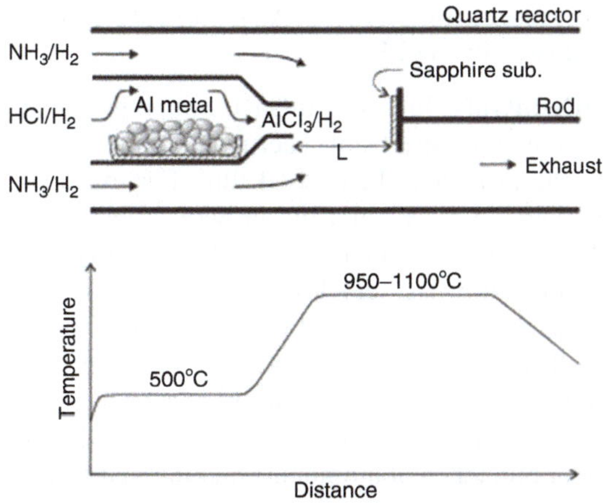

Fig. 1.15 Schematic diagram of horizontal HVPE for the growth of AlN substrate

Fig. 1.16 HVPE-AlN templates by TDI Inc. (2 in. AlN-on-sapphire, 3 in. AlN-on-sapphire, 4 in. GaN-on-sapphire, and 6 in. AlN-on-Si from left to right)

template refers to the deposited AlN layer on the heterogeneous substrates, such as Si or sapphire or SiC, for the further epitaxy of AlGaN devices.

The advantage is that the wafer size depends on that of the original substrate. But for the HVPE growth of AlN template compared with the growth by MOCVD and MBE, there is a same problem that serious cracks would be found while the thickness of AlN layer is up to several micrometers. So TDI developed a new stress control technology for the epitaxy of HVPE AlN template. By their self-owned multi-wafer HVPE growth system, they have achieved AlN templates of the thickness of 75 μm without cracks. The thickness of 2 in. AlN/SiC template without cracks is up to 10–30 μm, and the dislocation density is about 1×10^7 cm^{-2}, which is lower one order of magnitude than that of previous AlN/SiC template.

Research on the growth of AlN single crystals has been conducted for decades. Freestanding AlN substrates have been fabricated by sublimation, physical vapor transport (PVT), or sublimation-recombination technique, flux methods, and solution growth. The two most promising techniques are PVT and solution growth.

At present, freestanding AlN substrate with the diameter of more than 1 in. was prepared by HVPE.

InN-InGaN InN and InGaN compounds can extend the bandgap to the visible region of the spectrum. Thermodynamic instability of InN brings serious limitations to the growth of InN-based film. For example, the growth temperature of InGaN is below 800 °C, and the growth of InN is only at temperatures below 650 °C. The growth of stoichiometric InN is very difficult, because the ammonia at that temperature has very low cracking efficiency. Another important problem is that it is obviously impossible to use InCl to grow In-based compounds. Like the HVPE growth of GaN, $InCl_3$ is stable at low temperature. Presumably, at lower temperatures, an additional product of gas-phase $InCl_3{\cdot}NH_3$ plays a very key role for subsequent InN deposition. Even so, the thermodynamic instability of InN and poor efficiency of ammonia cracking make the growth of InN become very difficult.

1.4.4 The Main Difficulties of HVPE

As is known to all, HVPE-GaN growth has the very high growth rate, for the growth of almost of any GaN orientation can reach 100 μm/h in the order of magnitude, which means that HVPE is possible for the preparation of GaN substrates. But up to now, there are only a few research reports that the HVPE growth of large-scale GaN can be sustained for more than 10 h, and the thickness is rarely more than 10 mm, which is mainly caused by the technical characteristics of HVPE method. The main difficulties of HVPE are as follows:

- *Space parasitic reaction.* The parasitic reaction outside the substrate in HVPE system under atmospheric pressure is more outstanding than other growth method. A large number of parasitic GaN particles will be deposited on the exit of GaCl tube, the inner wall of quartz glass tube, and the substrate surface. GaN deposited on the exit of GaCl tube will consume GaCl so as to reduce the growth rate and cause the damage of the GaCl tube; GaN particles deposited on the growth surface forming defects will reduce the quality of the material, while GaN deposited on the wall of quartz reactor can cause the breakage of quartz reactor.
- *Consumption of liquid gallium source.* With the consumption of metal Ga source, the growth rate of GaN will be affected. In the growth process for a long time, the consumption of the gallium source cannot accurately be known, and it is also difficult to add during the growth process. Although we can increase the capacity of Ga container, the weight of Ga is limited at high temperature in quartz glass container. And while the temperature changes between high- and low-temperature changes, the stress from the liquid-solid gallium metal and quartz increases the chances of the breakdown of quartz reactor.
- *Ammonium chloride powder.* Another technical problem in HVPE method is that the growth reaction under atmospheric pressure will produce large amounts

of ammonium chloride powder, aggregating in the downstream blocking the exit of quartz reactor even to explode, which may affect the growth. The long continuous growth is needed for the growth of GaN boule. Although most of HVPE equipment claims to solve the problem, so far, the continuous growth of HVPE system still is little more than 10 h.

- *Brittle quartz reactor*. The HVPE reactor is usually made of quartz because of the strong corrosive gases. The high price of high-quality quartz material, the precision processing, and the difficulties in installation and maintenance greatly increase the cost, which affects the mass preparation of GaN material.

In spite of the disadvantages, with the deepening of the research, most of the problems can be solved with in-depth research.

1.4.5 Epitaxial Lateral Overgrowth by HVPE

The heterogeneous epitaxial growth of GaN will certainly cause high dislocation density in the epitaxial layer, typically up to 10^{10} cm^{-2}. A variety of methods have been used to reduce the dislocation density, achieving preliminary results. Compared with other growth technologies such as MOCVD and MBE, HVPE has the potential advantage in preparing large-scale GaN film with low dislocation density. In HVPE-GaN film, several microns of terrace structure and hundreds microns of honeycomb cell structure can often be observed, which all shows that HVPE-GaN growth has the very high ratio of horizontal/vertical growth rate, more suitable to the ELO technology. Using the ELO technology, Japan Nichia greatly improved the life of blue LDs, more than 10,000 h of continuous running.

The ELO process is to prepare the patterned masks on HVPE- or MOCVD- GaN layer by applying conventional photolithographic processing or wet etching to 130-nm-thick SiO_2 layers deposited by plasma-enhanced chemical vapor deposition. As a result of surface energy choice, GaN is only grown in the window part, and the nucleation on the covered SiO_2 parts is difficult [48, 49]. A variety of mask designs were investigated, including nonparallel stripes arranged in a "wagon-wheel" pattern, arrays of circular apertures, parallel stripes oriented along the <0001> direction, parallel stripes oriented along the <1–100> direction, and parallel stripes oriented along higher index directions. When the thickness of epitaxial GaN in the window is more than that of masked SiO_2 layer, the vertical and lateral growth of GaN will happen at the same time. When the lateral growth reaches a certain degree, full cover GaN epitaxial layers can be obtained. Because the growth is quasi-free and perpendicular to the direction of the climbing dislocation from original GaN layer, GaN in the overgrowth regions ("wings") has a high quality. ELO regrowth was carried out under similar growth conditions to those used for the GaN films by HVPE. It was found that masks consisting of parallel stripes oriented along the <1–100> direction exhibited reproducibly vertical stripe sidewalls and

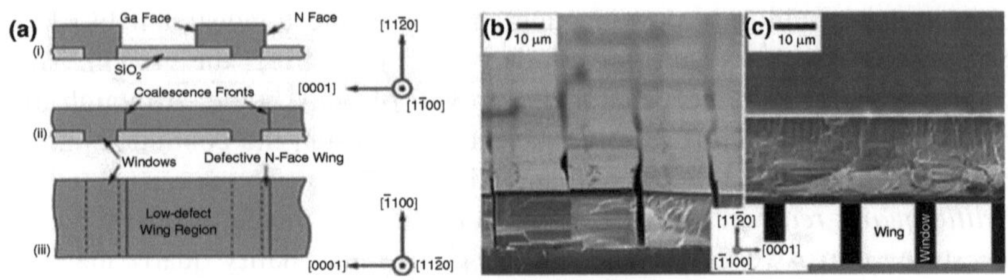

Fig. 1.17 (**a**) Schematics of HVPE LEO a-plane growth using mask stripes/openings oriented along the GaN <1–100> direction, (**b**) inclined cross-sectional SEM images of uncoalesced LEO stripes, (**c**) SEM cross-sectional image of coalesced a-plane LEO GaN film

Fig. 1.18 SEM images of GaN ELO growth in the hexagonal window and parallel window along the <11–20> direction

thus were most effective to lower the dislocations and stacking-fault densities in the overgrown regions ("wings"), as shown in Fig. 1.17 [50].

ELO is an effective, direct means of significantly reducing the propagation of such defects via blocking and redirection. In a manner analogous to c-plane ELO, the stripe direction, width, and period all play an important role in the morphology and defect reduction during overgrowth. Figure 1.18 is the morphology of ELOG GaN [51]. The experimental results show that the dislocation density of GaN on the SiO_2 mask is several orders of magnitude smaller than that of GaN on window part.

As above mentioned, the biggest characteristic of HVPE system is high growth rate, which is suitable for the growth of GaN thick film. But when the thickness of GaN on sapphire is more than 20 μm, the cracks will appear in GaN epitaxial layer. ELO technology can effectively alleviate the phenomenon, thus obtaining low-density, large-scale high-quality GaN thick film. In order to further improve the quality of ELO epitaxial layer, new methods have been tried. For example, two-step ELO growth of GaN is one of them, mainly considering that the lateral growth depends on the conditions, such as temperature, carrier gas flow rate, and V/III ratio. The ELO growth is first carried out at a relatively low V/III ratio in order to produce smooth vertical {11–20} interface and then at a higher V/III ratio to promote the lateral growth [52]. Another double-layer ELO technology is used to reduce the

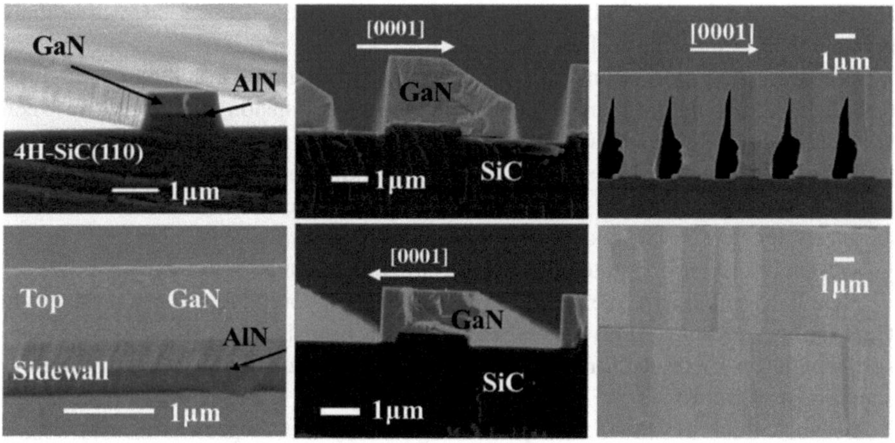

Fig. 1.19 SEM micrographs of GaN by pendeo-epitaxial regrowth

dislocations in the interface of lateral GaN on SiO_2 mask, by changing the direction of second ELO GaN after the first ELO.

In addition, the pendeo-epitaxy growth technology is also very attractive. It is a technique of lateral epitaxy without mask, which is usually on SiC or SiC/Si heterogeneous substrate, using AlN buffer layer as the mask [53]. After etching out part of AlN layer, GaN is not directly grown on SiC substrate but only selective epitaxial growth on the AlN layer. GaN grow up along the longitudinal and lateral directions at the same time, and with the increase of the thickness, the front of lateral epitaxial growth also laterally extends. Because GaN cannot be grown on SiC substrate between adjacent AlN films, it seems that the lateral spreading of GaN on AlN layer is impending, like GaN crystals hanging on both sides of AlN, so-called pendeo-epitaxy. Figure 1.19 shows SEM images of GaN by pendeo-epitaxy process. The characteristic of the pendeo-epitaxy technology avoids the effect of the mask on ELO GaN, especially the impurity and the interface stress, leading to a higher quality of GaN.

1.4.6 Freestanding HVPE-GaN Substrate

An alternative approach for GaN homo-substrates for a long time is the use of thick GaN layers grown by hydride vapor-phase epitaxy on sapphire. These thick layers are subsequently removed from the sapphire substrate by a suitable process and then can serve as quasi homo-substrates for device fabrication. Currently, hydride vapor-phase epitaxy (HVPE) provides the highest growth rate of GaN single crystals, reaching 1 mm/h or more. In addition, because the growth is usually performed at atmospheric pressure, large-scale HVPE reactors are available. For these reasons,

HVPE has considerable advantages for the mass production of large-area GaN crystals for freestanding wafers.

However, the process can suffer from serious problems of cracking of the thick GaN layer due to the large mismatch in the thermal expansion coefficients between GaN layer and the base substrate. Despite strenuous attempts to overcome this problem, difficulties still exist in reproducibly manufacturing GaN crystals that have a sufficiently large area for practical use. Reducing the density of dislocations is another important issue, because the HVPE method still involves hetero-epitaxial growth on a foreign substrate with a large lattice mismatch.

Here is to summarize the present state of research concerning such freestanding GaN homo-substrates, with special emphasis on laser-induced liftoff process and self-separation methods for the separation of the thick GaN layers from their sapphire substrates.

Because thick HVPE-GaN layers are generally hetero-grown on sapphire, a highly desirable process is required to enable a fast, reliable, and high-yield detachment. In practice, such a process will involve a sacrificial layer somewhere between the substrate and GaN layer, which is removed by a specific chemical or thermal treatment. One possibility would be to deposit a sacrificial layer which makes use of selective chemical etching, e.g., of AlN versus GaN in KOH. Another method is developed to cause GaN films to self-detach from its sapphire substrate during cooldown from growth temperature due to the accumulation of thermal stress and the lattice mismatch between GaN and sapphire. A third approach is the complete removal of the substrate by etching or polishing, which is quite time-consuming, especially in the case of sapphire or SiC substrates.

1.4.6.1 Laser Liftoff Process

Another useful process, much more flexible and significantly faster than the abovementioned methods, is the laser-induced delamination from a transparent substrate such as sapphire, or laser liftoff (LLO). In this process, the separation of a GaN layer from the substrate is achieved by irradiation of the substrate-film interface through the substrate with high-power excimer laser pulses at a wavelength which is transmitted by the substrate but is strongly absorbed in the GaN layer [54]. The absorption of such high-intensity laser pulses causes a rapid thermal decomposition of the irradiated GaN interfacial layer into metallic Ga and gaseous N_2 (Fig. 1.20) [55]. High-intensity laser pulses enter the sample via the sapphire substrate and thermally decompose a thin GaN layer at the substrate interface. The energy density of laser pulses is often in the range between 400 and 900 mJ/cm^2, and the threshold is approximately 250 mJ/cm^2. Below this critical absorbed pulse intensity threshold, no visible alteration of GaN layer can be seen. A hot plate can be used to raise the substrate temperature during the process, in order to relieve some of the accumulated thermal strain. The temperature of hot plate is usually above 730 °C but below 830 °C (decomposition temperature of GaN). At this temperature range, the strain and the bowing were visibly reduced and hardly noticeable. Ideally,

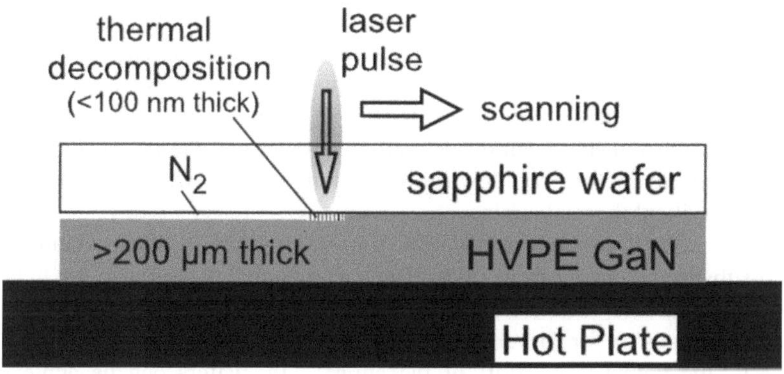

Fig. 1.20 Schematic view of the laser liftoff (LLO) process

Fig. 1.21 Photograph of a 275-μm-thick freestanding GaN film, after removal from 2 in. sapphire

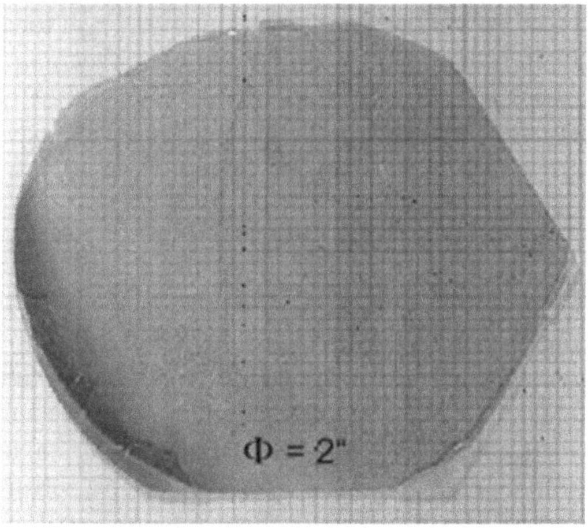

in situ laser liftoff in the deposition reactor close to growth temperature would be the best solution to avoid all bowing or cracking problems caused by the thermal expansion mismatch between GaN and the sapphire substrate.

A successful separation of thick HVPE-GaN layers from sapphire depends on several factors, including film thickness and thickness homogeneity across the wafer, the stresses, and stress gradients caused by the large thermal expansion coefficient mismatch. The differences of the thermal expansion coefficients result in substantial macroscopic wafer bowing, although most of the wafers exhibited very little or even no near-surface cracking across the whole wafer. But for the laser liftoff procedure, the bowing of thicker GaN films (>100 μm) often gave rise to extensive fracturing because of the strain inhomogeneity at the boundary of already released and thus relaxed areas and still attached portions of the wafer, as shown in Fig. 1.21 [55].

Such a LLO process has some intrinsic advantages compared with other possibilities abovementioned. First, sapphire as the commonly used substrates would be even possible to recycle the separated sapphire after LLO for further deposition runs. Second, the LLO process does not require any specific sacrificial layers in the growth sequence. Third, the LLO method is quite fast and can be scaled up easily. A 2 in. wafer in principle could be lifted off in a few seconds. Finally, the liftoff process could also in situ occur directly after deposition of the III-nitride film in the deposition reactor before cooldown to room temperature, which would avoid the build-up of critical thermal strain.

The laser liftoff process is not only for the production of freestanding GaN pseudo-substrates starting from thick HVPE-GaN epilayers but also the delamination of thin GaN device heterostructures for the purpose of wafer bonding onto foreign substrates or for flip-chip bonding in device technology. Using optimized processes, the defect-free liftoff of the entire 2 in. wafers can be achieved for GaN film thicknesses ranging from 3 to 300 μm. Separation of thin GaN layer from sapphire substrates opens up new possibilities for the formation of electrical contacts, for the extraction of photons, and for thermal management. The laser liftoff method has already become a major technique in III-nitride devices. Up to now, LLO separation process of thin GaN device heterostructures is widely applied for the vertical-structured GaN-based LED chip.

1.4.6.2 Self-Separation Methods

In addition to the LLO technology, some methods have been developed to realize the separation of GaN layer from sapphire substrate. The self-separation of all these methods is mostly based on the drastic change of thermal stress and the lattice mismatch between GaN layer and sapphire substrate.

In the early twenty-first century, an epitaxial procedure was developed to obtain self-separated GaN films from sapphire substrate during the cooling down to room temperature from the growth temperature [56]. In this example, thick GaN layers have been grown on particular sapphire substrates by hydride vapor-phase epitaxy (HVPE). Sapphire substrates have stripe-shaped GaN seeds (Fig. 1.22), which is formed by a rather complex pretreatment of the wafer including lithography. The separation of thick GaN layer took place during the growth process at the interface of GaN/sapphire, because of the lattice mismatch and the thermal stress between GaN and sapphire. The narrow stripe-shaped structure of the seeds makes the self-separation easy.

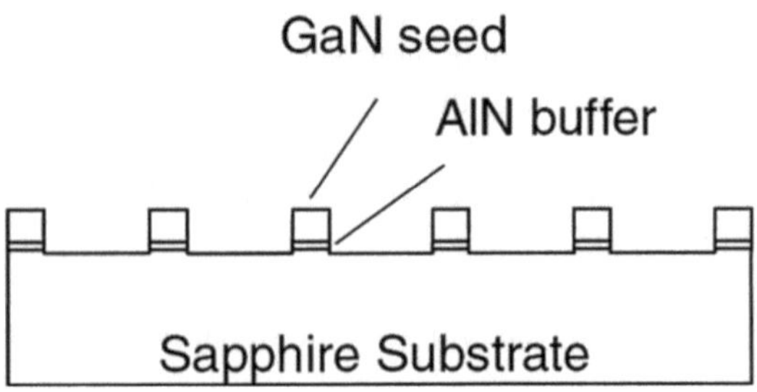

Fig. 1.22 Schematic diagram of substrate structure with GaN seeds

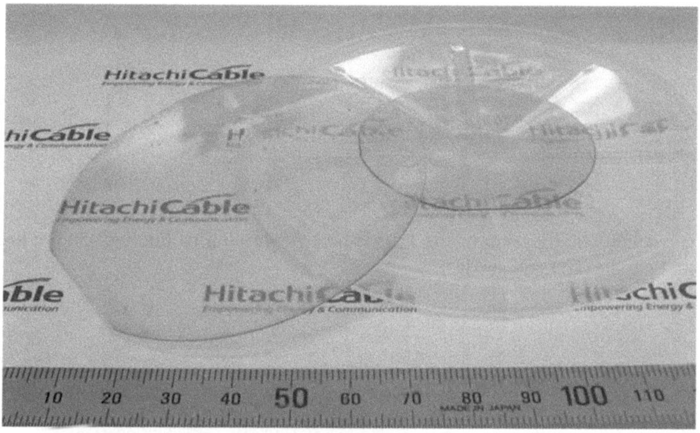

Fig. 1.23 Photograph of a 2 in. and a 3 in. VAS-GaN wafer

An original void-assisted separation (VAS) technology has been successfully established for the fabrication of large-area, high-quality GaN wafers with excellent reproducibility [57–59]. In this case, a fragile TiN layer that contains numerous small voids is formed at the interface between thick GaN layer and substrate. The thick GaN layer can be separated from the substrate by breakage of the fragile layer due to the thermal stress generated during the cooling process after the HVPE growth. The mechanical strength of the fragile layer can be controlled by changing the density of TiN voids. It is therefore possible to cause the separation by the application of a very small thermal stress. By using the HVPE-VAS method, high-quality GaN wafers with the diameter of over 3 in. have been prepared (Fig. 1.23).

Different from the abovementioned methods using a complex pretreatment including lithography, void-assisted separation technology enables highly reproducible large-area separation without producing any cracks. After that, a series of self-separation methods have been developed, by using a sacrificial layer (such as graphene, BN, nitride nanostructured interlayer, etc.) to form a weak interface layer close to substrate like the TiN voids.

All these self-separation approaches are not only to obtain freestanding GaN layer but also to reduce the threading dislocation density of GaN wafer to the order of 10^6 cm^{-2} due to the approximate ELO process.

1.4.7 Current Development Trend in HVPE-GaN Substrate Materials

Hydride vapor-phase epitaxy for the GaN growth is the most important purpose for preparing high-quality GaN substrates, because the high growth rate of HVPE provides the key technology of GaN quasi-substrate. By adopting the method of

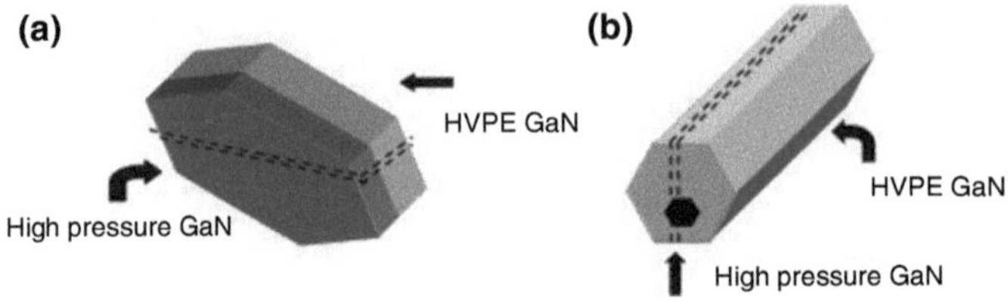

Fig. 1.24 The schematic diagram of the HVPE post-processing technology on the high-pressure platelet (**a**) and needle (**b**) GaN seeds

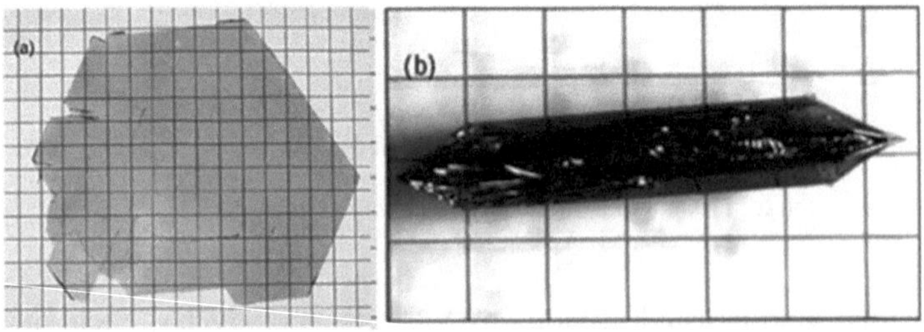

Fig. 1.25 The platelet (**a**) and needle (**b**) GaN single crystal by high-pressure solution method

atmospheric HVPE, a few companies and research institutes have prepared 2–6 in. GaN substrate materials (including GaN substrates and templates), but the price of freestanding GaN substrate is expensive, from $1000 to $5000 per piece of 2 in., which is mainly due to the technology of not fully mature and high cost, in the urgent need of improvement. In recent years, based on the combination of various substrate growth technologies, a variety of new composite technology for the growth of GaN substrates has been developed.

1.4.7.1 Combined GaN Crystal Growth

The combined GaN crystal growth refers to prepare a large-scale high-quality GaN crystal by the combination of different technologies, such as HVPE, HPNSG, ammonothermal growth, Na-flux method, etc.

The TopGaN Inc. claimed that they would expand the size of GaN regrown by HVPE on the platelet and needle GaN single crystal by high-pressure solution method, as shown in the schematic diagram in Fig. 1.24 [2], even though the dislocation density of GaN increased from almost zero to around 1×10^6 cm^{-2}. Two morphological GaN crystals grown under high N_2 pressure, hexagonal platelets and hexagonal needles are shown in Fig. 1.25.

The size of GaN crystal expanded by HVPE process is typically up to 100 mm, after getting rid of the original GaN seed crystal, as shown in Fig. 1.26. If the HVPE growth rate is lowered to 20 μm/h, 100 mm GaN single crystal almost without dislocations on n-type high-pressure GaN seed can be obtained.

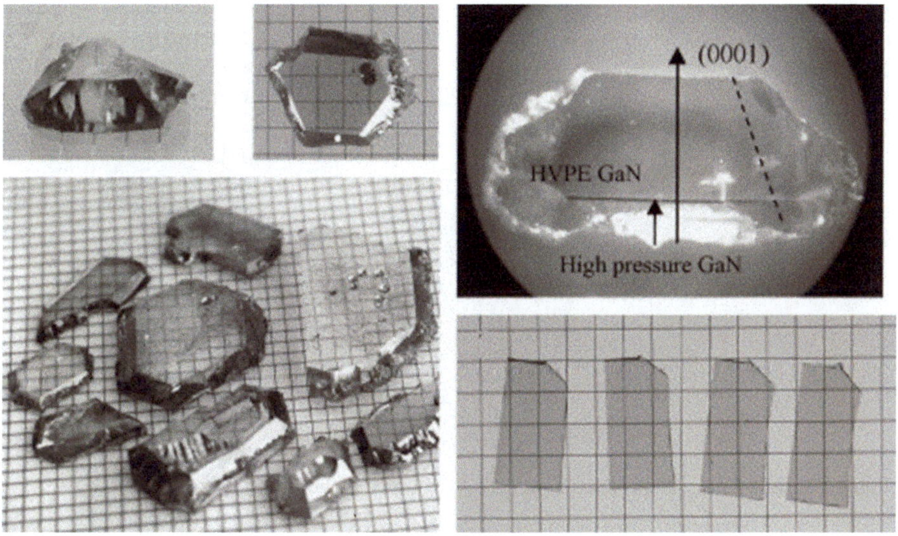

Fig. 1.26 The large-scale GaN crystal by the HVPE post-processing technology on the high-pressure platelet GaN seeds

Fig. 1.27 The large-scale GaN crystal by the HVPE post-processing technology on the high-pressure needle GaN seeds

Although the crystallographic plane of GaN needle seed by HPNSG is uncertain, transparent, or pale yellow, GaN crystal with stable crystal shape will be obtained by a series of the same process of HVPE, as shown in Fig. 1.27. Structure analysis showed that crystal quality of epitaxial GaN is very high. After the optimization growth by HVPE on almost dislocation-free high-pressure GaN seed, GaN crystal quality is not much degradation, even lower than 1×10^4 cm^{-2}. Of course, single HVPE process is preferable to expand the size of GaN, but that requires a new HVPE system to grow more than 100 h. This is the only obstacle for the HVPE post-process technology on the high-pressure GaN seeds. But now, only a few of HVPE equipment can continue to grow more than 10 h, which is caused by the characteristics of HVPE growth.

Fig. 1.28 2 in. bulk GaN crystal by Na-flux method on HVPE-GaN seeds

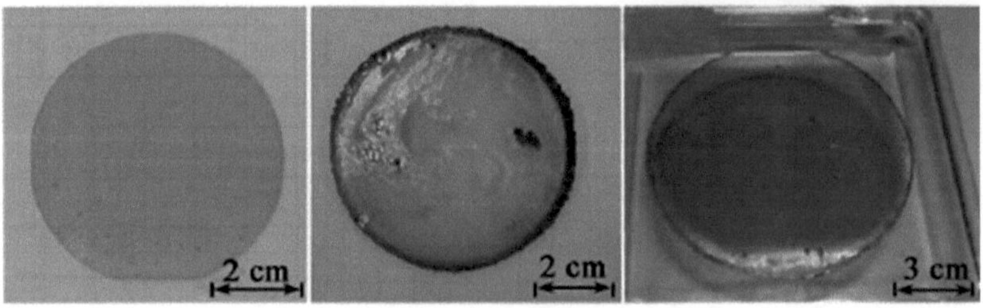

Fig. 1.29 Bulk GaN wafers of different scale by Na-flux method

By the HVPE post-processing technology on the high-pressure needle GaN seeds, a nonpolar GaN substrate can even be got. Nonpolar GaN substrate with high electrical conductivity has a special attraction in preparation of laser diode.

Different from the abovementioned growth technology of HVPE post-process on HPNSG GaN, some researchers studied the reverse growth of homogeneous GaN by high-pressure liquid-phase epitaxy technology on HVPE-GaN crystal. Osaka University finally got the GaN crystal with the thickness of 1.1 mm and diameter of 50 mm by molten Na-flux growth (atmospheric pressure, growth temperature from 40 to 860 °C), using HVPE-GaN chip with diameter of 45 mm as the seed crystal [13, 15], as shown in Fig. 1.28. Dislocation density is reduced from 1×10^8 cm^2 of seed crystal to 2.3×10^5 cm^2. Uniform GaN crystal of more than 2 in. with free of cracks has been already successful grown on the seed crystal, and the thickness is up to more than 750 μm, as shown in Fig. 1.29.

HVPE-GaN wafers can also be used as the seeds for ammonothermal growth of GaN. Poland Ammono Inc. [60, 61] has at present already succeeded in commercial mass production of polar, nonpolar, and semipolar GaN wafers with diameters of 1–2 in. and dislocation densities as low as 5×10^3 cm^{-2} (Fig. 1.30). But GaN single crystal of more than 2 in. in diameter is still in the development stage because of the lack of large-scale GaN seed. A large number of seeds can be placed in the ammonothermal reaction cavity at the same time, which is suitable for low-cost mass production of high-quality GaN.

1.4.7.2 GaN Boule by HVPE Technology

As described earlier, although ammonothermal growth and Na-flux growth for GaN single crystal seem to be a more productive potential, the small size by these

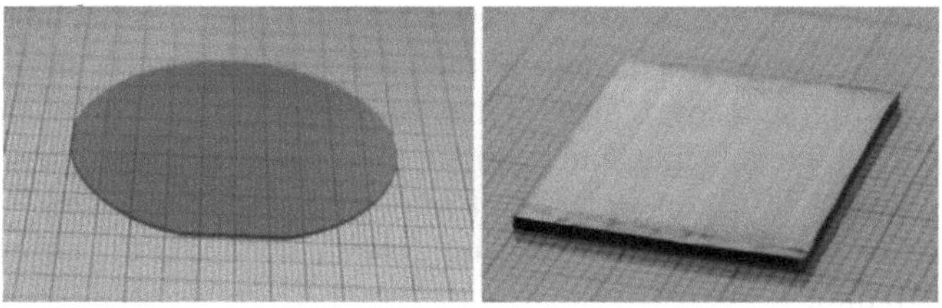

Fig. 1.30 A large dimension GaN single crystal by ammonothermal growth on HVPE-GaN seeds

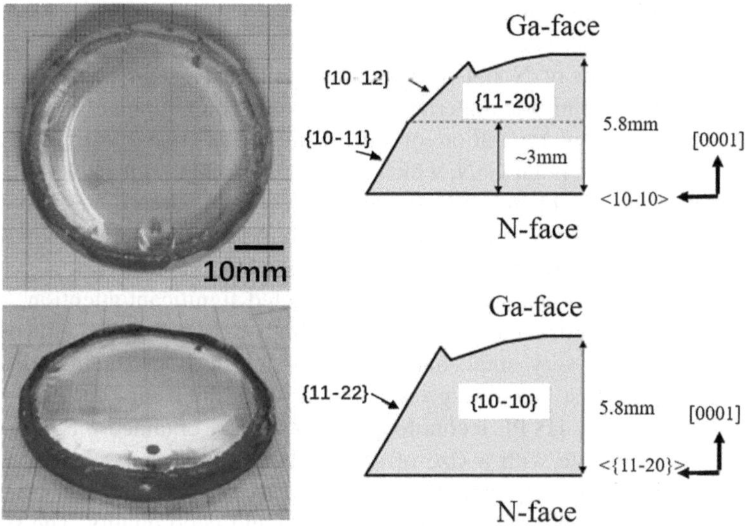

Fig. 1.31 Photograph of GaN boules with thickness of 5.8 mm by HVPE

two methods is still not competitive to the large-scale commercial production. Therefore, atmospheric hydride vapor-phase epitaxy is still a major means to obtain high-quality GaN with a large size. Wafers derived from boules produced by HVPE method will become the choice of substrates for GaN-based lasers and high-performance GaN-on-GaN light-emitting diodes.

Over the past few years, some companies, such as Sumitomo, TDI, Lumilog and Kyma Technologies, have increased the investment in this aspect and have been able to provide 2 in. GaN substrates of the dislocation density around 1×10^6 cm^2. The thickness of 2 in. GaN substrate is up to several millimeters, and a GaN boule grown via HVPE which is essentially the state of the art is shown in Fig. 1.31 [62]. It is said that the latest vertical HVPE system designed by AXITRON can be used for the growth of GaN boule with a length of 7 cm, and they also develop the planetary vertical HVPE system for the growth of multi-wafer GaN substrates at the same time.

If the size of the boules and wafers derived from this technique can be substantially increased in the manner of the evolution of the growth of ever-larger quartz crystals, sufficient cost reduction may be realized to meet the increasing need for chemically similar substrates that contain a very low density of dislocations. At present, the number of useful wafers that can be commonly gleaned from a single boule is in the range of 2–6, as there are multiple challenges in the growth of boules thicker than 10 mm.

1.4.7.3 Nonpolar GaN Substrate

Nakamura, the winner of Nobel physics prize, thinks that there are two ways to improve the quality of nitride material, a larger boule and nonpolar GaN. Nonpolar GaN will bring a new revolution of nitride materials. GaN-based devices are generally grown on the polar GaN, which leads to the strong built-in electric field in the active layer, the tilt of band gap, electron and lower luminous efficiency, etc. The power, efficiency, and life of the devices on the nonpolar substrate will significantly be improved, due to the reducing of the polarization effect and stress.

Nonpolar and semi polar substrates have attracted significant attention, but now the performance is restricted because of the high defect density. However, the substrate size is still very small and unsuitable for mass production. In China, Nanjing University first began to grow nonpolar GaN thick film on lithium meta-aluminate substrate by HVPE technology. In 2007, they obtained the freestanding nonpolar GaN substrate with a size of almost 2 in. by the self-separation due to thermal mismatch between nonpolar GaN and $LiAlO_2$ (Fig. 1.32). The sample, named as m-GaN, has a good (10–10) orientation and good quality. The stacking-fault density is about 10^4 cm^{-1}, while the thickness is only 90 μm. The full width at half maximum of m-GaN (110) by XRD analysis is about 773 arcs.

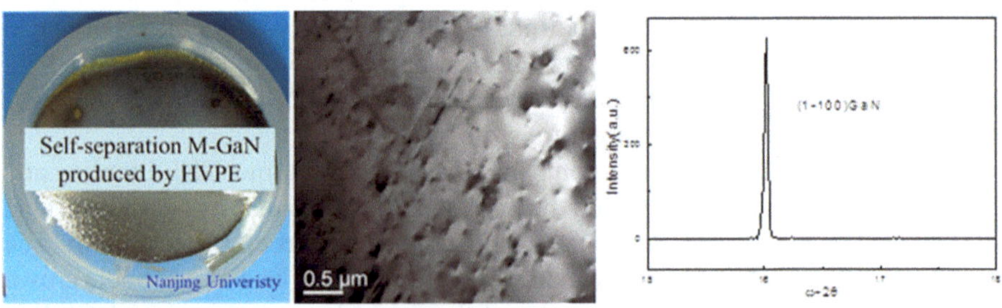

Fig. 1.32 Freestanding m-GaN substrate by HVPE growth (source: Nanjing University in 2007)

1.4.7.4 Low-Cost HVPE-GaN Templates on Sapphire

D. Martin et al. developed a low-cost process for the growth of GaN template directly only grown on sapphire substrate by HVPE method [63]. By this method, the thickness of high-quality GaN template is only 8 μm, far less than that of other high-quality GaN/sapphire template, 20–300 μm. Due to reducing the use of raw materials and simplified growth steps, this new process will greatly cut down the cost of the laser diodes and the transistor on GaN template.

Here, a two-step process similar to that of metal organic vapor-phase epitaxy was used to grow GaN template by Aixtron HVPE equipment. Some research shows that the quality and polarity of GaN are highly affected by the surface treatment of sapphire. First, after the cleaning in H_2 gas, sapphire substrate is exposed to 1050 °C of ammonia for the nitriding process about 4 min. Then the temperature is dropped to 600 °C to grow AlN layer on the substrate, and the temperature is increased to 1045 °C to continue the GaN growth. If the substrate is cooled to room temperature after the nitriding and exposed to the air, GaN layer is Ga-polar, and after the process again from cleaning to high-temperature growth, it will produce mixed-polar or N-polar GaN film. N-polar GaN film has more surface quality problems, and the doping is more difficult to control due to its higher impurity. Instead, Ga-polar GaN has a smooth surface, directly used for the subsequent growth without surface treatment. The surface roughness of GaN grown by this new process is 0.54 nm in the range of 5 × 5 μm by atomic force microscopy (AFM), and there is no crack in GaN template. The dislocation density after the acid etching process is only 1×10^8 cm^{-2}.

And the LED structure has also been grown on 8 μm HVPE-GaN template by MOCVD, whose properties are almost equal to that of the LED structure on MOCVD-GaN template. Some manufacturers of GaN template have interest in this growth technology.

1.5 Summary

The development of freestanding GaN substrate for homogeneous epitaxy is very important to high-power and high brightness LED, LDs and high-power microwave devices, etc. High-quality substrate materials provide excellent lattice-match and thermal-transportation characteristic for high-performance devices of high mass production. Through the use of the native substrate, nitride developers and device manufacturers will benefit its profit from epitaxial growth for the device. In this paper, we review the several methods and the latest progress in the area of the III-nitride substrates. Three main techniques (HVPE, ammonothermal, and Na-flux growth) are currently employed for the development and manufacturing of bulk GaN crystals for substrates. We have a special discussion on the technological aspects of HVPE for the production of GaN substrate materials, because HVPE is mostly suitable for obtaining the large-scale and low-cost GaN substrates at present.

In the near future, while exploring new methods for creating larger, high-quality crystals with the desired structural, optical, and electrical properties at an even lower cost, the size of nitride substrate materials will be continuously increased, and the defect density will reduce, reaching a level of device applications.

References

1. D. Hanser, M. Tutor, E. Preble, M. Williams, X. Xu, D. Tsvetkov, L. Liu, J. Cryst. Growth **305**, 372–376 (2007)
2. I. Grzegory, B. Ucznik, M. Bokowski, et al., J. Cryst. Growth **300**(1), 17–25 (2007)
3. J. Karpiski, S. Porowski, S. Miotkowska, J. Cryst. Growth **56**(1), 77–82 (1982)
4. M. Bokowski, M. Wroblewski, B. Lucznik, et al., Mater. Sci. Semicond. Process. **4**(6), 543–548 (2001)
5. D. Ehrentraut, R.T. Pakalapati, D.S. Kamber, W. Jiang, D.W. Pocius, B.C. Downey, M. McLaurin, M.P. D'Evelyn, Jpn. J. Appl. Phys. **52**, 08JA01 (2013)
6. W. Jiang, D. Ehrentraut, B.C. Downey, D.S. Kamber, R.T. Pakalapati, H.D. Yoo, M.P. D'Evelyn, J. Cryst. Growth **403**, 18 (2014)
7. H. Yamane, M. Shimada, S.J. Clarke, et al., Chem. Mater. **9**(2), 413–416 (1997)
8. T. Yamada, H. Yamane, Y. Yao, et al., Mater. Res. Bull. **44**(3), 594–599 (2009)
9. T. Yamada, H. Yamane, H. Iwata, et al., J. Cryst. Growth **286**(2), 494–497 (2006)
10. M. Morishita, F. Kawamura, M. Kawahara, et al., J. Cryst. Growth **284**(1), 91–99 (2005)
11. F. Kawamura, H. Umeda, M. Kawahara, M. Yoshimura, Y. Mori, T. Sasaki, H. Okado, K. Arakawa, H. Mori, Jpn. J. Appl. Phys. **45**, 2528–2530 (2006)
12. F. Kawamura, H. Umeda, M. Morishita, et al., Jpn. J. Appl. Phys. **2**(45), L1136–L1138 (2006)
13. Y. Mori, M. Imade, K. Murakami, et al., J. Cryst. Growth **350**(1), 72–74 (2012)
14. F. Kawamura, M. Morishita, M. Tanpo, et al., J. Cryst. Growth **310**(17), 3946–3949 (2008)
15. Y. Mori, M. Imade, M. Maruyama, et al., ECS J. Solid State Sci. Technol. **2**(8), N3068–N3071 (2013)
16. M. Imade, K. Murakami, D. Matsuo, et al., Cryst. Growth Des. **12**(7), 3799–3805 (2012)
17. M. Imanishi, K. Murakami, H. Imabayashi, et al., Phys. Status Solidi C **2012**, 1–5 (2012)
18. E. Maissner, B. Birkmann, S. Hussy, G. Sun, J. Friedrich, G. Mueller, Phys. Status Solidi **2**, 2040–2043 (2005)
19. B.N. Feigelson, R.M. Frazier, M. Gowda, J.A. Freitas, M. Fatemi, M.A. Mastro, J.G. Tischer, J. Cryst. Growth **310**, 3934–3940 (2008)
20. S. Fischer, C. Wetzel, W.L. Hansen, E.D. Bourret-Courchesne, B.K. Meyer, E.E. Haller, Appl. Phys. Lett. **69**, 2716 (1996)
21. H.P. Maruska, J.J. Tietjen, Appl. Phys. Lett. **15**, 327 (1969)
22. T.L. Chu, K. Ito, R.K. Smeltzer, S.S.C. Chu, J. Electrochem. Soc. **121**(1), 159–162 (1974)
23. M.J. Ilegems, J. Crystal Growth **13/14**, 360–364 (1972)
24. R.K. Crouch, W.J. Debnam, A.L. Fripp, J. Mater. Sci. **13**, 2358–2364 (1978)
25. T. Detchprohm, K. Hiramatsu, H. Amano, I. Akasaki, Appl. Phys. Lett. **61**, 2688 (1992)
26. A. Usui, H. Sunakawa, A. Sakai, A.A. Yamaguchi, Jpn. J. Appl. Phys. **36**, L899 (1997)
27. J.J. Nickl, W. Just, R. Bertinger, Mater. Res. Bull. **9**, 1413–1420 (1974)
28. H. Lee, M. Yuri, T. Ueda, J.S. Harris, Mater. Res. Soc. Symp. Proc. **423**, 233–238 (1996)
29. H. Tsuchiya, M. Akamatsu, M. Ishida, F. Hasegawa, Jpn. J. Appl. Phys. **36**, L748–L750 (1996)
30. Y. Morimoto, K. Uchiho, S. Ushio, J. Electrochem. Soc. **120**, 1783–1785 (1973)
31. P.J. Born, D.S. Robertson, J. Mater. Sci. **15**, 3003–2009 (1980)
32. V.S. Ban, J. Electrochem. Soc. **119**, 761 (1972)
33. D.K. Wickenden, K.R. Faulkner, R.W. Brander, B.J. Isherwood, J. Cryst. Growth **9**, 158–164 (1971)

34. G. Nataf, B. Beaumont, A. Bouille, S. Haffouz, M. Vaille, P. Gibart, J. Cryst. Growth **192**, 73 (1998)
35. W. Zhang, T. Riemann, H.R. Alves, M. Heuken, D. Meister, W. Kriegseis, D.M. Hofmann, J. Christen, A. Krost, B.K. Meyer, J. Cryst. Growth **234**, 616 (2002)
36. R.J. Monlnar, K.B. Nichols, P. Maki, E.R. Brown, I. Melngailis, Mater. Res. Soc. Symp. Proc. **378**, 479–484 (1995)
37. J.J. Naniwae, S. Itoh, H. Amano, K. Itoh, K. Hiramatsu, I. Akasaki, J. Cryst. Growth **99**, 381–384 (1990)
38. R.J. Monlnar, P. Maki, R. Aggarwal, Z.L. Liau, E.R. Brown, I. Melngailis, W. Götz, L.T. Romano, N.M. Johnson, Mater. Res. Soc. Symp. Proc. **423**, 221–226 (1996)
39. M. Sano, M. Aoki, Jpn. J. Appl. Phys. **15**, 1943–1950 (1976)
40. A. Nikolaev, Y. Melnik, N. Kuznetsov, A. Strelchuk, A. Kovarsky, K. Vassilevski, V. Dmitriev, Mater. Res. Soc. Symp. Proc. **482**, 251–256 (1998)
41. G. Jacob, M. Boulou, M. Furtado, J. Cryst. Growth **42**, 136–143 (1977)
42. H. Tsuchiya, K. Sunaba, S. Yonemura, T. Suemasu, F. Hasegawa, Jpn. J. Appl. Phys. **36**, L1–L3 (1997)
43. A. Yamaguchi, T. Manak, A. Sakai, H. Sunakawa, A. Kimura, M. Nido, A. Usui, Jpn. J. Appl. Phys. **35**, L873–L875 (1996)
44. H. Tsuchiya, T. Okahisa, F. Hasegawa, H. Okumura, S. Yoshida, Jpn. J. Appl. Phys. **33**, 1747–1752 (1994)
45. W.M. Yim, E.J. Stofko, P.J. Zanzucchi, J.I. Pankove, M. Ettenberg, S.L. Gibert, J. Appl. Phys. **44**, 292–296 (1973)
46. J. Hagen, R.D. Metcalfe, D. Wickenden, W. Clark, Solid State Phys. **11**, L143–L146 (1978)
47. B. Baranov, L. Däweritz, V.B. Gutan, G. Jungk, H. Neumann, H. Raidt, Phys. Status Solidi **49**, 629–636 (1978)
48. T.S. Zheleva, O.H. Nam, M.D. Bremser, R.F. Davis, Appl. Phys. **71**, 2472 (1997)
49. A. Sakai, H. Sunakawa, A. Usui, Appl. Phys. Lett. **71**, 2259 (1997)
50. M.D. Craven, S.H. Lim, et al., Appl. Phys. Lett. **81**(7), 1201 (2002)
51. B. Beaumont, P. Gibart, M. Vaille, S. Haouz, G. Nataf, A. Bouille, J. Cryst. Growth **189/190**, 97 (1998)
52. H. Marchand, J.P. Ibbetson, P.T. Fini, X.H. Wu, S. Keller, S.P. Denbaars, J.S. Speck, U.K. Mishra, MRS Int. J. Nitride Semicond. Res. **4S1**, G4.5 (1999)
53. B.P. Wagner, Z.J. Reitmeier, J.S. Park, D. Bachelor, D.N. Zakharov, Z. Liliental Weber, R.F. Davis, J. Cryst. Growth **290**(2), 504–512 (2006)
54. C.R. Miskys, M.K. Kelly, O. Ambacher, M. Stutzmann, Phys. Status Solidi **6**, 1627–1650 (2003)
55. M.K. Kelly, R.P. Vaudo, V.M. Phanse, L.G. gens, O. Ambacher, M. Stutzmann, Jpn. J. Appl. Phys. **38L**, 217 (1999)
56. K. Tomita, T. Kachi, S. Nagai, A. Kojima, S. Yamasaki, M. Koike, Phys. Status Solidi **194**, 563 (2002)
57. Y. Oshima, T. Eri, M. Shibata, H. Sunakawa, A. Usui, Phys. Stat. Sol. A **194**(2), 554 (2002)
58. A. Usui, T. Ichihashi, K. Kobayashi, H. Sunakawa, Y. Oshima, T. Eri, M. Shibata, Phys. Stat. Sol. A **194**(2), 572 (2002)
59. Y. Oshima, T. Eri, M. Shibata, H. Sunakawa, K. Kobayashi, T. Ichihashi, A. Usui, Jpn. J. Appl. Phys. **42**, L1 (2003)
60. R. Dwilinski, R. Doradzinski, J. Garczynski, L.P. Sierzputowski, A. Puchalski, Y. Kanbara, K. Yagi, H. Minakuchi, H. Hayashi, J. Cryst. Growth **310**, 3911 (2008)
61. R. Dwilinski, R. Doradzinski, J. Garczynski, L. Sierzputowski, R. Kucharski, M. Zajac, M. Rudzinski, R. Kudrawiec, J. Serafinczuk, W. Strupinski, J. Cryst. Growth **312**, 2499 (2010)
62. K. Fujito, S. Kubo, H. Nagaoka, T. Mochizuki, H. Namita, S. Nagao, J. Cryst. Growth **311**, 3011 (2009)
63. D. Martin, J. Napierala, M. Ilegems, R. Butté, N. Grandjean, Appl. Phys. Lett. **88**, 241914 (2006)

Chapter 2
SiC Single Crystal Growth and Substrate Processing

Xiangang Xu, Xiaobo Hu, and Xiufang Chen

2.1 Introduction for SiC Single Crystal Materials

Silicon carbide (SiC) is a IV-IV compound material. In the basic structure unit of SiC, the silicon atoms and carbon atoms are combined together by sp3 covalent bonding, where each carbon atom is surrounded by four silicon atoms and vice versa. The extremely strong Si-C bonds give it unique physical, chemical, and mechanical properties [1], such as high electric breakdown field, high thermal conductivity, and so on. In addition, SiC is well known for its polytypism. The polytypic nature of SiC is that the crystal structure of SiC owns different ordering sequences in one dimensional with the same stoichiometry. And up to now, more than 250 SiC polytypes have been identified [2]. Meanwhile, different SiC polytypes exhibit distinct properties. Currently, the main polytypes of SiC used for device fabrication are 4H-SiC and 6H-SiC. 4H-SiC is an ideal material for power electronic and high-frequency devices, while 6H-SiC is usually employed for optoelectronic device fabrication.

The study about SiC has a quite long history. SiC was first discovered in the process of synthesizing diamond by Berzelius in 1824 [3]. Later, in 1885, Acheson invented a method for SiC preparation in melting furnace. The synthesis process was named as "Acheson process" afterward [4]. In this process, silica and coke are used as the source materials, while salt is employed as additive. The main product of Acheson process is SiC powder, which is mainly used as an abrasive in the processing technologies such as cutting, grinding, and polishing. The by-products in this process are mainly small 6H-SiC plates. In 1905, French scientist Moissan discovered the natural SiC in meteorites, which inspired the intensive interests in

X. Xu (✉) · X. Hu · X. Chen
State Key Laboratory of Crystal Materials, Shandong University, Jinan, China
e-mail: xbhu@sdu.edu.cn

J. Li, G. Q. Zhang (eds.), *Light-Emitting Diodes*, Solid State Lighting Technology and Application Series 4, https://doi.org/10.1007/978-3-319-99211-2_2

the potential application of SiC [5]. After that, a series of simple SiC-based devices were fabricated. In 1907, the first SiC light-emitting diode (LED) was made by an engineer named Round [6]. In 1920, SiC-based sensor was applied in the radio receivers. However, at that time, there was no efficient method to synthesize SiC single crystals with high quality. As a result, the development of SiC growth once got stranded. Up to now, many methods have been invented for SiC single crystal growth, including vapor growth and solution growth.

2.1.1 Vapor Growth Method

In 1955, Jan Antony Lely proposed a sublimation technique to prepare SiC single crystals, which opened up a new era of SiC material and devices [7]. This method was name as Lely method later. The schematic diagram of Lely method is shown in Fig. 2.1. The industrial-grade SiC powder was used as source material and placed around the dense graphite crucible wall. The porous graphite was employed to separate the source material and the growth cavity. In the crucible, the temperature around the crucible wall was higher than in the center of the crucible. As a result, the growth was driven by the temperature gradient in the crucible. At first, the source material sublimated at high temperature (about 2500 °C); thus gas species was produced. After that, the vapors penetrated into the porous graphite to grow lamellar crystals on the inner surface. During the growth, argon was used as inert gas, and the pressure in the growth chamber was kept around 1 atm. The gas species randomly nucleated at the inner cavity wall. The defects and micropipe density in

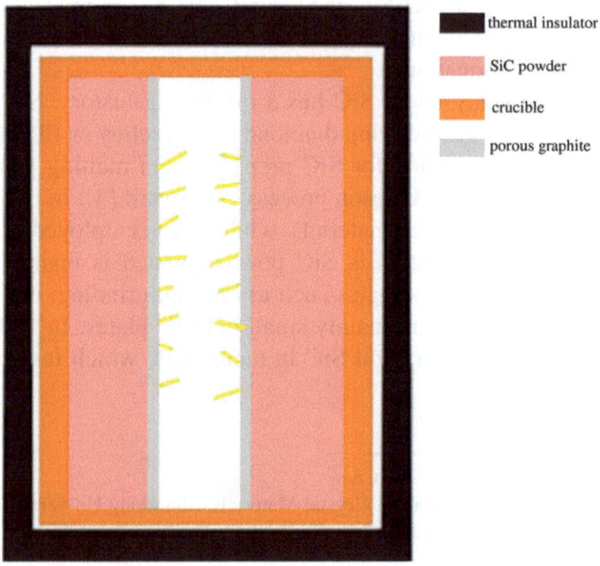

Fig. 2.1 Schematic diagram of Lely method for SiC crystal growth (Provided by Dr. Yan Peng at Shandong University)

the grown crystal are pretty low, which means this method can prepare 6H-SiC single crystals with high quality. However, crystals obtained by this method are in small size and irregular shape. Therefore, the crystals cannot be used for device fabrication. At the same time, due to the random nucleation at the inner cavity wall, 4H and 15R-SiC polytype inclusions often appeared in the grown crystals.

In order to solve the problem of polycrystalline nucleation, Tairov and Tsvetkov modified Lely method by an innovative use of SiC seed in 1978 [8]. The modified Lely method (also called "seed sublimation method" or "physical vapor transport method") was constituted of three steps: (1) sublimation of SiC source material, (2) mass transport of sublimed gas species, and (3) surface reaction and crystallization. The schematic diagram of growth furnace is shown in Fig. 2.2. In this method, SiC powder is put at the bottom of the crucible, and the seed is kept on the lid of the crucible. The distance of source-to-seed is in the range of 30–50 mm. Electromagnetic induction is employed to heat up the crucible in the temperature range of 2100–2400 °C. The graphite felt or porous graphite is used as thermal insulation. During the growth process, high purity argon is flowed as carrier gas, and the pressure in the growth chamber is kept at 20–60 mbar. The SiC powder was decomposed at high temperature, generating Si, Si2C, and SiC2 gas species. The gas species driven by the temperature gradient are transported to the seedand then

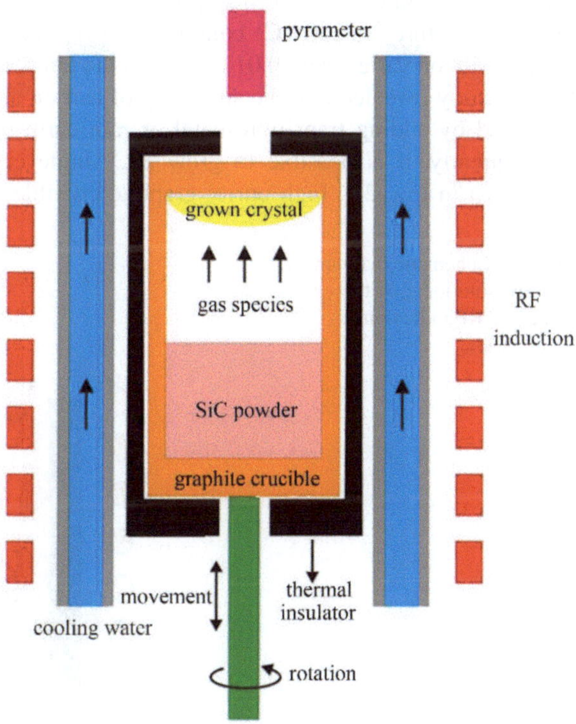

Fig. 2.2 Schematic diagram of a modified Lely method for SiC crystal growth (Provided by Dr. Xianglong Yang at Shandong University)

crystallize on the seed. During the growth, the main reactions were summarized as follows:

$$\begin{aligned} &Si_2C\,(g) + SiC_2\,(g) \leftrightarrow 3SiC\,(s) \\ &SiC_2\,(g) + 3Si\,(g) \leftrightarrow 2Si_2C\,(g) \\ &Si_2C\,(g) \leftrightarrow 2Si\,(g) + C\,(s) \end{aligned} \tag{2.1}$$

In this method, the crystal was grown approximately along isotherm lines. As a result, it is crucial to design the thermal field. Though micropipes and dislocations are commonly observed in the crystals grown by this method, the modified Lely method is currently the most mature method for growing bulk SiC single crystals. Using this method, 6H-SiC and 4H-SiC have been commercialized since 1991 and 1994, respectively. Recently, in Cree Research Inc., 4H-SiC single crystals with a diameter of 200 mm were successfully grown [9].

2.1.2 Solution Growth Method

Besides vapor growth, solution is another SiC single crystal growth method. Different from other semiconductor materials (such as Si and GaAs), SiC does not have any liquid phase in normal engineering conditions. According to theoretical analysis, the liquid phase of SiC appears only under high-pressure (>105 bar) and high-temperature (>3200 °C) conditions [10]. The solubility of carbon atoms in melting silicon varies from 0.01 to 19% in the temperature range of 1412–2830 °C. Recent study revealed that the solubility of carbon atoms in melting silicon can be increased by adding transition metal or rare earth metals to silicon solvent [11]. Consequently, it is possible to grow SiC single crystals by solution technique. As shown in Fig. 2.3, high purity graphite crucible is used as both container and

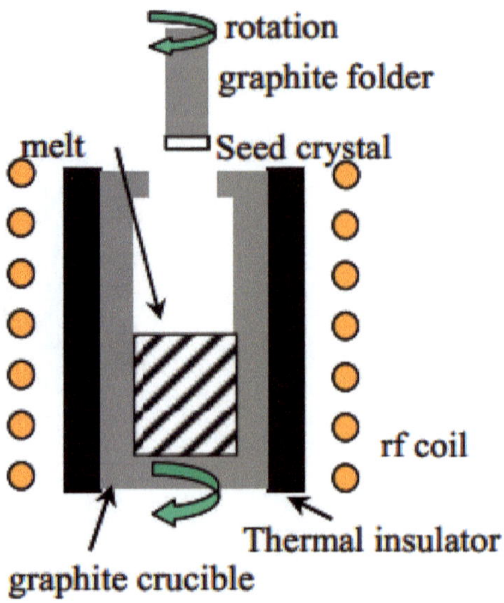

Fig. 2.3 Schematic diagram of a SiC solution growth heated by RF induction

carbon source. High purity Si and metals (such as Ti, Cr) are mixed and put in the crucible at argon atmosphere. SiC seed is mounted on a graphite folder. The typical growth temperature is in the range of 1600–1750 °C. During the growth, the seed is dipped into the ternary solution gradually, while the seed and crucible are rotated in opposite direction. Given that the crystals are grown under near thermal equilibrium condition, there are little micropipes and dislocations in SiC crystals grown by solution method. Therefore, high-quality SiC crystals can be grown by solution method.

2.1.3 High-Temperature Chemical Vapor Deposition (HTCVD) Method

HTCVD is the third method for bulk SiC single crystal growth. High growth rates and long SiC single crystals are expected by this method in comparison with PVT and solution methods. Figure 2.4 is a schematic illustration of a HTCVD SiC growth method [12]. A graphite heater is heated by induction coils and a cylindrical insulator is on the outside of the heater. A seed crystal was fixed on the seed holder. The Si and carbon sources are SiH4 and C3H8, respectively. H2 is used as carrier gas. At high temperature, the source gases were cracked and transformed into the reaction species such as Si, SiC2, Si2C, and C2H2. Then the species were deposited on the seed crystal. HTCVD method has the following advantages: one is the continuous gas source supply. This enables long-term growth which contributes to a long SiC crystal ingot; the other is the accurately controllability of the partial

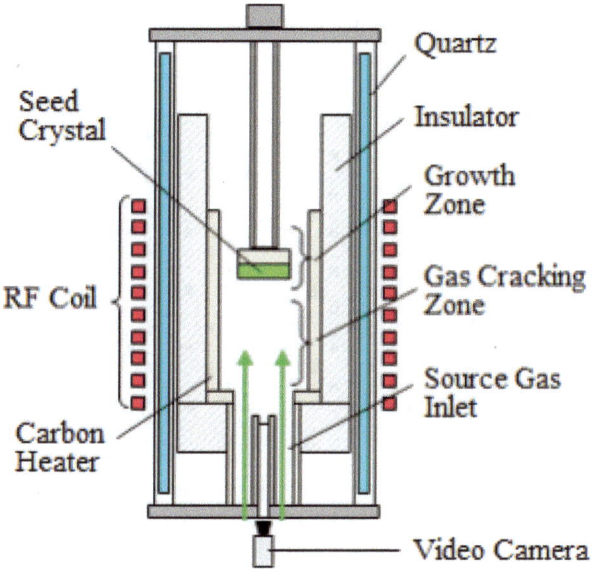

Fig. 2.4 Schematic illustration of the structure of a HTCVD SiC growth method

pressure, the gas flow, and the species ratio of the source gases which make the possibility to grow SiC crystal with Si/C stoichiometry; in addition, high purity semi-insulating SiC crystal can be grown by using high purity gas source.

In summary, SiC bulk single crystals can be grown by PVT, solution, and HTCVD. PVT is a conventional SiC crystal growth method and suitable for mass production. In comparison, the solution and HTCVD are suitable for the high structural quality SiC and the high purity semi-insulating SiC crystal growth, respectively.

2.2 Structure and Physical Properties of SiC

SiC possesses excellent mechanical, thermal, electrical, physical, and chemical properties which make it an ideal semiconductor material for the fabrication of the novel electronic and optoelectronic devices. From a viewpoint of crystallography, SiC is the best known example of polytypism. The variation in occupied sites along the c-axis in a hexagonal close-packed system brings about different crystal structures which were termed as polytypes. Figure 2.5 shows schematically the structures of typical SiC polytypes [13]. There are three possible sites, denoted by A, B, and C. Two layers cannot successively occupy the same site. For example, the next layer on the top of an "A" layer must occupy either "B" or "C" sites. In principle, there are infinite possibilities for the stacking sequence. However, for most materials, only some stacking structures are stable such as the zinc blende or wurtzite structure. For SiC, there are more than 200 polytypes. Polytypes are represented by the number of Si-C bilayers in the unit cell and the crystal system (C for cubic, H for hexagonal, and R for rhombohedral). 3C-SiC is often called β-SiC and other polytypes are termed as α-SiC. The structures of popular SiC polytypes

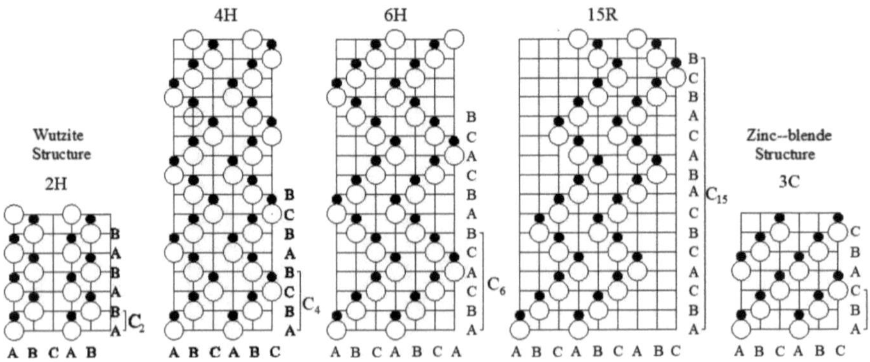

Fig. 2.5 Schematic diagram of SiC structure with different polytypes viewed along [11–20] direction in hexagonal system

Table 2.1 Comparison of basic properties of several semiconductor materials

Property/material	4H-SiC	6H-SiC	Si	GaAs
Thermal conductivity (W/cm K)	4.9	4.9	1.3	0.5
Bandgap (eV)	3.2	3.0	1.12	1.42
Intrinsic material transparent	Yes	Yes	No	No
Available doping	n, p	n, p	n, p	n, p
Saturated electron drift velocity (10^7 m/s)	2.0	2.0	1.0	2.0
Electron mobility (cm^2/Vs)	1000	600	1450	8500
Critical breakdown electrical field (MV/cm)	3	3.2	0.3	0.6
Lattice constant (a)	3.073	3.081	3.84	4.00
Lattice mismatch with GaN (%)	3.8	3.5	−17	−22
Thermal expansion mismatch with GaN (%)	−0.11	−0.12	−0.17	0.11

are 4H, 15R, 6H, and 3C as shown in Fig. 2.5 where open and closed circles donate Si and C atoms, respectively.

Considering the immediate neighbor site of Si-C bilayer stacking, the lattice sites with hexagonal-structured surroundings are termed as "hexagonal sites," and those with cubic-structured surroundings are called as "cubic sites." 4H has one hexagonal and one cubic site and 6H-SiC one hexagonal and two inequivalent cubic sites, while 3C-SiC contains only cubic sites. Hexagonal and cubic sites differ in the location of the second-nearest neighbor leading to different crystal fields. As a consequence, the energy levels of dopants, impurities, and point defects are affected by the lattice site or the hexagonal/cubic ratio. It has been found that the bandgap of SiC polytype increases monotonically with the increase of its hexagonality. The bandgap at room temperature is 2.36 eV for 3C-SiC, 3.26 eV for 4H-SiC, and 3.02 eV for 6H-SiC, respectively.

Compared with the first-generation semiconductor Si and the second-generation semiconductor GaAs, the third-generation semiconductor SiC possesses better physical and chemical properties. Table 2.1 lists the basic properties of several semiconductor materials [14]. From this table, we can see that SiC has higher thermal conductivity, wider bandgap, higher hardness, and higher chemical stability.

The basic properties of SiC material make the electronic devices the follow advantages:

1. The bandgap of SiC is nearly three times as that of Si. Wide bandgap can assure the long-term reliability of device which runs at high temperature. At high temperature, intrinsic carrier excitation will occur, which further causes the device failure. The wider the bandgap, the higher the device limit operating temperature. The limit operating temperature for Si device is lower than 200 °C. In contrast, the limit operation temperature for SiC device exceeds 400 °C.
2. The critical breakdown electrical field of SiC is ten times as that of Si. Thus, SiC can sustain much higher voltages than Si. Devices can be shrunk in size reducing the resistive losses. It reduces system complexity and cost while improving reliability.

3. SiC has a much higher thermal conductivity roughly three times as that of silicon. For SiC device, heat is conducted away much more effectively. It greatly reduces complexity of cooling systems.
4. SiC can carry much higher currents almost five times as that of silicon. It can reduce the area of devices and further reduce the parasitic capacitance.
5. Power devices made from SiC can switch roughly ten times faster than those made from silicon. Therefore, the power conversion circuits operate faster for SiC devices. It means much smaller energy storage capacitors and inductors can be made from SiC.

In addition to the above advantages, SiC is also an ideal substrate for the fabrication of GaN-based optoelectronics and radio frequency devices due to small lattice mismatch with GaN.

Therefore, SiC as the representative of third-generation semiconductor materials has extensive application prospects.

2.3 SiC Single Crystal Growth by PVT Method

From the phase diagram of the Si-C binary system, we know that it is very difficult to obtain stoichiometric SiC liquid phase. It is impossible to employ congruent melt method for SiC single crystal growth. Up to now, the most successful growth method for bulk SiC single crystal is PVT method which consists of the following procedures: (1) sublimation of SiC source, (2) mass transport of sublimed species, and (3) surface reaction and crystallization. Therefore, PVT method was also called as sublimation method.

The principle and growth furnace for the PVT method has been introduced. In order to grow high-quality SiC single crystal, both thermodynamic and kinetic factors must be considered. The process control to maintain optimal thermal and chemical conditions is very important so that an ideal growth interface can be obtained.

Figure 2.6 shows the partial pressures of SiC sublimed species from SiC-C(s) and SiC-Si(l) systems at high temperature [15]. In the gas phase, the dominant species are Si2C, SiC2, and Si which are not stoichiometric SiC.

In the procedure of SiC growing, SiC powder was decomposed at the temperature higher than 2300 °C; the gas phase in sublimation growth is usually Si-rich. This causes the source more and more rich in carbon, i.e., the graphitization of source. In order to avoid the formation of carbon inclusion in SiC single crystal, Si-rich SiC source is often used.

For the PVT growth of SiC, the growth occurred in a nearly closed graphite crucible. We can only measure the temperature at the upper surface of crucible lid or bottom of crucible by using a pyrometer. It is impossible to know the thermal field inside the crucible. Because growth experiment alone is not sufficient to develop a well-controlled PVT growth procedure, modeling and simulation of SiC PVT

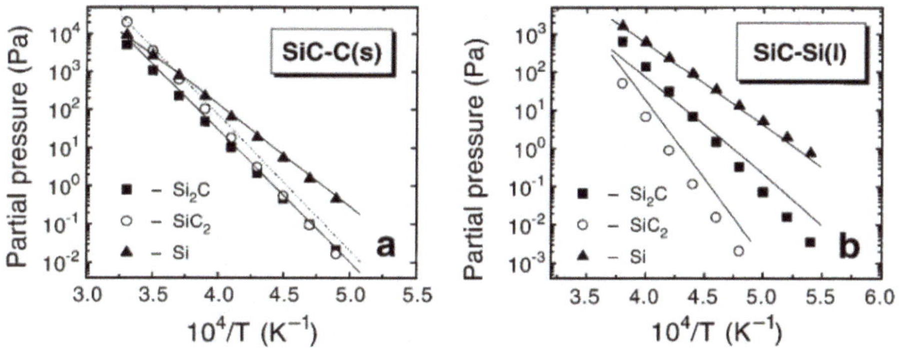

Fig. 2.6 Partial pressures of Si, Si_2C, and SiC_2 vs. temperature calculated for three-phase equilibria: (**a**) SiC-C(s) and (**b**) SiC-Si(l)

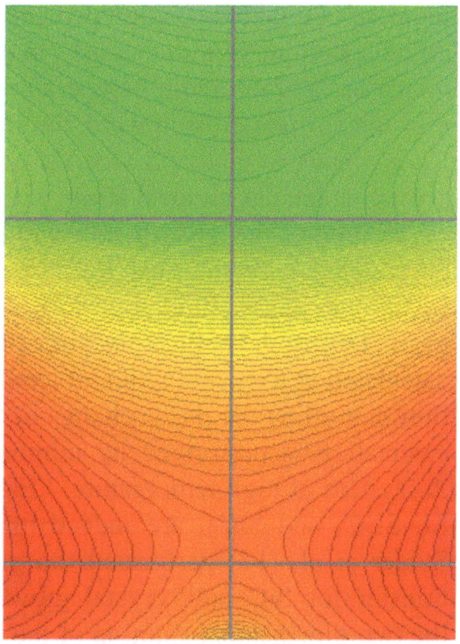

Fig. 2.7 Typical thermal field of SiC PVT growth (Provided by Dr. Yan Peng at Shandong University)

growth are necessary. In the simulation of thermal field, heat transfer by thermal conduction, gas-phase convection, and radiation should be considered. In the heat transfer calculation, the large crystallization energy of SiC must be considered. Heat transfer by radiation is a dominant process at very high temperature in the PVT growth.

Figure 2.7 shows a typical thermal field of SiC PVT growth which was simulated by using a virtual reactor (SiC) software exploited by the semiconductor technology research company. From this figure, the thermal field is described by multiple temperature contour lines. The SiC powder is located in the high-temperature zone of the lower part of the crucible, whereas SiC seed is placed in the low-temperature zone of the crucible lid. At the driving force of temperature gradient, gas species were transported from the source into the seed, and crystal growth occurred. It has been found that the thermal field was affected by the following parameters such as thickness of insulation shield, shape of crucible, diameter of temperature measurement hole, heating frequency and power, the relative position between crucible and induction coil, etc.

If the influence of mass transport on growth rate is neglected, the shape of SiC growth interface is approximately as same as that of temperature contour lines. For actual crystal growth, a slight convex thermal field is favorable for the thermal stress control and diameter enlargement. Sometimes, to obtain an ideal thermal field, the relative position between crucible and induction coil should be adjusted. Figure 2.8 shows the influence of the relative position between crucible and induction coil on thermal field in which crucible position percent denotes the deviation degree of real crucible position away from standard position.

From Fig. 2.8, we can see that as the crucible move upward corresponding to the increase of crucible position percent, the highest temperature zone shifts toward the crucible bottom. Simultaneously, in the growth cavity, the axial temperature gradient increases, and the radial temperature gradient decreases. Therefore, following the movement of crucible upward, the shape of growth interface will become gradually from convex to plan and vice versa. Figure 2.9 shows a typical as-grown 4H-SiC crystal which has an ideal shape with a slight convex interface.

To predict the growth rate and interface shape, a mass transfer model must be coupled with the heat transfer calculation and the thermodynamic database. In the gas phase, fluid transport is based on the low-pressure kinetic theory of gases. Diffusion coefficients, viscosity, conductivity, and specific heat of species are calculated as a local function of temperature, pressure, and composition. The Stefan flow caused by the phase change of SiC has to be considered. The thermodynamic calculation is performed by minimization of the total free energy of the Si-C-Ar system at high temperature. Calculations indicate that three particular species, Si, Si_2C, and SiC_2, are indeed important to describe sublimation growth [16]. Selder et al. [17] performed the global numerical simulation of heat and mass transfer for SiC bulk crystal growth by PVT. The physical model is based on the two-dimensional solution of the coupled differential equations describing mass conservation, momentum conservation, conjugate heat transfer including surface to surface radiation, multicomponent chemical species mass transfer, and advective flow. The model also includes the Joule volume heat sources induced by the electromagnetic field. Figure 2.10 shows the calculated temperature distribution inside the growth cell and the growing crystal for three different points in time. Due to the existence of radial temperature gradient, the growth front is getting convex. However, the shape of growth front is not only depending on the thermal field at the crystal/gas interface but also influenced by gas species mass transfer especially at the periphery of the growing crystal. As a consequence of two effects, the radial growth rate distribution is smoothed out as the crystal length increases.

Figure 2.11 shows the calculated radial growth rate distributions for the three growth stages. We can see that the growth rate at the crystal center is decreasing with the increase of crystal length, and the radial growth rate distribution is smoothed out

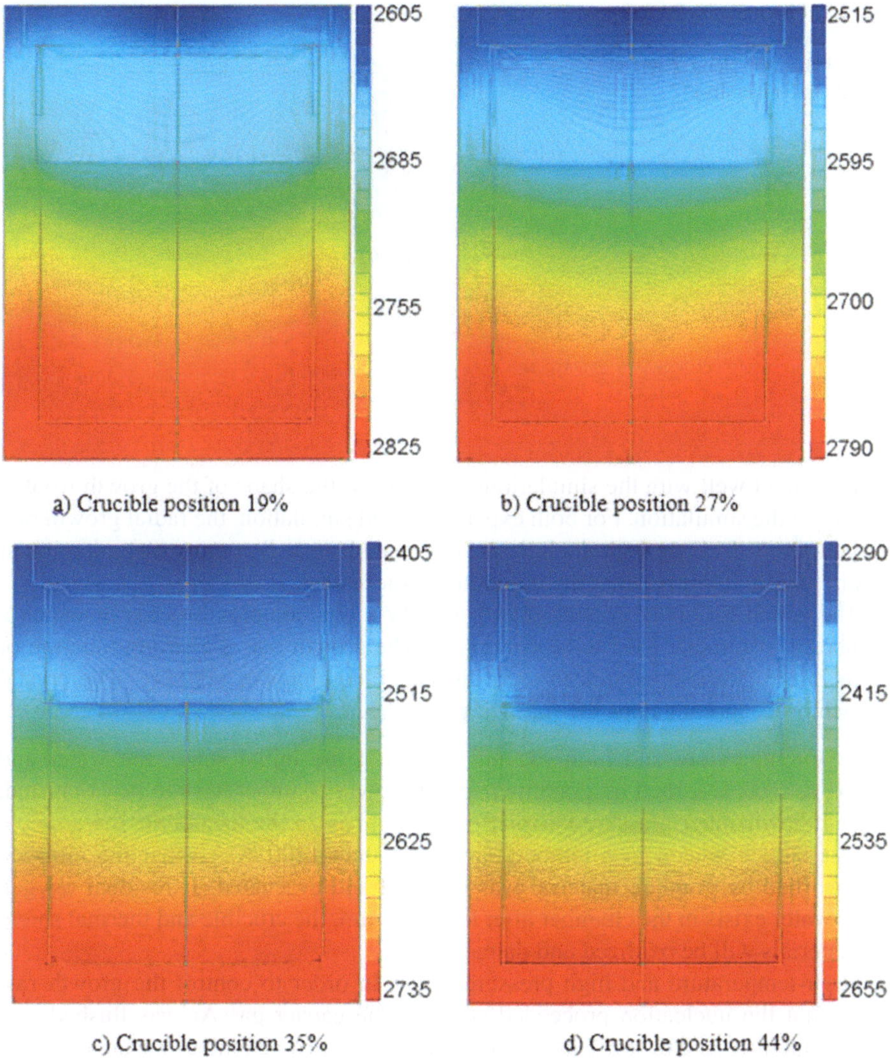

Fig. 2.8 Influence of relative position between crucible and induction coil on the thermal field (Provided by Dr. Peng Y at Shandong University). (**a–d**) Correspond to 19% 27%, 35% and 44% crucible position, respectively

with increasing growth time. The above simulation was confirmed by actual SiC single crystal growth experiment.

Figure 2.12 demonstrates a longitudinal cut slice of a SiC crystal. In this figure, dark regions are nitrogen doped, and bright regions correspond to nominally undoped SiC. The positions of the marked and unmarked regions provide the detailed information about the growth front evolution. The crystal growth process is

Fig. 2.9 Photograph of 4 in. 4H-SiC single crystal (Provided by Dr. Peng Y at Shandong University)

in agreement well with the simulations. Especially, the shape of the growth front is similar to the simulation. For both experiment and simulation, the radial growth rate variation is decreasing with the increase of crystal length. By comparing experiment to simulation, it implies that the modeling is successful.

The actual SiC bulk crystal growth via PVT is a complex process in which a number of parameter have to be controlled. The growth rate is a strong function of temperature gradient, source to seed distance, and chamber pressure. The crystal growth procedure includes the following six stages:

1. Low-temperature and high-vacuum stage: At the initial stage, the water and oxygen gas absorbed by graphite parts in growth chamber have to be exhausted, and the chamber pressure is lower than 100 nbar. In the meantime, the crucible was slowly heated to a temperature lower than 1400 °C so that the nitrogen absorbed by graphite material was released and evacuated. If residual oxygen or water exists in the chamber after high vacuum, the crucible and thermal shield materials will be oxidized and damaged.
2. High-temperature and high-pressure stage: In order to control the growth rate so that the nucleation proceeded steadily, the carrier gas Ar was flushed into the chamber, and the pressure in the chamber reached 800 mbar. Because the SiC powder was decomposed into Si, Si_2C, and Si_2C gas species at high temperature, the partial pressures of SiC gas species will be lowered at high background pressure. In this case, the growth rate was controlled at a very low level. It is favorable for SiC nucleation on seed.
3. SiC nucleation stage: At this stage, temperature and pressure were kept stably. SiC was nucleated on seed at a low growth rate. An ideal status is that a large area facet was formed without any structural defects.
4. Pressure decreasing and crystal growth stage: After SiC nucleation was completed, crystal growth rate was fastened by decreasing growth pressure. The growth pressure was decreased from present into a target pressure. Pressure decreasing rate can be expressed as follows:

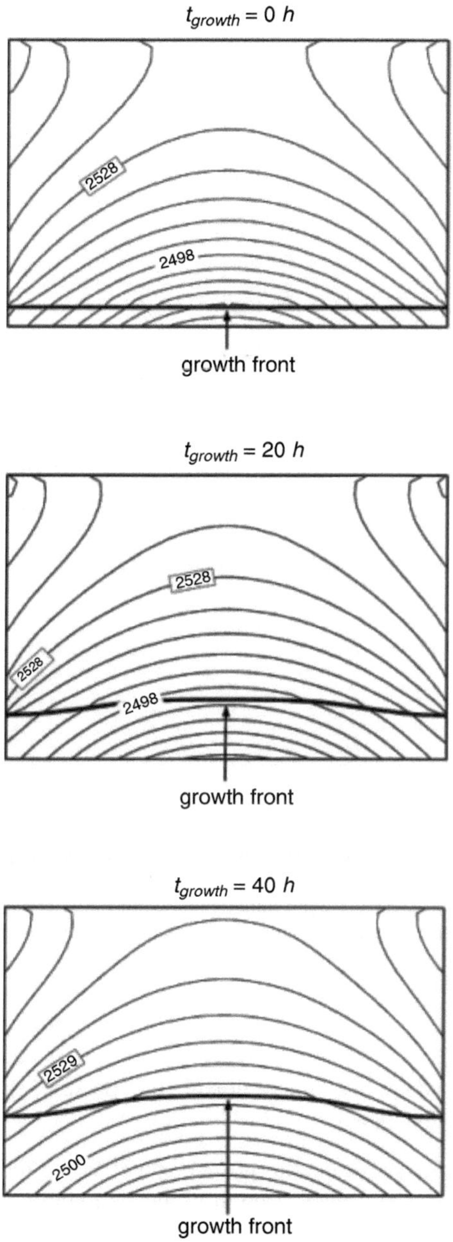

Fig. 2.10 Temperature distribution and growth front for three different points in time during SiC PVT growth. The seed is positioned at the bottom of the individual images. The SiC source is on the top

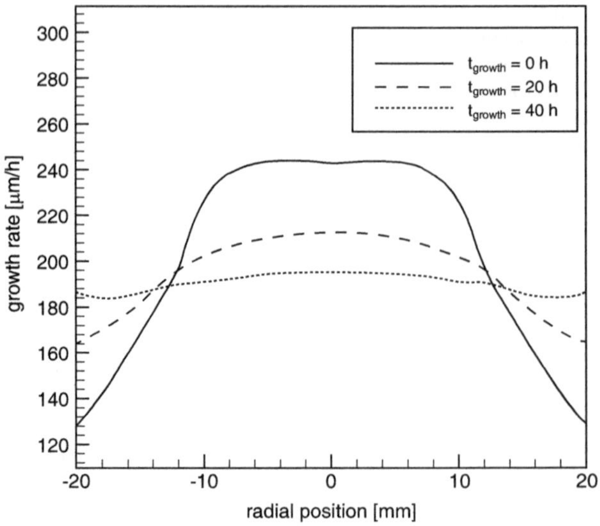

Fig. 2.11 Radial distribution of the calculated growth rate for three different points in time

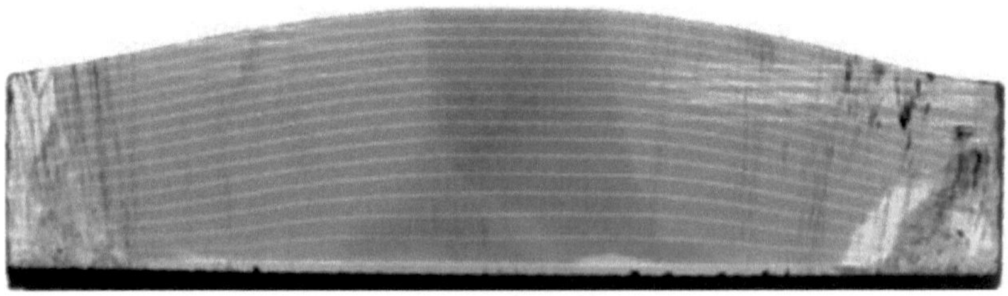

Fig. 2.12 Longitudinal cut slice of a SiC crystal in which the shape of growth front was marked by an interface demarcation technique. Undoped regions exhibit a bright contrast

$$p = p_0 e^{-t/\tau} \tag{2.2}$$

where p_0 and p donate the present and target pressure, respectively, and τ is constant.

5. Stable crystal growth stage: When the growth pressure reached the target value, the crystal growth proceeded stably at a certain temperature and pressure. The optimal growth rate was proximately 200–300 µm/h.
6. Pressure increasing and cooling stage: When the crystal ingot length was estimated to be 20–30 mm, crystal growth should be terminated by increasing pressure. Then the temperature was lowered slowly into room temperature. The whole growth procedure was ended.

2.4 The Formation and Control of Structural Defects in SiC Single Crystals

SiC single crystals contain various structural defects such as foreign polytypes, micropipes, dislocations, inclusions, point defects, etc. Defects in SiC wafers will deteriorate the device performance and shorten the device lifetime. So defects should be reduced or eliminated.

2.4.1 Micropipe Defects

A micropipe is actually a hollow core associated with a super-screw dislocation. It penetrates through the entire crystal along the c-axis with diameter from tens of nanometers to several micrometers. A SiC device with a micropipe in its active area cannot support significantly high electrical field. Therefore, micropipe is known as a “killer defect” [18]. According to the Frank theory [19], a dislocation whose Burgers vector exceeds a critical value, of the order of magnitude 10 Å, is only in equilibrium with an empty tube at its core. When a micropipe was formed, the elastic strain energy released by the formation of hollow core and surface free energy created by the interior cylindrical surface of hollow core reached a balance. The radius of a hollow core is given by

$$r = \mu b^2/8\pi^2\gamma \tag{2.3}$$

Here γ is the specific surface free energy of the material, and μ the rigidity modulus. The equation has been confirmed by different experiments for micropipes in SiC [20–22].

Figure 2.13 shows an atomic force microscopic (AFM) image of a micropipe in 6H-SiC. At the central region, a circular empty tube can be identified.

Huang et al. [23, 24] have taken the direct image of the super-screw in SiC by back-reflection white-beam synchrotron topography and simulated it by tracing ray method. Based on the dislocation theory, it was considered that the distorted region around a super-screw dislocation core consists of a large number of small crystallites which are orientated by the strain field and diffract X-rays independently according to their local lattice orientations. Under these conditions, calculating the traces of the diffracted X-rays enables one to map the spots where these rays intersect the recording plate, and the contrast intensity of the direct image is then determined by the density distribution of the intersection spots. Figure 2.14 shows the back-reflection synchrotron topograph of a 6c super-screw dislocation in 6H-SiC wafer. Comparison of synchrotron topograph of Fig. 2.6a to the kinematic contrast of Fig. 2.6b for a 6c super-screw dislocation in 6H-SiC wafer indicated that hollow-core micropipes in 6H-SiC the c-axis with Burgers vectors being multiples of the lattice constant *c*.

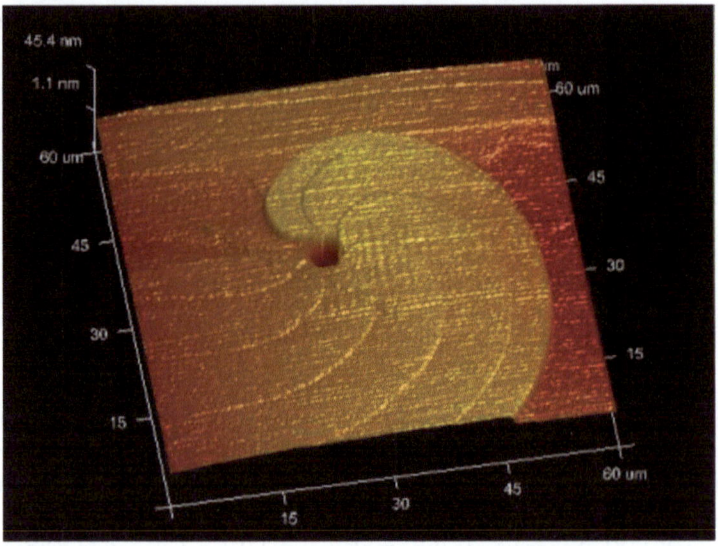

Fig. 2.13 An AFM image of a micropipe in 6H-SiC single crystal (Provided by Dr. Peng Y at Shandong University)

However, it is should be noted that Fig. 2.14a is actually the magnified image of the super-screw dislocation, and the diameter of the circular spot is the function of sample-recording film distance and Burgers vector. Figure 2.15 shows the principle for the contrast formation of a super-screw dislocation. From this figure the formation of the circular contrast feature on the film can be easily understood. From an overall viewpoint, only the different semi-apex angles of the twisted cones determine diffracted X-rays overlap in space and further determine the overall image features of the dislocation on topograph. Obviously, both the magnitude of the Burgers vector and the sample-film distance influence the topographic contrast, especially the diameter of the white-contrast region.

It was found that most of micropipes have the characteristic of mixed dislocations, i.e., their Burgers vectors have both edge and screw contents [23]. Heindl et al. investigated the structures of micropipes in SiC by using AFM and found that micropipes were hollow-core dislocations according to Frank's model but contain dislocations of mixed type [25].

The causes of micropipe generation during PVT growth of SiC include back-side bubble in seed, foreign polytype, inclusion, etc.

Kuhr et al. have observed the formation of micropipes caused by hexagonal void [26]. Figure 2.16a shows a back-reflection topograph of the area above a void, and Fig. 2.16b shows the same area below the void [26]. The area the void will pass through is outlined in white in Fig. 2.16a, and the hexagon of light and dark orientation contrast that occupies the center of Fig. 2.6b represents the trace of the

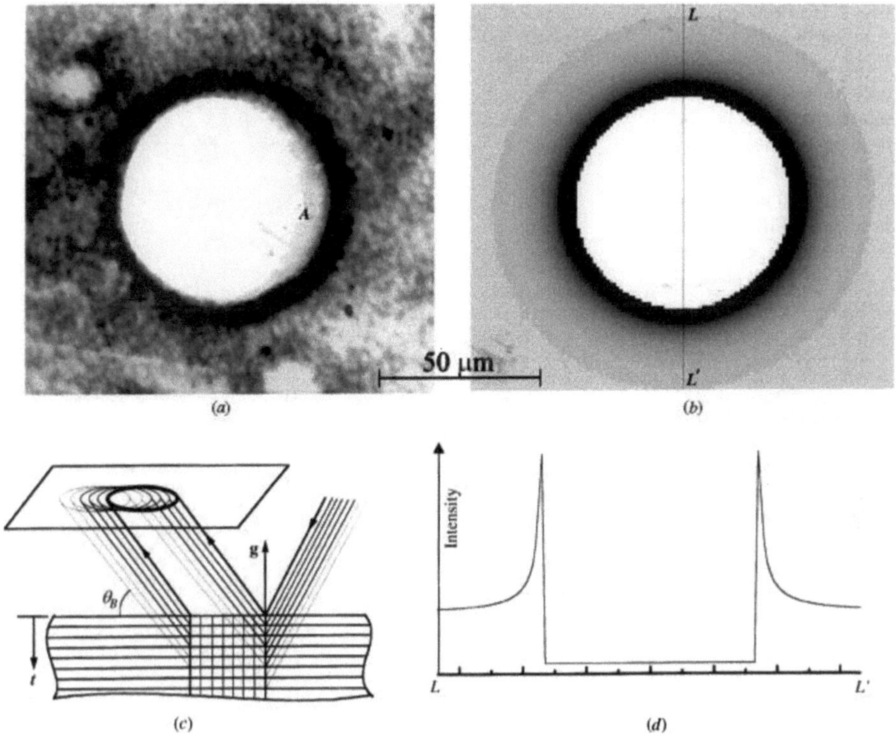

Fig. 2.14 Back-reflection synchrotron topograph of a 6c super-screw dislocation. (**a**) Synchrotron topograph of super-screw dislocation. (**b**) Kinematic contrast of a screw dislocation (Burgers vector magnitude b = 6c = 91.02 Å), simulated by tracing ray method. (**c**) Schematic representation of the displaced diffraction images associated with the deeper layers. (**d**) Radial intensity profile of the simulated image in (**c**)

void's path through the crystal during crystal growth. Large white spots on the left and lower right of the hexagon in both images are micropipes present in both wafers.

Figure 2.17 is a schematic illustration of the void and micropipe formation process. Voids are often formed at the interface between the seed crystal and crucible cap. Evaporation steps at the top of the void and growth steps on the bottom of the void provide strong evidence that SiC transport occurs within the void and that the void moves along the temperature gradient within the growing boule. Dislocations line up along the trace of the void path, and screw dislocations combine to form micropipes at the corners of the void, leaving an area directly beneath the void that is free of screw dislocations. This mechanism for micropipe formation is likely to be dominant in any crystal that contains hexagonal voids. Elimination of hexagonal voids by the use of a back-side evaporation barrier and proper seed mounting techniques was effective to reduce micropipe densities in PVT-grown SiC.

Figure 2.18 is a scanning image of a longitudinal cut 4H-SiC crystal. From this image, we can see that some foreign polytypes including 15R with yellow color and

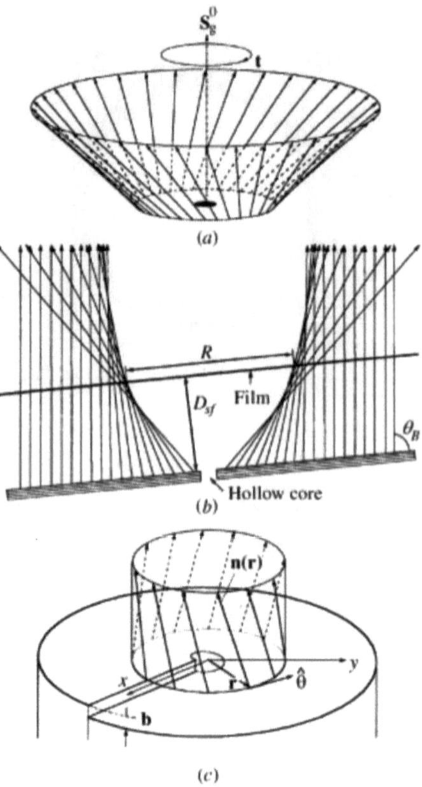

Fig. 2.15 Principle for the contrast formation of a super-screw dislocation in back-reflection synchrotron topograph. (**a**) Schematic drawing of one of the twisted cones. *t* twisting direction. (**b**) Overlap of the twisted cones in space. (**c**) Inclination of coordinate-dependent normal *n* causing the twist of diffracted X-ray

6H with green color appear in 4H-SiC crystal. As soon as the polytype phase transition occurred, a large number of micropipes were formed on the interface between the host polytype and other foreign polytype. Since the foreign polytype causes the serious mismatch in the stacking sequence as the foreign polytype nucleus meets the host polytype, the stacking mismatch and the associated large strains trigger micropipe formation. For 4H-SiC crystal growth, the temperature window for keeping polytype stability is very narrow. Any fluctuation in the temperature profile or pressure in growth cavity may cause unintentional supercooling and deviation of the C/Si ratio at growth front which further lead to the formation of foreign polytypes.

In addition, Si droplets and carbon inclusions in SiC crystals also cause the formation of micropipes.

Glass et al. [27] believed that the excess Si on the growth surface at the initial stage and the accumulation of free C after the source graphitization are the main causes for the formation of micropipes. Figure 2.19 shows an example of (a) a technological issue where a graphite particle has caused a micropipe and (b) a process instability issue where constitutional supercooling has produced Si droplets on the growth surface which have led to micropipe streaming.

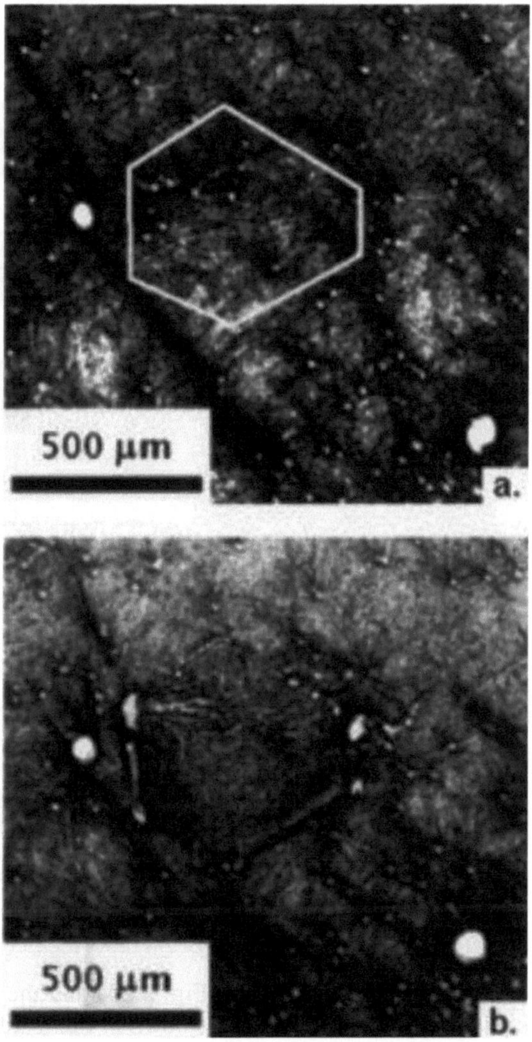

Fig. 2.16 Back-reflection X-ray topographs of the area above a void (**a**) and below a void (**b**) in 6H-SiC. The projected passage of the void has been traced in white. The screw dislocations are visible above the void in (**a**) and are not visible beneath the void in (**b**)

Origins of micropipes in SiC come from two sources. One kind of micropipes was generated from the instability growth conditions or other defects such as seed back-side void, foreign polytype, C or Si inclusion, etc. The others originated from the extension of micropipe in seed, i.e., micropipes have the heredity. For the former case, micropipes were controlled by optimization of the growth conditions such as the seed and the source temperature, the pressure of ambient gas, etc. For the latter case, many efforts have been made in the past decade to reduce or eliminate micropipes in SiC crystals.

To eliminate micropipes in SiC crystals, rhombohedral plane seeds were used to grow SiC crystals [28, 29]. Because the 4H-SiC {0 3−3 8} plane is inclined to the

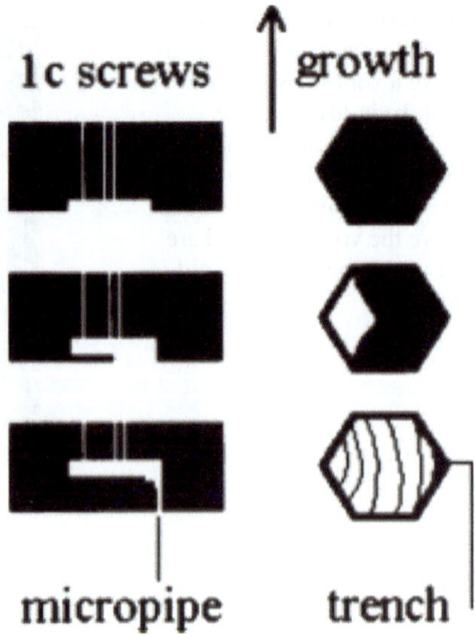

Fig. 2.17 Schematic diagram showing void formation and movement. The void begins as a cavity in the seed attachment layer, growth within the void occurs by lateral attachment and step flow, and the trench closes at the corners of the void, forming micropipes that extend beneath the void

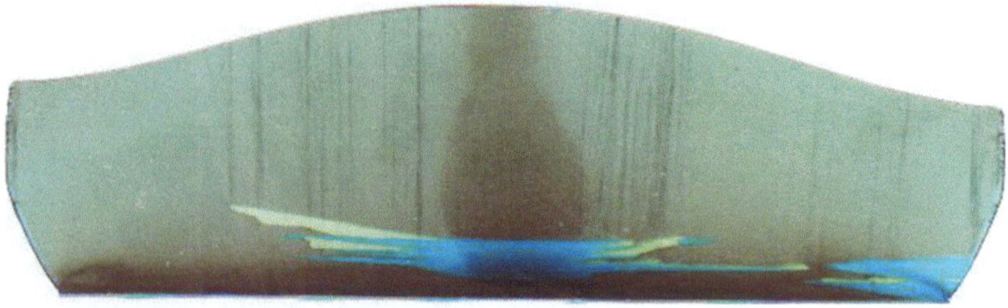

Fig. 2.18 Scanning image of a longitudinal cut 4 in. 4H-SiC crystal (Provided by Dr. Yang X L at Shandong University)

c-plane at 54.7° as shown in Fig. 2.20a and the 4H-SiC {0 3−3 8} plane is semi-equivalent to 3C-SiC {100} as shown in Fig. 2.20b, the 4H-SiC{0 3−3 8} plane has a low surface free energy and can be used as seed plane.

Shiomi et al. [28] found that a micropipe and stacking fault-free region was generated in crystals grown on the 4H-SiC {0 3−3 8} seed, as shown in Fig. 2.21. At the height of H′, there were two regions in the ingot, one was a perfect region without any micropipe and stacking fault (region A) and the other was a stacking fault region (region B). In other words, micropipe and stacking fault-free seed can be obtained by several repeating 4H-SiC crystal growth runs on {0 3−3 8} seed.

Figure 2.22a, b shows the surface morphology of the 4H-SiC {0 3−3 8} wafers after molten KOH etching. Many stripes due to the stacking faults are observed in the wafer from region B. In contrast, region A has high perfection without deep etch pits due to micropipes. It was obvious that the growth on the 4H-SiC {0 3−3 8} seed was suitable to achieve a SiC wafer without micropipes or stacking fault.

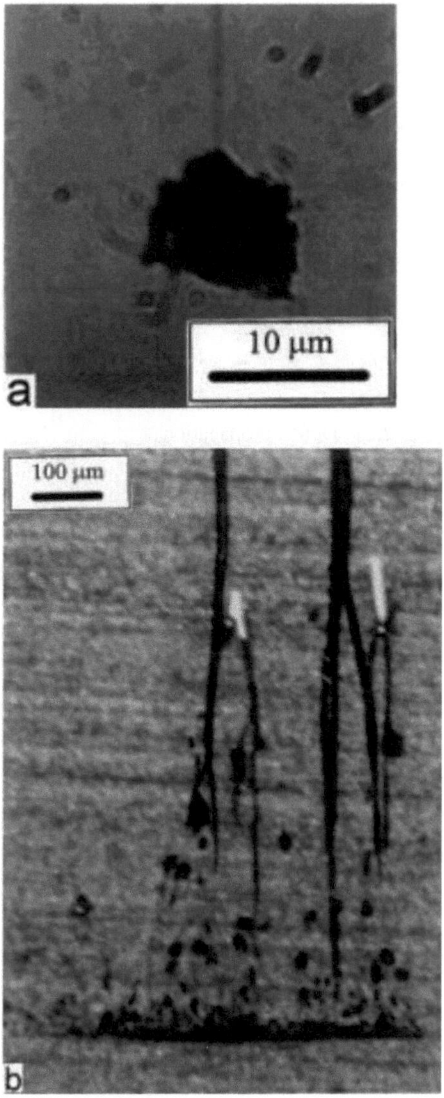

Fig. 2.19 (**a**) Graphite particle with resulting micropipe and (**b**) results of a process instability issue where constitutional supercooling has produced Si droplets on the growth surface leading to micropipe streaming

Li et al. [29] found that {0 1−1 2}, {0 1−1 3}, and {0 1−1 4} planes are the natural appearance faces for 4H-SiC. It implies that each of these planes has a low source free energy and can be used as seed face. By comparing the crystals grown on three seed faces, it was found that the quality of crystals grown on the {0 1−1 4} facet is much better than those of the crystal grown on other facets.

The above experiments confirmed that using non-*c*-face seed for SiC crystal growth is an effective method to eliminate micropipes in case of the existence of micropipes in seed. At present, micropipe-free wafers are commercially available from most suppliers.

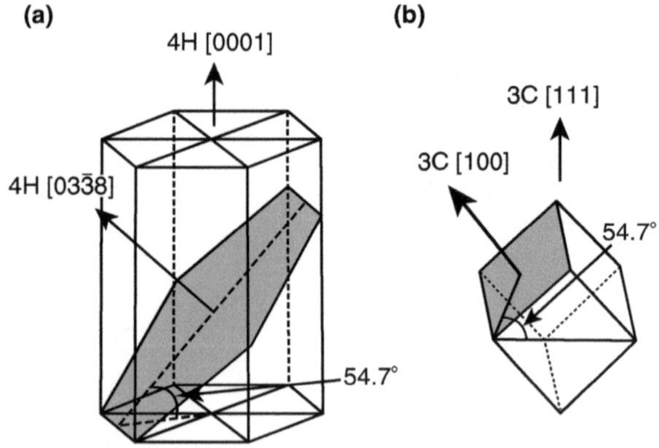

Fig. 2.20 Schematic illustration of unit cell: (**a**) 4H-SiC and (**b**) 3C-SiC

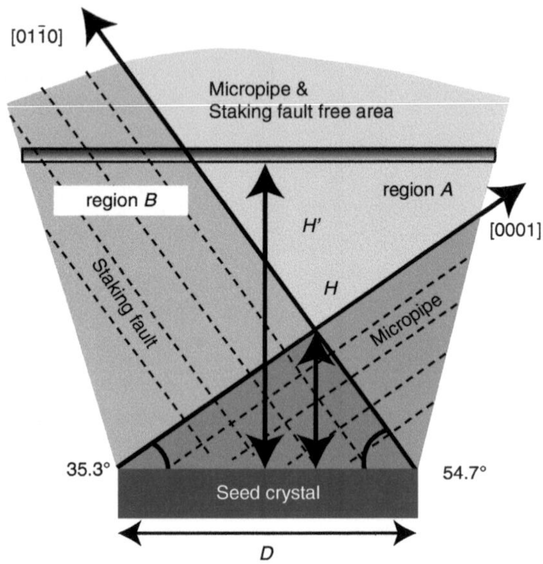

Fig. 2.21 Defect propagation on the 4H-SiC {0 3−3 8} seed

2.4.2 Foreign Polytypes

There are more than 200 polytypes for SiC materials. These polytypes have different characteristics. Because of nearly same stacking fault energy, several different polytypes such as 4H, 6H, and 15R are easily formed simultaneously in a crystal during crystal growth. Figure 2.23 shows a scanning image of a SiC wafer. In this image, we can see that 6H (green), 4H (yellow), and 15R (light yellow) coexist in a wafer. The mixing of the polytypes destabilizes growth and deteriorates the characteristic of SiC as an electronic material. Therefore, it is important to avoid the formation of foreign polytypes and keep the host polytype stability.

In the experiment, it was found that growth parameters such as the growth temperature, pressure in chamber, supersaturation, vapor-phase stoichiometry, impu-

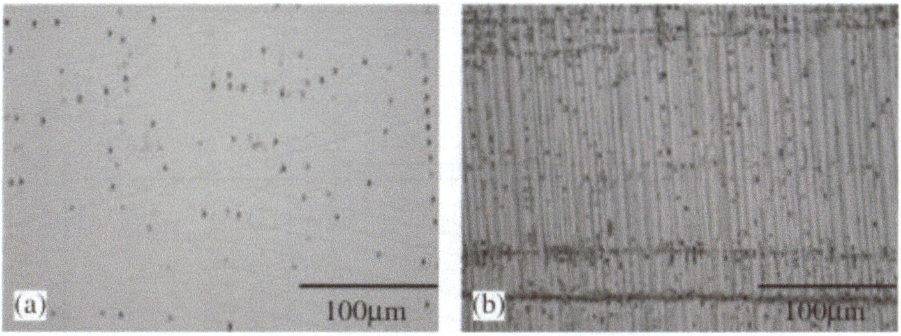

Fig. 2.22 Surface morphology of the 4H-SiC {0 3−3 8} wafer after molten KOH etching: (**a**) the micropipe and stacking fault-free area (region A) and (**b**) the micropipe-free area with stacking fault (region B)

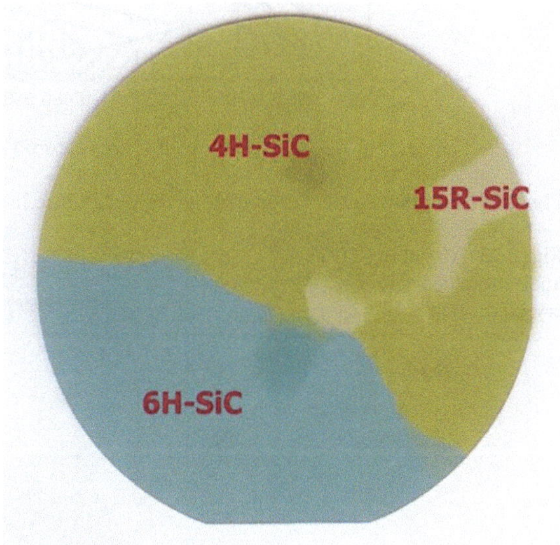

Fig. 2.23 Scanning image of a 4 in. SiC wafer in which 4H, 6H, and 15R polytypes coexist (Provided by Dr. Yang X L at Shandong University)

rities, and polarity of seed surface play important roles to influence the polytype stability. Therefore, the control of polytype is a complicated issue.

Knippenberg [30] reported empirical observation of the relative amount of individual polytype in SiC crystal growth as shown in Fig. 2.24. From this figure, we can see that the 3C is a metastable polytype, and 2H exists only at a low-temperature range. Above 2000 °C, 4H, 6H, and 15R are observed and sublimation growth can be performed. For SiC wafers to be used for electronic devices, 4H is the desired polytype. However, 4H is observed in a narrow temperature window and relative small amount. Thus, 4H-SiC crystal growth is more difficult than 6H.

Based on the classical thermodynamical analysis and crystal growth theory, Shiramomo et al. [31, 32] calculated the free energy of forming a critical nucleus in case of 2D nucleation as shown in Fig. 2.25. It can be expressed as follows:

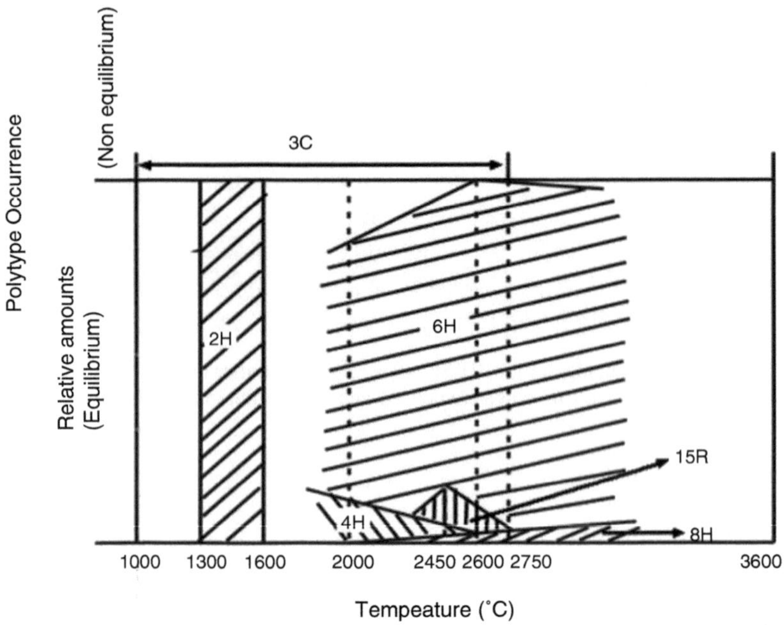

Fig. 2.24 Empirical observation of polytype occurrence in SiC crystal growth

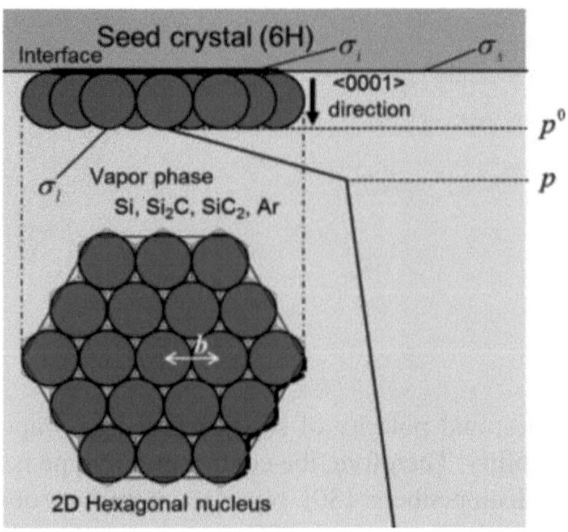

Fig. 2.25 Model of 2D nucleation theory for SiC crystal growth

$$\Delta G_2^* = \frac{b^4 \sigma_l^2}{\Delta\mu - \left(\frac{\sqrt{3}}{2}\right) b^2 (\sigma_l + \sigma_i - \sigma_S)} \tag{2.4}$$

where b is the first neighbor distance corresponding to the in-plane lattice constant. σ_s, σ_i, and σ_l are the surface energies of the substrate surface, the

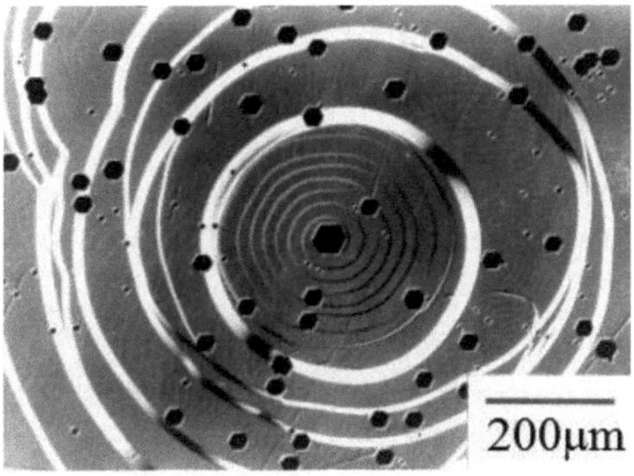

Fig. 2.26 Optical microscopic image of the growing surface of 6H-SiC after the chemical etching

layer/substrate interface, and the layer surface, respectively, as shown in Fig. 2.25. $\Delta\mu$ is the bulk chemical potential change associated with the molecular change from the vapor state to the condensed state.

To clarify which kind of polytype will be favored in the nucleation processes, the nucleation energies of 4H and 6H-SiC as a function of temperature and pressure which are the most easily generated polytypes in the sublimation method are calculated and compared. The calculation results indicated that when *c*-face was used as seed face, 4H-SiC 2D nuclei formation free energy is lower than that of 6H-SiC. This implies that the formation of 4H-SiC is more stable than that of 6H-SiC at nucleation stage in case of *c*-face seed. Additionally, the difference of nucleation energy between 4H-SiC and 6H-SiC decreased as the growth temperature increased and pressure decreased corresponding to increase of supersaturation of carbon species. This means that there is a large probability of the formation of a mixture of 4H and 6H polytypes in case of higher supersaturation, i.e., higher temperature of a seed and lower pressure in the growth chamber. In contrast, 6H-SiC is the most stable polytype for the growth on Si-face seed.

From the kinetic viewpoint, polytype replication through spiral growth around threading screw dislocations plays a critical role for the polytype stability. Figure 2.26 shows the typical surface morphology of a 6H-SiC crystal taken by optical microscope [33]. From this image, we can see that the outcrop of a threading screw dislocation on the surface provides an infinite step source. The spiral step will be kept through the whole growth procedure as long as the growth condition is stably maintained. At the step edges, the stacking information is provided, which ensures the replication of the polytype in the growth procedure.

2.4.3 Dislocations

SiC single crystal ingots and substrates contain a variety of dislocations. The Burgers vector, the extended direction, and the typical density are listed in Table 2.2.

Table 2.2 Main dislocation categories in SiC single crystals

Dislocation	Burgers vector	Extended direction	Typical density (cm^2)
Threading screw dislocation (TSD)	<0 0 0 1>	<0 0 0 1>	10^2–10^4
Threading edge dislocation (TED)	<1 1 −2 0>/3	<0 0 0 1>	10^2–10^4
Basal plane dislocation (BPD)	<1 1 −2 0>/3	In {0 0 0 1} plane	10^2–10^4

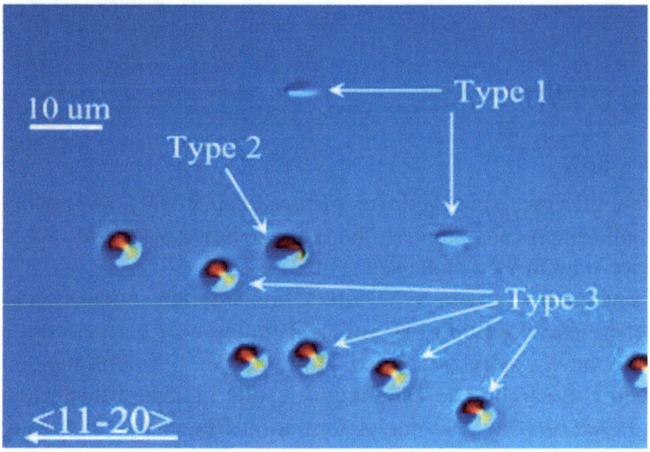

Fig. 2.27 The morphology of pits on SiC surface after oxidation

Figure 2.27 shows the typical surface morphology of SiC crystal after oxidation [34]. The oxidation pits are formed by oxidizing defects. Due to anisotropy of SiC structure, the oxidation rates of SiC are (0 0 0−1) > (1 1−2 0) > (0 0 0 1). Therefore, the oxidation pits were formed by the difference of the oxidation rate. In this figure, three kinds of pits were observed. Type 1, type 2, and type 3 correspond to basal plane dislocation, screw dislocation, and threading edge dislocation. The etch pits formed by molten KOH etching for SiC crystals have similar shapes.

As the device contains TSD, highly localized current may concentrate in the dislocation neighbor resulting in micro-plasmas. As a consequence, TSD causes the reduction of charge carrier lifetime [35]. BPDs are one of the troublesome defects in SiC materials. Stacking faults originating from the BPDs cause the increase of V_F under constant current for bipolar SiC devices, the increase of reverse-biased leakage, and the decrease of forward-biased current for unipolar SiC devices [36]. Diodes with higher TEDs densities have higher leakage current and slightly lower breakdown voltage compared to those without dislocations [37]. Among the three kinds of dislocations, the influence of TED on the performance of SiC device is relatively weak.

Since TED and BPD have the same Burgers vector, they can convert each other. To decrease the effect of dislocation on device performance, the conversions from BPD to TED are often controlled in CVD epitaxial growth by optimizing the growth parameters.

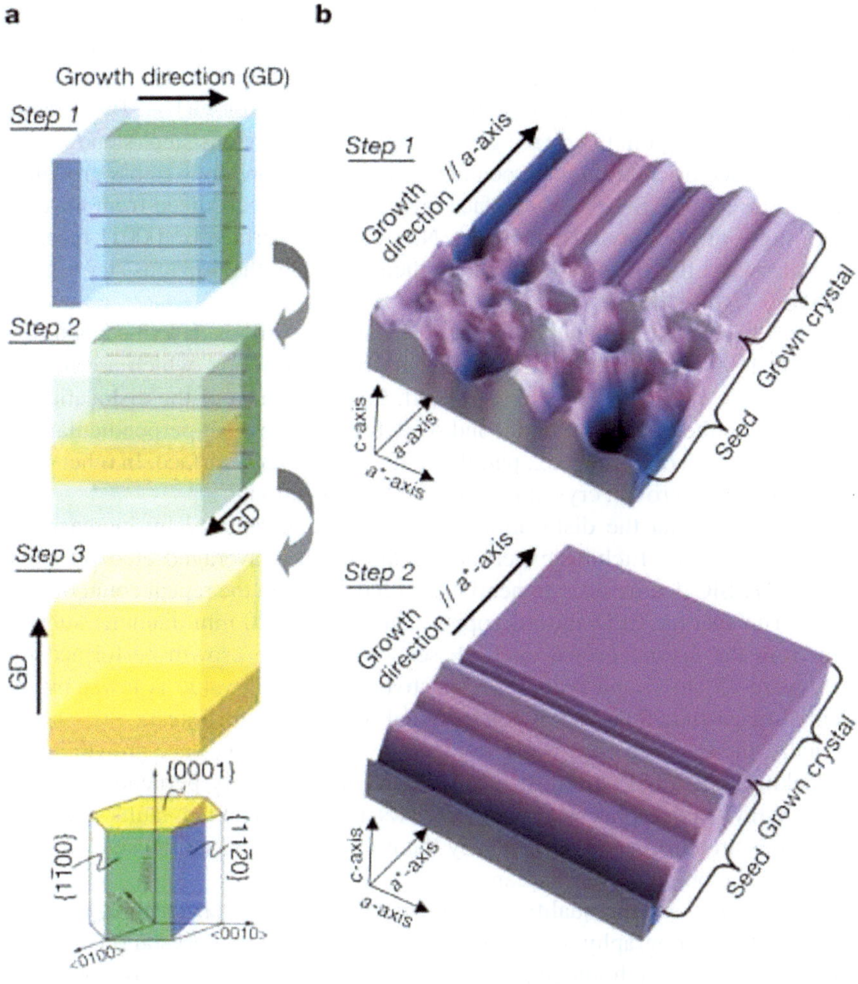

Fig. 2.28 Schematic diagram of the "repeated a-face" growth process. The growth sequences are as follows. Step 1: the first a-face growth. Step 2: the second a-face growth perpendicular to the first a-face (the seed was sliced from the first a-face growth crystal). Step 3: c-face growth with offset angle of several degrees (the seed was sliced from the second a-face growth crystal). (**a**) The first and second a-faces are {1 1−2 0} and {1−1 0 0}, respectively. (**b**) Top side view, showing {0 0 0 1} lattice plane irregularities of seed and grown crystal in steps 1 and 2. The a*-axis is perpendicular to both the a-axis and the c-axis

For conventional semiconductor materials such as Si and GaAs, dislocation-free crystals are generally grown from molten sources by means of the "necking" process. But "necking" process is not suitable for SiC crystals grown by PVT because rapid increase of crystal diameter is impossible. To reduce the dislocation density in SiC crystal, "repeated a-face" method was proposed by Nakamura et al. for the ultrahigh-quality SiC single crystal growth [38]. Figure 2.28 shows the "repeated a-face" growth process.

The process to eliminate dislocations is described as follows. Step 1: *a*-face ({1 1−2 0} or {1−1 0 0}) growth along the *a*-axis (<1 1 −2 0> or <1 −1 0 0>) direction, using a seed sliced from a *c*-face growth ingot. Step N ($N = 2, 3, 4, \ldots$): *a*-face ({1 −1 0 0} or {1 1 −2 0}) growth along the *a*-axis (<1 −1 0 0> or <1 1 −2 0>) direction, using a seed sliced from the *a*-face growth ingot of the previous step (step $N - 1$)—the seed surface orientation is perpendicular to both the previous step's growth and <0 0 0 1> directions. Step $N + 1$: *c*-face growth, using a seed sliced from the *a*-face growth ingot of the previous step (step N)—the seed surface is {0001} face, with several degrees off-axis toward the perpendicular to both the previous step's growth and <0 0 0 1> directions.

The principle for the reduction of the dislocation density is described as follows. In step 1, an *a*-face seed with a high density of dislocations, which are inherited from the crude *c*-axis SiC crystal, was used. In step 2, most of the dislocations are not exposed to the surface of the second seed, because they exit perpendicular to the first *a*-face growth direction, i.e., parallel to the second seed surface. In other words, the second *a*-face growth crystal inherits fewer dislocations.

It was found that the dislocation density could be reduced by increasing the repeating N counts. Etching experiment indicated that averaged etch pit density (EPD) in 4H-SiC decreased exponentially with increase in the repeat count of *a*-face growth. The average EPD and micropipe densities of a 20 mm diameter substrate, taken from the crystal grown on RAF seed with *a*-face growth performed three times, were 75 cm^{-2} and 0 cm^{-2}, respectively. The EPD value is lower by three orders of magnitude than that of conventional-grade SiC substrates.

Finally, the stacking fault was eliminated by the last step *c*-face growth because the stacking faults and partial dislocations are inherited only perpendicular to the *c*-axis. X-ray topographic observation shows that the stacking faults and partial dislocations are eliminated perfectly by the *c*-face growth since the stacking faults and partials lie on the {0 0 0 1} plane.

Figure 2.29 shows the quality assessment for the repeated *a*-face growth crystal by synchrotron topography. From Fig. 2.29a, we can see that crystal quality of the RAF substrate is very homogeneous, and there are very few macroscopic defects and dislocation. The long-range lattice warp is very small with the curvature radius of the lattice of about 800 m. In contrast, the conventional-grade substrate has many macroscopic defects and dislocation networks as shown in Fig. 2.29b.

Therefore, RAF is an effective method to eliminate dislocations and other defects in SiC single crystals.

2.5 Control of Electrical Characters of SiC Crystals Grown by Sublimation Growth

For the fabrication of vertical devices, low-resistivity wafers are needed to minimize the series resistance, while high-resistivity wafers are desired for the fabrication of lateral high-frequency devices, to reduce the parasitic impedance. In SiC

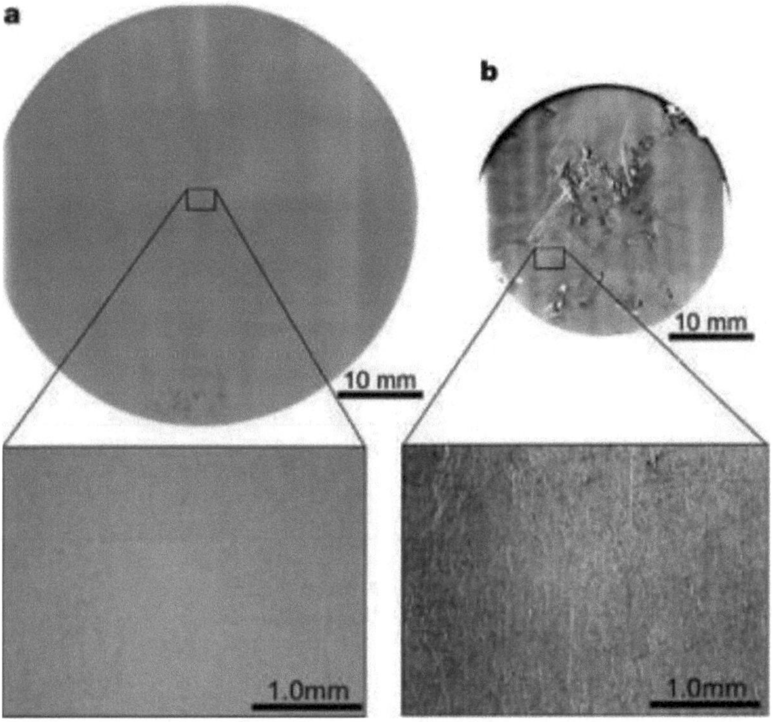

Fig. 2.29 Synchrotron monochromatic beam X-ray topographs. 4H-SiC (0001) 8° off-axis substrate, 2.0 in. in diameter, manufactured by the RAF process (**a**) and a 1.2 in. diameter specimen manufactured by the conventional process (only *c*-face growth (**b**). Averaged EPD and micropipe densities of the RAF growth crystal were about 250 cm^{-2} and 0 cm^{-2}, respectively, and those of crystal grown by the conventional method were about 3×10^4 cm^{-2} and 30 cm^{-2}, respectively. Magnified images are shown below. Dislocation networks are shown in the magnified image of (**b**)

single crystal, N and Al are generally the dopants for n-type and p-type boules, respectively. Semi-insulating characters of SiC crystals were realized by introducing deep levels compensating all shallow donors or acceptors.

2.5.1 n-Type Doping

In the procedure of SiC sublimation growth, nitrogen doping is generally conducted by introducing nitrogen gas into the growth chamber. In this case, carrier gases include Ar and N_2 with a suitable proportion. The nitrogen concentration in grown SiC crystal is approximately proportional to the square root of the nitrogen partial pressure during crystal growth and independent of growth rate. This implies that nitrogen incorporation is determined by the equilibrium between nitrogen in the gas phase and nitrogen adsorbed on the growing surface [39].

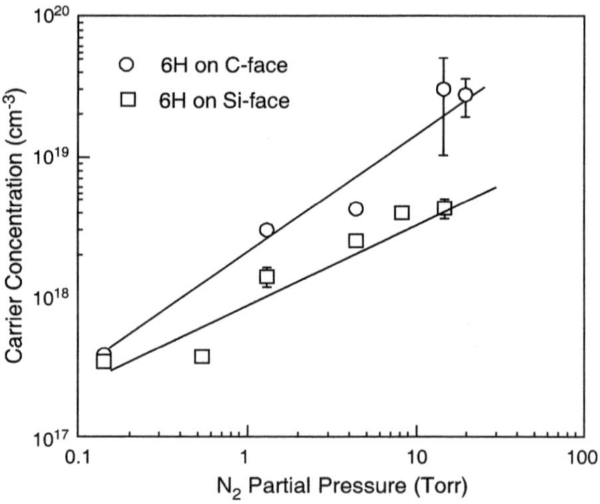

Fig. 2.30 N_2 partial pressure dependence of the carrier concentration in 6H-SiC grown on the (0 0 0 1) Si and the (0 0 0−1) *c*-faces

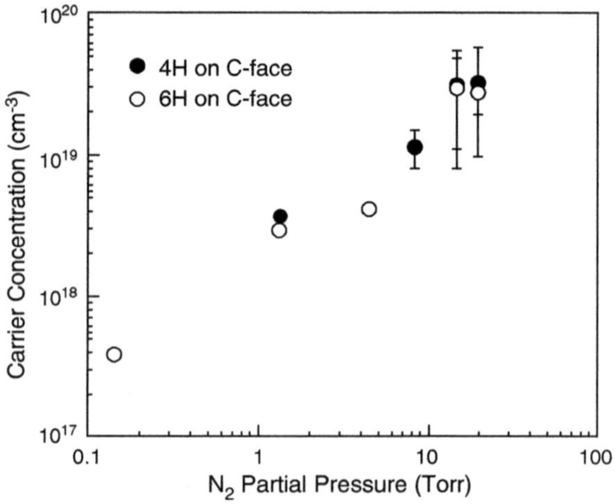

Fig. 2.31 N_2 partial pressure dependence of the carrier concentration in 4H and 6H-SiC grown on the *c*-face

Figure 2.30 shows the N_2 partial pressure dependence of the carrier concentration in 6H-SiC crystals grown on the (0 0 0−1) C and the (0 0 0 1) Si-faces [40]. The carrier concentration increased as the nitrogen partial pressure increased for both crystals. The crystals grown on the *c*-face always had higher carrier concentrations than those grown on the Si-face, and the difference became larger at higher carrier concentrations. The carrier concentration difference originated mainly from the large compensation of the impurity acceptor atoms in the low carrier concentration regime.

Figure 2.31 shows the N_2 partial pressure dependence of the carrier concentration in both 4H and 6H-SiC crystals grown on the *c*-face. Both crystals exhibited the similar concentration variation trend although 6H-SiC has larger nitrogen donor ionization.

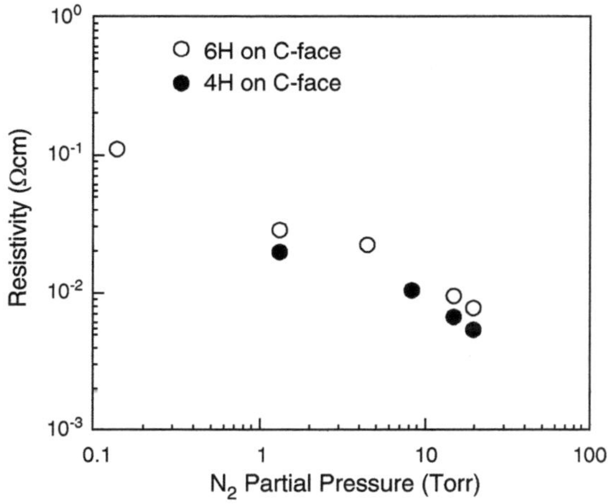

Fig. 2.32 N_2 partial pressure dependence of the resistivities of 4H and 6H-SiC crystals grown on the *c*-face

Figure 2.32 shows the resistivities of the 6H and 4H-SiC crystals as a function of the nitrogen partial pressure. The 4H-SiC crystal always exhibited lower resistivities at the same carrier concentration because of its higher electron mobility. At a nitrogen partial pressure of 20 Torr corresponding to the nitrogen concentration of 10^{20} cm^{-3}, bulk resistivities 7.6×10^{-3} Ω cm for 6H-SiC and 5.3×10^{-3} Ω cm for 4H-SiC were obtained. However, the resistivities of n-type 4H-SiC for the electronic devices were in the range of 0.01–0.03 Ω cm. Therefore, the electron mobility is rather low for 4H-SiC in the above experiment. In n-type SiC wafers, there are several kinds of deep levels or electron traps at a relatively high concentration of 10^{14}–10^{15} [41, 42].

It has been found that stacking faults were formed in highly nitrogen-doped 4H-SiC subjected to an oxidized or annealed in Ar at high temperature [43, 44]. When the nitrogen concentration exceeded 1×10^{9} cm^{-3}, the SF occurrence became distinct. Therefore, in order to decrease the resistivity of SiC, it is necessary to grow lightly nitrogen-doped 4H-SiC crystal and decrease the deep level impurity concentrations to very low level.

Recently, Kato et al. have grown low-resistivity n-type 4H-SiC crystals by sublimation using Al-N co-doping technique [45]. Figure 2.33 shows the schematic illustration of the Al-N co-doping 4H-SiC crystal growth. The furnace has two RF coils for heating SiC and Al source, respectively. N and Al doping levels were controlled in the range of 5×10^{18} cm^{-3} to 1×10^{20} cm^{-3} during growth.

It was found that the stacking faults (SFs) were suppressed in the n-type 4H-SiC grown by the co-doping technique. Figure 2.34 shows the etching morphologies of longitudinal cut SiC crystals after molten KOH etching treatment. In Fig. 2.34a, the etched surface shows morphology with line-shaped etch pits caused by a lot of SFs generated due to the high *N* doping of 3×10^{19} cm^{-3}. In contrast, the high N-Al co-doped crystal does not show any sign of SF generation in spite of the high *N* concentration of 8.8×10^{19} cm^{-3} as shown in Fig. 2.34b. Furthermore, the co-

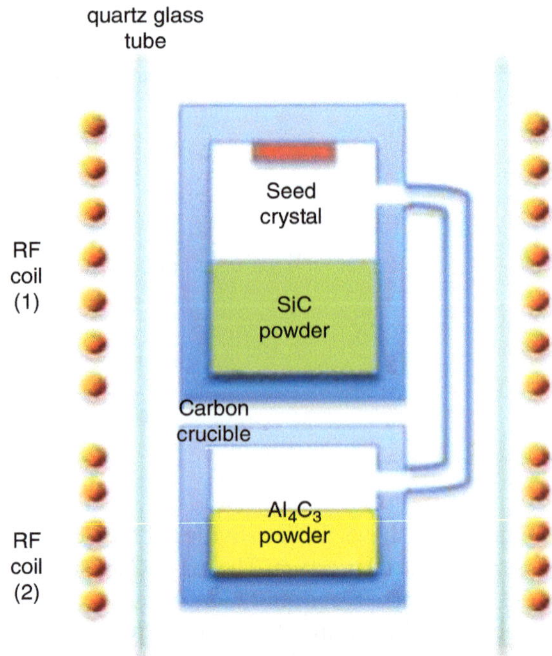

Fig. 2.33 Schematic illustration of Al-N co-doping 4H-SiC crystal growth

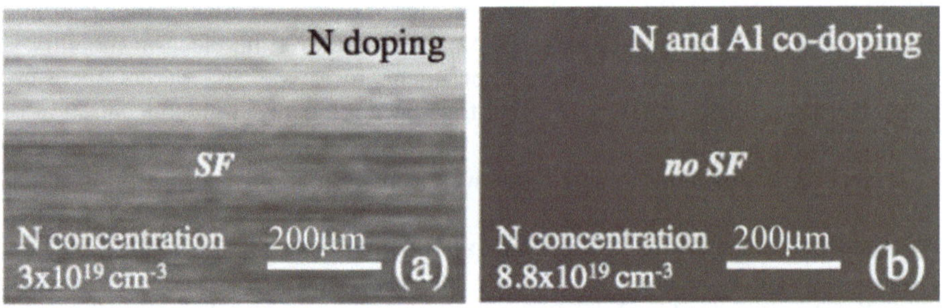

Fig. 2.34 The etching morphologies of longitudinal cut SiC crystals after molten KOH etching treatment. (**a**) Conventional sublimation-grown SiC with *N* concentration of 3×10^{19} cm^{-3}. (**b**) Co-doping-grown SiC with *N* concentration of 8.8×10^{19} cm^{-3} and Al concentration of 4.2×10^{19} cm^{-3}, respectively

doped 4H-SiC crystals show higher activation ratio of donor compared with the conventional nitrogen-doped crystals. When N and Al concentrations for co-doped 4H-SiC are 8.8×10^{19} cm^{-3} and 4.2×10^{19} cm^{-3}, respectively, the resistivity reaches 8.5 mΩ cm.

Therefore, low resistivity for n-type 4H-SiC could be achieved by N-Al co-doping technique. The SFs were suppressed until higher N concentration of 8.8×10^{19} cm^{-3} in the co-doped crystal. At the same time, the co-doped crystals showed higher activation ratio of donor compared with the conventional n-type 4H-SiC crystals.

2.5.2 p-Type Doping

For high-power SiC device application, substrates with low resistivity are required to reduce the unnecessary resistance of the devices. Especially for the fabrication of n-channel IGBTs, showing high performance compared to p-channel IGTs, p-type 4H-SiC substrates with low resistivity are required. Aluminum is often used as p-type dopant in SiC crystal. In general, aluminum doping is realized by adding aluminum or aluminum-containing compound into the SiC source. Aluminum doping is much more difficult than nitrogen doping during SiC crystal growth because severe depletion of the aluminum source occurs in the procedure of crystal growth at high temperature [46]. The Al concentration in crystal is almost proportional to aluminum vapor pressure in the growth chamber.

To realize the effective aluminum doping in 4H-SiC single crystals, Hens et al. have grown p-type SiC by using modified physical vapor transport (M-PVT) technique [47]. Figure 2.35 shows the schematic illustration of M-PVT SiC crystal growth method. In the left configuration of this figure, the Al source is located in the reservoir beneath crucible bottom, and the gas was flowed through the additional pipe only in small amount. If the additional flux is ten times larger than the conventional PVT, crystal growth will be disturbed in the center zone of the seed

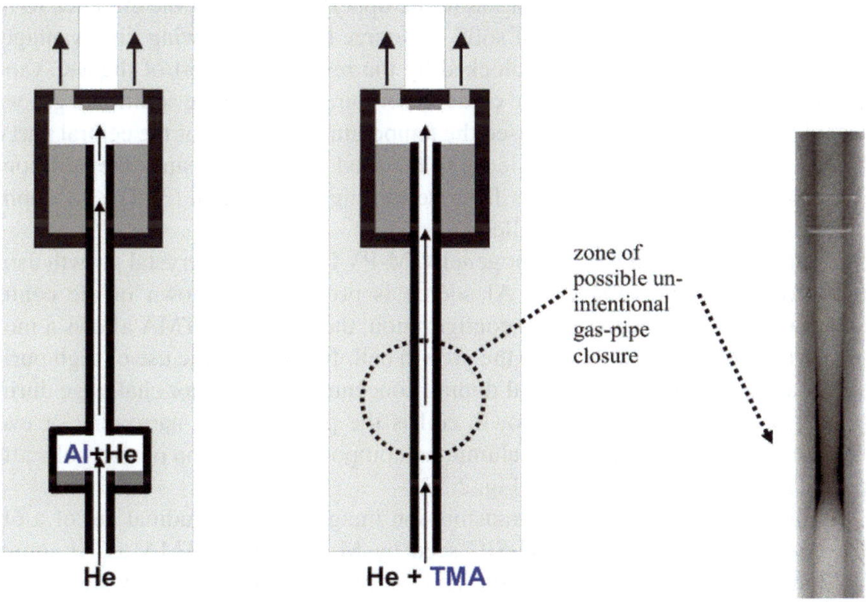

Fig. 2.35 Aluminum p-type doping SiC crystal growth by M-PVT technique. (Left) Regular setup with heated aluminum-containing reservoir. (Center) New setup with TMA source. (Right) X-ray image of gas pipe visible zone of unintentional deposition of carbon/graphite and related aluminum compounds due to decomposition of TMA

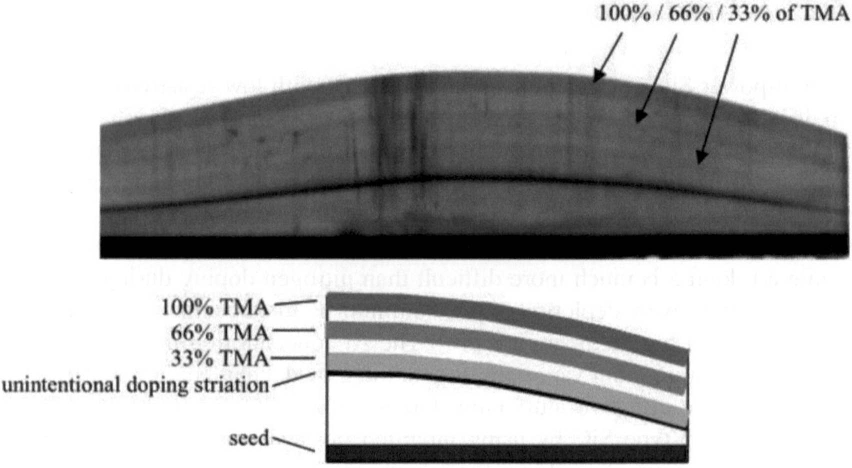

Fig. 2.36 Top: scanning image of longitudinal cut 6H-SiC crystal by M-PVT using TMA as Al source. Al doping concentration is indicated by the gray-level contrast. Bottom: schematic illustration of longitudinal cut of the highlighted doping stripes

crystal. If the additional flux is ten times smaller than the conventional PVT, the pipe will be closed by SiC deposition in the upper pipe zone at the entrance to the growth cell. Therefore, the use of solid Al source has the following disadvantages: (1) the additional pipe is easily blocked by the reverse transport of the SiC vapor species from the hot powder to the colder reservoir. (2) When the additional gas was flowed into the growth cell, it caused the temperature elevating of the central part of the gas room. It further possibly leads to the seed sublimation since the additional gas prevents the Si and C species from depositing on the seed. (3) The Al doping concentration is not easily controlled.

To improve the Al doping homogeneity, M-PVT p-type SiC crystal growth using tri-methyl-aluminum (TMA) as Al source is proposed as shown in the central configuration of Fig. 2.35. In this configuration, the feeding of TMA allows a more precise control of the Al flux into the growth cell. In addition, the use of high purity TMA can reduce the unintentional doping concentration. A major challenge during feeding of TMA into the hot growth cell is the pipe aperture narrowing or even blockade by graphite and related aluminum compound deposition on the walls after long run as shown in the right of Fig. 2.35.

Figure 2.36 showsan optical transmission image of a longitudinal cut of a 6H-SiC crystal grown on *c*-face 6H-SiC seed by M-PVT using TMA as Al source. The total gas flux was kept constant, and the relative values of TMA were varied in the following sequence: 0%, 33%, 0%, 66%, 0%, and 100%. The dark stripes correspond to 100%, 66%, and 33% of TMA, respectively. The black stripe is related to an unintentional high TMA flux when turning on the dopant supply for the first time. It was found that different Al incorporation levels from the low to

Fig. 2.37 Charge carrier mobility versus charge carrier concentration for a number of 6H-SiC crystals grown on *c*-face and Si-face SiC seeds

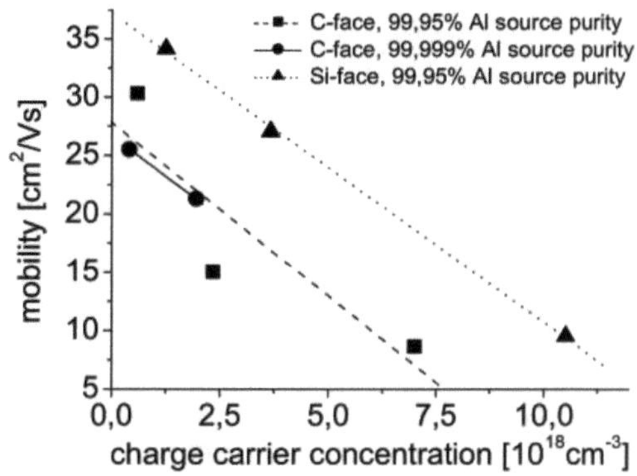

Table 2.3 Result of SIMS, Raman spectroscopy, and resistivity measurement of various crystals grown with Al-N co-doping

Growth no.	Al concentration (cm^{-3})	N concentration (cm^{-3})	Al/N	Dominant polytype	Carrier concentration	Resistivity (mΩ cm)
1	1.8×10^{-20}	1.5×10^{-20}	1.2	4H	3.0×10^{19}	86
2	3.1×10^{-20}	2.3×10^{-20}	1.3	4H	4.0×10^{19}	88
3	2.6×10^{-20}	1.8×10^{-20}	1.4	4H	4.0×10^{19}	97
4	1.5×10^{-20}	1.3×10^{-20}	1.5	6H	–	68

mid-10^{20} cm^{-3} could be achieved in SiC crystals by M-PVT using TMA as the Al source.

From a kinetic viewpoint, *c*-face is unfavorable for aluminum incorporation and favorable for nitrogen incorporation because the Al and N occupy the Si-lattice site and C-lattice site, respectively. The background N often causes the compensation of acceptors in p-type SiC. Even if the compensation is only in the range of 3–5%, it leads to a pronounced reduction of charge carrier mobility and conductivity. Figure 2.37 shows the influence of seed polarity on the compensation of acceptor. In case of Si-face seed, unintentional acceptor compensation of aluminum is much suppressed compared to the use of Si-face seed. From this figure, we can see clearly that the suppression of compensation causes the increase of the charge carrier mobility and further the conductivity increase by more than factor 2.

Recently, low-resistivity p-type 4H-SiC crystals were grown by sublimation using aluminum and nitrogen co-doping [48]. The crystal growth was performed by a two-zone heating furnace, which has two RF coils for heating high purity SiC source material and Al_4C_3. Table 2.3 shows representative results of SIMS, Raman spectroscopy, and resistivity measurement for the grown crystal. The lowest resistivities obtained for 4H and 6H-SiC were 86 mΩ cm and 68 mΩ cm, respectively.

The relation of Al and N concentrations of grown crystals is plotted in Fig. 2.38. It was found that Al/N = 1:1 line acted as a border of n-type and p-type. When the Al/N is larger than 1, the crystal exhibits the p-type electrical characteristics and vice versa.

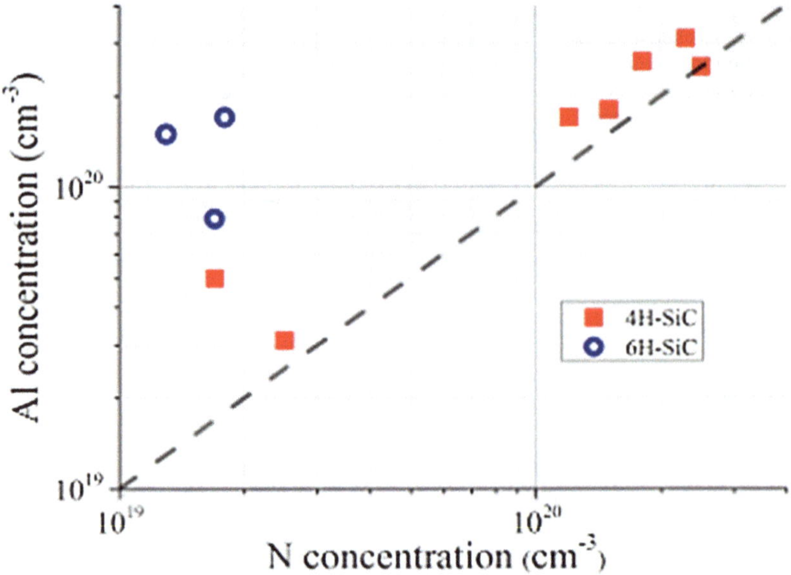

Fig. 2.38 Al and N concentrations measured by SIMS. The guide line show the Al/N = 1:1 line, as a border of n-type and p-type. Notation of 4H and 6H indicates the dominant polytypes of grown crystals

2.5.3 *Semi-insulating*

High-resistivity wafers are generally used fabrication of SiC- or GaN-based high-frequency devices to minimize the parasitic capacitances between the terminals, including the ground. Because it is very difficult to reduce the background dopant density below 10^{10} cm^{-3} by purification processes, compensation of dopants is often used to decrease the density of free carriers in the bands. In SiC, vanadium was the first element used as a compensation center to create semi-insulating wafers. Vanadium is an amphoteric impurity in SiC. It acts as an acceptor-like trap in n-type SiC and a donor-like trap in p-type SiC.

Vanadium deep levels in SiC have been identified by Schneider et al. [49]. Two deep levels are formed by vanadium in 6H-SiC: a V^{3+}/V^{4+} acceptor level about 0.8 eV below the conduction band and a V^{4+}/V^{5+} donor level approximately in the middle of the bandgap, i.e., about 1.5 eV.

Assuming that the donors (mainly nitrogen), acceptors (aluminum, boron, etc.), and vanadium are present simultaneously in the SiC crystals in electrically active concentrations N_D, N_A, and N_v, respectively, four compensation regimes can appear in 6H-SiC.

1. $N_V > N_D - N_A > 0$, vanadium compensates shallow donors which may be themselves partially compensated by acceptors.

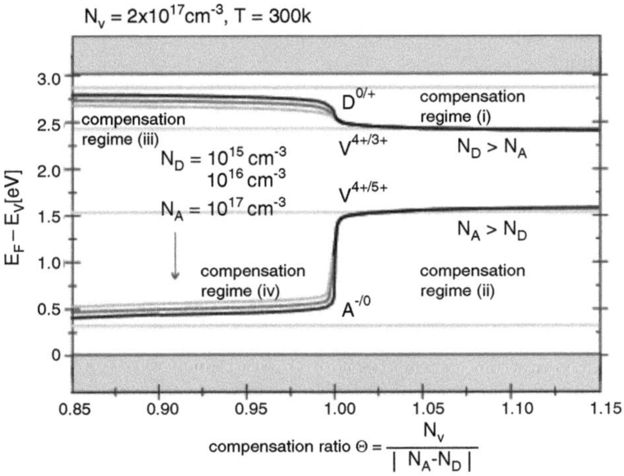

Fig. 2.39 Dependence of the Fermi level on the compensation ratio Θ for different shallow impurity concentrations N_A and N_D in vanadium-doped 6H-SiC

2. $N_V > N_A - N_D > 0$, vanadium compensates shallow acceptors which may be themselves partially compensated by donors.
3. $N_D > N_V + N_A$, domination of shallow donors lead to n-type conducting behavior, $\Theta < 1$.
4. $N_A > N_V + N_D$, domination of shallow acceptors leads to p-type conducting behavior, $\Theta < 1$.

Figure 2.39 shows the position of the Fermi level calculated from the charge carrier neutrality equation

$$n + N_A^- + N_{V(3+)} = p + N_D^+ N_{V(5+)} \tag{2.5}$$

as a function of $\Theta = N_V/|N_D - N_A|$ for different $N_D - N_A$ [50]. As long as vanadium compensates all other impurities as in regimes (1) and (2), semi-insulating SiC will be obtained. Another key condition for the preparation of semi-insulating SiC is that N_V cannot exceed the vanadium solubility limit.

V-doped semi-insulating 6H-SiC bulk single crystals have been grown by sublimation method from different research group. Bickermann et al. [50] found that if vanadium is added as a solid source to the SiC powder, it quickly exhausts during growth. Figure 2.40 is a scanning image of a longitudinal cut of a crystal grown with a vanadium content of 0.32 wt % in the source material. We can see that at the first stage, serious vanadium precipitation occurred. Following the crystal growth preceding, the vanadium concentration was gradually decreased, and the vanadium precipitation particles became smaller in size, but precipitation density does not change significantly. From GDMS measurements at the transition region

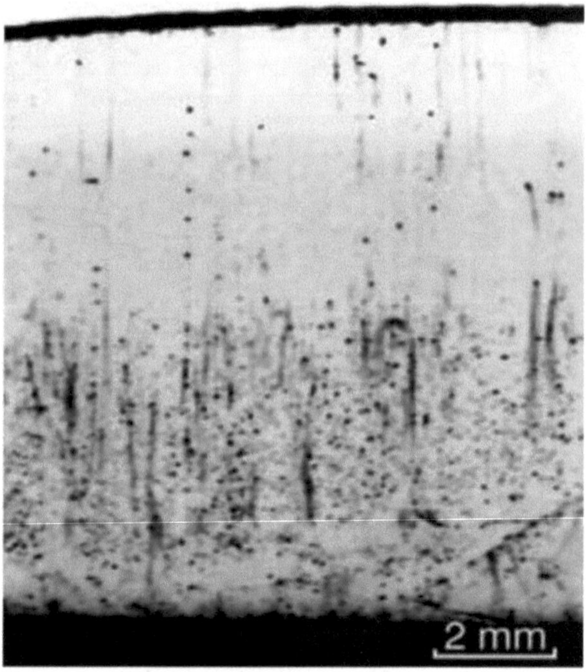

Fig. 2.40 The scanning image of a longitudinal cut 6H-SiC with vanadium doping. Semi-insulating 6H-SiC was grown by sublimation method using the solid vanadium source

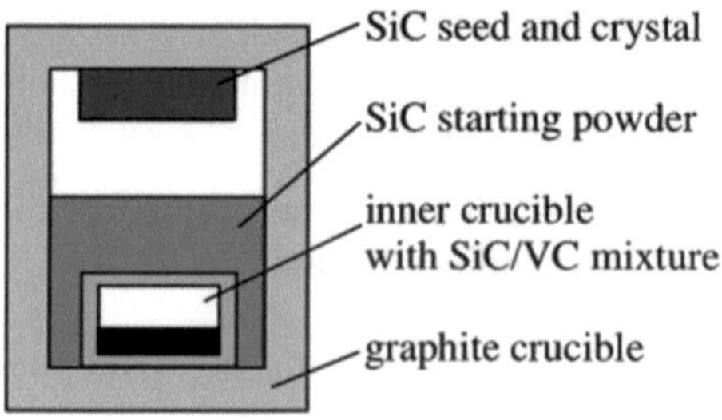

Fig. 2.41 Schematic illustration of the inner crucible located inside the growth container

from precipitated to non-precipitated areas, the vanadium solubility in SiC was confirmed to be about 4×10^{17} cm^{-3}.

In order to increase the semi-insulating yield of the grown crystals and avoid vanadium precipitation or depletion, vanadium concentrations in SiC crystals have to be lower than the solubility but higher than N_D in the whole crystal volume. To realize the vanadium uniform doping in SiC crystal, Bickermann et al. [50] used an inner crucible in which the SiC/VC mixture was filled as shown in Fig. 2.41. In this case, vanadium evaporation rate decreases as the container gets "sealed" in the early stages of growth due to SiC sublimation. A dense SiC/VC mixture sublimes at the top of the inner crucible and acts as an infinite vanadium source with nearly constant supply during SiC crystal growth. Therefore, semi-insulating, precipitate-free SiC crystals can be grown using the inner container doping method. Crystal volume yield is about 80%. The predominantly nitrogen-doped layer forming at the

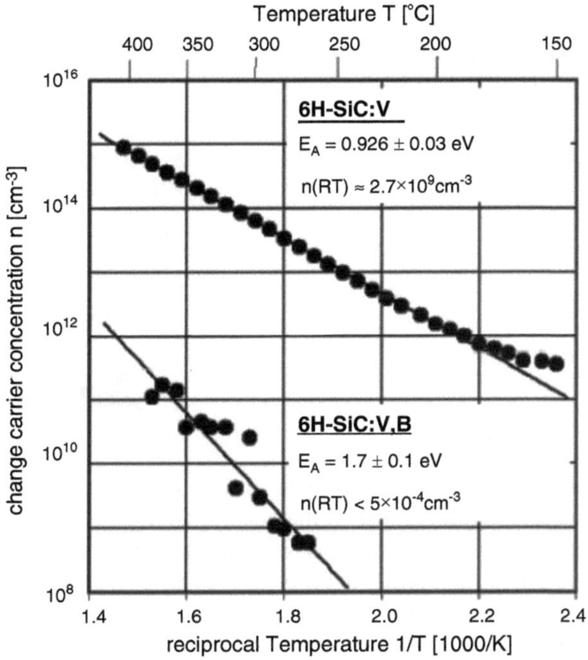

Fig. 2.42 Hall effect measurements of a V-doped and V-B co-doped 6H-SiC crystals

early stages of growth expanded to about 20% of the crystal volume. The vanadium concentration is about 5×10^{17} cm^{-3} near the seed and 2×10^{17} cm^{-3} at the end of growth, respectively. It implies that vanadium uniform doping has been realized. Similarly, V-B co-doping semi-insulating SiC crystals can be grown by using the inner container.

Figure 2.42 shows the Hall effect measurements of a V-doped and a V-B co-doped 6H-SiC crystals. For the V-doped crystal, the thermal activation energies are as high as 926 meV. It confirms that the vanadium compensates all shallow nitrogen donors. Extrapolation to room temperature gives the specific resistivities between 2×10^9 and 6×10^{10} Ω cm.

To overcome the difficulty in vanadium doping, semiconducting character could be also achieved by the use of intrinsic point defects. Therefore, high purity semi-insulating (HPSI) SiC was expected. The development of a semi-insulating SiC material requires two key components. First, a deep electronic level is necessary to trap either electrons or holes and to supply a large activation barrier to their subsequent release. Second, shallow levels are required to supply compensation to the deep electronic level so that the Fermi level can be pinned to the deep electronic level. In high purity SiC, the shallow electronic levels are mainly the boron and nitrogen. The deep electronic levels are vacancies.

The main difficulty for the growth of high purity SiC is the control of impurity incorporation in the crystal during growth. The metallic impurities can be easily eliminated by the high-temperature halogenation. But the nitrogen and boron are

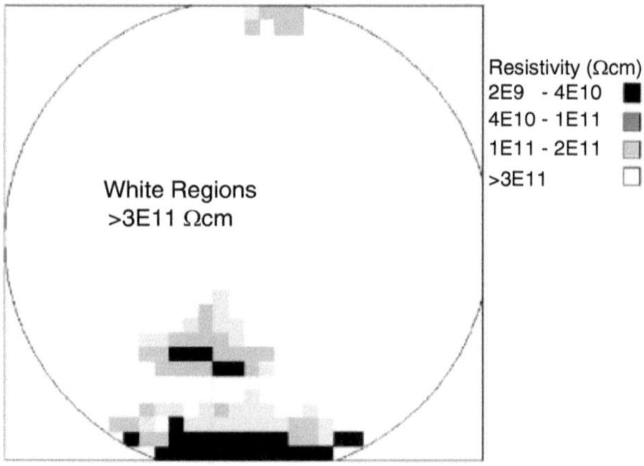

Fig. 2.43 High-resolution resistivity map of a 2″ diameter high purity semi-insulating 4H-SiC

difficult to reduce. Nitrogen comes from the absorbed air of the graphite parts and insulation. Boron is the accompanying product of the graphite. Therefore, the purification of graphite parts, insulation, and source is the critical technology for the growth of high purity SiC crystal. At early research work, Augustine et al. attempted to grow high purity SiC single crystals using sublimation method [51]. Unfortunately, the high purity SiC crystal did not exhibit semi-insulating electrical character due to the impurity contamination. Glow discharge mass spectroscopy (GDMS) measurements reveal that the metallic impurity concentrations are lower than the 1×10^{15} cm^{-3}, while the boron concentration is as high as 6.0×10^{16} cm^{-3}. Crystals exhibit resistivities in the 10^3 Ω cm range and are p-type due to residual high concentration of boron. It failed to obtain the high purity semi-insulating SiC crystals.

Jenny et al. [52, 53] reported the successful growth of high purity semiconducting 4H-SiC crystals by sublimation method. High-resolution resistivity map of a 2 in. diameter 4H-SiC was shown in Fig. 2.43. The entire wafer is semi-insulating with resistivity greater than 2×10^9 Ω cm. More than 80% of the wafer area has a resistivity $>3 \times 10^{11}$ Ω cm.

Secondary ion mass spectroscopy detection for the HPSI indicated that the only elements were nitrogen (2×10^{16} cm^{-3}) and boron (7×10^{15} cm^{-3}), and the concentrations of all other elements were below the SIMS detection limits or below the instrument background.

For examining electronic levels in semiconductors, Hall effect is an effect tool. Using the technique, we can determine several important electronic parameters, including the activation energy of the electronic level, the concentration of the principal defect, and the total net compensation. For the HPSI samples, the Hall effect is not suitable due to the extremely low mobility of the sample. Figure 2.44 shows the temperature-dependent resistivities of five HPSI samples. From the resistivity versus temperature, the obtained activation energies are 0.9, 1.1, 1.2,

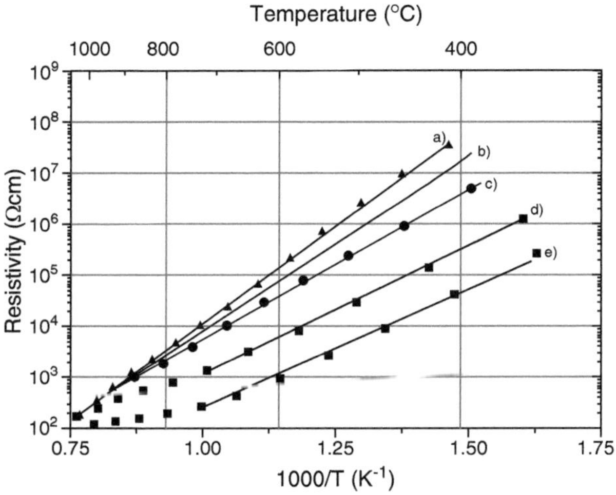

Fig. 2.44 Temperature-dependent resistivities for five HPSI samples whose activation energies are (**a**) 1.53 eV, (**b**) 1.33 eV, (**c**) 1.19 eV, (**d**) 1.06 eV, and (**e**) 0.90 eV

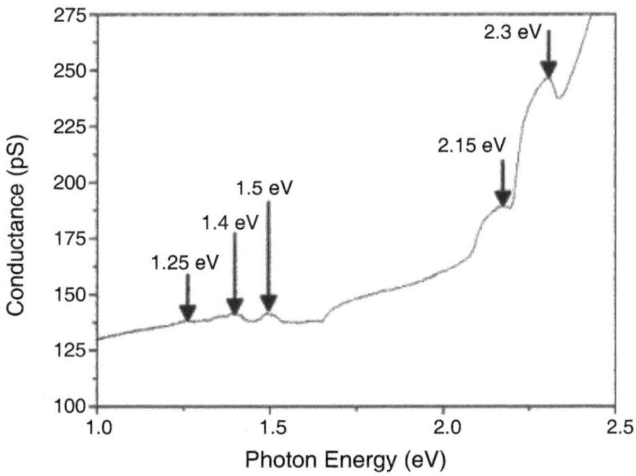

Fig. 2.45 Admittance versus photon energy plot that reveals the presence of five distinct energy levels

1.35, and 1.5 eV. Therefore, several defects are responsible for the semi-insulating character in the HPSI samples.

To further confirm the existence of multiple deep levels in HPSI, optical admittance spectroscopy (OAS) was used to measure the admittance of HPSI sample.

Figure 2.45 shows the OAS data of HPSI samples. It reveals that the activation energies for several deep levels are 1.25, 1.4, 1.5, 2.15, and 2.3 eV.

For activation energies greater than half of the bandgap, the bandgap energy (3.2 eV for 4H-SiC) has to be subtracted so that the data can be compared with the data obtained by other characterization techniques. Therefore, the actual activation

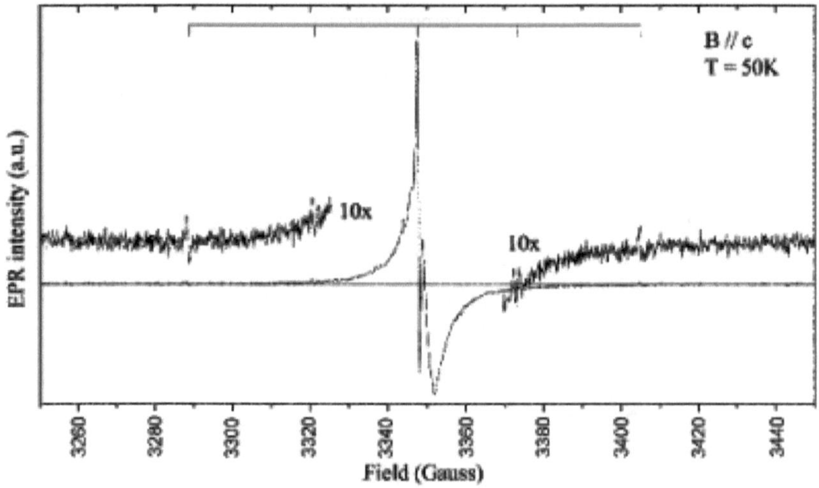

Fig. 2.46 The EPR spectrum of a 4H-SiC HPSI sample whose Fermi level was pinned to the 1.5 eV level and exhibits the carbon vacancy

energies are 0.9, 1.05, 1.25, 1.4, and 1.5 eV after the bandgap energy subtraction, which are very close to those obtained by temperature-dependent resistivity. The above two-group data proves that the source of the deep levels is probably to be intrinsic in nature.

Figure 2.46 is an electron paramagnetic resonance (EPR) spectrum of a sample exhibiting the activation energy of 1.5 eV. The large sharp peak with two hyperfine doublets is the main feature in the spectrum. It was identified as a carbon vacancy. The fact that defect is present in SI material suggests that it is electrically active and is responsible for at least one of the defects observed in the temperature-dependent resistivity measurements.

In addition, the SI material has a high thermal stability. After the HPSI wafers were subjected to 1650 °C annealing, for 90 min, no degradation in the SI character was observed in the annealed HPSI wafer. SI character preservation makes the HPSI wafers suitable for the application as ideal substrate for microwave devices.

2.6 Processing of Large-Diameter SiC Wafers

The processing from ingot to wafer for SiC is same as that for other semiconductors. SiC single crystal ingots are generally grown on {0 0 0 1} seed. The typical ingots are cylindrical in shape with a length of 20–30 mm. The crystallographic orientations of the ingots such as <0 0 0 1>, <1 1−2 0>, and <1 0−1 0> are determined by X-ray diffraction. Then the ingots are sliced into a number of wafers which were finally lapped, polished, and cleaned.

Fig. 2.47 As-grown SiC single crystal boules (Provided by Dr. Yan Peng at Shandong University)

SiC single crystal substrates are considered to be suitable for thin film growth of semiconductive GaN with wide energy bandgaps [54], because the lattice mismatch between SiC and GaN is quite small (3.5%) compared with sapphire substrates (16.1%) [55]. The quality of the deposition films is strongly dependent on the quality of substrates. Therefore, the processing of SiC wafers is extraordinarily important for film growth. However, the hardness of SiC is very closer to that of diamond, which makes it greatly difficult to process large-diameter SiC crystals. In addition, SiC substrate is very sensitive to surface defects resulting from lapping and polishing, which may propagate into an epitaxial layer and degrade device quality. Since MP achieves material removal through plastic deformation, scratches and subsurface damage layer containing dislocations are unavoidable [56]. Research on the epitaxial growth of 4H-SiC has shown that scratches are the primary contributors to polytype inclusions in the epilayers [57]. The polishing process of SiC substrates [51, 58–60] was investigated extensively. The removal mechanism of scratches in MP and physical and chemical process in CMP were analyzed. KOH etching and HRXRD were applied to evaluate the subsurface damage of SiC substrates. The effects of MP and CMP on the surface roughness were assessed. Three main procedures of SiC wafer processing were introduced. Variations of surface residual stresses during mechanical processing could be assessed by HRXRD.

SiC single crystals were grown by the sublimation method. Figure 2.47 is the image of as-grown SiC single crystals. The crystal boule was sliced into a number of wafers by diamond wires. Surface morphologies of polished wafers were observed by means of AFM and optical microscope. High-resolution X-ray diffractometer and KOH etching were used to assess the structural perfection of the wafer. Step profilometer and flatness measurement system were used to investigate the surface roughness and flatness. The SiC wafer processing includes the following steps.

2.6.1 Crystal Boule Slicing

It is very significant to slice high-quality SiC crystals into thin wafers with minimum warp, uniform thickness, and low kerf loss; otherwise it will bring great difficulty for succedent lapping and polishing. This has revitalized many interests in wire saw machining technology [58]. In general, diamond wire saw was used to slice SiC boule into wafers. The total thickness variation is less than 30 μm with warp about 20 μm for 2 in. wafers.

2.6.2 Lapping

Lapping is the second step for machining SiC in large-scale production. However, the lapping of silicon carbide is difficult because of its low fracture toughness, making it very sensitive to cracking [59]. Lapping process involves many parameters such as plate speed, pressure, grit size, abrasive density, etc. Thus, efficient lapping requires strict selection for operating parameters to maximize removal rate while controlling surface integrity.

Lapping plate is a cast-iron disk, its rotation speed is 40–70 rpm, and the applied pressure is 50–200 g/cm^2. The abrasive (B_4C) grain size is between 100 and 20 μm. The SiC material removal rate (MRR) is 5–20 μm/h owing to its high hardness and high wear resistance. MRR is increscent with the increasing of abrasive grit, density, plate speed, and pressure. However, a very high B_4C concentration is found to reduce the removal rate [60].

After lapping,the surface unflatness is 6 μm as shown in Fig. 2.48. Figure 2.49 shows the optical microscopic image of lapping surface. It can be seen that there exist lots of pits; the surface roughness R_a is about 100 nm after lapping. Due

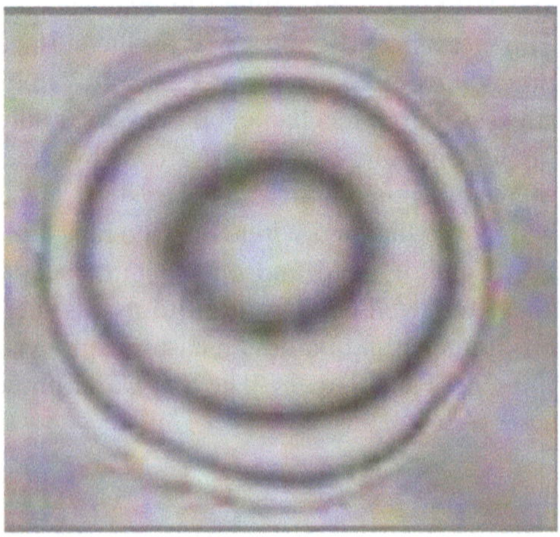

Fig. 2.48 Surface unflatness of SiC wafer after lapping (Provided by Dr. Yan Peng at Shandong University)

Fig. 2.49 Optical microscopic image of SiC wafer surface after lapping (Provided by Dr. Yan Peng at Shandong University)

to the great abrasive grit of lapping, it causes serious surface damage and large residual stresses. X-ray rocking curve measurement of (0 0 0 4) reflection for the wafer indicated that the full-width at the half-maximum (FWHM) is 122.4 arcs after lapping.

In the whole machining process, each step is very important for the successive processing. In order to reduce the total manufacturing time, it is preferable to obtain better ground surfaces, even if this takes longer lapping time [61].

2.6.3 Mechanical Polishing

MP is an important process aimed to remove the damaged layer caused by lapping to make a good surface and to amend the surface geometry exactly. The mechanism of SiC material removal by finer grit; harder abrasives, such as B_4C; and diamond in MP process actually includes the formation of mechanical microfractures and then the removal of these microfractures because of higher hardness of the abrasive and inherent hardness of the workmaterial [62].

The surface roughness after MP is $R_a = 3.1$ nm (measured by step profilometer). As shown in Fig. 2.50, the surface unflatness is 2 μm. Compared with surface quality after lapping, the surface unflatness decreased further after MP.

The polishing procedure was generally performed by diamond-based polishing slurries with decreasing grain size. A large number of scratches on the surface could be observed by AFM after MP. These scratches spread out in random directions with the depth in the range of 5–8 nm as shown in Fig. 2.51. This also indicates

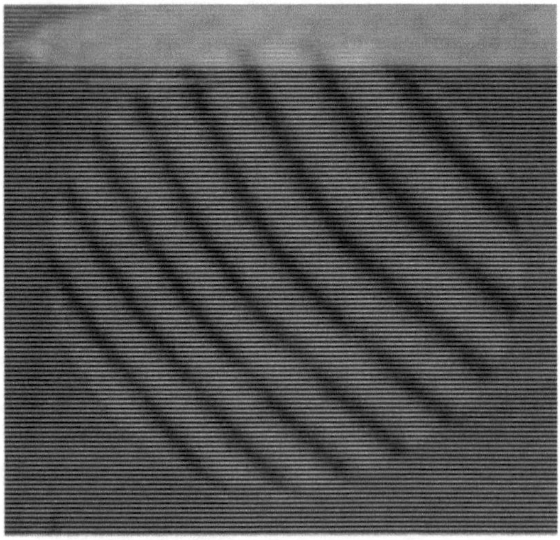

Fig. 2.50 Surface unflatness of SiC wafer after MP (Provided by Dr. Yan Peng at Shandong University)

that the material removal in MP procedure is dominant by the formation of brittle microfracture on a submicroscopic scale.

For both the mechanical lapping and polishing, subsurface damage was formed by harder abrasives. The subsurface damage has been found to extend in a depth range from one-half to one-seventy-fifth of the abrasive slurry particle size [63]. KOH etching was used to identify the subsurface damage as a part of the optimization of CMP process. Figure 2.52 displays the full-width at the half-maximum (FWHM) of (0004) reflection rocking curve of the SiC surface etched by KOH after MP with 2 μm diamond powder. It can be seen that when the etched depth reaches about 0.77 μm, the FWHM tends to keep constant. Thus, the damage depth of the surface can be considered to be about 0.77 μm.

2.6.4 Chemo-mechanical Polishing

CMP can be used to finish hard, brittle workmaterials with extremely smooth and damage-free surfaces. It depends on both chemical and mechanical effectiveness of the abrasive and the environment with respect to the workmaterial. In MP, diamond is harder than SiC; although the surface defects can be minimized by using very fine diamond particles, they cannot be completely eliminated [64]. Therefore, to produce extremely smooth and damage-free surface, further polishing process must be carried on. CMP was first demonstrated by Yasunaga et al. for polishing single crystal silicon using a soft abrasive ($BaCO_3$) [65]. Later on, this method was widely used to polish other crystal materials.

Theoretically any abrasive that can react with the workmaterial in a given environment and form a reaction product can be used for CMP [59]. For SiC CMP, the selected abrasive is SiO_2 which is much softer than SiC. The grit size of SiO_2 is 32 nm.

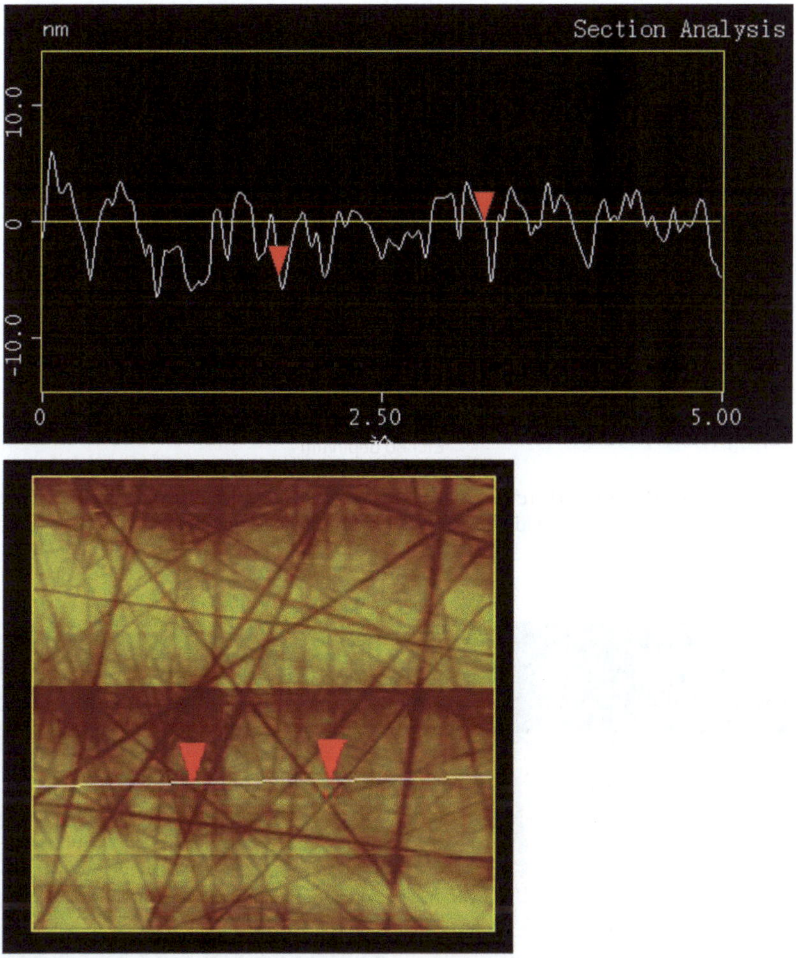

Fig. 2.51 AFM micrograph of the surface after MP ($5 \times 5\ \mu m^2$) (Provided by Dr. Yan Peng at Shandong University)

The wafer surface after CMP has a final surface roughness $R_a = 0.18$ nm (measured by AFM). Figure 2.53 is the AFM image of the surface finished by CMP, it reveals that the surface is extremely smooth, and there are no any scratches and other surface defects. Chemical reactions are performed between suspended silicon bonds in SiC and the alkaline (PH > 10) slurry of colloidal silica. The main chemical equation is as follows:

$$Si + 2\,NaOH + H_2O \rightarrow Na_2SiO_3 + 2\,H_2 \uparrow \tag{2.6}$$

The reaction product Na_2SiO_3 is soluble and can be removed easily from SiC surface by subsequent mechanical action by SiO_2. Chemical reaction will continue

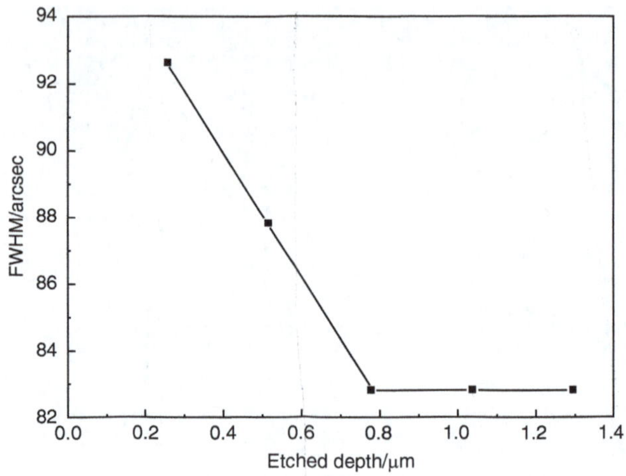

Fig. 2.52 The FWHM of (0004) reflection rocking curve of the surface etched by KOH after MP with 1 μm diamond powder (Provided by Dr. Yan Peng at Shandong University)

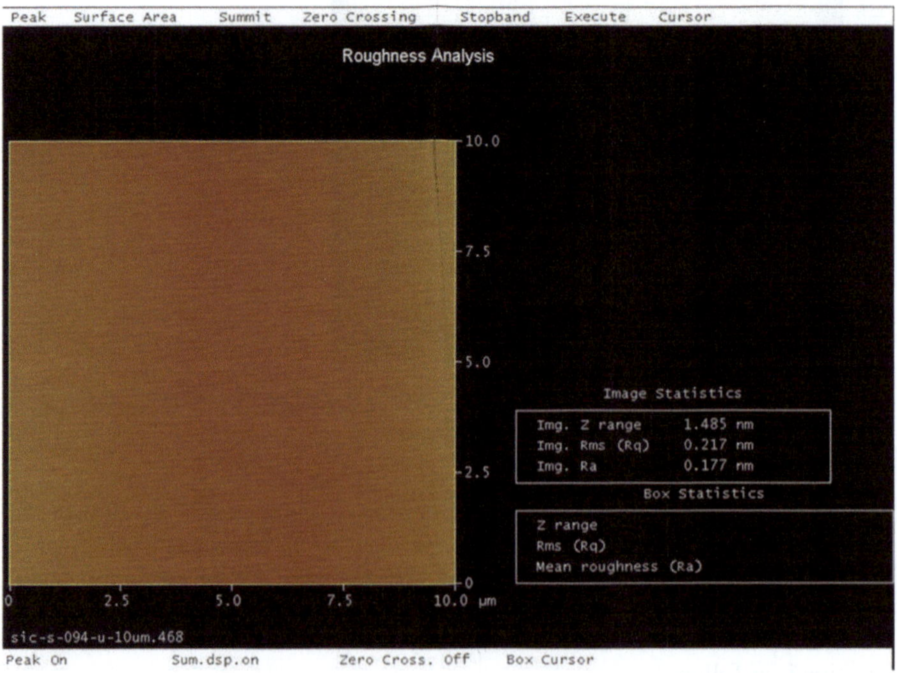

Fig. 2.53 AFM image of the SiC wafer surface finished by CMP (10 × 10 μm) (Provided by Dr. Yan Peng at Shandong University)

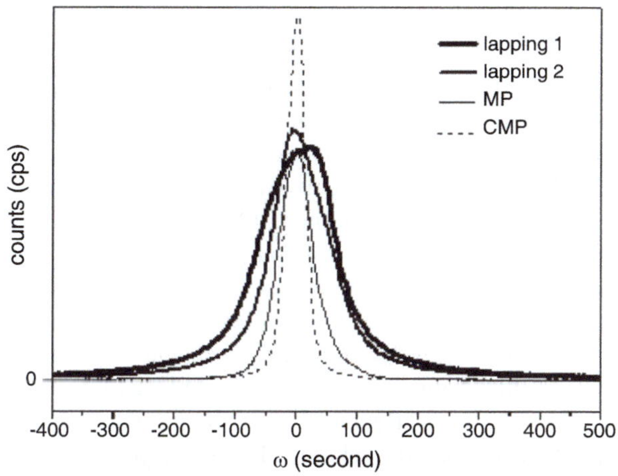

Fig. 2.54 Rocking curves of the SiC wafer surface in each machining process (Provided by Dr. Yan Peng at Shandong University)

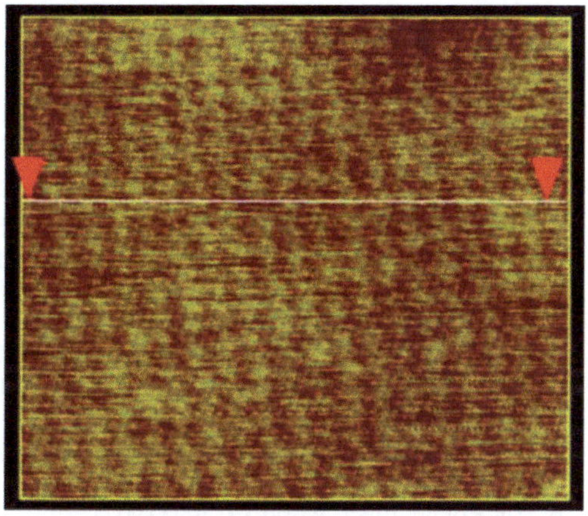

Fig. 2.55 AFM image of the CMP surface after KOH etching (Provided by Dr. Yan Peng at Shandong University)

after the passivating layers are removed [63]. Since material removal by this mechanism does not depend on the hardness but on the chemical potentials, it is possible to remove material by abrasives substantially softer than the workmaterial [63].

By HRXRD analysis shown in Fig. 2.54, the FWHM of (0004) reflection rocking curve of SiC wafer surface finished by CMP is 39.4 arcs which decreases much more than that after lapping. This indicates thin subsurface damage layer, good crystal lattice integrity, and low residual stresses in SiC wafer after CMP. From Fig. 2.54, it can be seen that FWHM falls down with each machining process. It proves that the damage layer and residual stresses have been minimized when the grain size of the abrasive decreases. The subsurface damage has been found to extend from depths ranging from one-half to one-seventy-fifth of the abrasive slurry size [64]. No scratches were visible on surfaces after KOH etching as shown in Fig. 2.55.

Three main machining processes of 2 in. SiC wafers were introduced. Slicing marks can be removed step by step by lapping and polishing. Lapping causes great residual stresses and deep damage layer which can be reduced gradually with subsequent polishing processes. Mechanical polishing produced smooth surface with a large number of scratches. These scratches can be effectively removed by CMP. After CMP, extremely smooth and low damage layer surface with roughness $R_a = 0.3$ nm was obtained. FWHM falls down with each machining process confirming that the surface damage layer and residual stresses minimize when the grain size of abrasive decreases. By the three main machining processes mentioned above, the final surface has low residual stresses and fine surface finish.

In summary, we introduce the SiC single crystal growth and substrate technologies in this chapter. It is difficult to reflect fully the status of SiC material research and development. Recently, new SiC growth technologies such as solution method and high-temperature chemical vapor deposition are well developed. However, we still hope that the contents in this chapter are helpful for SiC crystal grower and substrate processing technician.

References

1. J.B. Casady, R.W. Johnson, Status of silicon carbide (SiC) as a wide-bandgap semiconductor for high-temperature applications: a review. Solid State Electron. **39**, 1409 (1996)
2. R. Han, X. Xu, X. Hu, et al., Development of bulk SiC single crystal grown by physical vapor transport method. Opt. Mater. **23**, 415 (2003)
3. J.J. Berzelius, Ann. Phys. Chem. Lpz. **1**, 169 (1824)
4. E.G. Acheson Production of artificial crystalline carbonaceous materials, carborundum. English Patent 17911 (1892)
5. H. Moissan, Étude du siliciure de carbone de la météorite de cañon diablo. C. R. Acad. Sci. **140**, 405 (1905)
6. H.J. Round, Elect. World **19**, 309 (1907)
7. W.F. Kmippenberg, Growth phenomena in silicon carbide. Philips Res. Rep. **18**, 16 (1963)
8. Y.M. Tairov, V.F. Tsvetkov, General principles of growing large-size single crystals of various silicon carbide polytypes. J. Cryst. Growth **52**, 146 (1981)
9. A.R. Powell, J.J. Sumakeris, Y. Khlebnikov, et al., Bulk growth of large area SiC crystals. Mater. Sci. Forum **858**, 5 (2016)
10. D.H. Hofmann, M.H. Müller, Prospects of the use of liquid phase techniques for the growth of bulk silicon carbide crystals. Mater. Sci. Eng. B **61**, 29 (1999)
11. K. Danno, H. Saitoh, A. Seki, et al., High-speed growth of high-quality 4H-SiC bulk by solution growth using Si-Cr based melt. Mater. Sci. Forum **645**, 13 (2010)
12. J. Kojima, Y. Tokuda, E. Makino, N. Sugiyama, et al., Developing technologies of SiC gas source growth method. Mater. Sci. Forum **858**, 23 (2015)
13. G.R. Fisher, P. Barnes, Toward a unified view of polytypism in silicon carbide. Philos. Mag. B **61**, 217 (1990)
14. R.P. Adrian, B.R. Larry, SiC materials-progress, status, and potential roadblocks. Proc. IEEE **90**, 942 (2002)
15. S.Y. Kaprov, Y.N. Makarov, M.S. Ramm, Simulation of sublimation growth of SiC single crystals. Phys. Status Solidi B **202**, 201 (1997)
16. M. Pons, E. Blanquet, J.M. Dedulle, Thermodynamic heat transfer and mass transport modeling of the sublimation growth of silicon carbide crystals. J. Electrochem. Soc. **143**, 3727 (1996)

17. M. Selder, L. Kadinski, Y. Makarov, et al., Global numerical simulation of heat and mass transfer for SiC bulk crystal growth by PVT. J. Cryst. Growth **211**, 333 (2000)
18. P. Pirouz, On micropipes and nanopipes in SiC and GaN. Philos. Mag. A **78**, 727 (1998)
19. F.C. Frank, Capillary equilibria of dislocated crystals. Acta Cryst **4**, 407 (1951)
20. S.I. Maximenko, P. Pirouz, T.S. Sudarshan, Open core dislocations and surface energy of SiC. Mater. Sci. Forum **527-529**, 439 (2006)
21. S.I. Weimin, M. Dudley, R. Glass, V. Tsvetkov, C.H. Carter Jr., Hollow-core screw dislocations in 6H-SiC single crystals: a test of Frank's theory. J. Electron. Mater. **26**, 128 (1997)
22. P. Krishna, S.S. Jiang, A.R. Lang, An optical and X-ray topographic study of giant screw dislocations in silicon carbide. J. Cryst. Growth **71**, 41 (1985)
23. X.R. Huang, M. Dudley, V.M. Vetter, W. Huang, S. Wang, Direct evidence of micropipe-related pure superscrew dislocations in SiC. Appl. Phys. Lett. **74**, 355 (1999)
24. X.R. Huang, M. Dudley, W.M. Vetter, W. Huang, W. Si, C.H. Carter Jr., Superscrew dislocation contrast on synchrotron white-beam topographs: an accurate description of the direct dislocation image. J. Appl. Crystallogr. **32**, 516 (1999)
25. J. Heindl, W. Dorsch, H.P. Strunk, Dislocation content of micropipes in SiC. Phys. Rev. Lett. **80**, 740 (1998)
26. T.A. Kuhr, E.K. Sanchez, M. Skowronski, Hexagonal voids and the formation of micropipes during SiC sublimation growth. J. Appl. Phys. **89**, 4625 (2001)
27. R.C. Glass, D. Henshall, V.F. Tsvetkov, C.H. Carter Jr., SiC seeded crystal growth. Phys. Status Solidi A **202**, 149 (1997)
28. H. Shiomi, H. Kinoshita, T. Furusho, T. Hayashi, et al., Crystal growth of micropipe free 4H-SiC on 4H-SiC (0 3 -3 8) seed and high purity semi-insulating 6H-SiC. J. Cryst. Growth **292**, 188 (2006)
29. J. Li, O. Filip, B.M. Epelbaum, X. Xu, M. Bickermann, A. Winnacker, Growth of 4H-SiC on rhombohedral (0 1 -1 4) plane seeds. J. Cryst. Growth **308**, 41 (2007)
30. W.F. Knippenberg, Growth phenomena in silicon carbide. Philips Res. Rep. **18**, 161 (1963)
31. T. Shiramomo, B. Gao, F. Mercier, S. Nishizawa, et al., Thermodynamical analysis of polytype stability during PVT growth of SiC using 2D nucleation theory. J. Cryst. Growth **352**, 177 (2012)
32. K. Kakimoto, B. Gao, T. Shiramomo, S. Nakano, S.I. Nishizawa, Thermodynamic analysis of SiC polytype growth by physical vapor transport method. J. Cryst. Growth **324**, 78 (2011)
33. N. Sugiyama, A. Okamoto, K. Okumura, T. Tani, N. Kamiya, Step structures and dislocations of SiC single crystals grown by modified lely method. J. Cryst. Growth **191**, 84 (1998)
34. Y. Nakano, T. Nakamura, A. Kamisawa, H. Takasu, Investigation of pits formed at oxidation on 4H-SiC. Mater. Sci. Forum **600–630**, 377 (2009)
35. R. Singh, K.G. Irvine, D.C. Capell, J.T. Richmond, D. Berning, A.R. Hefner, Large area ultra-high voltage 4H-SiC p-i-n rectifiers. IEEE Trans. Electron. Devices **49**, 2308 (2002)
36. J. Zhang, P. Alexandrov, T. Burke, J.H. Zhao, 4H-SiC power bipolar junction transistor with a very low specific on-resistance of 2.9 $m\Omega\ cm^2$. IEEE Electron Device Lett. **27**, 368 (2006)
37. S. Ryu, A.K. Agarwal, R. Singh, J.W. Palmour, 1800V NPN bipolar junction transistors in 4H-SiC, 2001. IEEE Electron Device Lett. **22**, 124 (2001)
38. B. Nakamura, I. Tunjishima, S. Yamaguchi, T. Ito, et al., Ultrahigh quality silicon carbide single crystals. Nature **430**, 1009 (2004)
39. N. Ohtani, M. Katsuno, J. Takahashi, et al., Impurity incorporation kinetics during modified-lely growth of SiC. J. Appl. Phys. **83**, 4487 (1998)
40. K. Onoue, T. Nishikawa, M. Katsumo, et al., Nitrogen incorporation kinetics during the sublimation growth of 6H and 4H-SiC. Jpn. J. Appl. Phys. **35**, 2240 (1996)
41. S. Jang, T. Kimoto, H. Matsunami, Deep levels in 6H-SiC wafers and step controlled epitaxial layers. Appl. Phys. Lett. **65**, 581 (1994)
42. A.O. Evwaraye, S.R. Smith, W.C. Mitchel, Shallow and deep levels in n-type 4H-SiC. J. Appl. Phys. **79**, 7726 (1996)

43. M. Katsuno, M. Nakabayashi, T. Fujimoto, et al., Stacking fault formation in highly nitrogen-doped 4H-SiC substrates with different surface preparation conditions. Mater. Sci. Forum **600-630**, 341 (2009)
44. N. Ohtani, M. Katsuno, M. Nakabayashi, et al., Investigation of heavily nitrogen-doped n^+ 4H-SiC crystals grown by physical vapor transport. J. Cryst. Growth **311**, 1475 (2009)
45. T. Kato, K. Eto, S. Takagi, T. Miura, et al., Growth of low resistivity n-type 4H-SiC bulk crystals by sublimation method using co-doping technique. Mater. Sci. Forum **778-780**, 47 (2014)
46. N. Schulze, J. Gajowski, K. Semmelroth, M. Laube, G. Pensl, Growth of highly aluminum-doped p-type 6H-SiC single crystals by the modified lely method. Mater. Sci. Forum **353-356**, 45 (2001)
47. P. Hens, U. Kunecke, P. Wellmann, Aluminum p-type doping of bulk SiC single crystals by tri-methyl-aluminum. Mater. Sci. Forum **600-603**, 19 (2009)
48. K. Eto, H. Suo, T. Kato, H. Okumura, Growth of low resistivity p-type 4H-SiC crystals by sublimation with using aluminum and nitrogen co-doping. Mater. Sci. Forum **858**, 77 (2015)
49. J. Schneider, H.D. Muller, M. Maier, W. Wilkening, F. Fuchs, Infrared spectra and electron spin resonance of vanadium deep level impurities in silicon carbide. Appl. Phys. Lett. **56**, 1184 (1990)
50. M. Bickermann, R. Weingartner, A. Winnacker, On the preparation of vanadium doped PVT grown SiC boules with high semi-insulating yield. J. Cryst. Growth **254**, 390 (2003)
51. G. Augustine, V. Balakrishna, C.D. Brandt, Growth and characterization of high purity SiC single crystals. J. Cryst. Growth **211**, 339 (2000)
52. J.R. Jenny, S. Muller, A. Powell, V.F. Tsvetkov, et al., High purity semi-insulating 4H-SiC grown by the seeded sublimation method. J. Electron. Mater. **31**, 366 (2002)
53. J.R. Jenny, D.P. Malta, S. Muller, et al., High purity semi-insulating 4H-SiC for microwave device applications. J. Electron. Mater. **32**, 432 (2003)
54. T. Sasaki, T. Matsuoka, Substrate-polarity dependence of metal-organic vapor-phase epitaxy-grown GaN on SiC. J. Appl. Phys. **64**, 4531 (1988)
55. P. Kung, C.J. Sun, A. Saxler, H. Ohsato, M. Razeghi, Crystallography of epitaxial growth of wurtzite-type thin films on sapphire substrates. J. Appl. Phys. **75**, 4515 (1994)
56. S. Yu, S. Karpov, A.V. Kulik, I.A. Zhmakin, Y.N. Makarov, et al., Analysis of sublimation growth of bulk SiC crystals in tantalum container. J. Cryst. Growth **211**, 347 (2000)
57. N. Ohtani, T. Fujimoto, M. Katsuno, T. Aigo, et al., Growth of large high-quality SiC single crystals. J. Cryst. Growth **237–239**, 1180 (2002)
58. W.I. Clark, A.J. Shih, C.W. Hardin, R.I. Lemaster, et al., Fixed abrasive diamond wire machining—part I: process monitoring and wire tension force. Int J Mach Tool Manu **43**, 523 (2003)
59. H.K. Xu, S. Jahanmir, L.K. Ives, Material removal and damage formation mechanisms in grinding silicon nitride. J. Mater. Res. **11**, 1717 (1996)
60. M. Forsberg, N. Keskitalo, J. Olsson, Effect of dopants on chemical mechanical polishing of silicon. Microelectron. Eng. **60**, 149 (2002)
61. Z. Zhong, Surface finish of precision machined advanced materials. J. Mater. Process. Technol. **122**, 173 (2002)
62. M. Jiang, R. Komanduri, On the finishing of Si3N4 balls for bearing applications. Wear **215**, 267 (1998)
63. P. Vicente, D. David, J. Camassel, Raman scattering as a probing method of subsurface damage in SiC. Mater. Sci. Eng. B **80**, 348 (2001)
64. M. Jiang, N.O. Wood, R. Komanduri, On chemo-mechanical polishing (CMP) of silicon nitride (Si3N4) workmaterial with various abrasives. Wear **220**, 59 (1998)

Chapter 3
Homoepitaxy of GaN Light-Emitting Diodes

Ke Xu, Miao Wang, Taofei Zhou, and Jianfeng Wang

3.1 Bulk GaN Substrates for Light-Emitting Devices

Crystal growth technologies are widely developed for the realization of GaN substrate with large size and high quality [1]. There are mainly three routes. The first one is improved from high-pressure high-temperature growth method. By adding sodium into the Ga melt for the enhancement of N concentration, the sodium flux growth method can grow GaN under temperature about 800 °C and lower pressure about 5 MPa [2]. The second route is referenced from hydrogen-thermal method, which is very mature for the growth of quartz. It is named as ammonothermal method by using ammonia as the solvent and KNH_4 or NH_4Cl as mineralizer; GaN crystal can be grown under low temperature about 600 °C and high pressure about 400 MPa [3]. The third growth method is HVPE, which is initially used in the growth of GaAs and InP [4]. With the advantage of high growth rate as several hundred micrometers per hour by HVPE method, GaN can be grown up to several millimeters, which is thick enough for the fabrication of bulk substrates. However, since there is a lack of GaN seed crystal, thick GaN films have to be grown on foreign substrates, such as sapphire, GaAs, and silicon. The big lattice and thermal mismatch between GaN film and the starting foreign substrate bring big challenge of strain control and dislocation reduction. On the other hand, separation of GaN from the foreign substrate is another challenge after the HVPE growth. Self-separation and laser lift-off technique are well developed to obtain crack-free GaN films.

K. Xu (✉) · M. Wang · T. Zhou · J. Wang
Suzhou Institute of Nano-tech and Nano-bionics, Chinese Academy of Sciences, Suzhou, China
e-mail: kxu2006@sinano.ac.cn

J. Li, G. Q. Zhang (eds.), *Light-Emitting Diodes*, Solid State Lighting Technology and Application Series 4, https://doi.org/10.1007/978-3-319-99211-2_3

In this section, the progress in growth of bulk GaN by HVPE is introduced. The main challenges and solutions of HVPE growth method include dislocation reduction, strain control, and doping of GaN.

3.1.1 Growth Mechanism of HVPE System

Generally, HVPE reactor of GaN mainly includes two reacting zones. One is the source zone for providing chloride gas of gallium, and the other is deposition zone, in which the GaN is formed from the reaction of gallium source and nitride source (NH_3). Therefore, in the HVPE growth, main chemical reactions can be divided into two parts. One is in the source zone where the gallium is kept at a certain temperature and reacts with HCl gas, which is commonly used as reactive gas and introduced into the source zone over metal to form chloride gas of gallium. Then the gaseous species formed at the source zone are transported by carrier gas to deposition zone to further react with nitride source to form GaN. The carrier gases are generally H_2 and inert gas (N_2, He, Ar, etc.).

In the source zone, the gaseous species and their equilibrium partial pressures are very important for the formation of GaN in the deposition zone. Generally, the chemical reactions simultaneously happen in the source zone:

$$\mathrm{Ga\,(l) + HCl\,(g) = GaCl\,(g) + \frac{1}{2}H_2\,(g)} \tag{3.1}$$

$$\mathrm{Ga\,(l) + 2HCl\,(g) = GaCl_2\,(g) + H_2\,(g)} \tag{3.2}$$

$$\mathrm{Ga\,(l) + 3HCl\,(g) = GaCl_2\,(g) + \frac{3}{2}H_2\,(g)} \tag{3.3}$$

$$\mathrm{2GaCl_3\,(g) = (GaCl_3)_2\,(g)} \tag{3.4}$$

The equilibrium partial pressures in the source zone can be calculated by thermodynamic analysis [5]. In HVPE growth, the temperature of source zone is usually around 850 °C, at which the major gaseous species of gallium is formed by Eq. (3.1). It should be noticed that the input partial pressure of HCl should not be so high in order to make almost all the HCl introduced into the source zone react with gallium. Therefore, the equilibrium partial pressure of GaCl should be almost equal to that of HCl at source zone. From the thermodynamic analysis, if the temperature of the source zone is too low or the partial pressure of input HCl is too high, or the reaction area of gallium with HCl is too small, the reactions might be insufficient in the source zone, which means the reactions will be kinetically limited.

The GaCl formed at source zone and NH_3 are transported to the deposition zone separately by a carrier gas mixture of H_2 and inert gas. At the deposition zone, the source species are mixed, and the following chemical reactions occur simultaneously.

$$\mathrm{GaCl\,(g) + NH_3\,(g) = GaN\,(s) + HCl\,(g) + H_2\,(g)} \tag{3.5}$$

$$\mathrm{GaCl\,(g) + HCl\,(g) = GaCl_2\,(g) + \frac{1}{2}H_2\,(g)} \tag{3.6}$$

$$\mathrm{GaCl\,(g) + 2HCl\,(g) = GaCl_3\,(g) + H_2\,(g)} \tag{3.7}$$

$$\mathrm{2GaCl_3\,(g) = (GaCl_3)_2\,(g)} \tag{3.8}$$

Then, gaseous species at the growth zone include GaCl, $GaCl_2$, $GaCl_3$, $(GaCl_3)_2$, NH_3, HCl, H_2, and inert gas. The partial pressures of these gaseous species at the growth zone as a function of temperature at deposition zone have been calculated by Koukitu et al. [6]. The partial pressures of $(GaCl_3)_2$ and $GaCl_2$ are very small under typical growth conditions. The equilibrium constants for these reactions can be calculated from the equilibrium equations [7]. $GaCl_3$ also reacts with NH_3 to form GaN, whose equilibrium constant is close to zero at usual growth temperature. Though the equilibrium constants are close for both reactions using GaCl and $GaCl_3$, the partial pressure of $GaCl_3$ is far lower than that of GaCl. Therefore, Eq. (3.5) is the dominate reaction.

The driving force for the deposition can be obtained from the difference between the number of Ga atoms put in and the amount of Ga atoms remaining in the vapor phase, which can be written as [6]:

$$\Delta P = P^0_{\mathrm{GaCl}} - \left(P_{\mathrm{GaCl}} + P_{\mathrm{GaCl_3}}\right) \tag{3.9}$$

where P^0_{GaCl}, P_{GaCl}, and $P_{\mathrm{GaCl_3}}$ are the input partial pressure of GaCl, the partial pressure of GaCl, and $GaCl_3$ in the vapor phase in deposition zone, respectively. From the calculation [6], the growth temperature has an important influence on the driving force. One point should be noticed that the influence of H_2 on the decrease of the driving force is more significant at high growth temperatures and the driving force decreases with the increase of H_2 in the carrier gas [8].

3.1.2 Progress in HVPE Growth of GaN Substrate

3.1.2.1 Dislocation Reduction and Strain Control

To obtain high-quality and crack-free bulk GaN substrate, dislocation reduction and strain control are the most important issues during HVPE growth. Epitaxial lateral overgrowth (ELOG) is an effective and traditional method to bend the dislocation line and form void structure in GaN films simultaneously, which helps to enhance the dislocation annihilation for dislocation reduction and release the growth and thermal stress [9, 10].

A standard ELOG process is as below: (a) depositing mask film such as SiN, SiO_2, etc. on GaN film; (b) etching dielectric film to expose GaN as window area; and (c) growing GaN layer from the window and then laterally overgrowing to cover the mask. By carefully adjusting the mask/window shape, the growth parameters for the overgrowth, reasonable lateral overgrowth rate, and smooth surface can be achieved [10–13]. Other technologies base on ELO have been proposed like double-layer ELO [14, 15] and three-step ELO [16] in order to further eliminate the remaining TDs in the coalescence region.

By improvement of ELOG technology, some special technologies have been developed in HVPE system for the growth of high-quality and crack-free GaN layers, such as dislocation elimination by the epitaxial growth with inverse-pyramidal pits (DEEP), TiN-based nano-mask, and photoelectron-chemical-etched nanowires.

DEEP was reported by Kensaku Motoki from Sumitomo Electric [17]. During the growth of GaN layer, there are numerous large hexagonal inverse-pyramidal pits constructed mainly by {11 −22} facets appearing on the surface. While GaN growing, dislocations are collected to the center of the hexagonal pits parallel to (0001) in the <11 −20> or <1 −100> direction, and therefore dislocations are eliminated within the hexagonal pits except for its center. DEEP method can produce a high-quality GaN layer, but the dislocation distribution on surface of GaN is no uniform. Far away the center of pits possesses low DDs; on opposite, the center of pits gathering dislocation possesses very high DDs [18, 19].

Besides ELOG and DEEP, nano-mask is recognized as a good approach that not only reduces the DDs but also easily separates the freestanding GaN from sapphire. Yuichi Oshima et al. developed a novel technique for preparing large-scale freestanding GaN wafers called VAS (void-assisted separation) by thin TiN film [20].

ELOG, DEEP, and nano-mask are using dielectric film as mask for the overgrowth of GaN; another unique method without dielectric film as mask is developed in our institute [21]. By the electrode-less photoelectron-chemical (PEC), long and straight GaN nanowire arrays are obtained, which have a density about 10^7 cm^{-2}, diameters ranging from 150 to 500 nm, and corresponding lengths ranging from 10 to 20 μm. It is found that GaN nanowires are almost dislocation and strain free (Fig. 3.1a, b), because PEC etching is beginning from the dislocation core. Based on these GaN NWs, we can get high-quality and crack-free GaN layer (Fig. 3.1c, d), with thickness about 400 μm and dislocation density about 10^4–10^6 cm^{-2}.

In addition to the abovementioned methods, increasing thickness of GaN is another effective approach to decrease the defect density of GaN substrate. The dislocation density reduced to 10^6 cm^{-2} as the thickness increased to 5 mm [22]. According to the same research from our group, it is found that the dislocation density is declining sharply from 10^8 to 10^4 cm^{-2} when the thickness is from several micrometers increasing to several millimeters (Fig. 3.2a–h). The related panchromatic CL images of different thickness GaN are shown in Fig. 3.2i.

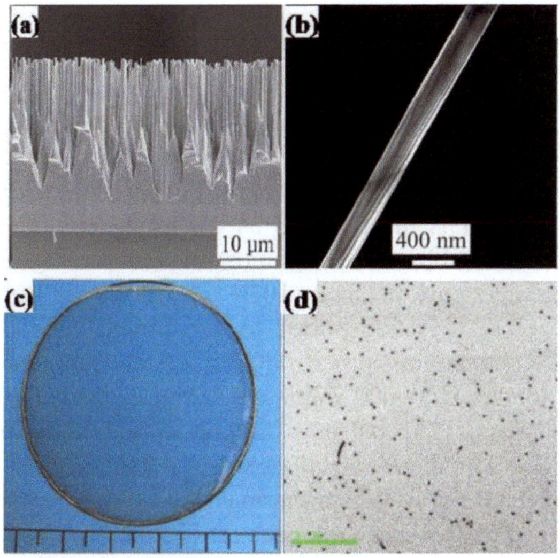

Fig. 3.1 (**a**) SEM image of GaN NW array [21]. (**b**) Weak-beam dark-field TEM image with $g = 11\text{–}22$, revealing that the GaN-NW does not possess dislocations [21]; Reproduced by permission of The Royal Society of Chemistry. (**c**) The photo of a 400 μm GaN base on the PEC GaN NW arrays template. (**d**) Panchromatic CL images show that the DDs of (**a**) are 3×10^6 cm^{-2}

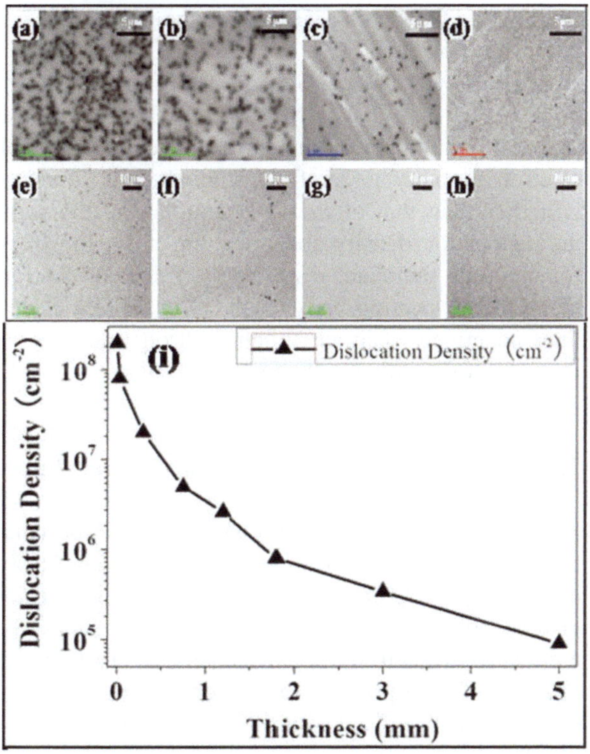

Fig. 3.2 Thickness from 5 μm to 5 mm and panchromatic CL images: (**a**) 5 μm, 2×10^8 cm^{-2}; (**b**) 20 μm, 5×10^7 cm^{-2}; (**c**) 200 μm, 2×10^7 cm^{-2}; (**d**) 0.7 mm, 5×10^6 cm^{-2}; (**e**) 1.2 mm, 3×10^6 cm^{-2}; (**f**) 1.8 mm, 8×10^5 cm^{-2}; (**g**) 3 mm, 3.4×10^5 cm^{-2}; (**h**) 5 mm, 8.0×10^4 cm^{-2}; (**i**) the relation between thickness and DDs

3.1.2.2 Si-Doping for n-GaN Substrate

The growth of *un*-doped GaN is simple but not suitable to make LED, LD, or high-power electronic devices, because its resistivity is relatively high with the typical value about 1 Ω cm. In order to control its electrical properties, a flexible and reproducible doping during HVPE growth of GaN is necessary and important.

SiH_4 is mostly used as the n-type doping source in MOCVD [23], but it is not suitable in HVPE, most of which is a hot-wall system with typical growth temperature of about 1040 °C. SiH_4 will decompose into silicon and hydrogen before it is transported to the growth zone of HVPE system, contributing less to the doping process. Great efforts have been made to find a proper doping source. Undoped single crystalline Si was used to react with HCl gas to produce dopants [24], but it is difficult to realize the precise control of the doping level. Dichlorosilane (SiH_2Cl_2) was a suitable choice for HVPE growth of Si-doped GaN because of its higher thermal stability.

The free carrier concentrations increase linearly in a semilog plot with SiH_2Cl_2 flow rates from 5 to 25 sccm [25]. This is different from the case in MOCVD growth of Si-doped GaN with SiH_4, where the free carrier concentrations increase linearly with SiH_4 flow rates [23].

The growth rate of GaN in HVPE is about tens to hundreds microns per hour, higher than that in MOCVD, so HVPE could easily realize growth of GaN with different dislocation density. The influence of dislocation density on the electron mobility is carefully studied. It is found that the mobility of FS GaN is higher than that of the GaN template at the same carrier concentration, with the dislocation density about 10^6 cm^{-2} for the freestanding GaN and 10^8 cm^{-2} for the GaN template, respectively. Edge dislocation introduces acceptor centers along the dislocation line, which could capture electrons and make the dislocation lines negatively charged. When electrons travel across the dislocations, it will be scattered, thus reducing the mobility [26].

3.1.2.3 Fe-Doping for High-Resistivity GaN Substrate

Semi-insulating (SI)-GaN substrate is very important for the performance of GaN-based HEMT device. However, *un*-doped GaN grown by HVPE in general shows n-type conductivity due to the residual donor impurities such as O and Si, which may give rise to the degradation of device performances. By compensation of the residual carriers from electron trapping centers, SI crystal is obtained by doping with Fe, Cr, or Zn [27–29]. Among these different doping elements, Fe-doping is wildly used for the reproducibility and controllability. With a Fe concentration above 10^{15} cm^{-3}, it is sufficient to compensate unintentionally incorporated donor impurities (oxygen and silicon) and native defects in GaN for the SI property [30, 31]. Although the formation energy and concentration of the point defect and/or the complex structure will be changed during the annealing process, the SI property of GaN:Fe bulk films grown by HVPE is thermal stability up to 1050 °C [32]. The resistivity decreases with the temperature with an activation energy about 0.5–0.6 eV which is attributed to Fe deep acceptor [30, 33]. Fe-doped bulk SI GaN (SI-GaN:Fe) grown by HVPE has been commercialized and resulted in a significant improvement of the device performance and reliability of the AlGaN/GaN heterostructure field effect transistors [34, 35]. Besides the SI property, the other properties of the GaN:Fe also have been widely interested, such as a diluted magnetic semiconductor to realize future spintronic applications [36, 37].

Compensation mechanisms of GaN:Fe have attracted much more attention, since these are important to the optical, electrical, and magnetic properties of the crystal [28, 38]. The Fe atoms incorporated in GaN matrix substitute Ga sites and introduce the charge transfer level ${Fe_{Ga}}^{3+/2+}$ in the midgap, which are crucial to carrier-mediated ferromagnetism and to predict band offsets in heterostructures on the basis of the internal reference rule [36, 39]. Since the charged state of ${Fe_{Ga}}^{3+/2+}$ is transferred from Fe^{3+} to Fe^{2+} by capturing an electron, the Fe atom acts as a compensating deep acceptor in GaN [28, 32]. The Fermi level is pinned approximately 0.5–0.6 eV below the conduction band minimum (CBM), which will change the formation energies of native point defects of GaN. The Fe^{3+} on Ga site will split into the ground state $^6A_1(S)$ and the excited states $^4T_1(G)$, $^4T_2(G)$, and $^4E(G)$ [28, 38]. A series of characteristic IR luminescence with a sharp zero-phonon line (ZPL) at 1.299 eV due to the spin-forbidden $^4T_1(G) \rightarrow {}^6A_1(S)$ transition will be observed at low temperature, as shown in Fig. 3.3a. The additional lines above ZPL are attributed to the splitting of the excited $^4T_1(G)$ state of Fe^{3+} under the combined effects of spin-orbit coupling, Jahn-Teller coupling, and the axial distortion of the trigonal crystal field in C_{3v} of the wurtzite lattice structure [28, 40]. The luminescence appears at energies less than 1.3 eV which are attributed to the local vibrational modes and phonon mode. Besides the IR part emissions, the blue and red broad bands which are attributed to the defect structures related to the doped Fe ion are generally observed, while the common yellow band is nearly unobserved due to the ionizing of the shallow donor involved in the related defect level. On the other hand, since the carrier-lifetime-killer actor of the doped transition metal impurities, quenching of the intensity of near band-edge excitonic emissions in the UV region is expected, as those shown in Fig. 3.3b [32, 38]. Raman investigation on the GaN:Fe bulk crystal confirms a strain-free incorporation of iron. Besides the normal Raman mode, an additional Raman mode at 670 cm^{-1} is usually observed in Fe-doped samples, which was assigned to a kind of phonon mode originated from VN. But, as those shown in Fig. 3.3c, unlike the familiar point-defects, this mode is thermal stability after annealing up to 1050 °C. Hence, a Fe-VN complex structure-related Raman mode is considered [41].

3.1.2.4 Minority Diffusion Lengths in Bulk GaN

The dislocations strongly affect the carrier properties in GaN including minority diffusion lengths and surface recombination velocities which are important for device performance. For example, in photovoltaic detectors, due to a large

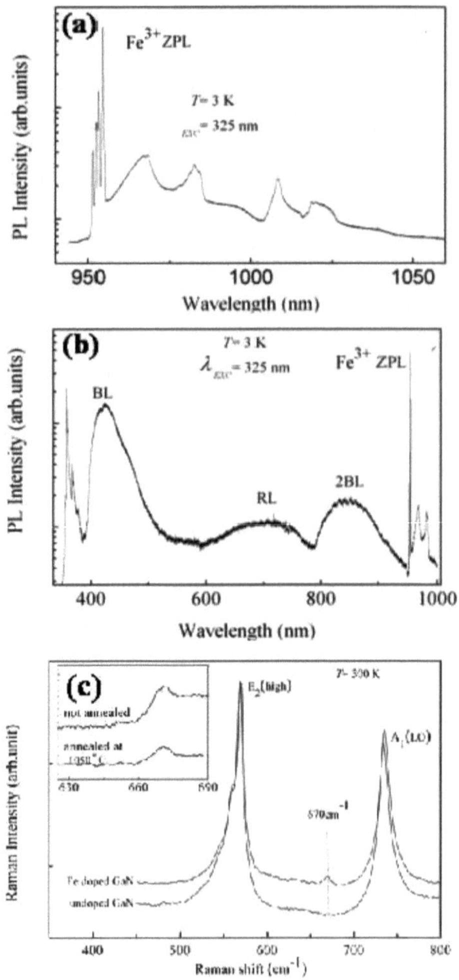

Fig. 3.3 (**a**) A series of characteristic IR luminescence with a sharp zero-phonon line (ZPL) at 1.299 eV at low temperature. (**b**) Quenching of the intensity of near band-edge excitonic emissions in the UV region. (**c**) A kind of phonon mode originated from VN

absorption coefficient of GaN, carriers are generated close to the surface and recombine. A sufficiently long minority diffusion length and the suppression of surface recombination velocity are helpful in the realization of high sensitivity [42]. However, for the most of characterization methods, such as photoluminescence [43], surface photovoltage [44], and photocurrent [45] measurements, the spatial resolution of as-measured carrier properties is low, which makes it difficult to reveal the relationship between the experimental results and the local dislocation structures. Electron-beam-induced current (EBIC) method is capable of achieving the inhomogeneity of minority diffusion length along a depth gradient, but a p-n junction or a Schottky barrier has to be made at cross section [46]. Recently, a combination of surface photovoltage spectroscopy (SPS) method and kelvin probe force microscopy (KPFM) was reported for the simultaneous measurement of the

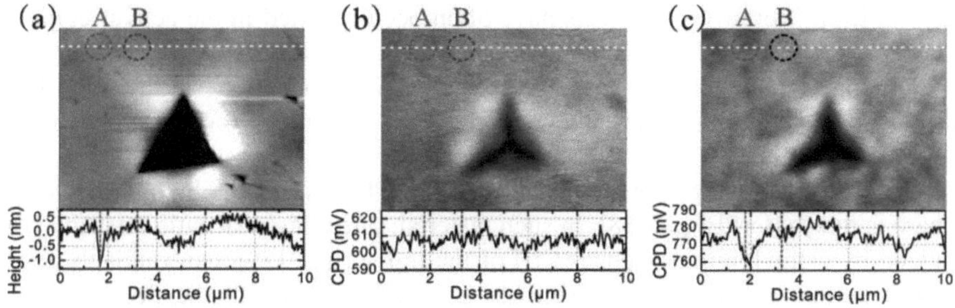

Fig. 3.4 (**a**) The topographic image around a nanoindentation with a scan area of $10 \times 10\,\mu m$. A V-pit of thread dislocation is marked with A, and a nearby plane position is marked with B. (**b**) and (**c**) are the CPD images of the same area acquired under dark condition and under UV illumination with wavelength of 360 nm, respectively. The curves in the images are the profiles along the white lines

topography, the local minority diffusion length, and the surface recombination velocity at a single thread dislocation [47]. The contact potential difference (CPD) at nanometer scale varying with the incident photon energy was measured with corresponding topography image. SPS responses at single thread dislocations near a nanoindentation on a HVPE-grown GaN surface can be distinguished. As shown in Fig. 3.4a–c, the thread dislocations introduced by a nanoindentation were observed as V-pits, where the photovoltage was lower than that on plane surface under ultraviolet illumination. Compared with those on plane surface, the calculated hole diffusion length is 90 nm shorter and the surface electron recombination velocity is 1.6 times higher at an individual thread dislocation.

3.2 Structural Characterization in Homoepitaxial GaInN/GaN Light-Emitting Diode Growth

Group III-nitride device development still suffers from mismatched heteroepitaxial growth. Mismatch in lattice constants and thermal expansion coefficients between substrate, mostly sapphire or SiC, and epitaxial layer inhibits perfect crystal formation, resulting in high densities of threading dislocations [48]. Low-dislocation-density bulk GaN has therefore long been sought for as the ideal substrate for homoepitaxial growth. It was found that the performance of LEDs could be improved by a reduction of the threading dislocation (TD) density, especially for green LEDs [49]. Different techniques have been used to reduce the TD density, such as lateral epitaxial overgrowth [50]. These methods, however, reduce the TD density only in parts of the wafer, leaving the rest useless. With the recent advances in hydride vapor-phase epitaxy (HVPE), bulk GaN substrates with a TD density as low as 10^5 cm^{-2} provide an ideal template for homoepitaxial growth [51]. Moreover, V-defects have often been reported in the active region of GaInN/GaN LEDs [52], especially with high In composition.

The main challenges are (1) to obtain epitaxial production worthy-sized bulk crystals of such high qualities and (2) to replicate the bulk performance in epitaxial overgrowth [53].

One very promising approach is to grow thick layers of GaN by hydride vapor-phase epitaxy (HVPE) [54–58]. In such material, threading dislocation densities, as judged by cathodoluminescence (CL) of the top surface [59], can routinely reach values as low 10^6 cm^{-2} and even 10^5 cm^{-2}. Wafers up to the dimensions of 2 inch are commercially available [60, 61]. The second challenge now calls for metalorganic vapor-phase epitaxy (MOVPE) to replicate the low dislocation density in homoepitaxial films on these substrates [62].

Here, two types of GaN templates were used for the analysis of the structural characterization of homoepitaxial GaN/GaN light-emitting diode: (1) 300 mm thick freestanding HVPE GaN and (2) 2 mm thick MOVPE GaN on 330 mm thick c-plane sapphire. HVPE GaN and MOVPE GaN templates were n-type doped with Si donor concentrations of 1×10^{18} and 1×10^{19} cm^{-2}, respectively. The surface of HVPE GaN was polished to atomic layer flat level via a proprietary chemical mechanical polishing (CMP) process. The root mean square (RMS) of the surface roughness of the HVPE GaN templates was 0.1–0.2 nm which was comparable to those of as-grown MOVPE GaN on sapphire. From (0004) X-ray rocking curve measurements, the full-widths at half-maximum were about 90 arcsec, while those for MOVPE-GaN were about 200–250 arcsec. An AIX-200RF MOCVD reactor was employed for homoepitaxy of GaInN-based light-emitting diodes (LED) consisting of (1) 0.5 mm thick n-GaN, (2) 5 periods of 3 nm thick GaInN quantum well and 6 nm thick GaN barrier, and (3) 0.2 mm thick *p*-GaN. Both templates were loaded into the growth chamber side by side. In order to prevent thermal decomposition of the templates, NH_3 was introduced into the chamber while raising the templates' temperature to growth temperature of *n*-GaN (1100 °C). The homoepitaxy growth proceeded immediately once the growth temperature was reached. Typical dopants of Si and Mg were utilized for the growth of *n*- and *p*-GaN with doping concentration of 3×10^{18} and 1×10^{20} cm^{-3}, respectively. Optimized growth conditions on the sapphire wafer were exploited for all epitaxial layers.

After the growth, both types of samples were evaluated for optical performance, crystalline quality, and surface morphology via photoluminescence (PL), X-ray diffraction (XRD), atomic force microscopy (AFM) techniques, and transmission electron microscopy (TEM), respectively. As shown in Table 3.1, the roughness of the *p*-layer due to heavy Mg doping was a factor of 2 higher on the sapphire-based structure than that of the GaN template.

Then typical LED fabrication processes such as reactive ion etching (RIE), e-beam evaporation, and photolithography were introduced to form mesa, transparent contacts, and electrodes on the LED devices. The size of the LED device was 350 μm × 350 μm with an effective mesa area of 300 μm × 300 μm. Transparent contacts on the mesa were formed by 5 nm of Ni followed by 5 nm of Au. A metal stack of Ti/Al/Ti/Au was formed on MOVPE-grown n-GaN layer for the electrical

Table 3.1 Reported performances of nonpolar and semipolar LEDs prepared on bulk-GaN substrates

Orientation	Structure	Wavelength (nm)	Radiant flux (mW)	Voltage (V)	Peak shift due to current increase	Reference
a	7-nm 5QW, no EBL[a]	450	...	4.4	None	[63]
a	3-nm 5QW	522	0.07	...	Red shift due to thermal effect	[64]
m	8-nm 6QW, 2.2-20 nm	407.4	23.7	5.2	Negligible blue shift	[65, 66]
m	3-nm 5QW	435	1.79	5	Blue shift	[67]
m	8-nm 6QW	468	8.9	6.4	Less than c-plane LEDs	[68]
m	2–4 nm 3QW	460–506	Max. 6	...	Red shift to be investigated	[69]
$(10\overline{1}\overline{1})$	2.5-nm 6QW	412	20.58	3.93	Minimal due to piezofield reduction	[70]
$(10\overline{1}\overline{1})$	3-nm 6QW	443.9	13.41	8.29	Minimal due to piezofield reduction	[71]
$(11\overline{2}2)$, $(11\overline{2}\overline{2})$	3-nm SQW	425, 530, 580	1.76, 1.91, 0.54	3.4, 3.8, 3.0	Blue shift due to state filling	[72]
$(11\overline{2}2)$	4-nm 6QW	522.4	5.0	6.9	Low-energy localized state filling	[73]
$(11\overline{2}2)$	3.5-nm SQW	562.7	5.9 (pulsed)	...	Localized energy state filling	[74]

Shown data were taken at 20 mA, unless mentioned otherwise. In the column for structures are the thickness and number of InGaN QWs [75]
[a]Electron-blocking layer

contact. Electroluminescence (EL) was employed to evaluate the optical output performance for LED on quasi-bulk HVPE GaN and LED on thin MOVPE GaN template on sapphire, named hereafter "LED on GaN" and "LED on sapphire," respectively. (0002) XRD curves confirmed the superlattice periods of 8.5 and 8.8 nm for GaInN/GaN active regions in LED on GaN and LED on sapphire as designed, respectively. A higher number of satellite peaks as well as one order of magnitude higher satellite peak intensities derived from the active region of the LED on GaN samples suggested superior crystalline quality and sharper interface of the GaInN quantum well (QW) and GaN barrier compared to those of the LED on sapphire samples. Fitting of the (0002) XRD curves assuming that the GaInN QW layers were pseudomorphically grown on GaN yielded indium contents of 13% and 9% for LED on GaN and sapphire, respectively. Though the difference in indium incorporation could be explained by compressive strain and/or template temperature, it will require further systematic study to clarify this discrepancy.

As shown in Fig. 3.5, PL peak wavelengths observed for as-grown LED on GaN and sapphire were 440 and 420 nm, respectively. The PL results agreed with (0002) XRD results and reconfirmed that indium incorporation was higher in GaInN QW layers in LED on GaN compared to that of the LED on sapphire. Moreover, the peak intensity of the LED on GaN was 20 times as strong as that that of the LED on sapphire. This big difference cannot be solely explained by the difference of the indium content.

The results of the optical output power of fully fabricated LED dies show that electroluminescence intensities of the LED on GaN were much stronger than those of the LED on sapphire when operating under the same driving current. At driving current of 0.5 A, the radiation power of the LED on GaN could reach 7 mW. Also, the LED on GaN's driving current could achieve values as high as 600 mA, which is much higher than the LED on sapphire. The latter reached the voltage limit of 7 V in the testing system at the driving current of only 80 mA. This result suggests the

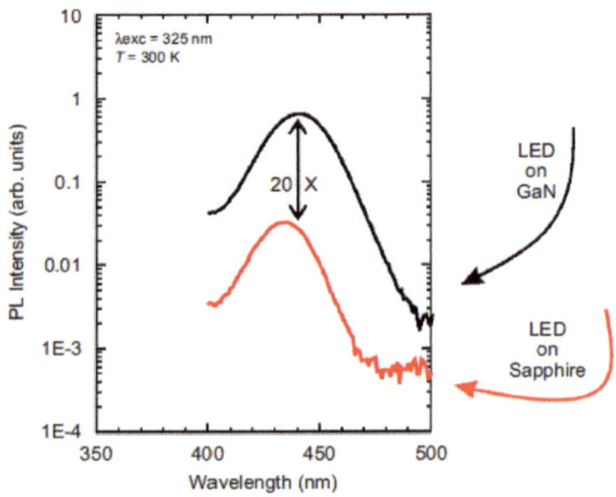

Fig. 3.5 Photoluminescence spectra of as-grown LED on GaN and LED on sapphire [53]

suitability of bulk GaN as a homoepitaxial template for high optical output power LEDs and hence solid-state lighting applications.

3.2.1 Evaluation of Threading Dislocation Density

Threading dislocation densities in the heteroepitaxial GaN on c-plane sapphire were as low as mid-10^8 cm^{-2} [6], while the threading dislocation densities on HVPE GaN were found to be about mid-10^6 to lower than 10^7 cm^{-2} [76]. After the homoepitaxial growth, threading dislocation densities in both samples increased to lower than 10^8 and lower than 10^9 cm^{-2} for the LEDs on GaN and sapphire, respectively. Jasinski et al. [77] reported that the threading dislocations on MOVPE-grown GaN and GaInN layers on quasi-bulk GaN templates were propagated directly from the template. The resulting threading dislocation densities of the homoepitaxial layers hence were of the same order of magnitude as those of the quasi-bulk GaN itself. The heavily doped Mg might initiate microstructural defects in p-GaN and lead to formation of additional dislocations. However, the process of newly formed dislocations which increased the threading dislocations densities in both samples by one order of magnitude was unknown. Investigations by transmission electron microscopy will be required to further investigate the nature of dislocations formed in the homoepitaxial layers. Nevertheless, this result suggests that existing dislocations could play a major role in the expansion of dislocation densities in the homoepitaxial growth.

Using scanning electron microscopy (SEM), the number of epitaxial defects, i.e., hexagonal pits, could be determined (Fig. 3.6). These defects appeared as dark spots in CL images taken at the peak wavelength of each sample. In the CL image of the LED on GaN, there were bands of low-intensity areas, while there were high densities of low-intensity spots scattered randomly over the mesa area. It was very difficult to identify dislocation-related dark spots on either sample. The pit densities were 7×10^4–3.6×10^5 cm^{-2} for the LED on GaN which were almost one order of magnitude smaller than those found on the LED on sapphire where defect densities were approximately 1–2 $\times$ 10^6 cm^{-2}.

However, when observing the LED on GaN at lower temperatures, i.e., 77–150 K, an additional number of dark spots besides those pits was observed as shown in Fig. 3.7. These additional dark spots represented dislocations formed in the regrown epitaxial layers. The corresponding dislocation densities were found to be about 2×10^6 and 5×10^6 cm^{-2} at 438 and 385 nm, respectively. At 385 nm, there was strong emission observed from pits. The emission at 385 nm was attributed to donor–acceptor pair recombination. On the other hand, it was difficult to estimate more accurate densities of the dark spots for the LED on sapphire due to its high dislocation density.

For comparison, the CL images at 367 nm were collected from an etched *n*-GaN area. For the LED on GaN sample, there were a number of dark spots that clearly related to threading dislocations. The density of those dark spots was

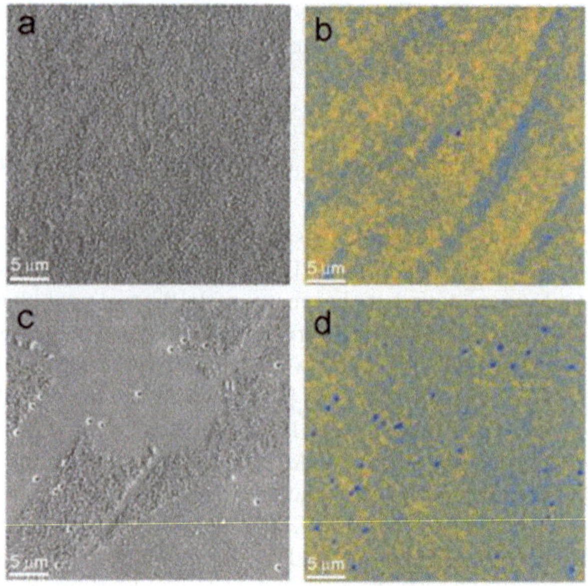

Fig. 3.6 SEM (a, c) and CL (b, d), micrographs of mesa area on the LED on GaN (a, b) and the LED on sapphire (c, d). The CL micrographs were collected at 440 and 420 nm for LED on GaN and sapphire, respectively. Markers represent 5 μm. Rugged surface of both samples were formed during LED fabrication process. Pits can be clearly identified as dark spots in CL images [53]

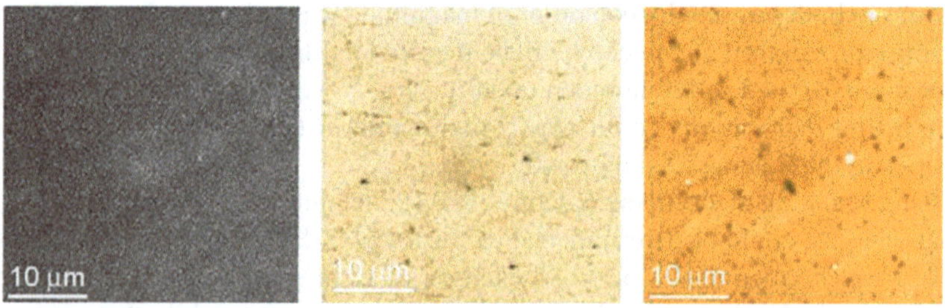

Fig. 3.7 SEM (left) and CL (middle and right) micrographs of mesa area on LED on GaN at 79 K. The central and right images were taken at 438 and 385 nm, respectively [53]

4.5×10^6 cm^{-2}. This number was similar to those found for the LED on GaN at low temperature. There are still a number of control parameters such as template cleaning procedure and growth conditions to optimize in order to further reduce the dislocation density in the homoepitaxial layer on such quasi-bulk HVPE GaN templates.

A cross-sectional TEM micrograph is shown in Fig. 3.8. It reveals the set of five QWs between n-type and p-type layers. No growth boundary is observable at the homointerface of the bulk GaN template and the epitaxial GaN layer. The contrast variations in the multiple quantum well (MQW) stack may be the result of radiation damage, ion-milling damage, or local charge or strain buildup. It was also suggested that phase separation in the GaInN QWs might be possible. Furthermore, no defects are observed along a lateral length of the QWs for over 1.2 μm. This suggests a low density of structural defects and good crystalline quality of the epitaxial material [78].

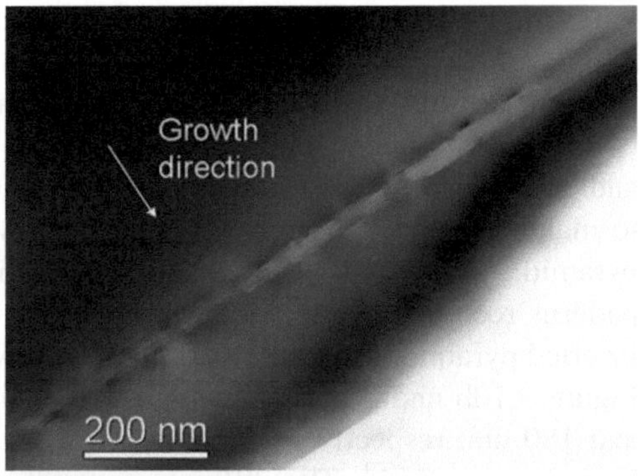

Fig. 3.8 Bright-field image of the QWs recorded along the $[1\bar{1}00]$ zone axis. The narrow darker lines are the QWs. Note that there is no defect along a lateral distance of more than 1.2 μm [78]

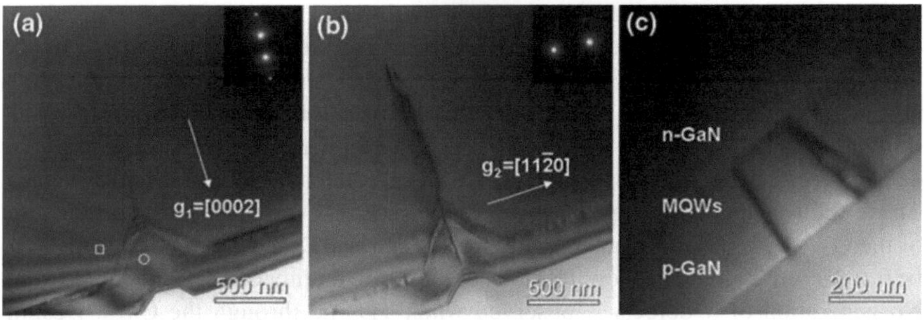

Fig. 3.9 Bright-field images at $g1 = [0002]$ (**a**) and $g2 = [11\bar{2}0]$ (**b**). The insets are the corresponding diffraction patterns. A large inverted pyramid with a diameter of 650 nm and depth of 480 nm is seen. The TD connected with the inverted pyramid is determined to be edge-type by the $g \cdot b = 0$ invisibility criterion. Subfigure (**c**) was taken along the $[1\bar{1}00]$ zone axis. It shows TDs that originate in the QWs [78]

It was also observed that large inverted pyramid defects were initiated by TDs, as shown in Fig. 3.9. The $g \cdot b = 0$ invisibility criterion was used to determine the defect type. In Fig. 3.9a, the TD reveals no contrast when the sample was tilted to $g1 = [0002]$. This indicates a Burgers vector of $b = 1/3\langle 11\bar{2}0\rangle$ and identifies it as an edge-type TD. The length of these TDs is usually 1.5 μm or more, while the epitaxial layer is only 0.75 μm thick. These TDs should therefore originate from the bulk GaN template and not from the epitaxial growth.

These large inverted pyramids have diameters from 600 to 650 nm at the top, and their depths range from 460 to 480 nm. Their origin beneath the QW region corresponds very well to the thickness of the overgrown n-layer. It can be concluded that they should form at the beginning of the homoepitaxial growth. During the growth of the n-type GaN, there must have been no or only very little growth of GaN around the vicinity of the TDs. However, after the deposition of five periods of QWs, the p-type material begins to fill in the V-shaped gap. Since the p-type GaN cannot fill in the gap entirely, pits with diameter from 300 to 500 nm remain at the surface and can be observed in the scanning electron microscopy (SEM) images

[53]. Thus the density of such large inverted pyramids was determined to be around $3 \times 10^5\ cm^{-2}$ by counting the pits in the SEM images. The TDs themselves continue to propagate until they reach the sample surface.

Inversion domains are known to exist in the homoepitaxial layer of GaN [79] and could lead to the formation of pits on the surface [80]. Therefore, in order to make sure that whether there are any inversion domain inside of the inverted pyramids, convergent beam electron diffraction (CEBD) is used to determine that patterns recorded inside (circle in Fig. 3.9a) and outside (square in Fig. 3.9a) the inverted pyramid, respectively. Both of them were taken along the $\left[1\overline{1}00\right]$ zone axis. Figure 3.10b and d are simulated patterns of Ga polarity at thickness of 210 nm and 180 nm, respectively. From left to right, the indices of each disc are [0002], [0000], and $\left[000\overline{2}\right]$. The different contrast within the [0002] and $\left[000\overline{2}\right]$ discs is caused by the polarity. By comparing the experimental patterns with the simulated ones for both polarities, we find that the material both inside and outside of the inverted pyramid has the same Ga polarity. Consequently, the inverted pyramids are not inversion domains. Besides TDs connected with inverted pyramids, TDs originating in the QWs are observed as well, as shown in Fig. 3.9c. The total TD density is estimated to be $2 \times 10^8\ cm^{-2}$ or less from cross-sectional TEM images. Those additional defects suggests that there is a need for further growth optimization of the QW region. For the heteroepitaxial LEDs, the TD density was determined to be $6 \times 10^9\ cm^{-2}$ by plan-view TEM. These TDs are mainly of edge and mixed type. No generation of V-defects is observed when TDs go through the QWs. Despite

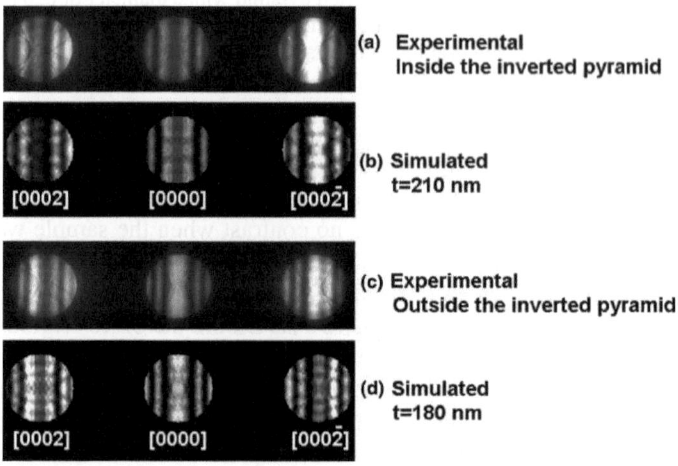

Fig. 3.10 CBED patterns of the GaN thin films recorded (**a**) inside the inverted pyramid (circle in Fig. 3.9); (**c**) outside the inverted pyramid (square in Fig. 3.9). Both of them were taken along the $\left[1\overline{1}00\right]$ zone axis. (**b**) and (**d**) show simulated CBED patterns for Ga polarity at thicknesses of 210 nm and 180 nm, respectively. The different contrast within [0002] and $\left[000\overline{2}\right]$ discs is due to the polarity. Comparisons between the experimental patterns and simulated ones suggest both materials have the same Ga polarity [78]

the existence of large inverted pyramids, the TD density has been reduced in the homoepitaxial LED to 1/30 of that in the heteroepitaxial one. This must be the reason for the 20-fold stronger photoluminescence intensity in the homoepitaxial material, as described in Ref. [53].

3.2.2 Electrical Characterization and Optical Characterization of Homoepitaxial InGaN/GaN Light-Emitting Diodes

InGaN/GaN-based light-emitting diodes (LEDs) grown heteroepitaxially on sapphire or SiC substrates contain a high density of threading dislocations ($\sim 10^8$–10^{10} cm^{-2}) due to large lattice and thermal expansion mismatch between the substrate and III-nitrides. It has been found that the high density of dislocations has limited adverse effects on the optical performance of the LEDs, especially at blue and green wavelengths, due to strong carrier localization effects in InGaN alloys [81, 82]. However, the dislocations, particularly those threading through the active region, may have a pronounced influence on the electrical characteristics of the LEDs by enhancing carrier tunneling [83, 84], dopant diffusion [85–87], and contact metal migration [88]. Anomalously high leakage currents are generally observed in commercially available GaN-based LEDs, which in many cases are grown on sapphire substrates. The high junction leakage raises concerns about device reliability, particularly under high-power operation conditions [89].

InGaN/GaN MQW LEDs on a freestanding GaN substrate can substantially reduce the defect density. As a consequence, the LEDs show a marked reduction in reverse leakage current, forward tunneling current, and improved injection efficiency under low bias compared to their counterparts on conventional sapphire substrates.

Cao et al. compared the electrical characteristics of InGaN/GaN multiple-quantum-well (MQW) LEDs grown on bulk GaN and sapphire substrates [90, 91]. The bulk GaN substrates with a thickness of 300 μm were produced by the HVPE technique [92, 93]. The materials were unintentionally doped and had a room-temperature resistivity of ~0.076 Ω cm. The dislocation density in the bulk GaN is $\sim 2 \times 10^7$ cm^{-2}, several orders of magnitude lower than those in the heteroepitaxial GaN. Homoepitaxial growth was conducted on the Ga-face of the GaN substrates by low-pressure metalorganic chemical vapor deposition at 1050 °C using trimethylgallium (TMGa) and ammonia precursors. Silane and bis-cyclopentadienyl-magnesium (Cp_2Mg) were used for n- and p-type doping, respectively. Hydrogen was used as the carrier gas.

LEDs with peak wavelengths ~405 nm were grown on both GaN and sapphire substrates in the same growth run. The LED structure consisted of a 10-period $In_xGa_{1-x}N$/GaN ($x = 0.1$–0.21) multiple-quantum-well (MQW) active region sandwiched between a 2 μm *n*-GaN layer (Si 5×10^{18} cm^{-3}) and a 0.1 mm p-type AlGaN cladding layer and a 0.2 μm *p*-GaN contact layer (Mg$\sim 1 \times 10^{19}$ cm^{-3}).

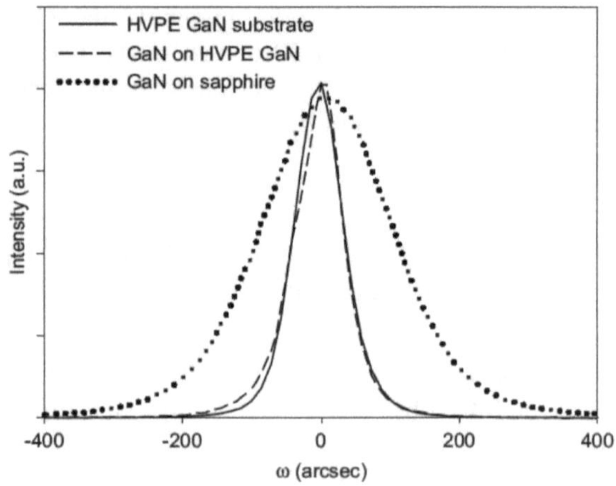

Fig. 3.11 HRXRD rocking curves of the (0002) reflection of an HVPE GaN and GaN epilayers grown on GaN and sapphire [91]

Top-emitting LEDs with a size of 300 μm × 300 μm were fabricated using standard photolithography and dry etch techniques [94]. For direct comparisons, identical devices were also grown on 2 in. sapphire substrates with a 100 nm GaN buffer layer under the same epitaxy conditions.

The homoepitaxial GaN epilayers exhibited a smooth, defect-free surface with a clear step structure, as revealed by AFM measurements, indicating good 2D step-flow growth. Second ion mass spectrometry (SIMS) measurements showed that the C, H, and O concentrations in the homoepitaxial structures were lower than those in heteroepitaxial GaN by a factor of 2–4 and are close to their SIMS detection limits. The reduced impurity incorporation in the homoepitaxial GaN is likely a result of reduced microstructural defects which may act as impurity segregation sites [95]. High-resolution X-ray diffraction (HRXRD) characterization showed that the rocking curve FWHM of the (0002) reflection of the GaN substrates, the epilayers on GaN, and the epilayers on sapphire are in the range of 85–129, 79–119, and 224–245 arcsec, respectively. Figure 3.11 compares the HRXRD rocking curves of a GaN substrate before and after homoepitaxial growth of a 3 mm GaN and a typical GaN epilayer on sapphire. The rocking curve FWHM of the homoepitaxial GaN is 79 arcsec, which is slightly smaller than that of the substrate (85 arcsec) and much smaller than that for GaN grown on sapphire (230 arcsec). Since the FWHM of the (0002) peak reflects the degree of lattice distortion from dislocations, the smaller FWHM of the homoepitaxial GaN confirms that the threading dislocation density is substantially reduced.

Figure 3.12 compares typical *I*–*V* characteristics of the LEDs measured at increasing temperatures. At low and moderate forward biases, two main exponential segments with different slopes can be distinguished for the LED on sapphire. The *I*–*V* characteristics can be represented by $I = I_0 \exp(qV/E)$. The temperature-independent energy parameter E has values of 190 and 70 meV in the voltage ranges 0–2.0 and 2.0–2.8 V, respectively. No realistic ideality factors can be extracted. These behaviors are characteristic of tunneling current in a semiconductor diode [83,

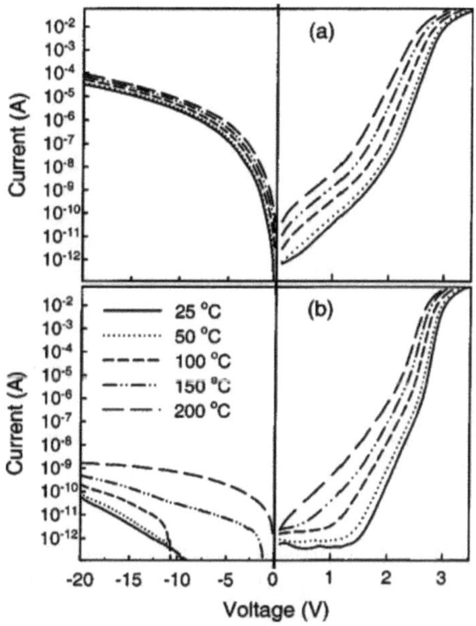

Fig. 3.12 Temperature-dependent $I-V$ characteristics of the LEDs grown on (**a**) sapphire and (**b**) GaN [90]

84]. The forward *I–V* characteristics of the LED on GaN also divide into two distinct linear sections with different slopes (see Fig. 3.12b). However, the slopes appear to be a function of temperature. At low injection levels, tunneling may still dominate, but the slope change suggests the involvement of thermally activated currents. As the forward current increases, diffusion and recombination currents start to dominate over the tunneling component. In fact, the forward current at bias >2.6 V can be described by the conventional drift-diffusion model as $I = I_0 \exp(qV/1.5kT)$ until series resistance in the diode dominates. The dominance of diffusion-recombination current reflects the high material quality of the homoepitaxial LED, where defect-assisted tunneling current is greatly suppressed.

The reverse-bias current in the LED on sapphire is much higher than conventional diffusion and generation currents, which are unobservably small in wide bandgap semiconductor diodes. The strong field dependence but low-temperature sensitivity of the leakage current is indicative of carrier tunneling. In sharp contrast, the LED on GaN shows a dramatic reduction in reverse current by more than six orders of magnitude. The remaining leakage current is a function of both applied bias and temperature, suggesting a combination of tunneling current and thermal generation current. In particular, as temperature increases from 100 to 150 °C, there is a sudden jump in the low-bias current, probably arising from thermal ionization of carriers from deep traps and trap-assisted tunneling process. An activation energy of ~0.5 eV is extracted from the $\log(I) - 1/T$ plot at -12.5 V where the change of the current as a function of temperature is roughly an exponential function.

The defect density in the homoepitaxially grown InGaN/GaN MQW LED is remarkably reduced compared to the similar device grown on sapphire, leading

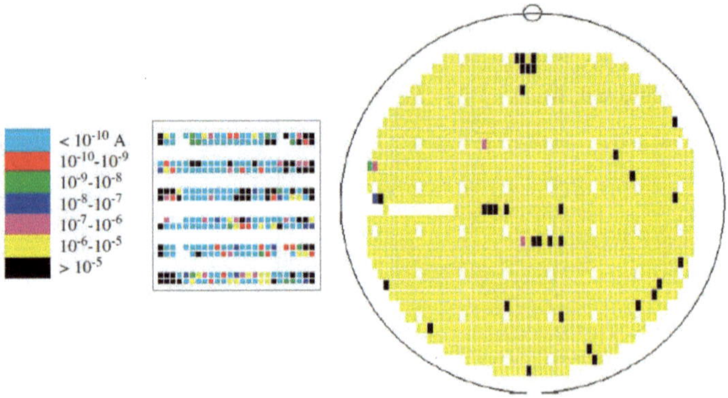

Fig. 3.13 The distribution of reverse currents at −10 V in LEDs grown on 1 × 1 cm^2 GaN and 2 in. sapphire substrates [91]

to much improved electrical characteristics. In contrast to the strong tunneling behaviors observed in the LED on sapphire, thermally activated currents were found to be dominant in the LED on GaN. An inhomogeneous distribution of reverse current in the LED on sapphire provides evidence that V-defects and the associated screw and mixed dislocations are electrically active and behave as leakage current pathways.

To investigate the distribution of leakage current in the LEDs, wafer mapping was carried out. Figure 3.13 shows the maps of leakage current measured at −10 V. The current is relatively high (~3 μA) in the LEDs on sapphire but very uniform across the 2 in. wafer. Only 2% of the LEDs have leakage current >10 μA at −10 V. While most LEDs on GaN have extremely low reverse leakage (<10^{-3} μA), ~20% of the LEDs are quite leaky. Interestingly, most leaky devices are located near the wafer edges, where the density of macroscopic defects in the GaN substrate is higher. This finding supports the above assumption that major defects originating from the substrate surface cause the low yield and poor scalability of the homoepitaxial diodes. Figure 3.14 shows the mapping results of the forward voltage at 50 mA. The average value for the LEDs on GaN is 3.9 V, lower than 4.2 V for the LEDs on sapphire. Approximately 5% of the LEDs on GaN, mainly along the wafer edges, have a forward voltage lower than 3 V. These devices are essentially shorted in both forward and reverse directions.

The superior electrical characteristics of most homoepitaxial LEDs are consistent with their improved optical properties. At 20 mA, the LEDs on GaN are twice more efficient than the LEDs on sapphire. The performance improvement is even more pronounced at low injection currents. This can be ascribed to the high structural quality of the MQW structures where nonradiative recombination rates are greatly reduced.

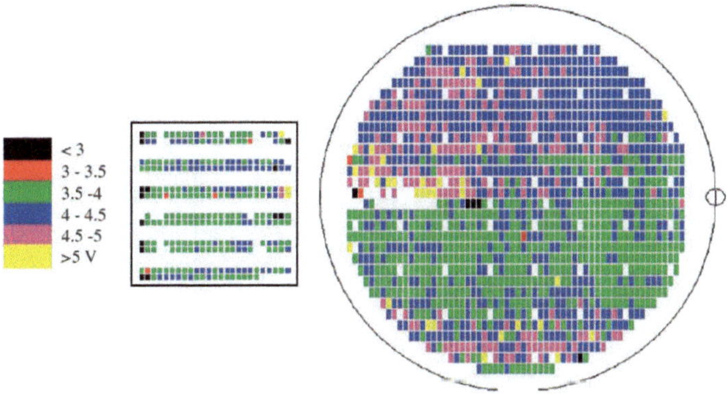

Fig. 3.14 The distribution of forward voltages at 50 mA of LEDs grown on 1×1 cm^2 GaN and 2 in. sapphire substrates [91]

3.3 Nonpolar and Semipolar Orientations GaN LED Grown on Bulk GaN Substrates

3.3.1 Problems with Conventional c-Plane LEDs and Motivation for Nonpolar and Semipolar Orientations

As the spectral range of light emission was extended from blue to green, we encountered a problem in luminescence efficiency, which is strongly related to InGaN material quality and QW structure. The material issue is that InGaN alloys tend to show spinodal decomposition [96, 97]. Furthermore, underlying GaN layers cause pseudomorphic growth of InGaN, thus, high In content is difficult and tends to introduce misfit dislocations and other types of defects. This problem is related to thermodynamics and growth kinetics; hence growth techniques and optimization will improve the materials. This aspect of the problem is common to all crystallographic orientations; nevertheless, different orientations may have different degrees of tolerances. The empirically found trend of high In incorporation into $(11\bar{2}2)$-oriented InGaN films may be related to this aspect. Further investigations are expected [75].

The QW structure-related challenge is due to the quantum-confined stark effect (QCSE, LEDs with thick active layers were dimmer than those with thin active layers) (Fig. 3.15). This effect is inevitable as long as strained QW structures are used in the polar orientation. In this case, a crystallographic solution may be given: employment of nonpolar orientation to have the polarization vectors lie in the plane, thus in the device direction, so no strain-induced electric fields appear. This is the main reason for the nonpolar orientation. Once we eliminate the QCSE, thicker active layers, as thick as the critical thickness, can be employed, which offer advantages, e.g., a larger active region volume. In addition to nonpolar orientations, it has been found by computation that semipolar orientations can provide similar

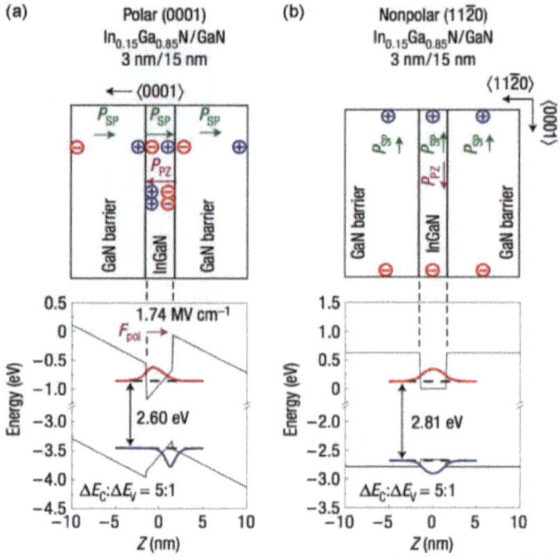

Fig. 3.15 QW structures on (**a**) polar and (**b**) nonpolar orientations and their band diagrams. Polarization charges appear at interfaces of the polar-oriented QW and induce electric fields that spatially separate electrons and holes in the QW. In nonpolar orientations, polarization charges do not affect the band structure. Because of the internal electric fields in the polar-oriented QW, transition energy occurs to be smaller than that of the nonpolar-oriented QW (QCSE). When the QW is embedded in the common +c-oriented LED structure, the internal electric fields are increased as LED positive bias is increased. After Chichibu et al. [98]

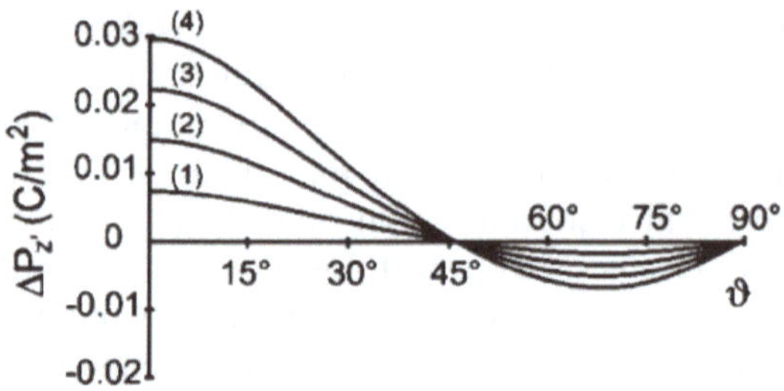

Fig. 3.16 Computed polarization charge (spontaneous and piezoelectric) density in InGaN/GaN QWs as a function of tilt angle of the growth plane with respect to the c-plane. Internal field strength becomes zero around $\theta = 45°$. After Romanov et al. [100]

effects in eliminating the QCSE. These calculations depend on material parameters such as elastic constants. Some of the necessary parameters are not well known to date, particularly those of InN [99]. These uncertainties introduce variations in the calculated results; Fig. 3.16 is an example result of these studies [100, 101]. Nevertheless, it has been believed that the advantage will be obtained for angles around 45°–60°. These angles include $(10\bar{1}1)$ and $(11\bar{2}2)$ orientations.

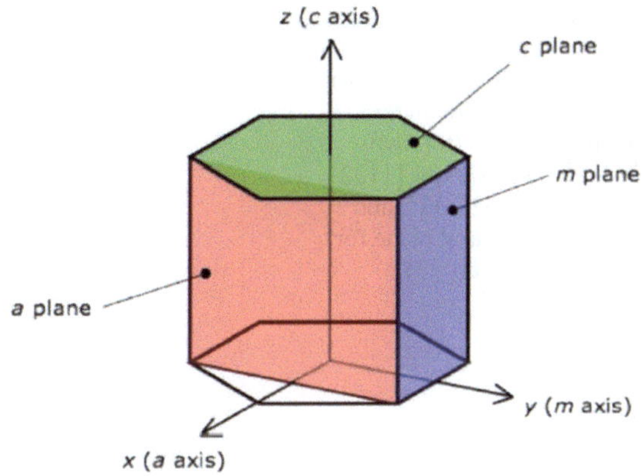

Fig. 3.17 Hexagonal prism representing a GaN crystal unit cell with common nomenclature of planes and axes. Coloring of the planes is for clarity and does not indicate any physical significance [75]

In addition, extra properties such as optical polarization are considered to be advantages over c-plane devices in some applications.

3.3.2 Crystallography and Piezoelectricity

A significant difference in nitride semiconductors from other conventional optoelectronic semiconductors is the crystal structure. Nitride semiconductor materials crystallize in wurtzite structures, while other typical optoelectronic semiconductor materials crystallize in zinc blende structures under normal conditions. Hexagonal lattices are expressed by using the Miller-Bravais index [102, 103]. Figure 3.17 shows a hexagonal prism representing a GaN crystal. The (0001) plane is the conventional growth plane and is called the $+c$ plane. All the current commercial products are fabricated via the c-orientation growth. The c axis [0001] is an electrically polar axis in wurtzite crystals. The $(000\overline{1})$ plane is often called the $-c$ plane; the zinc blende equivalence is considered to be the {111} B plane. Vertical planes are parallel to [0001], thus electrically nonpolar, and they are called nonpolar planes. Low-index nonpolar planes are $\{11\overline{2}0\}$ and $\{1\overline{1}00\}$, called a and m planes, respectively. It is a convention that the z axis of a 3-D spatial coordinate system is defined to be parallel to the c axis. It is a common practice that the x and y axes are defined to be parallel to the a and m axes, respectively. Note that Cartesian coordinates are conventionally used in wurtzite crystals. In some literature, the x and y axes are defined to be parallel to the m and a axes, respectively, as these two crystallographic axes are chemically different but equivalent in the sense of continuum mechanics [104]. Planes that make angles with respect to the c axis are called semipolar planes. Figure 3.18 shows three low-index semipolar planes in relation to the two nonpolar planes, which are important in InGaN LEDs. Semipolar planes are seldom atomically flat planes containing atomic steps. Detailed discussions are given in [105].

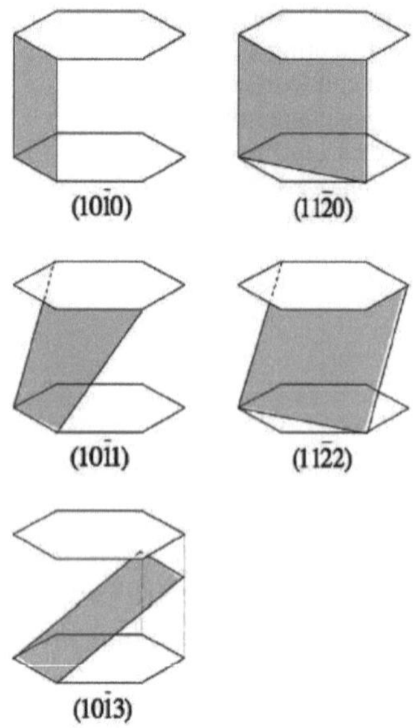

Fig. 3.18 Important semipolar planes (bottom three, painted in gray) shown along with the low-index nonpolar planes (top two). $\{10\bar{1}1\}$, $\{10\bar{1}3\}$, and $\{11\bar{2}2\}$ planes are at 62°, 32°, and 58° from the basal plane for GaN, respectively [75]

It is calculated from the closely packed structure of spheres that the ideal wurtzite structure possesses a *c/a* ratio of 1.633, where *c* and *a* are the conventional lattice constants in the *c* and *a* crystallographic directions, respectively. In reality, GaN, for example, has a ratio of 1.627, which implies that the crystal is compressed in the *c* direction under standard conditions (room temperature, 1 atm). Consequently, the centers of positive and negative charges from the ionicity of the Ga–N bond are displaced from each other, and electric charges theoretically appear at the opposite surfaces of the crystal. This is called spontaneous polarization. It is also possible to intentionally stress crystals (e.g., coherently grown heteroepitaxial films). The *c/a* ratio changes from the unstrained state, and additional polarization charges will result, which is called piezoelectric polarization. In any case, semiconductor crystals are not able to sustain voltages larger than their bandgaps, as carriers start shifting within the crystal to neutralize polarization charges (Fig. 3.19a), or ions in the atmosphere may adhere to the crystal surface to neutralize the polarization charges [106]. As a result, electric fields induced by spontaneous/piezoelectric polarization are irrelevant in large homogeneous materials (Fig. 3.19b). However, in heterostructures, this is not the case. There will be polarization-charge discontinuities at the interfaces, and net polarization charges induce internal electric fields. These electric fields are relevant in small-size heterostructures, e.g., QWs, resulting in the QCSE (Fig. 3.15). The strength of the internal electric fields is determined by the projection (i.e., cos θ) of the polarization vector (i.e., strain state, *c/a* deviation from 1.633) onto the plane of interest. Nonpolar orientations utilize cos $\theta = 0$, and semipolar orientations try to attain *c/a* $=$ 1.633 (although strained semipolar crystals do not necessarily remain wurtzite, rigorously speaking; see [100] for details).

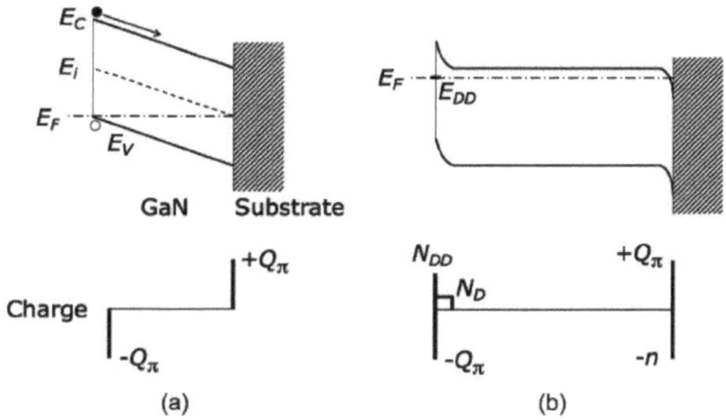

Fig. 3.19 Effects of polarization fields on carrier generation and field screening. (**a**) Idealized case of intrinsic semiconductors described by the band diagram and charge distribution. As a crystal grows on an insulating substrate, E_V meets E_F as a result of polarization fields. Holes are created at E_V, and electrons are swept toward the substrate by the polarization fields. Polarization charges ($+Q_\pi$ and $-Q_\pi$) are located at the interface and growth surface and are screened by carriers as the film grows. (**b**) More realistic case of thick n-type semiconductor films. Generated electrons have gathered at the substrate interface ($-n$) and neutralized the polarization charges. At the film surface, E_F is pinned at a deep donor level (E_{DD}), and a space charge layer is created where donors are charged (N_D). Polarization charges at the surface are neutralized by the charged donor and deep donors (N_{DD}) [75]

3.3.3 Performance of Nonpolar and Semipolar-Oriented LEDs Using Homoepitaxial Substrates

Several heteroepitaxial approaches implement *m*-axis or *a*-axis growth orientations [107, 108] yet result in very high densities of threading dislocations ($\sim 10^{10}$ cm^{-2}) and stacking faults ($\sim 10^5$ cm^{-1}) likely limiting LED performance [108]. The use of low-dislocation-density bulk GaN substrates sliced from boules recently resulted in better nonpolar homoepitaxial growth [72, 109]. UV and blue GaInN LEDs on the *m*-plane [67] and green and yellow LEDs on semipolar $(11\bar{2}2)$ substrate [73] have been demonstrated, the best of which show properties superior to those of c-plane growth [67, 72, 73, 110].

Bulk GaN substrates became available commercially in 2006. By that time, c-plane GaN wafers were commercially available but were limited in supply at a high cost. By making *c*-oriented hydride vapor-phase epitaxy (HVPE) GaN grow thick enough ($\sim$5 mm), it is possible to slice it in other orientations, thereby obtaining nonpolar and semipolar GaN substrates [111]. Hence, the size of the sliced wafers depends on how thick *c*-oriented GaN crystals were grown, and the current HVPE growth technology enables to provide approximately 10-mm-thick crystals. Although harvested nonpolar/semipolar substrates are small in view of commercial LED production, they are large enough to carry out basic research to investigate the potential of nonpolar- and semipolar-oriented LEDs. The performance of homoepitaxial LEDs is summarized in Table 3.1.

Most nonpolar LEDs reported to date have been fabricated on the m orientation. LED performances were drastically improved by two research groups via homoepitaxy [66, 67, 112]. Kim et al. [68] claimed an advantage of growing thick QWs due to the absence of the QCSE. The external quantum efficiency was determined to be ~40% in the early study [67]. The current desire is to achieve luminescence of longer wavelengths [69], as only one report can be found on LEDs beyond 500 nm with reasonable optical output power [113]. We may be able to have an optimistic view on achieving green light emission from m-plane LEDs, as a similar obstacle was confronted by c-plane LEDs during the 1990s. Other problems include hillock formation on epitaxial m-plane films. The mechanism of hillock formation and their effects on LED characteristics are still unknown. Hillock formation can be avoided by employing vicinal substrates [112, 114, 115]. Tsuda et al. [116] reported smooth film growth without hillocks. Although a study by Yamada et al. [63] indicated that the m-orientation had a higher potential for light-emitting devices than the a-orientation, recent accomplishments on a-plane LEDs emerged in 2008 have demonstrated competitive performances [64, 72]. Green light emission was achieved, yet output power was comparable to heteroepitaxial LEDs.

Despite the advantage of arbitrary slicing GaN crystals, experimental studies on semipolar LEDs have been limited to $(10\overline{1}\overline{1})$ and $(11\overline{2}2)$ orientations [117]. These two low-index planes share a tilt angle of approximately 60° with respect to the *c*-plane; thus, they are considered to be advantageous in terms of the QCSE. Green, yellow, and amber light emission has been realized on $(11\overline{2}2)$ with reasonable optical output [70, 73, 74]. The $(11\overline{2}2)$-oriented InGaN yellow LEDs were reported to have more stable performances against changes in ambient temperature than AlInGaP yellow LEDs [70]. $(10\overline{1}\overline{1})$-oriented LEDs have exhibited strong blue light emission [71, 118] with 34% external quantum efficiency, and longer wavelength emission has now been sought. As far as green emitters are concerned, the $(11\overline{2}2)$ orientation seems to have advantages among reported nonpolar/semipolar orientations to date. An important achievement on laser diodes to be mentioned here was reported by Enya et al. [119] in July 2009. Pure green laser diodes were demonstrated on the $(20\overline{2}1)$ plane, which has not been used for LEDs.

It is not intuitive that if the m-plane is preferred to the a-plane in device fabrication, why does the $(11\overline{2}2)$-plane (an inclined a-plane) seems to have advantages over the $(10\overline{1}\overline{1})$-plane (an inclined m-plane). We do not know yet whether these are inherent properties of these crystallographic planes or if it is only a matter of growth techniques and conditions.

3.4 Efficiency Droop and Efficiency Enhancement of Homoepitaxial InGaN/GaN Light-Emitting Diodes

The development of high-brightness LEDs suffers from the nonthermal rollover of internal quantum efficiency (IQE) at high current density known as the efficiency droop [120–122]. For the best commercial LEDs, the IQE peaks at a current density of a few A cm^{-2} but is reduced by ~25% at high current density J ~100 A cm^{-2}. Several mechanisms for efficiency droop have been proposed, including carrier leakage [120] and carrier delocalization from In-rich regions [121–124]. Great progresses have been made to mitigate this efficiency droop: semiempirical modeling [125], compressive stressed substrates [126], InGaN barriers for active region [127], etc.

Katsushi Akita et al. characterized InGaN-based blue LEDs on GaN substrates with low TDDs and on c-plane sapphire substrates and revealed the advantages of using GaN substrates [128]. GaN substrates were grown by HVPE, and their TDDs were less than 1×10^6 cm^{-2}. The detail fabrication flows of LEDs were provided in Ref. [128]. The LEDs with 3-nm-thick QWs and 5-nm-thick QWs were fabricated and characterized. The indium compositions of these two types of QWs were designed to be 14% and 10%, respectively, so as to obtain the same emission wavelength of 450 nm. The conditions for the carrier localization of the QWs are expected to be different between the two LEDs.

Figure 3.20a, b shows the output powers and the EQEs of the LEDs with 3-nm-thick QWs (sample (a)) and 5-nm-thick QWs (sample (b)) on sapphire substrates as a function of forward current. Figure 3.20c, d shows the output powers and the EQEs of the LEDs with 3-nm-thick QWs (sample (c)) and 5-nm-thick QWs (sample (d)) on GaN substrates as a function of forward current. As shown in Fig. 3.20a, the output power of sample (a) does not increase linearly with increasing forward current. The EQE of sample (a) was 8.8% at 20 mA but decreased with increasing forward current and dropped to 5.6% at 200 mA. This phenomenon is typically observed in blue LEDs on sapphire substrates and indicates that the carrier localization is in play. In contrast, the EQE of sample (b) dropped drastically at 20 mA. The EQE increased gradually with increasing forward current, but it was lower than that of sample (a) at 200 mA. As shown in Fig. 3.20c, the current dependence of the output power of sample c was almost the same as that of sample (a). The EQE of sample (c) was as high as 8.3% at 20 mA. It decreased with increasing forward current and dropped to 4.9% at 200 mA. This phenomenon indicates that the carrier localization is in play in LEDs also on GaN substrates with low TDDs as well as on sapphire substrates and that nonradiative recombination centers (NRCs) unrelated to threading dislocations (TDs) exist in the active layers. In contrast, the output power of sample (d) increased almost linearly with increasing forward current. The EQE of the LED was as high as 7.9% at 20 mA and diminished only slightly with increasing forward current to 6.6% at 200 mA. The decrease of the EQE of the LED was suppressed in the case of sample (d). This suggests that the concentrations of NRCs are lower for QWs with smaller indium compositions. This is supported by the fact that the large decrease of the EQE at high current densities is not observed in the near-UV LED with QWs with smaller indium compositions in Ref. [129]. The output powers were 17 mW at 20 mA (12.5 A/cm^2) and 148 mW at 200 mA (125 A/cm^2) when sample (d) was mounted p-side down and molded with epoxy. The EQEs were 30% at 20 mA and 26% at 200 mA. This value is considerably high at such a high current density. On the sapphire substrate, the

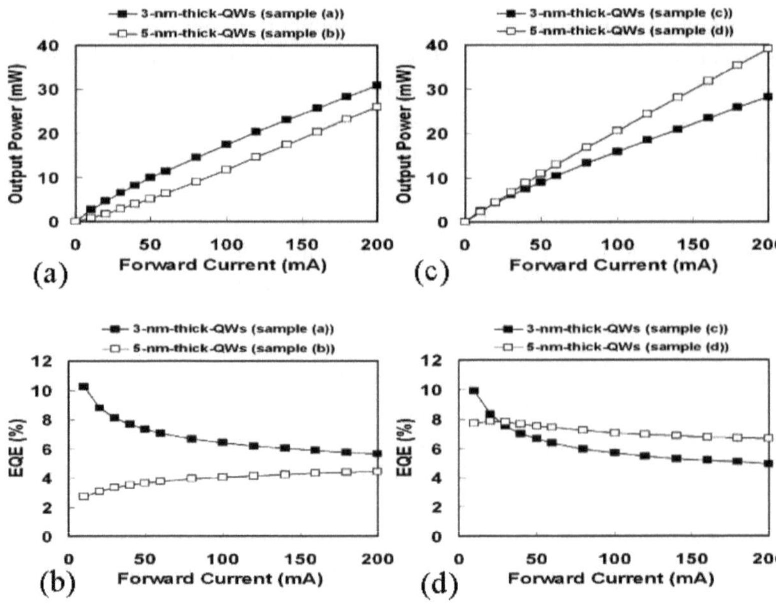

Fig. 3.20 (**a**) The output power and (**b**) the EQE of LEDs on sapphire substrates as a function of forward current; (**c**) the output power and (**d**) the EQE of LEDs on GaN substrates as a function of forward current [128]

EQEs were estimated 30% at 20 mA and 19% at 200 mA when sample (a) was mounted p-side down and molded with epoxy. Therefore, the use of GaN substrates in conjunction with thick QWs is a promising approach for fabrication of high power LEDs.

3.5 Light Efficiency Extraction

The total device efficiency of an LED is usually described as the external quantum efficiency (EQE). This figure of merit describes the efficiency of turning electrical carriers into photons. EQE can loosely be broken into internal quantum efficiency (IQE), injection efficiency, and extraction efficiency. The injection efficiency and IQE are largely a function of the crystal quality and electrical optimization of the LED structure. The extraction efficiency of a device describes the efficiency of the LED structure and packaging at removing photons generated in the active region of the device from the semiconductor material. It is known that the extraction efficiency of LEDs is limited by high refractive index contrast between the GaN (2.4) and air (1.0). The angle of the light escape cone is only 23°, which leads to a light extraction efficiency as low as 4.18% per crystal face [130]. For conventional LED on sapphire, smaller refractive index (1.78) and higher transparency of sapphire compared to bulk GaN are favor of light extraction. In contrast, a considerable amount of downward light in homoepitaxial LED is absorbed by the FS-GaN substrate. Thus, the problems about extraction efficiency for the LEDs on FS-GaN are more serious than the counterpart on sapphire. Currently, it has been demonstrated that several methods are used to improve extraction efficiency in GaN based on FS-GaN substrate, such as etching roughness of N-Face GaN [131, 132], geometric die shaping [133–135], and photonic crystal (PhC) structures [136].

3.5.1 Surface Treatment Methods

The critical angle for photons to escape from GaN film is determined by Snell's law. The angle is crucially important for the light-extraction efficiency of LEDs. It has been demonstrated that changing surface of the backside GaN can be used to improve light-extraction efficiency in InGaN-based LEDs on GaN substrate. This can be understood that the effective index of the GaN film has been decreased. It means more photons which should be total reflected can escape from GaN film for improving the total light efficiency.

Yi-Keng Fu et al. fabricated nitride-based LEDs with a roughed backside GaN substrate by chemical wet-etching process. The results show that they can enhance the 20 mA output power by 95% and 29% from the near-UV LED and blue LED, respectively, compared with standard LEDs [132]. Hong Zhong et al. used ICP etching to pattern the backside of semipolar LEDs, and a substantial increase (100% before packaging and 33% after) in output power was observed at 20 mA drive current in comparison with a smooth backside reference sample [137]. Yuji Zhao et al. fabricated the first 30-mW-class semipolar blue light-emitting diode (LED) on a freestanding $\left(10\overline{1}\overline{1}\right)$ GaN substrate by using microscale periodic backside structures [131]. The schematic structure of the device with the roughened backside is shown in Fig. 3.21a, while SEM images of the backside of the GaN substrate before (Fig. 3.21b) and after (Fig. 3.21c) backside roughening are also presented. A schematic graph (left) and an optical micrograph (right) of a working blue LED using this packaging method are shown as insets in Fig. 3.22. Room temperature (RT) electroluminescence (EL) measurements under pulsed conditions with a duty cycle of 1% were performed in an integrating sphere. Figure 3.22 shows the light output power vs current and external quantum efficiency vs current curves of the LED. At a forward current of 20 mA, the semipolar LED has an output power of 31.1 mW and an EQE of 54.7%. At 350 mA, the LED has a slightly lower EQE of 45.4% and an output power of 458 mW. The roughened devices demonstrated a better performance by having a sixfold increase of the output power after packaging, compared with a fourfold increase in the case of conventional devices mainly due to the dramatic enhancement of photon extraction from the backside of the substrate.

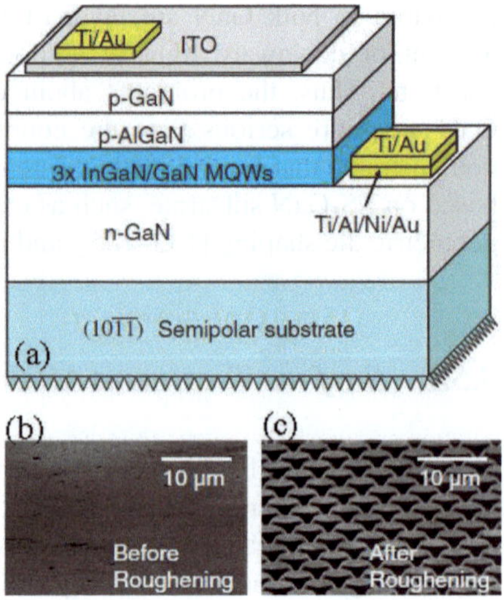

Fig. 3.21 (**a**) Schematic views of the semipolar $(10\bar{1}\bar{1})$ LED device with backside roughening structures. (**b**) SEM images of the backside of the GaN substrate before roughening and (**c**) after roughening, from a 10-tilted angle [131] (Copyright (2010) The Japan Society of Applied Physics)

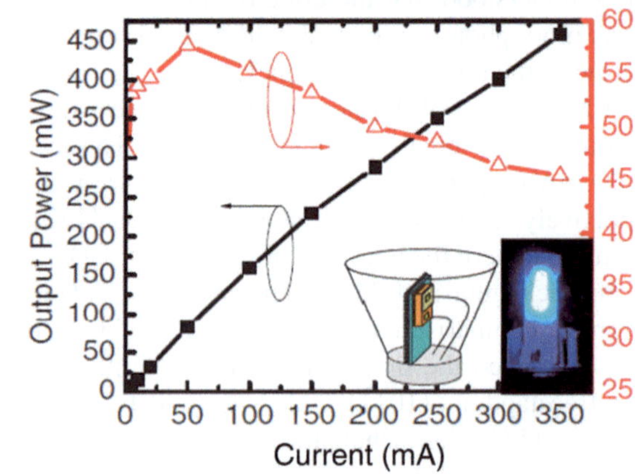

Fig. 3.22 Light output power vs current and external quantum efficiency vs current curves for a packaged $(10\bar{1}\bar{1})$ LED with backside roughening under pulsed conditions. Insets: Schematic graph (left) and optical micrograph (right) of a working blue LED using a transparent packaging method [131] (Copyright (2010) The Japan Society of Applied Physics)

3.5.2 Chip Shaping Method

Chip shaping is a method that has proven successful for material systems that are more fully mature than the nitrides [138]. Carefully designing the chip geometry of bulk GaN-based LEDs structure can improve the chances of light extraction for photons incident on device sidewalls. By breaking the lateral symmetry in a LED, conditions can be created such that each total internal reflection moves the photon path closer to being within the critical angle.

Bo Sun et al. realized a light extraction efficiency enhancement of bulk GaN light-emitting diodes (LEDs) in the shape of truncated pyramid. Compared with the reference LEDs, an enhancement of up to 46% on the light output power from rectangle-shaped LEDs chip with the inclination angle (~44°) has been observed. Compared with the common triangle-shaped and hexagon-shaped LEDs, large size

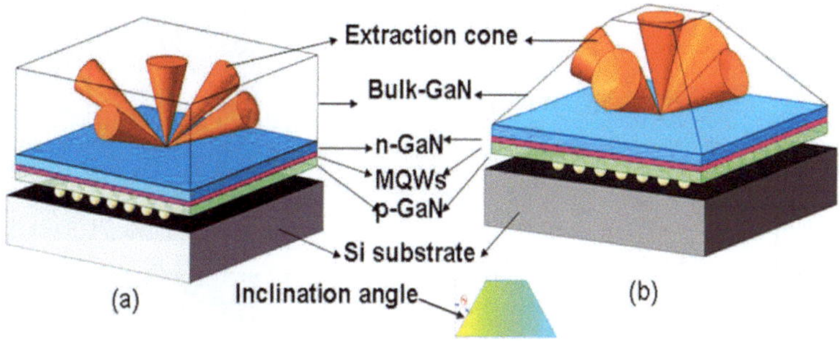

Fig. 3.23 Schematic diagrams of conventional flip-chip bulk-GaN LEDs (**a**) and truncated pyramidal-shaped flip-chip bulk-GaN LEDs (**b**) [134]

of conventional rectangular LEDs shaped with truncated pyramid shows more obvious enhancement in light extraction efficiency [134]. The schematic diagram of bulk-GaN LEDs with a truncated pyramidal shape was shown in Fig. 3.23.

Bulk-GaN LEDs have a relative large windows layer for light extraction in which the horizontal and vertical dimension is comparable. That means photons generated from multiple quantum wells have more chances to be scattered by sidewalls compared to conventional sapphire-based LEDs. However, for rectangular-shaped LED chip, the efficiency is still limited as light totally internally reflected at a semiconductor-to-ambient interface cannot change its incidence angle upon any of the facets. In case of a truncated pyramidal-shaped LED chip as shown in Fig. 3.23b, when light reflected from one facet to another, its incidence angle is changed. It means that light has several chances to escape from the LED chip which can then be extracted through top surface as well as the tilted sidewalls after considering Fresnel losses. The tilted sidewall also increases the light escape cones associated with the sidewalls and the same to the light escape cones related to the top surface. According the simulations, the light escape cones associated with the sidewall gradually enlarge with increasing the inclination angle. When the inclination angle was increased to 44°, the light escape cones associated with top surface and sidewalls start to overlap. Photo source emitted from QWs can be regarded as a dipole emitter, of which the normalized radiation patterns, given by the power per unit of solid angle, were a lambert-shaped (defined by $\sin^2\theta$ or $\cos^2\theta$). About 74% of dipole light power was confined in the 60° light cone, which means that the light extraction efficiency from sidewalls can be greatly enhanced if the light escape cone associated with the sidewall more merged in the dipole radiation patterns.

The L-I characteristics of truncated pyramidal-shaped bulk-GaN LEDs with different inclination angle and reference LEDs are shown in Fig. 3.24. The light output power of LEDs increased as the inclination angle enlarged from 0° to 30°; however, the light output power reached its maximum when the inclination angle increased to 44°; furthermore, a slight power reduced was found after the inclination angle enlarged to 50°. The experiment results are consistent with the simulation results. Compared with the reference LEDs, the LEDs with an inclination angle of

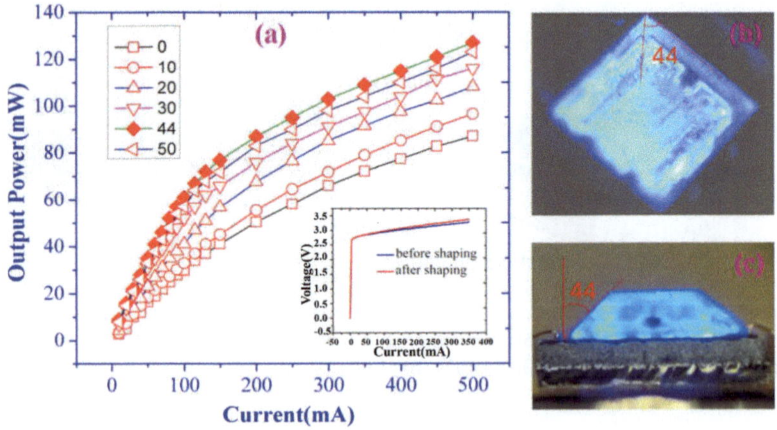

Fig. 3.24 (**a**) L-I characteristics of the truncated pyramidal-shaped LEDs (from inclination angle 0°–50°, respectively) and reference LEDs, and the insets show the $I-V$ curve of LEDs before and after laser shaping, (**b**) tilt view, and (**c**) cross-sectional view of optical micrographs of truncated pyramidal-shaped LEDs (with inclination angle 44°) at 5 mA injection current [134]

~44° have reached a larger enhancement of light extraction efficiency, up to 46%. The inset of Fig. 3.24 shows $I-V$ curve of LEDs before and after laser shaping. It can be found that the laser-shaping process does not cause undesirable damage to the LED chip. Figure 3.24b, c also shows the tilt view and cross-sectional view of optical microscopic images of the LED with an inclination angle of ~44° operated at an injection current of 5 mA. Light extracted among the four sidewalls can be easily observed.

In addition, it was found that compared to triangle-shaped and hexagon-shaped LEDs, large size of conventional rectangular LEDs shaped with truncated pyramid showed more significant improvement in light extraction efficiency. This could make the conventional rectangular bulk-GaN LEDs have a better improvement of external quantum efficiency even up to very high current densities just using the standard fabrication process.

3.5.3 *Photonic Crystal Method*

To enhance extraction of LEDs, photonic crystal (PhC) structures have drawn much attention, which could lead to efficiently coupling light from the dielectric-guided modes into air [139–142]. In addition to increasing the extraction efficiency of LEDs, periodic PhC structures have the ability to enhance the directionality, especially along the vertical orientation. Recently, Weisbuch et al. demonstrated two air-gap embedded PhCs, which created a waveguide with highly confined and

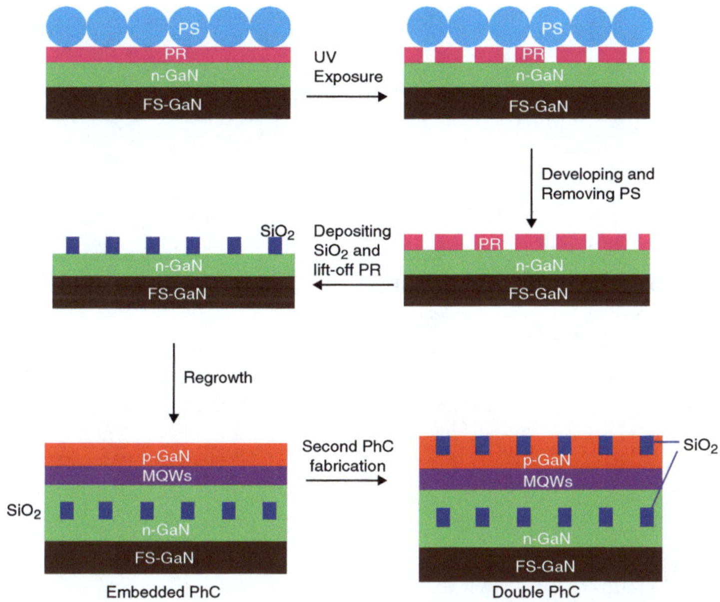

Fig. 3.25 Schematic illustration of the process for the fabrication of embedded PhC and double PhC LEDs. Reprinted with permission from Ref. [136], Optical Society of America

well-extracted mode while exhibiting no significant deleterious effects on the LEDs [143, 144].

Tongbo Wei et al. reported homoepitaxially grown InGaN/GaN light-emitting diodes (LEDs) with SiO_2 nanodisks embedded in *n*-GaN and *p*-GaN as photonic crystal (PhC) structures by nanospherical-lens photolithography. The introduction of SiO_2 nanodisks doesn't produce the new dislocations and doesn't also result in the electrical deterioration of PhC LEDs. The light output power of homoepitaxial LEDs with embedded PhC and double PhC at 350 mA current is increased by 29.9% and 47.2%, respectively, compared to that without PhC. The corresponding light radiation patterns in PhC LEDs on GaN substrate show a narrow beam shape due to strong guided light extraction, with a view angle reduction of about 30° [136].

Figure 3.25 shows the fabrication process flow for embedded and double SiO_2 PhC LEDs.

Figure 3.26a demonstrates the scanning electron microscope (SEM) image of cross-sectional view of the *n*-GaN laterally reovergrown over the SiO_2 nanodisks in the embedded PhC LED. It is noted that the SiO_2 nanodisks are fully surrounded by the GaN layer, without leaving the voids. The diameter, period and height of the embedded SiO_2 nanodisk are 400, 900 and 200 nm, respectively. According to the top view of PhC structure, there is a uniform hexagonal-lattice distribution of SiO_2 nanodisks as shown in Fig. 3.26b. Selectively regrowth of p-GaN with a thickness of 150 nm is carried out to fill the space between SiO_2 nanodisks. Furthermore, transmission electron microscopy (TEM) is employed to investigate

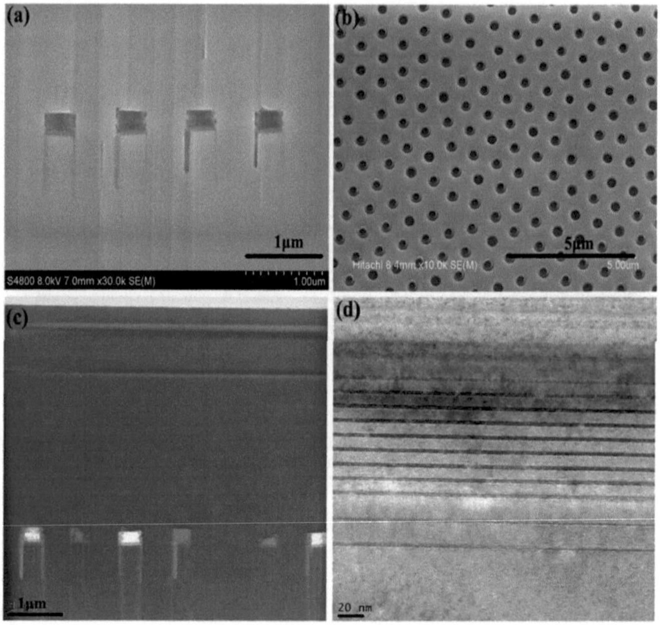

Fig. 3.26 (**a**) Cross-sectional view of embedded PhC of SiO_2 nanodisks surrounded by n-GaN and (**b**) tilted top view of p-GaN surface with top SiO_2 PhC structure. (**c**) Cross-sectional TEM image of the LED with embedded SiO_2 PhC (**d**) is the magnified region of MQWs from PhC LED in (**c**). Reprinted with permission from Ref. [136], Optical Society of America

the crystalline quality of GaN layers that are homoepitaxially grown on the SiO_2 PhC structure. As shown in Fig. 3.26c, d, almost no threading dislocations can be observed in both n-GaN and MQWs structures, implying the high-quality growth of GaN on FS-GaN substrate. In the epitaxial lateral overgrowth (ELOG) on sapphire, Wuu et al. reported the SiO_2 array could block the dislocation propagation, but new dislocations may be introduced in the lateral coalescence region during the second growth [145]. Unlike the heteroepitaxial growth, there are also no new dislocations formed on the SiO_2 nanodisks in the grown process.

The far-field emission patterns from the PhC LEDs show much smaller view angles but obvious enhancement in the overall integrated emission intensity. The light is further redirected to the top escape cone through the twice transmission of embedded and top SiO_2 arrays, resulting in the most significant focusing effect and more photon capable of escaping from the chips.

Bulk-GaN-based LEDs are attracting more and more attention since the first LEDs grown on bulk GaN substrates were reported by the UCSB group in 2005, due to the potential to fabricate high-power and high-efficiency devices. Over the past decade, many approaches to further improve the performance of native bulk-GaN-based LEDs have been demonstrated. Even though the developing technologies of GaN substrates are still on the way, their applications to LEDs have attracted much interest of community. Although these freestanding substrates still have relatively high densities of dislocations and stacking faults, GaN-based LEDs were successfully realized by homoepitaxy on polar, nonpolar, and semipolar substrates, with the progresses in the improvements of internal quantum efficiency, engineering of light extraction, and suppressing the efficiency droop. The achievements in GaN LEDs will more and more clearly prove the great advantages and irreplaceability of GaN substrates. Along with the advance of GaN-freestanding substrates in decreasing the density of defects and lowering the cost of production, wider applications of GaN substrates in high-efficiency and high-power LEDs for solid-state lighting can be desirable, on condition that homoepitaxy technology is well developed and widely used.

References

1. T. Paskova, D.A. Hanser, K.R. Evans, Proc. IEEE **98**, 1324 (2010)
2. H. Yamane, M. Shimada, S.J. Clarke, F.J. DiSalvo, Chem. Mater. **9**, 413 (1997)
3. M. Bockowski, Cryst. Res. Technol. **42**, 1162 (2007)
4. S. Lourdudoss, N. Gopalakrishnan, R. Holz, M. Deschler, R. Beccard, Value-Addition Metall. **177** (1998)
5. H. Murakami, J. Kikuchi, Y. Kumagai, A. Koukitu, Phys. Status Solidi C **3**, 1457 (2006)
6. A. Koukitu, S. Hama, T. Taki, H. Seki, Jpn. J. Appl. Phys. **37**, 762 (1998)
7. A. Koukitu, J. Kikuchi, Y. Kangawa, Y. Kumagai, J. Cryst. Growth **281**, 47 (2005)
8. Y. Kumagai, K. Takemoto, T. Hasegawa, A. Koukitu, H. Seki, J. Cryst. Growth **231**, 57 (2001)
9. A. Usui, H. Sunakawa, A. Sakai, A. Yamaguchi, Jpn. J. Appl. Phys. **36**, L899 (1997)
10. K. Hiramatsu, K. Nishiyama, A. Motogaito, H. Miyake, Y. Iyechika, T. Maeda, Phys. Status Solidi A **176**, 535 (1999)
11. Y. Kawaguchi, S. Nambu, H. Sone, M. Yamaguchi, H. Miyake, K. Hiramatsu, N. Sawaki, Y. Iyechika, T. Maeda, Mrs Internet, J. Nitride Semicond. Res. **4**, G4.1 (1999)
12. Y. Kawaguchi, S. Nambu, M. Yamaguchi, N. Sawaki, H. Miyake, K. Hiramatsu, K. Tsukamoto, N. Kuwano, K. Oki, Phys. Status Solidi A **176**, 561 (1999)
13. Y. Kawaguchi, Y. Honda, M. Yamaguchi, N. Sawaki, K. Hiramatsu, Phys. Status Solidi A **176**, 553 (1999)
14. Z. Yu, H.M.A.L. Johnson, J.D. Brown, N.A. El-Masry, J.F. Muth, J.W. Cook, J.F. Schetzina, K.W. Haberern, H.S. Kong, J.S. Edmond, Mrs Internet, J. Nitride Semicond. Res. **4**, G4.3 (1999)
15. F. Yun, Y.T. Moon, Y. Fu, K. Zhu, U. Ozgur, H. Morkoc, C.K. Inoki, T.S. Kuan, A. Sagar, R.M. Feenstra, J. Appl. Phys. **98**(8), 123502 (2005)
16. S. Nagahama, N. Iwasa, M. Senoh, T. Matsushita, Y. Sugimoto, H. Kiyoku, T. Kozaki, M. Sano, H. Matsumura, H. Umemoto, K. Chocho, T. Mukai, Jpn. J. Appl. Phys. **39**, L647 (2000)
17. K. Motoki, T. Okahisa, S. Nakahata, N. Matsumoto, H. Kimura, H. Kasai, K. Takemoto, K. Uematsu, M. Ueno, Y. Kumagai, A. Koukitu, H. Seki, J. Cryst. Growth **237**, 912 (2002)
18. K. Motoki, T. Okahisa, S. Nakahata, A. Matsumoto, H. Kimura, H. Kasai, K. Takemoto, K. Uematsu, M. Ueno, Y. Kumagai, A. Koukitu, H. Seki, Mater. Sci. Eng. B-Solid State Mater. Adv. Technol. **93**, 123 (2002)
19. K. Motoki, T. Okahisa, R. Hirota, S. Nakahata, K. Uematsu, N. Matsumoto, J. Cryst. Growth **305**, 377 (2007)
20. Y. Oshima, T. Eri, M. Shibata, H. Sunakawa, K. Kobayashi, T. Ichihashi, A. Usui, Jpn. J. Appl. Phys. **42**, L1 (2003)
21. J.Q. Liu, J. Huang, X.J. Gong, J.F. Wang, K. Xu, Y.X. Qiu, D.M. Cai, T.F. Zhou, G.Q. Ren, H. Yang, CrystEngComm **13**, 5929 (2011)

22. K. Fujito, S. Kubo, H. Nagaoka, T. Mochizuki, H. Namita, S. Nagao, J. Cryst. Growth **311**, 3011 (2009)
23. S. Nakamura, T. Mukai, M. Senoh, Jpn. J. Appl. Phys. **31**, 2883 (1992)
24. A.V. Fomin, A.E. Nikolaev, I.P. Nikitina, A.S. Zubrilov, M.G. Mynbaeva, N.I. Kuznetsov, A.P. Kovarsky, B.J. Ber, Phys. Status Solidi A **188**, 433 (2001)
25. E. Richter, C. Hennig, U. Zeimer, L. Wang, M. Weyers, G. Tränkle, Phys. Status Solidi (a) **203**, 1658 (2006)
26. N.G. Weimann, L.F. Eastman, D. Doppalapudi, H.M. Ng, T.D. Moustakas, J. Appl. Phys. **83**, 3656 (1998)
27. A.Y. Polyakov, N.B. Smirnov, A.V. Govorkov, N.Y. Pashkova, J. Kim, F. Ren, M.E. Overberg, G.T. Thaler, C.R. Abernathy, S.J. Pearton, R.G. Wilson, J. Appl. Phys. **92**, 3130 (2002)
28. E. Malguth, A. Hoffmann, W. Gehlhoff, O. Gelhausen, M.R. Phillips, X. Xu, Phys. Rev. B **74**, 165202 (2006)
29. B. Monemar, O. Lagerstedt, J. Appl. Phys. **50**, 6480 (1979)
30. R.P. Vaudo, X.P. Xu, A. Salant, J. Malcarne, G.R. Brandes, Phys. Status Solidi A **200**, 18 (2003)
31. A.Y. Polyakov, N.B. Smirnov, A.V. Govorkov, N.V. Pashkova, A.A. Shlensky, S.J. Pearton, M.E. Overberg, C.R. Abernathy, J.M. Zavada, R.G. Wilson, J. Appl. Phys. **93**, 5388 (2003)
32. M. Kubota, T. Onuma, Y. Ishihara, A. Usui, A. Uedono, S.F. Chichibu, J. Appl. Phys. **105**, 083542 (2009)
33. A.Y. Polyakov, N.B. Smirnov, A.V. Govorkov, S.J. Pearton, Appl. Phys. Lett. **83**, 3314 (2003)
34. Y.M. Fan, Z.H. Liu, G.Z. Xu, H.J. Zhong, Z.L. Huang, Y.M. Zhang, J.F. Wang, K. Xu, Appl. Phys. Lett. **105**, 062108 (2014)
35. A.Y. Polyakov, N.B. Smirnov, A.V. Govorkov, T.G. Yugova, A.V. Markov, A.M. Dabiran, A.M. Wowchak, B. Cui, J. Xie, A.V. Osinsky, P.P. Chow, S.J. Pearton, Appl. Phys. Lett. **92**, 042110 (2008)
36. A. Bonanni, M. Kiecana, C. Simbrunner, T. Li, M. Sawicki, M. Wegscheider, M. Quast, H. Przybylinska, A. Navarro-Quezada, R. Jakiela, A. Wolos, W. Jantsch, T. Dietl, Phys. Rev. B **75**, 125210 (2007)
37. A. Bonanni, A. Navarro-Quezada, T. Li, M. Wegscheider, Z. Matej, V. Holy, R.T. Lechner, G. Bauer, M. Rovezzi, F. D'Acapito, M. Kiecana, M. Sawicki, T. Dietl, Phys. Rev. Lett. **101**, 135502 (2008)
38. D.O. Dumcenco, S. Levcenco, Y.S. Huang, C.L. Reynolds, J.G. Reynolds, K.K. Tiong, T. Paskova, K.R. Evans, J. Appl. Phys. **109**, 123508 (2011)
39. J.M. Langer, H. Heinrich, Phys. Rev. Lett. **55**, 1414 (1985)
40. R. Heitz, P. Maxim, L. Eckey, P. Thurian, A. Hoffmann, I. Broser, K. Pressel, B.K. Meyer, Phys. Rev. B **55**, 4382 (1997)
41. M. Zhang, T.F. Zhou, Y.M. Zhang, B. Li, S.N. Zheng, J. Huang, Y.P. Sun, G.Q. Ren, J.F. Wang, K. Xu, H. Yang, Appl. Phys. Lett. **100**, 041904 (2012)
42. L. Chernyak, A. Osinsky, A. Schulte, Solid State Electron. **45**, 1687 (2001)
43. P. Scajev, K. Jarasiunas, S. Okur, U. Ozgur, H. Morkoc, J. Appl. Phys. **111**, 023702 (2012)
44. M. Reshchikov, A.M. Foussekis, A.A. Baski, J. Appl. Phys. **107**, 113535 (2010)
45. D. Wee, G. Parish, B. Nener, J. Appl. Phys. **111**, 074503 (2012)
46. A. Cavallini, L. Polenta, A. Castaldini, Microelectron. Reliab. **50**, 1398 (2010)
47. Z.H. Liu, K. Xu, Y.M. Fan, G.Z. Xu, Z.L. Huang, H.J. Zhong, J.F. Wang, H. Yang, Appl. Phys. Lett. **101**, 252107 (2012)
48. F.A. Ponce, D. Cherns, W.T. Young, J.W. Steeds, Appl. Phys. Lett. **69**, 770 (1996)
49. C. Wetzel, T. Salagaj, T. Detchprohm, P. Li, J.S. Nelson, Appl. Phys. Lett. **85**, 866 (2004)
50. O. Nam, M. Bremser, T. Zheleva, R. Davis, Appl. Phys. Lett. **71**, 2638 (1997)
51. C. Mion, J.F. Muth, E.A. Preble, D. Hanser, Appl. Phys. Lett. **89**, 092123 (2006)
52. X.H. Wu, C.R. Elsass, A. Abare, M. Mack, S. Keller, P.M. Petroff, S.P. DenBaars, J.S. Speck, S.J. Rosner, Appl. Phys. Lett. **72**, 692 (1998)
53. T. Detchprohm, Y. Xia, Y. Xi, M. Zhu, W. Zhao, Y. Li, E.F. Schubert, L. Liu, D. Tsvetkov, D. Hanser, C. Wetzel, J. Cryst. Growth **298**, 272 (2007)

54. A.D. Hanser, L. Liu, E.A. Preble, D. Tsvetkov, M. Tutor, N.M. Williams, K. Evans, Y. Zhou, D. Wang, C. Ahyi, C.-C. Tin, J. Williams, M. Park, D.F. Storm, D.S. Katzer, S.C. Binari, J.A. Roussos, J.A. Mittereder, Proceedings of the CS MANTECH Conference, April 24–27, Vancouver, BC, Canada, (2006)
55. M.K. Kelly, R.P. Vaudo, V.M. Phanse, L. Görgens, O. Ambacher, M. Stutzmann, Jpn. J. Appl. Phys. **38**, L217–L219 (1999)
56. J. Jasinski, W. Swider, Z. Liliental-Weber, P. Visconti, K.M. Jones, M.A. Reshchikov, F. Yun, H. Morkoç, S.S. Park, K.Y. Lee, Appl. Phys. Lett. **78**, 2297–2299 (2001)
57. C.L. Chao, C.H. Chiu, Y.J. Lee, H.C. Kuo, P.-C. Liu, J.D. Tsay, S.J. Cheng, Appl. Phys. Lett. **95**, 051905–051905–3 (2009)
58. T. Detchprohm, H. Amano, K. Hiramatsu, I. Akasaki, Appl. Phys. Lett. **61**, 2688 (1992)
59. S.J. Rosner, E.C. Carr, M.J. Ludowise, G. Girolami, H.I. Erikson, Appl. Phys. Lett. **70**, 420 (1997)
60. D. Hanser, L. Liu, E.A. Preble, D. Thomas, M. Williams, Mater. Res. Soc. **798**, Y2.1.1 (2004)
61. K. Lee, K. Auh, MRS Internet, J. Nitride Semicond. Res. **6**, 9 (2001)
62. C.R. Miskys, M.K. Kelly, O. Ambacher, G. Martínez-Criado, M. Stutzmann, Appl. Phys. Lett. **77**, 1858–1860 (2000)
63. H. Yamada, K. Iso, M. Saito, H. Hirasawa, N. Fellows, H. Masui, K. Fujito, J.S. Speck, S.P. DenBaars, S. Nakamura, Phys. Status Solidi RRL **2**(2), 89–91 (2008)
64. J.P. Liu, J.B. Limb, J.-H. Ryou, D. Yoo, C.A. Horne, R.D. Dupuis, Z.H. Wu, A.M. Fischer, F.A. Ponce, A.D. Hanser, L. Liu, E.A. Preble, K.R. Evans, Appl. Phys. Lett. **92**(1), 011123 (2008)
65. M. Funato, T. Kotani, T. Kondou, Y. Kawakami, Y. Narukawa, T. Mukai, Appl. Phys. Lett. **88**(26), 261920 (2006)
66. H. Yamada, K. Iso, M. Saito, H. Masui, K. Fujito, S.P. DenBaars, S. Nakamura, Appl. Phys. Express **1**(4), 041101 (2008)
67. M.C. Schmidt, K.C. Kim, H. Sato, N. Fellows, H. Masui, S. Nakamura, S.P. DenBaars, J.S. Speck, Jpn. J. Appl. Phys., Part 2 **46**, L126 (2007)
68. K.-C. Kim, M.C. Schmidt, H. Sato, F. Wu, N. Fellows, Z. Jia, M. Saito, S. Nakamura, S.P. DenBaars, J.S. Speck, K. Fujito, Appl. Phys. Lett. **91**(18), 181120 (2007)
69. K. Iso, H. Yamada, H. Hirasawa, N. Fellows, M. Saito, K. Fujito, S.P. DenBaars, J.S. Speck, S. Nakamura, Jpn. J. Appl. Phys. **46**(36–40), L960–L962 (2007)
70. H. Sato, R.B. Chung, H. Hirasawa, N. Fellows, H. Masui, F. Wu, M. Saito, K. Fujito, J.S. Speck, S.P. DenBaars, S. Nakamura, Appl. Phys. Lett. **92**(22), 221110 (2008)
71. A. Tyagi, H. Zhong, N.N. Fellows, M. Iza, J.S. Speck, S.P. DenBaars, S. Nakamura, Jpn. J. Appl. Phys. **46**(4–7), L129–L131 (2007)
72. T. Detchprohm, M. Zhu, Y. Li, Y. Xia, C. Wetzel, E.A. Preble, L. Liu, T. Paskova, D. Hanser, Appl. Phys. Lett. **92**, 241109 (2008)
73. M. Funato, M. Ueda, Y. Kawakami, Y. Narukawa, T. Kosugi, M. Takahashi, T. Mukai, Jpn. J. Appl. Phys., Part 2 **45**, L659 (2006)
74. H. Sato, A. Tyagi, H. Zhong, N. Fellows, R.B. Chung, M. Saito, K. Fujito, J.S. Speck, S.P. DenBaars, S. Nakamura, Phys. Status Solidi RRL **1**(4), 162–164 (2007)
75. H. Masui, S. Nakamura, S.P. DenBaars, U.K. Mishra, IEEE Trans. Electron Dev. **57**(1), 88–100 (2010)
76. M. Iwaya, T. Takeuchi, S. Yamaguchi, C. Wetzel, H. Amano, I. Akasaki, Jpn. J. Appl. Phys. **37**, L316 (1998)
77. J. Jasinski, Z. Liliental-Weber, D. Huang, M.A. Reshchikov, F. Yun, H. Morkoc, C. Sone, S.S. Park, K.Y. Lee, Mater. Res. Soc. **722**, K3.1.1 (2002)
78. M. Zhu, Y. Xia, W. Zhao, Y. Li, J. Senawiratne, T. Detchprohm, C. Wetzel, Journal of Elec Materi **37**(5), 641–645 (2008)
79. Z. Liliental-Weber, J. Washburn, K. Pakula, J. Baranowski, Microsc. Microanal. **3**, 436 (1997)
80. H.K. Cho, J.Y. Lee, G.M. Yang, Appl. Phys. Lett. **80**, 1370 (2002)
81. T. Mukai, S. Nakamura, Jpn. J. Appl. Phys., Part **1**(38), 5735 (1999)
82. X.A. Cao, S.F. LeBoeuf, L.B. Rowland, C.H. Yan, H. Liu, Appl. Phys. Lett. **82**, 3614 (2003)

83. P.G. Eliseev, P. Perlin, J. Furioli, P. Sartori, J. Mu, M. Osinski, J. Electron. Mater. **26**, 311 (1997)
84. X.A. Cao, K. Topol, F. Shahedipour, J. Teetsov, P.M. Sandvik, S.F. eBoeuf, A. Ebong, J. Kretchmer, E.B. Stokes, S. Arthur, A.E. Kaloyeros, n.D. Walker, Proc. SPIE **4776**, 105 (2002)
85. K. Leung, A.F. Wright, E.B. Stechel, Appl. Phys. Lett. **74**, 2495 (1999)
86. B. Monemar, B.E. Sernelius, Appl. Phys. Lett. **91**, 181103 (2007)
87. C. Sasaoka, H. Sunakawa, A. Kimura, M. Nido, A. Usui, A. Asakai, J. Ryst. Growth **189/190**, 61 (1998)
88. C.Y. Hsu, W.H. Lan, Y.S. Wu, Appl. Phys. Lett. **83**, 2447 (2003)
89. X.A. Cao, P.M. Sandvik, S.F. LeBoeuf, S.D. Arthur, Microelectron. Eliab. **43**, 1987 (2003)
90. X.A. Cao, J.M. Teetsov, M.P. D'Evelyn, D.W. Merfeld, C.H. Yan, Appl. Phys. Lett. **85**(1), 7–9 (2004)
91. X.A. Cao, H. Lu, E.B. Kaminsky, S.D. Arthur, J.R. Grandusky, F. Shahedipour-Sandvik, J. Cryst. Growth **300**(2), 382–386 (2007). ISSN 0022-0248
92. R.P. Vaudo, X. Xu, C. Lario, A.D. Salant, J.S. Flynn, G.R. Brandes, Phys. Status Solidi A **194**, 494 (2002)
93. X. Xu, R.P. Vaudo, C. Loria, A. Salant, G.R. Brandes, J. Chaudhuri, J. Crystal Growth **246**, 223 (2002)
94. X.A. Cao, E.B. Stokes, P. Sandvik, N. Taskar, J. Kretchmer, D. Walker, Solid State Electron. **46**, 1235 (2002)
95. I. Arslan, N.D. Browning, Phys. Rev. Lett. **91**, 165501 (2003)
96. T. Matsuoka, J. Cryst. Growth **189/190**, 19–23 (1998)
97. A. Koukitu, Y. Kumagai, J. Phys. Condens. Matter **13**(32), 6907–6934 (2001)
98. S.F. Chichibu, A. Uedono, T. Onuma, B.A. Haskell, A. Chakraborty, T. Koyama, P.T. Fini, S. Keller, S.P. DenBaars, J.S. Speck, U.K. Mishra, S. Nakamura, S. Yamaguchi, S. Kamiyama, H. Amano, I. Akasaki, J. Han, T. Sota, Nat. Mater. **5**(10), 810–816 (2006)
99. I. Vurgaftman, J.R. Meyer, J. Appl. Phys. **94**(6), 3675–3696 (2003)
100. A.E. Romanov, T.J. Baker, S. Nakamura, J.S. Speck, J. Appl. Phys. **100**(2), 023522 (2006)
101. T. Takeuchi, H. Amano, I. Akasaki, Jpn. J. Appl. Phys. **39**(2A), 413–416 (2000)
102. H.M. Otte, A.G. Crocker, Phys. Status Solidi B **9**(2), 441–450 (1965)
103. F.C. Frank, Acta Crystallogr. **18**, 862–866 (1965)
104. J.F. Nye, *Physical Properties of Crystals: Their Representation by Tensors and Matrices* (Oxford University Press, New York, 1985)
105. H. Masui, S.C. Cruz, S. Nakamura, S.P. DenBaars, J. Electron. Mater. **68**(6), 756–760 (2009). https://doi.org/10.1007/s11664-009-0777-4
106. U.K. Mishra, J. Singh, *Semiconductor Device Physics and Design* (Springer, Dordrecht, 2008), pp. 67–72
107. T. Takeuchi, S. Lester, D. Basile, G. Girolami, R. Twist, F. Mertz, M. Wong, R. Schneider, H. Amano, I. Akasaki, IPAP Conf. Ser. **1**, 137 (2000)
108. A. Chakraborty, T.J. Baker, B.A. Haskell, F. Wu, J.S. Speck, S.P. DenBaars, S. Nakamura, U.K. Mishra, Jpn. J. Appl. Phys., Part 2 **44**, L945 (2005)
109. A. Chakraborty, B.A. Haskell, S. Keller, J.S. Speck, S.P. DenBaars, S. Nakamura, U.K. Mishra, Jpn. J. Appl. Phys., Part 2 **44**, L173 (2005)
110. K. Okamoto, H. Ohta, S.F. Chichibu, J. Ichihara, H. Takasu, Jpn. J. Appl. Phys., Part 2 **46**, L187 (2007)
111. K. Fujito, K. Kiyomi, T. Mochizuki, H. Oota, H. Namita, S. Nagao, I. Fujimura, Phys. Status Solidi A **205**(5), 1056–1059 (2008)
112. K. Okamoto, H. Ohta, D. Nakagawa, M. Sonobe, J. Ichihara, H. Takasu, Jpn. J. Appl. Phys. **45**(45–45), L1 197–L1 199 (2006)
113. Y.-D. Lin, A. Chakraborty, S. Brinkley, H.C. Kuo, T. Melo, K. Fujito, J.S. Speck, S.P. DenBaars, S. Nakamura, Appl. Phys. Lett. **94**(26), 261108 (2009)
114. A. Hirai, Z. Jia, M.C. Schmidt, R.M. Farrell, S.P. DenBaars, S. Nakamura, J.S. Speck, K. Fujito, Appl. Phys. Lett. **91**(19), 191906 (2007)

115. H. Yamada, K. Iso, H. Masui, M. Saito, K. Fujito, S.P. DenBaars, S. Nakamura, J. Cryst. Growth **310**(23), 4968–4971 (2008)
116. Y. Tsuda, M. Ohta, P.O. Vaccaro, S. Ito, S. Hirukawa, Y. Kawaguchi, Y. Fujishiro, Y. Takahira, Y. Ueta, T. Takakura, T. Yuasa, Appl. Phys. Express **1**(1), 011104 (2008)
117. H. Masui, S. Nakamura, *Advances in Light Emitting Materials* (Trans Tech, Zurich, 2008), pp. 211–231
118. H. Zhong, A. Tyagi, N.N. Fellows, F. Wu, R.B. Chung, M. Saito, K. Fujito, J.S. Speck, S.P. DenBaars, S. Nakamura, Appl. Phys. Lett. **90**(23), 233504 (2007)
119. Y. Enya, Y. Yoshizumi, T. Kyono, K. Akita, M. Ueno, M. Adachi, T. Sumitomo, S. Tokuyama, T. Ikegami, K. Katayama, T. Nakamura, Appl. Phys. Express **2**(8), 082101 (2009)
120. M.H. Kim, M.F. Schubert, Q. Dai, J.K. Kim, E.F. Schubert, J. Piprek, Y. Park, Appl. Phys. Lett. **91**, 183507–183507-3 (2007)
121. S. Chichubu, T. Azuhata, T. Sota, S. Nakamura, Appl. Phys. Lett. **70**, 2822–2824 (1997)
122. S.F. Chichibu, T. Azuhata, M. Sugiyama, T. Kitamura, Y. Ishida, H. Okumura, H. Nakanishi, T. Sota, T. Mukai, J. Vac. Sci. Technol. **B 19**, 2177–2183 (2001)
123. Y.C. Shen, G.O. Mueller, S. Watanabe, N.F. Gardner, A. Munkholm, M.R. Krames, Appl. Phys. Lett. **91**, 141101 (2007)
124. X. Ni, J. Lee, S. Liu, V. Avrutin, Ü. Özgür, H. Morkoç, A. Matulionis, J. Appl. Phys. **108**, 033112–033112-13 (2010)
125. Q. Dai, Q. Shan, J. Wang, S. Chhajed, J. Cho, E.F. Schubert, M.H. Crawford, D.D. Koleske, M.-H. Kim, Y. Park, Appl. Phys. Lett. **97**, 133507–133507-3 (2010)
126. J.H. Son, J.L. Lee, Opt. Express **18**, 5466–5471 (2010)
127. Y.-K. Kuo, J.-Y. Chang, M.-C. Tsai, S.-H. Yen, Appl. Phys. Lett. **95**, 011116–011116-3 (2009)
128. K. Akita, T. Kyono, Y. Yoshizumi, H. Kitabayashi, K. Katayama, J. Appl. Phys. **101**, 033104–033104-5 (2007)
129. M. Yamada, T. Mitani, Y. Narukawa, S. Shioji, I. Niki, S. Sonobe, D. Kouichiro, S. Masahiko, T. Mukai, Jpn. J. Appl. Phys. **41**(12B), L1431 (2002)
130. E. Schubert, *Light Emitting Diodes*, 2nd edn. (Cambridge University Press, Cambridge, 2006), p. 93
131. Y.J. Zhao, J. Sonoda, C.-C. Pan, S. Brinkley, I. Koslow, K. Fujito, H. Ohta, S.P. DenBaars, S. Nakamura, Appl. Phys. Express **3**, 102101 (2010)
132. Y.-K. Fu, B.-C. Chen, Y.-H. Fang, R.-H. Jiang, Y.-H. Lu, R. Xuan, K.-F. Huang, C.-F. Lin, Y.-K. Su, J.-F. Chen, C.-Y. Chang, IEEE Photon. Technol. Lett. **23**(19), 1373–1375 (2011)
133. S.E. Brinkley, C.L. Keraly, J. Sonoda, C. Weisbuch, J.S. Speck, S. Nakamura, S.P. DenBaars, Appl. Phys. Express **5**(3), 032104 (2012)
134. B. Sun, L.X. Zhao, T.B. Wei, X.Y. Yi, Z.Q. Liu, G.H. Wang, J.M. Li, J. Appl. Phys. **113**(24), 243104 (2013)
135. T.B. Wei, K. Wu, Y. Chen, J. Yu, Q. Yan, Y.Y. Zhang, R. Duan, J. Wang, Y. Zeng, J.M. Li, IEEE Electron Device Lett. **33**(6), 857–859 (2012)
136. T. Wei, Z. Huo, Y. Zhang, H. Zheng, Y. Chen, J. Yang, Q. Hu, R. Duan, J. Wang, Y. Zeng, J. Li, Opt. Express **22**, A1093–A1100 (2014)
137. H. Zhong, A. Tyagi, N. Pfaff, M. Saito, K. Fujito, J.S. Speck, S. Nakamura, Jpn. J. Appl. Phys. **48**(3R), 030201 (2009)
138. M.R. Krames, M. Ochiai-Holcomb, G.E. Hofler, C. Carter-Coman, E.I. Chen, I.-H. Tan, P. Grillot, N.F. Gardner, H.C. Chui, J.-W. Huang, S.A. Stockman, F.A. Kish, M.G. Craford, T.S. Tan, C.P. Kocot, M. Hueschen, J. Posselt, B. Loh, G. Sasser, D. Collins, Appl. Phys. Lett. **75**, 2365 (1999)
139. C. Wiesmann, K. Bergenek, N. Linder, U.T. Schwarz, Laser Photon. Rev. **3**(3), 262–286 (2009)
140. Y.-J. Kim, M.-K. Kwon, K.-S. Lee, S.-J. Park, S.H. Kim, K.-D. Lee, Appl. Phys. Lett. **91**(18), 181109 (2007)
141. H.W. Huang, J.K. Huang, K.Y. Lee, C.F. Lin, H.C. Guo, IEEE Electron Device Lett. **31**(6), 573–575 (2010)

142. K.H. Li, H.W. Choi, J. Appl. Phys. **110**(5), 053104 (2011)
143. J. Jewell, D. Simeonov, S.-C. Huang, Y.-L. Hu, S. Nakamura, J. Speck, C. Weisbuch, Appl. Phys. Lett. **100**(17), 171105 (2012)
144. A. David, B. Moran, K. McGroddy, E. Matioli, E.L. Hu, S.P. DenBaars, S. Nakamura, C. Weisbuch, Appl. Phys. Lett. **92**(11), 113514 (2008)
145. D.S. Wuu, W.K. Wang, K.S. Wen, S.C. Huang, S.H. Lin, S.Y. Huang, C.F. Lin, R.H. Horng, Appl. Phys. Lett. **89**(16), 161105 (2006)

Chapter 4
GaN LEDs on Si Substrate

Fengyi Jiang, Jianli Zhang, Qian Sun, and Zhijue Quan

4.1 Epitaxy of GaN LED on Si Substrate

4.1.1 Overview of GaN Epitaxy on Si

Substrate is the basis of GaN epitaxial growth, which has great impact on the crystal quality, the strain, the luminescence behavior, and the light extraction mode. To some extent, Si is not considered to be an excellent choice as the substrate for GaN epitaxy. Such a sense is mainly attributed to the mismatch in the lattice constant and the thermal expansion coefficient. As shown in Fig. 4.1, the lattice mismatch and thermal mismatch between GaN and Si (111) substrates are 16.9% and 57%, respectively, which are much larger than that of GaN/sapphire and GaN/6H-SiC. These factors lead to high dislocation density, film cracking, and wafer bending, bringing great challenge to the epitaxial growth of GaN.

There is great challenge but also great potential to grow GaN on Si substrate. Leaving the weak point aside, Si substrate gets many other attractive advantages. Firstly, the availability of Si wafers makes the substrate relatively cheap, and the cost goes down when the diameter gets larger. Benefiting from the development of microelectronic in semiconductor industry, the processing technology of Si-related materials is very mature, which makes the fabrication of LED device very flexible. And the LEDs can be also integrated into the microelectronic devices. Besides, the Si substrate has good thermal and electric conductivity, which is good to the

F. Jiang (✉) · J. Zhang · Z. Quan
National Institute of LED on Si Substrate, Nanchang University, Nanchang, China
e-mail: jiangfy@ncu.edu.cn

Q. Sun
Suzhou Institute of Nano-Tech and Nano-Bionics, Suzhou, China

J. Li, G. Q. Zhang (eds.), *Light-Emitting Diodes*, Solid State Lighting Technology and Application Series 4, https://doi.org/10.1007/978-3-319-99211-2_4

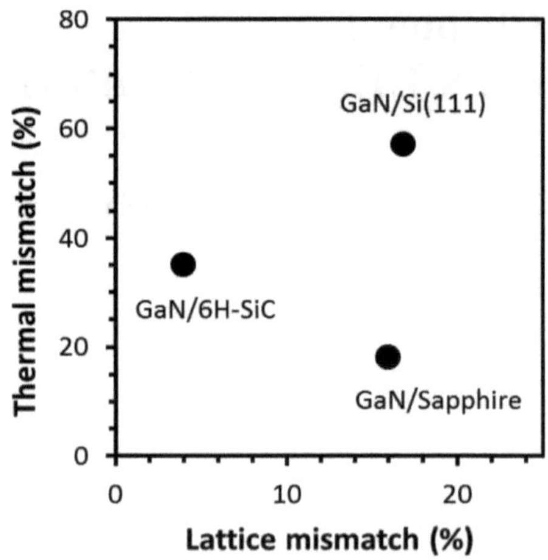

Fig. 4.1 The lattice and thermal mismatch of GaN epitaxy on different substrates

device reliability. Moreover, the large lattice and thermal mismatch between Si and GaN bring great opportunity to grow high-quality InGaN quantum wells. Once dislocation density and film cracking are under-controlled, the tensile-type thermal strain and lattice strain raised by the Si substrate will become positive factors for indium incorporation. Growth temperature can be enhanced to improve the quality of InGaN quantum wells (QWs).

Since the 1970s, when the first GaN on Si substrate was reported [1], great efforts on GaN/Si technology have been made by various researchers in order to industrialize the GaN LEDs on Si. But it took until 2009 for the first bright LED on Si substrate to be present [2]. And nowadays, the LEDs on Si substrate have already been commercially available and widely spread in our lives.

Then, what are the crucial technologies for GaN LEDs growing on Si substrates? Firstly, film-cracking problem must to be solved; otherwise the GaN epitaxy cannot be processed to devices. One of the approaches is to pattern the substrate into small grids via etching off grooves or growing dielectric strips. During the growth process, the region of the grooves or strips will not grow GaN, which acts as "cracks" on the wafer and helps to release stress. Moreover, the grid boundaries can localize the stress within small parts; even if there is a crack initiate at one of the grids, it will be isolated within the grid and not propagated to the entire wafer. Another approach to avoid cracking is to balance the thermal strain by lattice strain with the assistance of buffer. The buffer consists of multiple AlGaN layers with Al content gradually decreased from 100% to 0%. Lattice constant of AlGaN will increase as the Al content decreased, and compressive strain will accumulate within AlGaN film. Once the buffer layers grow thick enough, the compressive stain can compensate the tensile strain raised by the thermal mismatch and avoid film cracking.

Dislocation density in the GaN film also has great impact on the performance of LEDs. Direct growth of GaN on Si substrate results in high threading dislocation density up to 10^{10} cm^{-2}, which is useless for LED device. Epitaxial lateral overgrowth (ELOG) can be applied to reduce the dislocation density in GaN on Si substrate. An AlN layer was normally deposited on the Si substrate as the buffer. Upon that, island growth mode (3D) of GaN can be formed with the help of lattice strain between GaN and AlN. And then lateral overgrowth can be conducted to coalescent the islands. The 3D-ELOG growth mode can effectively reduce the dislocation density to a magnitude of 10^8 cm^{-2}, which is comparable to that of GaN grown on sapphire substrate.

In this chapter, we will focus on the technologies of GaN LEDs on Si substrate, including substrate preparation, epitaxial growth, device fabrication, and device characterization. The unique techniques of GaN on Si will be introduced in detail, whereas the generic techniques will be briefly mentioned.

4.1.2 Buffer Technology

Among the planes in the lattice of Si, (111) plane is most suitable for GaN film growth because of its hexagonal symmetry favoring epitaxial growth of the GaN (0001) plane with wurtzite structure. The epitaxial relationship is GaN (0001) parallel to Si (111), GaN [11−20] parallel to Si [−110], and GaN [−1100] parallel to Si [11−2] [3].

The large difference in the lattice constants of GaN (a_{GaN} = 0.3189 nm) and Si ($a_{Si(111)}$ = 0.3840 nm) yields a lattice parameter mismatch (16.9%) resulting in a high threading dislocation density (TDD) of 10^{10} cm^{-2}. Another severe problem is the large thermal dilatation mismatch between GaN and Si. The in-plane thermal expansion coefficient of GaN is 5.59×10^{-6} K^{-1} [4] as compared to 2.59×10^{-6} K^{-1} of Si, which leads to a large tensile stress during cooling from the growth temperature (about 1000 °C) to room temperature. The stress can be determined via the curvature of the sample which is proportional to the stress value [5]. Under typical MOCVD growth conditions, the stress amounts to 0.9 GPa μm^{-1} GaN [6]. The great tensile stress will cause cracking and a concave bending of the GaN/Si epi-wafer, which cause problems to device applications.

In order to obtain low TDD and crack-free GaN film on Si substrate, a suitable buffer layer is needed.

4.1.2.1 Thin AlN Buffer on Grid-Patterned Si Substrate

The challenges in using Si as the substrate are as great as the benefits. At the very beginning, it is necessary to appropriately treat the Si substrates to obtain the optimum state for growing high-quality GaN layer. The results presented in this section are based on Si (111) substrate.

The substrate orientation is usually oriented by an X-ray diffractometer; in the process of determining substrate orientation, cutting, and subsequent polishing, the error is introduced, so that the real surface of the Si substrate with the expected crystal surface has a certain deviation, and the degree of deviation is usually defined by the miscut angle. Research shows that the miscut angle of Si (111) substrate has a significant influence on the optical properties of GaN-based LED grown on it [7]. Of course, the miscut angle closer to 0° is better, but taking the control of the substrate production into account, it is usually controlled within 0.3° [7].

The thickness of substrates has evident effect on the growth of GaN film and LED structure on Si. Usually, the thicker the substrate, the smaller is the amount of deformation of the epitaxial wafer during the entire epitaxial growth and cooling process, which will favor the control of the wavelength uniformity of LED structure. However, the increase in substrate thickness will add to the cost of the Si substrates and at the same time is not conducive to lift-off of the Si substrate when the vertical structure chip is fabricated, and therefore, the thickness of the substrate should not be too thick. For a 2 in.-patterned Si (111) substrate with a thickness of 0.43–1 mm, an epitaxial wafer with good wavelength uniformity can be obtained, whereas for a 6 in.-patterned Si (111) substrate, the thickness is typically chosen to be 1–1.5 mm. For un-patterned Si substrates, thicker substrates are often required in order to obtain LED films with crack-free and good wavelength uniformity.

Grid-patterned Si substrate (GPSS) fabrication is one of the key technologies for the selective area growth (SAG) on GPSS. The size of the unit patterns usually depends on the size of the chips to be obtained. Figure 4.2 shows a schematic view of a GPSS, in which Fig. 4.2a is a vertical view, and it can be seen that the Si substrate is divided into separate unit patterns by pattern boundary. There are two different approaches for pattern boundary, one is obtained by growing layer of dielectric film (such as SiN_x, SiO_2) and the other by means of photolithography to obtain the dielectric film boundary, as shown in Fig. 4.2b, and another method is etching trenches on the Si substrate by photolithography as shown in the sectional view of Fig. 4.2c. Since such a boundary is amorphous, no oriented crystalline seed layer can be grown on it during the process of growing GaN; the GaN thin film is also divided into separate cell patterns, which greatly reduces the thermal stress between the GaN thin film and the substrate. By using the above two types of boundary, a high-quality GaN film can be obtained.

Before loading the substrates, it is necessary to clean the Si substrate, which typically consists of two parts. The first part refers to the wet cleaning prior to being placed in the MOCVD reaction chamber. This process is usually done using the typical Si substrate cleaning technology widely used in the IC industry and will not be repeated here. The second part refers to the dry cleaning in the MOCVD reaction chamber. The purpose of this process is to remove the native oxide layer on the surface of the Si substrates and obtain a flat surface suitable for epitaxial growth. Figure 4.3 shows two atomic force microscope (AFM) images of the Si (111) substrate after being polished, wet cleaned, and dry cleaned in the MOCVD reaction chamber. As can be seen from Fig. 4.3a, large amount of scratches present on the surface of the Si (111) substrate after polished and wet cleaned. Such a surface is not suitable for obtaining a high-quality epitaxial film. The scratches on the surface of the Si (111) substrates disappear, and the step flow occurs as a result of dry cleaning in the reaction chamber (usually at a substrate temperature (1100 ± 100) °C in H_2 atmosphere); the root mean square (RMS) of surface roughness is also reduced from 0.583 to 0.178 nm (10 μm $\times$ 10 μm range), which creates good surface conditions for subsequent epitaxial growth.

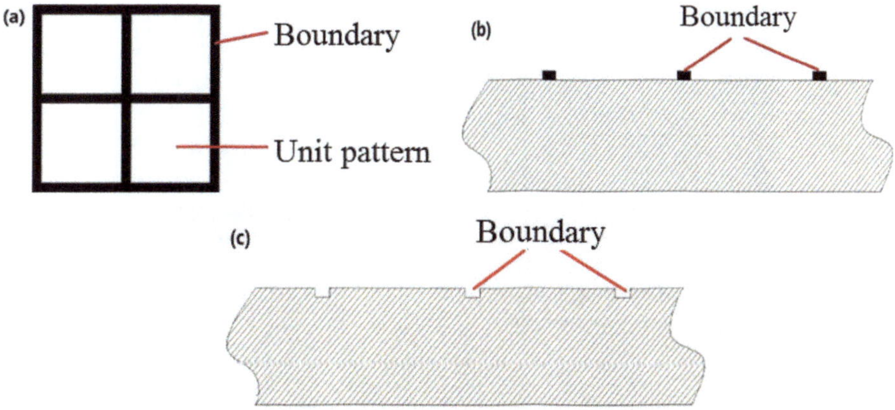

Fig. 4.2 Scheme for patterned Si substrate in (**a**) vertical view, (**b**) sectional view of Si substrate with dielectric film boundary, and (**c**) sectional view of Si substrate with trench boundary

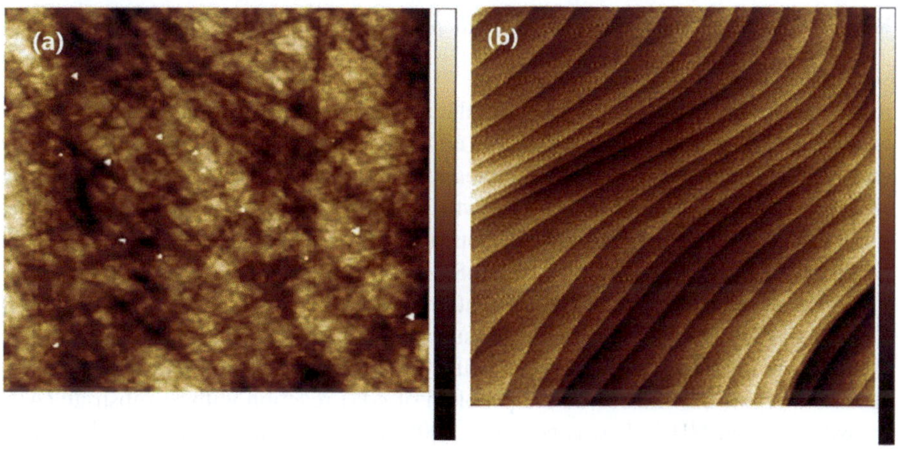

Fig. 4.3 AFM images of polished and wet-cleaned Si (111) substrate surface (10 μm × 10 μm scanning area). (**a**) Before dry cleaning in MOCVD, RMS = 0.583 nm; (**b**) after dry cleaning in MOCVD, RMS = 0.178 nm

When GaN is grown on Si substrates directly, Si surface easily reacts with NH_3 to form SiN_x. GaN single crystal cannot grow on this layer [8], and the Si substrate reacts with metal Ga to form so-called Ga melt-back etching phenomenon [9], resulting in macroscopic defects on the surface of the epitaxial wafer as shown in Fig. 4.4. In order to overcome the above problems, researchers introduced an intermediate layer or buffer layer between the GaN and the Si substrate to prevent the GaN from directly contacting with the Si substrate or to avoid the formation of SiN_x on the surface of the Si substrate by employing an intermediate layer without NH_3. These buffer layers include 3C-SiC [10], AlAs [11], γ-Al_2O_3 [12], BN [13], and so on. Although these buffer layers can solve the problems, these usually require

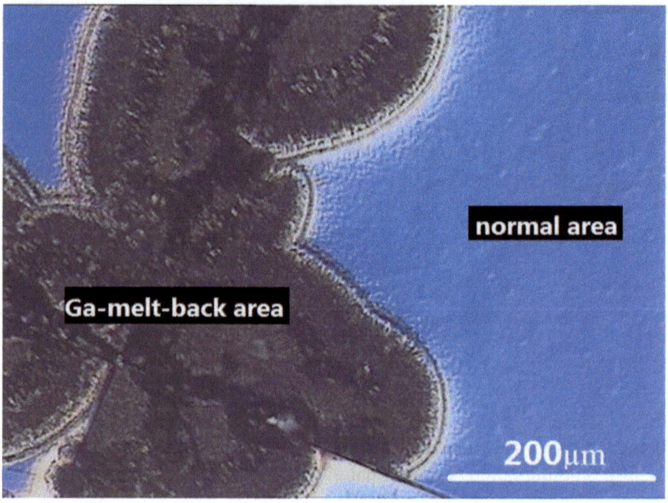

Fig. 4.4 Ga melt-back defects formed by reaction between Ga and Si substrate

a two-step epitaxy growth, which complicates the growth process. In contrast, AlN is an ideal intermediate material; a good coverage of AlN on the Si substrates surface cannot only serve as seed layer or nucleation layer to prevent the melt-back etching but also facilitate the subsequent growth of GaN [14]. Al and Ga belong to the group III, which can grow GaN without two-step epitaxial growth. Moreover, in comparison with the bond formation of Si–N, Al and N atoms tend to bond together much easier, which can avoid the formation of SiN_x polycrystalline layer.

As early as 1993, H. Amano et al. succeeded in using AlN as the intermediate layer to obtain a single crystal GaN film [15]. However, it was not mentioned in Amano's report how he solved the problem of NH_3 reaction with Si substrate (AlN growth required NH_3). It was not until 2000 that Khan clearly stated in his study that a few monolayers of metallic aluminum (Al) were deposited on the Si substrate prior to the growth of the AlN intermediate layer, which is believed to be useful for accelerating the subsequent AlN growth mode transferring from three-dimensional (3D) island growth to two-dimensional (2D) film growth [16, 17], thereby cleverly addressing the reaction of NH_3 with the surface of the Si substrate and Ga melt-back etching problem. This is very necessary to the following growth of high-quality GaN [14]. It was proved in both MBE [14, 17, 18] and MOCVD systems [16, 19, 20]. Much work had been paid in studying AlN growth temperature, V/III ratio, and suitable thickness [15, 21–23].

As mentioned above, GaN grown on Si shows a large mismatch dislocation density, usually $>10^{10}$ cm^{-2} (edge, screw, and mixture), due to the large lattice mismatch between the two materials. Therefore, reducing the dislocation density is a key technology for growing GaN films on Si substrates. Epitaxial lateral overgrowth (ELOG) is an important technique to reduce the dislocation density and

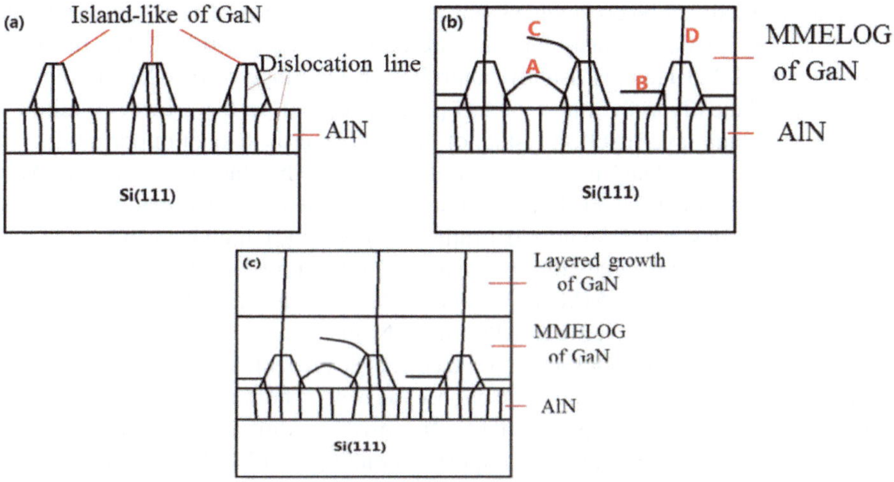

Fig. 4.5 Scheme for reducing the dislocation density of GaN film on Si substrate achieved by MMELOG in which the black lines represent dislocation lines. (**a**) Formed island-like GaN nucleation layer, (**b**) island-like GaN by means of MMELOG and then combined into a whole, (**c**) layered growth of GaN film with low dislocation density

is widely used in the growth of GaAs, InP, GaN, and other materials [24]. Threading dislocations can be blocked at the hetero interface. In 1999, Kung grew GaN on Si substrates by using ELOG technology [25, 26]. Honda et al. [26] and Dadgar et al. [27] also successfully grew GaN films on Si substrates by SAG technique and ELOG technique, respectively.

Jiang et al. [7] proposed a method to reduce the dislocation density of GaN films on Si substrates by referring to ELOG technology, which is called "maskless micro epitaxial lateral overgrowth technology (MMELOG)." Here, "maskless" refers to the fact that there is no need to growing masks (conventional ELOG needs to be fabricated before growth), while "micro" means a much smaller lateral epitaxial scale than the conventional ELOG. This method is described in detail below. Figure 4.5 shows a schematic representation of this technique: (1) growing AlN buffer layer on Si (111) substrate; (2) forming a "GaN island-like nucleation layer" on the AlN buffer layer; (3) the "nucleation layer" keeps on growing up, called the 3D GaN layer; and (4) growth combine layer, through the ELOG process 3D GaN layer, will be merged into a GaN layer with flat surface.

As shown in Fig. 4.5a, in the vertical direction, only a portion of the threading dislocations (TDs) extends from the AlN buffer layer to GaN island, thus reducing the dislocation density within GaN layer. Notably, the density of the GaN island (the number of islands per unit area) and the size (the area where the GaN island is in contact with AlN) have a significant effect on the TDs extending from the AlN buffer layer to the GaN island. The smaller the density and size of GaN is, the lower the dislocation density in the GaN island will be. However, the density of GaN is too small and will not be conducive to the subsequent merge of ELOG into a whole. Therefore, proper GaN density and size are needed to be controlled. The distance between the GaN islands is usually controlled in the range of 0.5–2 μm, and the size of the islands is in the range of 1–5 μm. In addition to the growth conditions of the GaN island layer itself, the state of the AlN layer is also an important factor in the control of the GaN island. Generally, the smoother the surface of the AlN is, the higher the qualities of the crystal are, and in the island-like pattern, GaN is easier to grow. In general, small V/III ratio (usually <500), high pressure, and high temperature easily make GaN to grow in accordance with island-like patterns.

It is possible to achieve MMELOG on the island of GaN, mainly due to the large lattice mismatch (2.47%) between AlN and GaN, just because the existence of large lattice mismatch that provides a high surface energy prevents GaN from growing on AlN. Therefore, subsequent GaN will choose to grow on homogeneous GaN islands, and then coupled with the appropriate GaN growth conditions, ELOG of GaN can be achieved. At this point, the most important condition for ELOG is the large V/III ratio, usually reaching more than 2000. In the process of ELOG, the dislocations in the island of GaN will change or interact with each other. For example, two dislocations react with each other and disappear (shown in red letter A in Fig. 4.5b), dislocation lines turn parallel to the growth surface (shown in red letter B in Fig. 4.5b), dislocation lines divert in a certain direction (shown in red letter C in Fig. 4.5b), and dislocation lines extend along the growth direction (shown in red letter D in Fig. 4.5b). In this case A, B, and C correspond to the decrease of the dislocation density in the subsequent layer, and as shown in Fig. 4.5c, only the dislocation of D case will remain in the GaN epitaxial layer. The evolution mechanism is based on the results observed by transmission electron microscope (TEM), and the results will be given later. With regard to the evolutionary mechanism of dislocations in the ELOG process, previous study of GaN ELOG has been reported [24–30], thus no longer go into much detail here.

Figure 4.6 shows TEM photographs of GaN film with crack-free and low dislocation density obtained on Si (111) substrate, using a 100-nm-thick AlN buffer layer. As can be seen from Fig. 4.6a, in the AlN layer, there exist a large number of dislocations where white lines in the figure represent dislocation lines (dislocation density $>10^{10}$ cm^{-2}), and after island-like growth and MMELOG, GaN dislocation density is reduced, 10^8 cm^{-2} order of magnitude. Figure 4.6b–d are partial enlarged views of Fig. 4.6a, where the red letters of A, B, and C in the figure represent three kinds of dislocation evolution of A, B, and C given in Fig. 4.5b, respectively. In order to show the reduction of GaN dislocation density by the island-like growth and MMELOG mode more intuitively, we have prepared a sample of GaN grown directly on AlN layer. The TEM photograph was shown in Fig. 4.6e. The dislocations in AlN buffer layer in Fig. 4.6e extend extensively to the upper GaN layer, so that the dislocation density within the GaN layer is quite high compared to the large drop in the AlN/GaN interface in Fig. 4.6a.

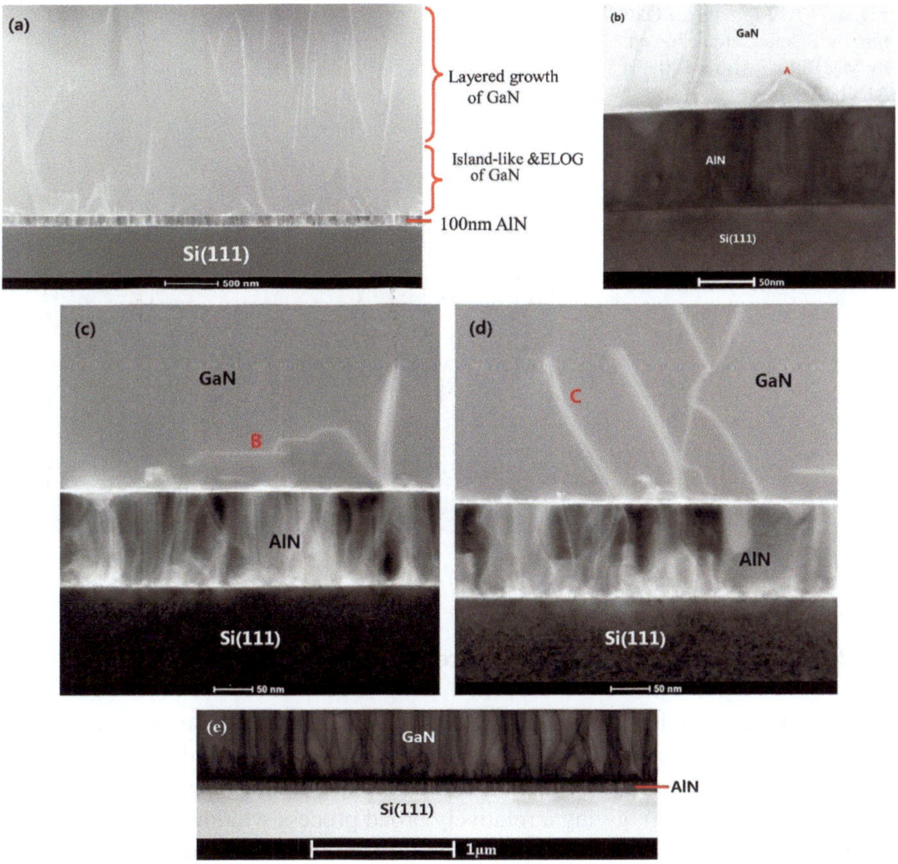

Fig. 4.6 TEM image of GaN film on Si substrate with a 100 nm AlN buffer layer. (**a**) High-quality, low dislocation density GaN film grown on the AlN buffer layer using island-like growth and MMELOG technology, (**b**) example of evolution mechanism of A class dislocation in Fig. 4.5, (**c**) example of evolution mechanism of B class dislocation in Fig. 4.5, (**d**) example of evolution mechanism of C class dislocation in Fig. 4.5, (**e**) high dislocation density GaN film grown directly on the AlN buffer layer

After the first reduction of GaN island-like growth model and the second reduction of MMELOG process, the dislocation density decreased by two orders of magnitude, from 10^{10} to 10^8 cm^{-2}. Using this technique, GaN with (002) and (102) plane double-crystal rocking curve are obtained at half peak width of 230 arcsec and 330 arcsec, respectively. The GaN film is further flattened by subsequent layered growth (also known as 2D growth), which provides a good surface condition for the subsequent growth of the LED structure. Figure 4.7 shows the AFM morphology of the GaN surface after the growth of the layered GaN. The surface roughness RMS was 0.148 nm.

Fig. 4.7 AFM image of GaN film on Si substrate achieved by MMELOG with a 100 nm AlN buffer layer RMS = 0.148 nm (2 μm × 2 μm scanning area); the thickness of GaN film is 3.2 μm

4.1.2.2 Graded AlGaN Buffer on Bare Si Substrate

Despite the lattice mismatch of GaN with sapphire substrates, the two-step growth technique [31] can realize a high-quality GaN film normally under a compressive strain because the coefficient of thermal expansion (CTE) of sapphire is larger than that of GaN (Fig. 4.8a), giving a relatively broad process window with little risk in micro-crack formation. Epitaxial growth of GaN-based LEDs on large diameter Si substrates has a great potential in significantly reducing the LED manufacturing cost for energy-efficient solid-state lighting. This epitaxial integration, however, was hindered by two major technological challenges. The 16.9% lattice mismatch between GaN and Si normally causes a high density (10^9–10^{10} cm^{-2}) of threading dislocations (TDs), often working as non-radiative recombination centers and hence deteriorating the LED efficacy [32]. Due to the huge CTE misfit (57%), Si substrates after the epitaxial growth lag behind GaN in wafer shrinking during the cooldown from growth temperature and effectively stretch the GaN film, which may lead to tensile stress, wafer bowing, and micro-crack network formation in the GaN layer (Fig. 4.8b) [33]. It should be mentioned that even the strong residual tensile stress in the nominally crack-free GaN-on-Si LED epitaxial wafers right after the growth still can often induce micro-cracks under device operation over time, causing carrier leakage, light output decay, and other reliability issues.

In order to avoid the formation of micro-crack network in GaN epitaxial film grown on bare Si substrates, various buffer and/or AlN interlayers have been inserted between GaN and Si. The positive lattice mismatch of +2.4% between GaN ($a = 0.3189$ nm) and AlN ($a = 0.3112$ nm) can be utilized in the Al-composition-graded AlGaN/AlN buffer layers to build up enough compressive strain in the GaN

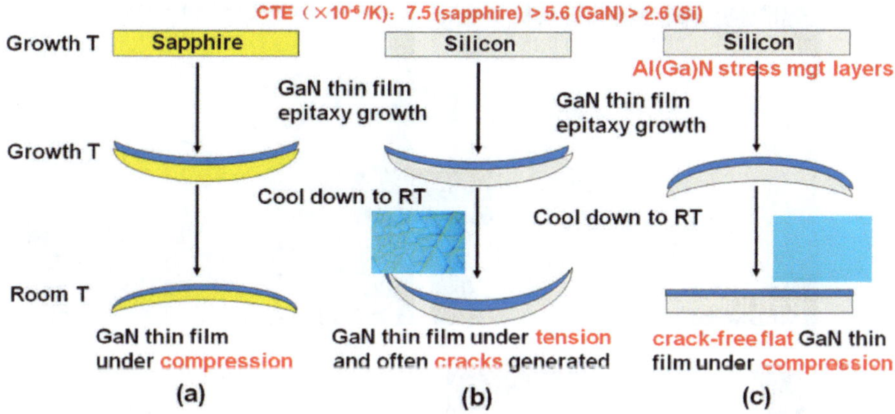

Fig. 4.8 Schematic diagram of the stress evolution during growth and cooldown of GaN film grown on sapphire (**a**), and Si substrates without (**b**) and with (**c**) Al-composition-graded AlGaN/AlN buffer layers

film at epitaxial growth temperature, for the compensation of the tensile stress induced by the CTE difference during the cooldown process [34, 35]. With a five-step-graded AlGaN/AlN multilayer buffer (Fig. 4.10a), crack-free GaN film (Fig. 4.9d) with an edge crack length shorter than 0.5 mm was obtained on Si (111) substrates.

It was found that the compressive strain built up within the Al-composition-graded AlGaN/AlN buffer layers could not only be used to compensate the tensile stress due to the CTE mismatch during the cooldown but also induce the inclination and annihilation of threading dislocations (TDs) at the interfaces according to the cross-sectional TEM observation (Fig. 4.9a, b). The full width at half maximum (FWHM) of GaN (0002) and (10-11) double-crystal X-ray rocking curves (XRCs) for the as-grown 2-μm-thick crack-free GaN film on Si (111) with a five-step-graded AlGaN/AlN multilayer buffer were 475 and 1147 arcsec, respectively (Fig. 4.10c, d), which, however, were still substantially larger than those of GaN grown on sapphire substrates. It should be mentioned that the large difference between the GaN (0002) and (10-11) XRCs FWHMs reflected a high density of edge-type dislocations, which are normally non-radiative recombination centers in LEDs. It is essential to reduce the edge-type TD density for the realization of high-efficacy LEDs grown on Si substrates.

Meanwhile, it was noted that there was a strong interplay between the defect engineering and the stress management for GaN growth on bare Si substrates. The inclination of existing edge TDs (Fig. 4.9c) effectively generates in-plane misfit dislocation segments, which result in a partial relaxation of the built-up compressive strain within the AlGaN/AlN multilayer buffer and the GaN overgrown layer [35–37]. A partial strain relaxation of the AlGaN/AlN multilayer buffer and the GaN epitaxial layer was clearly revealed by the displacement of their reciprocal lattice points (RLPs) along the Qx direction (not exactly aligned along the Qz

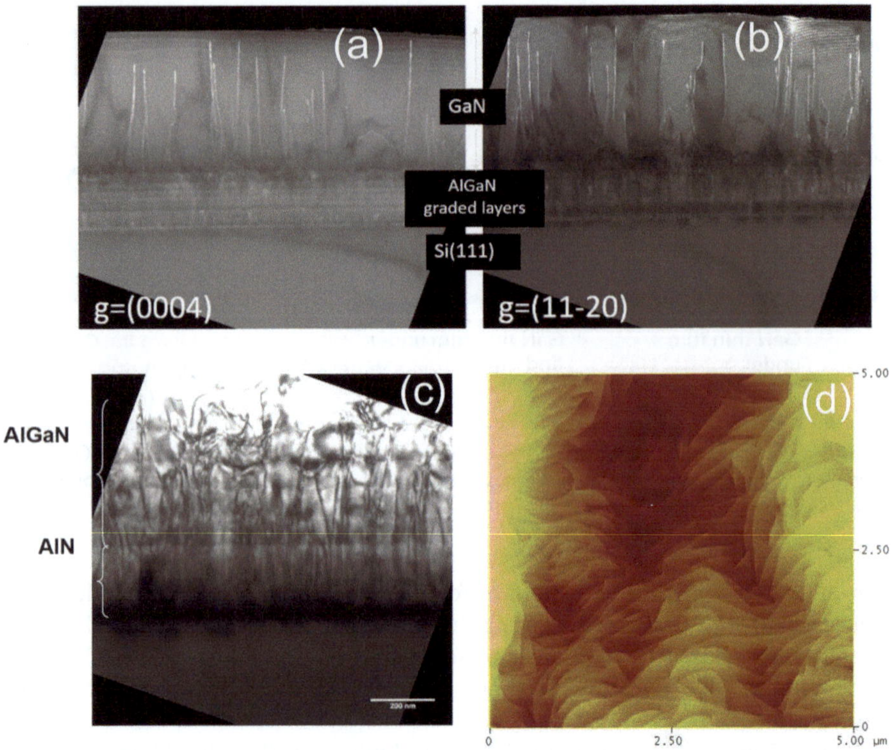

Fig. 4.9 Dark-field two-beam cross-sectional TEM images of GaN grown on Si (111) with a five-step-graded AlGaN/AlN buffer under a diffraction condition of (**a**) $g = 0004$ and (**b**) $g = 1120$, (**c**) bright-field TEM of the five-step-graded AlGaN buffer layers, (**d**) AFM image of the as-grown GaN surface with a root mean square roughness of 0.56 nm

direction) [35] in the (1124) X-ray reciprocal space mapping (RSM) (Fig. 4.10b). The inclination of TDs cancels out some compressive strain and may facilitate the formation of micro-cracks, which renders a limited thickness for crack-free GaN film grown on Si substrates. Therefore, it is also critical to reduce the edge-type TD density during the growth of GaN on Si for stress control.

It was observed that in AlGaN/AlN material system, greater lattice-mismatch compressive strain promotes larger angle bending of TDs [36, 37]. In the aforementioned five-step-graded AlGaN/AlN buffer structure [35] and other groups' reports [34, 38], typically a series of high Al-composition AlGaN layers (having a limited lattice mismatch with the underlying AlN layer) were firstly deposited on AlN/Si but gave little contribution to the TD bending/reduction and compressive strain accumulation. For instance, there is only 0.48% lattice mismatch between AlN and $Al_{0.80}Ga_{0.20}N$, which makes the TD bending at a small angle, and most of the TDs still propagate into the upper layers, contributing a broad XRC for GaN (10-11) diffraction.

Based on the understanding about the lattice mismatch, TD reduction, and strain relaxation, Sun et al. revised and simplified the design of the AlGaN/AlN

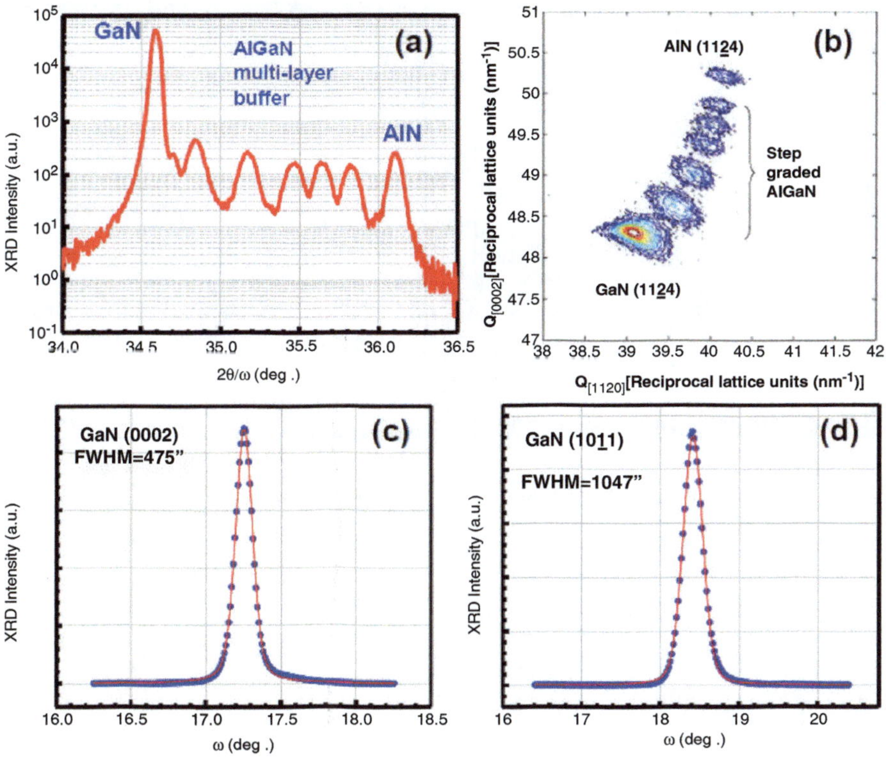

Fig. 4.10 (**a**) (0002) $2\theta/\omega$ scan of GaN grown on Si(111) with a five-step-graded AlGaN/AlN buffer, (**b**) reciprocal space mapping of the as-grown structure measured in an off-axis (1124) configuration, (**c**) double-crystal X-ray rocking curves of GaN (**c**) (0002) and (**d**) (10-11) diffractions

multilayer structure for the growth of crack-free high-quality GaN on Si substrates [39]. By eliminating the high Al-composition AlGaN layers, Sun et al. grew $Al_{0.35}Ga_{0.65}N$ as the first AlGaN layer in direct contact with AlN on Si. An increased compressive strain within the AlN/$Al_{0.35}Ga_{0.65}N$ structure, induced by the 1.6% lattice mismatch, facilitates the TDs bending at a larger angle, as well as their annihilation through dislocation interaction. With a greatly reduced TD density, compressive strain can be built up more effectively during the subsequent growth of $Al_{0.17}Ga_{0.83}N$/GaN for the compensation of tensile stress due to the CTE mismatch during the cooldown, resulting in a crack-free high-quality GaN film on Si. According to the TEM observations (Fig. 4.11a, b), from the AlN nucleation layer, through the $Al_{0.35}Ga_{0.65}N$/$Al_{0.17}Ga_{0.83}N$ buffer layers, to the GaN thick layer, a substantial portion of TDs were filtered out, and the crystalline quality was improved remarkably (Fig. 4.11c). And the as-grown high-quality GaN film on Si showed an atomically smooth surface featured with a step-flow morphology (Fig. 4.11d), which was distinctly different from that of the defective GaN grown on Si

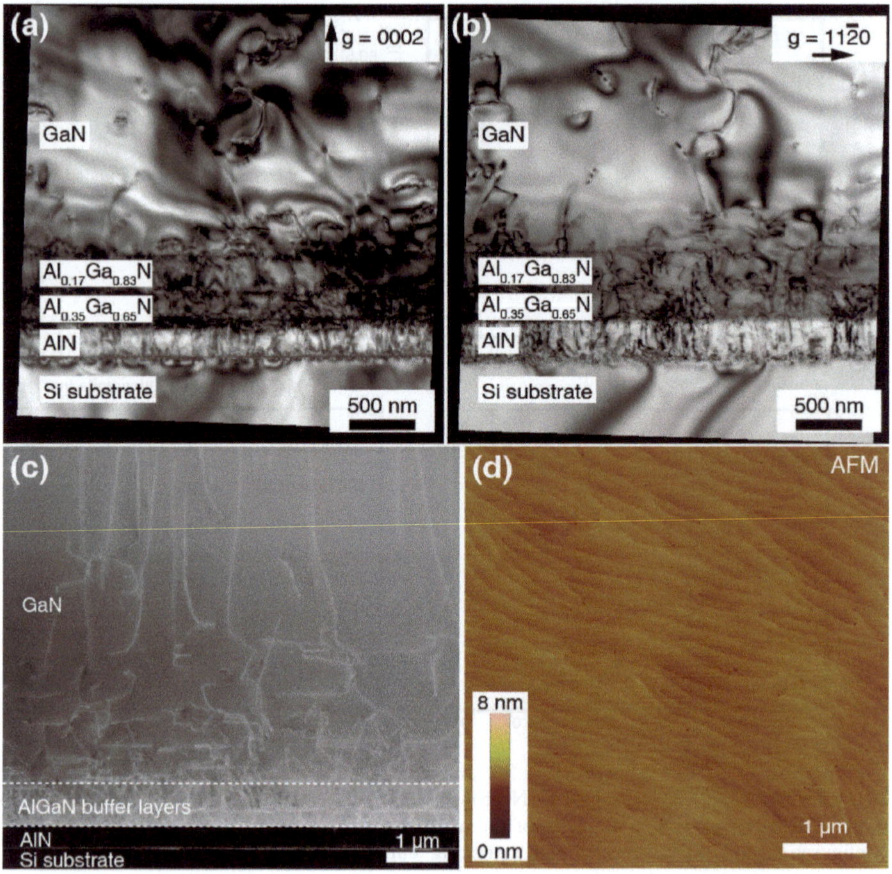

Fig. 4.11 Cross-sectional weak-beam bright-field TEM images of GaN grown on Si with an Al0.17Ga0.83N/Al0.35Ga0.65N/AlN buffer, obtained with (**a**) $g = 0002$ and (**b**) $g = 1120$, revealing TDs with screw and edge components, respectively. (**c**) Cross-sectional high-angle annular dark-field scanning TEM image of GaN grown on Si. (**d**) AFM image of the surface of this GaN film grown on Si, showing a step-flow surface morphology with a root mean square roughness of 0.19 nm

with a five-step-graded AlGaN/AlN buffer (Fig. 4.9d). The surface step pinning by the TDs was substantially suppressed for the high-quality GaN with a reduced TD density.

The crystalline quality of the as-grown GaN film on Si with an $Al_{0.17}Ga_{0.83}N/Al_{0.35}Ga_{0.65}N$/AlN buffer was also evaluated by DCXRC measurements in a skew symmetric geometry. The full width at half maximum (FWHM) of (0002), (2021), (2023), and (101 m) ($m = 1, 2, 3, 4,$ and 5) DCXRCs for 3.6-μm-thick n-type GaN (Si doping 8×10^{18} cm^{-3}) grown on Si was all below 300 arcsec (Fig. 4.12a, b), indicating a low density of edge TDs [40, 41]. It's noted that edge TDs as NRCs are more detrimental to IQE than screw and mixed ones [42]. The TD density

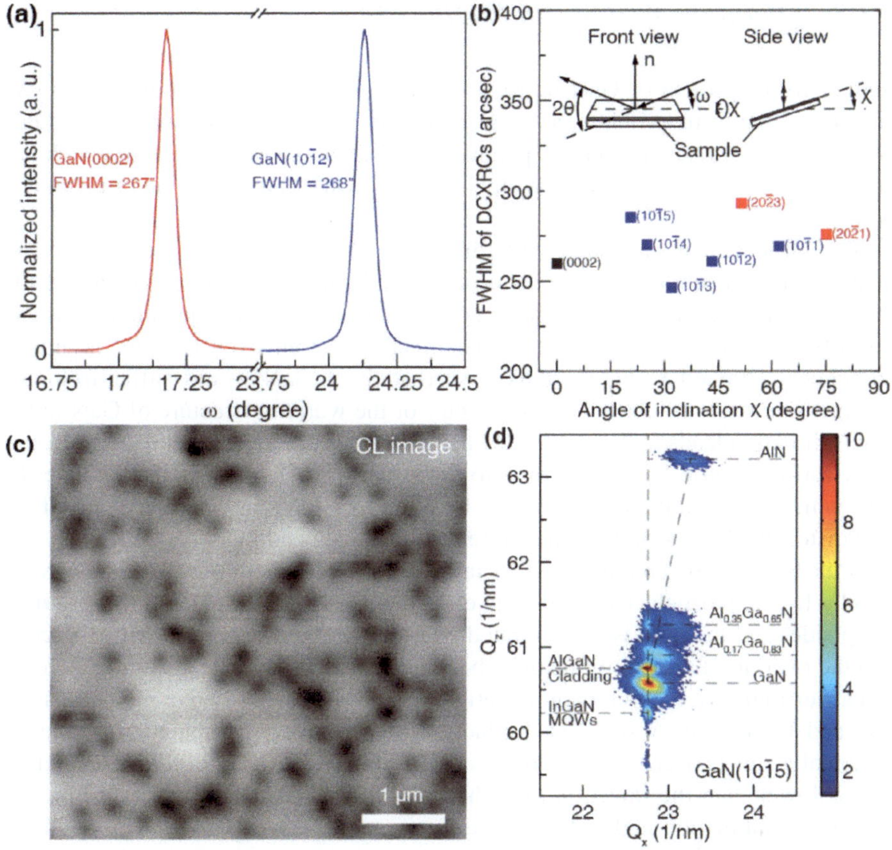

Fig. 4.12 (**a**) Double-crystal X-ray rocking curves of the high-quality GaN grown on Si around the (0002) and (10-12) diffractions. (**b**) FWHM of DCXRCs taken from a series of skew reflections at an increasing angle χ with respect to the (0001) plane for a 3.6-μm-thick n-type GaN (Si doping at 8×10^{18} cm^{-3}) grown on Si. The insets are schematic diagrams illustrating the inclination angle χ. (**c**) Panchromatic CL image of the high-quality GaN film grown on Si. The density of the dark spots, representing the TD density in the GaN film grown on Si, is about 5.8×10^8 cm^{-2}. (**d**) RSM of GaN heterostructure grown on Si measured around (1015) asymmetric reflection

in the as-grown GaN-on-Si film was estimated to be around 5.8×10^8 cm^{-2} based on the dark spot density in the panchromatic cathodoluminescence (CL) image (Fig. 4.12c). The as-grown GaN-on-Si template provides a high-quality material platform for the subsequent coherent growth of highly efficient InGaN/GaN multiple quantum well (MQW) active region, as evidenced by their vertically aligned RLPs along the Qz direction with respect to the GaN RLP in the (1015) RSM (Fig. 4.12d).

4.1.3 Quantum Well Strain Engineering

Similar to the growth of GaN on Si substrate, growth of the quantum well (QW) encounters the same problem, as there exists lattice mismatch between the InGaN well and GaN barrier. The lattice parameter of InN a-axis is 0.3548 nm, which has 10% lattice mismatch compared with GaN. For InGaN QW, the value of mismatch is 2–3%.

The large lattice mismatch will directly lead to the decrease of the crystal quality, such as producing new dislocations and causing indium segregation. On the other hand, it will introduce huge compressive stress to the quantum wells. X-ray diffraction measurement shows that InGaN quantum well (3 nm) is almost completely strained along the a-axis, so that the InGaN would suffer from the compressive stress of GaN. The asymmetry of the wurtzite structure of GaN in the c-axis direction leads to the existence of spontaneous polarization and piezoelectric polarization electric field; according to the strain of InGaN, one can calculate the QW which exists in the huge pressure field, up to 1 MV/cm, and the electric field will distort the band structure and cause serious band bending.

If the band is bent and the carrier is injected into the quantum well, it is separated rapidly by the piezoelectric field. The electrons are concentrated and trapped near the p side, and the hole is concentrated near the n side. Carrier separation occurs, and the overlap of the wave functions becomes smaller, which leads to the increase of carrier radiative recombination lifetime and lowers the radiative recombination rate and the efficiency of LED. Reduction of radiative recombination efficiency will make the carrier accumulation in quantum wells, as the carrier depletion rate is lower than the injection rate, thereby further causing carrier overflow and Auger recombination and reducing the luminous efficiency of LED. In addition, the presence of stress will also affect the indium incorporation during the epitaxial growth.

To summarize, the key issue to enhance the efficiency of LED is to reduce the stress in the QWs, which is beneficial to both material growth and device performance. One of the approaches is to grow AlInN or AlInGaN as the barrier to make its lattice parameter the same as InGaN well. As illustrated in Fig. 4.13, despite the bowing factors, for a blue LED with $In_{0.2}Ga_{0.8}N$ QW, if we grow $In_{0.2}Al_{0.8}N$ as the barrier, the lattice constant of a-axis is the same as the QW, while the barrier potential is higher than that of GaN as the barrier.

Using $Al_xGa_yIn_zN$ quaternary alloy as the barrier which can also achieve the effect, Table 4.1 lists the different composition and corresponding bandgap energy (without considering the bowing factors) of $Al_xGa_yIn_zN$ barrier matched with $In_{0.2}Ga_{0.8}N$ well to achieve the same lattice constant. Considering the carrier localization effect not worse than GaN, the Al composition must be set to at least 30%. Theoretically, these are good methods to release stress, but it is difficult for material growth, mainly attributed to opposite growth behavior of AlN and InN. For AlN which is easy to pre-react, the suitable growing conditions are low pressure, small V/III ratio, and high temperature, whereas InN, which is easy to decompose,

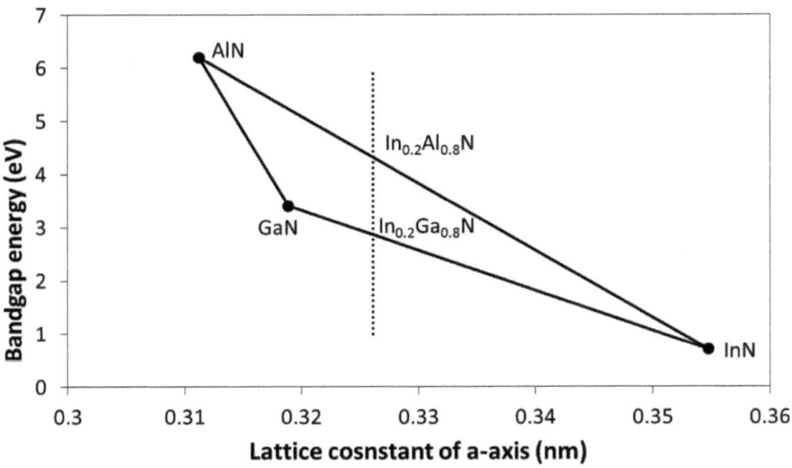

Fig. 4.13 Schematic diagram of lattice-matched InGaN and InAlN

Table 4.1 Calculated composition and bandgap energy of $Al_xGa_yIn_zN$ barrier matched with In0.2Ga0.8N well

x	y	z	Bandgap energy (eV)
0.10	0.68	0.22	3.08
0.20	0.56	0.24	3.30
0.30	0.44	0.26	3.53
0.40	0.31	0.29	3.75
0.50	0.19	0.31	3.97

is suitable to grow under high pressure, high V/III ratio, and low temperature. The antagonistic growth conditions of the two materials bring great difficulty in growing high-quality AlInN or AlGaInN materials. Thus to grow lattice-matched barrier is not a practical way to relax strain within the QWs.

A common way to relax strain of QW is to insert a layer or multiple layers of InGaN with relatively low indium content, after n-GaN and before the QWs. Such a layer is often called as pre-strain layer. The purpose of growing pre-strain layer is to partially relax the strain form n-GaN with the help of its indium content, which makes InGaN accommodate the lattice of QWs.

To grow the pre-strain layer, one should be concerned not only on the effect of strain relaxing but also on the quality of the material. If we bring too much negative effect to the QW crystal quality, then the strain relaxing becomes worthless. Thus the content of indium should be properly controlled. Too much indium will result in quality deterioration, and too little indium will not generate enough strain relaxation. Normally, the indium content in the pre-strain layer is set to be one third of that in the QW. Instead of growing a single layer of InGaN as the pre-strain layer, InGaN/GaN superlattices (SLs) are preferred when the pre-strain layer is thick. It is known that the crystal quality gets worse when growing InGaN: it is necessary to repair up the crystal with the help of GaN before InGaN grow too thick. And repeating of the InGaN/GaN SLs will form pre-strain layer which gives consideration to both

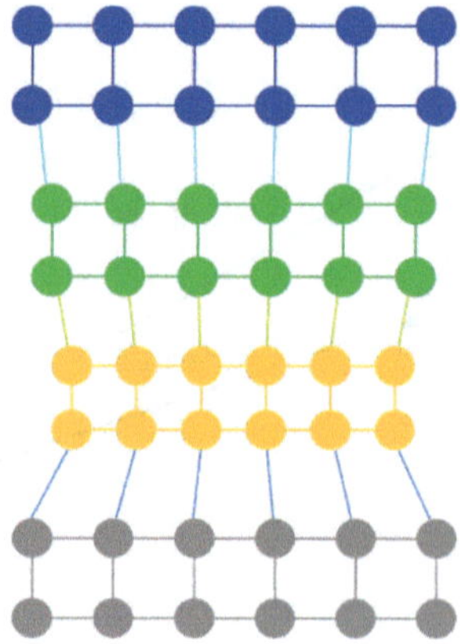

Fig. 4.14 Schematic illustration of strain evolution in a LED structure with pre-strain layer grown on Si substrate

strain relaxing and QW quality. As illustrated in Fig. 4.14, a pre-strain layer inserted between n-GaN and QWs makes the lattice evolved more gently, which is helpful to buffer the compressive stress in InGaN QWs.

The large lattice and thermal mismatch between Si and GaN bring great challenge in material growth, but also the opportunity to grow high-quality InGaN quantum wells. Once dislocation density and film cracking are under control, the tensile-type thermal strain and lattice strain raised by the Si substrate will become positive factors for indium incorporation. Growth temperature can be enhanced about 20 degrees to improve the quality of QWs.

4.1.4 V-Pits of GaN LED

There is a high density of threading dislocations (TDs) in InGaN/GaN LED grown on foreign substrates, due to the poor matching between the epi-layer and substrate in the lattice parameter and the thermal expansion coefficient. The density of TDs is about 10^8–10^9 cm^{-2} for LED grown on sapphire or SiC substrate and 10^9–10^{10} cm^{-2} for LED grown on Si substrate. The TDs are regarded as the non-radiative recombination centers [43] but seem to be insensitive to the luminous efficiency of GaN-based LEDs, unlike that of the traditional semiconductor materials. Two mechanisms have been proposed to explain this phenomenon. The traditional theory considers that most of carriers in the InGaN wells are localized in the In-rich regions and recombined before they reach the TDs [44, 45]. The recent theory proposed that the reason behind this phenomenon ascribes to the V-shaped pits (V-pits). The V-shaped pits with six {10-11}-oriented sidewalls, induced from dislocations, are the inverted hexagonal pits embedded in MQW structures [46]. Figure 4.15 shows the structure of V-shaped pits.

As a typical characteristic in GaN-based LEDs, the V-pits had been paid attention as early as 1998 [47, 48]. These works mainly focused on the formation mechanism

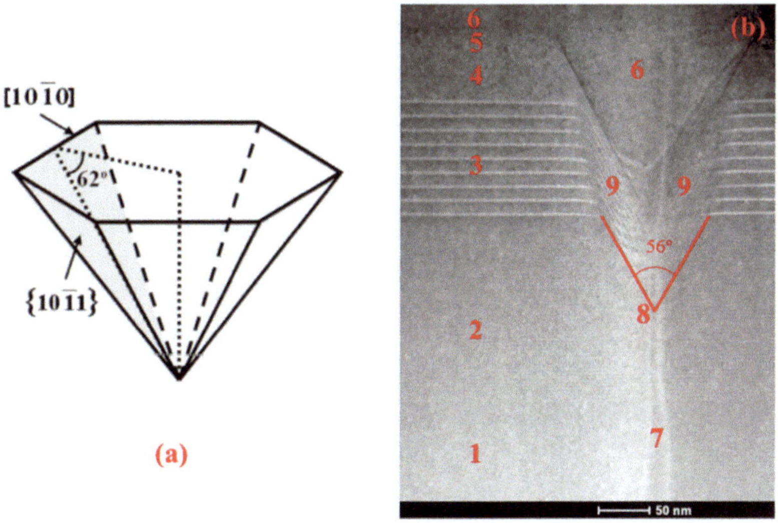

Fig. 4.15 Schematic structure of the V-shaped pits (**a**). STEM images of V-shaped pits in GaN-based LEDs grown on Si substrate (**b**). The numbers represent (1) n-type GaN, (2) InGaN/GaN SLs, (3) c-plane MQW, (4) HIL, (5) P-AlGaN EBL, (6) p-type GaN, (7) threading dislocation, (8) the position of pit generation, (9) the sidewall QW of V-shaped pits

of V-pits. Over time, the researchers began to follow the influence of the V-pits on the performance of LED. In 2000, Takahashi [49] examined the relations between the TD densities and the photoluminescence intensities of the SQW in which V-pits were intentionally formed, comparing them with the relations for a normal SQW under strong excitation conditions. The results indicated that the intentionally formed V-pits increase the luminescence intensity and reduce the dependence of photoluminescence intensity on the TD density of underlying GaN. In 2005, Hangleiter found that the TDs can be self-screened by the formation of V-shaped pits, and a physical model is proposed to explain this effect [50, 51]. The thickness and In concentrations in the sidewall structure of V-shaped pits were found to be much lower compared to those of the flat region, which provides an energy barrier around each dislocation. Consequently, V-shaped pits induced from dislocations can effectively screen the dislocations themselves and prevent carriers from non-radiative recombination. Here, we call this physical mode as "V pits screening TD" model.

Subsequently, tremendous efforts have been exerted to study the influences of V-shaped pits on the performance of InGaN/GaN MQW LEDs [52–60]. In our previous work [61], we observed the electroluminescence of sidewall MQWs at cryogenic temperatures. A broad short-wavelength EL peak with strong relative intensity to the main emission was observed at cryogenic temperatures in the LED with the unintentionally doped (UID) EBL but absent in the heavily doped (HD) EBL sample with identical epitaxial layer structure. Figure 4.16 shows the EL

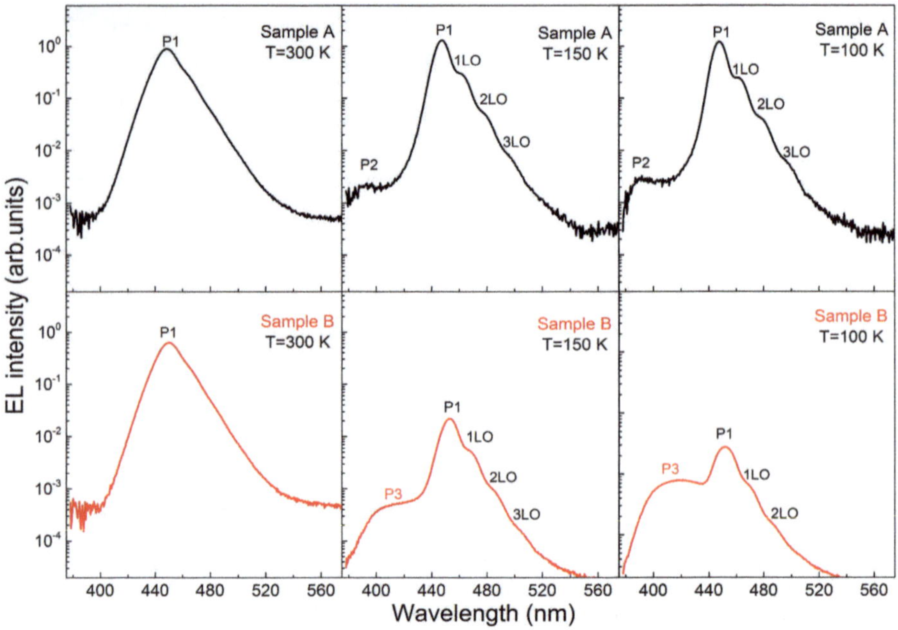

Fig. 4.16 EL spectra of the experimental samples at typically 35 A/cm^2. The spectra of the same sample are comparative in intensity. The emission intensity of Sample B decreases seriously as temperature drops, which is in sharp contrast to that of Sample A. In addition, P3 appears from 150 K in Sample B and increases in intensity as temperature further drops to 100 K

spectra of Sample A with the UID EBL and Sample B with the HD EBL at typically 35 A/cm^2. The main emission peak (P1) originates from the *c*-plane MQWs, and the emission peak (P2) involves to the Mg-related transition. After a careful analysis, we identify the sidewall MQWs in the V-pits as the emitter of P3. The peak wavelength of P3 is shorter than that of P1. It can lead to the conclusion that In concentrations in sidewall structure of V-shaped pits were much lower compared to those of the flat region. This conclusion is consistent with the "V pits screening TD" model and provides a solid proof of the existence of the self-screening effect.

In recent years, the "V pits screening TD" model is accepted by many researchers that the unexpectedly high emission efficiency of InGaN-based LEDs with high dislocation density attributes to the V-shaped pits. In the model, the role of V-pits is only screening TD. However, Li [62] reported an experimental phenomenon irrelevant to screening TD. In the research, it was demonstrated that the V-shaped pits have a significant influence on the hole injection depth.

In our previous work [63], a physical model, called the "V pits enhancing hole injection" model, has been established by using numerical simulation. In the model, screening dislocations is just one of the roles of V-shaped pits on the improvement of quantum efficiency, and V-shaped pits also play an important role on the hole injection into the MQWs. Due to the lower polarization charge densities in the

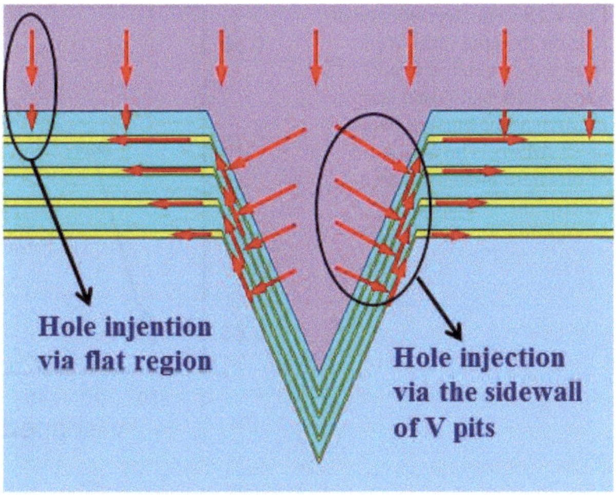

Fig. 4.17 The diagram about the process of hole injection into c-plane QW. The red arrows denote the direction of hole transport

sidewall structure with lower In concentration and {10-11}-oriented semi-polar facets, the injection of holes into the MQW via the sidewalls of the V-shaped pits is easier than via the flat region. It is helpful to reduce the efficiency droop and then improve the quantum efficiency of the LED. Therefore, InGaN/GaN MQW LEDs with high dislocation density still have high emission efficiency. Figure 4.17 illustrates clearly the process of hole injection into c-plane QW. There are two ways for hole injection into c-plane QW. One way is hole injection via flat region, in which holes are directly injected from p-type layer into c-plane QW. Another one is hole injection via the sidewall of V-pits, in which most of the holes in the V-shaped pit are injected into sidewalls QW laterally rather than vertically from p-type layer, run along the sidewall QW and then flow into c-plane QW.

Based on the "V pits enhancing hole injection" model, it can be well understood that the V-shaped pits have a significant influence on the hole injection depth [62]. The model is also used to explain the experimental results, in which LED with the larger size of V-shaped pit has higher quantum efficiency but lower forward voltage [64]. In the work, a careful simulation calculation is conducted, whose numerical models simultaneously include the screening dislocations and enhancing hole injection of V-shaped pits. It shows good agreement between the experimental data and the simulated results. The larger the V-shaped pit size is, the better it is for screening TDs and the injection of holes. It is revealed that screening TDs is the dominant role of V-shaped pits only at the small injection current density and has no effect on forward voltage. Yet the enhancing hole injection is the more significant function of V-shaped pits at the normal work current density and reduces the forward voltage. So, LED with the larger size of V-shaped pit having higher quantum efficiency but lower forward voltage is largely due to more hole injection via larger V-shaped pits.

In addition, we also experimentally demonstrate that the V-pits act as paths of hole injection into the c-plane MQWs at cryogenic temperatures [65]. It is evidenced

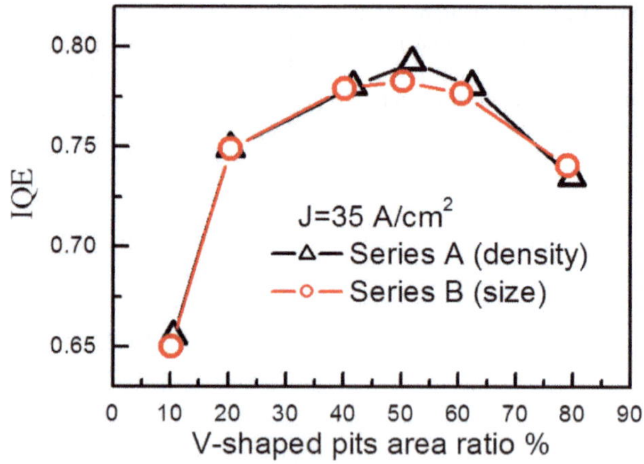

Fig. 4.18 The dependence of IQE on the area ratio of pits from the calculated results of Series A (density) and Series B (size) at the current density of 35 A/cm^2. The area ratio of pits here means the area ratio of the V-shaped pits accounting for the last QW

by the unusual broadening and blueshift of the emission of the c-plane MQWs, which occurs when the proportion of hole current flowing via the V-pits is increased.

It is confirmed that the V-shaped pits can promote the injection of holes in addition to screening the dislocations and then improve the quantum efficiency of the LED. According to this model, the density and size of pits are closely related with the hole injection, thereby affecting the quantum efficiency of LED. Hence, it is important to investigate the effect of the density and size of pits on improving the performance of LED. Two series of calculations are performed [66]. The one is the calculation for different density of pit with fixed size (Series A). Another is the calculation for different size of pit with fixed density (Series B). The calculated IQE curves show that the IQE is firstly enhanced and then reduced with an increase in both density and size of pits. The calculated results are in good agreement with the experimental results of the reported paper [67, 68]. The IQE displays the similar variation trend for different series of calculations. It implies that there is the same key factor for the dependence on both density and size of pits. The dependence of IQE on the area ratio of pits from the calculated results of Series A and Series B at the current density of 35 A/cm^2 is presented in Fig. 4.18. The area ratio of pits here means the area ratio of the V-shaped pits accounting for the last QW. As described in Fig. 4.18, the dependences of the calculated IQE on area ratio are almost the same for both two series of calculations. It indicates that the area ratio is the key factor. The calculated results show that the optimal value of area ratio is about 50% at the current density of 35 A/cm^2. For there lies an optimal value for pits density and the V-pits are induced from dislocations, there lies an optimal value for the density of dislocations, which is a new concept contrary to conventional viewpoints. It is the reason for the unexpectedly high emission efficiency of InGaN-based LEDs with high dislocation density, especially for LED grown on Si substrate.

It should be noted that the above results are obtained from the models based on two hypotheses in the simulation. The one is that the V-shaped pits are uniformly distributed. Another is that all of the pits have the same size. It is very difficult to realize the two hypotheses for a real LED. Therefore, the enhancing effect of V-pit on hole injection will be weakened in the real devices, which resulting in the real optimal data is much smaller than the theoretical value (<50%). However, the results will point out the future development direction for boosting the blue LED performance: how to get the same size and uniformly distributed V-pits in the device.

In conclusion, the V-shaped pits can promote the injection of holes in addition to screening the dislocations and then improve the quantum efficiency of the LED. At present, a lot of researches are focused on the role of V-shaped pits screening the dislocations, but the effect of V-shaped pits on hole injection has only recently begun to get attention. It is significant to the device performance, which deserves intensive study.

4.2 Device Processing of GaN LEDs on Si Substrate

Most GaN-based LEDs are grown on sapphire, SiC, or Si substrates. The substrates of sapphire and SiC are chemically stable and mechanically tough; thus they are difficult to be removed and often be processed to lateral structure. For GaN on Si substrate, as the substrate can be easily removed by wet etching, it can be processed to a vertical thin film structure.

The chip fabrication process of vertical LEDs grown on Si substrate can be simply described in Fig. 4.19. After the growth of (a) GaN LED epi-structure on Si substrate is finished, (b) we deposited metal layers on the p-surface of the wafer

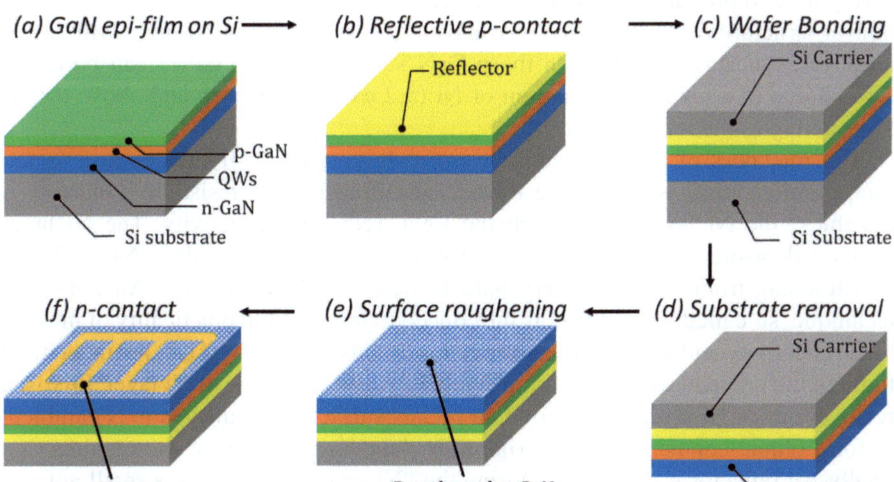

Fig. 4.19 Device fabrication processes of vertical LEDs grown on Si substrate, (**a**) growing Epi-film on GaN, (**b**) making reflective p-contact, (**c**) bonding the wafer to a new Si carrier, (**d**) removing the original Si substrate, (**e**) roughening n-GaN surface and (**f**) making n-contact

which act as p-contact and light reflector; (c) then the p-surface is bonded to a new electric conductive carrier via metal bonding technique; (d) the original Si substrate is etched off by wet etching, leaving the n-surface exposed to the air; (e) the n-surface is then roughed by wet etching to enhance light extraction; and finally, (f) n-contact is prepared [69], completing the chip process.

The chip fabrication process of vertical LEDs adopts many common techniques used in lateral LEDs fabrication, which are not needed to be described in detail. Some specific techniques used in vertical LED fabrication process will be focused below.

4.2.1 Reflective P-Type Ohmic Contact

The efficiency of a LED comprises not only internal quantum efficiency but also extraction efficiency. As Si can absorb a large amount of visible light emitted from the QWs, a reflective mirror is often placed between p-GaN and Si carrier to enhance the light extraction efficiency. In a vertical structure LED, current flows vertically through the Si carrier; thus the mirror layer between p-GaN and Si carrier functions not only as a reflecting layer but also as a p-contact layer.

Despite high reflectivity, it is difficult to make low resistivity p-contact as the hole concentration of p-GaN is low and the work function of p-GaN is high (7.1 eV). It becomes more difficult to select a suitable material as the reflective contact layer when we need to consider both reflectivity and resistivity. Ag is the most promising reflection metal for visible light, but its work function is only 4.3 eV, which makes it very hard to produce good ohmic contact to p-GaN. To overcome this problem, a bilayer structure of Ni/Ag reflector is developed. An ultrathin Ni layer with a few angstroms thick is deposited on the p-GaN surface; ohmic contact can be easily formed due to a high work function of Ni (5.2 eV). Ag layer is laid above the Ni layer acting as the reflection layer.

The thickness of Ni layer must be properly controlled. If the Ni layer is too thin, ohmic contact is hard to be formed, resulting in high resistivity; and if it is too thick, the Ni layer will absorb the light, reducing reflectivity. The Ni layer is often deposited by electron beam evaporation, and the required Ni thickness is a few angstroms which is very hard to be precisely controlled. An improved technique, so-called Ni sacrifice, is introduced to improve the practicality of making high reflectivity and low resistivity reflector [70]. The "sacrifice" can be understood through the procedures of depositing, annealing, and removing of Ni layer. Firstly, a relatively thick Ni layer is deposited on p-GaN surface and annealed in N_2 ambient to form ohmic contact; then the original Ni layer is removed by wet etching; and finally Ag reflector is deposited. During the "Ni sacrifice" process, a small amount of Ni atoms will be diffused into p-GaN and form ohmic contact during annealing process, and the extra Ni atoms are removed to ensure high reflectivity.

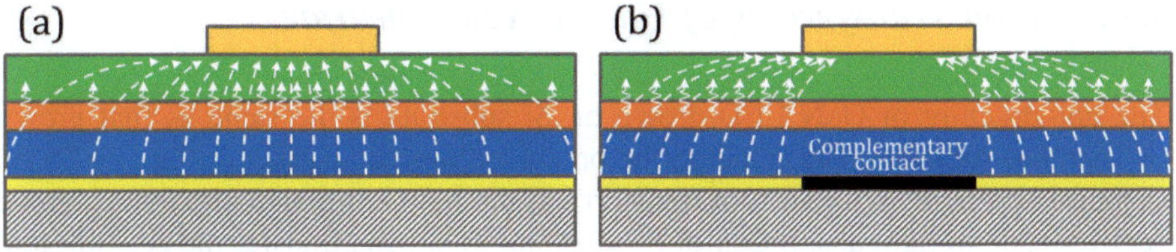

Fig. 4.20 Light extraction and current spreading in vertical LED (**a**) without complementary contact and (**b**) with complementary contact

4.2.2 Complementary Contact

Because of large tensile strain existing in GaN on Si, the thickness of epi-layer cannot be too thick, normally 2–3 μm. Compared with the width and length of LED chip of several hundred μm, the sidewall emission can be neglected. Majority of light will be emitted from the n-GaN surface. When n-contact is deposited on n-GaN surface, it will raise two problems: current crowding and light blocking. As illustrated in Fig. 4.20a, in a conventional LED structure, current prefers to flow following the path with lowest resistance; thus most of the current will be concentrated under n-contact, which is known as current crowding. Since most current crowd under n-contact, most light will emit from the region of QWs under n-contact, but the light from that region will be scattered by the n-contact, and this is called light blocking. Current crowding and light blocking will largely affect internal quantum efficiency as well as light extraction efficiency, distorting the performance of vertical LEDs.

We design a complementary contact structure to solve the problems of current crowding and light blocking [71]. As illustrated in Fig. 4.20b, we form a high-resistive contact on p-GaN surface with the same geometry of n-contact, which is the so-called complementary contact. The high-resistive contact can be achieved by deposition of dielectric materials like SiO_2 or Si_3N_4, forming a high-resistive media between p-GaN and p-contact, but not affecting light reflection. Current is forced to flow away from the region under n-contact and is distributed more uniformly across the junction, relieving current crowding. And because no current is flowing through the region under n-contact, no light will be emitted from the junction area under n-contact, and thus almost no light will be blocked by n-contact.

Experimental results of LED devices show that the complementary contact is effective to improve light extraction efficiency. With complementary contact, light extraction efficiency can be enhanced by 10–40%; the actual value depends on the coverage ratio of n-contact.

4.2.3 Film Transferring of GaN to New Substrate

The epi-structure of GaN LED grown on Si substrate is the same as that grown on sapphire substrate. But it cannot be processed to lateral structure like LEDs on sapphire substrate. Without film transferring, the light extraction efficiency can be very low due to light absorption of Si substrate. If we can transfer the epi-film to a new carrier with a reflector, the light extraction efficiency can be largely enhanced.

After the complementary p-contact with high reflectivity is prepared, film transferring process can be carried out. The film transferring comprises two main steps: firstly, the as-prepared wafer is attaching to a new carrier, and then, the original Si substrate is removed by wet etching.

The as-prepared wafer can be attached to a new carrier via thermal bonding or electrical plating. In thermal bonding, bonding media is deposited on both the wafer surface and carrier surface, and by applying high pressure under high temperature, bonding media diffuse into each side to form a continuous phase and join the wafer and carrier together. Both the bonding media and the carrier should have some special features such as good mechanical properties to sustain the structure, high thermal conductivity and electrical conductivity to conduct heat and electricity, good chemical stability to resist chemicals during chip fabrication, and matched thermal expansion coefficient with GaN films to lower thermal strain. And for the bonding media, good adhesivity is a must to attach the film to the carrier. Si is an outstanding carrier candidate with acceptable electrical conductivity, good thermal conductivity, high chemical stability, and matched thermal expansion coefficient. Metals are also used as carrier for LED transferring due to the excellent thermal conductivity and low price. Au is a promising bonding media with excellent properties in the aspects of adhesivity, thermal and electrical conductivity, and chemical stability. Indeed Au was widely used as the bonding media in semiconductor manufacturing, although the cost is high. Instead of pure Au, Au-Sn alloy with Au weight percentage from 10 to 90% is used as the bonding media to reduce the production cost. The Au-Sn alloy brings another advantage: by utilizing alloy composition to the eutectic point of Au-Sn system, bonding temperature can be dramatically reduced. Although Au-Sn alloy can reduce production cost to some extent, the price of it is still too high. A diffusion bonding technique is developed by using base metals as bonding media. The principle of diffusion bonding is based on diffusion of miscible metals: if two pieces of metals with smooth surface are laid together and sustain compressive force at high temperature, their atoms will diffuse into each other and form a continuous interface. As illustrated in Fig. 4.21, after Ag reflector is prepared, three layers of metals are deposited above the reflector sequentially, namely, diffusion barrier layer, bonding layer, and diffusion layer, respectively. The barrier layer above the Ag reflector is to prevent diffusion from the bonding layer to the Ag reflector. The bonding layer, normally in the form of high melting point metals like Cu, Mo, or Au, is acting as solvent; and the diffusion layer, normally in the form of low melting point metals like In or Sn, is acting as solute. A new Si carrier with surface covered by bonding layer is attached to as-prepared wafer. Under compressive stress and a

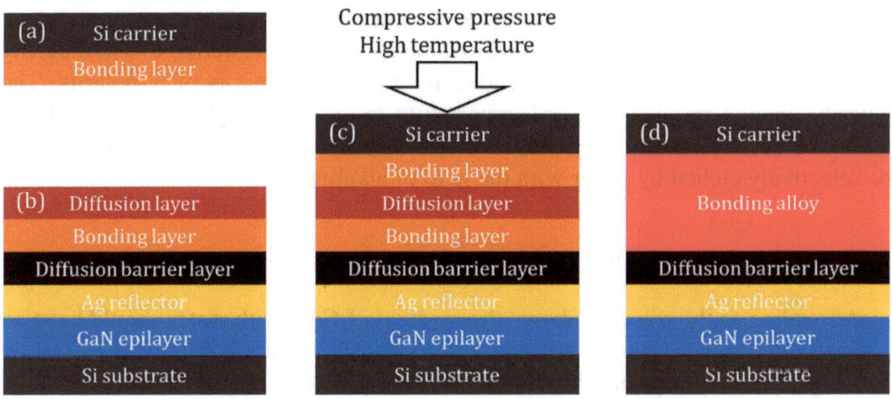

Fig. 4.21 Schematic illustration of diffusion bonding process, (**a**) depositing bonding layer on a Si carrier, (**b**) depositing reflector layer, diffusion barrier layer, bonding layer and diffusion layer on epi-wafer, (**c**) joining the two parts together under high pressure and high temperature and (**d**) the diffusion layer and the boding layers will form a continuous ally phase

proper high temperature, the atoms in the diffusion layer and bonding layers will diffuse into each other and finally form a continuous alloy phase.

After completing the bonding process, the original Si substrate can be removed by wet etching. Wet etching of Si is a very mature technology in IC industry; both anisotropic and isotropic etching modes have been developed. Anisotropic wet etching of Si is designed to create shapes like corners, grooves, or trenches, which will be used and discussed in later process of surface roughing. In the Si substrate removal process, isotropic etching of Si with aqueous HNO_3/HF is often used. The etching mechanism includes two steps: firstly oxidize Si to SiO_2 by HNO_3 and then dissolve SO_2 by HF. Related reactions are listed below:

$$Si + 4HNO_3 \rightarrow SiO_2 + 4NO_2 + 2H_2O$$
$$SiO_2 + 6HF \rightarrow H_2SiF_6 + 2H_2O$$

With surface covered by Au thin film, new Si carrier can be protected during etching, and the original Si substrate will be removed. The GaN film is then carried by the new Si substrate. The position of n- and p-GaN is upside down with p-GaN on the bottom and n-GaN exposed to the air.

4.2.4 Surface Roughening of N-Polar N-Type GaN

In GaN-based LEDs, light is very hard to escape from the bulk to the air as the total reflection angle of GaN is very small. Although a reflector is prepared at the backside of p-surface, it is still not enough to have high light extraction efficiency with a flat n-surface. A normal method is to make the n-surface rough to enhance the opportunity of light to escape [72, 73].

Normally, GaN epi-film grown on Si substrate is of Ga-polarity. As mentioned above, during the chip fabrication process, the epi-film is transferred to a new carrier with n-surface exposed to the air, indicating the top surface is of N-polarity. The polarity change provides a good opportunity to surface roughening. Ga-polarity surface is resistive to most chemicals at room temperature. But N-polar GaN can be selectively etched by water with the help of alkaline:

$$2GaN + 3H_2O \overset{OH^-}{\rightarrow} 2Ga_2O_3 + 2NH_3$$

According to the reaction above, the reaction of GaN with water should have no polarity dependence. But in fact, the truth is that Ga atoms are firstly attacked by hydroxyl (OH-) group and then reacted with water to form Ga_2O_3 and hydroxyl group. The polarity dependence of wet etching can be attributed to the difference in bond structure [74]. As illustrated in Fig. 4.22, in N-polar GaN, each nitrogen atom has one dangling bond perpendicular to the surface, and hydroxyl groups are free to attack Ga atoms, forming hexagonal pyramid-shaped etching surface, while in Ga-polar GaN, each nitrogen atom stretches out three dangling bonds covering

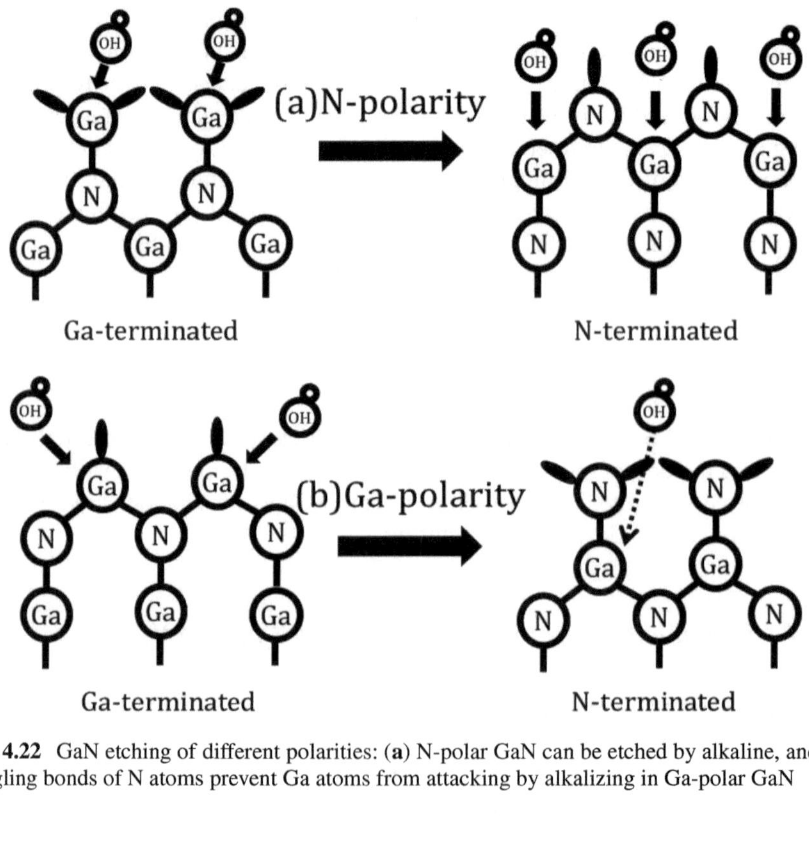

Fig. 4.22 GaN etching of different polarities: (**a**) N-polar GaN can be etched by alkaline, and (**b**) dangling bonds of N atoms prevent Ga atoms from attacking by alkalizing in Ga-polar GaN

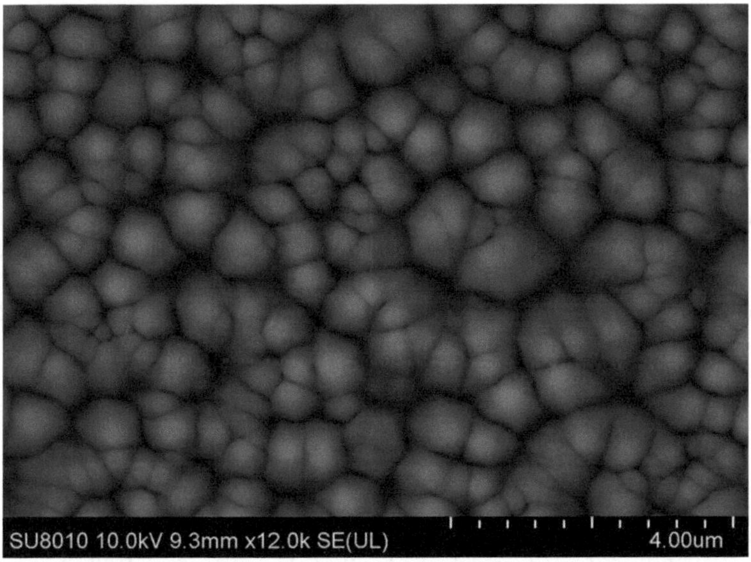

Fig. 4.23 SEM image of roughened N-polar surface by wet etching in KOH solution

the Ga atoms below them. Hydroxyl groups are repelled away from Ga atoms by the negatively charged dangling bonds; thus Ga-polar GaN is resistant to etching by alkaline.

As shown in Fig. 4.23, after wet etching in KOH solution, the N-polar GaN surface is roughened with hexagonal-shaped pyramids covering the whole surface. The facets of the hexagonal pyramids are of $\{10\bar{1}0\}$ orientation.

One can observe that the pyramids are not very uniform with different sizes. That can be attributed to the nonuniform defect distribution as the pyramids are initialed from dislocations.

With a roughened surface, the light can have a high probability to escape from the pyramid to the air after a few rounds of reflections. In vertical LEDs on Si substrate with a reflector, light extraction efficiency can be enhanced by 30–50% percent if surface roughening is applied; the value can be higher without a reflector.

4.2.5 Ohmic Contact on N-Polar N-GaN

The polarity change brings merit to surface roughening but also brings difficulty to n-contact making. In lateral LEDs, making ohmic contact on Ga-polar n-GaN is very easy using Al- or Ti-based metals, while in vertical LEDs, ohmic contact on N-polar n-GaN is difficult to obtain because of the formation of V_{Ga}-O_N complex on the surface of N-polar GaN [75]:

$$O + V_N \rightarrow O_N$$
$$V_{Ga} + O_N \rightarrow V_{Ga} - O_N$$

The formation of V_{Ga}-O_N not only consumes V_N but also acts as acceptors and compensates donor-like defects V_N, lowering electron concentration and leading to a poor ohmic contact. A surface treatment before n-contact deposition with Ar plasma can help to improve the stability of ohmic contact on N-polar n-GaN. With Ar plasma treatment, V_N concentration can be dramatically increased, and the existence of V_{Ga}-O_N will no more affect the electron concentration on the surface. Therefore, ohmic contact on N-polar n-GaN with high stability can be easily obtained with the help of Ar plasma treatment [76].

4.2.6 Device Passivation

Reliability is very important to a commercial LED. The most effective way to improve reliability of an electronic device is to do passivation. With the protection of passivation layer, LED can be protected from surface recombination and foreign-atom diffusion.

Figure 4.24 lists four vertical LEDs with different types of passivations. (a) Without passivation, surface recombination occurs when electrons and holes travel via the surface of the device, reducing the radiative recombination efficiency [77]; foreign ions such as Na + may diffuse into the active region, causing current leakage; and H atoms may diffuse into p-GaN [78], resulting in degradation of p-GaN. (b) Passivation of sidewalls with SiO_2 dielectric material can help to protect key region of the device – the QWs. (c) Passivation of sidewalls and n-surface and passivation of the entire surface can provide better protection. But the quality and the thickness of the passivation layer should be properly controlled to avoid light extraction impediment [79]. (d) Beyond n-surface and the sidewalls, p-surface is also partially passivated [80] near the edge. With such a design, current is forced to flow away from the surface of the sidewall, further preventing recombination of carriers at the interface of GaN/SiO_2. And leakage current can be dramatically reduced.

4.3 Device Characterization of Vertical Thin Film LEDs Based on GaN/Si Technology

For the LED grown on Si substrate and be fabricated into vertical thin film structure, it gets some unique features which are different from the conventional lateral LEDs. These features, specifically uniform current flow, excellent heat conduction, and single-face light emission, will bring many merits to the performance of LED.

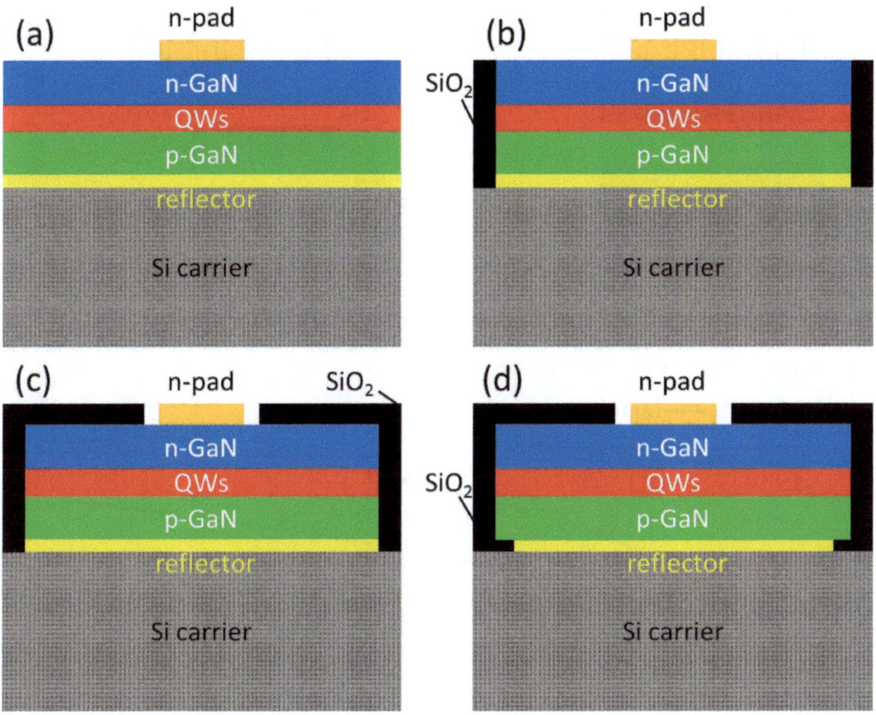

Fig. 4.24 Schema of a vertical LED (**a**) without passivation, (**b**) with sidewalls passivated, (**c**) with entire surface passivated, and (**d**) with double side passivated

As illustrated in Fig. 4.25a, for a typical lateral LED structure with both n-pad and p-pad on the same sides, the current flows downward through the QWs and then travels laterally to the n-side; the current is very crowded under the n-pad. Nonuniform current flowing will largely affect the performance of LED; it raises the operation voltage, causes regional over temperature, and reduces the reliability of LEDs. Under high current density, the status becomes even worse. In a vertical structure LED with n-pad and p-pad laid on two opposite sides, current flows directly through the QWs to n-pad. The current flowing path is symmetrical. With proper n-contact, current distribution can be very uniform across the whole QW region, and current crowding can be avoided. A simple way to justify the current spreading performance of a LED is to measure its I-V characteristic. Normally, a better current spreading will result in a lower forward voltage under high current density.

The dependences of voltage on current density of the two types of LEDs are plotted in Fig. 4.26. The solid line represents lateral LED on sapphire substrate, and the dash line represents vertical LED on Si substrate. Obviously, the vertical LED has a lower voltage at every current density, and the gap becomes bigger

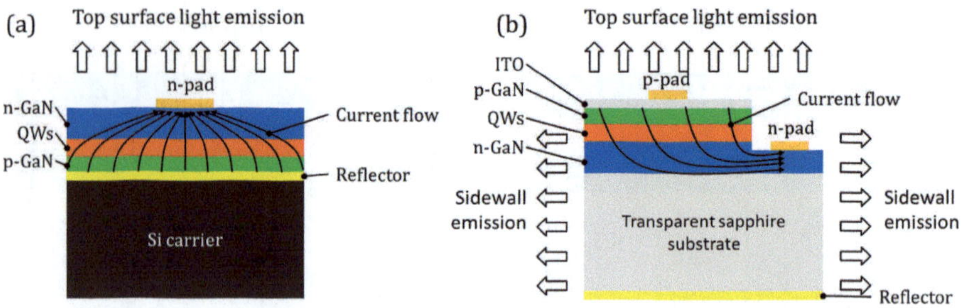

Fig. 4.25 Comparison of LED with (**a**) vertical structure and (**b**) lateral structure

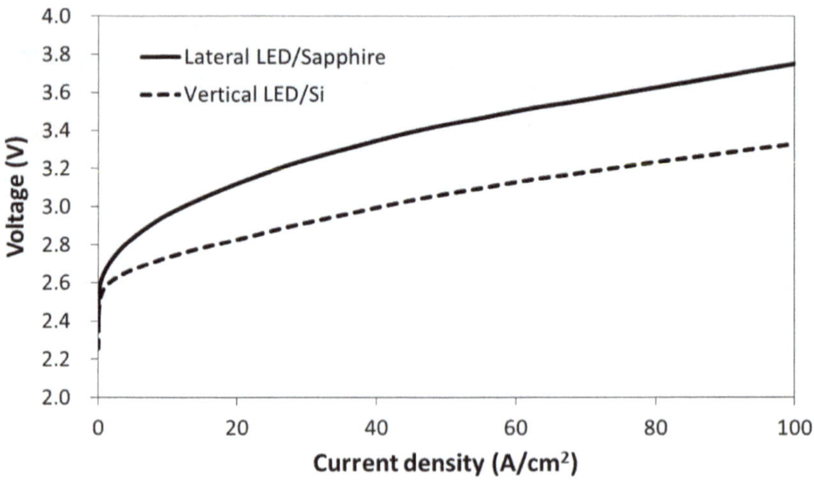

Fig. 4.26 Comparison of V-I characteristics between vertical thin film LED and conventional lateral LED

when current density increases. This phenomenon can be mainly attributed to the difference of current spreading. Commonly, the better current spreading of a LED, the lower the operation voltage will be.

The thermal conductivity is a very important feature for LED devices. Although the efficiency of LEDs is relatively high, still ten percent of the energy is lost in the form of heat. If the heat cannot be conducted away rapidly, it will accumulate inside the device and heat up the junction. Under a high temperature, the device will suffer thermal droop, that is, the efficiency of a LED drops down when temperature is increased. Besides, the reliability of the LED will also be largely affected; the LED devices will degrade faster under a higher temperature. Therefore, it is very important for a LED to maintain a relatively low temperature, which required good thermal conductivity. As in Fig. 4.25b, a conventional lateral LED has a thick sapphire substrate at the backside, and heat needs to be conducted away from the substrate. The thermal conductivity of sapphire is relatively low (45 W/mK). Under

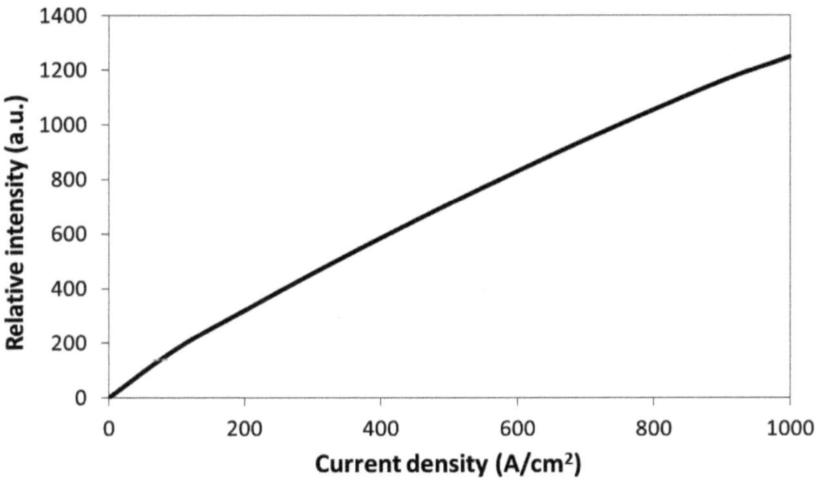

Fig. 4.27 Emission intensity versus current density of a vertical thin film LED

high current density, the junction temperature will rise up to very high level and lead to a very poor reliability. The nonuniform current distribution will aggravate the status. For the vertical structure, the heat is flowing through the Si carrier. Thus a conventional lateral LED structure is not suitable for high power application. A solution to this is to mount the top face instead of the bottom face on the sink, which is called as "flip chip" structure. The flip chip structure reduces the thermal diffusion length and directly contacts GaN to the sink, which can largely reduce the junction temperature. With flip chip structure, LEDs grown on sapphire can be used in high power application.

The Si carrier has a much better thermal conductivity (150 W/mK) which is two times higher than that of sapphire, resulting in a lower junction temperature. Combined with the good current spreading behavior, very high current density can be applied to the vertical thin film LEDs. Figure 4.27 demonstrates a vertical thin film LED chip in the size of 0.1 mm^2 working under various current densities. The emission intensity of the device shows no obvious decay up to a current density of 1000 A/cm^2.

The junction temperature is measured to compare the heat conduction capability between the lateral and vertical LEDs. The measuring is based on I-V testing method [81]. Figure 4.28 plots the junction temperatures of a lateral LED and a vertical LED at various current densities under a same ambient temperature of 30 °C. At 5 A/cm^2, the junction temperature of the vertical LED on Si is 40 °C, which is slightly lower than that of the lateral LED. As current density increases, the junction temperature of both samples increases. When current density increased to 100 A/cm^2, for vertical LED the junction temperature is 85 °C, whereas the junction temperature for lateral LED is 102 °C.

Another advantage for vertical thin film LED is on the light extraction mode. As illustrated in Fig. 4.25b, because the sapphire substrate is transparent and the

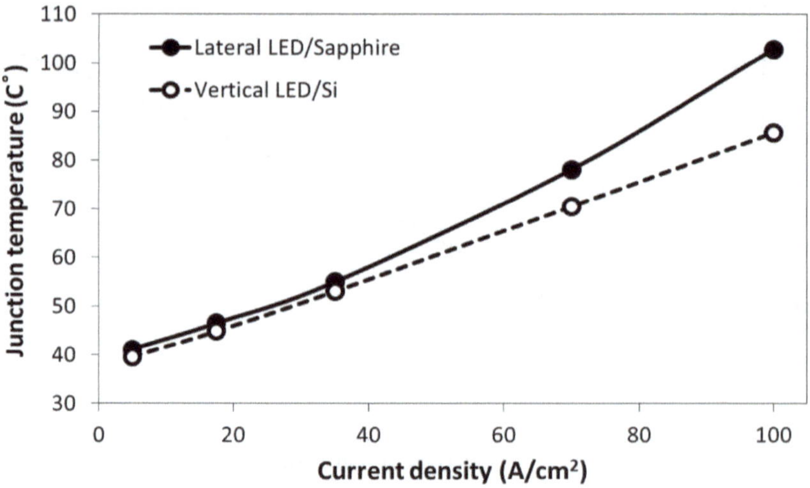

Fig. 4.28 Comparison of junction temperature between vertical thin film LED and conventional lateral LED

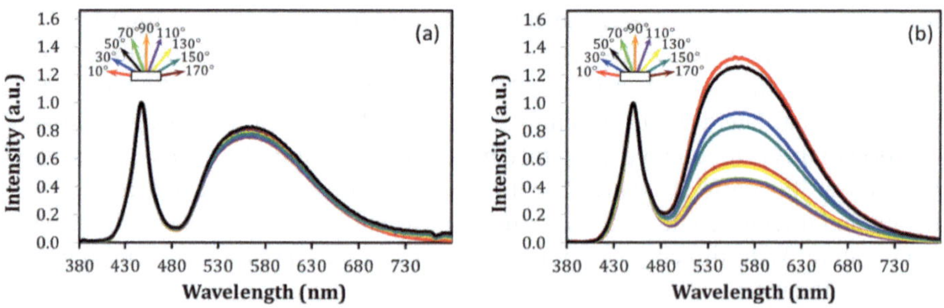

Fig. 4.29 The spectra of white LEDs of (**a**) vertical thin film structure on Si and (**b**) lateral structure on sapphire in different viewing angles

thickness is 70 μm, a large amount of light will be emitted from the sidewall of the sapphire. Excluding the bottom surface, there are five surfaces that can be the faces for light emission. It is difficult to manipulate the light emission, for instance, in white light application, as there are too many faces where light can come out, it is hard to apply the phosphor to the LED. As in Fig. 4.29b, the spectra of a phosphor which converted white LED based on lateral structure on sapphire substrate are measured in different viewing angles. The portion of yellow light is different in different viewing angles, which will lead to nonuniform color temperature. For the thin film vertical structure with epi-film thickness of only 2 μm, almost all the light comes out from the top surface. Thus it is very easy to control the light in LED applications. As shown in Fig. 4.29a, the light emitted from a vertical thin film phosphor-converted white LED has almost the same spectra in any direction.

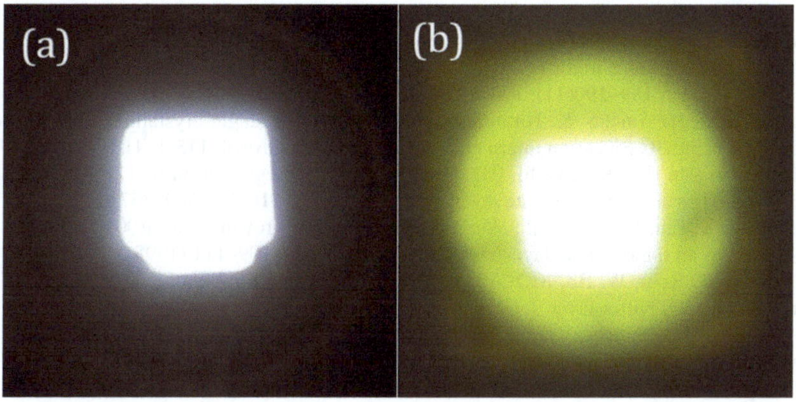

Fig. 4.30 Projection of white light of (**a**) vertical thin film LED on Si and (**b**) lateral LED on sapphire

The light emitted from the vertical thin film LED is highly directional. With simple light path design, it can propagate parallel in a long distance. As illustrated in Fig. 4.30a, the projection of a vertical thin film white LED in a distance of 5 m is recorded by a camera. It can be seen that the facula is very uniform, and the border of which is very sharp. By comparison, the facula of a lateral LED is less uniform with blurry border (Fig. 4.30b).

In conclusion, the thin film vertical LED on Si substrate has its unique features. In combination with good current spreading and excellent heat conductivity, LEDs on Si can be applied in high power application under high current densities. Because of one face emission behavior, LEDs on Si have high directional emission light, and it can be applied in directional lighting applications such as projectors, headlights, and car lamps.

References

1. T.L. Chu, Gallium nitride films. J. Electrochem. Soc. **118**(7), 1200–1203 (1971)
2. F. Jiang, L. Wang, X. Wang, et al., High power InGaN-based blue LEDs grown on Si substrates by MOCVD. Abstract Book: The 8th International Conference on Nitride Semiconductors, vol. 1 (2009), pp. 82–83
3. A. Krost, A. Dadgar, GaN-based optoelectronics on Si substrates. Mater. Sci. Eng. **B93**, 77–84 (2002)
4. H.P. Maruska, J.J. Tietjen, Preparation and properties of vapor-deposited single-crystalline GaN. Appl. Phys. Lett. **15**, 327 (1969)
5. T. Kozawa, T. Kachi, H. Kano, et al., Thermal-stress in GaN epitaxial layers grown on sapphire substrates. J. Appl. Phys. **77**, 4389–4392 (1995)
6. A. Dadgar, J. Bläsing, A. Diez, et al., Metalorganic chemical vapor phase epitaxy of crack-free GaN on Si (111) exceeding 1 μm in thickness. Jpn. J. Appl. Phys. **39**, L1183–L1185 (2000)
7. F. Jiang, J. Liu, L. Wang, et al., High optical efficiency GaN based blue LED on Si substrate. Sci. Sin. Phys. Mech. Astron. **45**, 067302 (2015)

8. E. Butter, G. Fitzl, D. Hirsch, et al., The deposition of group III nitrides on Si substrates. Thin Solid Films **59**, 25–31 (1979)
9. H. Ishikawa, K. Yamamoto, T. Egawa, et al., Thermal stability of GaN on (1 1 1) Si substrate. J. Cryst. Growth **189–190**(11), 178–182 (1998)
10. T. Takeuchi, H. Amano, K. Hiramatsu, et al., Growth of single crystalline GaN film on Si substrate using 3C-SiC as an intermediate layer. J. Cryst. Growth **115**(1–4), 634–638 (1991)
11. N.P. Kobayashi, J.T. Kobayashi, P.D. Dapkus, et al., GaN growth on Si(111) substrate using oxidized AlAs as an intermediate layer. Appl. Phys. Lett. **71**(24), 3569–3571 (1997)
12. L. Wang, X. Liu, Y. Zan, et al., Wurtzite GaN epitaxial growth on a Si(001) substrate using γ-Al2O3 as an intermediate layer. Appl. Phys. Lett. **72**(1), 109–111 (1998)
13. J.H. Boo, C. Rohr, W. Ho, MOCVD of BN and GaN thin films on Si: new attempt of GaN growth with BN buffer layer. J. Cryst. Growth **189–190**(1–2), 439–444 (1998)
14. E. Calleja, M.A. Sánchez-García, F.J. Sánchez, et al., Growth of III-nitrides on Si(1 1 1) by molecular beam epitaxy Doping, optical, and electrical properties. J. Cryst. Growth **201**(5), 296–317 (1999)
15. A. Watanabe, T. Takeuchi, K. Hirosawa, et al., The growth of single crystalline GaN on a Si substrate using AIN as an intermediate layer. J. Cryst. Growth **128**(1–4), 391–396 (1993)
16. J.W. Yang, A. Lunev, G. Simin, et al., Selective area deposited blue GaN–InGaN multiple-quantum well light emitting diodes over Si substrates. Appl. Phys. Lett. **76**(76), 273–275 (2000)
17. S.A. Nikishin, N.N. Faleev, V.G. Antipov, et al., High quality GaN grown on Si (111) by gas source molecular beam epitaxy with ammonia. Appl. Phys. Lett. **1999**(75), 2073–2075 (1999)
18. M.A. Sanchez-Garcia, E. Calleja, E. Monroy, et al., The effect of the III/V ratio and substrate temperature on the morphology and properties of GaN- and AlN-layers grown by molecular beam epitaxy on Si(1 1 1). J. Cryst. Growth **183**(1–2), 23–30 (1998)
19. P. Chen, R. Zhang, Z.M. Zhao, et al., Growth of high quality GaN layers with AlN buffer on Si(1 1 1) substrates. J. Cryst. Growth **225**(2–4), 150–154 (2001)
20. A. Dadgar, M. Poschenrieder, J. Bläsing, et al., MOVPE growth of GaN on Si(111) substrates. J. Cryst. Growth **248**(10), 556–562 (2003)
21. H. Lahrèche, P. Vennéguès, O. Tottereau, et al., Optimisation of AlN and GaN growth by metalorganic vapour-phase epitaxy (MOVPE) on Si (1 1 1). J. Cryst. Growth **217**(1–2), 13–25 (2000)
22. S. Zamir, B. Meyler, E. Zolotoyabko, et al., The effect of AlN buffer layer on GaN grown on (1 1 1)-oriented Si substrates by MOCVD. J. Cryst. Growth **218**(2–4), 181–190 (2000)
23. R. Liu, F.A. Ponce, A. Dadgar, A. Krost, Atomic arrangement at the AlN/Si (111) interface. Appl. Phys. Lett. **83**, 860 (2003)
24. A. Sakai, H. Sunakawa, A. Usui, Defect structure in selectively grown GaN films with low threading dislocation density. Appl. Phys. Lett. **71**(16), 2259–2261 (1997)
25. P. Kung, D. Walker, M. Hamilton, et al., Lateral epitaxial overgrowth of GaN films on sapphire and Si substrates. Appl. Phys. Lett. **74**(4), 570–572 (1999)
26. Y. Honda, Y. Kuroiwa, M. Yawaguchi, N. Sawaki, Growth of GaN free from cracks on a (111)Si substrate by selective metalorganic vapor-phase epitaxy. Appl. Phys. Lett. **80**, 222 (2002)
27. A. Dadgar, M. Poschenrieder, A. Reiher, et al., Reduction of stress at the initial stages of GaN growth on Si(111). Appl. Phys. Lett. **82**(1), 28–30 (2003)
28. T.S. Zheleva, O.H. Nam, M.D. Bremser, et al., Dislocation density reduction via lateral epitaxy in selectively grown GaN structures. Appl. Phys. Lett. **71**(17), 2472–2474 (1997)
29. O.H. Nam, M.D. Bremser, T.S. Zheleva, et al., Lateral epitaxy of low defect density GaN layers via organometallic vapor phase epitaxy. Appl. Phys. Lett. **71**(18), 2638–2640 (1997)
30. O. Contreras, F.A. Ponce, J. Christen, et al., Dislocation annihilation by Si delta-doping in GaN epitaxy on Si. Appl. Phys. Lett. **81**(25), 4712–4714 (2002)
31. Q. Sun, Y.S. Cho, I.H. Lee, J. Han, B.H. Kong, H.K. Cho, Appl. Phys. Lett. **93**, 131912 (2008)
32. D. Zhu et al., J. Appl. Phys. **109**, 014502 (2011)
33. A. Dadgar et al., Phys. Status Solidi C. 1583 (2003)
34. K. Cheng et al., J. Electron. Mater. **35**, 592 (2006)

35. B. Leung, J. Han, Q. Sun, Phys. Status Solidi C **11**, 437 (2014)
36. A.E. Romanov, J.S. Speck, Appl. Phys. Lett. **83**, 2569 (2003)
37. D.M. Follstaedt, S.R. Lee, A.A. Allerman, J.A. Floro, J. Appl. Phys. **105**, 083507 (2009)
38. B.A.B.A. Shuhaimi, H. Kawato, Y. Zhu, T. Egawa, J. Phys. Conf. Ser. **152**, 012007 (2009)
39. Y. Sun et al., Nat. Photon **10**, 595 (2016)
40. B. Heying, X.H. Wu, S. Keller, Y. Li, D. Kapolnek, B.P. Keller, S.P. DenBaars, J.S. Speck, Appl. Phys. Lett. **68**, 643 (1996)
41. R. Chierchia, T. Böttcher, H. Heinke, S. Einfeldt, S. Figge, D. Hommel, J. Appl. Phys. **93**, 8918 (2003)
42. D. Cherns, S.J. Henley, F.A. Ponce, Appl. Phys. Lett. **78**, 2691 (2001)
43. S.J. Rosner, E.C. Carr, M.J. Ludowise, G. Girolami, H.I. Erikson, Appl. Phys. Lett. **70**, 420 (1997)
44. S.F. Chichibu, A. Uedono, T. Onuma, B.A. Haskell, A. Chakraborty, T. Koyama, P.T. Fini, S. Keller, S.P. Denbaars, J.S. Speck, U.K. Mishra, S. Nakamura, S. Yamaguchi, S. Kamiyama, H. Amano, I. Akaki, J. Han, T. Sota, Nat. Mater. **5**, 810 (2006)
45. S. Nakamura, Science **281**(14), 956 (1998)
46. L.C. Le, D.G. Zhao, D.S. Jiang, L. Li, L.L. Wu, P. Chen, Z.S. Liu, Z.C. Li, Y.M. Fan, J.J. Zhu, Appl. Phys. Lett. **101**, 252110 (2012)
47. I.H. Kim, H.S. Park, Y.J. Park, et al., Formation of V-shaped pits in InGaN/GaN multiquantum wells and bulk InGaN films. Appl. Phys. Lett. **73**(12), 1634–1636 (1998)
48. Y. Chen, T. Takeuchi, H. Amano, et al., Pit formation in GaInN quantum wells. Appl. Phys. Lett. **72**(6), 710–712 (1998)
49. H. Takahashi, A. Ito, T. Tanaka, et al., Effect of intentionally formed V-defects' on the emission efficiency of GaInN single quantum well. Jpn. J. Appl. Phys. **39**(6B), L569–L571 (2000)
50. A. Hangleiter, F. Hitzel, C. Netzel, D. Fuhrmann, U. Rossow, G. Ade, P. Hinze, Phys. Rev. Lett. **95**, 127402 (2005)
51. C. Netzel, H. Bremers, L. Hoffmann, D. Fuhrmann, U. Rossow, A. Hangleiter, Phys. Rev. B **76**, 155322 (2007)
52. S. Tomiya, Y. Kanitani, S. Tanaka, T. Ohkubo, K. Hono, Appl. Phys. Lett. **98**, 181904 (2011)
53. J. Abell, T.D. Moustakas, Appl. Phys. Lett. **92**, 091901 (2008)
54. L.C. Le, D.G. Zhao, D.S. Jiang, L. Li, L.L. Wu, P. Chen, Z.S. Liu, J. Yang, X.J. Li, X.G. He, J.J. Zhu, H. Wang, S.M. Zhang, H. Yang, J. Appl. Phys. **114**, 143706 (2013)
55. J. Kim, Y.-H. Cho, D.-S. Ko, X.-S. Li, J.-Y. Won, E. Lee, S.-H. Park, J.-Y. Kim, S. Kim, Opt. Express **22**(S3), A857–A866 (2014)
56. Z. Fang, J. Appl. Phys. **106**, 023517 (2009)
57. J. Kim, J. Kim, Y. Tak, S. Chae, J.-Y. Kim, Y. Park, IEEE Electron Device Lett. **34**(11), 1409 (2013)
58. P.H. Weidlich, M. Schnedler, H. Eisele, R.E. Dunin-Borkowski, P. Ebert, Appl. Phys. Lett. **103**, 142105 (2013)
59. S.-H. Han, D.-Y. Lee, H.-W. Shim, J.W. Lee, D.-J. Kim, S. Yoon, Y.S. Kim, S.-T. Kim, Appl. Phys. Lett. **102**, 251123 (2013)
60. Y.-H. Cho, J.-Y. Kim, J. Kim, M.-B. Shim, S. Hwang, S.-H. Park, Y.-S. Park, S. Kim, Appl. Phys. Lett. **103**, 261101 (2013)
61. X. Wu, J. Liu, Z. Quan, C. Xiong, C. Zheng, J. Zhang, Q. Mao, F. Jiang, Appl. Phys. Lett. **104**, 221101 (2014)
62. Y. Li, F. Yun, X. Su, S. Liu, W. Ding, X. Hou, J. Appl. Phys. **116**, 123101 (2014)
63. Z. Quan, L. Wang, C. Zheng, J. Liu, F. Jiang, J. Appl. Phys. **116**, 183107 (2014)
64. Z. Quan, J. Liu, F. Fang, G. Wang, F. Jiang, A new interpretation for performance improvement of high-efficiency vertical blue light-emitting diodes by InGaN/GaN superlattices. J. Appl. Phys. **118**, 193102 (2015)
65. X. Wu, J. Liu, F. Jiang, Hole injection from the sidewall of V-shaped pits into c-plane multiple quantum wells in InGaN light emitting diodes. J. Appl. Phys. **118**, 164504 (2015)
66. Z.J. Quan, J.L. Liu, F. Fang, et al., Opt. Quant. Electron. **48**(3), 1–8 (2016)

67. C.Y. Chang, H. Li, Y.T. Shih, T.C. Lu, Manipulation of nanoscale V-pits to optimize internal quantum efficiency of InGaN multiple quantum wells. Appl. Phys. Lett. **106**, 091104 (2015)
68. N. Okada, H. Kashihara, K. Sugimoto, Y. Yamada, K. Tadatomo, Controlling potential barrier height by changing V-shaped pit size and the effect on optical and electrical properties for InGaN/GaN based light-emitting diodes. J. Appl. Phys. **117**, 025708 (2015)
69. C. Xiong, F. Jiang, W. Fang, L. Wang, C. Mo, H. Liu, The characteristics of Gan-based blue LED on Si substrate. J. Lumin. **122**, 185–187 (2007)
70. G. Wang, X. Tao, F. Feng, C. Xiong, J. Liu, M. Zhang, F. Jiang, Effects of Ni-assisted annealing on P-type contact resistivity of Gan-based LED films grown on Si(111) substrates. Acta Phys. Sin. **60**(7), 078503 (2011)
71. F. Jiang, L. Wang, W. Fang, Semiconductor light-emitting device and method for making same. US patent. US7919784B1, Filed 26 Sept 2006 and Issued 5 Apr 2007 (2007)
72. T. Fujii, Y. Gao, R. Sharma, E.L. Hu, S.P. DenBaars, S. Nakamura, Increase in the extraction efficiency of Gan-based light-emitting diodes via surface roughening. Appl. Phys. Lett. **84**(6), 855–857 (2004)
73. Z. Yinhua, T. Yingwen, R. Jianping, J. Fengyi, Improvement for extraction efficiency of vertical Gan-based LED on Si substrate by photo-enhanced wet etching. Acta Opt. Sin. **29**(1), 252–255 (2009)
74. D. Li, M. Sumiya, S. Fuke, D. Yang, Q. Duanlin, Y. Suzuki, Y. Fukuda, Selective etching of Gan polar surface in potassium hydroxide solution studied by X-ray photoelectron spectroscopy. J. Appl. Phys. **90**(8), 4219–4223 (2001)
75. J.-O. Song, D.-S. Leem, S.-H. Kim, T.-Y. Seong, Formation of vanadium-based ohmic contacts to N-Gan. Korean J. Mater. Res. **13**(9), 567–571 (2003)
76. J. Liu, F. Feng, Y. Zhou, J. Zhang, F. Jiang, Stability of Al/Ti/Au contacts to N-polar N-Gan of Gan based vertical light emitting diode on Si substrate. Appl. Phys. Lett. **99**(11), 111112 (2011)
77. G.L. Martinez, M.R. Curiel, B.J. Skromme, R.J. Molnar, Surface recombination and sulfide passivation of Gan. J. Electron. Mater. **29**(29), 325–331 (2000)
78. M. Meneghini, L.R. Trevisanello, U. Zehnder, T. Zahner, U. Strauss, G. Meneghesso, E. Zanoni, High-temperature degradation of Gan LEDs related to passivation. IEEE Trans. Electron Dev. **53**(12), 2981–2987 (2007)
79. J. Liu, C. Qiu, Research of passivation and anti reflecting layer on Gan based blue LED on Si substrate. Acta Optica Sinica **30**(10), 2978–2982 (2010)
80. F. Jiang, J. Liu, L. Wang, Semiconductor light-emitting device with double-sided passivation. US Patent. US 2011/0001120 A1, Filed 25 Mar 2008 and Issued 13 May 2008 (2008)
81. F. Jiang, W. Liu, Y. Li, et al., Research on the junction-temperature characteristic of GaN light-emitting diodes on Si substrate. J. Lumin. **122–123**(1), 693–695 (2007)

Chapter 5
The AlGaInP/AlGaAs Material System and Red/Yellow LED

Guohong Wang, Xiaoyan Yi, Teng Zhan, and Yang Huang

5.1 AlGaInP/AlGaAs Material System Lattice and Bandgap Energy

AlGaInP/AlGaAs is the material of choice for the long-wavelength part of the visible spectrum, namely, for red, orange, yellow, and yellow-green wavelengths. Both of them are zinc blende structure as shown in Fig. 5.1.

$Al_xGa_{1-x}As$ for Al mole fractions $x < 0.45$ are direct-gap semiconductors. For Al mole fractions $x < 45\%$, the Γ conduction-band valley is the lowest minimum, and the semiconductor has a direct gap. For $x > 45\%$, the X valleys are the lowest conduction-band minimum, and the semiconductor becomes indirect.

AlGaAs lattice nearly matches with GaAs for whole Al content [1]. And AlGaAs is a high-efficiency material suitable for infrared and red wavelengths. However it is unsuitable for orange and shorter wavelengths due to the direct-indirect transition of the $Al_xGa_{1-x}As$ bandgap at x of 0.45 corresponding to a wavelength of 624 nm [2].

G. Wang · X. Yi · T. Zhan · Y. Huang
Semiconductor Lighting Technology Research and Development Center, Institute of Semiconductors, Chinese Academy of Sciences, Beijing, China

State Key Laboratory of Solid State Lighting, Beijing, China

Beijing Engineering Research Center for the 3rd Generation Semiconductor Materials and Application, Beijing, China
e-mail: ghwang@semi.ac.cn; spring@semi.ac.cn

J. Li, G. Q. Zhang (eds.), *Light-Emitting Diodes*, Solid State Lighting Technology and Application Series 4, https://doi.org/10.1007/978-3-319-99211-2_5

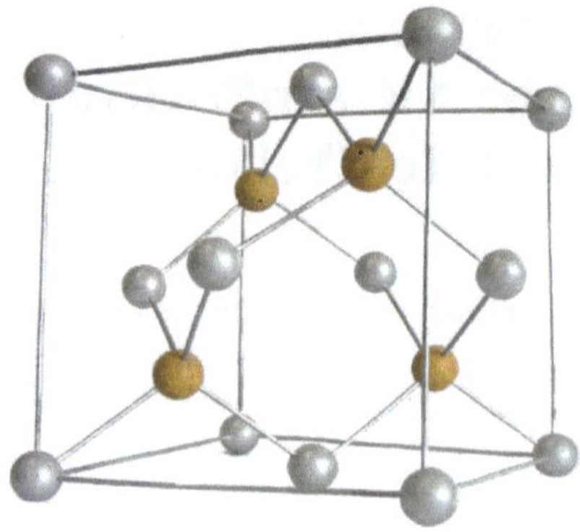

Fig. 5.1 AlGaInP zinc blende structure: gray ball, Al, Ga, In; yellow ball, As, P

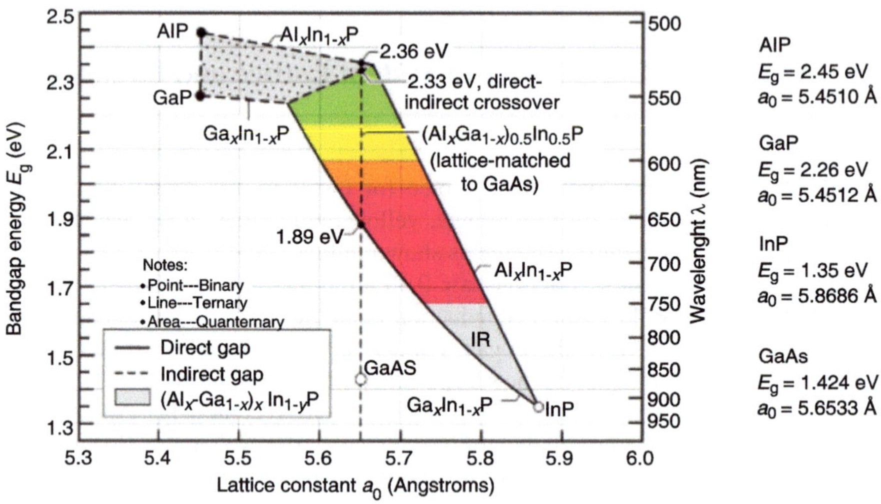

Fig. 5.2 AlGaInP zinc blende structure, gray ball: Al, Ga, In and yellow ball: As, P

$(Al_xGa_{1-x})_{1-y}In_yP$ alloy lattice constant varies with indium content linearly but changes very slightly as shown by Eq. (5.1), and the lattice mismatches with GaAs as shown by Eq. (5.2).

$$a_{(Al_xGa_{1-x})_{1-y}In_yP} = 5.45 \times (1 - y) + 5.8688 \times y \tag{5.1}$$

$$\frac{\Delta a}{a_{GaAs}} = -0.03567 + 0.0379 \times y \tag{5.2}$$

At the molar indium composition of 50%, $(Al_xGa_{1-x})_{0.5}In_{0.5}P$ is lattice matched to GaAs as shown in Fig. 5.2 [3], which can be calculated by Vegard's law [4, 5].

Such long-wavelength visible-spectrum devices will play an important role in solid-state lighting applications. In Fig. 5.2, the least squares fit to this data (indicated by the dashed lines) and give us the bandgap of $(Al_xGa_{1-x})_{0.5}In_{0.5}P$ alloy range at 2 K by Eqs. (5.3) and (5.4), and at 300 K by Eqs. (5.5) and (5.6).

$$Eg\,(\Gamma) = 1.985 + 0.610 \times x \tag{5.3}$$

$$Eg(X) = 2.282 + 0.0852 \times x \tag{5.4}$$

$$Eg\,(\Gamma) = 1.900 + 0.610 \times x \tag{5.5}$$

$$Eg(X) = 2.204 + 0.085 \times x. \tag{5.6}$$

Similar to AlGaAs, AlGaInP suffers from a direct-indirect transition of the bandgap. However, the direct-indirect transition in AlGaInP occurs at a higher energy compared with AlGaAs, which occurs at the energy of 2.23–2.33 eV corresponding to 556–532 nm.

Besides direct-indirect transition, AlGaInP exact bandgap energy suffers from order and disorder effect with In atoms and Ga and Al atoms, at which the direct-indirect transition occurs depending on the degree of randomness of the quaternary alloy semiconductor; generally it is lower for AlGaInP with a high degree of order [6].

The first-principles pseudopotential method with the virtual crystal approximation is used to calculate the band offsets at (001) $Ga_{0.5}In_{0.5}P/Al_xGa_{0.52-x}In_{0.5}P$ heterostructures. The valence-band offset VBO varies with x as VBO = 0.433x eV, while the inferred conduction-band offset CBO at Γ minima varies in x as CBO = $0.787x$ eV. The values for $CBO^{\Gamma}/\Delta Eg$ are around 0.65. And as shown in Fig. 5.3, the results are in very good agreement with the experimental data [7].

5.2 AlGaInP/AlGaAs Material Epitaxy by MOCVD

AlGaInP materials and devices have led to major advances in high-brightness LEDs. The fabrication of AlGaInP-based light emitters dates back to the mid-1980s. It is closely related to the development of metal-organic vapor phase epitaxy (MOVPE) because the standard growth technologies for conventional LEDs like liquid-phase epitaxy (LPE) or hydride vapor phase epitaxy (HVPE) were not suitable for the growth of AlGaInP. The first light-emitting AlGaInP-based devices were semiconductor lasers developed in Japan [8]. High-efficiency LEDs grown by MOVPE were reported a few years later [9, 10]. Despite the difficulties in epitaxial growth, the new material system was attractive, because it combined the possibility to achieve high efficiency with the flexibility to tune the emission wavelength over half of the visible spectrum from green to red. Continuous progress in epitaxial growth, processing technology, and the design of structures for effective

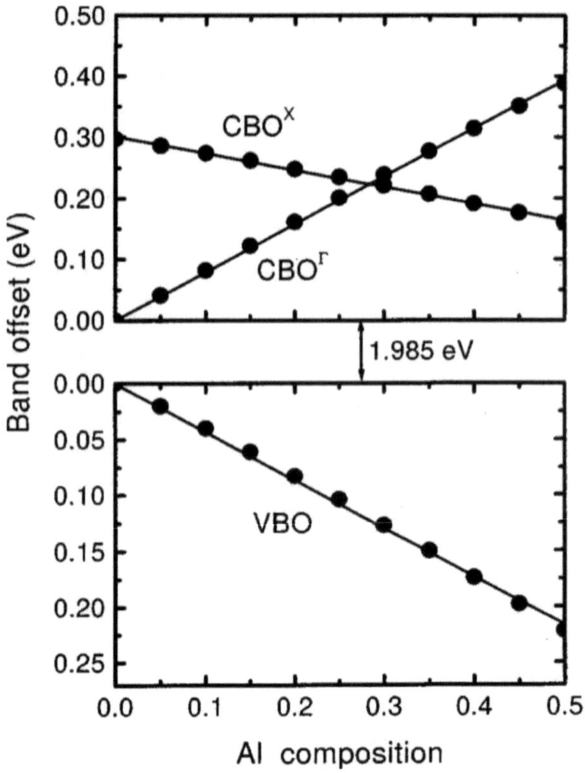

Fig. 5.3 Composition dependence of the valence- and conduction-band offsets at $Ga_{0.5}In_{0.5}P/Al_x Ga_{0.52-x}In_{0.5}P$ (001) heterostructures

light extraction enabled the fabrication of AlGaInP LEDs with record high external efficiencies above 50%. The possibility to efficiently generate light in frequently used colors like yellow, orange, or red accelerated the use of AlGaInP LEDs in many applications such as interior and exterior automotive lighting, traffic lights, full-color displays, or all kinds of indoor and outdoor signs.

As a metastable phase material without its own substrate, AlGaInP usually has been grown on the GaAs substrate. Thus the lattice matching between AIGaInP and GaAs substrate is the most important. Commonly before the epitaxy of AlGaInP layer, GaInP and AlInP materials were firstly deposited on GaAs substrates, respectively, according to the linear interpolation method. At the molar In composition of 50%, $(Al_xGa_{1-x})_{0.5}In_{0.5}P$ lattice matched with GaAs at room temperature; however, the difference of thermal expansion coefficient between AlGalnP and GaAs should be also considered. And the molar In composition x is usually set slightly larger than 0.5, which benefits for better crystal quality.

P-type and n-type doping are crucial for AlGaInP/AlGaAs material. And the control of doping concentration, uniformity, and repeatability plays a key role in the improvement of yield and performance of devices. Si_2H_6 and Cp_2Mg are commonly used as n-type and p-type dopants, and the doping concentration is controlled by adjusting the dopant flux. In addition, the doping of Mg is more complicated than that of Si. From the larger range of growth conditions, the hole concentration of materials is nonlinear with the flow of Cp_2Mg and is affected by temperature and growth speed, so the significant change of growth conditions should be avoided.

High-quality A1GaInP/AlGaAs bulk layers as well as quantum well (QW) structures have been grown by LP-MOVPE on (100) offset 6°–15° toward (111)B-oriented GaAs substrates. TMGa, TMAl, and TMIn were used as MO sources, and Cp_2Mg and SiH4 were used as p-type and n-type doping precursors. 100% arsine and phosphine as the group V element sources, the growth temperature was around 700 °C, with the reactor pressure of 50 mbar and the V/Ill ratio in the gas phase around 300 in hydrogen atmosphere. Especially emission wavelength blue shifting of AlGaInP epitaxial layers, including MQW structure, were observed on GaAs substrate orientation offset from (100) toward (311)B, this phenomenon is caused by the (111) ordering on the group III sublattice [11]. Now by the state of the art (311)B or (511)B orientation GaAs substrate were used widely for AlGaInP LED.

5.3 AlGaInP/AlGaAs LED Structure Design and Manufacture

Owing to the high refractive index of semiconductors, light incident on a planar semiconductor-air interface is totally reflected, if the angle of incidence is sufficiently large. As a result of total internal reflection, light can be "trapped" inside the semiconductor. Light trapped in the semiconductor will eventually be reabsorbed, e.g., by the substrate, active region, cladding layer, or metallic contact.

The external quantum efficiency (EQE) of an LED is the product of the internal quantum efficiency (IQE), $\eta_{\rm iqe}$, and the light extraction efficiency (LEE), $\eta_{\rm lee}$, i.e.,

$$\eta_{\rm eqe} = \eta_{\rm iqe} \times \eta_{\rm lee}$$

The LEE plays an important role in increasing the power efficiency of LEDs.

5.3.1 Bragg Reflector and Textured Chip Surfaces

Subsequent to the AlGaInP laser development that occurred in the early 1980s, AlGaInP LED development started at the end of the 1980s and early 1990s [9, 10, 12, 13]. These LEDs were grown in a p-side-up configuration on conductive n-type GaAs substrates. Except to the AlGaInP lasers, LED structures employ current-spreading layers. The effect of the current-spreading or light attractive window layer is shown in Fig. 5.4a [14].

Without a current-spreading layer, the current is concentrated under the center of the top contact. Because most of the light is generated in the region below the opaque

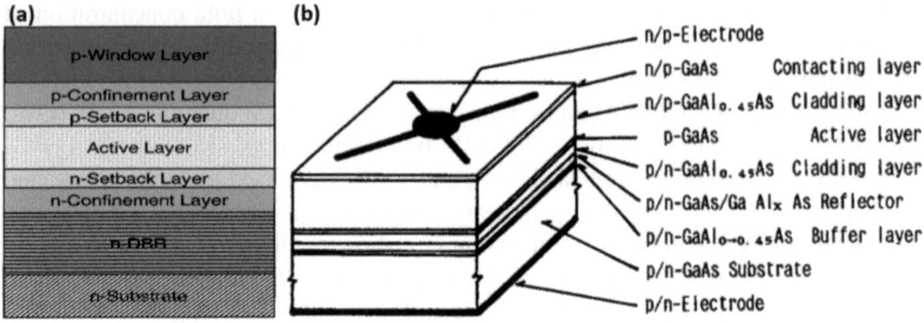

Fig. 5.4 (**a**) Schematic drawing of the layer structure of a typical high-brightness LED, (**b**) structure of DH-LED with Bragg reflector [14]

top contact, the light extraction efficiency of AlGaInP LEDs without a current-spreading layer is very low. For a sufficient thick and conductive current-spreading layer, the entire p-n junction plane of the LED chip lights up and not just the region below the top ohmic contact. It is desirable to spread the current beyond the contact area by a distance larger than the contact radius. However, spreading the current all the way to the edge of the chip could result in unwanted surface recombination.

In order to improve the extraction efficiency of AlGaInP LED based on absorbing GaAs substrate (referred as AS-substrate), a distributed Bragg reflector DBR is included between the substrate and the lower cladding layer shown as in Fig. 5.4b [15].

As shown in Fig. 5.5, a new structure of high-brightness light-emitting diodes (LED) is experimentally demonstrated. The thin window layer is composed of a 300-nm-thick indium-tin-oxide layer and a 500-nm-thick GaP layer for both current spreading and light antireflection. The two coupled distributed Bragg reflectors (DBRs) with one for reflecting normal incidence light and the other for reflecting inclined incidence light which is emitted to the GaAs substrate are employed in the LED fabrication. The coupled DBRs in the LED can provide high reflectivity with wide-angle reflection. With the injection current increasing, the luminance intensity of device A shows the substantial superiorities over device B; the luminous flux of device A is about 1.79 lm and of device B is 1.32 lm with the saturation currents 130 mA and 110 mA, respectively. The luminance efficiency of device A is larger than that of device B, and the curve of device A is slowly declining with the current increasing, which indicates that light generated inside device A would emit outside much more than device B. This is mainly attributed to the thin window layers and the coupled DBRs, which have the advantages of less optical absorption, better current spreading, and larger antireflection for the light emitting outside [16].

As shown in Fig. 5.6, metal-organic chemical vapor deposition (MOCVD)-based aluminum gallium arsenide (AlGaAs) used as the bottom window (BW), which was inserted between the light-emitting diode (LED) structure and the absorbing substrate, has been proposed to improve the extraction efficiency of 630 nm AlGaInP LEDs. In an AlGaInP LED with this AlGaAs BW, enhanced light

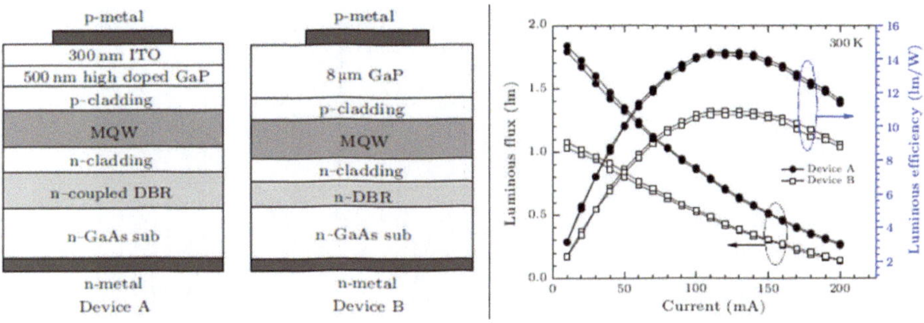

Fig. 5.5 Schematic view of cross sections of the AlGaInP LED studied, luminance intensity and luminance efficiency versus dc current for the LEDs studied [16]

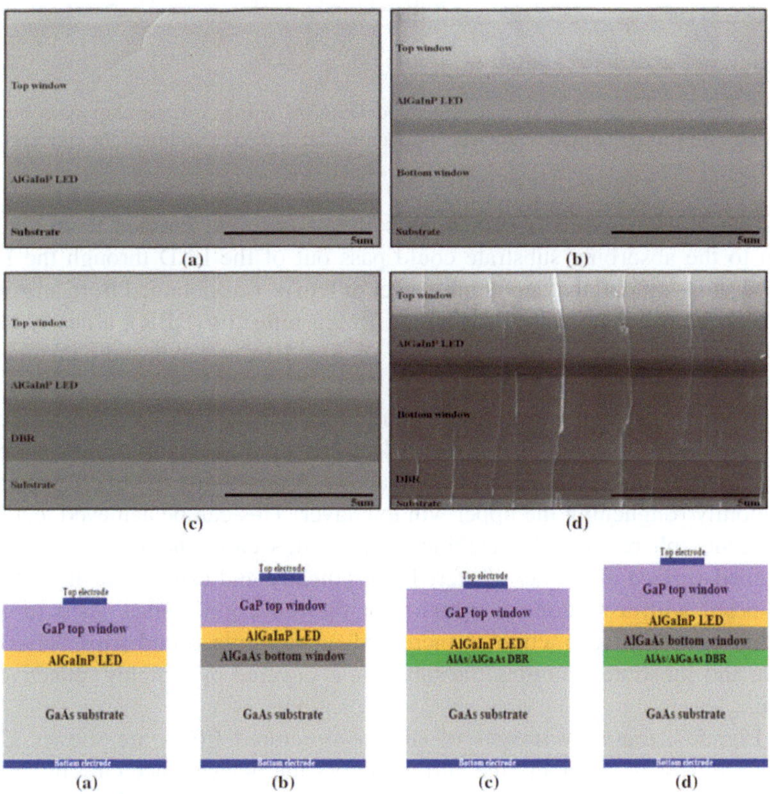

Fig. 5.6 (Color online) SEM images (upper panels) and schematics (lower panels) of structures of (**a**) conventional AlGaInP LED and AlGaInP LEDs with (**b**) BW only, (**c**) DBR only, and (**d**) both BW and DBR [17]

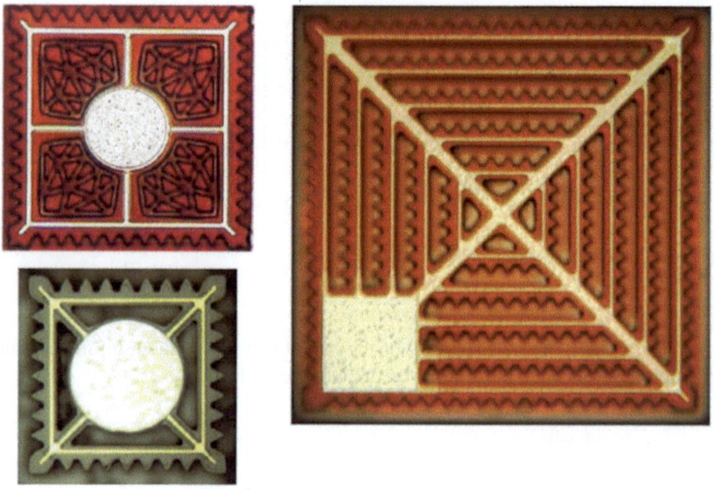

Fig. 5.7 Top-view images of surface-structured OSRAM chips. Various chip sizes have been fabricated with similar luminous efficiencies, demonstrating the scalability of the concept [19]

extraction efficiency was observed, as some of the light emitted from the active region to the absorbing substrate could pass out of the LED through the BW. In addition, it was found that an output power of 8 mW was obtained from an AlGaInP LED with both a BW and a distribution Bragg reflector (DBR), a nearly twofold improvement of over 4.2 mW that was obtained from a conventional one at an injection current of 80 mA [17].

Chip surfaces may be textured in order to increase the light extraction efficiency. An advantage of this approach is that it leads to a scalable chip design as opposed to the chip-shaping techniques discussed above. A possibility of structuring consists of randomly roughening the upper window layer. This can be achieved, e.g., using polystyrene spheres and dry etching [18]. In this case, incident photons at the roughened surface are either scattered isotropically and can escape from the chip or reflected with a random distribution of the reflection angles. As a result, the probability of light extraction is increased on multiple round trips between the surface and substrate assuming the substrate is covered by a highly reflective DBR mirror.

In Fig. 5.7, top-view images of surface-structured LEDs are shown. Contact frames are used to distribute the current across the device since current spreading in the window layer is strongly suppressed underneath the etched structures. Also, current densities are increased in the light extraction structures, enhancing the efficiency further. One advantage of the concept is that it leads to a completely scalable chip design, whereas other advanced extraction mechanisms require certain ratios between chip size and thickness for maximum efficiency. Larger chip sizes can simply be obtained by adding more contact frames with attached light extraction structures.

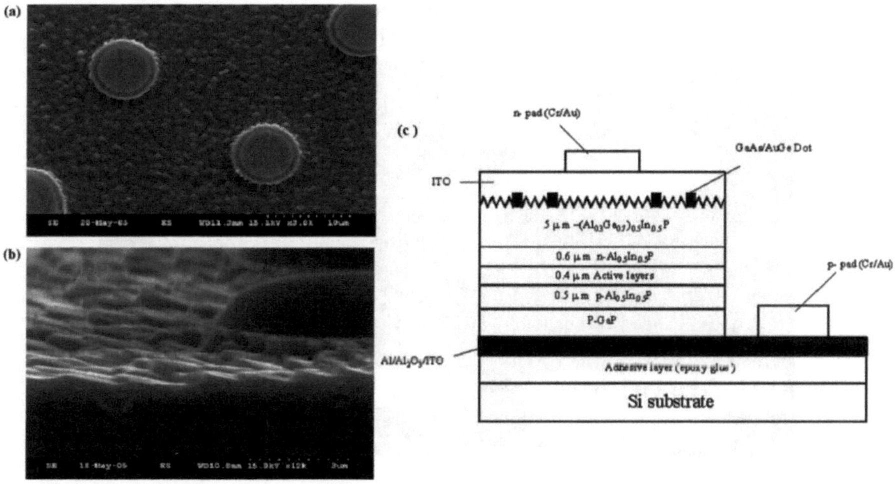

Fig. 5.8 SEM images of an n-side roughened-surface AlGaInP LED. (**a**) Top view and (**b**) cross-sectional side views. (**c**) Schematic cross section of a AlGaInP LED with n-side surface roughening

An n-side-up AlGaInP-based light-emitting diode (LED) with a triangle-like surface morphology was fabricated using the adhesive layer bonding technique, followed by wet etching to roughen the surface. The light output power of the roughened-surface LED was 1.6 times higher than that of a flat-surface LED at an injection current of 20 mA, i.e., a significant improvement attributed to the ability of the roughened surface to not only reduce the internal reflection between the rear mirror system and the semiconductor-air interface but also to effectively scatter the light outside the LED device. Figure 5.8c schematically depicts a cross-sectional image of AlGaInP LED with n-side surface roughening. Figure 5.8a shows scanning electron micrograph (SEM) images of the top and cross-sectional side views of the AlGaInP LED after H_3PO_4:HCl = 5:1 for 40 s of etching time. According to Fig. 5.8b, the roughened-surface morphology of the n-side-up AlGaInP-based LED displays a triangle-like feature, which tilts toward a specific direction associated with the lattice orientation. This feature could be related to the surface polarity of AlGaInP since this roughening feature can only be observed through wet etching only on the n-side-up surface of AlGaInP material after the GaAs substrate is removed. Our laboratory is currently investigating the detailed mechanism of forming the roughened surface [20].

5.3.2 *Transparent Substrate*

As shown in Fig. 5.9, to further eliminate the absorbed GaAs substrate affect, the transparent substrate (TS) was employed [21]. In this letter, they report the develop-

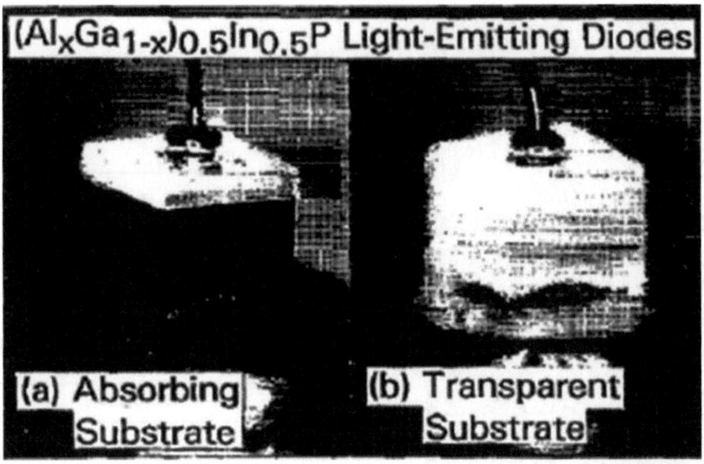

Fig. 5.9 Photomicrographs of (**a**) absorbing-substrate (GaAs) and (**b**) transparent-substrate (GaP) AlGaInP LED chips operating at 50 mA (dc) with an emission wavelength of λ ~ 600 nm [21]

ment of a new family of transparent-substrate (TS) $(Al_xGa_{1-x})_{0.5}In_{0.5}P/GaP$ LEDs whose luminous performance exceeds that of all other current LED technologies in the green to red (560–630 nm) spectral regime by at least a factor of 2. The maximum luminous efficiency of these devices is 41.5 lm/W at λ ~ 604 nm (20 mA, direct current). These devices are fabricated by semiconductor wafer bonding a "transparent" n-type GaP substrate to a p-n $(Al_xGa_{1-x})_{0.5}In_{0.5}P$ double heterostructure (DH) LED after selective removal of the n-type GaAs-absorbing substrate employed for lattice-matched growth. And in 1996, visible-spectrum transparent-substrate GaP-AlGaInP/GaP light-emitting diodes fabricated with large-area wafer-bonding technology operating at 635.6 nm exhibit low voltage (<2.1 V at 20 mA) and high external quantum efficiency (23.7%) [22].

In a conventional transparent-substrate chip, external quantum efficiency is limited by internal optical losses within the LED structure [23]. These losses include the finite reflectivity of ohmic contact metallization, reabsorption by the active layer, and free-carrier absorption. For AlGaInP LEDs, a trade-off exists between active layer reabsorption and electron confinement which results in an optimum active layer thickness depending on the emission wavelength [24]. This effect is pronounced for shorter-wavelength devices ($\lambda_p < 630$ nm) wherein the required thickness for sufficient electron confinement results in appreciable photon reabsorption in the active region. Consequently, light extraction techniques that require long photon path lengths may be inferior in realizing high-efficiency devices for such material systems where significant internal loss mechanisms are present. In order to maximize extraction efficiency in such structures, techniques are required which reduce the photon path length for extraction. These improvements can be realized by modifying the geometry of the LED chip. M. R. Krames et al. reported a truncated-

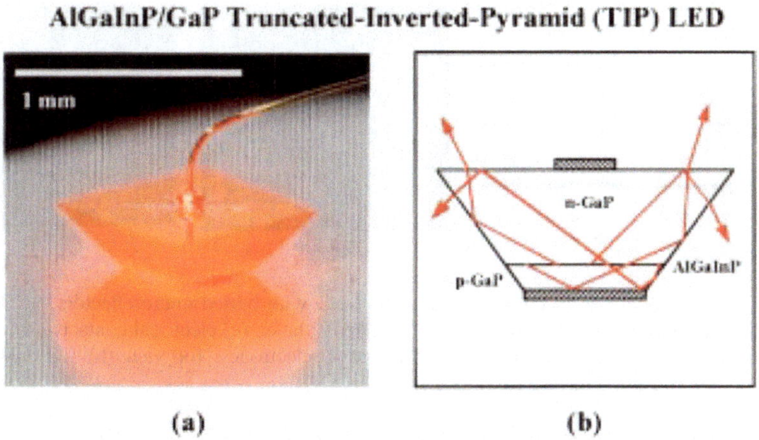

Fig. 5.10 Geometry of the truncated-inverted-pyramid LED: (**a**) photomicrograph of a TIP LED under forward bias; (**b**) schematic cross section showing means by which photons are redirected by total internal reflection [25]

inverted-pyramid (TIP) chip geometry, which provides substantial improvement in light extraction efficiency over conventional AlGaInP/GaP chips. The TIP geometry decreases the mean photon path length within the crystal and thus reduces the effects of internal loss mechanisms. In orange-emitting ($\lambda_p < 610$ nm) devices, peak efficiencies exceed 100 lm/W. And in the red ($\lambda_p < 650$ nm) wavelength regime, peak EQE of 55 and 60.9% are measured under direct current and pulsed operation, respectively. Photographs and schematic structures of the device are shown in Fig. 5.10a, b [25].

5.3.3 *Thin-Film Structure*

For traditional LEDs with AlGaInP structure, the epilayer is at first deposited on GaAs wafers, but the absorbing GaAs is subsequently removed in the fabrication process. Transfer to a new carrier can be facilitated either on chip or wafer level and usually involves an intermediate metal layer for soldering. Since the final result is a device in which the thin film of AlGaInP active material has been transferred from one wafer to another, the devices are frequently named "thin-film" LEDs. Compared to TS-LEDs with a basically isotropic emission pattern, thin-film LEDs are more directional top emitters which might be an attractive feature for many applications.

S. Illek et al. demonstrated a TF AlGaInP LED with an array of buried cone-shaped microreflectors. The LED structure is shown schematically in Fig. 5.11a–f. The cones are etched through the active layer and covered by a metal and a thin dielectric layer. Small openings in the dielectric provide for electrical conductivity

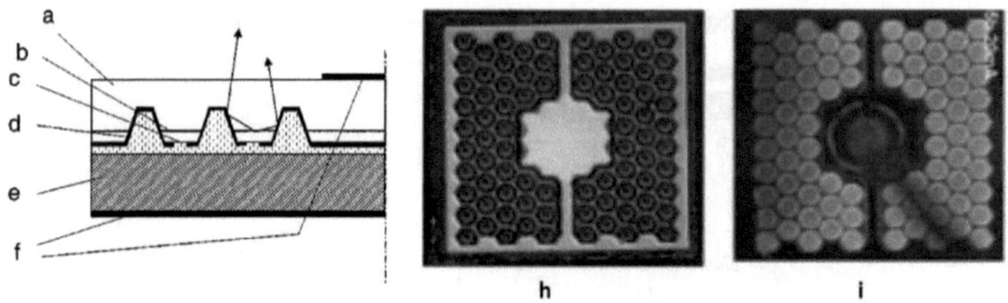

Fig. 5.11 Schematic cross section of a thin-film LED with buried microreflectors, (**a**) AlGaInP LED structure with removed GaAs substrate, (**b**) active layer, (**c**) electrical contacts, (**d**) microreflector with dielectric/metal mirror, (**e**) carrier, and (**f**) electrodes. Top view (**h**) and illumination pattern (**i**) of a 615 nm thin-film LED

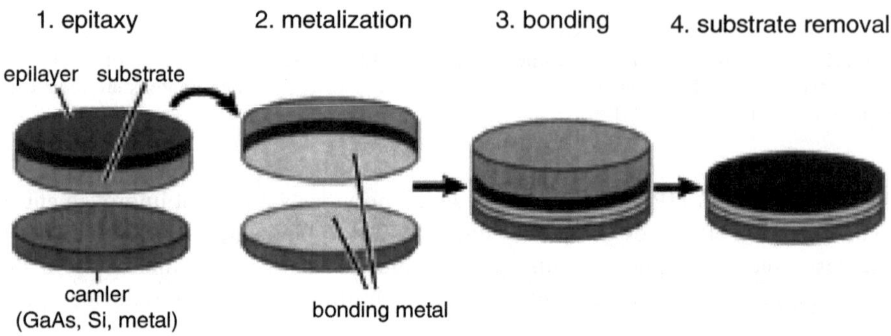

Fig. 5.12 Wafer-bonding processing steps for thin-film LED technology utilizing metal-to-metal bonding by means of soldering

and localize current injection and light generation to the center of the cones. The shape of the microreflectors is designed such that totally reflected light is guided upwards. As a result, the TF-LED with buried microreflector is predominantly top emitting. In contrast to TS-LEDs, very thick window layers in order to increase sidelight extraction are not required. Figure 5.11h shows the top view of a 615-nm device with a structured interface. Although the array of microreflectors is located underneath the planar top layers, it is clearly visible from the top. As shown in Fig. 5.11, the illumination pattern light is generated and extracted around the microreflectors, confirming the principle of operation.

Thin-film LEDs (TF-LEDs) as an alternative to TS-LEDs, offering high extraction efficiency, are made possible by TF technology. Here the epitaxial layer is removed from the original GaAs substrate and transferred to another carrier by means of metal-to-metal bonding. Stringent semiconductor-to-semiconductor wafer-bonding process parameters such as ultraflat surfaces and crystallographic orientation matching are not required for TF-LEDs shown as in Fig. 5.12 [14].

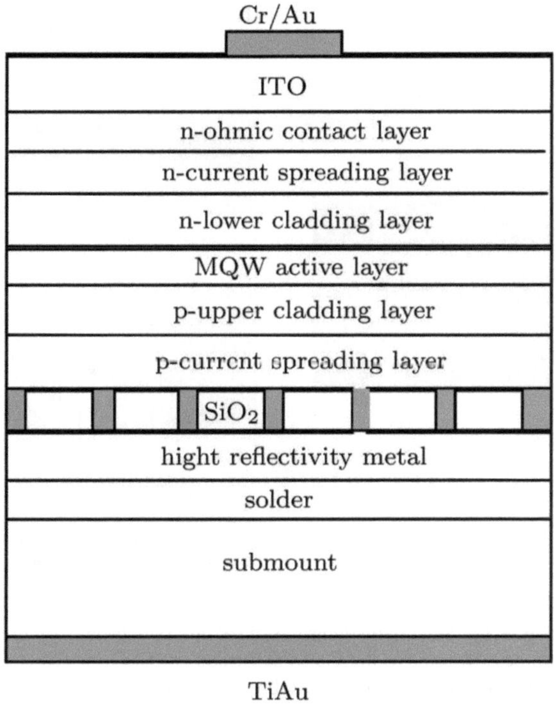

Fig. 5.13 Schematic structure of the ODR-LED [26]

High-brightness AlGaInP thin-film LED with ODR- and transparent-conducting ITO n-type contact has been fabricated by [26]. The schematic structure of the ODR-LED is shown in Fig. 5.13. The normal incidence reflectivity of ODR is 92.7% at $\lambda = 630$ nm (above 92.7% for 620–770 nm). It is found that the sheet resistance of the ITO films (95 nm) is of the order 23.5/□ with up to 92% transparency for 590–770 nm of the spectrum. ITO n-type contact can improve to spread current uniformly and enhance light output power for antireflection. The light output from the ODR-LED with ITO at forward current 20 mA exceeds that of AS-LEDs and ODR-LED without ITO by about a factor of 1.63 and 0.16, respectively. A favorable luminous intensity of 218.3 mcd from the ODR-LED with ITO could be obtained under 20 mA injection, which is 2.63 and 1.21 times higher than that of AS-LED and ODR-LED without ITO, respectively.

Bergenek et al. investigate the use of photonic crystals for light extraction from high-brightness thin-film AlGaInP light-emitting diodes with different etch depths, lattice constants, and two types of lattices (hexagonal and Archimedean). Both simulations and experimental results show that the extraction of high-order modes with a low effective index n_{eff} is most efficient. The highest external quantum efficiency without encapsulation is 19% with an Archimedean A7 lattice with reciprocal lattice constant $G = 1.5\ \text{K}_0$, which is 47% better than an unstructured reference device [27].

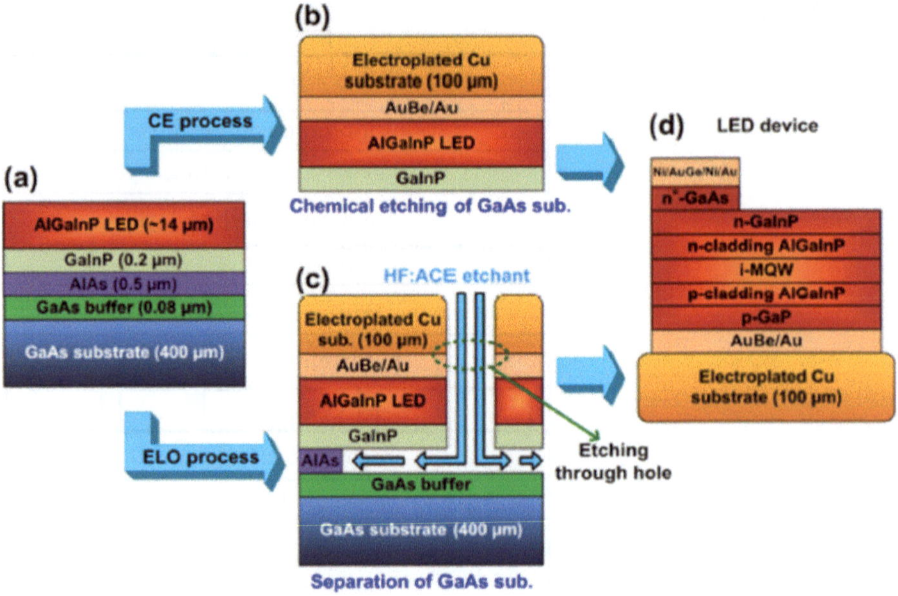

Fig. 5.14 Schematic diagrams of (**a**) epitaxial growth on the GaAs substrate, (**b**) chemical etching process to remove the GaAs substrate, (**c**) ELO process to separate the GaAs substrate, and (**d**) vertical-type AlGaInP LED fabricated on the Cu substrate [28]

Fan-Lei Wu et al. fabricated a thin-film vertical-type AlGaInP LEDs on Cu substrates. By performing the epitaxial lift-off (ELO) process, the LED device can be transferred from GaAs to Cu substrate, as shown in Fig. 5.14. It confirms that the design of patterned Cu substrate is very helpful to obtain the thin-film vertical-type AlGaInP LEDs. Additionally, via the ELO process, the separated GaAs substrate can be reused for production cost down [28].

A twice wafer-transfer technique can be used to fabricate high-brightness p-side-up thin-film AlGaInP-based light-emitting diodes (LEDs) with an indium-tin-oxide (ITO) transparent conductive layer directly deposited on a GaP window layer, without using postannealing. The ITO layer can be used to improve light extraction, which enhances light output power. The p-side-up thin-film AlGaInP LED with an ITO layer exhibited excellent performance stability (e.g., emission wavelength and output power) as the injection current increased. This stability can be attributed to the following factors: (1) refractive index matching, performed by introducing ITO between the epoxy and the GaP window layer enhances light extraction; and (2) the ITO layer is used as the current spreading layer to reduce the thermal accumulation in the epilayers. Figure 5.15 presents a flowchart illustrating the fabrication of thin-film p-side-up AlGaInP-based LEDs by using twice wafer-transfer technology [29]. A substantially higher output power of up to 181 mW at 1000 mA and high wall plug efficiency are obtained.

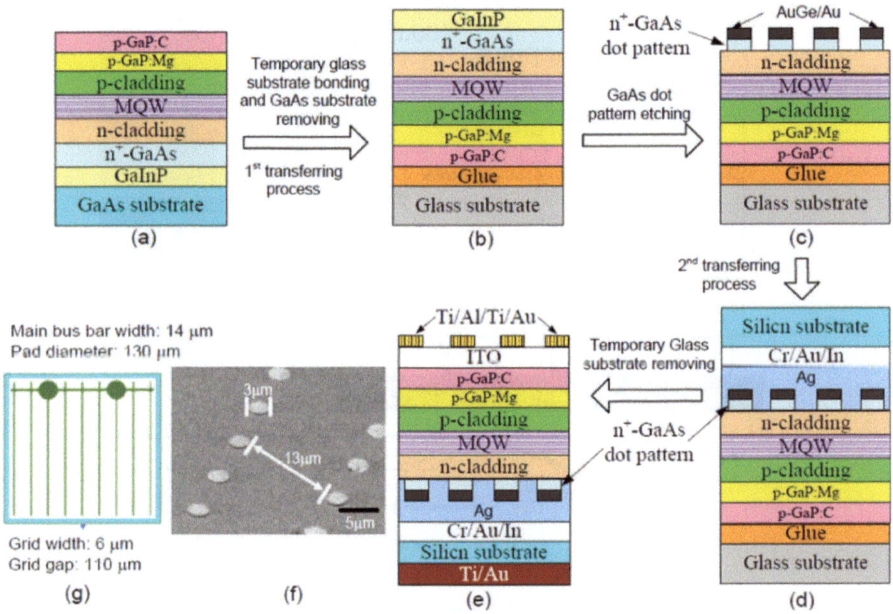

Fig. 5.15 (**a**)–(**e**) Schematic diagrams of fabrication process for p-side-up thin-film AlGaInP-based LED with ITO contact layer, (**f**) top-view image of dot-patterned GaAs on n-cladding surface for the thin AlGaInP-based LED after removing GaAs substrate, (**g**) overall macro schematic of device with detail layout

5.3.4 *Flip-Chip Structure*

Flip-chip light-emitting diode with a geometric sapphire shaping structure (GSSFC) [30] also was employed to eliminate absorbed GaAs substrate affect shown as in Fig. 5.16.

5.4 AlGaInP/AlGaAs Red/Yellow LED Application in Solid-State Lighting, Display, and Communication

5.4.1 *Application in Solid-State Lighting*

There are three common architectures for generating white light; the simulated optical power spectra for these three architectures are shown in Fig. 5.17. The phosphor-converted LED (pc-LED) is based on a blue LED to pump yellow-green and red wavelength optical downconverters (typically phosphors), thus producing white light. The hybrid LED (hy-LED) is based on a blue LED used to pump a yellow-green wavelength downconverter, and then the blue and yellow-green light

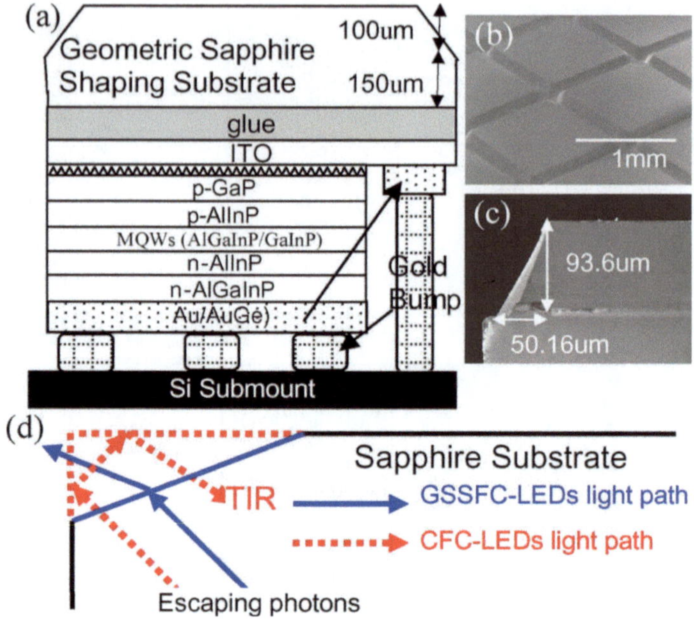

Fig. 5.16 (**a**) Schematic diagram of the AlGaInP GSSFC-LEDs structure. (**b**) SEM figures of the top view of geometric sapphire shaping substrate profile and (**c**) cross-sectional profile. (**d**) Described with the possible photons paths inside the structure of the GSSFC-LEDs and the CFC-LEDs

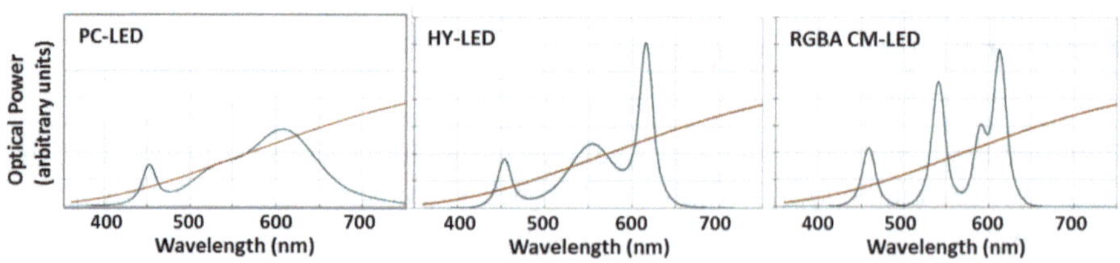

Fig. 5.17 Typical simulated optical power spectra for the three white light LED package architectures considered

is mixed with light from a red LED to again produce white light. The primary colors that compose a red, green, blue, and amber (RGBA) color-mixed LED combine to produce white light.

The current state-of-the art commercial pc-LED, with a luminous efficacy of approximately 137 lm/W, is about 33% efficient. Because of the fundamental Stokes efficiency loss associated with this architecture, 100% efficiency is not possible. Even if all other losses were eliminated, the current pc-LED with its current spectral distribution of optical power can at most have luminous efficacy of approximately 220 lm/W (LER × Stokes losses). This upper pc-LED potential can be considered approximately equal for warm and cool white. The current 100 nm FWHM wide red phosphor emission linewidth causes a significant spillover of light into the deeper red, where the human eye is less sensitive, and is a significant contributor to the spectral inefficiency of current pc-LED white light.

Optimization of the lighting quality was aimed at maximizing either luminous efficacy of radiation (LER) or color rendering index (CRI). In this case, people operated with the emission spectra only and did not account for the efficiencies of individual emitters, which could depend strongly on their emission wavelengths. Already in earlier studies, a trade-off between LER and CRI has been revealed, i.e., LER could be maximized at the expense of CRI and vice versa shown as in Fig. 5.17 [31, 32].

Figure 5.18a shows that the white light source comprising of two LEDs and the YAG-phosphor enables achieving the efficacy of ~215 lm/W at CCT = 2400 K and CRI 70–75. At higher CRI values of ~90–95, the efficacy slightly decreases down to ~195–200 lm/W. It should be noted that a very similar solution for the color mixing but with the use of a different phosphor has been suggested earlier at Osram within the Brilliant Mix concept. So high efficacy predicted for the warm white light is achieved due to a rather high "phosphor efficiency" of ~55% compared to WPE of LEDs emitting at the wavelengths close to 560 nm, i.e., deep in the "green gap." At that, the optimal emission wavelengths of two LEDs lie beyond the "green gap" of the LED efficiency (see Fig. 5.18c). Similar phosphor-converted sources, being optimized for generation of cool white light (CCT > 5000 K), provide remarkably lower values of efficacy and CRI (see Fig. 5.18a). In the case of white light source consisting of LEDs only, the transition from three optimized LED emitters (CRI < 92) to four ones (CRI > 92) was found to occur at CRI $\approx$ 92 irrespective of the correlated color temperature (CCT) of white light [33].

Optimal number of four for individual emitters with almost monochromatic emission spectra and their optimal emission wavelengths was recommended [34, 35] to maximize LER of white light sources at rather acceptable CRI values of ~80–90. Similarly to optimal LER, there is a distinct transition from the optimal number of LEDs of three (two nitride and one phosphide LEDs) to four (three nitride LEDs and one phosphor) occurring at CRI $\approx$ 92. However, the general behavior of efficacy difference considerably from that of LER is there is a maximum of the efficacy achieved at CRI of ~88–91. At CRI > 92, where four LEDs come into play, the efficacy drops down dramatically. The reason for the drop can been understood from the SPD spectra of the white light sources (Fig. 5.18c).

Quite different results have been obtained by optimization of the light sources consisting of three LEDs and the YAG-phosphor (Fig. 5.18b). In the case of a warm white light (CCT = 2400–3000 K), the maximum efficacy of ~180 lm/W is lower than in combination of two LEDs and phosphor. The reason for this is a lower WPE of the additional LED having the emission wavelength situated inside the "green gap." In the case of the neutral white light (CCT = 4000 K), however, the efficacy of ~190–195 lm/W at extremely high CRI of ~97 can be obtained. In this case, the optimal wavelength of the additional LED moves to the periphery of the "green gap" (Fig. 5.18d), thus increasing the efficacy of the white light source. Comparison of the optimization results obtained for the white light sources consisting entirely of

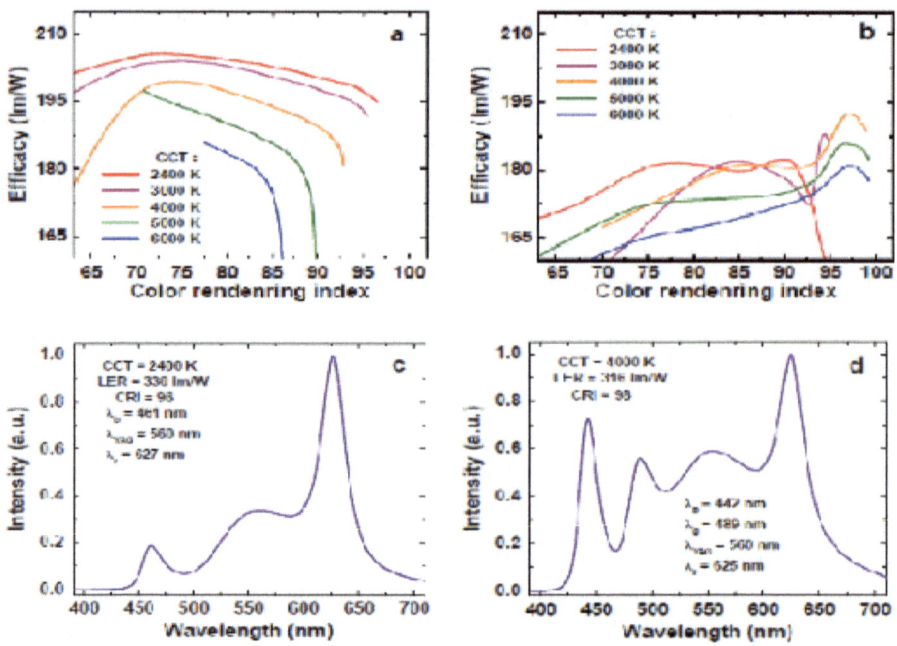

Fig. 5.18 Optimized efficacy of a light source comprising of a YAG:Ce^{3+} phosphor pumped by a blue LED and two (**a**) or three (**b**) LEDs as a function of CRI. Total emission spectra of optimized white light sources comprising of two LEDs and the YAG-phosphor (**c**) or three LEDs and the YAG-phosphor (**d**)

LEDs and those utilizing partial light conversion by a YAG:Ce^{3+} phosphor enables one to conclude that the latter scheme of the color mixing provides at the moment a systematically higher, by ~10%, efficacy at higher CRI values (~90–97 versus ~86–90). The major reason for this benefit is a high overall "phosphor efficiency," which exceeds remarkably WPEs of both nitride and phosphide LEDs in the "green gap" shown as in Fig. 5.19a [33]. In Fig. 5.19b, dashed lines are guides to the eye, illustrating the "green gap": the decrease in efficiency from the blue to the green-yellow and from the red to the green-yellow [36].

One can see that in the case of three LEDs, only the optimal wavelength $\lambda_g = 532$ nm gets deep into the "green gap" of LED efficiency, whereas in the case of four LEDs, both $\lambda_g = 514$ nm and $\lambda_y = 566$ nm are situated inside the gap. Because of a low efficiency of the green-gap LEDs, the latter leads to a dramatic efficacy reduction when four LEDs are utilized in the white light source.

Therefore, the phosphor-converted LEDs will be advantageous over the light sources mixing light from LEDs only, until the "green gap" problem is cardinally resolved. One more conclusion following from the optimization results is that the efficacy of ~200 lm/W at an excellent color rendition (CRI ~ 95) is feasible for phosphor-converted LEDs. So for the high-quality lighting source, AlGaInP LED now gives the most important contribution of low CCT application at the wavelength

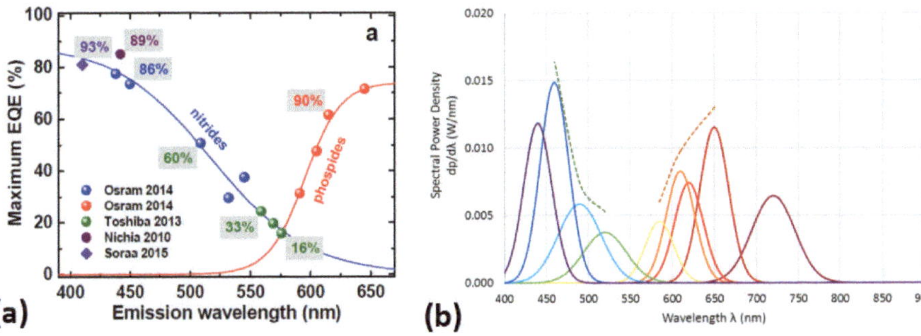

Fig. 5.19 (**a**) Green gap between the GaN and AlGaInP base LEDs as the state of the art, (**b**) spectral power densities of state-of-the-art commercial LEDs vs. wavelength

of 630 nm and for the excellent color rendition (CRI) to solve the green gap problem with the wavelength 566 nm.

As shown in Fig. 5.20a, relative LER is higher with narrower red linewidth, increasing by 15%, from 80% to 95%, as the linewidth decreases from the current 100 nm FWHM to 35 nm FWHM. It is important to note that the improvement continues as linewidth continues to narrow to even less than 35 nm, with no penalty in color-rendering quality. The challenge is thus to develop new red downconverters—phosphors, quantum dots, etc.—with narrower emission linewidth while maintaining high (greater than 90%) internal radiative quantum efficiency. As narrower linewidth red wavelength downconverters are explored, their center emission wavelength is also important. As can be seen in Fig. 5.20b, relative LER is higher the closer the center emission wavelength is to 614 nm. A center wavelength of 623.5 nm would incur a 5% efficiency penalty, and a center wavelength of 630 nm would incur a 10% efficiency penalty.

The hy-LED has a significant efficiency advantage over the more standard pc-LED architecture because the red LED incurs no Stokes deficit in generating red light, and red LEDs have intrinsically narrow linewidths with little spillover into the deep red where the human eye is relatively insensitive. A luminous efficacy of about 280 lm/W, or an efficiency of around 68%, is considered to be the hy-LED upper potential. In contrast, the pc-LED upper potential is only 255 lm/W and could be closer to 220 lm/W if narrower red-phosphor linewidths are not achieved. However, the hy-LED also has two major disadvantages, both associated with the AlInGaP technology used for the red LED. First, the thermal efficiency droop associated with these AlInGaP-based red LEDs is much greater than that associated with InGaN-based blue LEDs. Their different thermal behavior requires a control system to maintain a consistent color point, which adds complexity and cost to the lighting system. Second, AlInGaP-based red LED efficiencies decrease the shorter their red wavelengths, as illustrated in the spectral power densities for various LEDs in Fig. 5.20b. At 614 nm, which can be considered the ideal red peak for lighting as it is just long enough to provide good color-rendering quality but just short enough for good

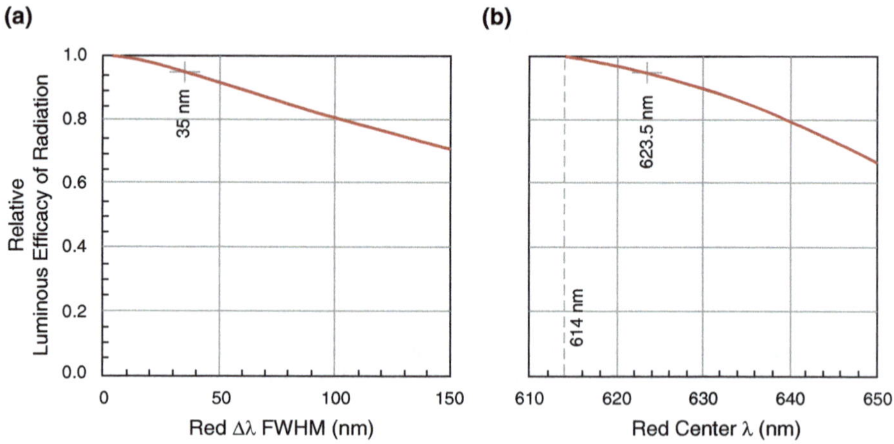

Fig. 5.20 Relative white light luminous efficacy of radiation (**a**) as the FWHM linewidth of the red phosphor increases, for a given red center wavelength of $\lambda = 614$ nm, and (**b**) as the red center wavelength increases, for a given FWHM linewidth of $\Delta\lambda = 7$ nm

sensitivity by the human eye (reasonably high LER), state-of-the-art research LED external quantum efficiencies (EQE) are only about 25%. This architecture faces the challenge of improving the red LED.

Red LED efficiency improvement is an important but tough challenge. Replacing even an ideal narrow linewidth red phosphor in a pc-LED architecture with high-efficiency red LED at 614 nm in a hy-LED architecture would enable approximately 10% improvement, and replacing a non-ideal wide-linewidth red phosphor in a pc-LED architecture would enable approximately 25% improvement. The key challenge is to overcome what appear to be fundamental limits associated with AlInGaP materials: an unfavorable band structure in the shallow red both for carrier transport/confinement and radiative carrier recombination (due to a direct to indirect bandgap crossover). A novel variant of AlInGaP, or a different material system entirely (e.g., InGaN), may provide a solution. The full exploitation of composition and band-structure engineering in semiconductor materials is often limited by strain issues associated with lattice mismatches to common substrates; however, recent research breakthroughs may have overcome these issues including metamorphic epitaxy and nano-compliancy. The development of novel substrates that are lattice-matched to material compositions of interest for 614 nm red LED emission may also reap benefits [38].

5.4.2 Application in Display

The first true all-LED flat panel television screen was possibly developed, demonstrated, and documented by James P. Mitchell in 1977. Inorganic light-emitting

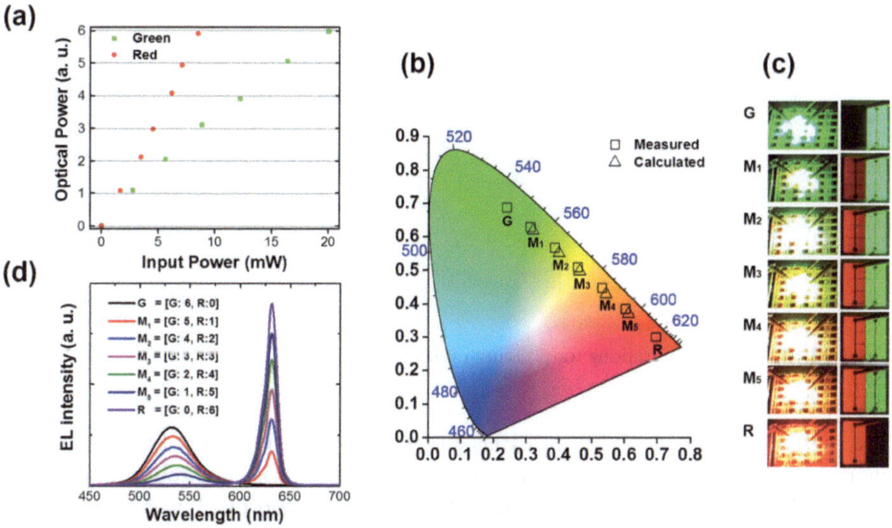

Fig. 5.21 Performance of additive color mixing on the dual color LEDs having a LAS-type array structure realized by adhesive bonding: (**a**) optical output and input powers tuned in the green and red LED subpixels, and (**b**) CIE color coordinates, (**c**) photographs (left) and microscope images (right), and (**d**) EL spectra of the dual color LEDs for the seven color modes. For reference, the CIE coordinates for the seven color modes were calculated using equation

diodes (LEDs) are the brightest and most efficient and stable light source for displays [39–41]. In the current display industry, inorganic LEDs are mainly used as backlights for thin-film-transistor liquid-crystal displays or red-green-blue (RGB) subpixels for outdoor LED displays. From 2016, as the interest in microdisplays such as the ones used in smartphones, smart watches, and head-mounted displays (HMDs) has increased, much effort has been made to apply highly efficient inorganic LEDs as their micro-display light source.

Chang-Mo Kang et al. introduced adhesive bonding and a chemical wet etching process to monolithically integrate two materials with different bandgap energies for green and red light emissions. As shown in Fig. 5.21, the structure can emit various colors by integrating both AlGaInP-based and InGaN-based LEDs onto one substrate which could be a solution to achieve full color with high resolution; the dual-color LEDs integrated by the bonding technique were tunable from the green to red color regions (530–630 nm) as intended [42].

5.4.3 Application in Communication

Visible light is the more suitable for indoor optical wireless communication (OWC) compared to the popular IR band which is preferable for long-haul communication.

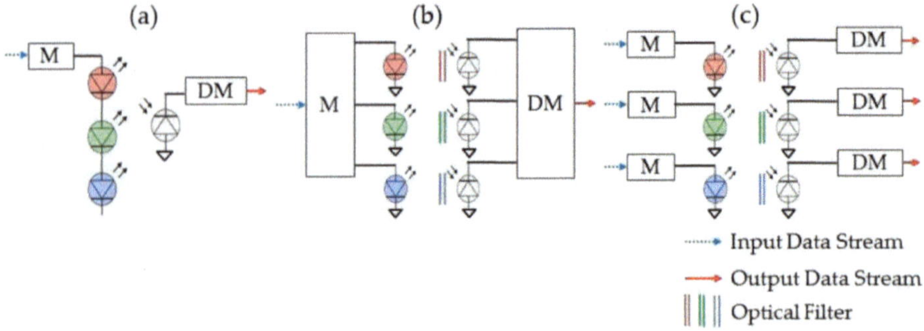

Fig. 5.22 (**a**)–(**c**) Configurations for utilizing multiple light sources in VLC

The high switching capability of LEDs along with other important features such as energy efficiency and longer lifetime makes them the most favorable light source that can be incorporated into visible light communication (VLC). At present, the arrangement of various types of LED is available, each having its own characteristics and unique properties. Based on the features, they can be used in different types of lighting applications. The potential market growth of LED lighting technology is foreseen to be very strong in the coming years, which creates a strong case for this lighting technology to be integrated into VLC [43].

The "multi-chip" approach utilizes three or more LED chips emitting different colors, typically red, green, and blue (RGB), to produce white light. Depending on the light intensities of the different chips, color control can be achieved. Multi-chip WLEDs have lower color rendering index (CRI) than pc-LEDs. pc-LEDs are cheaper and less complex compared to these multi-chip LEDs; however, they have a bandwidth limitation due to the low phosphor conversion efficiency. When using multi-chip LEDs in VLC, there are three ways to utilize the chips: connecting the pins in series to modulate all sources at the same time (Fig. 5.22a), independently modulating each chip for a single communication channel (Fig. 5.22b), and independently modulating each chip enabling parallel communication channels (Fig. 5.22c).

These methods are depicted in Fig. 5.22; the figure "M" denotes modulation and "DM" denotes demodulation. In the first method, we see a considerable bandwidth enhancement compared to pc-LEDs, since there is no slow phosphor component in the generated light. Optical filtering is used in pc-LED systems to suppress the slow phosphor decay and extract the much faster blue response, but this comes at the cost of losing a considerable portion of the received signal power. However, in multicolored sources, almost the entire optical power of the emitted spectrum for each channel can be harnessed by optical filtering. This is because no phosphor wavelength converters are present. The second method is utilized in the newly introduced color shift keying (CSK) modulation by IEEE. The intensity of each color channel generates a specific color point in the CIE 1931 color coordinates. This can be utilized as constellation points for modulation purposes. Optical filters are required at the receiver to extract the intensities of each color channel to generate

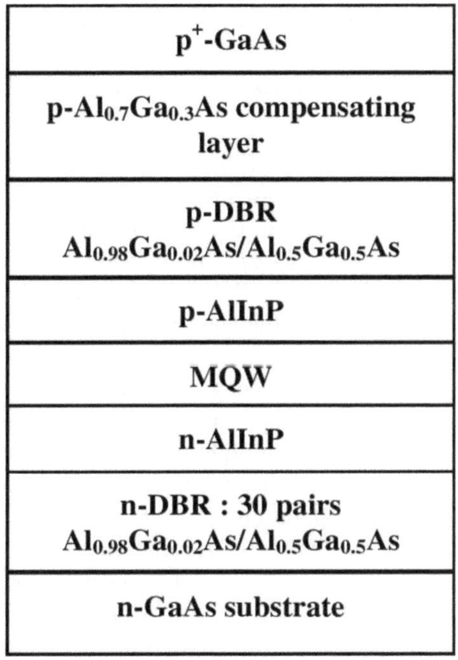

Fig. 5.23 Schematic diagram of AlGaInP-based LEDs for VLCs [46]

the color coordinate. This method is discussed elaborately in Section IV-F. In the third method, each color can be modulated independently, provided white color balancing is maintained. At the receiver, optical filtering is used to extract data from each color channel. This system has the potential for wavelength division multiplexing (WDM) [44, 45].

In 2008, Oh et al. investigated the effects of reflectivity of p-DBR and the number of QWs in active layers on the spectral characteristics and optical power of RCLEDs, in order to develop high-power RCLEDs appropriate for VLCs. The schematic diagram for the AlGaInP-based LED structures for VLC application is shown in Fig. 5.23. As the reflectivity of p-DBR increased, the FWHM of EL spectrum was reduced from 12.3 to 3.6 nm, whereas the relative integrated intensity decreased from 1.0 to 0.37, which was attributed to the improvement of spectral purity of the peaks with the inphase condition. As the number of QWs decreased, the optical power increased owing to the reduction of the optical loss of the recycling light in the active region. Using the optimized structural conditions, we demonstrated RCLEDs having a modulation speed up to 130 MHz in free space, which clearly showed that the optimized RCLED structure is a promising candidate for VLCs [46].

5.5 III-Nitrides Red/Yellow LED

5.5.1 GaN-Based Yellow Light-Emitting Diodes

AlGaInN-based LED, the wavelength of the light, can cover up all visible range. Same for GaN-based LED, GaN-based LED can basically cover all the visible light wavelength range. But there is a huge deficiency for this kind of material which is for GaN-based LED, the lighting efficiency is extremely unsatisfactory when the wavelength of the light is in long range (yellow light or red light). GaN can reach an acceptable efficiency when the wavelength of the light is short. For example, the efficiency of GaN-based blue light emitter can reach 70% [47]; for green light emitter, the efficiency can also reach 30% and above [48]. This phenomenon is called "green gap," in which the efficiency of LEDs has a huge drop when the wavelength of the light emitted is beyond the green light wavelength (550 nm to 600 nm especially) [49]. That does not only happen with GaN-based LED; that phenomenon has also happened with AlGaInN-based LED.

Unlike GaP or GaAs materials, the efficiency of the GaN-based light emitter has less sensitivity to slab staggering [50]. The increase of the GaN-based yellow light emitter efficiency can bring us significant improvement in many ways. The first improvement is white light illumination industry. This improvement is revolutionary which means fluorescent powder can be removed from the lighting system by using true yellow light which can create real white GaN-based light. On the other hand, the light quality of display and white illumination can be greatly enhanced by improving the efficiency of yellow light to overcome the "green gap". Also this can be used as the reference for blue and green light emitter to boost the efficiency even further. Furthermore, the highly efficient yellow light emitter can be used as a bedding for GaN-based red light emitter to accomplish full visible light spectrum in GaN LED [51].

Figure 5.24 shows that the efficiency varies with light wavelength changes, in which there is a gap showing when the wavelength is at green light [52].

LED efficiency reduces as current increases. This can be explained by several different reasons. The first reason is that when the density of the current pass through the diode gets increased, charge carriers will leak out from the active area (lighting area) [53]. This will lower the working efficiency. The second reason is auger recombination, which is a non-radiation recombination. The rate of recombination is proportional to the cube of the charge carrier concentration. When the charge carrier becomes more concentrated, the auger recombination will happen more frequently even in a domination condition; in that case the LED efficiency will drop [54].

LED efficiency reduces as temperature rises. When the temperature of the LED increases, the internal quantum efficiency will drop. Some studies claim that the reason for the efficiency drop as temperature rises is quite complicated, which include many factor and reasons such as defect, charging the carrier, thermal stress, heat dissipation, etc. [55]. Currently the drop of efficiency as temperature of the device rises has been treated as normalcy, but GaN-based yellow light emitter has

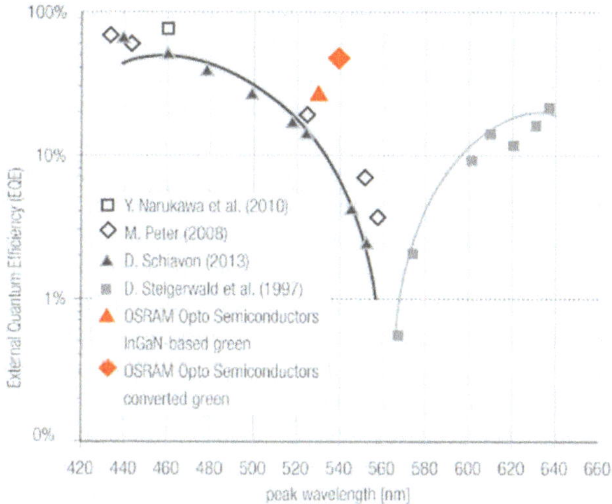

Fig. 5.24 Green gap, there are several factors that will reduce the efficiency of the GaN-based LED [52]

relatively lower temperature sensitivity compared with AlGaInP-based LED, which means the efficiency drop for GaN-based diode is less than AlGaInP-based diode if they have the same amount of temperature increase [56].

LED efficiency reduces as wavelength increases [57]. The wavelength of GaN-based LEDs can be changed by using different indium composition in the InGaN/GaN multiple-quantum-well (MQW). The higher the indium level, the longer the wavelength will be. But the problem is that the structure of InN and GaN has a significant difference in lattice constant, which means when the concentration level of indium increases, the lattice mismatch of quantum well of InGaN and GaN will be greater; this will cause an increase in stress. In this case, the piezoelectric effect of InGaN will reduce the overall efficiency. On the other hand, since both InN and GaN are non-dissolvable solids, when the concentration of In excesses a certain amount, InN will separate out that will reduce the crystalline quality and LED working efficiency. So the key point to optimize the efficiency of the long-wavelength light-emitting diodes is to reduce the stress force acting on quantum well and optimize the crystalline quality [58]. This makes the study on efficiency with wavelength droop considerably valuable and meaningful.

5.5.2 Progress on GaN-Based Yellow LED

The research of GaN yellow LED has begun very early by Nakamura in the 1990s [59]. And he successfully produced GaN-based yellow LED. But the epitaxial structure of the yellow LED is the same as both blue and green LEDs, in which yellow light is accomplished by reducing the growing temperature to increase the mixture concentration of indium. So the product he got was extremely inefficient in which the efficiency is only about 1%. And the efficiency for blue and green LEDs is 7.3% and 2.1%, respectively. For the next several decades, the efficiency of both blue and green LEDs has increased more than ten times. But the development of yellow LED technology is very limited. Also the facilities that yellow LED needs are not developed [56]. At present there are two directions to study GaN-based yellow LED. The first one is based on semipolarized or nonpolarized GaN. The main idea is to reduce piezoelectric field to increase lighting efficiency. The indium composition is relatively high in InGaN QW of Yellow LED. However, there is larger lattice mismatch between high indium composition InGaN QW and GaN carrier, which will bring compression stress to QW and leads to piezoelectric field. And piezoelectric field will separate carrier and make carrier overflow, which will reduce the efficiency. The second theory is based on low-dimensional structure of InGaN-like quantum dot or quantum wire. Using quantum dot of InGaN creates concentrated indium localized state, in which the localized state can restrict carrier effectively and polarization inside the localized state is comparatively less.

For the first theory, Funato and his colleague have successfully produced GaN-based yellow LED on (11–22) semipolar plane template in 2006 [60]. They believe that the mismatch between GaN well and GaN base should be very small, which can strongly reduce the stress inside the quantum well in order to reduce the negative effect created by piezoelectric field. But sadly the result is unsatisfactory, in which the efficiency of the new LED does not have much improvement compared with the previous LED. Then in 2008, Sato and his colleagues improved that technology by increasing the crystalline quality and reducing the dislocation density inside the quantum well [61]. The high power yellow InGaN LEDs with a peak emission wavelength of 562.7nm grown on low extended defect density semipolar (11–22) bulk GaN substrates by metal organic chemical vapor deposition. The report says when the LED is under a pulse current at 200 mA, the light intensity reaches 30 mW, which means the external quantum efficiency is nearly 6%. That is a significant improvement comparing with the previous result (1% efficiency). But the problem is that the wavelength for the light is only about 550 nm which is a bit short comparing with proper yellow light. After that, Yamamoto and his colleagues produced GaN green-yellow LED with a wavelength of 552 nm that grew at GaN (20–21) semipolarized substrate in 2010. The external efficiency for the new product can reach 12.6% in low current density condition [62]. There are several advantages for this kind of theory (produce yellow LED by using semipolarized or nonpolarized substrate). First is that the product given by homoepitaxy substrate usually has better crystalline quality. Second, the stress can be decreased since quantum well reduced when LED grows. Also the piezoelectric on the quantum well p-n junction is nearly zero. Those advantages can help boost LED efficiency. But the drawback is that the substrate is very difficult to get and the problem can be hardly avoided. The polarized c-GaN substrate is very hard to produce already, for semipolarized or nonpolarized substrate is even harder. This drawback makes the theory only be used in laboratory study. Applying the theory into scale production is hardly going to happen [56].

Since the limitation of the first theory, people are trying to find some solution by the second theory, by using quantum dot or quantum wire. In 2005, Seong-Ju Park et al. investigated the enhancement of phase separation in the InGan layer grown on a GaN layer with a rough surface, in order to form self-assembled In-rich quantum dots in the InGaN layer [63]. InGaN films were grown on both smooth and rough GaN surfaces with a root mean square (rms) roughness of 2.5 and 23.5 Å, respectively. The cross-sectional transmission electron microscopy (TEM) images of InGaN layers grown on smooth and rough GaN surfaces was analyzed by authors. The images show that the rough GaN layer induces phase separation in InGaN film even though the thickness of the InGaN layer is less than the critical layer thickness and the InGaN with a low In composition. For the sample that grew from rough surface substrate, obvious segregation of quantum well can be easily observed. By using that character, they can produce LED with obvious segregation indium in InGaN quantum well. This is a valuable beginning of the quantum dot technology; the theory provides both technological means and direction for further work.

With the progress before, Soh and his colleagues create nanometer-scaled pattern on a GaN template by etching method; then they produce InGaN/GaN quantum well on the GaN template [64]. The produced quantum well has lower edge dislocation density and less stress; also the InGaN quantum dot density can reach 4.5E9. LED with this specific structure can emit yellow light, and comparing with ordinary quantum well structure, the new structure can provide higher light intensity and less wavelength drift. After they observe the phenomenon, they create a structure to describe and explain the benefit (as shown in Fig. 5.25), which when the concentration of the InGaN is not equally distributed inside the quantum well, the area with more concentrated indium will have less energy gap in which charge carriers will be trapped in, also since the density of indium is relatively higher around the quantum well, which can create a good buffer against stress. So quantum well is a qualified lighting area.

By using this idea, Lv and his colleagues optimized the growing method by creating better growing condition for the quantum base [65], which includes using higher temperature when the base is growing and using hydrogen instead of nitrogen that can produce better base quality. That method successfully produced InGaN quantum well-based green-yellow LED with longer wavelength. But the efficiency for the device was not clearly recorded.

From previous papers, the most difficult part of InGaN quantum dot-based LED is that quantum dots need to be distributed uniformly inside the material and the crystalline quality needs to be ensured at the same time. On the other hand, the size of lighting quantum dot is limited which can hardly suffer huge current density. Also the material-generating craft is hard to control with bad stability.

The development of GaN-based yellow LED needs to utilize well-developed machines, tools, and growing crafts and use blue LED and green LED as references to innovate new material and crafts. This is a very feasible way to increase the efficiency of GaN-based yellow LED in the future [56].

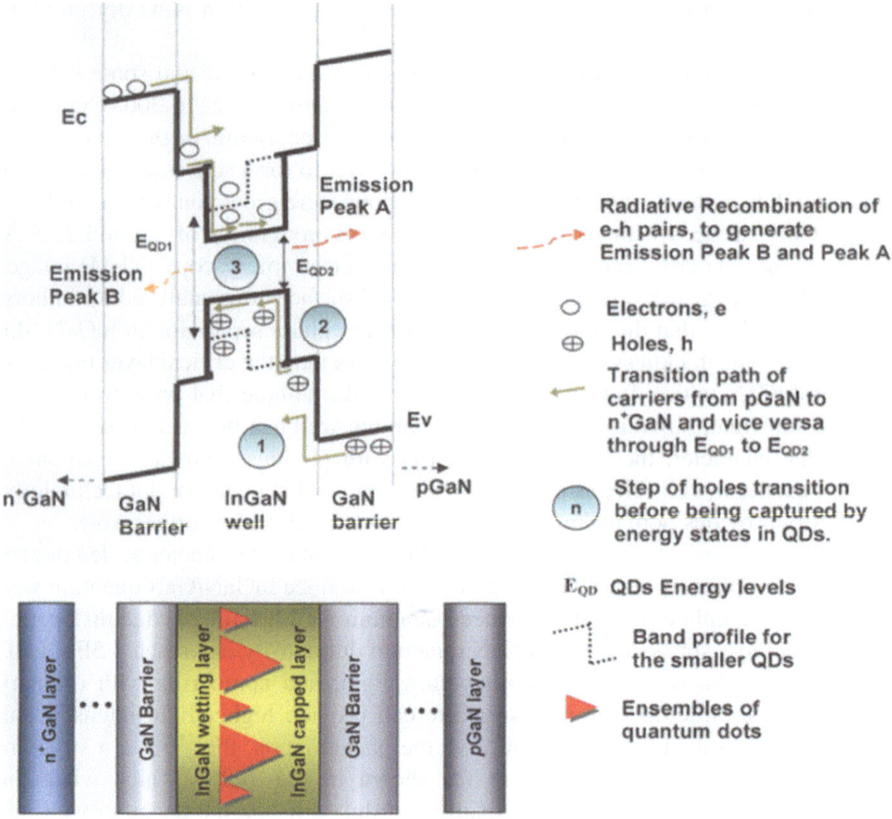

Fig. 5.25 Color online schematic band diagram of bimodal distribution of QDs embedded in the InGaN well layer of LEDs with illustration of their carrier transport kinetic [64]

5.5.3 Substrate Technique

Substrate is the fundamental of LED epitaxy growth, which will influence the quality of the crystalline quality, stress situation, and lighting method. There are several factors that need to be considered: the lattice mismatch between substrate material and epitaxy material, the chemical stability of substrate material, and the cost of substrate material. Currently the substrate material can be chosen from sapphire (Al2O3), silicon carbide (SiC), and silicon itself (Si). Also there are some other choices like GaN and AlN, but those materials are rare and have not been used in scaled production [66].

Currently the most common material for substrate is sapphire; most GaN-based LEDs on the mark were using sapphire as substrate. This is because sapphire has been used as substrate for a very long time which makes the craft technique well developed. Also sapphire has very good chemical stability. But the problem for this material is that the heat conduction is comparatively lower.

As for SiC, the lattice constant and the dilatation coefficient for SiC are similar with GaN, and also SiC contains good heat conduction and good optical properties (index of refraction is high); those advantages make SiC a quite competitive material for making substrate for GaN-based LED. But the problem is SiC is hard to produce which means the cost for SiC is comparatively higher. So this material is not currently widely used [67].

Si is a substrate material with huge potential; this is because silicon substrate is easy to produce. Due to the development of semiconductor, silicon-related material craft technique is quite mature (cheap). Also, silicon substrate is good at both heat and electric conduction and easy to assemble to electronic devices. But the problem is silicon has significant lattice mismatch and heat mismatch with GaN. At present, people are finding solutions to solve those problems. Some researchers use AlN buffer layer technology to reduce the negative influence caused by lattice mismatch [68]. As for heat mismatch problem, some designed pattern will help reduce epitaxy layer fracturing pheromone caused by heat mismatch between those two materials [69].

Since silicon as substrate has been successfully used to produce high-quality blue and green light LEDs, if the white light technique wants to be accomplished, using silicon as substrate for yellow LED seems the most suitable choice [70].

References

1. F.M. Steranka, AlGaAs red light-emitting diodes, in *High Brightness Light Emitting Diodes*, Semiconductors and Semimetals, vol. 48, (Academic, Cambridge, 1997)
2. H.C. Casey Jr., M.B. Panish, *Heterostructure Lasers, Part A and Heterostructure Lasers, Part B* (Academic, Cambridge, 1978)
3. C.H. Chen, S.A. Stockman, M.J. Peanasky, C.P. Kuo, OMVPE growth of AlGaInP for high efficiency visible light-emitting diodes, in *High Brightness Light Emitting Diodes*, Semiconductors and Semimetals, vol. 48, (Academic, Cambridge, 1997)
4. L. Vegard, Die konstitution der mischkristalle und die raumfüllung der atome. Z. Phys. **5**(1), 17–26 (1921)
5. A.R. Denton, N.W. Ashcroft, Vegard's law. Phys. Rev. A At. Mol. Opt. Phys. **43**(6), 3161–3164 (1991)
6. F.A. Kish, R.M. Fletcher, AlGaInP light-emitting diodes, in *High Brightness Light Emitting Diodes*, Semiconductors and Semimetals, vol. 48, (Academic, Cambridge, 1997)
7. X.H. Zhang, S.J. Chua, W.J. Fan, Band offsets at GaInP/AlGaInP(001) heterostructures lattice matched to GaAs. Appl. Phys. Lett. **73**(8), 1098–1100 (1998)
8. K. Kobayashi, Room-temperature CW operation of AlGaInP double-heterostructure visible lasers. Electron. Lett. **21**, 931–932 (1985)
9. C.P. Kuo, R.M. Fletcher, T.D. Osentowski, M.C. Lardizabel, M.G. Craford, V.M. Robbins, Appl. Phys. Lett. **57**, 2937 (1990)
10. H. Sugawara, M. Ishikawa, G. Hatakoshi, Appl. Phys. Lett. **58**, 1010 (1991)
11. M. Shigekazu, K. Masahiko, Characterization of OMVPE—grown A1GaInP by optical spectroscopy. SPIE Modulation Spectroscopy **1286**, 74–84 (1990)
12. R.M. Fletcher, C.P. Kuo, T.D. Osentowski, V.M. Robbins, US Patent No. 5,008,718 (1991)

13. H. Sugawara, M. Ishikawa, Y. Kokubun, Y. Nishikawa, S. Naritsuka, US Patent No. 5,048,035 (1991)
14. S. Illek, U. Jacob, A. Ploessl, P. Strauss, K. Streubel, W. Wegleiter, R. Wirth, Compound Semicond. **8**, 39 (2002)
15. T. Kato, H. Susawa, M. Hirotani, T. Saka, Y. Ohashi, E. Shichi, S. Shibata, J. Cryst. Growth **107**, 832 (1991)
16. C.H.E.N. Yi-Xin, S.H.E.N. Guang-Di, et al., Efficiency-enhanced AlGaInP light-emitting diodes with thin window layers and coupled distributed bragg reflectors. Chin. Phys. Lett. **28**(6), 067806 (2011)
17. H.J. Lee, Y.J. Kim, S.U. Kim, et al., Efficiency improvement of 630 nm AlGaInP light-emitting diodes based on AlGaAs bottom window. Jpn. J. Appl. Phys. **52**, 102101 (2013)
18. H.W. Deckman, J.H. Dunsmuir, Appl. Phys. Lett. **41**, 377 (1982)
19. N. Linder, S. Kugler, P. Strauss, R. Wirth, H. Zull, K.P. Streubel, High-brightness AlGaInP light emitting diodes using surface texturing. Proc. SPIE **4278**, 19–25 (2001)
20. Y.J. Lee, H.C. Kuo, S.C. Wang, T.C. Hsu, M.H. Hsieh, M.J. Jou, B.J. Lee, Increasing the extraction efficiency of AlGaInP LEDs via n-side surface roughening. IEEE Photon. Technol. Lett. **17**(11), 2289–2291 (2005)
21. F.A. Kish et al., Appl. Phys. Lett. **64**, 2839 (1994)
22. G.E. Hoefler, D. Vanderwater, D.C. DeFevere, F.A. Kish, M. Camras, F. Steranka, I.-H. Tan, Appl. Phys. Lett. **69**, 803 (1996)
23. I. Schnitzer, E. Yablonovitch, C. Caneau, T.J. Gmitter, Appl. Phys. Lett. **62**, 131 (1993)
24. N.F. Gardner, H.C. Chui, E.I. Chen, M.R. Krames, J.-W. Huang, F.A. Kish, S.A. Stockman, C.P. Kocot, T.S. Tan, N. Moll, Appl. Phys. Lett. **74**, 2230 (1999)
25. M.R. Krames, M. Ochiai-Holcomb, G.E. Höfler, et al., High-power truncated-inverted-pyramid $(Al_\chi Ga_{1-\chi})_{0.5}In_{0.5}P$ light-emitting diodes exhibiting >50% external quantum efficiency. Appl. Phys. Lett. **75**(16), 2365–2367 (1999)
26. Z. Jian-Ming, Z. De-Shu, X. Chen, et al., AlGaInP thin-film LED with omni-directionally reflector and ITO transparent conducting n-type contact. Chin. Phys. Soc. **16**(11), 3498 (2007)
27. K. Bergenek et al., Enhanced light extraction efficiency from AlGaInP thin-film light-emitting diodes with photonic crystals. Appl. Phys. Lett. **93**, 041105 (2008)
28. W. Fan-Lei, O. Sin-Liang, Y.-C. Kao, et al., Thin-film vertical-type AlGaInP LEDs fabricated by epitaxial lift-off process via the patterned design of Cu substrate. Opt. Express **23**(14), 18156–18165 (2015)
29. M.-C. Tseng, C.-L. Chen, N.-K. Lai, et al., P-side-up thin-film AlGaInP-based light emitting diodes with direct ohmic contact of an ITO layer with a GaP window layer. Opt. Express **22**(S7), A1862–A1867 (2014)
30. Y.C. Lee, H.C. Kuo, C.E. Lee, T.C. Lu, S.C. Wang, IEEE Photon. Technol. Lett. **20**, 23 (2008)
31. A. Žukauskas, R. Vaicekauskas, F. Ivanauskas, R. Gaska, M.S. Shur, Optimization of white polychromatic semiconductor lamps. Appl. Phys. Lett. **80**, 234–236 (2002)
32. Y. Ohno, Spectral design considerations for white LED color rendering. Opt. Eng. **44**, 111302 (2005)
33. K.A. Bulashevich, A.V. Kulik, S.Y. Karpov, Optimal ways of colour mixing for high quality white light LED sources. Phys. Status Solidi A **212**, 914–919 (2015)
34. J.M. Phillips, M.E. Coltrin, M.H. Crawford, A.J. Fischer, M.R. Krames, R. Mueller-Mach, G.O. Mueller, Y. Ohno, L.E.S. Rohwer, J.A. Simmons, J.Y. Tsao, Research challenges to ultra-efficient inorganic solid-state lighting. Laser Photonics Rev. **1**, 307–333 (2007)
35. J.Y. Tsao, M.E. Coltrin, M.H. Crawford, J.A. Simmons, Solid-state lighting: an integrated human factors, technology and economic perspective. Proc. IEEE **98**, 1162–1179 (2009)
36. Lumileds, *LUXEON Rebel Color Line Datasheet* (2017). http://www.lumileds.com/uploads/265/DS68-pdf. Accessed 31 May 2017
37. Y. Ohno, in *Color rendering and luminous efficacy of white LED spectra*. Calculations based on white light simulator (2004)
38. DOE SSL Program, in *Suggested Research Topics Supplement: Technology and Market Context*, ed. by J. Brodrick (2017)

39. T. Gessmann, E.F. Schubert, High-efficiency AlGaInP light-emitting diodes for solid-state lighting applications. J. Appl. Phys. **95**, 2203–2216 (2004)
40. G. Chen et al., Performance of high-power III-nitride light emitting diodes. Phys. Status Solidi A **205**, 1086–1092 (2008)
41. J. Day et al., III-nitride full-scale high-resolution microdisplays. Appl. Phys. Lett. **99**, 031116 (2011)
42. C.-M. Kang, S.-J. Kang, S.-H. Mun, et al., Monolithic integration of AlGaInP-based red and InGaN-based green LEDs via adhesive bonding for multicolor emission. Sci. Rep. **7**(1), 10333 (2017)
43. D. Karunatilaka, F. Zafar, V. Kalavally, R. Parthiban, LED based indoor visible light communications: state of the art. IEEE Commun. Surv. Tutorials **17**(3), 1649–1678 (2015)
44. Y. Tanaka, T. Komine, S. Haruyama, M. Nakagawa, Indoor visible light data transmission system utilizing white LED lights. IEICE Trans. Commun. **86**(8), 2440–2454 (2003)
45. J. Vucic, C. Kottke, K. Habel, K.D. Langer, in *Proc. OFC/NFOEC, Los Angeles, CA, USA*. 803 mbit/s visible light WDM link based on DMT modulation of a single RGB LED luminary (2011), pp. 1–3
46. O.H. Hwa Sub, J.O.O. Jee Hue, L.E.E. Jin Hong, et al., Structural optimization of high-power AlGaInP resonant cavity light-emitting diodes for visible light communications. Jpn. J. Appl. Phys. **47**(8), 6214 (2008)
47. C. Xinlian, K. Fanmin, k. Li, et al., Study of light extraction efficiency of flip-chip GaN based LEDs with different periodic arrays [J]. Opt. Commun. **314**, 90–96 (2014)
48. S. Shinji, H. Rei, H. Jongil, et al., InGaN light emitting diodes on c-face sapphire substrates in green gap spectral range [J]. Appl. Phys. Express **6**(11), 111004 (2013)
49. O. Jeong Rok, C. Sang-Hwan, O. Ji Hye, et al., The realization of a whole palette of colors in a green gap by monochromatic phosphor-converted light-emitting diodes [J]. Opt. Express **19**(5), 4188–4198 (2011)
50. J. Davies Matthew, D. Philip, C.-P. Massabuau Fabien, et al., The effects of varying threading dislocation density on the optical properties of InGaN/GaN quantum wells [J]. Phys. Status Solidi C **11**, 750–753 (2014)
51. J.L. Zhang, F.Y. Jiang, J.L. Liu, et al., *Study on Epitaxial Growth and Device Characterization of GaN Based Yellow Light-Emitting Diodes on Si Substrate* (Nanchang University, Nanchang, 2014), pp. 3–11
52. Steve Bush, Osram starts to bridge green gap (2015). http://en.ofweek.com/news/Osram-starts-to-bridge-green-gap-23824
53. C.H. Wang, C.C. Ke, C.Y. Lee, et al., Hole injection and efficiency droop improvement in InGaN/GaN light-emitting diodes by band-engineered electron blocking layer [J]. Appl. Phys. Lett. **97**(26), 261103 (2010)
54. G. Wei, Z. Fan, M. Morteza, et al., Auger recombination in III-nitride nanowires and its effect on nanowire light-emitting diode characteristics [J]. Nano Lett. **11**(4), 1434–1438 (2011)
55. C.H. Wang, J.R. Chen, C.H. Chiu, et al., Temperature-dependent electroluminescence efficiency in blue InGaN-GaN light-emitting diodes with different well widths [J]. IEEE Photon. Technol. Lett. **22**(4), 236–238 (2010)
56. J.L. Zhang, F.Y. Jiang, J.L. Liu, et al., *Study on Epitaxial Growth and Device Characterization of GaN Based Yellow Light-Emitting Diodes on Si Substrate* (Nanchang University, Nanchang, 2014), p. 15
57. J.R. Chen, Y.C. Wu, S.C. Ling, et al., Investigation of wavelength dependent efficiency droop in InGaN light emitting diodes [J]. Appl. Phys. B **98**(4), 779–789 (2010)
58. S. Dong-Soo, H. Dong-Pyo, O. Ji-Yeon, et al., Study of droop phenomena in InGaN-based blue and green light-emitting diodes by temperature-dependent electroluminescence [J]. Appl. Phys. Lett. **100**(15), 153506 (2012)
59. N. Shuji, S. Masayuki, I. Naruhito, et al., High-brightness InGaN blue, green and yellow light-emitting diodes with quantum well structures [J]. Jpn. J. Appl. Phys. **34**(Part 2, 7A), L797 (1995)

60. F. Mitsuru, U. Masaya, K. Yoichi, et al., Blue, green and amber InGaN/GaN light-emitting diodes on semi polar {11-22} GaN bulk substrates [J]. Jpn. J. Appl. Phys. **45**(26), L659 (2006)
61. S. Hitoshi, B. Chung Roy, H. Hirohiko, et al., Optical properties of yellow light-emitting diodes grown on semi polar (11-22) bulk GaN substrates [J]. Appl. Phys. Lett. **92**(22), 221110–221113 (2008)
62. Y. Shuichiro, Z. Yuji, P. Chih-Chien, et al., High-efficiency single-quantum-well green and yellow-green light-emitting diodes on semi polar (20-21) GaN substrates [J]. Appl. Phys. Express **3**(12), 122102 (2010)
63. P. Il-Kyu, K. Min-Ki, B. Sung-Ho, et al., Enhancement of phase separation in the InGaN layer for self-assembled in-rich quantum dots [J]. Appl. Phys. Lett. **87**(6), 061906 (2005)
64. C.B. Soh, W. Liu, H. Hartono, et al., Enhanced optical performance of amber emitting quantum dots incorporated InGaN/GaN light-emitting diodes with growth on UV-enhanced electrochemically etched nanoporous GaN [J]. Appl. Phys. Lett. **98**(19), 191906 (2011)
65. L. Wenbin, W. Lai, W. Jiaxing, et al., InGaN/GaN multilayer quantum dots yellow-green light-emitting diode with optimized GaN barriers [J]. Nanoscale Res. Lett. **7**(1), 1–8 (2012)
66. S.L. Delage, D. Christian, Wide band gap semiconductor reliability: status and trends [J]. Microelectron. Reliab. **43**(9–11), 1705–1712 (2003)
67. K. Ryosuke, K. Toshiyuki, S. Atushi, et al., Realization of extreme light extraction efficiency for moth-eye LEDs on SiC substrate using high-reflection electrode [J]. Phys. Status Solidi C **7**(7–8), 2180–2182 (2010)
68. P. Chen, R. Zhang, Z.M. Zhao, et al., Growth of high quality GaN layers with AlN buffer on Si (111) substrates [J]. J. Cryst. Growth **225**(2), 150–154 (2001)
69. L. Junlin, z. Jianli, M. Qinghua, et al., Effects of AlN interlayer on growth of GaN-based LED on patterned silicon substrate [J]. CrystEngComm **15**(17), 3372–3376 (2013)
70. J.L. Zhang, F.Y. Jiang, J.L. Liu, et al., *Study on Epitaxial Growth and Device Characterization of GaN Based Yellow Light-Emitting Diodes on Si Substrate* (Nanchang University, Nanchang, 2014), p. 16

Chapter 6
The InGaN Material System and Blue/Green Emitters

Ning Zhang and Zhiqiang Liu

6.1 Blue LEDs

6.1.1 Buffer Layer for the Growth of GaN and Growth of High-Quality GaN Materials

III-nitrides can be of wurtzite (Wz), zincblende (ZB), and rocksalt structure. For bulk AlN, GaN, and InN, wurtzite is the thermodynamically stable structure under ambient conditions, which has a hexagonal unit cell and thus two lattice constants c and a. The space grouping for the wurtzite structure is $P6_3mc$ in the Hermann-Mauguin notation. The Wz structure consists of two interpenetrating hexagonal close-packed sublattices with different types of atom, offset along the c-axis by 5/8 of the cell height (5c/8), (0001) planes of Ga and N pairs stacked in an ABABAB sequence. Atoms in the first and third layers are directly aligned with each other. Three surfaces are of special importance in nitrides, which are (0001) c-planes, (11–20) a-planes, and (1–100) m-planes. Other surfaces could also be observed in nitrides nanostructures, i.e., (1–102) r-planes and (11–23) n-planes. For epitaxy GaN thin film or single crystals, determining the orientation is not straightforward and will depend on growth conditions, substrates, or doping.

Because bulk gallium nitride crystals are still not commercially available now, gallium nitride materials are mainly grown on substrates of another material such

N. Zhang · Z. Liu (✉)
Institute of Semiconductors, Chinese Academy of Sciences, Beijing, China

State Key Laboratory of Solid State Lighting, Beijing, China

Beijing Engineering Research Center for the 3rd Generation Semiconductor Materials and Application, Beijing, China
e-mail: lzq@semi.ac.cn

J. Li, G. Q. Zhang (eds.), *Light-Emitting Diodes*, Solid State Lighting Technology and Application Series 4, https://doi.org/10.1007/978-3-319-99211-2_6

as sapphire, silicon carbide, silicon, etc. Normally, the lattice constant mismatch is the most important factor for a material as a suitable substrate for gallium nitride crystals heteroepitaxy. Moreover, properties other than the lattice constants including the material's crystal structure, surface finish, composition, reactivity, chemical, thermal, and electrical properties can also influence quality of the gallium nitride epitaxial layer; those properties of a substrate for gallium nitrate crystals should be considered roundly.

Sapphire, single-crystal aluminum oxide, was the original substrate used in Maruskas and Tietjen's pioneering study of GaN epitaxy by hydride vapor-phase epitaxy (HVPE) in 1969 [1]. GaN epitaxy on c-plane sapphire result in c-plane oriented films but with a 30° rotation of the in-plane GaN crystal directions with respect to the same directions in the sapphire. The rotation of the (0001) nitride plane with respect to the sapphire (0001) occurs to reduce the lattice constant mismatch, and the mismatch would be 30% without this rotation. However, the large lattice constant mismatch still reaches 16% [2] between the substrate and the epilayer to lead to high dislocation density in the GaN epitaxial film [2]. In order to improve the surface morphology and crystalline quality of epitaxial film, two-step growth process is widely adopted to reduce problems caused by lattice mismatch in growing gallium nitride (GaN) with metal-organic chemical vapor deposition (MOCVD) on a sapphire substrate. In 1986, Amano et al. used low-temperature AlN buffer layer to significantly improve the optical and electrical properties of the GaN layers grown on sapphire [3, 4]. And later, Wickenden et al. [5] and Nakamura [6] independently developed the use of low-temperature GaN buffer layer. Usually, before the growth of a LT buffer layer, the nitridation of sapphire is carried out to reduce the defect density, enhancing the electron mobility, and reducing the yellow luminescence in subsequently deposited films. Sapphire is nitridated by exposure to ammonia/hydrogen gas mixtures in MOCVD. After nitridation, a LT GaN or AlN buffer (usually about 500–550 °C for MOCVD) dramatically improves the surface morphology and crystalline quality of GaN sequentially deposited at HT (usually about 1100 °C for MOCVD) on sapphire. Annealing at a high growth temperature crystallizes the amorphous buffer layers into films with preferential orientation ([0001] for c-sapphire) and reduces the density of low-angle grain boundaries.

Generally, sapphire is commonly used as the substrate for III-nitride LEDs. However, the large lattice mismatch and thermal expansion coefficient mismatch between GaN and sapphire substrate cause high TDD in the range of 10^8 to 10^{10} cm^{-2} [7]. TDs are very harmful to electronic and optoelectronic devices because they act as non-radiative electron-hole recombination centers [8]. Thus, it is always desirable to reduce the number of TDs in GaN. There are several different growth approaches to reduce TDD in GaN films recently. Epitaxial lateral overgrowth (ELOG) [9] and its derivatives can significantly reduce TDD, but it inevitably needs to interrupt the growth process. Meanwhile, growing on PSS is another alternative way, and it can also reduce TDD efficiently [10]. Different ELOG techniques with ex situ substrate patterning are widely used to reduce the TDD and grow high-quality GaN layers [11–13] which have also been adapted to AlGaN layers [14, 15]. But, there are some drawbacks for ex situ ELOG because of long ex situ masking procedures and the presence of localized TTDs in the window regions. Moreover, a relatively thick overgrowth of the mask is essential to coalesce the wing area. In order to overcome the drawbacks of ex situ ELOG, in situ ELO techniques are developed, especially in small scales. In situ ELOG techniques are effective to improve the quality of GaN epilayers comparing to the ex situ ELOG techniques. Tanaka' work [16] shows that SiNx intermediate layers deposited in situ are an effective tool to grow high-quality GaNe pi layers. However, Engl et al. [17] observed that SiNx interlayers do not reveal any visible improvement in crystal quality of AlGaN layers, since AlGaN does not grow as selective as GaN [18]. Forghani et al. use the implementation of SiNx for AlGaN layers with 20% Al content and developed an appropriate model for the effect of SiNx interlayers on the reduction of edge-type TD in AlGaN [19].

Extensive research on growth conditions for high-quality gallium nitride crystals growth on sapphire has been done to reduce TDD, 3D-2D growth process [20] was applied, and cone-shaped patterned sapphire substrates were used as substrates. During the 3D growth process, growth condition was to adjust to control the facets including {1$\underline{1}$01} inclined facets and top (0001) plane. The thickness of 3D layer should be near to the height of cone-shaped patterns on which there were little GaN deposited. At the end of the process, cone-shaped patterns were surrounded by 3D layer. In the following 2D growth process, inclined side facets coalesce quickly, and the interaction of TDs with the side facets causes the TDs to bend over. Some samples of GaN films on PSS were grown by a Thomas Swan close-coupled showerhead (CCS) MOCVD where a 633 nm laser interferometer was used to monitor the growth. Trimethylgallium (TMGa) and ammonia (NH3) were used as precursors, and pure hydrogen was used as the carrier gas. At the onset, the PSSs whose specification was 2 um (bottom) × 1 um (space) × 1.5 um (height) were annealed at 1060 °C in H_2 atmosphere, then cooling down the reactor to 530 °C, and an about 30-nm-thick GaN nucleation layer was deposited. Three-dimensional layer was grown on the nucleation layer which is recrystallized by ramping up the temperature to 1045 °C with the rate of 1.5 °C/s and holding the temperature for about 90 s. During this period, low temperature (about 1010 °C) and V/III ratio (about 670) were set to encourage the 3D growth. Besides, in order to optimize the parameters of 3D layer growth, the pressure was varied from 4.5×10^4 Pa to 6.5×10^4 Pa, and the growth time was adjusted from 1000 s to 850 s and 750 s. After the 3D layer, 2D growth process for merging quickly was recommended. Then raising the temperature (1060 °C) and V/III ratio (about 1200), a pressure was set to 2.0×10^4 Pa. Furthermore, another sample was grown with normal process to compare with the 3D-2D process. After growth, morphologies were examined by SEM at different stages of sample A. Moreover, TDD was assessed by etch-pit density (EPD) measurements, and XRD was used to support the results.

Morphological evolution of GaN layer growth with the 3D-2D process can be obviously observed in Fig. 6.1. The growth temperature and reflectance transient are taken from the sample and the morphology at different stages of growth. As shown in inset (1), amorphous GaN is deposited uniformly on the PSS after the nucleation. And the thickness of nucleation layer is about 30 nm. In the duration,

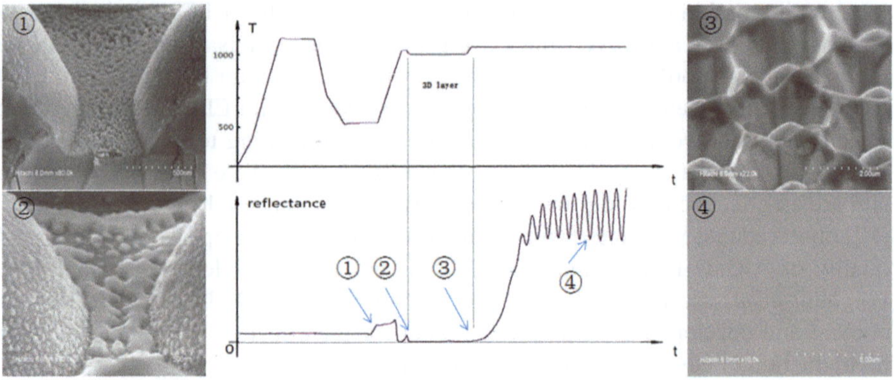

Fig. 6.1 3D-2D growth temperature and reflectance transient taken from sample (4.5 × 10^4 Pa, 1000 s 3D layer growth time). SEM insets show the morphology at different stages of growth as the nucleation before (1) and after recrystallization (2), at the end of 3D layer (3), and after coalescence (4) [12]

the reflectivity increases with the nucleation layer growing because the reflectivity is in proportion to the thickness of nucleation layer. As nucleation layer recrystallize under a high temperature, the materials are redistributed. GaN grains are formed, and then they gain or lose particles from one grain and transport to another; thus smaller grain shrinks, while larger grain grows. Finally the platelet-shaped islands are mainly left on the space of PSS, as can be seen in the inset (2). In the 3D layer growth, growth conditions are adjusted to achieve that vertical growth rate is much higher than lateral growth rate. 3D layer surrounds the cone-shaped pattern, and the thickness of 3D layer is near to the height of cone-shaped pattern. Furthermore, the hexagonal recesses have $\{1\bar{1}01\}$ incline facets and narrow (0001) plane. In the following 2D process, lateral growth rate is much higher than vertical growth rate; thus the incline facets coalesce quickly, and the interaction of TDs with the side facets causes the TDs to bend over. However, TDs in the (0001) surface will not bend over. The phenomenon can be observed in Fig. 6.2, and dislocations above the patterns bend over (in circle 1), while those among the patterns extend to the surface (in circle 2).

The condition of 3D layer is critical to reduce the TDD. Pressure and growth time are changed for the samples. All samples are etched in hot H_3PO_4/H_2SO_4 (3/1) solution, and then EPD is examined by the AFM. And XRD is used to calculate TDDs. For XRD, the measurements of rocking curves of (002) and (102) are carried out, and the XRD (002) and (102) rocking curves of sample (4.5 × 10^4 Pa, 1000 s 3D layer growth time) reach 211 arcsec and 219 arcsec, respectively.

And the densities of screw-type dislocation D_S and edge-type dislocation D_E can be estimated from the XRD FWHM values based on the following formulas [21]:

$$Ds = \frac{\beta_S^2}{4.35 \times |b_S|^2} = \frac{\beta_{002}^2}{4.35 \times (b_s \cdot \cos\alpha)^2} \tag{6.1}$$

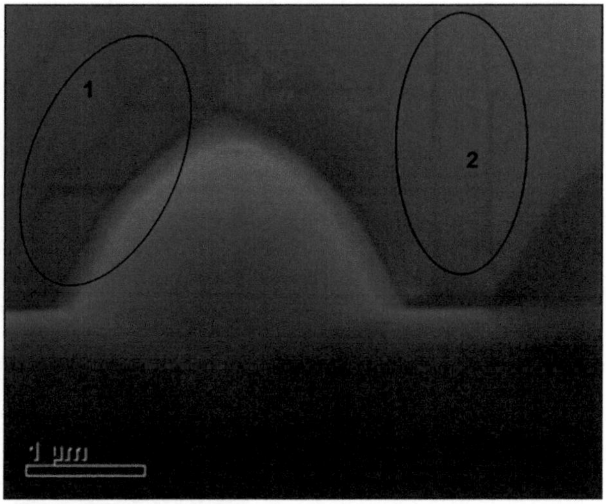

Fig. 6.2 Bending-over TDs and unbending-over TDs [12]

$$D_E = \frac{\beta_e^2}{4.35 \times |b_e|^2} = \frac{\beta_{102}^2 - \beta_{002}^2}{4.35 \times (b_e \cdot \sin\alpha)^2} \tag{6.2}$$

where $|b_S|$ and $|b_e|$ are the Burgers vector sizes of the screw-type dislocation ($|b_S| = 0.5185$ nm) and edge-type dislocation ($|b_e| = 0.3189$ nm), respectively, and β_{002} and β_{102} are XRD FWHM values of (002) and (102). And α is the angle between the reciprocal lattice vector (K_{hkl}) and the (001) surface normal. With the FWHM of (002) and (102) rocking curves of sample (4.5×10^4 Pa, 1000 s 3D layer growth time), the dislocation density is 1.685×10^8 by calculation, and it meet the EPD 9.80×10^7 of the sample, which is shown in Fig. 6.3, AFM image (5 um $\times$ 5 um) of sample (4.5×10^4 Pa, 1000 s 3D layer growth time) after wet chemical etching.

TDDs of the samples which grew with the 3D-2D process are much less than that of the sample of normal process which grew with normal process. Besides, TDDs increase as the pressure increases. And TDDS increase as the growth time decreases. Since the pressure also effects on the growth rate, both pressure and growth time relate to the thickness of 3D layer. Thus, the thickness is critical parameter for reducing the TDDs. Three-dimensional layer has the incline facets $\{1\underline{1}01\}$ and (0001) plane; TDs in the (0001) surface will not bend over, so small area of (0001) plane is good for reducing TDDs. Generally, the thickness of 3D layer is recommended to be near to the height of cone-shaped pattern to further reduce TDDs.

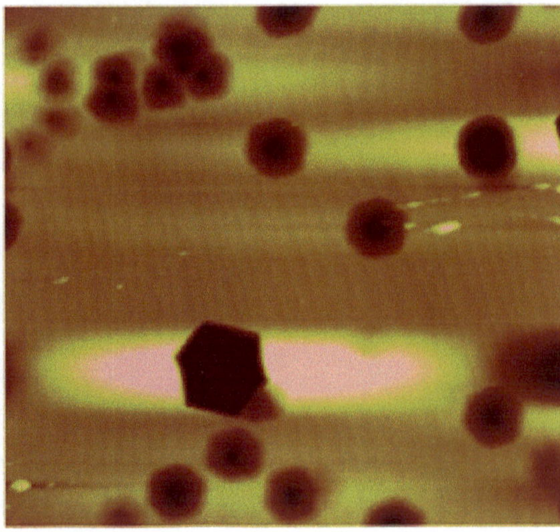

Fig. 6.3 AFM image (5 um × 5 um) of sample (4.5 × 10^4 Pa, 1000 s 3D layer growth time) after wet chemical etching [12]

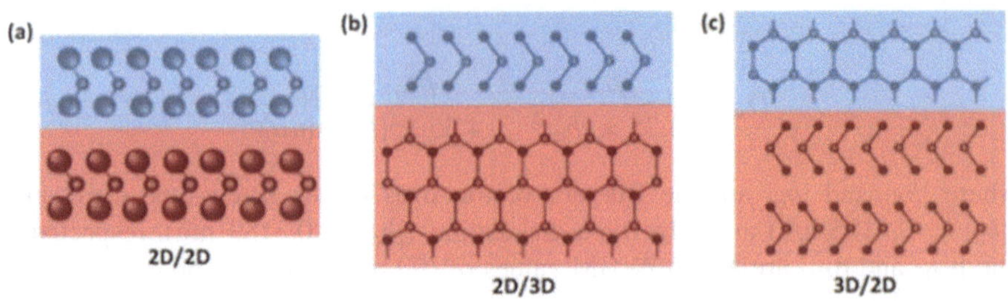

Fig. 6.4 The principle of vdWE. (**a**) 2D material grown on 2D material. (**b**) 2D material grown on 3D material. (**c**) 3D material grown on 2D material

6.1.2 New Buffer Layer for High-Quality GaN Materials

Since the native substrate of GaN is very expensive, we usually use Si [22], sapphire [23], and SiC [24] as the substrate to grow GaN films, which is called heteroepitaxial growth. In this way, there are many strong chemical bonds at the interface to connect the epilayer and the substrate, so that it can lead to many defects and dislocations during the epitaxial process because of the mismatch, so the quality of the GaN films reduced. To solve this problem, a buffer layer like AlN or GaN grown at a lower temperature was introduced, and it can decrease the defects to below 10^8 cm^{-2}. Recently, another growth mechanism was reported to improve the heteroepitaxial growth of GaN. That is van der Waals epitaxy (vdWE) [25], which means the connection between the substrate and the epilayer is weak in van der Waals interactions rather than strong chemical bonds, as shown in Fig. 6.4, so we can ignore the mismatch in this way.

For vdWE, graphene is a perfect candidate, for it is a 2D material without darling bonds on the surface, and it is hexagonal structure like GaN in plane. Besides, it is van der Waals interactions between different graphene layers, so it is easy to separate them. Since then, what can we do by using graphene?

Firstly, we can use graphene as a buffer layer.

As we said before, graphene acting as a buffer layer can prevent the dislocations caused by mismatch and then improve the crystal quality. However, there is an obstacle we need to face with; it is the nucleation on the graphene surface. Indeed, without the dangling bonds, there is less mismatch, and also it cannot provide the sites for Ga or N atom to nucleate. So we need to adopt some other methods to promote the nucleation process.

The first way is ZnO nanowalls, which is mainly adopted by Chung et al. in 2010 [26]; they grew GaN films and LED structure on graphene/SiC substrate using this method. And then in 2012 [27], GaN grown on graphene/SiO_2 was achieved successfully, with the flat surface. It is a big progress in nitride growth since we can grow single crystal on amorphous substrates.

Another way is using the steps on SiC substrates. It is known that we can get graphene on SiC directly by a high-temperature sublimation process, and the step on it can act as the nucleation sites. Balushi et al. [28] studied the nitride nucleation on such substrate. It is found that GaN nucleated at the step-edge priorly and little at the terrace, the nucleation sizes will become larger as the temperature rises, and it is easier to nucleate at few-layer graphene surface. Jee et al. [29] grew GaN films using this method and got good-quality films with the RMS for only 3 Å, and the full-width half maximum (FWHM) of the GaN (0002) peak was 0.06°.

The third way is plasma treatment. We destroyed the graphene surface in this way indeed, even though we can get lots of nucleation sites. Chae et al. [30] grew GaN films on graphene/SiO_2 using this method, and the film quality is good without yellow luminescence band.

And also, we can use some other ways, like pulsed sputtering. Shon et al. [31] used this method to grow GaN films on graphene/amorphous SiO_2 substrate; they got the good-quality film and LED successfully.

Secondly, we can use graphene as a release layer.

Graphene is a 2D material without dangling bonds on the surface, so the connection between graphene layers is very weak, so that it is easy to release the epitaxy layer on graphene. In this way, we can transfer the grown material onto other substrates and make more kinds of devices. Exactly, the first GaN film grown on graphene was transferred to foreign substrates successfully. In 2010 [26], Chung et al. reported that they grew GaN LED on ZnO-coated graphene/SiC substrate using ZnO nanowalls as an intermediate layer, and then the LED was transferred to glass, metal, and plastic substrates, which was found that they all exhibit good performance. In 2014 [29], Jee et al. grew graphene on SiC directly by a high-temperature sublimation process and then grew GaN LED on it. They use the kind of SiC substrate with vicinal steps which can facilitate nucleation. It is needed to mention that they release the GaN films by a 2-mm-thick Ni stressor, and after transferring the released GaN/Ni/tape stack on a host substrate, the SiC substrate

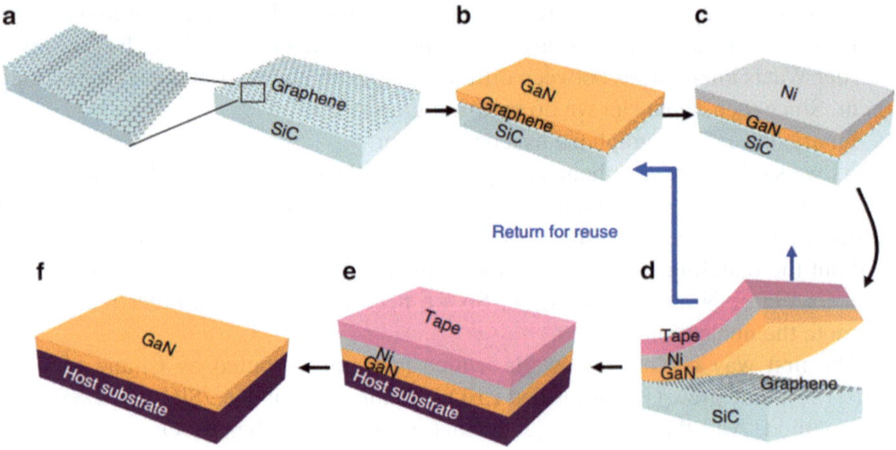

Fig. 6.5 Schematic method for growing/transferring single-crystalline thin films on/from epitaxial graphene

can be used again, as shown in Fig. 6.5. And also, there are many other groups who are focusing on the epitaxy release, like Qi et al. [32] in Soochow University, they grew crack-free gallium nitride microrods by HVPE and then achieved the release.

Thirdly we can use graphene as the heat dissipation layer.

Another advantage of graphene is the high thermal conductivity, so that we can use it as the dissipation layer. Han et al. [33] used the reduced graphene oxide (rGO) on sapphire substrate as such layer, and they found that the average temperature distribution on the chip surface is decreased from 51.4 °C to 47.06 °C, as shown in Fig. 6.5. Then, they adopted this method to grow GaN film on patterned sapphire substrate [34], and the averages of the surface temperature distribution decreased from 47.9 °C to 42.3 °C; the output power is 7% higher conventional counterpart.

It is obvious that we can improve the crystalline quality and even the device performance. However, there are still many questions we need to study, like the nucleation mechanism on graphene, the large-size graphene growth and transfer, and even the high-quality graphene grown on dielectric substrate to grow nitrides directly. There are much more work we need to do in the future.

Despite graphene, other 2D materials have also become a subject of great recent interest. Among these 2D materials, WS_2 and MoS_2 2D films whose lattice constants ($a = 3.15$ Å and 3.16 Å, respectively) [35] are remarkably close to those of AlN ($a = 3.11$ Å) and GaN ($a = 3.18$ Å) have lead to investigations to use relevant layered TMDCs as potential substrates for the epitaxy of AlN and further GaN films.

In 1997, Ohuchi and Chung [36] preliminary reported an attempt to grow AlN thin film on WS_2 films fabricated on Si(100) substrate. In the first part, they deposited highly oriented WS_2 films with WS_2 basal planes parallel to the substrate by MOCVD. Then, by a regular atomic layered growth (ALD) process, AlN films were grown in the order of several hundred Å and with the FWHM of approximately 0.25°. Two years later, Yamada et al. [37] had an investigation comparing the structural and optical properties of GaN films grown on MoS_2 substrate and with those on Al_2O_3 substrate. The MoS_2 substrate used in the study was cut from native mineral rock, and 0.8 μm GaN films were grown by radio-frequency MBE with a two-step process: low temperature in 200 °C and high temperature in 800 °C. Through a detailed characterization, they verified the fact that flatter surfaces, lower stress, and dislocations in GaN on MoS_2 compared to those on Al_2O_3. While lacing deepen investigations about the nucleation mechanism, both the two groups were all proved the feasibility of high crystalline quality III-V nitride thin films on alterative near-lattice-match substrates—MoS_2 and WS_2.

Given the slow development of III-V nitrides at that time and the limited advance of 2D materials, the near-lattice-match epitaxy investment is in a longtime state of stagnation. However, in the past few years, with the frenzy of activity in research in III-V nitrides and 2D materials and motivated by the advantages in overcoming the existing large lattice mismatch problems among III-V nitrides, this epitaxy gradually comes back into our sight again. In 2016, Gupta et al. [37] reported the MOCVD growth strain-free, single-crystal islands of GaN on the mechanically exfoliated flakes of several layered WS_2 and MoS_2, as shown in Fig. 6.6. In the same year, Ooi et al. [38, 39] had a further investigation about an alignment parameters of GaN/single-layer MoS_2 and WSe_2 heterostructure and opened up a way to integrate III-nitrides with 2D TMDCs for designing and modeling their heterojunction-based electronic and photonic devices.

Our research center also draw a wide attention about this field and have achieved a rudimental result of GaN films grown on large-area quartz substrates by inserting MoS_2 and WS_2 interlayers. Even though, we still face many new challenges, such as the difficulty in grown high-oriented crystal surface of these 2D materials.

6.1.3 Design and Growth of High-Efficiency Blue LEDs

Light-emitting diodes (LEDs) are used in a very broad range of applications, from displaying information, sensing, communications, to lighting and illumination. Generally, developing the high-efficiency blue LEDs should be conducted from two aspects: internal quantum efficiency and light extraction efficiency (LEE). The light extraction efficiency has seen remarkable progress in the past decade, increasing for nitride-based LEDs from roughly 20% to 85%. However, there is one of the most significant and enduring challenges facing high-power GaN-based LEDs, called efficiency droop, or the decrease in external quantum efficiency (EQE) of an LED with increasing drive current. The efficiency droop in GaN-based light-emitting diodes has been extensively studied, and many physical origins of the droop effect have been proposed, including poor hole-injection efficiency, polarization charges,

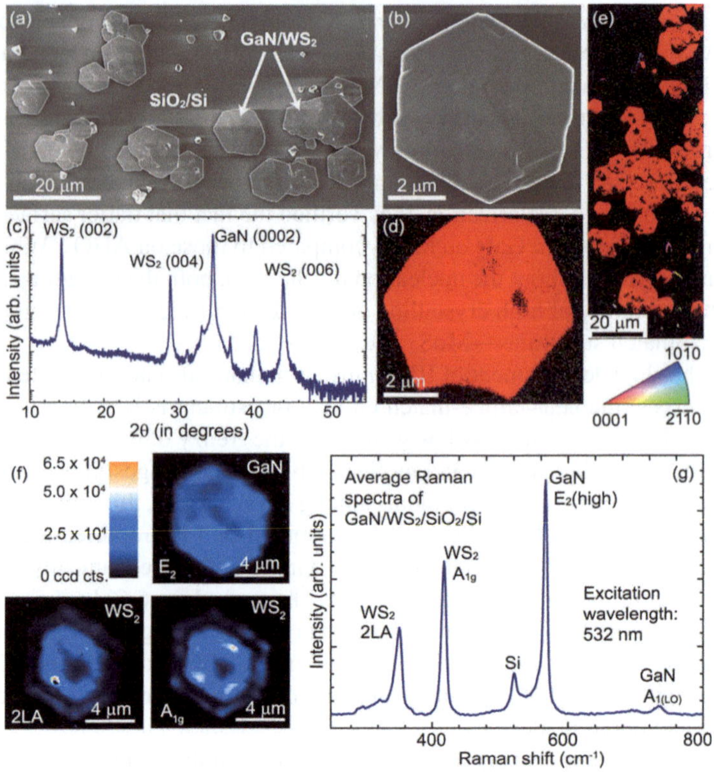

Fig. 6.6 Growth of GaN on exfoliated WS2 flakes. (**a**) Scanning electron microscopy confirms near-hexagonal crystals of GaN growing only in the region covered by the WS2 flakes. (**b**) Micrograph showing single hexagonal crystal of GaN grown on WS2. (**c**) X-ray diffraction profile of GaN on WS2 shows preferential (0002) orientation. (**d**) EBSD maps of GaN grown on exfoliated WS2 clearly show that the grown GaN layer is single crystal and (0002) oriented

Auger recombination, and carrier leakage, around which various design strategies and growth methods of high-efficiency blue LEDs are performed.

To improve the internal quantum efficiency of blue LEDs, several schemes have been put forward from different points of view, including improving hole injection, reducing polarization charge, and eliminating the carrier leakage.

Apart from the effect of electron-blocking layer (EBL) on the transport of holes, hole injection is further hindered as compared to electron injection due to the fact that there is a large disparity between electron and hole mobility. Some researchers aimed to enhance the hole injection by improving the hole concentration of p-GaN, but obtaining the high-quality and carrier concentration p-GaN is still a huge challenge.

Hwang et al. [40] introduced a new device named light-emitting triodes which have two anodes for promoting the injection of holes into the active region shown

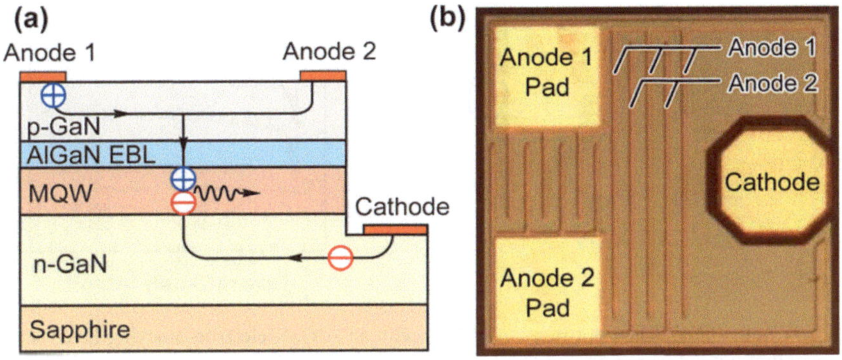

Fig. 6.7 (Color online). (**a**) Schematic drawing of the operation of an LET showing enhanced hole injection into active region by an anode-to-anode bias. (**b**) Photo of an LET with interdigitated anode fingers

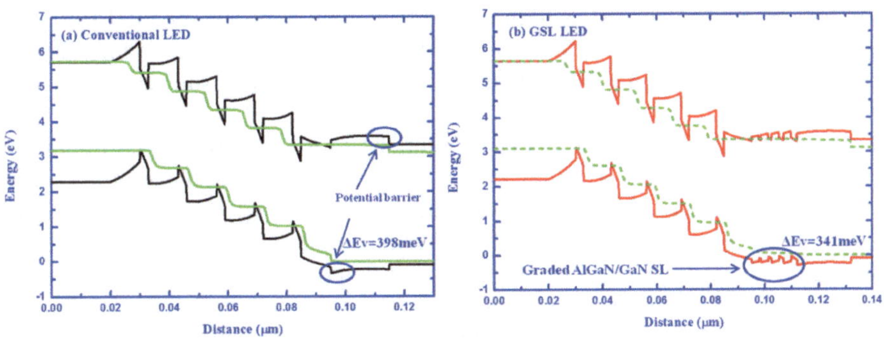

Fig. 6.8 Calculated energy-band diagrams of the conventional (**a**) and GSL (**b**) LEDs at current density of 100 A/cm^2

in Fig. 6.7. As the anode-to-anode bias increases, the efficiency at the same current density increases, whereas the efficiency droop decreases substantially.

Kang et al. [41] introduced a graded superlattice (GSL) AlGaN/GaN inserting layer to decrease the effective barrier height of holes, as presented in Fig. 6.8. The piezoelectric polarization field near the last barrier is suppressed effectively by introducing the GSL inserting layer. As a result, the efficiency droop radio is improved from 35.8% to 19.4% at current density of 100 A/cm^2.

Current commercially available InGaN-based LEDs grown on the "polar" c-plane of the crystal suffer from internal polarization-related electric fields that separate the electron and hole wave functions in the quantum wells and limit the radiative recombination rate. On the other hand, devices grown on nonpolar or semipolar orientations have been demonstrated with eliminated or reduced polarization fields and are theoretically predicted to have higher radiative recombination rates than c-plane devices.

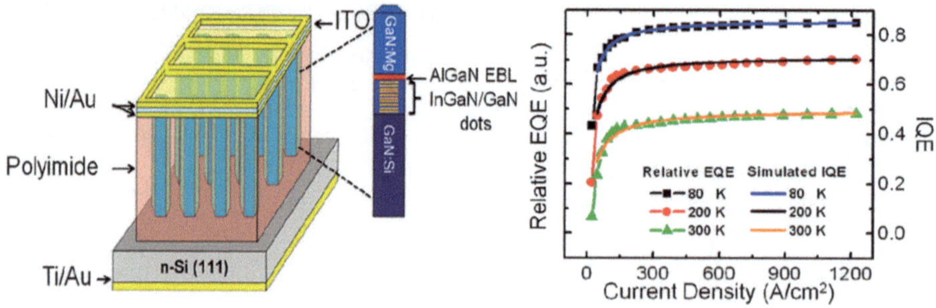

Fig. 6.9 Left: InGaN/GaN dot-in-a-wire LEDs with an AlGaN electron-blocking layer between the quantum dot active region and the p-GaN section. Right: Variations of the measured relative external quantum efficiencies with injection current at 80 K, 200 K, and 300 K. Variations of the simulated IQE with current using the ABF model show a good agreement with experimental results

Zhao et al. [42] report a high-power blue light-emitting diode (LED) with a high external quantum efficiency and low droop on a free-standing (20-2-1) GaN substrate; in higher current density regions, the LED showed outstanding performance.

Chung et al. [43] have introduced multilayer QBs (MLBs) as an alternative approach to reduce polarization effects, and time-resolved photoluminescence measurement showed that polarization field was reduced by 19% in the multilayer barrier light-emitting diodes structures. Besides, optical power measurements on packaged devices showed overall increase of external quantum efficiency for all currents up to the current density of 150 A/cm^2. The flow of energetic electrons flying over the active region (without being captured) to recombine with holes in p-type GaN or at the p-type contact electrode, i.e., electron leakage, is known as a common problem in GaN-based LEDs, and it is the reason why an AlGaN electron-blocking layer (EBL) is implemented on the p-side of the active region. However, the EBL in GaN-based LEDs is often unable to completely block election leakage.

Recently, Nguyen et al. [44] have investigated for the first time the impact of electron overflow on the performance of nanowire light-emitting diodes (LEDs), as showed in Fig. 6.9, and further demonstrated that electron overflow in nanowire LEDs can be effectively prevented with the incorporation of a p-doped AlGaN electron-blocking layer, leading to the achievement of phosphor-free white light-emitting diodes that can exhibit for the first time virtually zero efficiency droop for injection currents up to ~2200 A/cm^2.

Ni et al. [45] introduced an InGaN staircase electron injector with step-like increased In composition, an "*electron cooler*," proposed for an enhanced thermalization of the injected hot electrons to reduce the overflow and mitigate the efficiency droop, which is illustrated in Fig. 6.10.

To improve the light extraction efficiency, various texturing techniques have been used [46]. The two main ones have been surface roughening and substrate patterning, while side shaping can lead to sizeable low loss side extraction when

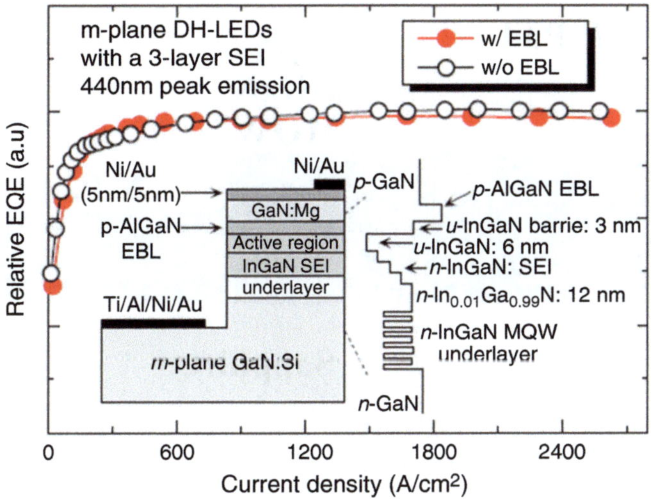

Fig. 6.10 (Color online). Relative EQE of two m-plane LEDs grown on freestanding m-plane (1-100) GaN substrates with a three-layer SEI: one with and one without the EBL. The inset shows the schematic for the LED with a 10 nm EBL (p-$Al_{0.15}Ga_{0.85}N$)

using transparent index-matched substrates. Photonic crystals also lead to remarkable LEEs, but their main advantage is emission directionality, which does not seem yet a large enough advantage to offset the demanding fabrication technology.

6.1.4 Device of High-Efficiency InGaN/GaN LEDs

There are three basic chip structures for InGaN/GaN LEDs: lateral LED, flip-chip LED, and vertical LED. The main light escaping surface of lateral LED and flip-chip LED are top surface and back sapphire surface, respectively. The p and n electrodes of the lateral LED and flip-chip LED are on the same side. However, the p electrode of the lateral LED is transparent or translucence, while that of the flip-chip LED has large reflectivity. The vertical LED has p and n electrodes on top and backside, respectively, so the substrate of the vertical LED is conductively.

6.1.4.1 InGaN/GaN Lateral LED

The chip structure of the lateral LED is the most easily implemented LED chip structure for InGaN/GaN LED on sapphire epitaxial wafers which are the most common used InGaN/GaN LED wafers and is a common form of structure. There are other kinds of lateral LEDs which have different substrates, for example, LEDs

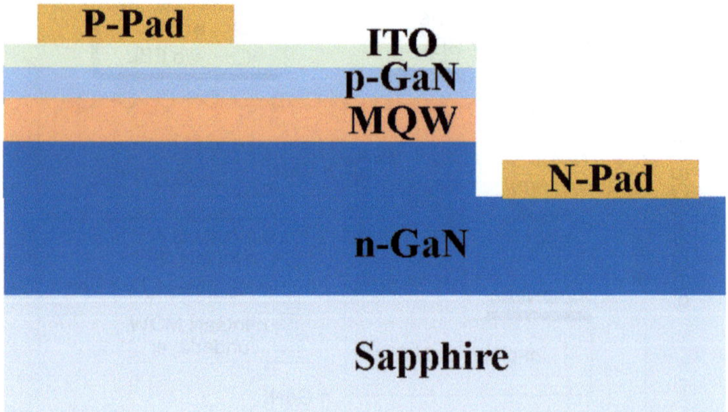

Fig. 6.11 The structure diagram of the InGaN/GaN lateral LEDs on sapphire substrate

with silicon substrate or silicon carbide substrate. Silicon absorbs blue light severely, and high-quality silicon carbide for GaN epitaxy is too expensive, so those LEDs are not common.

Figure 6.11 shows the LED for this structure. Since sapphire substrate is elecinsulating, the n-type GaN materials need to be exposed, which can be achieved by etching process. Because p-GaN is very thin, the current expansion capability is poor, and the p-GaN surface should be used as the light escaping surface, so it needs to deposit a layer of transparent conductive material on its surface, such as ITO, NiAu, ZnO, etc.

In early lateral LED chip, the p transparent electrode is made of NiAu thin film which has several nanometers thick. However, the light absorption of the NiAu thin film is severe, and the transmittance of it is about 70% for 460 nm blue light. The ITO application on InGaN/GaN LED is a big improvement which is used to replace NiAu thin film as p transparent electrode. ITO is indium tin oxide film which has a relative low absorption for blue light compared to NiAu thin film, and the light extraction efficacy of the LED is much improved when ITO is used to replaced NiAu thin film. Another big improvement for lateral LED is the application of PSS. PSS is patterned sapphire substrate, and it has aligned micrometer scale circular cone structures covered densely up its top surface which InGaN/GaN LED was grown on. The light extraction of the lateral LED with PSS is much better than lateral LED with flat sapphire substrate. Nichia is first to combine ITO and PSS technologies to improve the efficacy the InGaN/GaN blue LED. Figure 6.12 shows the structure of this kinds of high efficacy blue LED. The back surface of the sapphire is designed to reflect light, and the DBR (distributed Bragg reflector), metal reflector, or DBR metal combined ODR (omnidirectional reflector) reflector is deposited on the back sapphire surface to improve light extraction efficacy.

Because of the sapphire substrate, thermal conductivity is only 40 w/(m °C), only one over ten of copper, so a major defect of the lateral LED is a poor heat

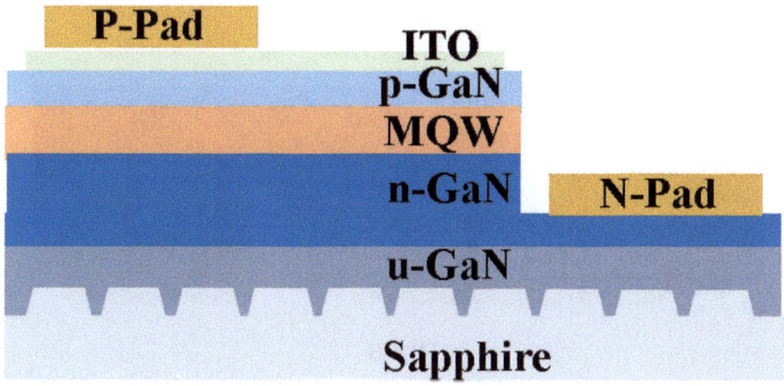

Fig. 6.12 The structure diagram of Nichia high efficacy InGaN/GaN lateral LED with ITO and PSS

dissipation performance. In addition, if the arrangement of electrode wire and the chip size do not match appropriately, it easily led to cause the edge-crowding effect of the current, leading to the local overheating of the chip. These problems will degrade the performance of the LED chip and shorten its lifetime.

Although it has some disadvantages, the lateral LED is easily fabricated and is cheaper compared to other LED chip form. The lateral LED has been widely used in indoor illumination, backlighting, decoration, display, and other lighting fields.

6.1.4.2 InGaN/GaN Flip-Chip LED

The light escape direction of the LED flip chip is the opposite of the lateral LED, so the flip-chip LED is inverted, and the p and n electrode are both down and have the ability to reflect light. Early flip-chip LED commonly has a silicon or ceramic substrate which has good thermal conductivity. The advantage for silicon substrate is that the electrostatic protection circuit is integrated in silicon substrates to protect LED chip. The p and the n electrodes of LED chip are connected to p and n electrodes on silicon substrate with little golden ball through flip-chip bonding technology. Its structure is shown in Fig. 6.13. The heat generated by such high-power LEDs does not need to pass through a chip's sapphire substrate but rather directly to a silicon or ceramic substrate with a higher thermal conductivity and then to the metal base.

The abovementioned flip-chip LED technology provides a solution to improving LED performance in thermal dissipation and light extraction when the lateral LED chips with NiAu thin and flat sapphire substrate are still mainstream chips. However, this kind of flip-chip LED has little advantages in thermal dissipation in practical application because the total cross area of the gold ball which is part of the heat dissipation channel is much smaller than the whole chip area though it has better thermal conductivity than sapphire. When the high efficacy lateral LED with ITO

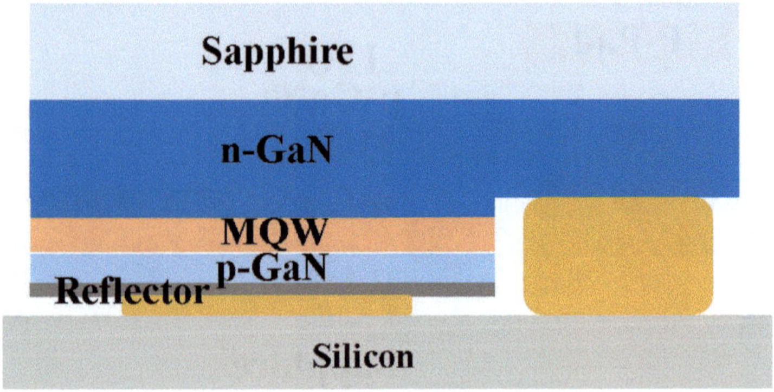

Fig. 6.13 The structure diagram of the InGaN/GaN flip-chip LED

and PSS became widely used and show better performance, this early type flip-chip LEDs were abandoned gradually due to its higher manufacture cost.

The flip-chip LED did not disappear. A new type flip-chip technology combined CSP (chip-scale packaging) was introduced to reduce the LED package cost in the 2010s. Chip-scale packaging without using gold line, based on the flip-chip bonded technology, on the basis of the traditional LED chip packaging, reduced the gold thread encapsulation process and save wire rack, Au wire bonding, leaving only the chip tie-in phosphors, and encapsulation adhesive is used. Figure 6.14 shows the structure of this new type flip-chip LED. There is no need for silicon or ceramic substrate. n pad is enlarged, and its area is as large as p pad. So the n large pad is overlapping with p electrode, and they were separated by passive layer such as dioxide silicon and nitride silicon thin film. The distance between p and n large pad is about 200 μm to insure soldering easily and successfully. After CSP, the top and four sides of the flip chip were covered by solidified encapsulated silica gel which is mixed with YAG phosphors. The CSP packaged LED is compact with relatively small size and can be directly used on PCB (printed circuit board) by SMD (surface mounting technology). The thermal resistance of the chip is much lower than that of the lateral LED, and the operation current is improved.

As a new packaging technology product, the non-gold-line chip level light source (CSP flip-chip LED) is completely free of the problems such as the lack of light, blinking, and light attenuation caused by the welding or contact of the gold thread. Compared with the traditional packaging process, the packaging density of the chip level light source has increased 16 times; the packaging volume has shrunk by 80%, and the design space of the lamp is larger. CSP flip chips with more stable performance, better heat dissipation, and more uniform light distribution, smaller volume, have favored by more and more LED lighting enterprises and end-product application enterprises.

Another type flip-chip LED is called thin-film flip-chip LED (TFFC LED). The top sapphire substrate was moved by laser lift off in TFFC LED. The InGaN/GaN LED thin film was transfer to a new substrate commonly on silicon substrate, and

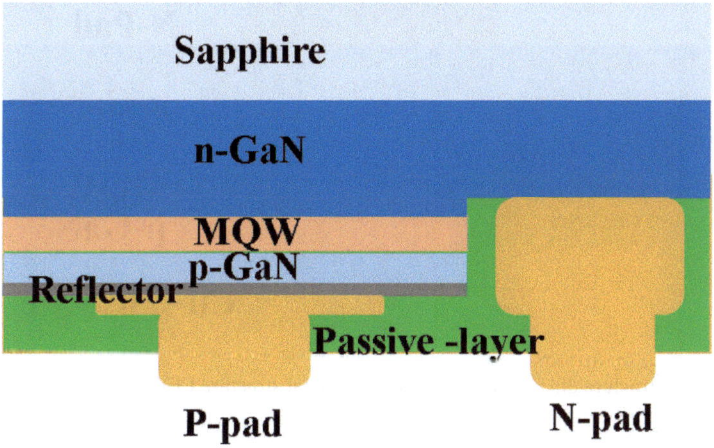

Fig. 6.14 The structure diagram of CSP flip-chip LED

the p and n electrodes of the InGaN/GaN LED are connected to p and n electrodes on silicon substrate. The total InGaN/GaN LED thin-film bottom surface with electrodes is cohered firmly to silicon substrate to avoid film cracking, while the top surface is roughened to improving light extraction. It is reported that the heat dissipation ability and light extraction efficacy of TFFC LED are excellent. The most severe problems of the TFFC LED are its low yield and high cost during manufacture process.

6.1.4.3 InGaN/GaN Vertical Structure LED

The p electrode and the n electrode of the vertical structure LED (VLED) are, respectively, on the top and bottom of the chip or on the contrary. The supporting substrate usually acts as heat dissipation path, while the substrate also acts as an electrode. Therefore, the LED of this structure overcomes the problem of current set edge effect and poor heat dissipation ability. Figure 6.15 is a typical structure diagram of the InGaN/GaN vertical structure LED. The vertical structure LED has many advantages, but it is difficult for GaN to grow directly on this new substrate. Sapphire and silicon are most commonly used substrates in nitride material growth by metal organic chemical vapor deposition (MOCVD). The sapphire substrate is insulating and has low heat conductivity, while the silicon severe absorb blue light, so these two kinds of material are not suitable to act as the vertical structure LEDs. So the vertical structure LED needs to remove the old substrate and attach the new substrate. This greatly increases the complexity and difficulty of the process. Still, vertical structural LEDs are favored by research institutions both at home and abroad. In particular, the news of the excellent high-power vertical structure LEDs, which has kept the most efficacy LED record for a long time, reported by Cree

Fig. 6.15 The structure diagram of InGaN/GaN vertical structure LED

has enhanced the attention and confidence of the VLED. With the development of technology, the vertical structure LED may be the best solution to realize the general lighting of the semiconductor.

There are several schemes for implementing InGaN/GaN vertical structure LED chip technology. For the InGaN/GaN LED epitaxy material grown on sapphire substrate, to realize this structure, there are two key processes, one is to attach the InGaN/GaN LED epitaxy material to the new thermally conductive substrate, and the second is to remove the old substrate. As for the remaining chip technology, it is compatible with the chip technology of the lateral LED. These two key technologies are not only important but also difficult to achieve. The new conductive thermal substrate attachments include wafer bonding and electroplating. The methods of removing the old substrate include laser lift off, substrate grinding, and wet drilling [47–49]. The most commonly used method is laser lift off. The two steps combine to achieve the vertical structure LED with n-type GaN on the top and p-type GaN at the bottom. There are also twice substrate transfer method, which can realize the vertical structure LED with the p-type GaN on the top and the n-type GaN at the bottom. Electroplated copper alloy is used as the substrate in the vertical structured LED chip produced by SemiLEDs, ISCAS, which is shown in Fig. 6.16. During this kind of vertical structure LED manufacture process, the sapphire substrate is removed by the laser lift off and recycled.

For InGaN/GaN LED epitaxial wafer grown on a silicon substrate, in the production of vertical structure LED, the different process is substrate removal method. It is easier to find quick corrosion methods compared to the sapphire wafers. So, wet etching method is often used to remove the silicon substrate. If the epitaxy substrate is conductive silicon, the substrate is retained to act as the negative electrode, and it can also form a vertical structure LED. However, the visible light reflectivity of the silicon substrate is very low, and the absorption of the light is severe. The growth of the GaN base LED on silicon carbide is similar to that in silicon.

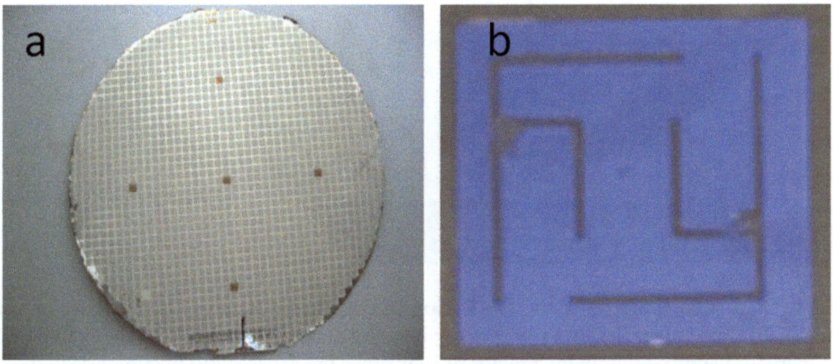

Fig. 6.16 VLED wafer and VLED chip photos with metal substrate produced by ISCAS

6.1.5 Tendency of Novel LEDs Structure and Application

Recently, the nanowire LED (Fig. 6.17) [50–54], mainly core/shell LED, has attracted large attention because of their potential advantages. In comparison to the conventional thin-film LEDs, core/shell nanorod structure LEDs have higher aspect ratio and larger active regions. Besides, the InGaN/GaN multi-quantum wells (MQWs) grown on nonpolar or semipolar GaN facets are suggested to be able to effectively increase the light extraction efficiency and reduce the quantum-confined stark effect (QCSE), which increases the internal quantum efficiency of LEDs through enhancing the radiative recombination rate in MQWs [55–57]. What's more, nanowires could improve the material quality with respect to thin films, a better handling of the strain induced by thermal expansion mismatch. Thus, the high-quality axial NW LEDs also have been reported a lot and proved to have some special performance [58, 59].

6.1.5.1 Synthesis

For the structure of nanowire LED, the growth of high-density ordered GaN nanowires is necessary for the follow-up metal contacting. The top-down etching, bottom-up selective vapor-liquid-solid (VLS) growth, and area MOCVD growth are three prevailing GaN nanowire growth methods. GaN dry plasma etching is widely used in the III-nitrides industry. However, reactive-ion etching (RIE) suffers from relatively low etch rates and plasma damage in devices. Compared with RIE, wet etching has the advantages of reducing surface damage and low cost and providing a simple method for device fabrication. Figure 6.18a shows the nanostructure mainly developed by wet etching. In our work [60–62], we found the suppression of the etch rate in the vertical direction could help to form the uniform nanowire. A thin AlGaN which was stable in the KOH solution was used to suppress the vertical direction. The wetting nanowires were surrounded by six $\{1\bar{1}00\}$ m plane. On the

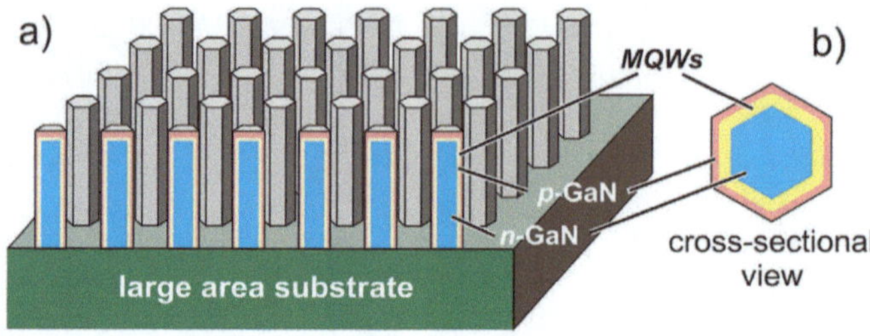

Fig. 6.17 Core-shell strategy for the fabrication of GaN-based nanoLEDs. Sketches of (**a**) core-shell nanoLED ensemble. (**b**) Cross-sectional view of a core-shell nanoLED. The active LED area can be increased approximately by a factor of four times the aspect ratio in comparison to a planar LED [57]

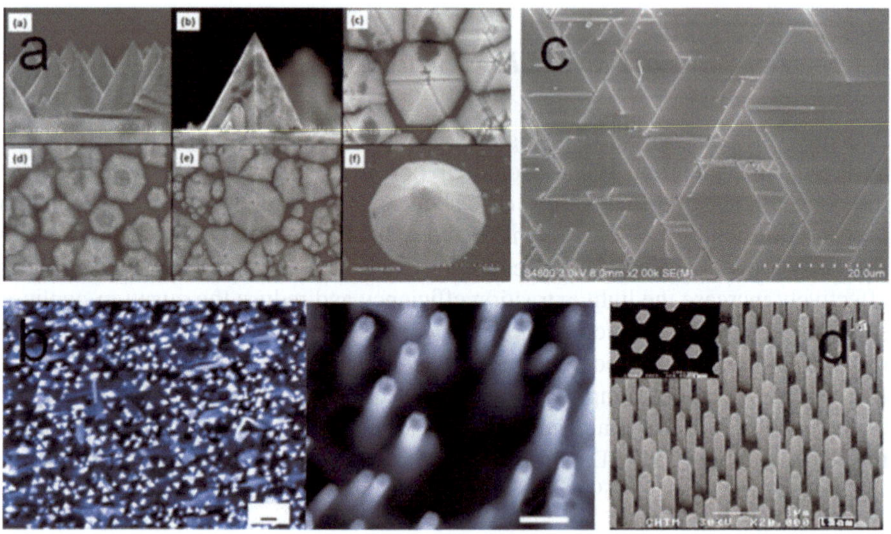

Fig. 6.18 Synthesis of GaN NWs by (**a**) top-down etching [61], (**b**, **c**) bottom-up selective vapor-liquid-solid (VLS) growth, and (**d**) area MOCVD growth [64, 68]

other hand, the existed large vertical etching could develop the pyramid structure with six $\{11\overline{2}1\}$ facets.

The top down may be an effective method to obtain the array nanowires. However, the nanowire crystal quality was poor because of the film growth and wetting. Since the discovery of the VLS process in 1964 by Wagner and Ellis [63], there have been many seminal works advancing NW synthesis and applications. To achieve GaN NWs with controlled orientation (or say, to obtain GaN NWs array), the most effective ways to control the growth direction of GaN nanowires is the epitaxial growth. Kuykendall et al. [64] have shown that hexagonal cross section, c-axis NWs grew perpendicular to (111) MgO, and m-axis NWs with a triangular cross section grow on (100) γ-LiAlO2 in MOVPE (Fig. 6.18b). Due to the epitaxial relationship, horizontal GaN NWs with various growth directions and cross sections have been obtained on sapphire substrate with various facet orientations from our group (Fig. 6.25c). Recently, our group proposed the fast and controlled growth of high-quality GaN nanowires by VLS-HVPE [65–67]. We also achieved the first growing of horizontal GaN nanowires by HVPE with the growth rate more than 400 μm/h (Fig. 6.18c) [67].

The selective area growth (SAG) is the most practical GaN nanowire growth methods because the high crystal quality has no introduction of foreign metals. Hersee et al. [68] reported for the first time the successful MOCVD growth of GaN nanorods on patterned SiO_2/GaN using the pulsed growth mode (Fig. 6.18d). During the vertical growth phase, the Ga and NH3 flows were alternatively pulsed starting after the GaN nanorods extrude out the holes. This is claimed to be very critical since if the growth conditions for "filling the holes" were maintained, then as soon as the nanorods emerged from the growth mask, lateral growth occurred, and the nanorod geometry was lost. With the pulsed growth mode, GaN nanorods with good homogeneity in size and morphology were obtained. Our team also systematically studied the growth technology [69, 70]. We found the hydrogen content has a large impact on the morphology of the nanostructure. Interestingly, the results were similar to the wet etching. The high hydrogen content benefited to the development of long one-dimensional nanorods, while lack of or no hydrogen carrier gas leads to form the pyramid structure.

6.1.5.2 Nanowire LED

Nanowire-based LEDs were first reported in the early 2000s by Harvard University from liaber's group. However, in the early work, GaN core-shell NW LEDs were already fabricated in a single NW scale and are not ideal for mass production. The meticulous processes are needed to define the location of the nanomaterial and attach electrical contacts with nanoscale precision. The problems inherent to single nanowire LEDs can be resolved using the aligned nanowires (as shown in the front part), which is highly desirable for scalable LED fabrication. Corresponding to the synthesis methods of GaN NWs, there are also three types of NWs LEDs: top-down nanoLED, bottom-up axial nanoLEDs, and core-shell nanoLED.

A straightforward method to fabricate GaN nanoLEDs is to etch planar GaN LEDs with nanomasks. The GaN nanorods were defined by various methods like conventional photolithography, e-beam lithography, and nanoimprint lithography, or other low-cost methods, for example, nanoscale self-organized nickel islands as etching mask or SiO_2 nanosphere lithography. However, using a top-down method, a large portion of GaN material is etched away and does not contribute to light emission any more. This will add additional cost during device fabrication compared to the bottom-up (growth) method. Besides, other advantages of GaN nanorods, for example, defect reduction and tuning the indium concentration by the nanorod size will not be applicable. On the other hand, plasma-assisted MBE was also employed to fabricate LEDs based on arrays of GaN nanocolumns in an axial configuration, which contain n-GaN/MQWs (InGaN/GaN)/p-GaN, on a Si substrate [57, 58]. By tuning the indium content in the MQWs, the emission color of the nanocolumn LED could be varied from violet to red [59, 71]. Moreover, white-light emission was achieved in these LEDs by gradually varying the indium composition within the InGaN segment [72]. However, for commercial application, these novel axial nanowires LED may face the challenge of high cost and inefficient.

Radial nanowire LED heterostructures generally consist of a nanowire core region, surrounding by an active region and coaxial shell layers [73]. Such radial nanowire LED structures are often grown by selective area growth using MOCVD, and the active region typically contains nonpolar (m-plane) and/or semipolar multi-quantum wells. Several fabrication methods have been developed to realize large area radial nanowire LEDs [50, 73, 74]. For example, the p-GaN shell can be overgrown to form a coalesced layer to effectively connect all the p-type segments of individual nanowires [73]. A relatively low turn-on voltage of ~2.65 V was measured, which is better than that of a reference planar LED device. It was also measured that the peak emission wavelength stayed nearly constant with increasing injection current, due to the absence of quantum-confined stark effect in the nonpolar m-plane InGaN quantum well active region. Other approaches for realizing large area core-shell nanowire LEDs are to fill the gap between individual nanowires and directly cover a graphene layer as the p-electrode contact [50, 74]. The core-shell nanowire LED has been considered the most likely to replace conventional thin-film LEDs. However, though several companies and numbers of research institutes have studied it a lot, the effectiveness of the NW LED is still low.

6.2 Green LEDs

This chapter describes a number of factors that affect the efficiency of green LED and analyzes the effects of polarization, carrier transport, carrier localization, current expansion, epitaxial structure design, Auger recombination and light extraction, and luminous efficiency. Aiming at the different influence mechanism, the droop effect of reducing the green LED is discussed, and the method of improving the quantum efficiency is discussed.

6.2.1 Polarization Fields in the InGaN-Based LEDs

The nitride materials have a wurtzite structure, which has the point group of C6V (6 mm). The C6V point group is a hexagonal structure with a single axis of symmetry or a polar axis. The centers of the positive and negative charge

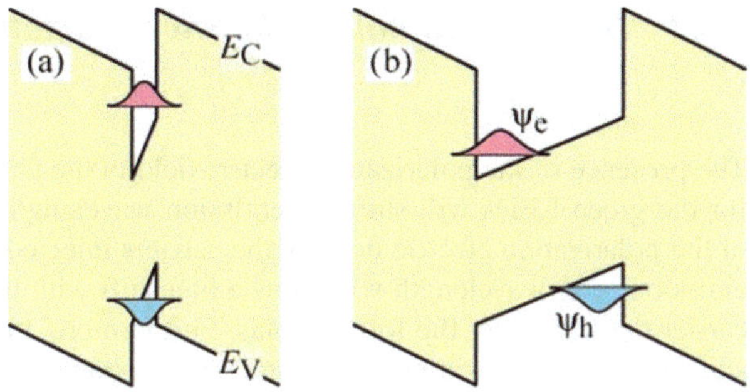

Fig. 6.19 The separation of the electrons and holes wave functions due to the band bending caused by the polarization electric field for the InGaN green LED

within the unit cells do not coincide, forming the torque and then resulting in spontaneous polarization effect. Spontaneous polarization field is determined by the nature of the material itself, regardless of whether it is subjected to the external stress. Piezoelectric polarization field is related to the stress applied on the nitride materials, which changed the position of the positive and negative charge centers in the unit cell, then resulting in a polarized electric field inside the material [75–77]. Unlike the III-arsenide compounds, an important property of the group III-nitride semiconductor material is its strong polarization effect, including spontaneous polarization and piezoelectric polarization. The total polarization in the nitride material is the sum of the piezoelectric polarization electrical field and the spontaneous polarization electric field [78].

This result shows that the lattice-mismatching between the InGaN layer and the GaN layer in the InGaN-based green LED produces a piezoelectric electric field on the order of MV/cm and bound sheet charges with a density of up to 10^{13} cm^{-2} at the interface of the heterostructure [75]. The polarization electric field will have a great influence on the band twist, the carrier distribution, and the photoelectric properties of the green LED [79].

As is shown in Fig. 6.19, the band bending caused by the polarization electric field in the quantum well leads to the separation of the electrons and holes wave functions. Thus, the effective band gap of the quantum wells and the recombination probability of the carriers will be reduced, which results in the decreasing of the luminous efficiency of the green LEDs. As the In composition in the multi-quantum well increases, that problem becomes more and more serious in the InGaN/GaN heterostructures [80]. The piezoelectric polarization along the [0001] crystal axis direction is at least one order of magnitude larger than spontaneous polarization [76]. For the In-rich green LEDs, the higher polarization filed in the InGaN/GaN multiple quantum wells has a great effect on the radiative recombination and efficiency droop. Various methods have been adopted to control the polarization fields and related QCSE in the green LED to weaken its negative effects and improve the luminous efficiency.

6.2.2 Internal Quantum Efficiency Promotion in the Green LEDs

The presence of the polarization electric field in the Ga-face InGaN quantum wells for the green LEDs will shift the emission wavelength [81]. Due to the screening of the polarization electric field by the carriers injected into the quantum wells, the emission peak wavelength will show a blueshift with the increasing of the injected carrier density under the forward bias. Furthermore, the peak wavelength will also shift with the increasing of the reverse bias voltage due to the reverse electric field which is opposite to the direction of the polarized electric field. The applied reverse fields flatten the energy band within the QW and then widen the band gap [82].

Controlling the stress in the epitaxial wafer can also effectively reduce the polarization electrical field in the active region of the green LED. It's reported that the stress alleviation in the high In composition quantum wells by pre-growing a low In composition of InGaN layer or InGaN/GaN superlattices layers between n-GaN and high In composition InGaN/GaN green light quantum wells (shown in Fig. 6.20) effectively reduced the polarization electric field within the green LED quantum well and then improved the wavelength stability and the luminous efficiency for green LED [83–87]. In order to reduce the wavelength blueshift in the green LEDs, Huang et al. inserted an InGaN layer between the n-GaN and the MQWs used as the pre-strained layer [83]. The growth temperature of the emitting QWs can be raised by 30 °C while keeping about the same emission wavelength in green region. The blueshift of the PL feature for the pre-strained sample grown under the same temperature with the compared sample indicated the reduction of the strength of the QCSE in the MQWs. The spectral blueshift in increasing injection current in the range of 45 mA was decreased by 46%. Similar results have been found by Park et al. [84]. They found that an employed graded superlattice (GSL) consisting of 12-stacked InGaN/GaN layers in the green LEDs showed a 4.7 nm blueshift and an enhanced output power under the photoexcitation. They attributed it to the increase of electron injection efficiency and the decrease of electron overflow to the p-GaN reduction in non-radiative defect sites due to the inserted GSL.

The transport of carriers in the Ga-face multiple quantum wells of the LEDs will be affected by the polarization fields [88–90]. Due to the strong polarization mismatch between the InGaN well and GaN barrier in the green LED, there is strongly polarization electrical fields with the direction from the p-GaN side toward the n-GaN side, resulting in band tilt. Then, a triangular barrier is formed in both the conduction band and valence band. As a result, the band bending of the quantum well reduces the confinement ability of carriers, making it easy for the electrons escaping from the quantum well and forming electrons leakage. Schubert argues that the presence of polarization fields reduces the carriers capturing probability in the quantum wells.

Effective dopant doping within the active regions can provide free electrons that shield the polarized sheet charges at the interfaces of the heterostructures in the active region and thus reduce the polarized electric field. The methods of providing

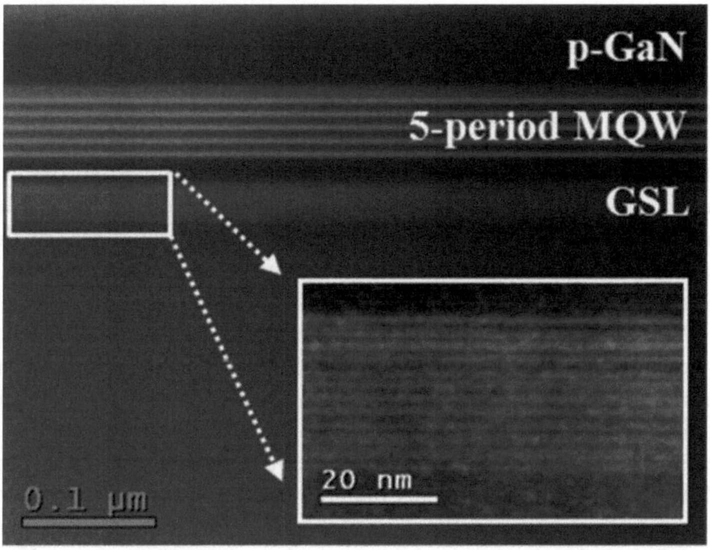

Fig. 6.20 TEM image with an inset of the inserted pre-strained structure for the LED sample. The inset shows the magnified image for the pre-strained superlattice structure

free carriers in MQWs by dopants doping mainly involve p-type or n-type impurity doping in the quantum wells or in the quantum barriers [91–95]. It has been reported that the donor impurity (Si) doping in the quantum barrier effectively shields the polarization electrical field and then improves the photoelectric performance [94]. Lin's group [94] reported that the Si doping in barrier layers with a dopants density of 3.4×10^{16} cm^{-3} for green LEDs not only screens the electrostatic field but also effectively reduces the roughness at the InGaN/GaN heterostructure interface. What's more, the current spreading could be enhanced by the Si doping in the barrier layers. The current crowding alleviation reduced the local heating and suppressing the efficiency degradation. But Ryou et al. [95] found that the Si doping concentration in the GaN barriers increases the barrier height for holes would block the holes transportation, thus decreasing the EL intensity. Zhang et al. employed the Mg doping in barriers in the InGaN-based green light-emitting diodes. The holes generated by the Mg impurity atoms in the barriers effectively screen the polarization fields and reduce the efficiency degradation of 12.4% at a high injection current. What's more, that research group also investigated the effect of the Mg back diffusion into the MQWs on the carrier recombination. The results show that the back-diffused Mg acceptors in the QWs closed to the p-layer could improve the hole injection efficiency and reduce the leakage current.

The nitride material with cubic structure has symmetry and no spontaneous polarization filed. However, that kind of nitride crystal for the green LEDs is difficult to be obtained by MOCVD. For the nitride materials with hexagonal phase, the intensity of the polarized electric field in the a-axis or m-axis directions is smaller than that along the c-axis [96, 97]. A LED with a small polarization effect in the direction perpendicular to the multi-quantum well is obtained by growing a hexagonal phase nitride material on a semipolar or nonpolar substrate. For example, in the m-plane (1-100) and a-plane (11-20) InGaN quantum wells, since the polarized electric field is parallel to the InGaN/GaN interface, the energy band of the quantum well remain flat. However, due to the limitations of the epitaxial growth techniques, the nitride material having a nonpolar or semipolar surface has a large number of defects [98]. What's more, the surface roughness of the film is relatively high. Thus, there are many non-radiative recombination defects in the semipolar or nonpolar LED, which reduces the radiation recombination efficiency of the MQWs. Therefore, the increasing of the external quantum efficiency for the nonpolar or semipolar LED is difficult. In addition, nonpolar or semipolar substrates for the nitride materials are expensive, which are not suitable for the large-scale commercial production.

The energy-band engineering of the active region though reducing the lattice mismatch between the In-rich InGaN quantum well and the GaN quantum barrier can also improve the luminous efficiency of the green LEDs. Some ways, such as the AlInGaN or InGaN barrier, were adopted to reduce the lattice-mismatching-related polarization field in the active region, improving the carriers coincidence rate and finally improving higher carrier recombination efficiency [99–105]. For example, in order to suppress the efficiency droop in the green LEDs through energy-band engineering, Lee et al. [106] employed the gradual In-content InGaN QW to reduce the valence-band bending and enhance the hole injection. The IQE in the In-graded QW LED was increased by 45.5% and 55.7% under the injection current of 20 and 100 mA, respectively. Furthermore, the In-graded QWs also decreased the electron leakage, suppressing the efficiency droop.

The electrons could participate in the radiation recombination only after they were captured by the In-rich quantum wells in the green LEDs. Electrons leakage means that the electrons transported though the active region to the p-type GaN layer, participating in the non-radiative recombination [107]. In order to suppress the electrons leakage, an AlGaN electron barrier layer (EBL) is inserted between the p-type layer and the active region for the green LEDs [108]. However, due to the polarization effect, the barrier height of the EBL conduction band for the electrons is reduced. As a result, the ability of the EBL to block the electrons is limited, reducing the radiative recombination efficiency. Therefore, electrons leakages are also considered to be one of the causes of efficiency droops in the green LEDs.

Vampola et al. [109] demonstrated the electrons leakages experimentally by inserting a p-doped, low Indium InGaN quantum well in the p-type EBL. Under the lower injected current, the inserted quantum well shows no light emitting, which means that the electrons leakage could be ignored. As the current increases, electrons transport through the active region and inject into the inserted low In-content quantum well and then recombine with the hole. The emitting light shows a shorter wavelength. The light with short wavelength, which is related to the electron leakage, appears when the external quantum efficiency is about to reach its peak. Then, the external quantum efficiency decreases when the intensity of the shortwave light increases.

The simulation results, shown in Fig. 6.21, by Chang et al. [110] show that the electrons leakage greatly reduces the peak EQE and moves the peak efficiency to a higher current density for the green LEDs with a 3.5-nm-thick undoped $In_{0.25}Ga_{0.75}N$ test well. In addition, some temperature-related EL testing results also show that electrons leakage may lead to efficiency droop. Nguyen et al. [44] reported that in the GaN-based nanowires LEDs, the most important factor limiting the quantum efficiency is the electron leakage, rather than the auger recombination. They demonstrated that droop was not present at a current density of 2200 A/cm^2 in a GaN-based nanowire LED with a p-doped AlGaN EBL layer.

The polarized sheet charges at the interface between the EBL and spacer layer are considered to be the fundamental factor in reducing the EBL limitation and the electrons leakage to the p-GaN layer [107, 111]. The Al composition in the AlGaN EBL is generally set from 5% to 30% in the green LEDs. However, with the increasing of Al composition in AlGaN, the polarization mismatch between AlGaN and adjacent GaN barrier layer also becomes more serious, and the effective barrier height of conduction band for the electrons in EBL is further reduced [107]. Thus, the electron-blocking effect of high Al component EBL is hard to achieve the desired target. Kim et al. showed that 60% of the electrons injected into the active region would overflow to the p-type layer [112]. In order to reduce the stress between the EBL and the barrier in the green LEDs, Lin et al. [113] adopted a quaternary InAlGaN/GaN superlattice electron-blocking layer (SL-EBL) (shown in Fig. 6.22) and improved the efficiency droop phenomena. With that new EBL structure, the optical output power of the green LED was enhanced by 57%, and the efficiency droop was reduced by 30%. Similar structure was also investigated by Tzou et al., and they found that the light output power for the green LEDs was enhanced by 53%.

The hole in the nitride LED has a large effective mass and a smaller mobility, which severely limits the transportation of holes in the active region, restricting the uniform distribution of holes in the multiple quantum wells. The inefficiency injection for the holes limits the external quantum efficiency of the green LED [114]. In addition, the existence of the EBL blocks the transport of holes into the In-rich MQWs. The high Al component also raises the valence-band barrier in EBL, then affecting the injection of holes into the active region. Hwang et al. introduced a new device called light-emitting triode (LET), which has two positive electrodes that can increase hole injection. With the LET, they demonstrated that low hole injection efficiency is an important cause of efficiency droop. To achieve a better hole injection, some methods were introduced. Lin et al. [115] coated Mg-modulation-doped InGaN/GaN superlattices on the In-rich MQWs. Due to the better conductivity, the light output power was enhanced by 100%, compared to that of the LEDs with a high-temperature grown p-GaN layer. It's known that Mg dopants react with the hydrogens in the GaN, resulting in the passivation of the Mg-related acceptor and lower activation efficiency of holes. Furthermore, the higher

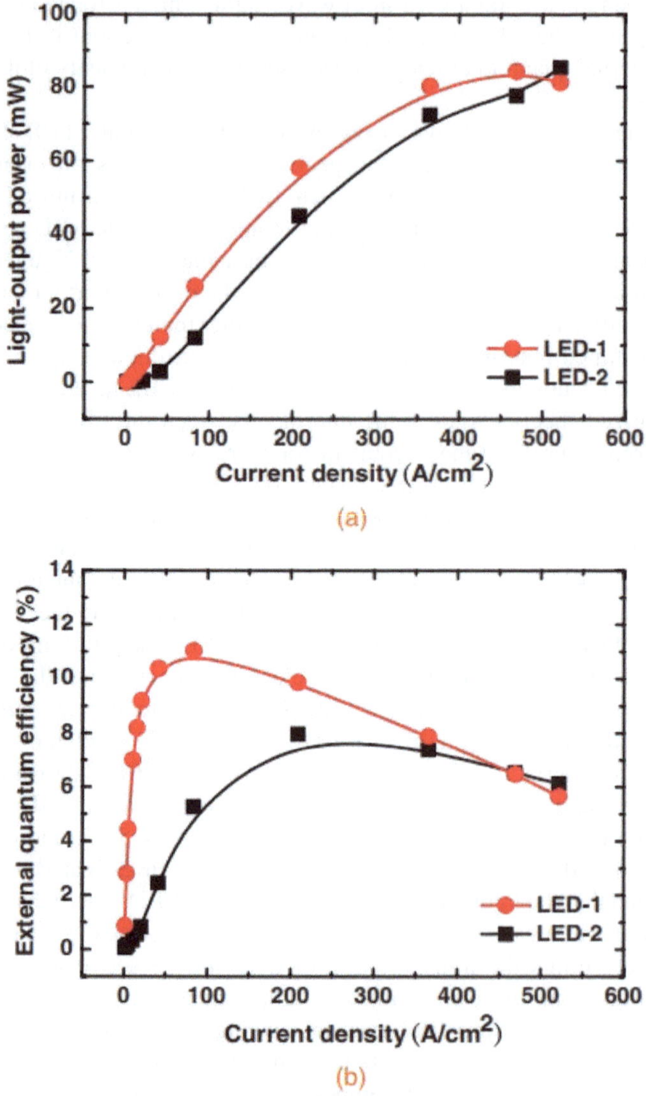

Fig. 6.21 (**a**) Light-output power of samples LED-1 (compared LED) and LED-2 (LED with a 3.5-nm-thick undoped $In_{0.25}Ga_{0.75}N$ test well) as a function of forward current density. (**b**) EQE of samples LED-1 and LED-2 as a function of forward current density

temperature to grow the p-GaN also causes some crystal damage to the InGaN layer, which is under the unstable state. To remove the hydrogen in p-GaN layer and suppress the thermal damage to In-rich MQWs pf the green LEDs, Kim et al. [116] proposed to anneal the p-GaN layer at a low activation temperature (600 °C) by coating a PdZn film on the top of the wafer and remove the hydrogen in the p-GaN

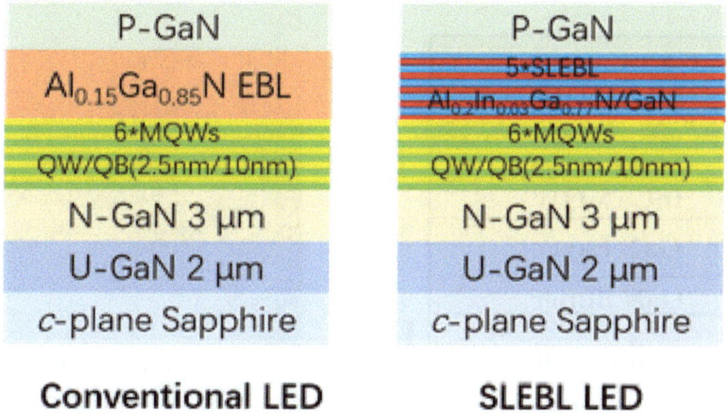

Fig. 6.22 Schematic diagrams of the conventional structure and SL-EBL LEDs

Fig. 6.23 Schematic diagram of the concept of band engineering at EBL

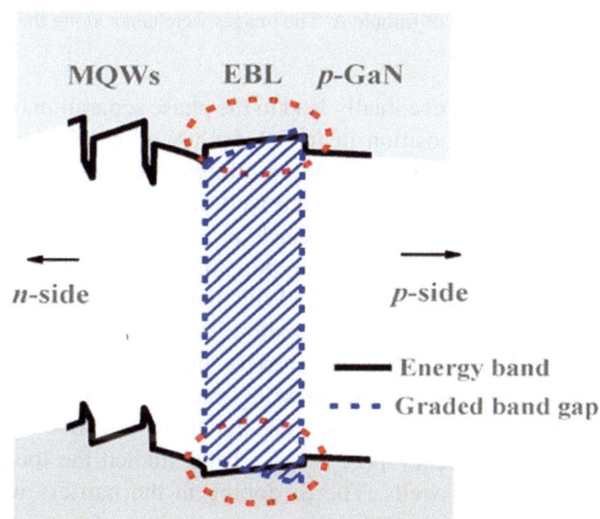

under the annealing temperature of 600 °C. This new annealing method helped to improve the EL intensity by 33% at 20 mA. Researchers also studied the effect of the graded-composition electron-blocking layer (GEBL) on the holes transportation in the green LEDs [117]. Results showed that the GEBL with aluminum composition increasing along the [0001] direction could level down the valence band (schematic diagram of the concept of band engineering at EBL is shown in Fig. 6.23) and then improve hole transportation across the EBL.

InGaN ternary alloy with high In composition is in the thermodynamically non-equilibrium state [77, 118, 119]. This problem is more serious for the green LEDs. In-rich InGaN in the metastable phase is prone to undergo phase separation, resulting in uneven composition distribution within the film. Furthermore, the change of the temperature, the rotating speed, inadequate atomic migration, and

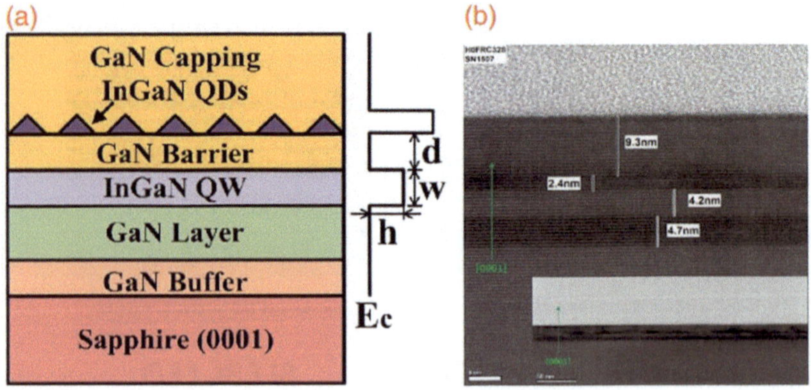

Fig. 6.24 (**a**) Schematic structure of the coupled QW-QD samples' epitaxial structure and conduction band. (**b**) High-resolution (HR) TEM image of sample A. The inset is bright-field (BF) TEM image of sample A. The images were taken along the GaN [11–20]

stress also will eventually lead to the phase separation of the InGaN. The fluctuation of the In composition in In-rich InGaN alloy will lead to the randomness of the potential energy distribution, resulting in the localization of the carriers [120, 121]. For the In-rich green quantum well, the shallow localization centers are related to the low In composition InGaN, in which the potential energy fluctuates around 150 meV. In comparison, the potential energy fluctuation of the deep localization center can reach 300 meV. With the increasing of injected carrier density, the carriers flow between the shallow local centers and the deep local centers. The localized center limits the carriers and reduces the transporting of electrons and holes to defects, thus suppressing the non-radiative recombination. Therefore, the localization of carriers is considered an effective way to increase the internal quantum efficiency [122–124]. Cheng studied the localization behaviors in green light quantum wells. The Si doping in the barriers will result in serious indium clusters, which is attributed to the strain.

Some groups tried to grow the self-organized quantum dots (QD) through phase separation of In-rich InGaN to realize the localization of the carriers for the green LEDs [122, 125–128]. With the $In_{0.25}Ga_{0.75}N$ QDs in the In-rich InGaN quantum wells, Zhang et al. [126] reduced the green LEDs efficiency droop by 32%. The recombination lifetime determined by the temperature-dependent photoluminescence measurements is 0.57 ns, which is due to the localization effect and the smaller piezoelectric field in the QDs. Yu et al. [128] employed the coupled InGaN/GaN QW and QDs structure (shown in the Fig. 6.24) in the green LEDs to enhance the internal quantum efficiency. The results showed that the carriers could tunnel from shallow QWs to deep QDs nearby. Compared with the conventional single QD layer, the IQE of the QDs was enhanced by more than two times, arriving at 45%.

It has been reported that internal non-radiative losses are associated with carrier delocalization in the InGaN-based LEDs [129]. Under low injection current densities, carriers are confined to the QD-like centers with the lowest potential in the quantum wells. As a result, in these localization regions, the defect-related non-radiative recombination efficiency is on the order of 10^7 s^{-1}. At high injection current densities, the carriers exhibit delocalization behavior and diffuse into other areas in the active region. The non-radiation recombination defects in these areas increase the SRH coefficient, making the droop more serious.

Auger recombination is the most widely debated issue in the study of the efficiency of nitride-based green LEDs [130–132]. Auger recombination means that electrons recombine with holes without emitting photons but rather passing energy to the third carrier. The Auger recombination efficiency is related to the Auger coefficient and is proportional to n^3, where n is the carrier concentration [133]. As the injected carrier density is enhanced, the Auger recombination increases, and the luminous efficiency is reduced. It is reported that only the Auger recombination coefficient greater than 10^{-31} cm^6 s^{-1} will significantly affect the green LED droop. In theory, for a single-crystal material, the Auger coefficient C is generally lower than 10^{-32} cm s^{-1}, while for the active region, C is 10^{-31} to 10^{-29} cm^6 s^{-1}. It is reported that when the Auger recombination coefficient is greater than 10^{-31} cm^6 s^{-1}, the effect of the Auger recombination on the droop is more obvious, which is in good agreement with the experimental results. Felix Nippert et al. found that the IQE losses are related to the temperature dependence of the radiative and Auger processes. While the radiative process increases only weakly with temperature, non-radiative Auger recombination increases steeply, becoming a significant contributor to the green gap [134].

6.2.3 *Light Extraction*

In the GaN-based green LED, the light extraction efficiency seriously restricts the improvement of external quantum efficiency. For the green LEDs, if the incident angle of the photon is less than the critical angle, the photon can escape from the materials; otherwise the photon is difficult to escape the semiconductor [135]. However, due to the high refractive index of the nitride semiconductor material and the absorption of photons in the material, the photon-extraction efficiency is very low.

To reduce the light absorption by the metal pads and enhance light extraction efficiency, the transparent highly conductive electrodes are employed in the LEDs. The transparent highly conductive electrode is used to increase the lateral current spread and reduce the photons absorption by the metal electrodes. Due to the low p-type carrier concentration in GaN, the lateral spread of current in p-type GaN is relatively poor. Therefore, indium tin oxide (ITO) [136] is widely used as a current spreading layer for the GaN-based LEDs. In addition, semiconductor materials such as ZnO [137], TiO2 [136], MgO [138], and graphene[139, 140], which are preferable in terms of lateral conductivity, are also used as the transparent electrodes.

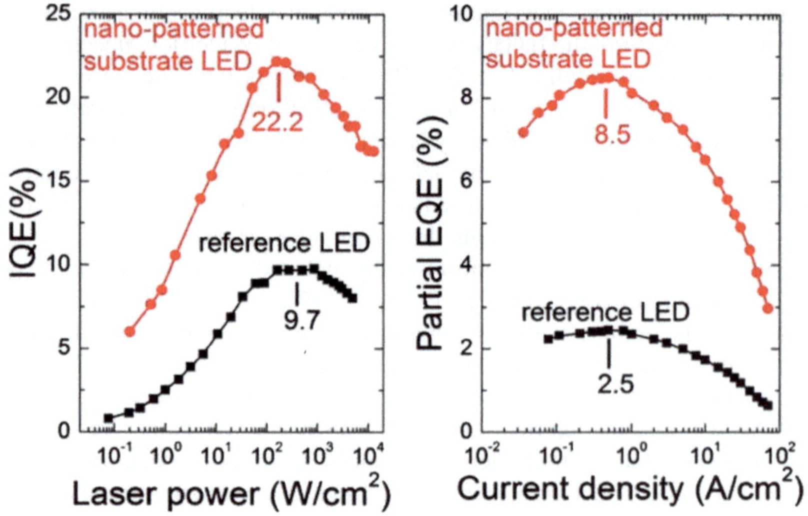

Fig. 6.25 (**a**) IQE as a function of excitation power density, (**b**) EQE as a function of current density for the nanopatterned substrate circles and the reference squares LED

The metal reflector is deposited on the back of the LED to improve the reflection of light, improving the efficiency of photons extraction [141, 142]. The most commonly used metal reflector for GaN-based LEDs is Ni/Al/Pt/Au. In addition, a distributed Bragg reflector (DBR) with multiple layers of different refractive index materials can also be used as the reflector [143, 144].

In order to improve the efficiency of photon extraction, new chip shapes are proposed [145–149]. Since the LED with a planar structure has a small critical angle, it is difficult for the photon to escape the semiconductor. Therefore, it is necessary to design a new chip shape structure to improve the probability of photon escape. For example, the hemispherical shape design can better improve the critical angle of the outgoing photon in the device. The small lens prepared by the etching method can greatly improve the light output power. Omron uses a combination of two lenses with different sizes and doubles the efficiency of LED light extraction. In addition, change in the shape of the LED chip, such as quadrilateral, triangular, round, pyramid, funnel, inverted trapezoid, etc., can increase the LED light area and finally improve the LED light extraction efficiency.

Patterned sapphire substrate (PSS) technology is also used to improve the efficiency of LED light extraction [150–153]. As is shown in Fig. 6.25, for the LED with sapphire substrate, the spectral interference and emission patterns reveal a 58% enhanced light extraction, while photoluminescence reveals a doubling of the internal quantum efficiency [152]. The nanoscale or microscale pattern on the substrate can effectively improve the photon reflection, change the emitting direction of the photons, and ultimately achieve the promotion of LED light extraction. Furthermore, PSS also can solve the problem of epitaxial growth quality.

Surface roughening technology can also improve the efficiency of LED light [154–156]. There are a variety of surface roughening techniques, such as laser bombardment of LED material surfaces, wet etching, secondary epitaxial growth of materials, or the combination of the above techniques. These surface roughening techniques are relatively simple and practical and can achieve better photon escape efficiency.

References

1. H.P. Maruskas, J.J. Tietjen, The preparation and properties of vapor deposited single-crystalline GaN. Appl. Phys. Lett. **15**(10), 327 (1969)
2. A.G. Bhuiyan, A. Hashimoto, A. Yamamoto, Indium nitride (Inn): a review on growth, characterization, and properties. J. Appl. Phys. **94**(5), 2779–2808 (2003)
3. H. Amano, N. Sawaki, I. Akasaki, Y. Toyoda, Metalorganic vapor phase epitaxial growth of a high quality GaN film using an Aln buffer layer. Appl. Phys. Lett. **48**(5), 353–355 (1986)
4. S. Yoshida, S. Misawa, S. Gonda, Improvements on the electrical and luminescent properties of reactive molecular beam epitaxially grown GaN films by using Aln-coated sapphire substrates. Appl. Phys. Lett. **42**(5), 427–429 (1983)
5. D.K. Wickenden, T.J. Kistenmacher, W.A. Bryden, J.S. Morgan, A.E. Wickenden, The effect of self nucleation layers on the MOCVD growth of gallium nitride on sapphire. Mater. Res. Soc. **221**, 167–172 (1991)
6. S. Nakamura, GaN growth using GaN buffer layer. Jpn. J. Appl. Phys. **30**(4A), L1705–L1707 (1991)
7. S. Nakamura, The roles of structural imperfections in Ingan-based blue light-emitting diodes and laser diodes. Science **281**(5379), 956–961 (1998)
8. S. Tomoya, S. Hisao, H. Maosheng, N. Yoshiki, K. Satoshi, T. Satoru, Y. Kenji, N. Katsushi, T.R. Linda, S. Shiro, Direct evidence that dislocations are non-radiative recombination centers in GaN. Jpn. J. Appl. Phys. **37**(4A), L398–L400 (1998)
9. K. Forghani, M. Klein, F. Lipski, S. Schwaiger, J. Hertkorn, R.A.R. Leute, F. Scholz, M. Feneberg, B. Neuschl, K. Thonke, O. Klein, U. Kaiser, R. Gutt, T. Passow, High quality AlGaN epilayers grown on sapphire using SiNx interlayers. J. Cryst. Growth **315**(1), 216–219 (2011)
10. H.-Y. Shin, S.K. Kwon, Y.I. Chang, M.J. Cho, K.H. Park, Reducing dislocation density in GaN films using a cone-shaped patterned sapphire substrate. J. Cryst. Growth **311**(17), 4167–4170 (2009)
11. C.I.H. Ashby, C.C. Mitchell, J. Han, N.A. Missert, P.P. Provencio, D.M. Follstaedt, G.M. Peake, L. Griego, Low-dislocation-density GaN from a single growth on a textured substrate. Appl. Phys. Lett. **77**(20), 3233–3235 (2000)
12. L. Meng, W. Guohong, L. Hongjian, L. Zhicong, Y. Ran, L. Panpan, L. Jing, Y. Xiaoyan, W. Junx, L. Jinmin, Low threading dislocation density in GaN films grown on patterned sapphire substrates. J. Semicond. **33**, 113002 (2012)
13. H.K.N. Kazumasa, O. Masaru, M. Hiromitsu, N. Mitsuhisa, M. Atsushi, M. Hideto, I. Yasushi, M. Takayoshi, Fabrication and characterization of low defect density GaN using facet-controlled epitaxial lateral overgrowth (facelo). J. Cryst. Growth **221**(1-4), 316–326 (2000)
14. K. Iida, T. Kawashima, M. Iwaya, S. Kamiyama, H. Amano, I. Akasaki, A. Bandoh, Epitaxial lateral overgrowth of Alxga1−Xn (X>0.2) on sapphire and its application to Uv-B-light-emitting devices. J. Cryst. Growth **298**, 265–267 (2007)

15. R. Matsuoka, T. Okimoto, K. Nishino, Y. Naoi, S. Sakai, Algan epitaxial lateral overgrowth on Ti-evaporated GaN/sapphire substrate. J. Cryst. Growth **311**(10), 2847–2849 (2009)
16. T. Satoru, T. Misaichi, A. Yoshinobu, Anti-surfactant in Iii-nitride epitaxy quantum dot formation and dislocation termination. Jpn. J. Appl. Phys. **39**(8B), L831–L834 (2000)
17. K. Engl, M. Beer, N. Gmeinwieser, U.T. Schwarz, J. Zweck, W. Wegscheider, S. Miller, A. Miler, H.J. Lugauer, G. Brüderl, A. Lell, V. Härle, Influence of an in situ-deposited intermediate layer inside GaN and AlGaN layers on SiC substrates. J. Cryst. Growth **289**(1), 6–13 (2006)
18. K. Yoshiki, K. Shota, H. Kazumasa, S. Nobuhiko, Selective growth of wurtzite GaN and AlGaN on GaN/sapphire substrates by metalorganic vapor phase epitaxy. J. Cryst. Growth **144**(3–4), 133–140 (1994)
19. O. Klein, J. Biskupek, U. Kaiser, K. Forghani, S.B. Thapa, F. Scholz, Simulation supported analysis of the effect of sinxinterlayers in AlGaN on the dislocation density reduction. J. Phys. Conf. Ser. **209**(1), 012018 (2010)
20. S. Haffouz, H. Lahrèche, P. Vennéguès, P. de Mierry, B. Beaumont, F. Omnès, P. Gibart, The effect of the Si/N treatment of a nitridated sapphire surface on the growth mode of GaN in low-pressure metalorganic vapor phase epitaxy. Appl. Phys. Lett. **73**(9), 1278–1280 (1998)
21. V. Srikant, J.S. Speck, D.R. Clarke, Mosaic structure in epitaxial thin films having large lattice mismatch. J. Appl. Phys. **82**(9), 4286–4295 (1997)
22. K.K. Ansah Antwi, C.B. Soh, Q. Wee, R.J.N. Tan, P. Yang, H.R. Tan, L.F. Sun, Z.X. Shen, S.J. Chua, Crystallographically tilted and partially strain relaxed GaN grown on inclined {111} facets etched on Si(100) substrate. J. Appl. Phys. **114**(24), 243512 (2013)
23. L. Liu, J.H. Edgar, Substrates for gallium nitride epitaxy. Mater. Sci. Eng. **37**(3), 61–127 (2002)
24. Z.Y. Xie, C.H. Wei, S.F. Chen, S.Y. Jiang, J.H. Edgar, Surface etching of 6h-Sic (0001) and surface morphology of the subsequently grown GaN via MOCVD. J. Electron. Mater. **29**(4), 411 (2000)
25. M.I. Utama, Q. Zhang, J. Zhang, Y. Yuan, F.J. Belarre, J. Arbiol, Q. Xiong, Recent developments and future directions in the growth of nanostructures by Van Der Waals epitaxy. Nanoscale **5**(9), 3570–3588 (2013)
26. K. Chung, C.H. Lee, G.C. Yi, Transferable GaN layers grown on ZNO-coated graphene layers for optoelectronic devices. Science **330**(6004), 655–657 (2010)
27. K. Chung, S. In Park, H. Baek, J.-S. Chung, G.-C. Yi, High-quality GaN films grown on chemical vapor-deposited graphene films. NPG Asia Mater. **4**(9), e24 (2012)
28. Z.Y. Al Balushi, T. Miyagi, Y.-C. Lin, K. Wang, L. Calderin, G. Bhimanapati, J.M. Redwing, J.A. Robinson, The impact of graphene properties on GaN and Aln nucleation. Surf. Sci. **634**, 81–88 (2015)
29. J. Kim, C. Bayram, H. Park, C.W. Cheng, C. Dimitrakopoulos, J.A. Ott, K.B. Reuter, S.W. Bedell, D.K. Sadana, Principle of direct Van Der Waals epitaxy of single-crystalline films on epitaxial graphene. Nat. Commun. **5**, 4836 (2014)
30. S.J. Chae, Y.H. Kim, T.H. Seo, D.L. Duong, S.M. Lee, M.H. Park, E.K. Suh, Direct growth of etch pit-free GaN crystals on few-layer graphene. RSC Adv. **5**(2), 1343–1349 (2012)
31. J.W. Shon, J. Ohta, K. Ueno, A. Kobayashi, H. Fujioka, Fabrication of full-color InGaN-based light-emitting diodes on amorphous substrates by pulsed sputtering. Sci. Rep. **4**, 5325 (2014)
32. L. Qi, Y. Xu, Z. Li, E. Zhao, S. Yang, B. Cao, J. Zhang, J. Wang, K. Xu, Stress analysis of transferable crack-free gallium nitride microrods grown on graphene/Sic substrate. Mater. Lett. **185**, 315–318 (2016)
33. N. Han, T.V. Cuong, M. Han, B.D. Ryu, S. Chandramohan, J.B. Park, J.H. Kang, Y.J. Park, K.B. Ko, H.Y. Kim, H.K. Kim, J.H. Ryu, Y.S. Katharria, C.J. Choi, C.H. Hong, Improved heat dissipation in gallium nitride light-emitting diodes with embedded graphene oxide pattern. Nat. Commun. **4**, 1452 (2013)
34. M. Han, N. Han, E. Jung, B.D. Ryu, K.B. Ko, T.v. Cuong, H. Kim, J.K. Kim, C.-H. Hong, Effect of curved graphene oxide in a GaN light-emitting-diode for improving heat dissipation with a patterned sapphire substrate. Semicond. Sci. Technol. **31**(8), 085010 (2016)

35. J.W. Chung, F.S. Ohuchi, Deposition of AlN on WS2 (0001) substrate by atomic layer growth process. MRS Online Proc. Library Arch. **449**, 379–384 (1996)
36. A. Yamada, K.P. Ho, T. Maruyama, K. Akimoto, Molecular beam epitaxy of GaN on a substrate of Mos2 layered compound. Appl. Phys. A **69**(1), 89–92 (1999)
37. P. Gupta, A.A. Rahman, S. Subramanian, S. Gupta, A. Thamizhavel, T. Orlova, S. Rouvimov, S. Vishwanath, V. Protasenko, M.R. Laskar, H.G. Xing, D. Jena, A. Bhattacharya, Layered transition metal dichalcogenides: promising near-lattice-matched substrates for GaN growth. Sci. Rep. **6**, 23708 (2016)
38. M. Tangi, P. Mishra, C.C. Tseng, T.K. Ng, M.N. Hedhili, D.H. Anjum, M.S. Alias, N. Wei, L.J. Li, B.S. Ooi, Band alignment at GaN/single-layer WSe2 interface. ACS Appl. Mater. Interfaces **9**(10), 9110–9117 (2017)
39. M. Tangi, P. Mishra, T.K. Ng, M.N. Hedhili, B. Janjua, M.S. Alias, D.H. Anjum, C.-C. Tseng, Y. Shi, H.J. Joyce, L.-J. Li, B.S. Ooi, Determination of band offsets at GaN/single-layer MoS2 heterojunction. Appl. Phys. Lett. **109**(3), 032104 (2016)
40. S. Hwang, W. Jin Ha, J. Kyu Kim, J. Xu, J. Cho, E. Fred Schubert, Promotion of hole injection enabled by Gainn/GaN light-emitting triodes and its effect on the efficiency droop. Appl. Phys. Lett. **99**(18), 181115 (2011)
41. J. Kang, H. Li, Z. Li, Z. Liu, P. Ma, X. Yi, G. Wang, Enhancing the performance of green GaN-based light-emitting diodes with graded superlattice AlGaN/GaN inserting layer. Appl. Phys. Lett. **103**(10), 102104 (2013)
42. Y. Zhao, S. Tanaka, C.-C. Pan, K. Fujito, D. Feezell, J.S. Speck, S.P. DenBaars, S. Nakamura, High-power blue-violet semipolar ($20\bar{2}\bar{1}$) InGaN/GaN light-emitting diodes with low efficiency droop at 200 A/cm^{2}. Appl. Phys. Express **4**(8), 082104 (2011)
43. H.J. Chung, R.J. Choi, M.H. Kim, J.W. Han, Y.M. Park, Y.S. Kim, H.S. Paek, C.S. Sone, Y.J. Park, J.K. Kim, E.F. Schubert, Improved performance of GaN-based blue light emitting diodes with InGaN/GaN multilayer barriers. Appl. Phys. Lett. **95**(24), 241109 (2009)
44. H.P. Nguyen, K. Cui, S. Zhang, M. Djavid, A. Korinek, G.A. Botton, Z. Mi, Controlling electron overflow in phosphor-free InGaN/GaN nanowire white light-emitting diodes. Nano Lett. **12**(3), 1317–1323 (2012)
45. X. Ni, X. Li, J. Lee, S. Liu, V. Avrutin, Ü. Özgür, H. Morkoç, A. Matulionis, T. Paskova, G. Mulholland, K.R. Evans, InGaN staircase electron injector for reduction of electron overflow in InGaN light emitting diodes. Appl. Phys. Lett. **97**(3), 031110 (2010)
46. C. Weisbuch, M. Piccardo, L. Martinelli, J. Iveland, J. Peretti, J.S. Speck, The efficiency challenge of nitride light-emitting diodes for lighting. Phys. Status Solidi A **212**(5), 899–913 (2015)
47. T. Bret, V. Wagner, D. Martin, P. Hoffmann, M. Ilegems, A mechanistic study of GaN laser lift-off. Phys. Stat. Sol. A **194**(2), 559–562 (2002)
48. J.S. Ha, S.W. Lee, H.J. Lee, H.J. Lee, S.H. Lee, H. Goto, T. Kato, K. Fujii, M.W. Cho, T. Yao, The fabrication of vertical light-emitting diodes using chemical lift-off process. IEEE Photon. Technol. Lett. **20**(1-4), 175–177 (2008)
49. Y.H. Zhou, Y.W. Tang, J.P. Rao, F. Jiang, Improvement for extraction efficiency of vertical GaN-based LED on si substrate by photo-enhanced wet etching. Acta Opt. Sin. **29**(1), 252–255 (2009)
50. M. Tchernycheva, P. Lavenus, H. Zhang, A.V. Babichev, G. Jacopin, M. Shahmohammadi, F.H. Julien, R. Ciechonski, G. Vescovi, O. Kryliouk, InGaN/GaN core-shell single nanowire light emitting diodes with graphene-based p-contact. Nano Lett. **14**(5), 2456–2465 (2014)
51. J.R. Riley, S. Padalkar, Q. Li, P. Lu, D.D. Koleske, J.J. Wierer, G.T. Wang, L.J. Lauhon, Three-dimensional mapping of quantum wells in a GaN/InGaN core-shell nanowire light-emitting diode array. Nano Lett. **13**(9), 4317–4325 (2013)
52. H.P. Nguyen, S. Zhang, A.T. Connie, M.G. Kibria, Q. Wang, I. Shih, Z. Mi, Breaking the carrier injection bottleneck of phosphor-free nanowire white light-emitting diodes. Nano Lett. **13**(11), 5437–5442 (2013)

53. R. Koester, J.S. Hwang, D. Salomon, X. Chen, C. Bougerol, J.P. Barnes, S. Dang Dle, L. Rigutti, A. de Luna Bugallo, G. Jacopin, M. Tchernycheva, C. Durand, J. Eymery, M-plane core-shell InGaN/GaN multiple-quantum-wells on GaN wires for electroluminescent devices. Nano Lett. **11**(11), 4839–4845 (2011)
54. F. Qian, S. Gradečak, Y. Li, C. Wen, C.M. Lieber, Core/multishell nanowire heterostructures as multicolor, high-efficiency light-emitting diodes. Nano Lett. **5**(11), 2287–2291 (2005)
55. P. Waltereit, O. Brandt, A. Trampert, H.T. Grahn, J. Menniger, M. Ramsteiner, M. Ramsteiner, M. Reiche, K.H. Ploog, Nitride semiconductors free of electrostatic fields for efficient white light-emitting diodes. Nature **406**(24), 865–868 (2000)
56. S.P. Chang, Y.C. Chen, J.K. Huang, Y.J. Cheng, J.R. Chang, K.P. Sou, Y.T. Kang, H.C. Yang, T.C. Hsu, H.C. Kuo, C.Y. Chang, Electrically driven nanopyramid green light emitting diode. Appl. Phys. Lett. **100**(6), 061106 (2012)
57. S. Li, A. Waag, GaN based nanorods for solid state lighting. J. Appl. Phys. **111**(7), 071101 (2012)
58. W. Guo, M. Zhang, A. Banerjee, P. Bhattacharya, Catalyst-free InGaN/GaN nanowire light emitting diodes grown on (001) silicon by molecular beam epitaxy. Nano Lett. **10**(9), 3355–3359 (2010)
59. R. Armitage, K. Tsubaki, Multicolour luminescence from InGaN quantum wells grown over GaN nanowire arrays by molecular-beam epitaxy. Nanotechnology **21**(19), 195202 (2010)
60. J. Ma, L. Wang, Z. Liu, G. Yuan, X. Ji, P. Ma, J. Wang, X. Yi, G. Wang, J. Li, Nitride-based micron-scale hexagonal pyramids array vertical light emitting diodes by N-polar wet etching. Opt. Express **21**(3), 3547–3556 (2013)
61. L. Wang, J. Ma, Z. Liu, X. Yi, G. Yuan, G. Wang, N-polar GaN etching and approaches to quasi-perfect micro-scale pyramid vertical light-emitting diodes array. J. Appl. Phys. **114**(13), 133101 (2013)
62. J. Kang, Z. Li, Z. Liu, H. Li, Y. Zhao, Y. Tian, P. Ma, X. Yi, G. Wang, Investigation of the wet-etching mechanism of Ga-polar AlGaN/GaN micro-pillars. J. Cryst. Growth **386**, 175–178 (2014)
63. R.S. Wagner, W.C. Ellis, Vapor-liquid-solid mechanism of single crystal growth. Appl. Phys. Lett. **4**(5), 89–90 (1964)
64. T. Kuykendall, P.J. Pauzauskie, Y. Zhang, J. Goldberger, D. Sirbuly, J. Denlinger, P. Yang, Crystallographic alignment of high-density gallium nitride nanowire arrays. Nat. Mater. **3**(8), 524–528 (2004)
65. S. Wu, L. Wang, X. Yi, Z. Liu, T. Wei, G. Yuan, J. Wang, J. Li, Influence of lateral growth on the optical properties of GaN nanowires grown by hydride vapor phase epitaxy. J. Appl. Phys. **122**(20), 205302 (2017)
66. S. Wu, L. Wang, X. Yi, Z. Liu, J. Yan, G. Yuan, T. Wei, J. Wang, J. Li, Crystallographic orientation control and optical properties of GaN nanowires. RSC Adv. **8**(4), 2181–2187 (2018)
67. S. Wu, L. Wang, Z. Liu, X. Yi, Y. Huang, C. Yang, T. Wei, J. Yan, G. Yuan, J. Wang, J. Li, Ultrafast growth of horizontal GaN nanowires by HVPE through flipping the substrate. Nanoscale **10**(13), 5888–5896 (2018)
68. D. Stephen, X.S. Hersee, X. Wang, The controlled growth of GaN nanowires. Nano Lett. **6**(8), 1808–1811 (2006)
69. P. Ren, G. Han, B.-L. Fu, B. Xue, N. Zhang, Z. Liu, L.-X. Zhao, J.-X. Wang, J.-M. Li, Selective area growth and characterization of GaN nanorods fabricated by adjusting the hydrogen flow rate and growth temperature with metal organic chemical vapor deposition. Chin. Phys. Lett. **33**(6), 068101 (2016)
70. K. Wu, T. Wei, D. Lan, X. Wei, H. Zheng, Y. Chen, H. Lu, K. Huang, J. Wang, Y. Luo, J. Li, Phosphor-free nanopyramid white light-emitting diodes grown on $\{10\bar{1}1\}$ planes using nanospherical-lens photolithography. Appl. Phys. Lett. **103**(24), 241107 (2013)
71. H. Sekiguchi, K. Kishino, A. Kikuchi, Emission color control from blue to red with nanocolumn diameter of InGaN/GaN nanocolumn arrays grown on same substrate. Appl. Phys. Lett. **96**(23), 231104 (2010)

72. H. Lin, Y. Lu, H. Chen, H. Lee, S. Gwo, InGaN/GaN nanorod array white light-emitting diode. Appl. Phys. Lett. **97**(7), 073101 (2010)
73. Y.H. Ra, R. Navamathavan, J.H. Park, C.R. Lee, Coaxial In(x)Ga(1-x)N/GaN multiple quantum well nanowire arrays on Si(111) substrate for high-performance light-emitting diodes. Nano Lett. **13**(8), 3506–3516 (2013)
74. Y.J. Hong, C.H. Lee, A. Yoon, M. Kim, H.K. Seong, H.J. Chung, C. Sone, Y.J. Park, G.C. Yi, Visible-color-tunable light-emitting diodes. Adv. Mater. **23**(29), 3284–3288 (2011)
75. V. Fiorentini, F. Bernardini, F. Della Sala, A. Di Carlo, P. Lugli, Effects of macroscopic polarization in III-V nitride multiple quantum wells. Phys. Rev. B **60**(12), 8849 (1999)
76. F. Bernardini, V. Fiorentini, in *Spontaneous vs. piezoelectric polarization in III-V nitrides: conceptual aspects and practical consequences*. arXiv preprint cond-mat/9908087, 1999
77. D. Doppalapudi, S. Basu, K. Ludwig Jr., T. Moustakas, Phase separation and ordering in InGaN alloys grown by molecular beam epitaxy. J. Appl. Phys. **84**(3), 1389–1395 (1998)
78. F. Bernardini, V. Fiorentini, D. Vanderbilt, Spontaneous polarization and piezoelectric constants of III-V nitrides. Phys. Rev. B **56**(16), R10024 (1997)
79. S. Watanabe, N. Yamada, M. Nagashima, Y. Ueki, C. Sasaki, Y. Yamada, T. Taguchi, K. Tadatomo, H. Okagawa, H. Kudo, Internal quantum efficiency of highly-efficient In x Ga 1−x N-based near-ultraviolet light-emitting diodes. Appl. Phys. Lett. **83**(24), 4906–4908 (2003)
80. A. Hangleiter, F. Hitzel, S. Lahmann, U. Rossow, Composition dependence of polarization fields in GaInN/GaN quantum wells. Appl. Phys. Lett. **83**(6), 1169–1171 (2003)
81. J. Xu, M.F. Schubert, A.N. Noemaun, D. Zhu, J.K. Kim, E.F. Schubert, M.H. Kim, H.J. Chung, S. Yoon, C. Sone, Reduction in efficiency droop, forward voltage, ideality factor, and wavelength shift in polarization-matched GaInN/GaInN multi-quantum-well light-emitting diodes. Appl. Phys. Lett. **94**(1), 011113 (2009)
82. T. Takeuchi, S. Sota, M. Katsuragawa, M. Komori, H. Takeuchi, H. Amano, I. Akasaki, Quantum-confined Stark effect due to piezoelectric fields in GaInN strained quantum wells. Jpn. J. Appl. Phys. **36**(4A), L382 (1997)
83. C.-F. Huang, T.-C. Liu, Y.-C. Lu, W.-Y. Shiao, Y.-S. Chen, J.-K. Wang, C.-F. Lu, C. Yang, Enhanced efficiency and reduced spectral shift of green light-emitting-diode epitaxial structure with prestrained growth. J. Appl. Phys. **104**(12), 123106 (2008)
84. J.Y. Park, J.H. Lee, S. Jung, T. Ji, InGaN/GaN-based green-light-emitting diodes with an inserted InGaN/GaN-graded superlattice layer. Phys. Status Solidi A **213**(6), 1610–1614 (2016)
85. W. Lundin, A. Nikolaev, A. Sakharov, E. Zavarin, G. Valkovskiy, M. Yagovkina, S. Usov, N. Kryzhanovskaya, V. Sizov, P. Brunkov, Single quantum well deep-green LEDs with buried InGaN/GaN short-period superlattice. J. Cryst. Growth **315**(1), 267–271 (2011)
86. K. Lee, C.-R. Lee, J.H. Lee, T.-H. Chung, M.-Y. Ryu, K.-U. Jeong, J.-Y. Leem, J.S. Kim, Influences of Si-doped graded short-period superlattice on green InGaN/GaN light-emitting diodes. Opt. Express **24**(7), 7743–7751 (2016)
87. Y. Xia, W. Hou, L. Zhao, M. Zhu, T. Detchprohm, C. Wetzel, Boosting green GaInN/GaN light-emitting diode performance by a GaInN underlying layer. IEEE Trans. Electron. Dev. **57**(10), 2639–2643 (2010)
88. L. Hsu, W. Walukiewicz, Effect of polarization fields on transport properties in AlGaN/GaN heterostructures. J. Appl. Phys. **89**(3), 1783–1789 (2001)
89. A. Hangleiter, J.S. Im, H. Kollmer, S. Heppel, J. Off, F. Scholz, The role of piezoelectric fields in GaN-based quantum wells. MRS Internet J. Nitride Semicond. Res. **3**, e15 (1998)
90. C. Wood, D. Jena, *Polarization Effects in Semiconductors: From Ab Initio Theory to Device Applications* (Springer, New York, 2007)
91. Z. Ning, L. Zhe, S. Zhao, R. Peng, W. Xiao-Dong, F. Xiang-Xu, D. Peng, D. Cheng-Xiao, Z. Shao-Xin, F. Bing-Lei, Reduction of efficiency droop and modification of polarization fields of InGaN-based green light-emitting diodes via Mg-doping in the barriers. Chin. Phys. Lett. **30**(8), 087101 (2013)

92. C.-Y. Huang, Q. Yan, Y. Zhao, K. Fujito, D. Feezell, C.G. Van de Walle, J.S. Speck, S.P. DenBaars, S. Nakamura, Influence of Mg-doped barriers on semipolar (20 2¯ 1) multiple-quantum-well green light-emitting diodes. Appl. Phys. Lett. **99**(14), 141114 (2011)
93. J. Zhang, X.-J. Zhuo, D.-W. Li, L. Yu, K. Li, Y.-W. Zhang, J.-S. Diao, X.-F. Wang, S.-T. Li, Effect of Mg doping in GaN interlayer on the performance of green light-emitting diodes. IEEE Photon. Technol. Lett. **27**(2), 117–120 (2015)
94. Z. Lin, R. Hao, G. Li, S. Zhang, Effect of Si doping in barriers of InGaN/GaN multiple quantum wells on the performance of green light-emitting diodes. Jpn. J. Appl. Phys. **54**(2), 022102 (2015)
95. J.-H. Ryou, J. Limb, W. Lee, J. Liu, Z. Lochner, D. Yoo, R.D. Dupuis, Effect of silicon doping in the quantum-well barriers on the electrical and optical properties of visible green light-emitting diodes. IEEE Photon. Technol. Lett. **20**(21), 1769–1771 (2008)
96. F. Scholz, Semipolar GaN grown on foreign substrates: a review. Semicond. Sci. Technol. **27**(2), 024002 (2012)
97. Z. Wu, A. Fischer, F. Ponce, B. Bastek, J. Christen, T. Wernicke, M. Weyers, M. Kneissl, Structural and optical properties of nonpolar GaN thin films. Appl. Phys. Lett. **92**(17), 171904 (2008)
98. N. Kriouche, P. Vennéguès, M. Nemoz, G. Nataf, P. De Mierry, Stacking faults blocking process in (11− 22) semipolar GaN growth on sapphire using asymmetric lateral epitaxy. J. Cryst. Growth **312**(19), 2625–2630 (2010)
99. H. Zhao, G. Liu, X.-H. Li, R. Arif, G. Huang, J. Poplawsky, S.T. Penn, V. Dierolf, N. Tansu, Design and characteristics of staggered InGaN quantum-well light-emitting diodes in the green spectral regime. IET Optoelectron. **3**(6), 283–295 (2009)
100. H. Zhao, G. Liu, J. Zhang, J.D. Poplawsky, V. Dierolf, N. Tansu, Approaches for high internal quantum efficiency green InGaN light-emitting diodes with large overlap quantum wells. Opt. Express **19**(104), A991–A1007 (2011)
101. Y.-A. Chang, Y.-T. Kuo, J.-Y. Chang, Y.-K. Kuo, Investigation of InGaN green light-emitting diodes with chirped multiple quantum well structures. Opt. Lett. **37**(12), 2205–2207 (2012)
102. H. Zhao, G. Liu, N. Tansu, Analysis of InGaN-delta-InN quantum wells for light-emitting diodes. Appl. Phys. Lett. **97**(13), 131114 (2010)
103. C. Bayram, F.H. Teherani, D. Rogers, M. Razeghi, A hybrid green light-emitting diode comprised of n-ZnO/(InGaN/GaN) multi-quantum-wells/p-GaN. Appl. Phys. Lett. **93**(8), 081111 (2008)
104. S. Verma, S.K. Pandey, S.K. Pandey, S. Mukherjee, Theoretical simulation of Hybrid II-O/III-N green light-emitting diode with MgZnO/InGaN/MgZnO heterojunction. Mater. Sci. Semicond. Process. **31**, 340–350 (2015)
105. L. Han, K. Kash, H. Zhao, High-efficiency green light-emitting diodes based on InGaN-ZnGeN 2 type-II quantum wells. Proc. SPIE **2014**, 90030 (2014)
106. Y.-J. Lee, C.-H. Chen, C.-J. Lee, Reduction in the efficiency-droop effect of InGaN green light-emitting diodes using gradual quantum wells. IEEE Photon. Technol. Lett. **22**(20), 1506–1508 (2010)
107. J. Piprek, S. Li, Electron leakage effects on GaN-based light-emitting diodes. Opt. Quant. Electron. **42**(2), 89–95 (2010)
108. N. Zhang, Z. Liu, T. Wei, L. Zhang, X. Wei, X. Wang, H. Lu, J. Li, J. Wang, Effect of the graded electron blocking layer on the emission properties of GaN-based green light-emitting diodes. Appl. Phys. Lett. **100**(5), 053504 (2012)
109. K.J. Vampola, M. Iza, S. Keller, S.P. DenBaars, S. Nakamura, Measurement of electron overflow in 450 nm InGaN light-emitting diode structures. Appl. Phys. Lett. **94**(6), 061116 (2009)
110. L.-B. Chang, M.-J. Lai, R.-M. Lin, C.-H. Huang, Effect of electron leakage on efficiency droop in wide-well InGaN-based light-emitting diodes. Appl. Phys. Express **4**(1), 012106 (2011)
111. J. Piprek, Z. Simon Li, Sensitivity analysis of electron leakage in III-nitride light-emitting diodes. Appl. Phys. Lett. **102**(13), 131103 (2013)

112. H.J. Kim, S. Choi, S.-S. Kim, J.-H. Ryou, P.D. Yoder, R.D. Dupuis, A.M. Fischer, K. Sun, F.A. Ponce, Improvement of quantum efficiency by employing active-layer-friendly lattice-matched InAlN electron blocking layer in green light-emitting diodes. Appl. Phys. Lett. **96**(10), 101102 (2010)
113. D.-W. Lin, A.-J. Tzou, J.-K. Huang, B.-C. Lin, C.-Y. Chang, H.-C. Kuo, in Greatly improved efficiency droop for InGaN-based green light emitting diodes by quaternary content superlattice electron blocking layer. *2015 International Conference on Numerical Simulation of Optoelectronic Devices (NUSOD)* (IEEE, 2015). pp. 15–16
114. C. Wang, S. Chang, P. Ku, J. Li, Y. Lan, C. Lin, H. Yang, H. Kuo, T. Lu, S. Wang, Hole transport improvement in InGaN/GaN light-emitting diodes by graded-composition multiple quantum barriers. Appl. Phys. Lett. **99**(17), 171106 (2011)
115. H.C. Lin, G.Y. Lee, H.H. Liu, N.W. Hsu; C.C. Wu, J.I. Chyi, in Polarization-enhanced Mg doping in InGaN/GaN superlattice for green light-emitting diodes. *Conference on Lasers and Electro-Optics, Optical Society of America*, 2009, p. CMM4
116. J.-Y. Kim, M.-K. Kwon, S.-J. Park, S. Kim, J.W. Kim, Y.C. Kim, Improving the performance of green LEDs by low-temperature annealing of p-GaN with PdZn. Electrochem. Solid-State Lett. **12**(5), H185–H187 (2009)
117. C. Wang, C. Ke, C. Lee, S. Chang, W. Chang, J. Li, Z. Li, H. Yang, H. Kuo, T. Lu, Hole injection and efficiency droop improvement in InGaN/GaN light-emitting diodes by band-engineered electron blocking layer. Appl. Phys. Lett. **97**(26), 261103 (2010)
118. N. El-Masry, E. Piner, S. Liu, S. Bedair, Phase separation in InGaN grown by metalorganic chemical vapor deposition. Appl. Phys. Lett. **72**(1), 40–42 (1998)
119. C. Tran, R. Karlicek, M. Schurman, A. Osinsky, V. Merai, Y. Li, I. Eliashevich, M. Brown, J. Nering, I. Ferguson, Phase separation in InGaN/GaN multiple quantum wells and its relation to brightness of blue and green LEDs. J. Cryst. Growth **195**(1), 397–400 (1998)
120. J. Wang, L. Wang, W. Zhao, Z. Hao, Y. Luo, Understanding efficiency droop effect in InGaN/GaN multiple-quantum-well blue light-emitting diodes with different degree of carrier localization. Appl. Phys. Lett. **97**(20), 201112 (2010)
121. B. Monemar, B. Sernelius, Defect related issues in the "current roll-off" in InGaN based light emitting diodes. Appl. Phys. Lett. **91**(18), 181103 (2007)
122. I.-K. Park, M.-K. Kwon, J.-O. Kim, S.-B. Seo, J.-Y. Kim, J.-H. Lim, S.-J. Park, Y.-S. Kim, Green light-emitting diodes with self-assembled In-rich InGaN quantum dots. Appl. Phys. Lett. **91**(13), 133105 (2007)
123. H. Jeong, H.J. Jeong, H.M. Oh, C.-H. Hong, E.-K. Suh, G. Lerondel, M.S. Jeong, Carrier localization in In-rich InGaN/GaN multiple quantum wells for green light-emitting diodes. Sci. Rep. **5**, 9373 (2015)
124. A. Hori, D. Yasunaga, A. Satake, K. Fujiwara, Temperature dependence of electroluminescence intensity of green and blue InGaN single-quantum-well light-emitting diodes. Appl. Phys. Lett. **79**(22), 3723–3725 (2001)
125. C.-H. Lu, Y.-C. Li, Y.-H. Chen, S.-C. Tsai, Y.-L. Lai, Y.-L. Li, C.-P. Liu, Output power enhancement of InGaN/GaN based green light-emitting diodes with high-density ultra-small In-rich quantum dots. J. Alloys Compd. **555**, 250–254 (2013)
126. M. Zhang, P. Bhattacharya, W. Guo, InGaN/GaN self-organized quantum dot green light emitting diodes with reduced efficiency droop. Appl. Phys. Lett. **97**(1), 011103 (2010)
127. W. Lv, L. Wang, L. Wang, Y. Xing, D. Yang, Z. Hao, Y. Luo, InGaN quantum dot green light-emitting diodes with negligible blue shift of electroluminescence peak wavelength. Appl. Phys. Express **7**(2), 025203 (2014)
128. J. Yu, L. Wang, D. Yang, Z. Hao, Y. Luo, C. Sun, Y. Han, B. Xiong, J. Wang, H. Li, Improving the internal quantum efficiency of green InGaN quantum dots through coupled InGaN/GaN quantum well and quantum dot structure. Appl. Phys. Express **8**(9), 094001 (2015)
129. J. Rong, L. Hai, C. Dun-Jun, R. Fang-Fang, Y. Da-Wei, Z. Rong, Z. You-Dou, Temperature-dependent efficiency droop behaviors of GaN-based green light-emitting diodes. Chin. Phys. B **22**(4), 047805 (2013)

130. K. Bulashevich, S.Y. Karpov, Is auger recombination responsible for the efficiency rollover in III-nitride light-emitting diodes? Phys. Status Solidi C **5**(6), 2066–2069 (2008)
131. Y. Shen, G. Mueller, S. Watanabe, N. Gardner, A. Munkholm, M. Krames, Auger recombination in InGaN measured by photoluminescence. Appl. Phys. Lett. **91**(14), 141101 (2007)
132. E. Kioupakis, P. Rinke, K.T. Delaney, C.G. Van de Walle, Indirect auger recombination as a cause of efficiency droop in nitride light-emitting diodes. Appl. Phys. Lett. **98**(16), 161107 (2011)
133. S. Karpov, ABC-model for interpretation of internal quantum efficiency and its droop in III-nitride LEDs: a review. Opt. Quant. Electron. **47**(6), 1293–1303 (2015)
134. F. Nippert, S.Y. Karpov, G. Callsen, B. Galler, T. Kure, C. Nenstiel, M.R. Wagner, M. Straßburg, H.-J. Lugauer, A. Hoffmann, Temperature-dependent recombination coefficients in InGaN light-emitting diodes: hole localization, auger processes, and the green gap. Appl. Phys. Lett. **109**(16), 161103 (2016)
135. A.I. Zhmakin, Enhancement of light extraction from light emitting diodes. Phys. Rep. **498**(4), 189–241 (2011)
136. H. Liu, V. Avrutin, N. Izyumskaya, Ü. Özgür, H. Morkoç, Transparent conducting oxides for electrode applications in light emitting and absorbing devices. Superlattice. Microst. **48**(5), 458–484 (2010)
137. K. Nakahara, K. Tamura, M. Sakai, D. Nakagawa, N. Ito, M. Sonobe, H. Takasu, H. Tampo, P. Fons, K. Matsubara, Improved external efficiency InGaN-based light-emitting diodes with transparent conductive Ga-doped ZnO as p-electrodes. Jpn. J. Appl. Phys. **43**(2A), L180 (2004)
138. T. Minami, Transparent conducting oxide semiconductors for transparent electrodes. Semicond. Sci. Technol. **20**(4), S35 (2005)
139. B.-J. Kim, C. Lee, Y. Jung, K. Hyeon Baik, M.A. Mastro, J.K. Hite, C.R. Eddy Jr., J. Kim, Large-area transparent conductive few-layer graphene electrode in GaN-based ultra-violet light-emitting diodes. Appl. Phys. Lett. **99**(14), 143101 (2011)
140. Y. Zhang, L. Wang, X. Li, X. Yi, N. Zhang, J. Li, H. Zhu, G. Wang, Annealed InGaN green light-emitting diodes with graphene transparent conductive electrodes. J. Appl. Phys. **111**(11), 114501 (2012)
141. J.-K. Sheu, I.-H. Hung, W.-C. Lai, S.-C. Shei, M. Lee, Enhancement in output power of blue gallium nitride-based light-emitting diodes with omnidirectional metal reflector under electrode pads. Appl. Phys. Lett. **93**(10), 103507 (2008)
142. J.K. Kim, T. Gessmann, H. Luo, E.F. Schubert, GaInN light-emitting diodes with RuO2/SiO2/Ag omni-directional reflector. Appl. Phys. Lett. **84**(22), 4508–4510 (2004)
143. Y. Zhao, D. Hibbard, H. Lee, K. Ma, W. So, H. Liu, Efficiency enhancement of InGaN/GaN light-emitting diodes with a back-surface distributed Bragg reflector. J. Electron. Mater. **32**(12), 1523–1526 (2003)
144. N. Nakada, M. Nakaji, H. Ishikawa, T. Egawa, M. Umeno, T. Jimbo, Improved characteristics of InGaN multiple-quantum-well light-emitting diode by GaN/AlGaN distributed Bragg reflector grown on sapphire. Appl. Phys. Lett. **76**(14), 1804–1806 (2000)
145. C.O. Egalon, R.S. Rogowski, in Increased efficiency LED. Google Patents, 1998
146. J.H. Lee; B.W. Oh; H.S. Choi; J.T. Oh; S.B. Choi, S.Y. Lee, in Vertical GaN-based LED and method of manufacturing the same. Google Patents, 2008
147. J.-Y. Kim, M.-K. Kwon, J.-P. Kim, S.-J. Park, Enhanced light extraction from triangular GaN-based light-emitting diodes. IEEE Photon. Technol. Lett. **19**(23), 1865–1867 (2007)
148. J.-W. Pan, C.-S. Wang, Light extraction efficiency of GaN-based LED with pyramid texture by using ray path analysis. Opt. Express **20**(105), A630–A640 (2012)
149. D.-W. Jeon, S.-J. Lee, T. Jeong, J.H. Baek, J.-W. Park, L.-W. Jang, M. Kim, I.-H. Lee, J.-W. Ju, Chemical lift-off of (11–22) semipolar GaN using periodic triangular cavities. J. Cryst. Growth **338**(1), 134–138 (2012)
150. Y.-J. Lee, J. Hwang, T. Hsu, M. Hsieh, M. Jou, B. Lee, T. Lu, H. Kuo, S. Wang, Enhancing the output power of GaN-based LEDs grown on wet-etched patterned sapphire substrates. IEEE Photon. Technol. Lett. **18**(10), 1152–1154 (2006)

151. J.-H. Lee, J. Oh, Y. Kim, J.-H. Lee, Stress reduction and enhanced extraction efficiency of GaN-based LED grown on cone-shape-patterned sapphire. IEEE Photon. Technol. Lett. **20**(18), 1563–1565 (2008)
152. Y. Li, S. You, M. Zhu, W. Hou, Defect-reduced green GaInN/GaN light-emitting diode on nanopatterned sapphire. Appl. Phys. Lett. **98**(15), 151102 (2011)
153. S.-J. Chang, Y. Lin, Y.-K. Su, C. Chang, T.-C. Wen, S.-C. Shei, J. Ke, C. Kuo, S. Chen, C. Liu, Nitride-based LEDs fabricated on patterned sapphire substrates. Solid State Electron. **47**(9), 1539–1542 (2003)
154. H. Huang, C. Lin, C. Yu, B. Lee, C. Chiu, C. Lai, H. Kuo, K. Leung, T. Lu, S. Wang, Enhanced light output from a nitride-based power chip of green light-emitting diodes with nano-rough surface using nanoimprint lithography. Nanotechnology **19**(18), 185301 (2008)
155. W.C. Peng, Y.C.S. Wu, Improved luminance intensity of InGaN–GaN light-emitting diode by roughening both the p-Ga N surface and the undoped-GaN surface. Appl. Phys. Lett. **89**(4), 041116 (2006)
156. L. Hui, L. Peixian, S. Huifang, Z. Guangcai, Enhancement of extraction efficiency of green LED via surface roughening. Electron. Sci. Technol. **6**, 11 (2010)

Chapter 7
Al-Rich III-Nitride Materials and Ultraviolet Light-Emitting Diodes

Jianchang Yan, Junxi Wang, Yuhuai Liu, and Jinmin Li

7.1 Heteroepitaxy of AlN Material by MOVPE

7.1.1 Al Precursor-Related Pre-reaction Issues in AlN MOVPE

High-quality AlN materials are the key element in obtaining efficient DUV LEDs. However, the chemical reactions between the Al related metal-organic precursors (trimethylaluminum, TMAl) and ammonia (NH_3) are very complicated in the MOCVD chamber and depend on the different growth atmosphere. The chemical equation for the AlN growth can be written as:

$$Al(CH_3)_3 + NH_3 = AlN\,(s) + 3CH_4 \tag{7.1}$$

Except for the formations of AlN, many parasitic products may form due to the parasitic gas reactions [1–5]. These products, such as Lewis acid-base adducts (TMAl:NH_3), polymers ($[DMAlNH_2]_n$, $[NH_3{:}DMAlNH_2]_n$, and so on),

J. Yan · J. Wang (✉)
Institute of Semiconductors, Chinese Academy of Sciences, Beijing, China

State Key Laboratory of Solid State Lighting, Beijing, China

Beijing Engineering Research Center for the 3rd Generation Semiconductor Materials and Application, Beijing, China
e-mail: jxwang@semi.ac.cn; yanjc@semi.ac.cn

J. Li
Chinese Academy of Sciences, State Key Laboratory of Solid State Lighting, Solid State Lighting R&D Center, Beijing, China

Y. Liu
Zhengzhou University, Zhengzhou, China

Nagoya University, Nagoya, Japan

J. Li, G. Q. Zhang (eds.), *Light-Emitting Diodes*, Solid State Lighting Technology and Application Series 4, https://doi.org/10.1007/978-3-319-99211-2_7

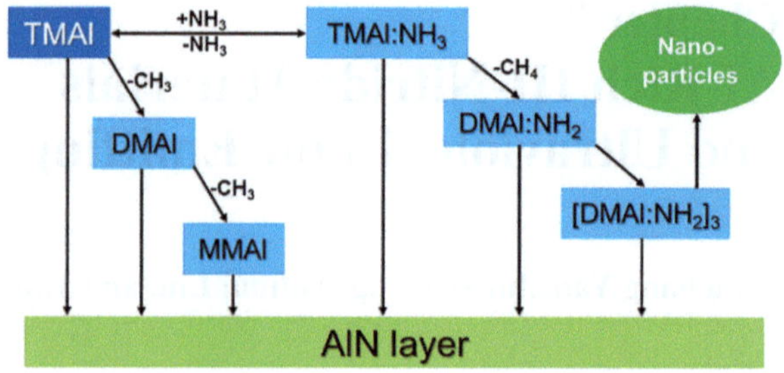

Fig. 7.1 Schematic of the main gas reaction paths in III-nitride MOVPE growth

and nano-sized particles, will interfere the growth process severely [6, 7]. The parasitic reactions will cause the contamination on the reactor walls and on the growing films. Furthermore, the surface morphology and the quality will be deteriorated [8, 9]. Besides, the growth efficiency will be decelerated owing to the waste of sources [8]. Thus, there are two essential aspects for improving the material quality and increasing the growth efficiency in AlN MOVPE growth. Firstly, the reaction mechanism including thermodynamic and kinetic theory should be studied. After that, the approaches to reduce the parasitic reactions should be found out.

7.1.1.1 Reaction Mechanism (Thermodynamics and Kinetics)

To date, many researches have been done to understand the gas reaction paths in AlN MOVPE growth. Generally, there are two competing paths in the gas reactions between TMAl and NH_3 which are called the adduct reaction path and the pyrolysis path [6, 9–11], as shown in Fig. 7.1. In the adduct reaction path, TMAl and NH_3 produce the donor-acceptor complex adduct, TMAl:NH_3, at room temperature upon mixing. When the temperature rose up, the adducts may decompose into amides, $DMAlNH_2$ (or $Al(CH_3)_2NH_2$), with the release of CH_4. The $DMAlNH_2$ adduct often can be observed near the nozzle [10]. After that, the amides can easily oligomerize into oligomers, $[DMAlNH_2]_x$ ($x = 2$ or 3), due to the high reactivity of amides. The oligomers are the main source of the nano-sized particles, which are the main contamination on the reactor walls and on the epilayers. Instead of contributing to the growing crystal, white powders are usually deposited on the reactor walls of the growth chamber. Additionally, it's possible that the second NH participates in the adduct reaction path to form the two-NH_3 adduct, NH_3:TMAl:NH_3, or the amide-adduct, NH_3:$DMAlNH_2$ for the large quantity of the NH_3 gas in the MOVPE chamber comparing with the quantity of the TMAl [7, 12–15]. In the pyrolysis path, TMAl directly pyrolyzes into DMAl, MMAl, and/or Al successively near the hot substrate region with the release of CH_3 [16].

The gas reaction paths as abovementioned have been proved by huge experimental results. At low temperature, the experimental mass spectrometry results confirmed the formation of cluster compounds in the gas phase upon laser irradiation of TMAl-NH_3 mixtures [16]. The TMAl:NH_3 adducts were structurally characterized in the solid state by X-ray diffraction (XRD) [17]. Its pyrolysis in the condensed phase yields the amido- and imidoalanes and AlN [18]. Sauls and Interrante also studied the detailed formation and thermal decomposition of the $TMAlNH_3$ in solution phase [19, 20]. The dissociation enthalpy of the TMAl-NH_3 bond between donor and acceptor was evaluated by solution calorimetry in benzene (95 $\pm$ 5 kJ/mol) [20] and in hexane (115 $\pm$ 1 kJ/mol) [21]. The Fourier transform infrared spectroscopic studies revealed the activation energy of methane elimination considering the kinetics of the gas reaction paths between TMAl and NH_3 is 107.5 kJ/mol [6]. Amato proposed the existence of $DMAlNH_2$ in the gas phase at MOCVD conditions through the observation of the intermediate in the decomposition of $TMAlNH_3$ in solution [22]. Müller has found the monomeric $DMAlNH_2$ as a production of UV irradiation of matrix-isolated TMAl-NH_3 [23]. These experimental results can prove the reasonability of the gas adduct reaction paths. Also, there are many studies about the pyrolysis reaction paths [24–26]. For the reason that these reactions won't produce the parasitic productions, we don't talk more about the experimental results about that.

In order to understand the gas-phase chemistry, it's necessary to study the reaction mechanisms by the thermodynamics and kinetics theory. The density functional theory (DFT) method is often applied to study the theoretical mechanisms. All calculations are often done in Gaussian 03 software package, Gaussian 09 software package, the Vienna ab initio simulation package (VASP) code, and so on [18, 25, 27, 28]. The atomic structures should be optimized and the varied transition states should be induced to verify the connections between reactants and products. After the detailed calculation, the changes of enthalpies and Gibbs free energies of the reactions at different growth parameter can be used to postulate the detailed mechanisms and further useful for the achievement of the high-quality AlN films. Here, we mainly paid our attention on the formation of the parasitic productions which are usually formed by the adduct reactions. Thus, among the gas reactions, we'll introduce them in four aspects: the initial process of the adduct reactions, the adduct reactions as the functions of temperature, the adsorption of the adducts on the surface, and the decomposition processes of the TMAl precursor.

7.1.1.2 The Initial Process of Adduct Reactions

Lisovenko et al. have explored the mechanisms of initial stages of gas-phase reactions between TMAl and NH_3 by DFT studies [18]. They considered the subsequent substitution of CH_3 groups in TMAl by NH_3 groups and substitution of hydrogen atoms in ammonia by TMAl groups. So the intermediates include $TMAlNH_3$, $Al(CH_3)_x(NH_2)_{3-x}$, $NH_x(Al(CH_3)_2)_{3-x}$ ($x = 0$–3), and the transition states. There are 13 kinds of transition states that have been induced and calculated. Here they just considered the $[DMAlNH_2]_2$ dimer production which is formed by the subsequent reactions. According to these results, the initial corresponding reaction paths have been obtained, as shown in Fig. 7.2. According to the thermodynamic characteristics of the reactions, the first methane elimination from $TMAlNH_3$ is the highest point

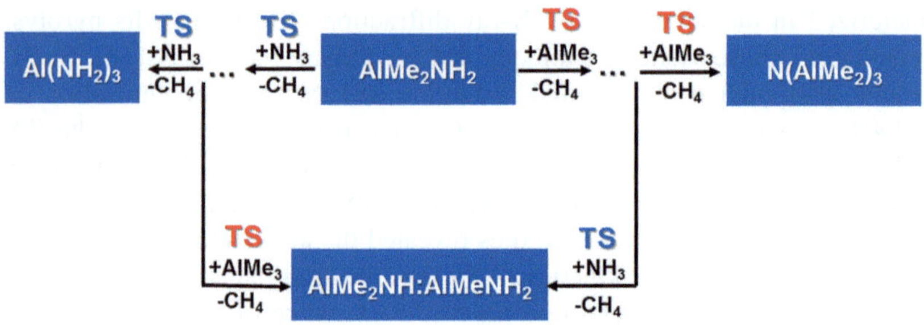

Fig. 7.2 The initial reaction pathways and corresponding compounds

on the reaction paths and the only one lying above the isolated TMAl and NH_3. It's an exothermic process for the subsequent substitution of methyl groups in TMAl by amido groups, but the intermediate steps make the activation energies of the reaction larger, even though the overall reaction profile is rather downhill. These results support that the first methane elimination is a rate-limiting step for the whole reactions. In addition, the reaction leading to $[DMAlNH_2]_2$ dimer and the reactions of subsequent substitution are competitive. Also, these processes may operate at the same time. Besides, they found higher vapor pressure is helpful for the dimerization reactions, while lower vapor pressures and excess of one of the reagents will favor the substitution pathways. According these results, the lower vapor pressure and excess of one of the reagents may suppress the formation of dimerization productions.

7.1.1.3 The Adduct Reactions as the Functions of Temperature

Ran Zuo calculated the changes of enthalpy ΔH and Gibbs free energy ΔG as the functions of temperature to predict the most probable reaction paths [27]. Their results and some literature values at room temperature are shown in Table 7.1, which indicates the accuracy of Zuo's quantum chemistry calculation. As shown in Fig. 7.3, there exists a critical temperature, T_{eq} ($T_{eq} \approx 480$ K), for the formation of $TMAl{:}NH_3$. When the $T < T_{eq}$, $\Delta G < 0$, and there is no energy barrier and the reaction will occur spontaneously; when $T > T_{eq}$, $\Delta G > 0$, and the adduct may dissociate back into TMAl and NH_3 with a small energy barrier equivalent to the heat release in the adduct formation; and when $T = T_{eq}$, $\Delta G = 0$, and the adduct formation and dissociation are in equilibrium. After the formation of $TMAl{:}NH_3$, there are two paths and corresponding two translate states: the traditional formation of $DMAlNH_2$ and the second participates of NH_3, which is the formation of the two-NH_3 adduct, $NH_3{:}TMAl{:}NH_3$. The changes of ΔH and ΔG at different temperatures in one-NH_3 involved adduct reaction path were shown in Fig. 7.4a, b. The changes of ΔH and ΔG at different temperatures in two-NH_3 involved adduct reaction path were shown in Fig. 7.4c, d. The two-NH_3 involved adduct reactions

Table 7.1 Comparisons of calculated results of ΔH and ΔG with literature values in the adduct reactions of TMX + NH3 (X = Ga and Al) at room temperature (ΔG* denotes values calculated from ΔG = ΔH − TΔS) [27]

Reactions	Calculated value (kcal/mol)	Literature value (kcal/mol)	Methods	Reference
$TMAl + NH_3 \rightarrow TMAl{:}NH_3$	ΔH: −21.56	ΔH: −19.06	B3LYP/def2-TZVPP	Lisovenko et al. [18]
		ΔH: −20.4	B3LYP/6-311++G(d,p)	Wang and Creighton [14]
		ΔH: −23.17	B3LYP/LanL2DZ*	Nakamura et al. [15]
		ΔH: −27.2	B3LYP/6-31G(d,p)	Simka et al. [12]
	ΔG: −8.27	ΔG: −7.21	B3LYP/def2-TZVPP	Lisovenko et al. [18]
		ΔG: −16.0	B3LYP/6-31G(d,p)	Simka et al. [12]

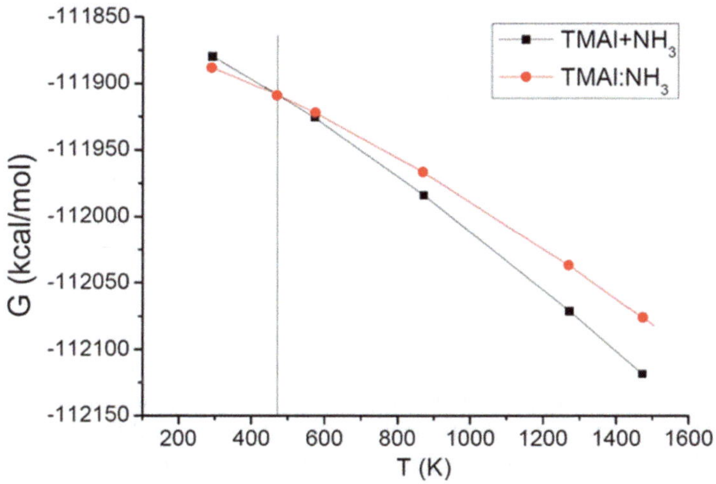

Fig. 7.3 Changes of Gibbs free energies with temperature for TMAl adduct reactions [27]

compared to the one-NH_3 involved adduct reactions are disfavored and the most probable adduct reaction path should be the one-NH_3 involved process. Their study indicated the importance of the temperature for the different reaction paths. The proper temperature should be chosen to obtain the high-quality AlN films.

7.1.1.4 The Surface Reaction of the Adducts

Yumi Inagaki studied the adsorption of the adducts on the surface of the epilayers [28]. They found $DMAlNH_2$ is one of the major reactive involved in the AlN

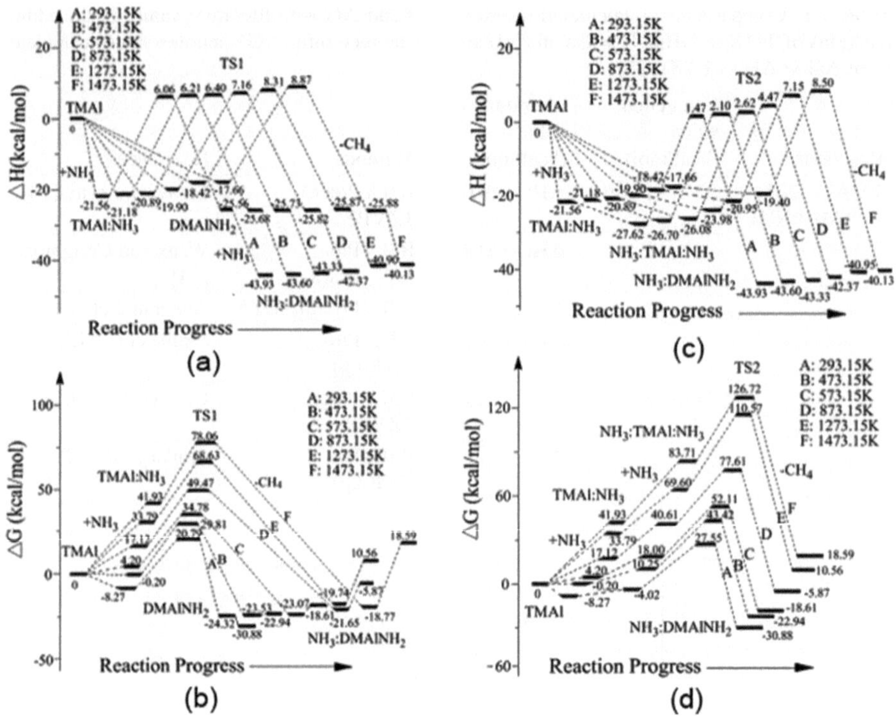

Fig. 7.4 Changes of ΔH (**a**) and ΔG (**b**) in one NH_3 involved adduct reaction path at different temperatures and changes of ΔH (**c**) and ΔG (**d**) in two NH_3 involved adduct reaction path at different temperatures [27]

growth according to the surface reactions between $DMAlNH_2$ and AlN, as shown in Fig. 7.5. The $DMAlNH_2$ can be adsorbed on the surface owing to their simulation results. Then the methane can eliminate from the surface, and finally the Al atoms of $DMAlNH_2$ have been incorporated into the lattice of the AlN epilayers with no energy barrier. So $DMAlNH_2$ is also one of the primary reactive species for AlN growth. This result shows the adduct reaction paths are also very important for the growth of AlN.

7.1.1.5 The Decomposition Processes of the TMAl

Obviously, the adduct reaction paths are the key element for the study of the parasitic reactions and parasitic products. However, the decomposition of the metal-organic precursor TMAl sometimes is also important for the design of the MOVPE. Kazuki Sekiguchi studied the decomposition processes of the TMAl at finite temperature [25]. The detailed chemical reaction formulas as shown follow and the calculated ΔG result is shown in Fig. 7.6.

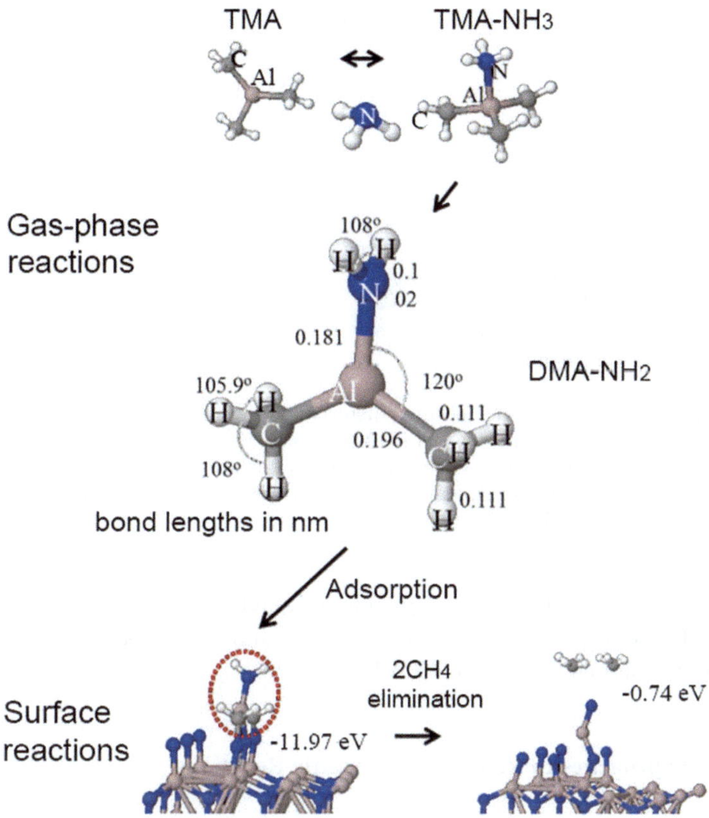

Fig. 7.5 Growth process, including gas-phase and surface reaction of AlN through DMAl-NH_2 (including the molecular structure of DMA-NH2 and adsorption and elimination energies at the surface) [28]

$$2Al(CH_3)_3\,(g) + H_2\,(g) = 2Al(CH_3)_2\,(g) + 2CH_4\,(g) \tag{7.2}$$

$$Al(CH_3)_3\,(g) + H_2\,(g) = Al\,(CH_3)\,(g) + 2CH_4\,(g) \tag{7.3}$$

$$2Al(CH_3)_3\,(g) + 3H_2\,(g) = 2Al\,(g) + 6CH_4\,(g) \tag{7.4}$$

$$Al(CH_3)_3\,(g) + NH_3\,(g) = (CH_3)_2\,AlNH_2\,(g) + CH_4\,(g) \tag{7.5}$$

The adduct reaction is the easiest to occur below 1500 K. TMAl almost completely reacts with not H_2 but NH_3, leading to the formation of the $(CH_3)_2AlNH_2$ adduct at 700 K. This is consistent with the AlN materials observed near the nozzle. Thus, it's important to design the nozzle to inhibit the reaction between precursors.

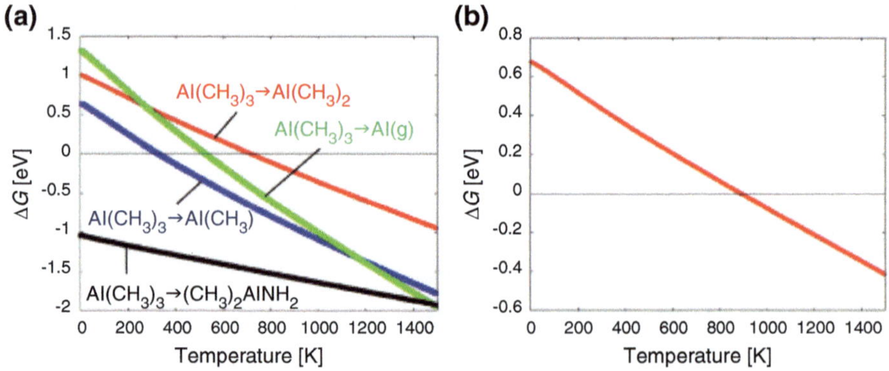

Fig. 7.6 (**a**) Calculated Gibbs free energy difference ΔG. The red, blue, green, and black lines show ΔG values for the reactions in which TMA becomes DMA, MMA, Al, and the $(CH_3)_2AlNH_2$ adduct, respectively. (**b**) Calculated Gibbs free energy difference ΔG. The red line shows ΔG for the reaction in which MMA decomposes into gas-phase Al Atoms [25]. Copyright (2017) The Japan Society of Applied Physics

TMAl was found to decompose into monomethylaluminum ($Al(CH_3)$) between 330 K and 1200 K and into gas-phase Al atoms above 1200 K. It's crucial for the design of the thermal field in the MOVPE chamber.

7.1.1.6 Approaches to Reduce the Parasitic Reactions

So far, many approaches have been used to reduce the parasitic reactions and further improve the AlN quality and growth efficiency. It does reduce the parasitic reactions to lower the chamber pressure or increase the growth temperature due to the study of the reaction mechanism mentioned above [4, 29–31]. However, the raising of the temperature is limited by the uniformity and stability of the chamber's thermal field. Moreover, the descending of the chamber pressure and the ascending of the growth temperature will impede the p-type doping as nitrogen vacancies acting as shallow donors in III-nitride materials are increasingly generated.

In recent years, the pulsed deposition method has been developed to inhibit the parasitic issue [32–34]. Through a pulsed and independent supply of the metal-organic gas precursors from NH_3, the pre-mixture of the gas reactant can be controlled and at last the AlN quality can be improved. What's more, the migration of Al atoms on the surface of the epilayers can be enhanced. Nevertheless, this method will waste huge time for waiting as the ordinary valve is difficult to switch at high frequency. So the growth rate has been decreased by a factor of two or three times owing to the limitation of the pulsed time. These inferiors make the pulsed method less attractive for production.

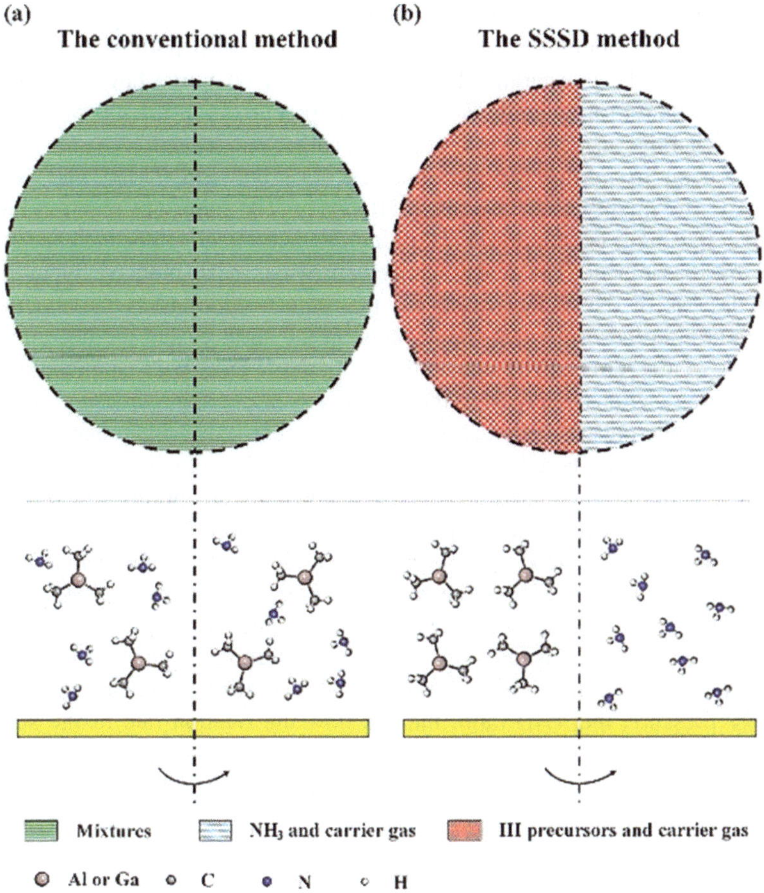

Fig. 7.7 (**a**) The conventional and (**b**) the spatial separated source delivery (SSSD) method of the MOCVD system [8]

Except for these conventional methods, some special approaches have been adopted [8, 35–37]. Lai et al. deposited an AlN epilayer with suppressed parasitic reactions by raising the hydrogen flow rate (20 slm) under a higher chamber pressure (100 torr) [37]. Yang et al. proposed a spatial separated source delivery (SSSD) method to alleviate the parasitic reactions between group III precursor and NH_3 [8]. The gas injection method of the MOCVD system has been modified, as shown in Fig. 7.7a. This design of the gas injection system can spatially separate group III precursor from NH_3 in gas phase. The separation of precursors near the injection system makes the combination reactions occur only on the surface of the substrates. This physical separation method directly prevents any reactions in the gas phase. Additionally, the growth rate has been promoted and the sources consumption has been reduced apparently. The uniformity of the epilayers has been improved. So the

SSSD method is a promising high-efficiency method to solve the parasitic reaction issue. Similarly, Chung et al. used the N_2 purge line to spatially separate the AlN precursors [36]. They separated TMAl and NH_3 by N_2 purge flow. It's similar as the SSSD method for spatially separating the precursors in the gas phase. So the parasitic reactions also can be inhibited efficiently.

It's crucial to inhibit the parasitic reactions between TMAl and NH_3 for improving the quality and the growth efficiency. The main gas reaction paths of AlN materials have been induced according the theoretical works and experimental works by researchers. The reaction mechanism has been induced, including the initial process of the adduct reactions, the adduct reactions as the functions of temperature, the adsorption of the adducts on the surface, and the decomposition processes of the TMAl. The conventional methods have been applied to induce the parasitic reactions: decreasing the chamber pressure and increasing the growth temperature, the pulsed method, and so on. In recent years, many new approaches have been employed: increasing the H_2 flow, the SSSD method through optimizing the gas injection system, and so on. These methods are promising to solve the parasitic reaction issue and do not decrease the growth efficiency. However, there are also many works need to be done about the parasitic reaction issues. The detailed chemical mechanism of the gas reaction pathways between TMAl and NH_3 gas should be unified and proved by more experimental results: translate states, intermediates, detailed gas reaction, and so on. Additionally, the relations between enthalpy ΔH, Gibbs free energy, ΔG and other growth parameters (such as the chamber pressure, the V/III ratio, the gas flow the rotation rate of the substrate, and so on) should be studied. The approaches to reduce the parasitic include the optimization of the growth parameter, the design of the MOVPE system, and so on.

7.1.2 Defects and Stress Control of AlN Epitaxy on Sapphire

The defects caused by parasitic reactions can be released by some methods as mentioned above. Additionally, the lattice mismatch and thermal mismatch between AlN and hetero-substrate owing to lack of native substrate bring the severe defects and stress issues [38]. Sapphire materials are usually chosen as the substrates caring about the crystal lattice and price factor [2]. There are still 13.3% mismatch existence between AlN and sapphire even after the two lattices rotate by a 30° angle [39, 40]. Moreover, the Al atomic surface mobility is very low comparing with Ga atom. It's difficult to form the 2D growth model for AlN resulting in the low-quality materials. The generation of threading dislocations along the island boundaries may decrease the internal quantum efficiency (IQE) of DUV LEDs due to the non-radiative recombination of excited carriers at these dislocations. The threading dislocation density (TDD) should be controlled in the range of 10^7 cm^{-2} to keep IQE close to unity according to Martin Guttmann's report [41]. Except for the defect issues, stress in AlN films also affects the property of DUV LEDs. Huge stress may result in cracks, wafer bowling, and high defect densities [42–44]. Cracks in epitaxial layer may cause electric leakage. Wafer bowling results in the difficult control of the wafer uniformity. Thus, the defects and stress control of AlN grown by MOCVD on sapphire really matters. The convenient techniques will be introduced as follows.

The quality of GaN has been improved prominently due to the development of two-step method. The two-step method also has been used in the growth of AlN [38]: after the H_2 flowing the sapphire substrate at high temperature for a short time to clean the surface, the low-temperature AlN (LT-AlN) layers are usually grown as the buffer layer at low to medium temperatures of about 600–1000 °C with high V/III ratio (>1000), then the high-temperature AlN layers are deposited on the LT-AlN buffer. The growth atmosphere usually set at low power. However, the quality of AlN grown by two-step methods still needs to be improved. From now on, many methods have been developed to improve the AlN quality. Thanks to those methods, the external quantum efficiency (EQE) of DUV LEDs has been improved. Comparing with the EQEs of visible-wavelength LEDs, it's no doubt the efficiencies and output powers of DUV LEDs will continue to develop. Here, we divide the conventional methods of the AlN heteroepitaxy into four kinds of techniques: epitaxial lateral overgrowth (ELOG), buffer-assisted technique, different interlayers, and special growth process control.

7.1.2.1 Epitaxial Lateral Overgrowth

ELOG on masked AlN template or patterned sapphire substrate has been proved to be an effective way to further reduce the TDDs in GaN epitaxy [45, 46]. Researches about AlN ELOG also have been studied in recent years [47, 48]. Researchers have found many threading dislocations would generate from the AlN/sapphire interface and then spread upward in the support mesa regions. After that, some dislocations over the mesa region and near the air gap would be looped, bend, or get annihilated. Thus, the TDDs in AlN films deceased apparently. Khan et al. reported 290 nm emission AlGaN-based DUV LEDs exhibiting stable cw-powers in excess of 2 mW over pulsed laterally overgrown (PLOG) AlN [47]. AlN with 15–10-μm-thick was grown using a pulsed growth mode on the micro-sized air-bridge AlN template. The achieved AlN films had significantly reduced number of threading dislocations ($\sim 10^7$ cm^{-2}) and showed a smooth surface in the lateral overgrowth area. The schematic diagram of the LED structure and the cross-sectional SEM image of PLOG AlN were shown in Fig. 7.8. Comparing with the two-step method, the TDD of AlN grown by PLOG decreased 10^2–10^3 cm^{-2}. Except for the air-bridge AlN template, there are many other methods to be used for the ELOG method, such as self-assembled patterned AlN nanorods [49], nano-sized trench-patterned AlN [50], and so on.

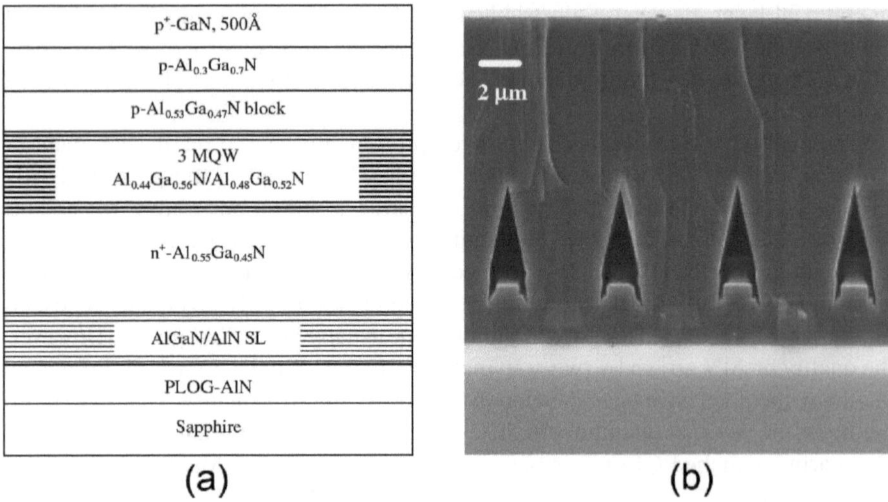

Fig. 7.8 (**a**) The schematic diagram of the LED structure and (**b**) the cross-sectional SEM image of PLOG AlN [47]. Copyright (2007) The Japan Society of Applied Physics

7.1.2.2 Buffer-Assisted Technique

Buffer plays a crucial role on the quality of epilayers [51]. There are mainly three effects for the growth of AlN. Firstly, buffer provides nucleation sites whose crystal orientation is similar with substrate for AlN epilayers. Secondly, the buffer layer can release some stress that comes from the lattice mismatch and thermal mismatch between the epilayer and foreign substrate. Thirdly, the buffer with the smooth and flat surface can help the nucleation of AlN for the reason of decreasing the growth contact angle of AlN crystal grains at high temperature. Thus, the AlN islands can coalesce in the range of smaller thickness and the growth model can easily transfer from 3D mode to quasi-2D mode.

The typical buffer of AlN films is AlN layer grown at low temperature (600–1000 °C). Recently, there are some advanced buffer layers that have been developed. Here we mainly introduce four kinds of advanced methods based on buffer-assisted technique: annealing of an AlN buffer at high temperature, sputtering AlN buffer, buffer with Mg-Si codoping pair, and dual buffer.

Miyake, Hideto et al. reported the annealing of an AlN buffer in a carbon-saturated N_2-CO gas AT 1650–1700 °C for 1 h [52]. The 2-μm-thick AlN on the annealed AlN buffer showed high quality by MOCVD. The FWHM values of the (002)- and (102)-plane XRCs were 16 and 154 arcsec, respectively, and the corresponding threading dislocation density was only 4.7×10^8 cm^{-2}. This annealing AlN buffer method is very significant to improve the quality of AlN films. Recently, researchers found the sputtering present a promising candidate as a buffer for AlN films owing to its low price and time-saving. Ohtsuka et al. improved the crystal quality of AlN by increasing the mobility of the Al sputtered atoms on the

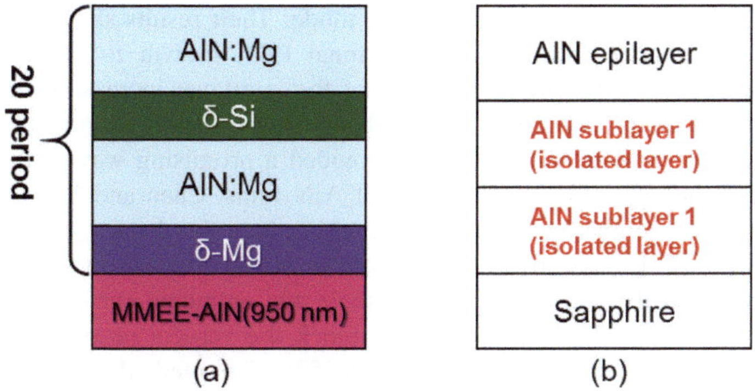

Fig. 7.9 (**a**) Schematic diagram of the Mg-Si codoping pair structure and (**b**) the schematic diagram of the dual buffer layer structure

surface through lowering sputtering pressure [53]. They investigated the effect of sputtering pressure on the morphology, crystal quality, and stress of AlN films by pulsed DC reactive sputtering. It was found that increasing pressure could cause the surface damage of the AlN sputtered films because of arcing. Meanwhile, the sputtering pressure would affect the crystal quality and residual stress of AlN films due to a change in the number and energy of Ar^+ ions and Al sputtered atoms. Soomro et al. reported a Mg-Si codoping pair for the quality improvement and misfit strain release of AlN films grown by MOCVD [54]. The schematic diagram of the Mg-Si codoping pair structure is shown in Fig. 7.9a. The achieved as-grown AlN:Mg-Si shows the release of misfit strain, whose stress was close to the stress-free statue. Zhao reported a dual buffer layer structure, which is shown in Fig. 7.9b [55]. This approach can weaken the negative nitridation effect and improve lateral growth condition in the initial growth stage. Their experiment results suggest the thicker the dual AlN buffer, the rougher the AlN surface and the higher the AlN quality. These advanced buffer-assisted techniques mentioned above were proved to improve the AlN quality through changing the growth process of structure of buffer layers.

7.1.2.3 Interlayers

In order to block the extend of the edge dislocations in the AlN epilayers, a few kinds of interlayers have been developed. Moreover, the interlayers can release residual stress in AlN films. The Mg-Si codoping pair structure mentioned above also can be regarded as an interlayer.

Four kinds of interlayers were introduced as follows: intermediate-temperature interlayers, in situ SiN_x masks, superlattice structure, and Si-doping AlN interlayers. Chen et al. reduced the TDD in AlN layers through adopting AlN intermediate-temperature interlayers [56]. They found the growth temperature and the thickness of the AlN interlayer affected the growth mode. Their results showed the growth mode would change from three-dimensional (3D) growth to two-dimensional (2D) growth mode with the temperature of the interlayer increasing from 470 to 670 °C. Then the growth mode would change into 3D growth mode again when the temperature rose to 870 °C. They provided a promising way to improve the surface morphology and crystal quality of AlN films. Chen and Yan also proved the intermediate-temperature interlayer could reduce the density of defect in the reference [56, 57], respectively. Vennegues et al. reported an in situ SiN_x mask originating from the decomposition of silane and ammonia can reduce the TDD in GaN films [58]. This SiN_x mask can block the vertical propagation of some dislocations by lateral epitaxial overgrowth [59]. This method can be used in the UV range with AlGaN compositions [60]. The superlattice has been developed as another usual interlayer of reducing TDD and release stress in AlN films. It has been proved the effectiveness of the short-period superlattice of alternating Al(Ga)N/(Al)GaN as strain-releasing and dislocation-annihilation layers [61, 62]. Streubel et al. reported Si-doped (10^{19} cm^{-3}) AlN interlayer was another approach of reducing TDD [63].

7.1.2.4 Special Growth Process Control

Growth process plays an important role in the growth mode of AlN epilayers. Some methods based on the control of special growth process have been developed recently, such as changing V/III ratios [64–66], NH_3 pulsed flow [32, 67], migration enhanced epitaxy (MEE) [68, 69], and mixture carrier gases [70].

V/III ratio is an important parameter for the growth of AlN films. It affects the growth mode, lateral growth rate, and crystal quality. Usually, V/III ratio is constant during the whole growth. Imura et al. controlled V/III ratio into changeable during the growth of AlN films [64, 65]. The multiple modulation of the V/III ratio resulted in higher-quality AlN layers than growth with an unchangeable V/III ratio. The dislocation density of the AlN films was $<3 \times 10^8$ cm^{-2}. They found different V/III ratio led to different growth rate for each facet. Thus, the macroscopic form of grain would change at the transition V/III ratio. The threading dislocations were annihilated with the formation of dislocation loops at the changing of the macroscopic grain [65]. Wang et al. reported the annihilation of defects with alternating high and low V/III ratios [66]. They found this structure was suitable to fabricate the stacking structure and superlattice layers. This method could help achieve an AlN films with optimal quality at a low temperature.

Oshika and Shatalov et al. developed the NH_3 pulsed-flow multilayer growth method and reduced the TDD by a factor of 100 [32]. Hu et al. investigated the effect of $H_2 + N_2$ mixture carrier gases on the AlN films [70]. They found the size of the initial AlN islands and the stress statue in AlN films could be controlled by the different carrier gases. Especially, the in-plain stress is only 0.1 GPa with the 0.5 slm N_2 + 2.5 slm H_2 mixture gas. Moreover, the AlN films had good crystal quality under this condition.

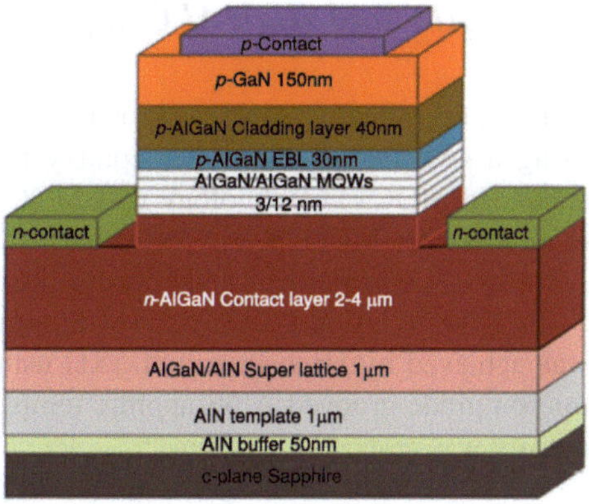

Fig. 7.10 Schematic diagram of a typical AlGaN-based DUV LED epitaxial structure

In conclusion, many methods have been developed to reduce the density of defects and release the stress in AlN films. These significant works paved the way on the commercial application of AlGaN-based DUV LEDs.

7.2 Structural Design for Efficient DUV LEDs

The typical epitaxial structure of a 280-nm AlGaN-based DUV LED is shown in Fig. 7.10, including high-temperature AlN template layer grown on planar sapphire substrate, AlGaN/AlN superlattice stress release layer, Si-doped n-AlGaN contact layer, $Al_xGa_{1--x}N/Al_yGa_{1-y}N$ MQWs active region, Mg doped p-AlGaN electron blocking layer, p-AlGaN/p+-GaN hole injection layer.

The n, p, and active regions of the UV LED are all AlGaN materials; the quality of AlGaN material is closely related to the efficiency of DUV LED. The relationship between the internal quantum efficiency (IQE) and dislocation density of the 280 nm UV LED was simulated by Kneissl et al. [71]. Ban et al. [72] also reported the relationship between the IQE and dislocation density under weak excitation and 1×10^{18} carrier concentration. According to the researches, when the dislocation density is $>10^{10}$ cm^{-2}, the IQE is only a few percent. When the dislocation density is reduced to 10^9 cm^{-2} level, the IQE of the 280 nm UV LED will be increased to 10–40%. If the IQE is increased to more than 40%, the dislocation density is required to reduce to 10^8 cm^{-2} lower level. Therefore, obtaining high-quality AlGaN epitaxial material is the basis of preparation of high-efficiency DUV LED.

The quality of AlGaN based on sapphire substrate depends on the quality of AlN template. Therefore, it is the key to improve the quality of AlN template. In our research group, a crack-free AlGaN epitaxial layer was successfully obtained by using high-temperature AlN intercalation [73]. Compared to the low-temperature AlN and AlGaN intercalation techniques, respectively, reported by Kamiyama et al. [74] and Amano et al. [75], the high-temperature AlN intercalation layer is more favorable to the quality of the subsequent epitaxial AlGaN material.

7.2.1 AlN and High Al Component AlGaN Epitaxy Technology

In 1992, high-quality AlN was grown on sapphire substrates by Khan et al. [76–78] using a switched atomic layer epitaxy (SALE) technology. In 2005, migration-enhanced MOCVD technology (MEMOCVD) was developed by SETi and Khan et al. [79]; the MEMOCVD technology can grow high-quality AlN templates and AlN/AlGaN superlattices and improve the quality of AlGaN material [80].

In 2009, a 270 nm UVC LED with a maximum output of 2.7 mW light output was achieved by Hirayama et al. [81] using the NH_3 pulse access method on the micron grade grooved AlN/sapphire template. Moreover, the threading dislocation density was reduced to 10^8 cm^{-2}.

In 2013, our research group realized epitaxy AlN templates on a flat sapphire substrate (FSS) [82] and further improved the crystal quality of the AlN template by the nano-patterned sapphire substrate (NPSS) [83]. The results show that the NPSS can effectively alleviate the stress in the AlN epitaxial layer.

7.2.2 Study on N-Type Doping of AlGaN Materials

For n-type-doped AlGaN, great progress has been made at present. Several study groups showed that n-type doping efficiency of AlGaN can be improved significantly by restraining the formation of self-compensating defects, and even n-type conductive Si-doped AlN was obtained [84–89]. In 2002, Hwang et al. [90] prepared Si-doped high Al (50%) component AlGaN by molecular beam epitaxy (MBE). The AlGaN has an electron concentration up to 1.25×10^{20} cm^{-3}. But the electron mobility is low. In 2011, AlGaN material with an Al component of 85% and the resistivity of 0.1 Ω cm was achieved based on AlN single crystal substrate by Collazo et al. [91]. Moreover, the doping concentration, carrier concentration, and carrier mobility of $Al_{0.8}Ga_{0.2}N$ material are 6×10^{18} cm^{-3}, 1×10^{18} cm^{-3}, and 40 cm^2 V^{-1} s^{-1}, respectively. Therefore, n-type conductance can be provided for DUV LED with wavelengths as short as 250–260 nm.

7.2.3 Study on P-Type Doping of AlGaN Materials

The problem of doping efficiency of p-type AlGaN materials is more prominent. The activation energy of Mg acceptors in p-type GaN is 160–200 meV [92]. In AlGaN materials, the activation energy of Mg can be as high as 510–600 meV. Therefore, the activation efficiency of Mg is very low, causing low hole concentration of p-type AlGaN (far below n type AlGaN) and low conductivity. Research on p-doped of AlGaN has been studied, such as the common Mg doping, Mg-δ doping, superlattice doping, codoping, and polarization-induced doping.

Jeon et al. [93] studied the influence of Mg concentration on the resistivity of 0.5-μm-thick $Al_xGa_{1-x}N$. In 2003, Nakarmi et al. [94] reported Mg-δ doping. Compared with the homogeneous p-doped AlGaN material, the Mg-δ doping leads to an increase in hole concentration. The horizontal and vertical conductivities are increased by two times and five times, respectively. In 2009, Simon et al. [95] proposed a new type of p-doping method, polarization-induced doping. They grew the Mg-doping AlGaN monolayer on the nitrogen surface GaN, and the Al component was graded from 0 linearly to 0.3. The hole concentration is up to 1×10^{18} cm^{-3}.

7.2.4 *Quantum Efficiency Study of UV LED Structure*

The metal polarity of AlGaN material has a strong spontaneous polar effect. Its spontaneous polarization is [000-1]. The spontaneous polarization increases with the increase of Al component. Thus, a strongly polarized electric field is generated along the c axis of the heterojunction interface or AlGaN MQWs region, leading to band bending and the spatial separation of the electron and hole wave functions (quantum-confined Stark effect, QCSE). The QCSE leads to an increase in carrier lifetime, a decrease in radiative recombination efficiency, and red shift of luminous wavelength. The electric field induced by spontaneous polarization and piezoelectric polarization is as high as MV/cm, which reduces the internal quantum efficiency of UV LEDs.

Marcinkevicius et al. [96] studied the internal polarization field of $Al_xGa_{1-x}N/Al_yGa_{1-y}N$ MQW with different components. The x value is 0.05–0.35, and the y value is 0.23–0.50. The result shows that, for low Al components, the measured polarization field values agree well with those calculated by the first principles; when the Al component is higher, the experimental values are contrary to the theoretical predictions. With the increase of applied bias, the electric field strength of the quantum well is decreased.

Fujioka [97] and Sumiya [98] et al. inserted a 1-nm AlN between AlGaN-MQWs and p-AlGaN electron barrier layers to suppress carrier overflow and the resulting parasitic luminescence and to block the spread of Mg to MQWs. Hirayama et al. [99] increased the effective barrier height by using multi-quantum barrier electron barriers to suppress the electronic leakage. As a result, current injection efficiency and external quantum efficiency were increased. The external quantum efficiency of 250 nm DUV LED was increased by 2.7 times. In order to improve the hole injection efficiency of UV LEDs, a gradual structure of p-AlGaN component was adopted in our research group [100], showing that the quantum efficiency of 297–299 nm UV LEDs was increased by 50%. A $AlN/Al_{0.7}Ga_{0.3}N$ electron barrier heterojunction was used in the 234–263 nm UVC LEDs by Mehnke et al. [101]. By optimizing the thickness of the AlN layer, the parasitic light was suppressed effectively, and the luminous efficiency of QW is improved.

7.3 Homoepitaxy of DUV LEDs on AlN Substrate

Rapid progress has been made recently in the development of AlGaN-based DUV-LEDs. The highest external quantum efficiency (EQE) of DUV-LEDs is higher than 20% at an emission wavelength of 275 nm [102]. However, DUV LEDs continue to have much lower output power from single chips than the more widely used blue LEDs, which is caused by high dislocation density in the AlGaN-based active layers and the extremely low light extraction efficiency of these devices. Traditionally, AlGaN-based DUV LEDs are grown on sapphire, SiC, or Si substrates. The large lattice and thermal expansion coefficient mismatches between these substrates and the epitaxial layers cause large numbers of dislocations and poor device reliability. The density of threading dislocations in nitride films grown on foreign substrates is high, typically $>10^8$ cm^{-2} even when special growth techniques are employed. In order to overcome the dislocation problem, AlN substrates with significantly lower dislocation density ($<10^4$ cm^{-2}) have been developed as a promising candidate due to similar thermal expansion coefficients and relatively small lattice mismatches.

7.3.1 Homoepitaxy on AlN Substrates

An aluminum hydroxide layer with the thickness of 5–10 nm forms on the surface when the AlN substrate is exposed to moist air or water, which consists of aluminum oxide hydroxide (AlOOH) or a mixture of AlOOH and aluminum trihydroxide [$Al(OH)_3$] [103, 104]. Nikishin et al. demonstrated that the aluminum hydroxide layer can result in increased dislocation density in homoepilayers [105]. In addition, surface scratches are incompletely removed after mechanically polishing (MP), which also may lead to strained AlN and the increased surface roughness due to re-nucleation and grain coalescence on these surfaces [106]. Therefore, the surface treatments of AlN substrates prior to epitaxy are critical to achieve high-quality homoepitaxy. The surface treatment schemes, including both ex situ and in situ processes, are reported to achieve suitably terminated surfaces. Ex situ surface treatments involve the removal of surface contamination with solvents and acids. A process of chemo-mechanical polishing (CMP) can obtain scratch-free substrates with the reduction of roughness, which is typically subjected to polishing in the alkaline slurry of submicron sized abrasives following the MP procedure. Figure 7.11 shows 2×2 μm^2 AFM images of MP and CMP AlN substrate surfaces. MP substrates had numerous polishing scratches, while CMP substrates were featureless, and root-mean-square (RMS) roughness is found to be 0.5 nm and 0.1 nm, respectively. These results suggest that work damage introduced during mechanical polishing is reduced by chemo-mechanical polishing. Then, wet etching with sulfuric and phosphoric acid mixtures is conducted to reduce the amount of surface hydroxide. In situ surface treatments include AlN substrates nitridation by

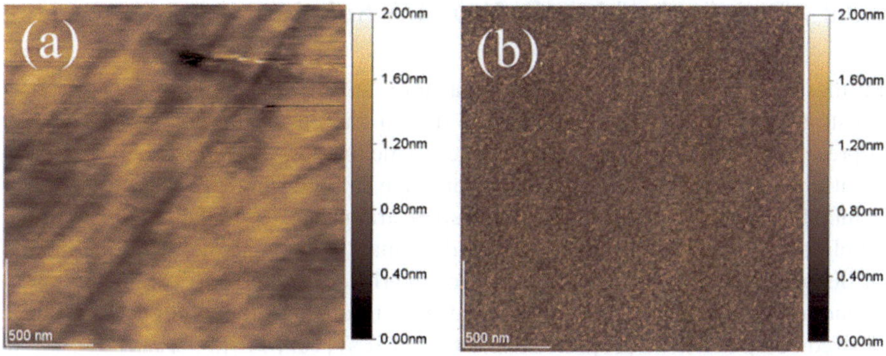

Fig. 7.11 $2 \times 2\ \mu m^2$ AFM images of (**a**) MP and (**b**) CMP AlN substrate surfaces [106]

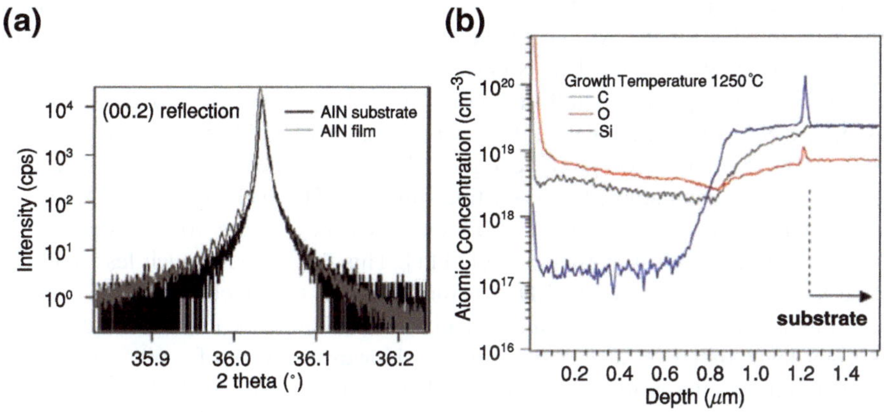

Fig. 7.12 HRXRD triple axis 2θ-ω scans of AlN substrates after CMP and after deposition of AlN homoepitaxial films grown at 1250 °C. (**b**) Calibrated SIMS depth profiles for C, O, and Si impurities in homoepitaxial AlN films grown at 1250 °C [107]

exposure to ammonia at high temperature. Generally, ammonia annealing at 1250 °C may convert the substrate hydroxide layer to AlN.

Once the AlN substrates surface is suitably prepared, high-quality homoepitaxial AlN may be achieved. Figure 7.12a shows HRXRD triple axis 2θ-ω scans of AlN substrates after CMP and after deposition of AlN homoepitaxial films grown at 1250 °C [107]. The lack of any peak shifts or shoulders in the films scan indicated that the film is epitaxial and strain-free. Kim et al. demonstrate modulated precursor flow epitaxial growth (MPEG) is an effective method to achieve high-quality homoepitaxial AlN with narrow X-ray rocking curve peak linewidths of 36 and 61 arcsec for the (002) and (102) diffraction conditions, respectively [108]. Note that homoepitaxial AlN films have low concentrations of background impurities, as shown in Fig. 7.12b.

7.3.2 Pseudomorphic AlGaN on AlN Substrates

Furthermore, the studies of strain relaxation in AlGaN alloys grown on AlN substrate are necessary to guide device design and ensure the low defect density is preserved in heterostructure active layers. Due to a low density of preexisting dislocations in the AlN substrate, the compressive strain during AlGaN heteroepitaxy cannot be relieved effectively. The built-up of strain energy eventually induces either an elastic surface roughening or plastic deformation via generation and inclination of dislocations, depending on the stressor interlayers and growth parameters used. Pseudomorphic AlGaN can be achieved by introducing graded alloy layer to the desired composition or properly superlattices (SLs) [109, 110].

The pseudomorphic growth has many advantages. At first, there are no misfit dislocations and new threading dislocations in the pseudomorphic AlN layer. For the pseudomorphic sample, the (0002) rocking curve width increases from 64 s for the AlN to 81 s for the $Al_xGa_{1-x}N$ layer, while the (10–12) rocking curve width increases from 89 to 104 s. This is in sharp contrast to the relaxed sample in which the (0002) rocking curve width increases from 49 s for the AlN to 239 s for the $Al_xGa_{1-x}N$ layer, while for the (10–12), rocking curve width increases from 30 to 302 s [109]. Secondly, it is possible to grow thick layers with a threading dislocation density (TDD) comparable to the starting substrate. The critical thickness for an n-type AlGaN contact layer with $x = 0.6$ on AlN is only about 40 nm as predicted by the Matthews and Blakeslee model [111]. This thickness is much less than the practical contact layer in an LED structure. Beyond this thickness, it is expected that misfit dislocations will form and lead to threading dislocations throughout the AlGaN layer based on energy equilibrium. However, layers of ~60% Al can be grown fully pseudomorphic up to a thickness of 0.5 μm, while layers of 70% Al can be grown nearly pseudomorphic up to a thickness of 1 μm, as shown in Fig. 7.13a. Thirdly, pseudomorphic AlGaN films exhibit smooth surfaces. It is found that layers begin relaxing by a surface roughening mechanism. Initially the surface appears to buckle to reduce the compressive strain in the layers. Upon continued growth, very rough surfaces were obtained for the $Al_xGa_{1-x}N$ layers with 60% Al at 0.5 mm thick. This is similar to what was seen previously in the growth of lower composition $Al_xGa_{1-x}N$ on low dislocation density bulk AlN substrates in which plateaus develop on the surface [110]. However the pseudomorphic $Al_xGa_{1-x}N$ layers with 70% Al can be grown with very smooth surfaces as seen in Fig. 7.13b. This step-flow growth pattern with atomically smooth surfaces is typical, and the RMS roughness is 0.1 nm in 2×2 μm^2 scan range.

Si-doping of pseudomorphic AlGaN films grown on AlN substrates is necessary to achieve DUV LEDs. Electrical resistivity below 0.1 Ω·cm with Al content below 85% is demonstrated [91]. The electrical resistivity of AlGaN films increases with the Al mole fraction, which results from the reduction of n-type carrier concentration due to the increased donor activation energy or compensating point defects. On the one hand, a significant increase in the activation energy is demonstrated for

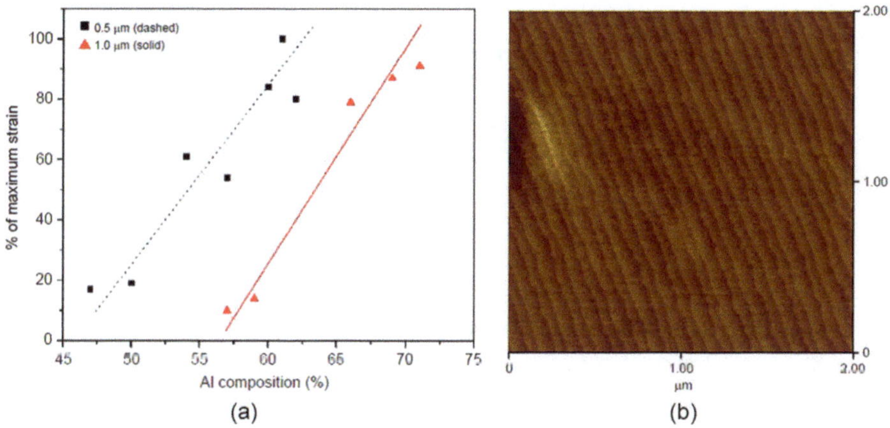

Fig. 7.13 (**a**) Percent of maximum (pseudomorphic) strain versus Al composition for different thickness layers (dashed line is 0.5 μm and solid line is 1.0 μm); (**b**) 2 × 2 μm^2 AFM scans showing the step-flow growth of the pseudomorphic n-type $Al_xGa_{1-x}N$ layer with a composition of 70% Al [109]

films with Al content greater than 80%. On the other hand, electrical compensation is another factor identified as causing a reduction in the carrier concentration.

7.3.3 Pseudomorphic DUV LEDs on AlN Substrates

The next step in obtaining a full LED structure is the growth of the multi-quantum wells (MQWs). The step-flow growth mode and atomically smooth surface continue during growth of the MQWs. Finally, pseudomorphic DUV LEDs on AlN substrates are fabricated, typically consisting of an Al-rich electron blocking layer, a p-type $Al_xGa_{1-x}N$ hole injection layer, and a p-type GaN contact following the MQWs growth. Figure 7.14 shows cross-sectional transmission electron microscopy (TEM) image in the active region and corresponding schematic of the device structure. No dislocations are generated in the whole LED structure except for p-GaN layer [112]. The dislocation density in the MQWs estimated is less than 10^6 cm^{-2}. In contrast, misfit and threading dislocations are observed at the interface between the p-GaN and p-AlGaN, which propagates through the p-GaN layer due to the large lattice mismatch (~2%).

Pseudomorphic DUV LEDs on AlN substrates exhibit noticeable improvement over those on sapphire in device efficiency due to the reduction of dislocation defects in the active region. An output power of over 60 mW has been demonstrated for pseudomorphic LEDs emitting at a peak wavelength of about 270 nm [33]. Especially, Bryan et al. report a record high IQE of ~80% by using $Al_{0.55}Ga_{0.45}N/Al_{0.85}Ga_{0.15}N$ MQW structures emitting at ~258 nm [113]. Another

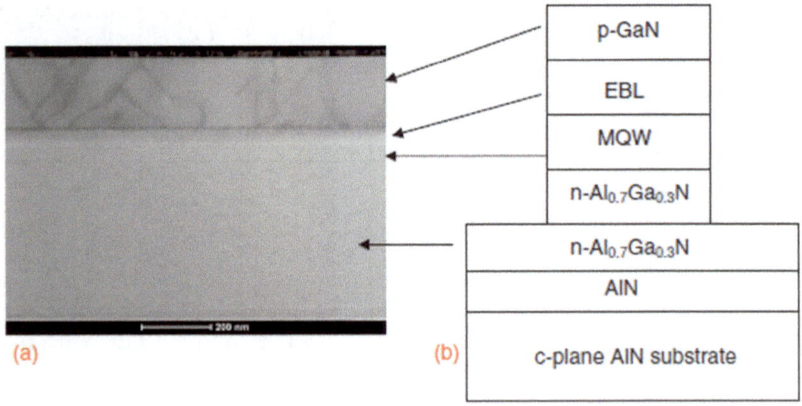

Fig. 7.14 (**a**) Cross-sectional TEM image showing low dislocation density in the active region and (**b**) corresponding schematic of the device structure [112]

important feature of the pseudomorphic DUV LEDs is the improved reliability and lifetime due to the high-quality pseudomorphic growth and high thermal conductivity of the AlN substrate. Along with advanced chip package technology, L50 lifetimes of LEDs are well in excess of 1000 h [114].

7.3.4 Light Extraction Efficiency for Pseudomorphic DUV LEDs on AlN Substrates

The pseudomorphic DUV LEDs are successful in reducing the dislocation density of devices. However, these devices are still challenged by the extremely low light extraction efficiency, which is mainly caused by total internal reflection and the absorption resulting from p-GaN contact layer and AlN substrates. While p-GaN layer strongly absorbs DUV light, a p-GaN contact layer is required to provide a good ohmic contact and to reduce the series resistance for reliable high-power LED operation. Therefore, flip-chip designs are generally used for DUV LEDs, and the light emission is thus extracted through the AlN substrate. The AlN substrate is generally prepared by physical vapor transport (PVT) with significant decrease of optical transparency at photon energies below the band gap, which is thought to be caused by the presence of Al vacancies, substitution impurities, and their complexes. Typically, PVT-AlN substrates with an absorption coefficient of 35 cm^{-1} may absorb approximately 50% of the light. Kinoshita et al. demonstrate that thick AlN grown by hydride vapor phase epitaxy (HVPE) on PVT-AlN substrates have both low dislocation density and high DUV optical transparency [115]. Figure 7.15 shows that external optical transmission spectra of PVT- and HVPE-AlN substrates measured at room temperature in air. The external optical

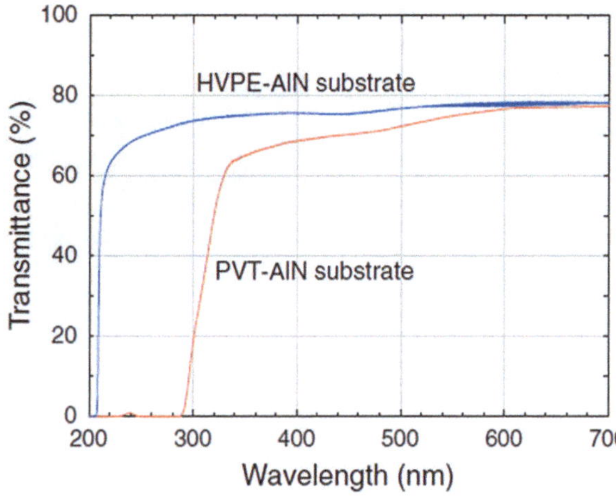

Fig. 7.15 External optical transmission spectra of PVT-AlN and HVPE-AlN substrates measured at room temperature at air [115]

transmittance of HVPE-AlN substrates in the deep-UV range of 220–300 nm was above 63%. DUV-LEDs fabricated on these substrates exhibit a single emission peak at 268 nm with the output power (LOP) and EQE of 28 mW and 2.4%, respectively [116]. Note that thinning AlN substrate may relieve the absorption of AlN substrate. For typical values of absorption coefficients in the AlN substrate (35 cm^{-1}), this reduces the light absorption in the substrate by an order of magnitude from the starting thickness of 425–200 μm [33].

Total internal reflection between AlN substrate and the ambient medium also results in low light extraction efficiency for the pseudomorphic DUV LEDs. The light extraction efficiency from a flat bottom side surface of the transparent AlN substrate of a DUV-LED at 265 nm is estimated to be <4% by three-dimensional (3D) finite difference time-domain (FDTD) calculations. Therefore, it is important to find ways to enlarge the light escape from AlN substrate surface. Many approaches, such as surface roughening and photonic crystal, have been proposed to increase the photon extraction efficiency. Shin-ichiro Inoue et al. reported the highest output power to date for DUV-LEDs with emission wavelengths shorter than 280 nm during CW operation by roughed AlN surfaces. Output power of more than 150 mW is observed at an injection current of 850 mA at a peak emission wavelength of 265 nm [117]. Figure 7.16 shows the schematic of DUV-LEDs on transparent AlN substrates with nanophotonic light extraction structures. The AlN bottom side surface configuration, which is composed of a hybrid structure of photonic crystals and sub-wavelength nanostructures, has been designed using FDTD calculations to enhance light extraction [118]. The LED with the nanophotonic light extraction structure shows considerably wider and stronger light emission in its both near-field and far-field patterns, very low efficiency droop, and an approximately 20-fold increase in the output power over that of a conventional flat surface LED.

With the large-size and high-quality AlN substrates available, the high-power and high-efficiency pseudomorphic DUV LEDs on these substrates will be achieved and thus allow replacement of traditional high-power mercury lamps.

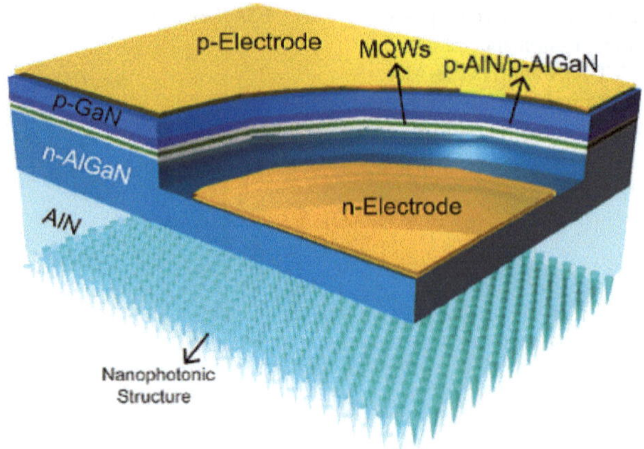

Fig. 7.16 Schematic of DUV-LEDs on transparent AlN substrates with nanophotonic light extraction structures [118]

7.4 Light Exaction Issues of DUV LEDs

Another challenge in obtaining highly efficient AlGaN-based DUV LEDs is the poor light extraction efficiency (LEE). DUV LEDs usually adopt flip-chip configuration for the light-absorbing p-GaN layer on the top of the LED structure. Serious total internal reflection (TIR) occurs at the semiconductor layer/substrate interface and the sapphire substrate/ambient interface due to the large refractive index difference ($n_{\mathrm{AlN}}{\sim}2.3$, $n_{\mathrm{sapphire}}{\sim}1.8$, $n_{\mathrm{air}} = 1$). According to the Snell's law, the critical angle of TIR on the interface of sapphire substrate to air is only about 33° at a wavelength of 275 nm. Furthermore, the intrinsic anisotropic optical polarization properties of Al-rich AlGaN materials also affect the LEE significantly [119]. Light can be emitted from c-plane AlGaN-MQWs as either transverse electric (TE) mode with electrical field vector **E** perpendicular to **c**-axis or transverse magnetic (TM) mode with **E** parallel to **c**-axis. And the TM mode light comes to predominate over the TE mode light when the emission wavelength gets shorter than 300 nm [120]. The TE mode light mainly propagates in the vertical direction, while the TM mode mainly transmits in the horizontal direction, which is difficult to extract due to large incident angle on the surface. Ryu et al.'s simulation showed that the TM mode LEE is much lower than the TE mode LEE, which is partly responsible for the decrease of LEE in DUV LEDs with shorter wavelength [121].

7.4.1 Al-Rich-Induced Optical Polarization Effect in DUV LEDs

As shown in Fig. 7.17, the wurtzite GaN or AlN alloy has three valence subbands, including the heavy hole (HH) band, the light hole (LH) band, and the crystal field split-off hole (CH) band. The topmost valence subband depends on the crystal

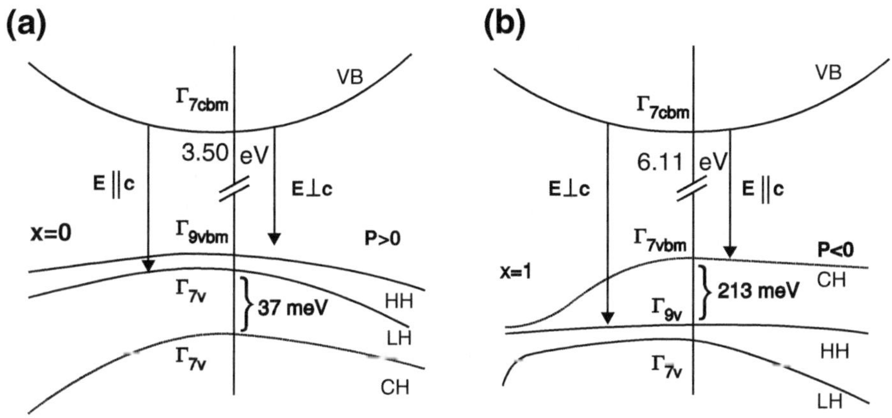

Fig. 7.17 The band structure near the Γ point of (**a**) GaN and (**b**) AlN. E is the electric field vector of the emission light and P is the degree of polarization [123]

field split-off energy (Δ_{CF}). GaN has a positive Δ_{CF} of 38 meV [122], making the HH band as the topmost subband. In contrast, AlN has a large negative Δ_{CF} of −219 meV, leading to the CH band as the topmost subband. Therefore, light emission due to the recombination between the conduction band and the topmost hole subband near the Γ point is mainly polarized with **E** perpendicular to c-axis (TE polarization) in GaN and with **E** parallel to **c**-axis (TM polarization) in AlN.

The $Al_xGa_{1-x}N$ alloy has an intermediate property, depending on the Al composition x. Nam et al. reported at $x = 0.25$, the three valence subbands near the Γ point become degenerated, and the degree of polarization is zero [123]. Thus $Al_xGa_{1-x}N$ generates predominant TM-polarized light when $x > 0.25$.

For the AlGaN multiple quantum well (MQW) DUV LEDs, the switch occurs at higher x due to other parameters, such as the strain in the QWs [124–126], the QW thickness [127–129], and the temperature and carrier density [127, 130, 131] within the QWs. Kolbe et al. [120] showed that the intensity of the TE-polarized light relative to the TM-polarized light decreases with decreasing emission wavelength from the AlGaN active layers. The degree of polarization is ~0 for LEDs with a wavelength of around 300 nm. They also found that the TM-polarized light becomes more dominant relative to the TE-polarized light for AlGaN MQWs with decreasing tensile in-plane strain [132]. Banal et al. [129] reported polarization switching up to x~0.80 in AlGaN/AlN (1.5 nm/13.5 nm) MQWs on sapphire substrates due to the in-plane compressive strain and the quantum confinement due to narrow quantum wells at 8.5 K. Bryan et al. [126] reported the polarization of the emitted light transitioned from TE mode to TM mode at 245 nm in pseudomorphically strained AlGaN grown on AlN template. Reich et al. [133] demonstrated strongly TE-polarized light emission as short as 239 nm from AlGaN MQW DUV LEDs on pseudomorphically strained epitaxial lateral overgrowth AlN/sapphire based on k·p perturbation theory.

The polarization property of the emitted light from the AlGaN MQWs influences the light extraction efficiency of DUV LEDs significantly. Since the TM-polarized light mainly propagates along the c-plane, it suffers more serious TIR issue at the back surface of the sapphire substrate and lower light extraction efficiency than the TE-polarized light. Therefore, to obtain high-efficiency DUV LEDs, it's necessary to either enhance the intensity of the TE-polarized light relative to the TM-polarized light emitted at a certain wavelength using narrow compressive QWs or to develop new strategies to efficiently extract TM-polarized light.

7.4.2 Surface Patterning and High Reflective Techniques for DUV LEDs

7.4.2.1 Surface Patterning

Surface pattering is one promising approach to reduce the internal reflections of light. Khizar et al. [134] reported a 55% enhancement of the light output power in 280-nm AlGaN based DUV LEDs with an integrated microlens array on the sapphire substrate. Pernot et al. [135] reported the average light output power of 270-nm DUV LED with moth-eyes structure on the sapphire substrate is improved by a factor of 1.5. Inoue et al. [117] showed high-power 265-nm DUV LEDs with large-area nanophotonic light extraction structure on the HVPE-AlN substrate, exhibiting a ~20-fold increase in the output power compared with a flat surface DUV LED. Dong et al. [83] reported higher internal quantum efficiency and also higher light extraction efficiency 282-nm AlGaN-based DUV LEDs on nano-patterned sapphire substrate (NPSS). The light scattering at the AlN/NPSS interface and AlN/air voids interface can decrease the internal reflection and thus increase the photon's escape opportunity from the LED structure [136].

Since the TM-polarized light mainly propagates within the c-plane and is extracted from the device sidewalls, disrupting the internal reflection on the substrate sidewalls would be beneficial to improve the extraction efficiency of TE-polarized light. Substrate sidewall roughening can be implemented by ultra-short pulse laser machining during the chip separation process and has been proven to be advantageous for enhanced light extraction efficiency in visible or UV LEDs [137–139]. Lee et al. [139] reported the roughened sidewalls of thick sapphire substrate using stealth laser dicing can promote the extraction probability of photons outside the DUV LED chip in the lateral directions.

We investigated the influence of the roughening position at the sapphire substrate sidewall on the light extraction of AlGaN-based 275-nm UV LEDs and found that effective regions for roughening exist on the substrate sidewalls for light extraction enhancement. Roughening outside the effective roughening region will decrease the sidewall light extraction efficiency due to the undesirable inward photon scattering at the substrate sidewall.

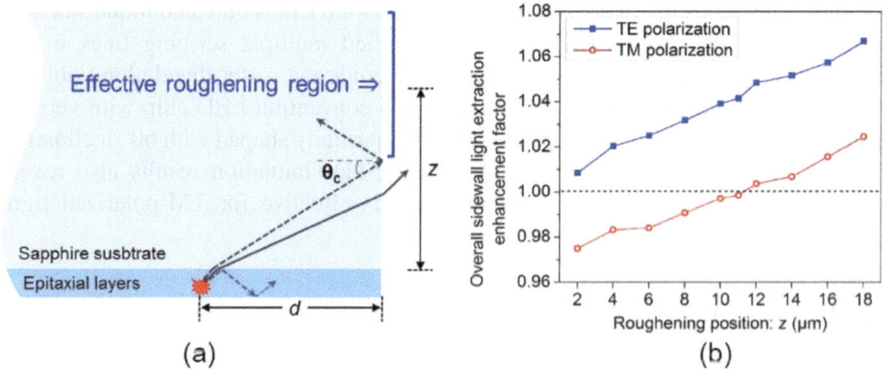

Fig. 7.18 (**a**) Schematic of the effective roughening region on the reference substrate sidewall; (**b**) the dependence of overall sidewall light extraction enhancement factor on the roughening position z simulated by the finite-difference time-domain (FDTD) method [140]

As shown in Fig. 7.18, for a TE- or TM-polarized dipole in LED's active layer, the effective roughening region is from $d \times \tan(\theta_c)$ away from the epitaxial layers to the substrate backside surface, where d is the distance from the dipole to the roughened sidewall and θ_c is the critical angle. Considering multiple TM-polarized dipoles in a real deep-UV LED, the effective roughening region is found equal to that of the point dipole in the central portion of the active layer. Therefore, in one LED with a side length of L, the effective roughening region of TM polarization is deduced to be the area from the sapphire substrate's backside to $L/2 \times \tan(\theta_c)$. On the other hand, the overall sidewall light extraction enhancement factor of TE polarization remains higher than one wherever the roughening position is, revealing that the whole sapphire substrate's sidewall is suitable for roughening for TE-polarized light extraction in DUV LEDs. The experimental results show that compared to the counterpart with two roughening layers, the TM polarization-dominant DUV LED with three roughening layers has 13.2% higher average light output power (LOP) at 20 mA thanks to the larger roughening area in the effective roughening region. However, when further applied additional roughening layers beyond the effective roughening region, the LOP decreases, which is inconsistent with the simulation.

Modifying the geometry of the LED chip can also help outcouple the light trapping in the structure. Krames et al. [141] reported 40% EQE enhancement of truncated-inverted-pyramid AlGaInP-based LEDs sawn by a beveled blade. Lee et al. [142] reported 55% light output enhancement in flip-chip GaN-based LEDs with oblique sapphire substrate sidewalls using a wet etching technique. Laser micromachining has been developed to shape the sapphire substrate, such as nanosecond (ns) laser dicing either with a laser-beam turning mirror [143] or with an oblique sample stage [144] to fabricate oblique substrate sidewalls, and shifted picosecond (ps) laser stealth dicing to form wavy substrate sidewalls [145]. We report a convenient ps laser multiple scribing method to shape the sapphire substrates of AlGaN-based flip-chip 276-nm LEDs into an oblique sidewall geometry for enhanced LEE [146]. The applied multiple scribing lines in the sapphire substrate are intentionally aligned to guide the wafer diced along oblique sidewalls with designed angles. Compared to a convention LED chip with vertical sidewalls, the LED with two opposite sidewalls partially shaped with 60° inclination shows 13.8% higher LOP at 50 mA. The FDTD simulation results also reveal that the oblique substrate sidewall geometry is effective for TM-polarized light extraction.

7.4.2.2 High Reflective Techniques

By introducing reflective structures in the DUV LEDs, part of the emitted light can be reflected back and the probability of light escaping through the substrate is improved. A transparent p-type contact layer, such as p-AlGaN layer [67, 147] or AlGaN/AlGaN short-period superlattice [148], and a high reflectivity p-metallization stack in the DUV spectrum can reduce the light absorption loss and improve the LEE. Shatalov et al. [67] reported a twofold increase of output power from the LED chip. Maeda and Hirayama [147] reported the external quantum efficiency of 287-nm DUV LED increasing from 2% to 5.5% by suing transparent p-AlGaN contact layer and highly reflective Ni(1 nm)/Al layers as the p-type electrode.

On the other hand, reflective structure at the slant mesa sidewalls can efficiently deflect the photons guided laterally along the c-plane and into the vertical direction and thus enhance the LEE. Lee et al. reported sidewall-emission-enhanced DUV LED with active mesa stripes [149] or arrays of truncated cone-shaped active mesas [150] to effectively extract TM-polarized photons down through the sapphire substrate. We reported a sidewall reflection method consisting of microarray mesas with a hexagonal shape and on-chip high reflective metal pads for DUV light [151]. The LOP of 278-nm LEDs was improved by 30.1% compared with that of LEDs without high reflective metal pads. The LOP can be further increased when reducing the side length of the hexagonal mesa thanks to the larger sidewall area for DUV reflection. Wierer et al. [119] demonstrated the enhancement is greater when there is a larger faction of in-plane light.

Acknowledgments The author acknowledges contributions of Qingqing Wu and Jiankun Yang for AlN material growth, Lili Sun for structural design for efficient DUV LEDs, and Yanan Guo for the light exaction issues.

References

1. C.H. Chen et al., A study of parasitic reactions between NH3 and TMGa or TMAl. J. Electron. Mater. **25**(6), 1004–1008 (1996)
2. M.A. Khan et al., III-nitride UV devices. Jpn. J. Appl. Phys. **44**(10), 7191–7206 (2005)
3. T. Uchida, K. Kusakabe, K. Ohkawa, Influence of polymer formation on metalorganic vapor-phase epitaxial growth of AlN. J. Cryst. Growth **304**(1), 133–140 (2007)

4. D.G. Zhao et al., Parasitic reaction and its effect on the growth rate of AlN by metalorganic chemical vapor deposition. J. Cryst. Growth **289**(1), 72–75 (2006)
5. S. Kim et al., Growth of AlGaN epilayers related gas-phase reactions using TPIS-MOCVD. J. Cryst. Growth **245**(3–4), 247–253 (2002)
6. J.R. Creighton, G.T. Wang, Kinetics of metal organic-ammonia adduct decomposition: Implications for group-III nitride MOCVD. J. Phys. Chem. A **109**(46), 10554–10562 (2005)
7. J.R. Creighton, G.T. Wang, M.E. Coltrin, Fundamental chemistry and modeling of group-III nitride MOVPE. J. Cryst. Growth **298**, 2–7 (2007)
8. H. Yang et al., Alleviation of parasitic reactions for III-nitride epitaxy in MOCVD with a spatial separated source delivery method by controlling the main reaction type. J. Cryst. Growth **465**, 1–5 (2017)
9. R. Zuo et al., Influence of gas mixing and heating on gas-phase reactions in GaN MOCVD growth. ECS J. Solid State Sci. Technol. **1**(1), P46–P53 (2012)
10. T.G. Mihopoulos, V. Gupta, K.F. Jensen, A reaction transport model for AlGaN MOVPE growth. J. Cryst. Growth **195**(1–4), 733–739 (1998)
11. R.P. Parikh, R.A. Adomaitis, An overview of gallium nitride growth chemistry and its effect on reactor design: Application to a planetary radial-flow CVD system. J. Cryst. Growth **286**(2), 259–278 (2006)
12. H. Simka et al., Computational chemistry predictions of reaction processes in organometallic vapor phase epitaxy. Prog. Cryst. Growth Charact. Mater. **35**(2–4), 117–149 (1997)
13. R.M. Watwe, J.A. Dumesic, T.F. Kuech, Gas-phase chemistry of metalorganic and nitrogen-bearing compounds. J. Cryst. Growth **221**, 751–757 (2000)
14. G.T. Wang, J.R. Creighton, Complex formation of trimethylaluminum and trimethylgallium with ammonia: Evidence for a hydrogen-bonded adduct. J. Phys. Chem. A **110**(3), 1094–1099 (2006)
15. K. Nakamura et al., Quantum chemical study of parasitic reaction in III-V nitride semiconductor crystal growth. J. Organomet. Chem. **611**(1–2), 514–524 (2000)
16. A. Demchuk, S. Simpson, B. Koplitz, Exploration of the laser-assisted clustering and reactivity of trimethylaluminum with and without NH3. Chem. A Eur. J. **107**(11), 1727–1733 (2003)
17. J. Müller et al., Structure of ammonia trimethylalane (Me3Al-NH3): Microwave spectroscopy, x-ray powder diffraction, and ab initio calculations. J. Am. Chem. Soc. **121**(19), 4647–4652 (1999)
18. A.S. Lisovenko, K. Morokuma, A.Y. Timoshkin, Initial gas phase reactions between Al(CH3)3/AlH3 and ammonia: Theoretical study. J. Phys. Chem. A **119**(4), 744–751 (2015)
19. F.C. Sauls, L.V. Interrante, Coordination compounds of aluminum as precursors to aluminum nitride. Coord. Chem. Rev. **128**(1–2), 193–207 (1993)
20. F.C. Sauls, L.V. Interrante, Z.P. Jiang, ME3AL.NH3 formation and pyrolytic methane loss - thermodynamics, kinetics, and mechanism. Inorg. Chem. **29**(16), 2989–2996 (1990)
21. C.H. Henricks, D. Duffy, D.P. Eyman, Lewis acidity of alanes. Interactions of trimethylalane with amines ethers and phosphines. Inorg. Chem. **7**(6), 1047–1051 (1968)
22. C.C. Amato, J.B. Hudson, L.V. Interrante, Identification of the gas-phase products which occur during the deposition of AIN using the organometallic percursor: [(CH3)2AINH2]3. Appl. Surf. Sci. **54**, 18–24 (1992)
23. J. Müller, Aminodimethylalane (Me2AlNH2): Matrix isolation andab InitioCalculations. J. Am. Chem. Soc. **118**(27), 6370–6376 (1996)
24. Y.S. Hiraoka, M. Mashita, Ab initio study on the dimer structures of trimethylaluminum and dimethylaluminumhydride. J. Cryst. Growth **145**(1–4), 473–477 (1994)
25. K. Sekiguchi et al., Thermodynamic considerations of the vapor phase reactions in III-nitride metal organic vapor phase epitaxy. Jpn. J. Appl. Phys. **56**(4S), 04CJ04 (2017)
26. D. Sengupta et al., Combined ab initio quantum chemistry and computational fluid dynamics calculations for prediction of gallium nitride growth. J. Cryst. Growth **279**(3–4), 369–382 (2005)

27. R. Zuo et al., Quantum chemistry study on the adduct reaction paths as functions of temperature in GaN/AlN MOVPE growth. ECS J. Solid State Sci. Technol. **5**(12), P667–P673 (2016)
28. Y. Inagaki, T. Kozawa, Chemical reaction pathways for MOVPE growth of aluminum nitride. ECS J. Solid State Sci. Technol. **5**(2), P73–P75 (2016)
29. A.V. Lobanova et al., Growth conditions and surface morphology of AlN MOVPE. J. Cryst. Growth **310**(23), 4935–4938 (2008)
30. R. Bouveyron, M.B. Charles, Growth by MOCVD of In(Ga)AIN alloys, and a study of gallium contamination in these layers under nitrogen and hydrogen carrier gas. J. Cryst. Growth **464**, 105–111 (2017)
31. J. Stellmach et al., High aluminium content and high growth rates of AlGaN in a close-coupled showerhead MOVPE reactor. J. Cryst. Growth **315**(1), 229–232 (2011)
32. H. Hirayama, S. Fujikawa, N. Kamata, Recent progress in AlGaN-based deep-UV LEDs. Electron. Commun. Jpn. **98**(5), 1–8 (2015)
33. J.R. Grandusky et al., 270 nm pseudomorphic ultraviolet light-emitting diodes with over 60 mW continuous wave output power. Appl. Phys. Express **6**(3), 032101 (2013)
34. J.P. Zhang et al., Pulsed atomic-layer epitaxy of ultrahigh-quality AlxGa1-xN structures for deep ultraviolet emissions below 230 nm. Appl. Phys. Lett. **81**(23), 4392–4394 (2002)
35. H.M. Foronda et al., Improving source efficiency for aluminum nitride grown by metal organic chemical vapor deposition. Semicond. Sci. Technol. **31**(8), 085003 (2016)
36. R.B. Chung et al., Growth and impurity characterization of AlN on (0001) sapphire grown by spatially pulsed MOCVD. Phys. Status Solidi A Appl. Mater. Sci. **213**(4), 851–855 (2016)
37. M.J. Lai et al., Improvement of crystal quality of AlN grown on sapphire substrate by MOCVD. Cryst. Res. Technol. **45**(7), 703–706 (2010)
38. M. Kneissl, *III-Nitride Ultraviolet Emitters: Technology and Applications* (Springer, Berlin, 2017)
39. H.-C. Seo, I. Petrov, K. Kim, Structural properties of AlN grown on sapphire at plasma self-heating conditions using reactive magnetron sputter deposition. J. Electron. Mater. **39**(8), 1146–1151 (2010)
40. Y.R. Lin, S.T. Wu, Growth of aluminum nitride films at low temperature. J. Cryst. Growth **252**(1–3), 433–439 (2003)
41. S.Y. Karpov, Y.N. Makarov, Dislocation effect on light emission efficiency in gallium nitride. Appl. Phys. Lett. **81**(25), 4721–4723 (2002)
42. A. Severino, F. Iucolano, Impact of growth conditions on stress and quality of aluminum nitride (AlN) thin buffer layers. Phys. Status Solidi B Basic Solid State Phys. **253**(5), 801–808 (2016)
43. X. Rong et al., Residual stress in AlN films grown on sapphire substrates by molecular beam epitaxy. Superlattice. Microst. **93**, 27–31 (2016)
44. P. Dong et al., AlGaN-based deep ultraviolet light-emitting diodes grown on nano-patterned sapphire substrates with significant improvement in internal quantum efficiency. J. Cryst. Growth **395**, 9–13 (2014)
45. Y. Li et al., Defect-reduced green GaInN/GaN light-emitting diode on nanopatterned sapphire. Appl. Phys. Lett. **98**(15), 151102 (2011)
46. Y. Zhang et al., Defect reduction in overgrown semi-polar (11-22) GaN on a regularly arrayed micro-rod array template. AIP Adv. **6**(2), 025201 (2016)
47. V. Adivarahan et al., Robust 290 nm emission light emitting diodes over pulsed laterally overgrown AlN. Jpn. J. Appl. Phys. Part 2 Lett. Express Lett. **46**(36–40), L877–L879 (2007)
48. M. Imura et al., Microstructure of epitaxial lateral overgrown AlN on trench-patterned AlN template by high-temperature metal-organic vapor phase epitaxy. Appl. Phys. Lett. **89**(22), 221901 (2006)
49. M. Conroy et al., Epitaxial lateral overgrowth of AlN on self-assembled patterned nanorods. J. Mater. Chem. C **3**(2), 431–437 (2015)
50. C. Xiang et al., Improved crystalline quality of AlN by epitaxial lateral overgrowth using two-phase growth method for deep-ultraviolet stimulated emission. IEEE Photon. J. **8**(5), 2300211 (2016)

51. M.I. Nathan, The blue laser diode. GaN based light emitters and lasers. Science **277**(5322), 46–47 (1997)
52. H. Miyake et al., Annealing of an AlN buffer layer in N2–CO for growth of a high-quality AlN film on sapphire. Appl. Phys. Express **9**(2), 025501 (2016)
53. M. Ohtsuka, H. Takeuchi, H. Fukuyama, Effect of sputtering pressure on crystalline quality and residual stress of AlN films deposited at 823K on nitrided sapphire substrates by pulsed DC reactive sputtering. Jpn. J. Appl. Phys. **55**(5S), 05FD08 (2016)
54. A.M. Soomro et al., Modified pulse growth and misfit strain release of an AlN heteroepilayer with a mg–Si codoping pair by MOCVD. J. Phys. D. Appl. Phys. **49**(11), 115110 (2016)
55. D.G. Zhao et al., Effect of dual buffer layer structure on the epitaxial growth of AlN on sapphire. J. Alloys Compd. **544**, 94–98 (2012)
56. S.C. Chen et al., Defect reduction in AlN epilayers grown by MOCVD via intermediate-temperature interlayers. J. Electron. Mater. **44**(1), 217–221 (2015)
57. J. Yan et al., AlGaN-based deep-ultraviolet light-emitting diodes grown on high-quality AlN template using MOVPE. J. Cryst. Growth **414**, 254–257 (2015)
58. P. Vennegues et al., Influence of in situ sapphire surface preparation and carrier gas on the growth mode of GaN in MOVPE. J. Cryst. Growth **187**(2), 167–177 (1998)
59. O. Klein et al., TEM investigations on growth interrupted samples for the correlation of the dislocation propagation and growth mode variations in AlGaN deposited on SiNx interlayers. J. Cryst. Growth **324**(1), 63–72 (2011)
60. K. Forghani et al., High quality AlGaN epilayers grown on sapphire using SiNx interlayers. J. Cryst. Growth **315**(1), 216–219 (2011)
61. H.-M. Wang et al., AlN/AlGaN superlattices as dislocation filter for low-threading-dislocation thick AlGaN layers on sapphire. Appl. Phys. Lett. **81**(4), 604 (2002)
62. J.P. Zhang et al., Crack-free thick AlGaN grown on sapphire using AlN/AlGaN superlattices for strain management. Appl. Phys. Lett. **80**(19), 3542 (2002)
63. K.P. Streubel et al., MOVPE growth for UV-LEDs. Proc. SPIE **7231**, 72310G (2009)
64. M. Imura et al., High-temperature metal-organic vapor phase epitaxial growth of AlN on sapphire by multi transition growth mode method varying V/III ratio. Jpn. J.Appl. Phys. Part 1 Regular Papers Brief. Commun. Rev. Papers **45**(11), 8639–8643 (2006)
65. M. Imura et al., Annihilation mechanism of threading dislocations in AlN grown by growth form modification method using V/III ratio. J. Cryst. Growth **300**(1), 136–140 (2007)
66. T.Y. Wang et al., Defect annihilation mechanism of AlN buffer structures with alternating high and low V/III ratios grown by MOCVD. CrystEngComm **18**(47), 9152–9159 (2016)
67. M. Shatalov et al., AlGaN deep-ultraviolet light-emitting diodes with external quantum efficiency above 10%. Appl. Phys. Express **5**(8), 082101 (2012)
68. R.G. Banal, M. Funato, Y. Kawakamia, Initial nucleation of AlN grown directly on sapphire substrates by metal-organic vapor phase epitaxy. Appl. Phys. Lett. **92**(24), 241905 (2008)
69. R.G. Banal, M. Funato, Y. Kawakami, Characteristics of high Al-content AlGaN/AlN quantum wells fabricated by modified migration enhanced epitaxy. Phys. Status Solidi C Curr. Topics Solid State Phys. **7**(7–8), 2111–2114 (2010)
70. W.G. Hu et al., Using different carrier gases to control AlN film stress and the effect on morphology, structural properties and optical properties. J. Phys. D Appl. Phys. **40**(23), 7462–7466 (2007)
71. M. Kneissl et al., Advances in group III-nitride-based deep UV light-emitting diode technology. Semicond. Sci. Technol. **26**(1), 014036 (2011)
72. K. Ban et al., Internal quantum efficiency of whole-composition-range AlGaN multiquantum wells. Appl. Phys. Express **4**(5), 052101 (2011)
73. Y. Zhao et al., Characterization of AlGaN on GaN template grown by MOCVD. Proc. SPIE **6841**, 68410K (2007)
74. S. Kamiyama et al., Low-temperature-deposited AlGaN interlayer for improvement of AlGaN_GaN heterostructure. J. Cryst. Growth **223**, 83–91 (2001)
75. H. Amano et al., Metalorganic vapor phase epitaxial growth of a high quality GaN film using an AlN buffer layer. Appl. Phys. Lett. **48**(5), 353–355 (1986)

76. M. Asif Khan et al., Low pressure metalorganic chemical vapor deposition of AlN over sapphire substrates. Appl. Phys. Lett. **61**(21), 2539–2541 (1992)
77. M. Asif Khan et al., Atomic layer epitaxy of GaN over sapphire using switched metalorganic chemical vapor deposition. Appl. Phys. Lett. **60**(11), 1366–1368 (1992)
78. M. Asif Khan et al., GaN/AlN digital alloy short-period superlattices by switched atomic layer metalorganic chemical vapor deposition. Appl. Phys. Lett. **63**(25), 3470–3472 (1993)
79. O. Ambacher, Growth and applications of Group III-nitrides. J. Phys. D. Appl. Phys. **31**(20), 2653–2170 (1998)
80. J. Zhang et al., AlGaN deep-ultraviolet light-emitting diodes. Jpn. J. Appl. Phys. **44**(10), 7250–7253 (2005)
81. H. Hirayama et al., Milliwatt power 270 nm-band AlGaN deep-UV LEDs fabricated on ELO-AlN templates. Phys. Status Solidi C **6**(S2), S474–S477 (2009)
82. Y. Jianchang et al., High quality AlGaN grown on a high temperature AIN template by MOCVD. J. Semicond. **30**(10), 103001 (2009)
83. P. Dong et al., 282-nm AlGaN-based deep ultraviolet light-emitting diodes with improved performance on nano-patterned sapphire substrates. Appl. Phys. Lett. **102**(24), 241113 (2013)
84. V. Adivarahan et al., Sub-milliwatt power III-N light emitting diodes at 285 nm. Jpn. J. Appl. Phys. **41**(Part 2, No. 4B), L435–L436 (2002)
85. J.P. Zhang et al., High-quality AlGaN layers over pulsed atomic-layer epitaxially grown AlN templates for deep ultraviolet light-emitting diodes. J. Electron. Mater. **32**(5), 364–370 (2003)
86. P. Cantu et al., Metalorganic chemical vapor deposition of highly conductive Al0.65Ga0.35N films. Appl. Phys. Lett. **82**(21), 3683–3685 (2003)
87. M.L. Nakarmi et al., Transport properties of highly conductive n-type Al-rich AlxGa1−xN(x⩾0.7). Appl. Phys. Lett. **85**(17), 3769–3771 (2004)
88. K. Zhu et al., Silicon doping dependence of highly conductive n-type Al0.7Ga0.3N. Appl. Phys. Lett. **85**(20), 4669–4671 (2004)
89. Y. Taniyasu, M. Kasu, T. Makimoto, Electrical conduction properties of n-type Si-doped AlN with high electron mobility (>100cm2V−1s−1). Appl. Phys. Lett. **85**(20), 4672–4674 (2004)
90. J. Hwang et al., Si doping of high-Al-mole fraction AlxGa1−xN alloys with rf plasma-induced molecular-beam-epitaxy. Appl. Phys. Lett. **81**(27), 5192–5194 (2002)
91. R. Collazo et al., Progress on n-type doping of AlGaN alloys on AlN single crystal substrates for UV optoelectronic applications. Phys. Status Solidi C **8**(7–8), 2031–2033 (2011)
92. M. Katsuragawa et al., Thermal ionization energy of Si and Mg in AlGaN. J. Cryst. Growth **189/190**, 528–531 (1998)
93. S.R. Jeon et al., Investigation of Mg doping in high-Al content p-type AlxGa1−xN (0.3<x<0.5). Appl. Phys. Lett. **86**(8), 082107 (2005)
94. M.L. Nakarmi et al., Enhanced p-type conduction in GaN and AlGaN by Mg-δ-doping. Appl. Phys. Lett. **82**(18), 3041 (2003)
95. J. Simon et al., Polarization-induced hole doping in wide-band-gap uniaxial semiconductor heterostructures. Science **327**(5961), 60–64 (2010)
96. S. Marcinkevičius et al., Intrinsic electric fields in AlGaN quantum wells. Appl. Phys. Lett. **90**(8), 081914 (2007)
97. A. Fujioka et al., Improvement in output power of 280-nm deep ultraviolet light-emitting diode by using AlGaN multi quantum wells. Appl. Phys. Express **3**(4), 041001 (2010)
98. S. Sumiya et al., AlGaN-based deep ultraviolet light-emitting diodes grown on epitaxial AlN/sapphire templates. Jpn. J. Appl. Phys. **47**(1), 43–46 (2008)
99. H. Hirayama et al., Marked enhancement in the efficiency of deep-ultraviolet AlGaN light-emitting diodes by using a multiquantum-barrier Electron blocking layer. Appl. Phys. Express **3**(3), 031002 (2010)
100. J. Yan et al., Improved performance of UV-LED by p-AlGaN with graded composition. Phys. Status Solidi C **8**(2), 461–463 (2011)
101. F. Mehnke et al., Efficient charge carrier injection into sub-250 nm AlGaN multiple quantum well light emitting diodes. Appl. Phys. Lett. **105**(5), 051113 (2014)

102. T. Takano et al., Deep-ultraviolet light-emitting diodes with external quantum efficiency higher than 20% at 275 nm achieved by improving light-extraction efficiency. Appl. Phys. Express **10**(3), 031002 (2017)
103. L.M. Svedberg, K.C. Arndt, M.J. Cima, Corrosion of aluminum nitride (AlN) in aqueous cleaning solutions. J. Am. Ceram. Soc. **83**(1), 41–46 (2000)
104. R. Dalmau et al., X-ray photoelectron spectroscopy characterization of aluminum nitride surface oxides: Thermal and hydrothermal evolution. J. Electron. Mater. **36**(4), 414–419 (2007)
105. S.A. Nikishin et al., Short-period superlattices of AlN/Al[sub 0.08]Ga[sub 0.92]N grown on AlN substrates. Appl. Phys. Lett. **85**(19), 4355 (2004)
106. A. Rice et al., Surface preparation and homoepitaxial deposition of AlN on (0001)-oriented AlN substrates by metalorganic chemical vapor deposition. J. Appl. Phys. **108**(4), 043510 (2010)
107. R. Dalmau et al., Growth and characterization of AlN and AlGaN epitaxial films on AlN single crystal substrates. J. Electrochem. Soc. **158**(5), H530 (2011)
108. H.J. Kim et al., Modulated precursor flow epitaxial growth of AlN layers on native AlN substrates by metal-organic chemical vapor deposition. Appl. Phys. Lett. **93**(2), 022103 (2008)
109. J.R. Grandusky et al., Pseudomorphic growth of thick n-type AlxGa1-xN layers on low-defect-density bulk AlN substrates for UV LED applications. J. Cryst. Growth **311**(10), 2864–2866 (2009)
110. Z. Ren et al., Heteroepitaxy of AlGaN on bulk AlN substrates for deep ultraviolet light emitting diodes. Appl. Phys. Lett. **91**(5), 051116 (2007)
111. J.W. Matthews, A.E. Blakeslee, Defects in epitaxial multilayers. I. Misfit dislocations. J. Cryst. Growth **27**, 118–125 (1974)
112. J.R. Grandusky et al., Properties of mid-ultraviolet light emitting diodes fabricated from Pseudomorphic layers on bulk Aluminum nitride substrates. Appl. Phys. Express **3**(7), 072103 (2010)
113. Z. Bryan et al., High internal quantum efficiency in AlGaN multiple quantum wells grown on bulk AlN substrates. Appl. Phys. Lett. **106**(14), 142107 (2015)
114. C.G. Moe et al., High-power pseudomorphic mid-ultraviolet light-emitting diodes with improved efficiency and lifetime. Proc. SPIE **8986**, 89861V (2014)
115. Y. Kumagai et al., Preparation of a freestanding AlN substrate from a thick AlN layer grown by hydride vapor phase epitaxy on a bulk AlN substrate prepared by physical vapor transport. Appl. Phys. Express **5**(5), 055504 (2012)
116. T. Kinoshita et al., Deep-ultraviolet light-emitting diodes fabricated on AlN substrates prepared by hydride vapor phase epitaxy. Appl. Phys. Express **5**(12), 122101 (2012)
117. S.-i. Inoue, N. Tamari, M. Taniguchi, 150 mW deep-ultraviolet light-emitting diodes with large-area AlN nanophotonic light-extraction structure emitting at 265 nm. Appl. Phys. Lett. **110**(14), 141106 (2017)
118. S.-i. Inoue et al., Light extraction enhancement of 265 nm deep-ultraviolet light-emitting diodes with over 90 mW output power via an AlN hybrid nanostructure. Appl. Phys. Lett. **106**(13), 131104 (2015)
119. J.J. Wierer et al., Influence of optical polarization on the improvement of light extraction efficiency from reflective scattering structures in AlGaN ultraviolet light-emitting diodes. Appl. Phys. Lett. **105**(6), 061106 (2014)
120. T. Kolbe et al., Optical polarization characteristics of ultraviolet (In)(Al)GaN multiple quantum well light emitting diodes. Appl. Phys. Lett. **97**, 171105 (2010)
121. H.Y. Ryu et al., Investigation of light extraction efficiency in AlGaN deep-ultraviolet light-emitting diodes. Appl. Phys. Express **6**(6), 062101 (2013)
122. J. Li et al., Band structure and fundamental optical transitions in wurtzite AlN. Appl. Phys. Lett. **83**(25), 5163 (2003)
123. K.B. Nam et al., Unique optical properties of AlGaN alloys and related ultraviolet emitters. Appl. Phys. Lett. **84**(25), 5264 (2004)

124. T.K. Sharma, D. Naveh, E. Towe, Strain-driven light-polarization switching in deep ultraviolet nitride emitters. Phys. Rev. B **84**(3), 035305 (2011)
125. J.E. Northrup et al., Effect of strain and barrier composition on the polarization of light emission from AlGaN/AlN quantum wells. Appl. Phys. Lett. **100**(2), 021101 (2012)
126. Z. Bryan et al., Strain dependence on polarization properties of AlGaN and AlGaN-based ultraviolet lasers grown on AlN substrates. Appl. Phys. Lett. **106**(23), 232101 (2015)
127. J.J. Wierer et al., Effect of thickness and carrier density on the optical polarization of $Al_{0.44}Ga_{0.56}N/Al_{0.55}Ga_{0.45}N$ quantum well layers. J. Appl. Phys. **115**(17), 174501 (2014)
128. T.M. Al Tahtamouni, J.Y. Lin, H.X. Jiang, Optical polarization in c-plane Al-rich AlN/AlxGa1-xN single quantum wells. Appl. Phys. Lett. **101**(4), 042103 (2012)
129. R. Banal, M. Funato, Y. Kawakami, Optical anisotropy in [0001]-oriented $Al_xGa_{1-x}N/AlN$ quantum wells (x>0.69). Phys. Rev. B **79**(12), 121308(R) (2009)
130. M. Hou et al., Effect of injection current on the optical polarization of AlGaN-based ultraviolet light-emitting diodes. Opt. Express **22**(16), 19589 (2014)
131. S.-H. Park, J.-I. Shim, Carrier density dependence of polarization switching characteristics of light emission in deep-ultraviolet AlGaN/AlN quantum well structures. Appl. Phys. Lett. **102**(22), 221109 (2013)
132. T. Kolbe et al., Effect of temperature and strain on the optical polarization of (In)(Al)GaN ultraviolet light emitting diodes. Appl. Phys. Lett. **99**(26), 261105 (2011)
133. C. Reich et al., Strongly transverse-electric-polarized emission from deep ultraviolet AlGaN quantum well light emitting diodes. Appl. Phys. Lett. **107**(14), 142101 (2015)
134. M. Khizar et al., Nitride deep-ultraviolet light-emitting diodes with microlens array. Appl. Phys. Lett. **86**(17), 173504 (2005)
135. C. Pernot et al., Improved efficiency of 255–280 nm AlGaN-based light-emitting diodes. Appl. Phys. Express **3**(6), 061004 (2010)
136. C.-H. Chan et al., Improved output power of GaN-based light-emitting diodes grown on a nanopatterned sapphire substrate. Appl. Phys. Lett. **95**(1), 011110 (2009)
137. J.H. Lee et al., Enhanced extraction efficiency of In GaN-based light-emitting diodes using 100-kHz femtosecond-laser-scribing technology. IEEE Electron Device Lett. **31**(3), 213–215 (2010)
138. Y.Y. Zhang et al., Light extraction efficiency improvement by multiple laser stealth dicing in In GaN-based blue light-emitting diodes. Opt. Express **20**(6), 6808–6815 (2012)
139. K.H. Lee et al., Light-extraction efficiency control in AlGaN-based deep-ultraviolet flip-chip light-emitting diodes: A comparison to In GaN-based visible flip-chip light-emitting diodes. Opt. Express **23**(16), 20340–20349 (2015)
140. Y. Guo et al., Light extraction enhancement of AlGaN-based ultraviolet light-emitting diodes by substrate sidewall roughening. Appl. Phys. Lett. **111**(1), 011102 (2017)
141. M.R. Krames et al., High-power truncated-inverted-pyramid $(Al_{1x}Ga_{1-x})_{0.5}In_{0.5}P/GaP$ light-emitting diodes exhibiting >50% external quantum efficiency. Appl. Phys. Lett. **75**(16), 2365 (1999)
142. C.E. Lee et al., Luminance enhancement of flip-chip light-emitting diodes by geometric sapphire shaping structure. IEEE Photon. Technol. Lett. **20**(1–4), 184–186 (2008)
143. X.H. Wang, P.T. Lai, H.W. Choi, Laser micromachining of optical microstructures with inclined sidewall profile. J. Vac. Sci. Technol. B Microelectron. Nanometer Struct. **27**(3), 1048 (2009)
144. B. Sun et al., Shape designing for light extraction enhancement bulk-GaN light-emitting diodes. J. Appl. Phys. **113**(24), 243104 (2013)
145. S.-J. Chang et al., GaN-based light-emitting diodes prepared with shifted laser stealth dicing. J. Disp. Technol. **12**(2), 1 (2015)
146. Y. Guo et al., Sapphire substrate sidewall shaping of deep ultraviolet light-emitting diodes by picosecond laser multiple scribing. Appl. Phys. Express **10**(6), 062101 (2017)
147. N. Maeda, H. Hirayama, Realization of high-efficiency deep-UV LEDs using transparent p-AlGaN contact layer. Phys. Status Solidi C **10**(11), 1521–1524 (2013)

148. M. Shatalov et al., High power AlGaN ultraviolet light emitters. Semicond. Sci. Technol. **29**(8), 084007 (2014)
149. J.W. Lee et al., An elegant route to overcome fundamentally-limited light extraction in AlGaN deep-ultraviolet light-emitting diodes: Preferential outcoupling of strong in-plane emission. Sci. Rep. **6**, 22537 (2016)
150. J.W. Lee et al., Arrays of truncated cone AlGaN deep-ultraviolet light-emitting diodes facilitating efficient outcoupling of in-plane emission. ACS Photon. **3**(11), 2030–2034 (2016)
151. Y. Guo et al., Enhancement of light extraction on AlGaN-based deep-ultraviolet light-emitting diodes using a sidewall reflection method, in *Wide Bandgap Semiconductors China (SSLChina: IFWS), 2016 13th China International Forum on Solid State Lighting: International Forum on*, (IEEE, New York, 2016), pp. 127–130

Chapter 8
Technology and Droop Study for High Internal Quantum Efficiency

Bo Shen and Zhizhong Chen

8.1 Introduction

Recently, the highest record of the luminous efficiency of white light-emitting diode (LED) was reported as high as 303 lm/W at the current of 350 mA by Cree Co (http://www.cree.com/News-and-Events/Cree-News/Press-Releases/2014/March/300LPW-LED-barrier). It indicates a lower cost and smaller size for lighting source. The internal quantum efficiency (IQE) has been approached to 90% for blue LED [1]. However, the actual luminous efficiency is about 150 lm/W for white LED in the market at present. There still exist many theoretical and technique problems preventing to reach the ultimate values of the luminous efficiency. The IQE of a LED is defined as the generated photon numbers divided by the number of pairs of injected carriers in the pn junction. Except for the radiative recombination, the non-radiative recombination also occurs in the semiconductor. If the radiative one dominates in the operation, the IQE should be high. The luminous efficiency will be high too upon the same light extraction efficiency (LEE) and serial resistance.

IQE is strongly dependent on the crystalline quality and device structure. Although the IQE of the blue LED is very high at normal injection level, those for green LED and UV LED are still rather low. The well-known "green gap" has become a big issue for some years [2]. One major obscure is the increased quantum-confined Stark effect (QCSE) when the indium content increases in InGaN/GaN multiple quantum wells (MQWs). The other cause is the large number of defects because of the phase separation in InGaN layers and large misfit between GaN barriers and InGaN quantum wells (QWs). The wave function overlapping of electron and hole in QWs reduces, and more non-radiative recombination centers

B. Shen · Z. Chen (✉)
Peking University, Beijing, China
e-mail: bshen@pku.edu.cn; zzchen@pku.edu.cn

J. Li, G. Q. Zhang (eds.), *Light-Emitting Diodes*, Solid State Lighting Technology and Application Series 4, https://doi.org/10.1007/978-3-319-99211-2_8

appear with the indium content increasing. On the other hand, when indium disappeared in QWs, the IQE could also be rather small, especially for the deep ultraviolet LED made by GaN/AlGaN MQWs [3]. Without carrier localization to the indium content fluctuation, the high-density threading dislocations (TDs) will become detrimental for light emission.

The IQE in III-nitride materials is usually estimated from the temperature dependence of the photoluminescence (PL) intensity [4]. The IQE at low temperature (below 10 K) is assumed as unity, and the IQE at room temperature (RT) is estimated as the ratio of the PL intensity at RT to that at low temperature. However, the non-radiative recombination and radiative recombination of the defects are not zero at low temperature. Moreover, the recombination process for optical excitation is different from that for electrical excitation. So the validity of the conventional IQE measurement should be carefully checked.

Efficiency droop (hereafter droop, for brevity) is a phenomenon that the luminous efficiency or IQE will reduce gradually with the current density increasing [5]. Even with good thermal dissipation, the droop exists and takes an important role under high injection level. It is ineffective to reduce the lighting cost by cutting down the chip size or increasing the operation current. The droop may be due to Auger recombination, current leakage, carrier delocalization or defect activated, and so on. The high carrier concentration in the active layer is a serious problem when the TD density is high and polarization field is strong. A lot of efforts have been made to alleviate the droop. These routes include carrier concentration reduction, uniform and symmetry injection for both electron and hole, low TDD GaN substrate, etc. Under high injection level, the many body effect [6] and energy bandgap renormalization (BGR) effect [7] should also be considered. However, droop is still an unresolved problem and required more efforts to dig out its origin and find some alleviation methods.

In this chapter, we will briefly introduce the technique progress for IQE promotion of blue LED. The techniques for high IQE green and UV LEDs will be stressed, respectively. Then the new methods for IQE measurement are given. The main content will be focused on the droop and high injection performance of LED. The high radiative recombination rate and low loss will be demonstrated as a result of alleviating the droop effect.

8.2 Techniques for High Internal Quantum Efficiency

The first blue LED with high brightness of 1 Cd was commercialized in Nichia Co. in 1993. It is amazing that the TDD is about 10^{10} cm^{-2} in the GaN material of the high-brightness blue LED [8]. The defects seem not affecting on the efficiency of the GaN-based LED. However, Tsao et al. reported that the defects may influence the performance significantly at high injection level, high temperature, or with the short wavelength UV emission, as shown in Fig. 5 of the Ref. [9]. It indicates that the IQE can be improved by TDD reduction. Soraa Co. reported their UV LED

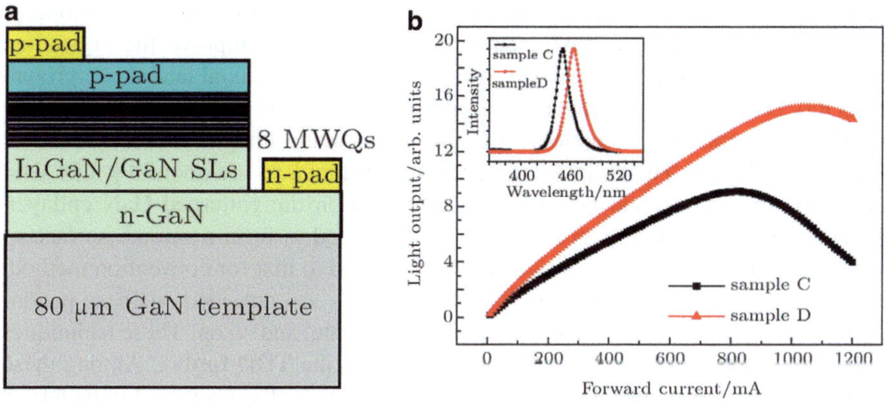

Fig. 8.1 (**a**) Structure of quasi-homoepitaxial LED, (**b**) the L–I curves under cw measurement mode of samples C (conventional) and D (quasi-homoepitaxial). The inset shows the normalized EL spectra [15]

on freestanding GaN substrate with external quantum efficiency (EQE) of 68% at 180 Acm^{-2} [10]. The TDD of GaN substrate is about $10^{5–6}$ cm^{-2}. All performances are maintained at J: 100∼1000 Acm^{-2}, as is device reliability.

The GaN substrate is very expensive compared to the popular sapphire substrate or Si substrate. The laser lift-off (LLO) and quasi-homoepitaxy techniques are developed to enhance the LED efficiency. Thirty-three micrometer freestanding thick film vertical structure LED (VSLED) was made by LLO and hydride vapor phase epitaxy (HVPE) technique [11]. Compared to the thin film (3–4 μm) VSLED on Cu or Si substrate, their electrical and optical performance is high because of less damage on the quantum wells caused by the LLO process. The thick film devices also have a low thermal resistance about 0.15–0.2 K/W. With the current of 500 mA, the 33 μm freestanding thick film VSLEDs obtain over 1.3 times light output power (LOP) than that of the 4 μm thin film VSLEDs on Cu substrates. Speck et al. found that the TDD in the GaN layers is inversely proportional to the film thickness [12]. The TDD of 80-μm-thick GaN template corresponds to 10^7 cm^{-2}, which is two orders lower than that of several-micron-thin GaN film on sapphire substrate. According to Reynolds' results, the surface is strain-free at the thickness [13]. Gorgen et al. proposed that a fully relaxed sample is able to incorporate more indium during its process of growth than a strained sample [14]. We fabricated GaN-based blue LED on 80-μm-thick GaN template with emission wavelength of 464 nm successfully by MOCVD, HVPE, and LLO techniques, as shown in Fig. 8.1 [15]. The light output at 350 mA from the thick GaN template LED is doubled compared to traditional LEDs, and the forward bias is also substantially reduced. The high performance of 80-μm-thick GaN template LED depends on the high crystal quality.

There is no doubt that the sapphire substrate is the most popular substrate for epitaxy of GaN-based material. Two-step growth method on the sapphire substrate opens a new era of nitride [16, 17]. Akasaki, Amano, and Nakamura won the 2014

Nobel Prize for their invention of the first GaN-based blue LED. Using the concept of selective area epitaxy (SAE), an effective method to improve the crystalline quality of GaN epilayer is provided, which is named as epitaxial lateral overgrowth (ELOG) [18]. The 2-μm-thick GaN epilayer was firstly grown on the c-plane sapphire substrate. Then a periodic SiO_2 mask was made along ⟨10–10⟩or⟨11–20⟩of the GaN epilayers. The stripe width, period, and thickness are about 3 μm, 6 μm, and 100 nm. The second growth was performed on the patterned GaN epilayer. Lateral overgrowth wing regions would be coalesced to form a smooth surface of GaN. The TDD reduced about two orders compared to that for convention method. After the basic ELOG technique, many evolutions are provided, such as pendo-epitaxy, air-bridge epitaxy, grooved sapphire substrate, and so on. These techniques solve the problems of the basic ELOG and reduce the TDD further. Among these methods, the grooved sapphire substrate has been gradually developed to patterned sapphire substrate (PSS) [19, 20], which becomes the most popular substrate for commercial GaN epitaxy now. Around 2005, many types of the PSS patterns appeared, including circular hole, truncated circular cone, circular cone, inverted circular cone, pyramid, inverted pyramid, hemisphere, and so on. PSS technology can not only improve the crystalline quality by TDD reduction and modify the strain but also enhance the light extraction efficiency by changing the propagation direction of the total inner reflection (TIR).

PSS technology is a mask-free method and does not require second growth. It is economic and acceptable for commercialization. In 2011, Wang et al. fabricated different period and shape of PSS by photoresist reflow technique [21]. They grew GaN films on these PSS by MOCVD. The TEM results show that the cone-shaped PSS is more effective to reduce the TDD in comparison to the truncated cone-shaped PSS. The short-period PSS makes the GaN epilayer more strain relaxed than the long period one. The epilayer on the cone-shaped PSS also shows more strain relaxation than that on the truncated cone-shaped PSS. Some researchers changed the tilt angle of the cone of PSS. They found that when the angle was about 30°, the crystalline quality was better than those of other tilt angles [22]. Huang et al. found the LEE is also enhanced for the LED grown on the 30° tilt PSS [23].

In 2000 Ashby et al. predicted that the TDD in GaN epilayer would reduce when the pattern size of the PSS decreased to nanoscale. Nano PSS (NPSS) has been applied in LED fabrication since 2007. Wang et al. thought that nanoscale pattern made the number of the pattern in per area increases [24]. The light would be more easily scattered and reflected out from the semiconductor. Su reported their 1.5 times light emission enhancement results by NPSS. They used the nanosphere lithography and ICP dry etching. The size of the sphere is about 450 nm. The spacing is about 50 nm with depth about 150 nm. They reported that the crystalline quality of GaN on NPSS was not improved compared to that on conventional micro-PSS. Moreover, the air gap in their epilayer was thought as the barrier in the thermal path, which led to the reduction of the saturation current of the LED. In 2011 Wetzel et al. fabricated NPSS by nanoimprint technology [25]. The period and depth of the NPSS are about 450 and 125 nm, respectively. Compared to the planar sapphire substrate, the LEDs grown on that NPSS achieved a three times light emission enhancement. The IQE is

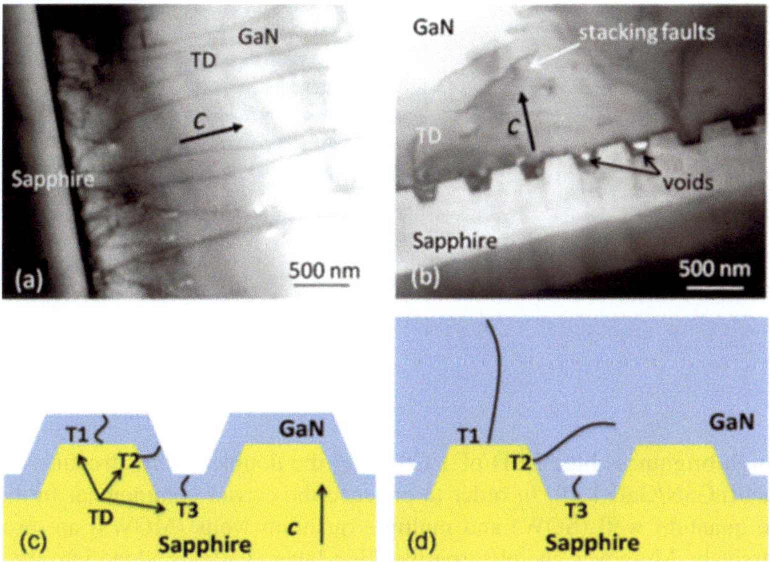

Fig. 8.2 Cross-sectional TEM of the epitaxial GaN/sapphire interface on (**a**) planar and (**b**) nano-patterned sapphire. Schematics showing early (**c**) and late (**d**) phases of the TD (T1, T2, T3) formation on the nano-patterned substrate [25]

enhanced two times, and the LEE is enhanced 58%. They observed 44% reduction of TDD in TEM images, as shown in Fig. 8.2.

Different views for the crystalline quality improvements of GaN on NPSS are reported by different research groups. It is accepted that the size and shape of PSS will determinate the growth route. The species transport processes on the NPSS is quite different to those on conventional PSS or planar substrate. Ee et al. found that the surface energy of the sapphire substrates was altered by nano-patterns [26]. We used silane in the process of the nucleation island growth and coalescence. Silicon, acting as a surfactant, will change the species mobility and surface energy [27]. An extremely smooth surface was achieved with an average roughness of 0.10 nm over $3 \times 3\ \mu m^2$ for a sample with 600 s 3D growth and 2400 s lateral growth on the NPSS. The silane addition causes the effective 3D growth. The size, homogeneity, and faceted sidewalls of islands by the 3D growth lead to a high crystalline quality, much strain relaxation, and a specular surface for the GaN epilayers. The growth time could be shortened more by further reducing the pattern size and optimizing the 3D growth conditions.

Besides the low TDD and strain-controlled substrate, the IQE improvement of LED mostly depends on the structure of the active layer. The first GaN-based LED was fabricated using metal-insulator-semiconductor (MIS) [28]. The first GaN-based pn junction LED was achieved by Amano et al. after p-type GaN was successfully obtained [29]. Both of the two LEDs are not very bright because of the lower carrier confinement in the junction area. S. Nakamura reported the

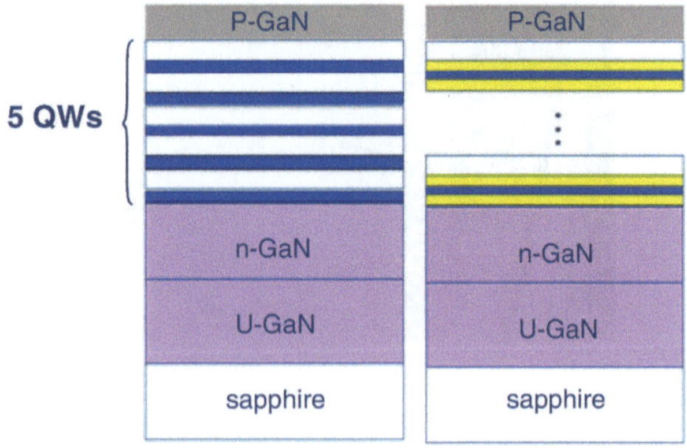

Fig. 8.3 The schematic structures of (**a**) the conventional LED, (**b**) the sandwiched LED

first high-brightness blue LED of 1 Cd using the double hetero-structure (DH) of AlGaN/InGaN/GaN [30]. In order to enhance the carrier confinement further, the single quantum well (SQW) and multiple quantum wells (MQWs) are proposed successively. Moreover, the electron-blocking layer (EBL) is added into the active layer after the MQWs growth because the mobility of electron is much higher than that of holes. Meanwhile, the electron reservoir layer (ERL) is also added before the MQWs growth to improve the carrier capture ability in the active layer [31].

Since the group III cation does not locate at the center of the nitrogen tetrahedron, the group III nitrides show very strong polarization, including spontaneous and piezoelectric ones. The polarization plays an important role on the IQE of almost all nitride LED on c-plane nitride template by bending the energy band. The QCSE makes the separation between the wave functions of electron and hole, which leads to the reduction of the radiative recombination possibility. The trapezoid QW shows better performance than the rectangular one [32]. The trapezoid QW is made by a graded indium content of InGaN QW and GaN barrier. The overlapping of the wave functions in QWs is increasing, and the homogeneous quantum dot-like attributes the high performance of the trapezoid QW [32]. We have developed a sandwiched QW structure to modify the polarization field [33]. The schematic structures are shown in Fig. 8.3. LEDs using InGaN/GaN MQWs with thin low-temperature GaN (LT-GaN) layers bounding each InGaN layer are grown by MOVPE. It is observed that the effect of polarization modification by LT-GaN is more significant. The LT-GaN layers also improve the crystal quality of the sandwiched InGaN. This in turn leads to lower defect-related recombination in the sandwiched QW, accompanied by higher radiative recombination efficiency and smaller FWHM of the emission peak. Doping in the barrier is an effective method to screen the polarization field [34]. With the doping concentration increasing, the Stokes shift becomes small. The Si concentration reaches $1 \times 10^{19}/cm^3$ so that the piezoelectric field can be fully screened. Furthermore, the interface of GaN/InGaN becomes smoother, and the radiative recombination lifetime reduces.

Kim et al. calculated the band diagram of InGaN/AlInGaN MQW LED structures, as shown in Fig. 8.4 [35]. When the total spontaneous and piezoelectric

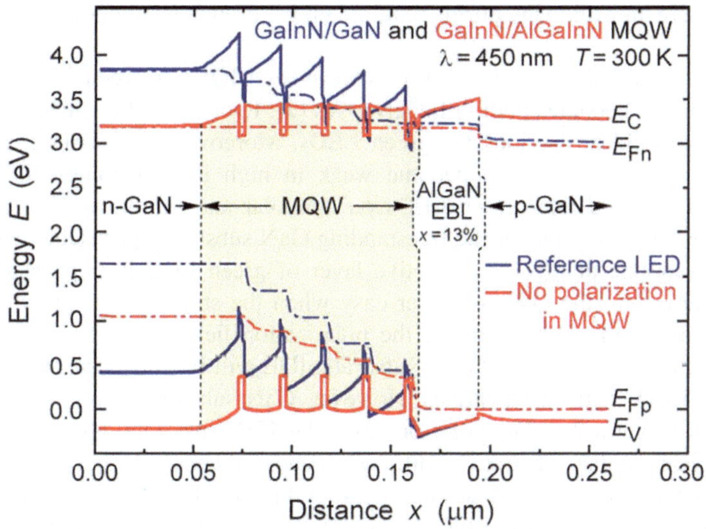

Fig. 8.4 Calculated band diagram of reference GaInN/GaNLED as well as GaInN/AlGaInN LED structure with polarization-matched MQW under a forward bias condition [35]

polarization charges in InGaN QWs are matched by equal polarization charges in AlGaInN quantum barriers, and the remaining polarization charges at the bottom and top of the MQW are respectively compensated by n- and p-type delta doping, the triangular barriers disappear. Besides the barrier, using a certain content of AlInGaN, the EBL can also be polarization matched to GaN. When the polarizations in barriers and EBL are matched to the GaN layer, the band becomes flat and the conduction band on the p-side is now higher than on the n-side making it much more difficult for electrons to reach the p-side. So the light output increases by 138% at 350 mA [35]. Some researchers found that the AlInGaN barriers could lead to interband transition instead of recombination by localization states or quantum dots. It shows high performance in high injection level [36]. In conventional injection level, the IQE is lower than that for GaN barriers.

Due to the direction of the net dipole moment along c-axis, the nonpolar or semipolar plane will show zero or small polarization in the growth direction. The less polarization field in QWs will enhance the light emission of the LEDs. The available substrates include m-, a-, and r-plane GaN; r- and m-plane sapphire; (100) planeγ-$LiAlO_2$; (100) and (110) plane $MgAl_2O_4$; and so on [37–39]. Researchers have achieved many excellent performances for nonpolar or semipolar LEDs, such as higher radiative recombination rate, high mobility, high In incorporation, polarized light, and so on. S. Nakamura reported their hole concentration of a-plane p-GaN as $2 \times 10^{18}/cm^3$. However, the luminous efficiency of the nonpolar or semipolar LEDs is not superior to that of polar ones. The growth of the GaN epilayer is very difficult on nonpolar and semipolar substrates because pure group III or N atom monolayer cannot be formed likely and stack fault densities become higher. There are still many efforts required on the growth on nonpolar or semipolar substrates.

As mentioned above, it is hard to achieve high IQE for green LED when the indium content becomes high in InGaN QWs [2]. Poor crystalline quality and high polarization field are serious for green LEDs. Moreover, the carrier localization around the In-rich area will become weak in high indium content QWs [40]. Nakamura reported their green LEDs on nonpolar and semipolar GaN substrate [41–43]. Firstly, the TDD of the freestanding GaN substrate approaches 10^5 cm^{-2}, so the crystalline quality of the active layer of green LED is improved. The In incorporation into the QWs is rather easy when the strain on the growth surface become small [14]. What's more, the polarization field is reduced or alleviated by the semipolar or nonpolar GaN substrate. IQE will be enhanced by less QCSE for green LED on these substrates. However, GaN substrate is very expensive. It is difficult to get a 2 in. nonpolar or semipolar GaN substrate. High-quality GaN growth on nonpolar substrate is difficult because there are two types of atoms, both nitrogen and group III atoms on the nonpolar plane, which will influence the mobility and anisotropy on the growth surface. Using semipolar substrate, the crystalline quality of LED wafer will be improved. It is achieved an external quantum efficiency (EQE) of 13.4% with a peak emission wavelength of 563 nm [44]. The EQE of the semipolar LED is much higher than that of AlInGaP LED (Fig. 8.5).

It is obvious that the nonpolar and semipolar substrates are very effective to enhance the IQE of green LED. The GaN substrate is too expensive to use for cheap LED fabrication. Heteroepitaxially grown (11–22) semipolar LED on sapphire substrates are continuously paid attention to due to the lower cost and larger sizes of sapphire substrate [45–47]. High-quality (11–22) GaN templates are grown on (10–12) r-plane PSS by MOCVD [45]. Recently, Wang group reported (11–22) plane InGaN LEDs from green to amber (605) on their overgrown GaN on micro-rod templates [46]. They used m-plane sapphire substrate. They provided many evidences of polarization reduction and obtained a polarization ratio of about 25% for their (11–22) LEDs. Nakamura et al. also obtained efficient semipolar (11–22) 550 nm yellow/green InGaN LEDs on low TDD (11–22) GaN/PSS [47]. It is hopeful to fabricate high efficiency semipolar InGaN LEDs with long emission wavelength on large area sapphire substrate (Fig. 8.6).

Besides the large area growth of (11–22) semipolar LED, the core-shell LEDs were reported to obtain high IQE of green LED. The m-plane of GaN is easy to grow or etched by induced coupling plasma (ICP) etching. The m-plane InGaN LED shell can be grown on the m-plane n-GaN core [48]. The cost is much lower than that using am-plane freestanding GaN substrate. The large surface of the core-shell LED induces new optical and electrical properties. The emission wavelength can be tuned to green or amber by changing the diameter of the nanorods [49]. In May 2012, Samsung Co. reported their core-shell green LED in the XVI International Conference on MOVPE (ICMOVPE-XVI). The IQE is as high as 75% at the wavelength of 550 nm. The value is five times of the highest IQE by other techniques. Glo Co. also reported their green LED based on nanorod array

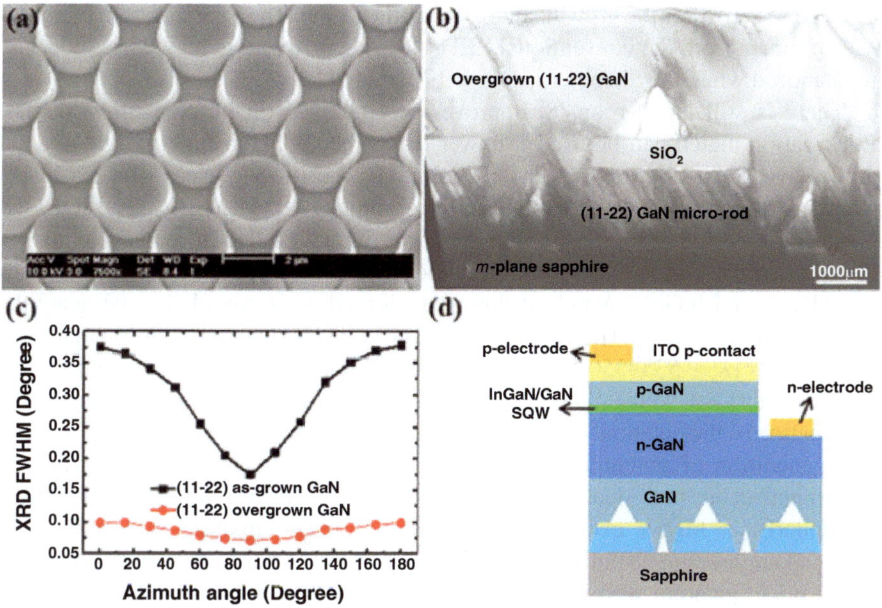

Fig. 8.5 (**a**) SEM image of a micro-rod GaN template; (**b**) TEM image of the (11–22) GaN overgrown on micro-rods; (**c**) XRD FWHMs of the overgrown (11–22) GaN and (11–22) GaN template, measured at azimuth angles ranging from 0° to 180°; (**d**) schematic of (11–22) InGaN SQW LED [46]

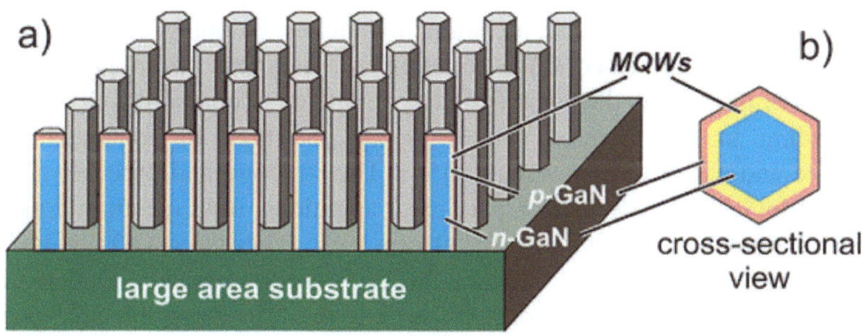

Fig. 8.6 Core-shell strategy for the fabrication of GaN-based nanoLEDs. Sketches of (**a**) core-shell nanoLED ensemble. (**b**) Cross-sectional view of a core-shell nanoLED. The active LED area can be increased approximately by a factor of four times the aspect ratio in comparison to a planar LED [48]

in the 10th International Conference on Nitride Semiconductor (ICNS-10). The SMASH project in the Europe Union organized Osram and Obducat companies and the research institutes in Germany, the UK, France, etc. to study the nanorod LEDs. It is much superior for nanorod LED arrays. The growth of nanorod avoided the effects of lattice and thermal misfit in heteroepitaxy by compliance "bridges" from the mismatched epilayers to the nano-patterned substrates [50]. The TDD even reduced to zero after the growth condition optimization. The emission area increases several times from the conventional LED [48]. There are no guided wave modes in the nanorod when the diameter is below 170 nm [51].

There is still much room to improve the efficiency of the polar green LEDs. Polarization match can be used to lower down the QCSE in high indium content active layer [35]. The Rensselaer Polytechnic Institute group used AlInGaN layers as polarization-matched barriers and electron-blocking layer (EBL). The IQE was enhanced 1.38 times compared with the conventional LEDs. However, the growth of high-quality AlInGaN is very difficult because of the opposite growth conditions for AlGaN and InGaN. A pulsed-flow growth method is provided to fabricate the AlInGaN barriers [52]. The electron overflow is much suppressed in N-polar GaN green LEDs ($\lambda \sim 535$–550 nm) [53]. The LEDs were found to operate with a low ($\sim$2.3 V) turn-on voltage. The reversed (N-face) polarization charges benefit the elimination of the potential barriers against carrier injection (outside the QW region) and generation of potential barriers to confine carriers (inside the QW region). O'Donnell et al. provided a nonradiative shunt for electron-hole recombination, particularly at higher excitation densities. Solid-state lighting is redefined by bandgap engineering techniques (SORBET). A novel bandgap engineering strategy based upon quantum well intermixing (QWIM) offers solutions to IQE green gap [2]. All graded SQW show a significant increase in IQE compared to a rectangular SQW for a wide range of current densities. A tapered AlGaN EBL were used in green LED structure, which improved the hole injection and electron overflow [54]. A charge asymmetric resonance tunneling (CART) structure was applied to nitride-based green LEDs to enhance their output efficiency [55]. The structures of barrier and wells can also be optimized, including the thickness, doping, content gradation, and so on. The sandwiched QWs structure is a potential technique because of prevention of the indium diffusion from low-temperature quantum wells [33].

According to the thermodynamics theory, the high In content InGaN is metastable and easy to phase separate [56]. The structure and processing of high In content InGaN QWs should be paid great attentions. InAs quantum dots (QDs) are put forward by Arakawa and Sakaki in 1982. They used QDs to lower the current density threshold and enhance the differential gain [57]. Due to its low defect density and high confinement of carriers, the QDs structures are hopeful to act as the active layer for green LEDs. Generally, the InGaN phase separation is very little when the size of QDs is small. In early works, InGaN QDs are obtained by self-assembling method [58]. In the self-assembling method, it is hard to control the size and density of the InGaN QDs. Arakawa et al. achieved uniform InGaN QDs by selective growth [59]. Mi et al. reported a high efficiency green, yellow, and amber emission from InGaN/GaN dot-in-a-wire hetero-structures [60]. The IQE is about 45% at 570 nm. The value is the highest one in InGaN-based green LED.

Although most green LEDs are grown on sapphire and freestanding GaN substrate, green LEDs on Si substrate is being industrialized [61–64]. Both the lattice mismatch and thermal mismatch between GaN and Si cause the tensile strain generated in GaN epilayer. The tensile strains benefit for In incorporation into InGaN QWs. It indicates that the QW growth temperature of LEDs on Si substrate can be set 20° higher than that of sapphire, which is helpful for improving the quality of QWs [61]. V-pits is also paid attention to in the green LED on Si substrate. The prestrained layers consisted of a low-temperature (LT)-GaN layer, InGaN/GaN SLs, and blue QWs. LT-GaN is used to generate V-pits. The V-pits can divide the QW plane into small areas, which lead to strain relaxation in QW. The V-pits may also form high potential barrier preventing carriers from recombination at TD. Due to the thin InGaN/GaN layers on the sidewall, the hole tunneling into the planar QWs becomes easier [62, 63]. In the longer wavelength at 566 nm, the EQE of high-power LEDs on Si substrate has been achieved to 9.4% [64].

As an alternative route, the surface plasmon (SP) are paid much attention to obtain high-efficiency green LEDs. SP mode is a collective electron oscillation on the surface by the external excitations [65]. Using Ag, Au, and Al films on the InGaN QW, the PL intensity was enhanced 17 times at the resonant wavelength [66]. The GaN spacer is about 10 nm thick between the InGaN QW and metal film. SP modes include SP polaritons (SPP) and localized SP (LSP). To realize the coupling of photons and SPPs, wave vector matching is essentially required, while LSP coupling does not require the matching. LSP coupled with QWs for green LED emission enhancement is of great interest for many researchers because of zero momentum-matching and adjustable resonance energy [67–70]. The metal material, size, shape, covering dielectric layer, and spacer thickness of the metallic nanoparticles (NPs) are studied to obtain high-efficiency green LED and to study its coupling mechanisms. Compared to that in the randomly distributed metallic NPs, LSP in the periodical ones shows many advantages for light extraction and spontaneous emission enhancement [69, 70]. The NPs with rigid period and the specific size of Ag can diffract and effectively scatter more light from the semiconductor and metal. We successfully fabricate Ag NPs array in photonic crystal (PhCs) holes of p-GaN by soft-stamp nanoimprinting and lift-off techniques [71].

Figure 8.7 shows the scanning electron microscope (SEM) images of Ag NPs array embedded into p-GaN layer of green LEDs. The PL intensity is enhanced to 4.5 times and the PL decay time reduced 3.8 times by the Ag NPs array. We also reveal that Ag NPs, with a diameter over a large range from 90 nm to 200 nm and α larger than 1.5, would help to get a significant emission enhancement on green LEDs.

The improvement of luminous efficiency for green LED is urgently required to promote the quality of semiconductor lighting. In a near future, polarization engineering, strain engineering, and MOCVD processing optimization can be carried out. With the development on nanoscience and technology, LSP, PhC, QDs, and core-shell structures can play important role in the green LED fabrication. The high-efficiency, low-cost green LED will be the key device to promote the development of semiconductor lighting.

Recently, AlGaN-based deep ultraviolet light-emitting diodes (DUV-LEDs) have received considerable attentions due to their promising applications, such as purification, sterilization, biomedicine, UV curing, gas sensing, optical data storage, and non-line-of-sight communication [72]. Compared with the EXE of the visible

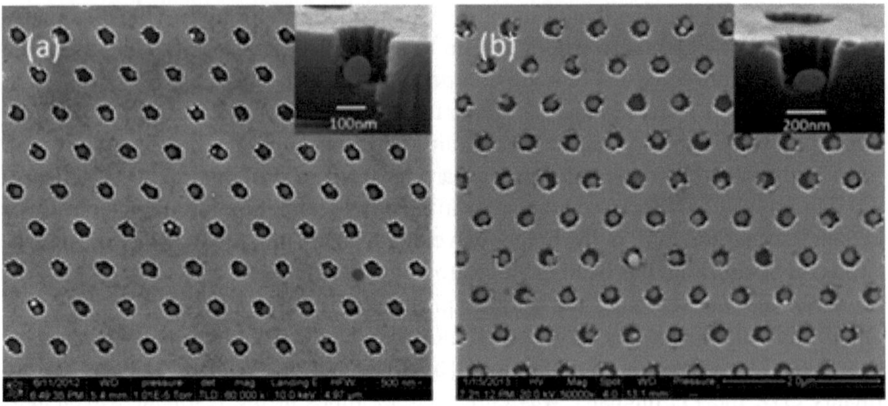

Fig. 8.7 Top view and cross-view (inset) SEM images for the Ag arrayed LED (**a**) sample 1 and (**b**) sample 2 on the green LED epitaxial wafer, for the hexagonal arrayed nanoholes into the p-GaN cladding layer; every hole is filled with single Ag NP [71]

LED usually exceeding 50%, the EXE of AlGaN-based DUV LED is hard to be beyond 10% [73]. There are many problems in substrate, high Al content AlGaN growth, polarization, p-doping, and so on.

I. Akasaki et al. analyzed the IQE of whole-composition-range AlGaN multi-quantum wells (MQWs) on AlGaN with various dislocation densities (DDs) by excitation-density-dependent PL measurement [74]. They found that with an excess carrier density of 1×10^{18} cm^{-3}, the IQE changes from 4 to 64% with the DD changing from $6 \times 10^{9}$to 2×10^{8} cm^{-2} for DUV MQWs with the emission wavelength from 230 to 350 nm. Because aluminum adatoms have a much larger sticking coefficient and much lower surface mobility than gallium adatoms, the UV-transparent low DD AlN and AlGaN is difficult to grow. A low V/III ratio (about 1:4) and high growth temperature in excess of 1200 °C were used to achieve two-dimensional growth mode for AlN on sapphire. A. Khan et al. suggested that a pulsed atomic-layer epitaxy (PALE) approach could be used to grow high-quality AlN and AlGaN films [75]. In PALE the flow rates of group III and group V precursors are sequentially modulated, thus enhancing the surface migration of aluminum and gallium adatoms. In our study the AlN buffer was fabricated by the PALE method, and then AlN epilayer was continuously grown on it [76]. TEM images show that PALE-AlN buffer is a region with high-density TDs, and the subinterface can play an obvious role for blocking TDs. Initially, dislocations are generated at the points of grain boundaries in the PALE buffer. With increase in the lateral growth rate, small islands are incorporated by bigger islands, and they became coalescence. During the coalescence of these islands, the dislocation line will be bended due to their image force. When encountering another dislocation, a dipole half loop will be formed. If the two dislocation lines have opposite Burger vectors, they will be annihilated by each other. Otherwise, the large kinks form and TDs reorient back into approximately [0001] direction [76].

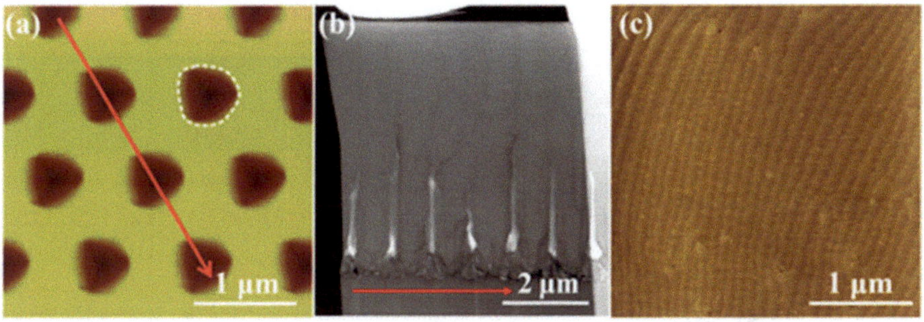

Fig. 8.8 (**a**) AFM image of a typical NPSS (3 × 3 μm^2). (**b**) The cross-sectional STEM image for this chosen sample fabricated by focused ion beam. (**c**) A typical AFM image of the surface morphology of the AlN sample on NPSS with 650 nm hole patterns (3 × 3 μm^2) [79]

PSS used for AlN growth are also paid much attention on. However, large spacing between micro-patterns usually requires a great coalescence thickness close to 10 μm, implying long growth time and high cost. To obtain a thinner coalescence thickness, AlN epitaxy on NPSS and nano-patterned AlN/sapphire template are considered [77, 78]. Although their results are exciting, the growth mechanism and dislocation evolution are not clear for AlN epitaxy on NPSS. We performed a series of AlN/NPSS experiments by changing the size of the pattern and V/III ratio [79]. Figure 8.8 shows the AFM image of the prepared NPSS with hexagonal hole patterns and STEM and AFM images of AlN epilayer on 650 nm hole diameter NPSS. The period is about 1 μm. The crystalline quality can be greatly improved by using optimized 1 μm period, 650 nm diameter NPSS. The XRD ω-scan FWHMs for (0002) and (10–12) reflections are 171 and 205 arcsec, respectively. The optimized NPSS contribute to eliminating almost entirely the threading dislocations (TDs) originating from the AlN/sapphire interface via bending the dislocations by image force from the void sidewalls before coalescence.

In addition, reducing the misorientations of the adjacent regions during coalescence adopting the low lateral growth rate is also essential for decreasing TDs in the upper AlN epilayer.

Stress in the AlGaN growth should be well controlled on the above AlN/sapphire template. Interlayers of AlN and AlGaN and its SLs have been grown at low temperatures to avoid cracking of AlGaN [80–82]. The use of AlN/AlGaN SLs to alleviate the cracking of thick AlGaN films over sapphire substrates has also been reported [82]. X-ray measurements show that these SLs were efficient in enhancing the AlGaN mosaic block dimensions. Development of the AlGaN/AlN SL buffer layers enabled growth of highly doped n-AlGaN to a thickness >2.0 μm [82], enabling development of UV-B and UV-C LED devices first with submilliwatt- and later with milliwatt-level output powers. Further optimization of the AlN buffer layer quality by means of PALE enabled improvements in the material quality of the n-AlGaN bottom-cladding layer, as well as the multiple QW layers, and resulted in the first milliwatt level UV-C LEDs [83]. We studied the influence of growth

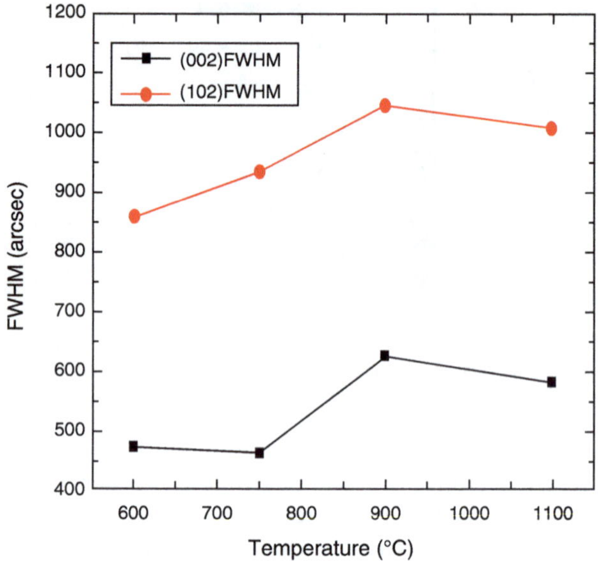

Fig. 8.9 (002)- and (102)-plane FWHMs of ω-scan XRD for $Al_{0.65}Ga_{0.35}N$ epilayers on AlN/sapphire template with GaN interlayer grown at different temperatures [84]

temperature for the low-temperature (LT) GaN interlayer on the crystalline quality of AlGaN. Figure 8.9 shows a suitable temperature as 600 °C for the growth of LT GaN interlayer.

Doping in high Al content AlGaN is difficult for both n- and p-type. The bandgap of the semiconductor increases, and the ionization energies for silicon (n-type dopant) and magnesium (p-type dopant) increase, resulting in a lower ionization efficiency. P-type doping of high-aluminum-content AlGaN is even more challenging. In order to increase the doping efficiency of Mg-doped AlGaN, indium (In) ambient assistant growth was used for Mg doping of AlGaN. It was found that In ambient assistant growth can decrease the acceptor activation energy significantly [84]. The LED device with p-type AlGaN grown under In ambient exhibit better electric properties than that without In ambient assistant growth. Surfactant In-assisted Mg delta doping in AlGaN is successful to obtain high p-type concentration of 4×10^{18} cm^{-3} [85]. A novel multidimensional Mg-doped SLs is proposed to enhance the vertical hole conductivity in conventional Mg-doped AlGaN SL which generally suffers from large potential barrier for holes [86]. The hole concentration reaches up to 3.5×10^{18} cm^{-3}, while the corresponding resistivity reduces to 0.7 Ω cm at room temperature. The hole in 3D SL are likely to be more delocalized rather than concentrated in the well compared with that in the conventional SL. High hole concentration can improve not only the lateral current spreading but also the hole injection efficiency into the active layer [87]. The IQE will be enhanced significantly with the carrier concentration increases in high Al content AlGaN.

For high IQE DUV-LEDs, the polarization effect should be concerned just like that in the GaN-based visible LEDs. Lin et al. show that through the employment of properly designed quaternary AlInGaN materials in the active region of UV LEDs, the polarization mismatch between hetero-layers can be efficiently mitigated [88]. Owing to the reduced built-in polarization field, not only the hole injection efficiency is improved, but also the electron leakage can be suppressed due to the effective potential height for electrons being increased. By tailoring the polarization in the EBL, the light output power of the AlGaN-based DUV-LED is improved significantly [89]. It indicates that the polarization effect contributes to both the suppression of the electron leakage and increase of the hole injection efficiency. The structures of the barrier and well are also designed to obtain better confinement of carriers. Chen et al. used staggered quantum wells to improve the DUV-LEDs performance [90]. It is due to the diminished piezoelectric polarization fields in QWs which can increase the density of electron and hole and the overlap of the electron and hole wave functions and thus enhance the radiative recombination rate. Wang et al. investigated AlGaN-based DUV-LEDs with composition-varying AlGaN multilayer barriers [91]. The better performance is attributed to the markedly reduced polarization-induced electrostatic field within the QWs. We improve the performance of DUV-LEDs by inserting single spike barriers [92]. The IQE and LOP were remarkably improved compared to those with conventional AlGaN MQWs structures. The spike barriers modulate the carrier distribution and increase the overlapping between electron and hole wave functions in MQWs.

8.3 Characterization for Internal Quantum Efficiency

Nowadays, most IQE values of LED or quantum wells are obtained by temperature-dependent PL (TDPL) [4, 93]. As shown in Fig. 8.10, the integrated PL intensity for the InGaN UL is almost constant at 17–150 K, and it falls gradually with further increase of temperature. Even at RT, however, it remains 71% of that at 17 K [93]. The non-radiative recombination centers near the MQW were greatly reduced by incorporating indium atoms into the underlying layer. Martinez et al. reported that IQE changed with the excitation power density [94]. The IQE increases to a maximum, and then droop appears with the injection level increases. They also performed TRPL as a complementary technique to study the effect of free-carrier screening on the radiative and nonradiative processes in the QWs. Schubert et al. reported a similar work and combined the ABC model to give the exact IQE. The IQE is as high as 64% at a carrier concentration of 1×10^{18} cm^{-3} for the sample with the lowest dislocation density (5.3×10^{8} cm^{-2}) [95]. The three main carrier recombination mechanisms in a bulk semiconductor are Shockley-Read-Hall non-radiative recombination, expressed as An, bimolecular radiative recombination Bn^2, and Auger non-radiative recombination Cn^3, where A, B, and C are the respective recombination coefficients, and n is the carrier concentration. Auger recombination affects LED efficiency only at very high excitation; thus, the generation rate G at

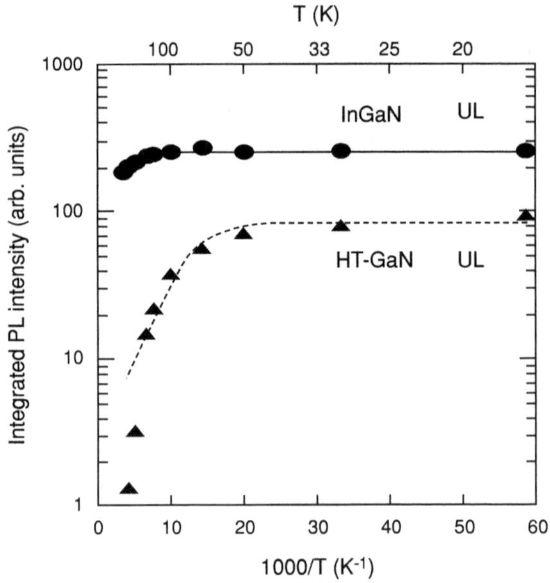

Fig. 8.10 Arrhenius plots of integrated PL intensity for InGaN MQWs prepared by using an InGaN UL (closed circles) and an HT-GaN UL (closed triangles). The solid line and dotted line are the fitting results obtained by thermal quenching equation for integrated PL intensity for InGaN MQWs prepared by using an InGaN UL and an HT-GaN UL, respectively [93]

steady state can be expressed as

$$G = \frac{A}{\sqrt{B\eta}}\sqrt{I_{PL}} + \frac{1}{\eta}I_{PL} \tag{8.1}$$

where η is a constant determined by the volume of the excited active region and the total collection efficiency of the luminescence. I_{PL} is the integrated PL intensity. According to the experimental PL integrated intensity, laser spot size, QW thickness, absorption coefficient, and so on, the IQE could be fitted exactly. However, the leakage of carriers from GaInN QWs may occur even when the energy of the optical excitation is much less than the bandgap energy of the GaN quantum barrier [96]. So the IQE results need be further modified by the effect of carrier leakage.

Considering the difference of electrical and optical excitation, we measured electroluminescence (EL) spectra under different injection level [97]. Then we also used ABC model to obtain the IQE. For the GaN-based LEDs, the recombination rate is given by

$$R = An + Bn^2 + Cn^3 + f(n), \tag{8.2}$$

where $f(n)$ denotes the carrier leakage. The light output power (P) is related to Bn^2. By defining

$$x = \sqrt{P} = n\sqrt{B} \tag{8.3}$$

Fig. 8.11 (a) Fitting process for EQE to x, (b) efficiencies as functions of injection current [97]

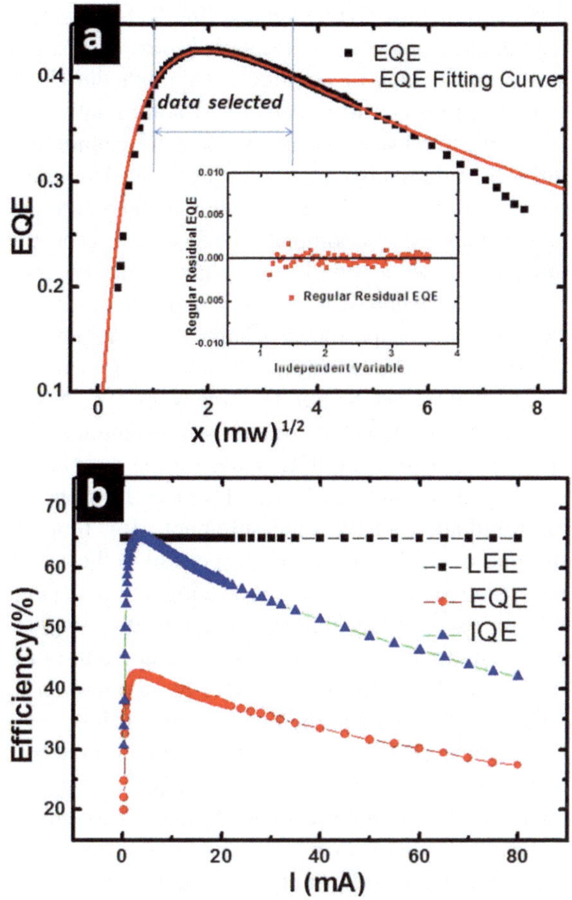

At low and moderate current injection levels, carrier leakage is negligible. When the IQE maximum approaches,

$$\mathrm{IQE}\,(n_0) = \frac{B{n_0}^2}{An_0 + B{n_0}^2 + C{n_0}^3} = \frac{a}{a+2} \tag{8.4}$$

where $a = {}^{B}\big/_{\sqrt{AC}}$, n_0 is carrier concentration at the IQE maximum.

Normally, the LEE is considered to be constant under varied current injections. With EQE derived from optical power versus injection current curve, IQE can be obtained by dividing EQE by the constant LEE. Figure 8.11 shows atypical fitting analyses process. By fitting the EL spectra under different injection level, IQE and LEE for the blue LED on PSS are 54.6% and 74.0%, respectively.

The assumption for IQE at cryogenic temperature as unity is not reasonable. Kawakami et al. found the evidences from simultaneous measurements of PL and photo-acoustic (PA) signals [98, 99]. PA signal caused by heat generation in nonradiative processes in the sample was detected as sound wave by a microphone and a signal lock-in amplifier. Although the PL and PA signals are considered to be proportional to radiative and non-radiative recombination rate, respectively, while the intensity of the two signals cannot be compared directly, and thus, the IQE values also cannot be derived. The energy supplied to the sample by optical excitation is emitted either in the form of "light" or "heat" by radiative or non-radiative recombination processes, respectively. Therefore, in case the PL intensity shows a superlinear behavior, the PA intensity should show a sublinear behavior. Conversely, in case PL intensity shows a sublinear behavior, PA intensity should show a superlinear behavior. This complementary relationship enables the comparison of the two signals' intensities to determine the IQE values [98].

They measured the IQE by three types of GaN samples by the PL/PA method and compared with these measured by conventional TDPL [98]. PL/PA measurements show quite reasonable IQE values for the samples.

They also estimate accurate lifetimes from the reliable IQE values estimated by the simultaneous PL/PA measurements. It is found that the radiative recombination lifetime in GaN increases in proportion to the 1.5th power of temperature and that nonradiative lifetime shows little temperature dependence although the nonradiative lifetime itself largely depends on sample quality [99]. It is concluded that decrease in emission efficiency with increasing temperature in GaN is not due to enhancement of nonradiative processes but due to suppression of radiative recombination rate, which is conformed to their early results [100].

Chichibu et al. thought that the IQE measured by the conventional TDPL method is potentially larger than the true IQE in the low-quality samples [101]. They presented the omnidirectional photoluminescence (ODPL) spectroscopy using an integrating sphere, in which both irradiating photons from the excitation source and emitted photons from the sample are omnidirectionally integrated and detected. The IQE in a material can be accurately measured. The ODPL spectra commonly exhibited a characteristic two-peak structure, as shown in Fig. 8.12. They assume the origin of the two-peak structure found in $I_{ODPL}(\omega)$ is the traveling of photons in nearly transparent energy regime below the absorption edge energy (E_{abs}). The other expression is that the traveling photons with the energy higher than E_{abs} of $I_{ODPL}(\omega)$ are completely reabsorbed before escaping from the sample [101]. They also give the expression of IQE:

$$\eta_{\text{int}} = \frac{\eta'}{\eta' + \eta_{\text{ext}}^{\text{direct}} - \eta' \eta_{\text{ext}}^{\text{direct}}} \tag{8.5}$$

where $\eta_{\text{ext}}^{\text{direct}}$ is the direct LEE from surface and η' is the EQE. At last they give a typical IQE as high as 15% for high-quality GaN single crystals grown by HVPE on a GaN seed crystal prepared by the ammonothermal method using acidic mineralizers [101]. They also use ODPL to obtain the absolute quantum efficiency (AQE) and IQE for HVPE growth GaN wafer [102]. The high value of $\eta_{\text{int}} = 70.9\%$

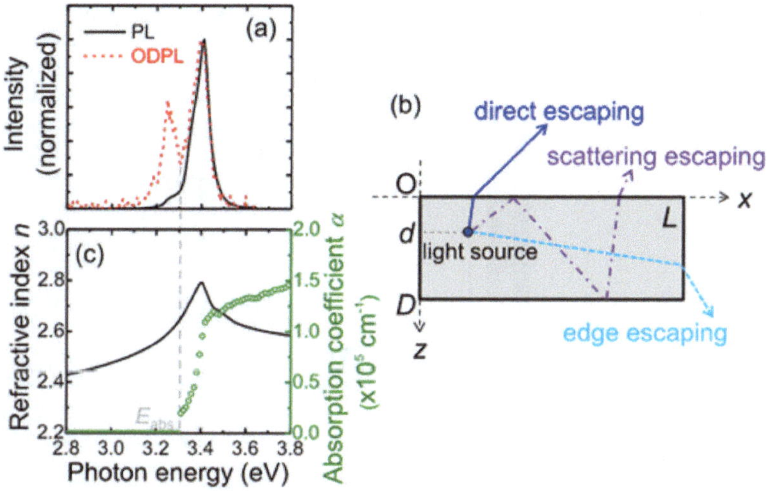

Fig. 8.12 (**a**) PL and ODPL spectra of the HVPE/AT-GaN crystal and (**b**) detectable light-traveling passes considered in the simulation of light extraction: (1) direct escaping from the surface; (2) scattered at the bottom and escaping from the surface; and (3) direct escaping from the edge. (**c**) Refractive index and absorption spectra of HVPE/AT-GaN [101]

may truly represent a considerable decrease in N_{NRC}, which corresponds to the decrease in [V_{Ga}X] of HVPE/AT-GaN in comparison with HVPE/GaN. They note that η_{int} determined in this study is insensitive to the surface quality, because both crystals were prepared using the same polishing process. Therefore, it can be a scale to evaluate the bulk quality.

8.4 Origins of Efficiency Droop

For general lighting, high luminous efficiency and high current density operated LEDs are absolutely the most valuable ones. However, the IQE typically peaks at a few A/cm^2 and then drops at increasing current density, down to values that can be as low as 50% of the peak value for current densities above hundreds of A/cm^2. Even if the measurement is performed under short pulse and low filling factor of current, the efficiency droop (hereafter droop, for brevity) exists too. Verzellesi et al. summarized the possible origins of the droop [103], as shown in Fig. 8.13. There are two types of mechanisms, namely, the internal losses and carrier leakage. Auger recombination is a three-particle process where an electron from the conduction band recombines with a hole in the valence band, and the excess energy resulting from this recombination is transferred to a third electron (eeh) or hole (hhe) by carrier-carrier interaction. The Auger coefficient C determines whether Auger recombination is the major cause for the droop. Previous measurements

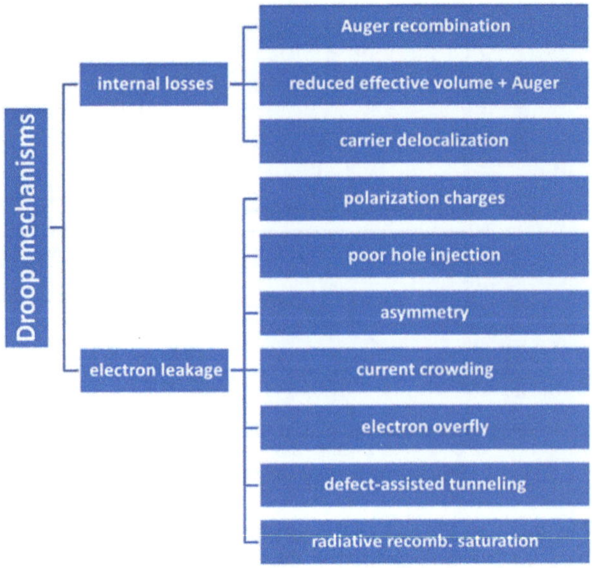

Fig. 8.13 Classification of droop mechanisms [103]

and estimations of the Auger coefficient in nitrides are ranged from 1×10^{-34} to 5.37×10^{-28} cm^6 s^{-1} [104]. One ab initio calculation suggested C values in the order of 10^{-30} cm^6 s^{-1} for a relatively narrow InGaN bandgap range around 2.5 eV, due to resonance between the lowest and the immediately upper conduction bands [105]. The coefficient is associated with indirect Auger processes in bulk InGaN, where the C values are about 3×10^{-31} cm^6 $s^{-1}$and 10^{-31} cm^6 $s^{-1}$for phonon-assisted and alloy-disorder-assisted Auger process, respectively [106]. Auger rates may be significantly enhanced in QWs with respect to the bulk case [107]. The direct measurement of the generation of Auger electrons is carried out by electron emission spectra after anode contact Cs [108]. For Auger being the sole cause for large droop effects that are observed in many reported devices, C should exceed 10^{-30} cm^6 s^{-1}.

Another major origin of the droop is the carrier leakage. Kim et al. reported [35] that droop is unrelated to junction temperature; it is not related to MQW efficiency but rather to the recombination of carriers outside the MQW region. Their simulations show that polarization fields in the MQW and electron-blocking layer enable the escape of electrons from the MQW region and thus are the physical origin of the droop. When polarization-matched AlGaInN quantum barriers and EBL are used, the droop at 350 mA decreases from 49.5% to only 5% [35]. The carrier leakage is well proved by Vampola et al. [109]. They grew a QW more with emission wavelength of 405 nm after the 450 nm blue MQWs and EBL. They found that the 405 nm emission did not appear under low injection level. It appeared at high injection level. This means the existence of carrier leakage from MQWs.

Besides the two major origins, there are many other factors causing droop. Carrier delocalization under high injection level is one of the important droop mechanisms [110]. It is observed a reduction in the localization induced S-shape temperature dependence of the peak photoluminescence energy with increasing excitation power density. This S-shape dependence is a key fingerprint of carrier localization. The onset of efficiency droop occurs at the same carrier density as the onset of the carrier delocalization. It is concluded that the saturation of the localized states is a contributory factor to efficiency droop whereby nonlocalized carriers recombine nonradiatively. Similar to the carrier delocalization under high injection, the extra, defect-related process has been modeled in Ref. [111] and termed density-activated defect recombination (DADR). Experimental droop data have been fitted in Ref. [111], by assuming a recombination rate proportional to the second power of the carrier density. With the carrier concentration increases, the defects become electrically and optically active. They turn out to be non-radiative recombination centers.

Except for the above polarization-induced carrier leakage, there are many other factors for carrier leakage, such as poor hole injection, current crowding, electron overfly, defect-assisted tunneling, radiative recombination saturation, and so on [103]. Both the hole concentration and mobility of Mg doping in p-GaN and AlGaN EBL layers are small [112]. The valence band offset and polarization charges at the EBL/spacer interface prevent the hole injection. Asymmetry in the electron and hole active doping concentration and in carrier transport parameters (mobility and related diffusivities) has been explicitly indicated as a droop cause [113]. Current crowding affects the droop through Auger effect or leakage in high current density area and defect-assisted tunneling in the low current density area. Electron overfly might be exacerbated by nonequilibrium electron transport effects induced by extra kinetic energy gained by electrons as they are injected from GaN into InGaN due to conduction-band discontinuity, since these "hot electrons" are assumed to have an enhanced overfly probability [114, 115]. Bandgap renormalization (BGR) and band-filling should occur when the carrier concentration is higher than mid 10^{18} cm^{-3} [116]. The BGR effect can build an additional potential barrier for carriers to inhibit their transferring from high to low energy state, thus leading to efficiency saturation [117]. Defect-assisted tunneling from the QW into defects in the p-side QB and "under" the QW through deep states in the QBs attributes to the carrier leakage and resulting droop. These defect-assisted tunneling effects have been correlated with the observed decrease in IQE in the low-energy part of the emission spectrum in a single QW LED [118]. The IQE and EL spectra of blue and green LEDs make the researchers conclude that a major role in the enhanced droop observed at low temperatures lies in the electron leakage promoted by the saturation of the radiative recombination rate [119].

There are still other explanations for the droop. Using fully microscopic many-body theory, Hader et al. investigate the droop causing mechanisms in GaN-based LEDs [6]. The most plausible droop model appears to be a combination of carrier delocalization at very low temperatures and pump powers, density-activated defect recombination at low to medium pumping, and injection/escape losses at strong pumping. Some groups found that the radiative recombination rate is saturated while the nonradiative one is increased with current density increases [119–121].

Lu et al. studied the droop by aging experiment on GaN-based LEDs [122]. They found that with increasing stressing time, the overall quantum efficiency of the aged LEDs drops while the peak-efficiency-current shifts toward higher magnitude. They thought that the sample variations were the root cause of the differences in opinions over the origin of efficiency droop. The absolute efficiency of LEDs should be stressed when comparing different device structures and material qualities. Temperature-dependent droops in LEDs are also studied [123]. In the low current density region, the temperature dependence of the EQE is mainly dominated by the Shockley-Read-Hall recombination and hence the dislocation density of the device. In the high current density region, carrier leakage from the quantum wells is the main factor influencing the decrease in the EQE with ambient temperature, and in this region, the LEDs become less temperature sensitive with increasing quantum well barrier height.

8.5 Some Remedies to Alleviate the Efficiency Droop

Nearly all of the reports on droop give their remedies to improve droop characteristics. However, there are no general methods to deal with all the droops effectively. Verzellesi et al. also propose a classification of the droop remedies based on their primary goal, whether it is (a) reducing the QW carrier densities, (b) improving the electron confinement within the active region, or (c) enhancing the hole injection into the active region [103]. Reducing the QW carrier densities is a direct method to resolve many problems that lead to droop, both for Auger recombination and carrier leakages. Lumileds Co. reported their InGaN/GaN double hetero-structure (DH) LED [124] (Fig. 8.14).

By increasing the thickness (~9–12 nm) of InGaN layer, the EQE reaches to its maximum at the current density of 200 A/cm^2. The spectral width of the DH LED is also sharper than that of the MQW LED. The shift of the peak wavelength is much smaller too. Their epi-structures are not clear but quite different from the DH LED reported in 1994 [30]. The earlier DH LED is GaN/InGaN/AlGaN structure. The thickness of InGaN is much high as about 50 nm. The FWHM of the spectrum is about 70 nm. DH structure reduces the droop significantly. However, with the thickness of InGaN increasing, the difficulties for InGaN growth and its large QCSE should be well dealt with.

QW number increase is also an easy approach to be considered. It can improve the lateral current spreading and lower the peak current densities for the same driving current [125]. With the QW number increases, electron leakage reduces and the droop mitigates. The actual carrier crowding may be due to the carrier density nonuniformity in the lateral and vertical directions. The effective area becomes small when the polarization field exists in QWs. So the polarization match by polarization engineering will increase the effective area of QWs, which causes the carrier density to be reduced. Chang et al. found that thick InGaN wells and polarization-matched AlGaInN barriers could achieve low-droop LEDs

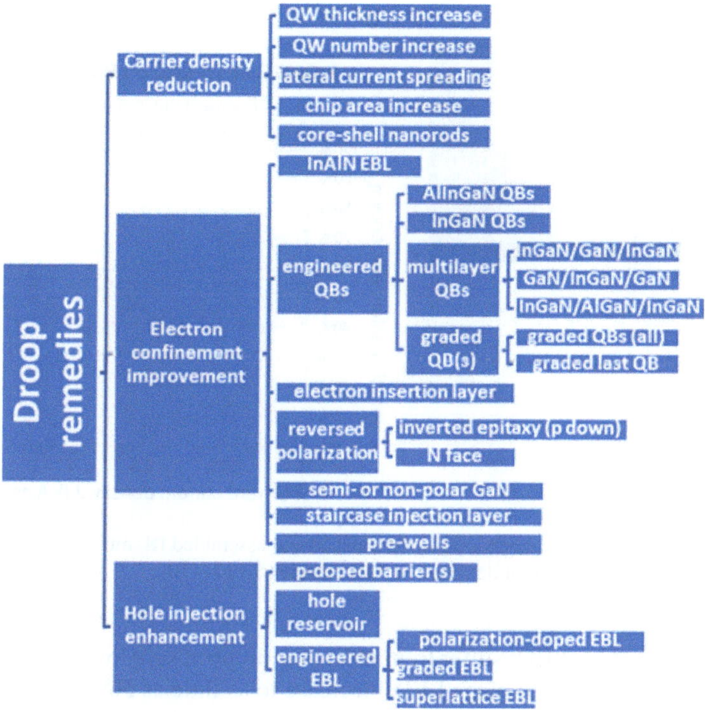

Fig. 8.14 Classification of droop remedies [103]

[126]. However, the number of QWs is reduced for the purpose of mitigating the additional compressive strain in AlGaInN barriers. Due to low activation efficiency of p-type, the current spreading and hole transportation are serious problems for carrier uniformity [112, 113, 127]. The improvement in lateral current uniformity is considered a key droop remedy in Ref. [128] on the basis of a simulation study providing optimized vertical LED structures with interdigital n-region contacts. Rozhansky et al. designed a p-type MQW and npn active structure to improve the asymmetry injection of hole and electron [129]. They found that the EQE did not reduce when the current density was above 60 A/cm^2. Because the radiative recombination occurred at the bottom QWs, their crystalline quality is poor, and the maximum EQE is only 1%. Then ap-doped In 0.05 Ga 0.95 N "hole reservoir" layer was inserted between the MQW and EBL regions [130]. They predicted a droop reduction from >35% to <12%.

The polarization match reduces not only the carrier density in QWs but also the QCSE in the active layer [126]. Another route reducing the QCSE is using nonpolar or semipolar structures [131, 132]. The reduction of the polarization field not only made the band diagram smoother but also restricted electron overflow to the p-GaN layer as shown in the simulations. With increasing the injection current density to 100 A/cm^2, the m-plane LEDs exhibit only 13% efficiency droop, whereas

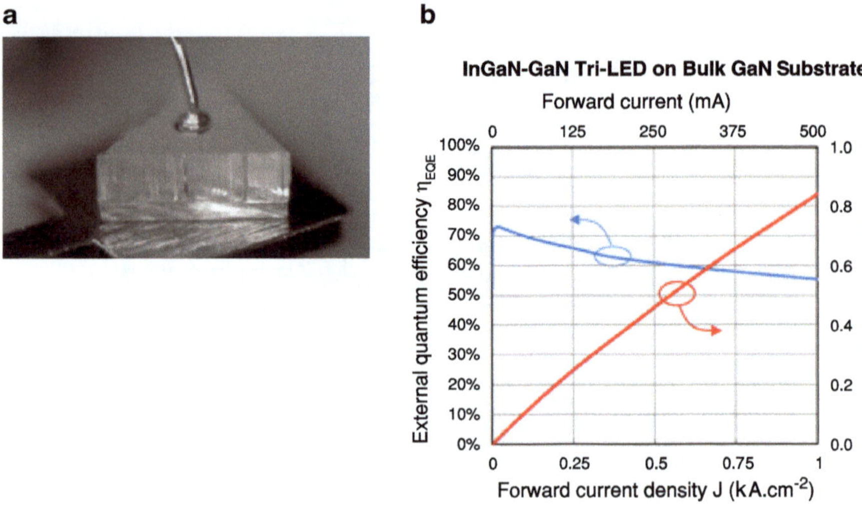

Fig. 8.15 (**a**) Optical microscope image of a finished and assembled III–nitride. (**b**) Pulsed, room temperature EQE versus current density J of a III–nitride tri-LED [10]

conventional c-plane LEDs suffer from efficiency droop at very low injection current density, and the EQE of c-plane LEDs decrease to as little as 50% of its maximum value. The nonpolar, semipolar InGaN-based LEDs can possess a high radiative recombination rate and low efficiency droop at a high injection current.

LED on freestanding GaN substrate shows high performance, especially for droop under high injection [10]. Figure 8.15 shows the InGaN/GaN tri-LED photograph and EQE-J curves. The EQE continues to be high even at current densities up to 1 kA cm^{-2}, at which point EQE = 55.3%, only a 24% drop from the peak value. The high performance in high injection level is due to (1) high LEE, (2) homoepitaxial technology for low current droop structures, and (3) uniform current spreading. We thought that the low TDD and no phonon barriers between the epilayer and substrate should also play important roles in the droop behavior.

Recently, much attention has been paid to specific small LEDs with chip size of tens microns (μ-LEDs), which can effectively reduce the droop phenomenon and endure high current density of even several kA/cm^2 [7, 133–138]. The distinct performances of tens microns μ-LEDs are attributed to uniform current spreading and low junction temperature (Tj) by some groups [136–138]. However, the maximum endured current density still increases as the size of the μ-LEDs reducing down below the current spreading length and operating under low-duty pulse current [7]. BGR and band-filling should occur when the carrier concentration is higher than mid 10^{18} cm^{-3} [116]. The BGR effect can build an additional potential barrier for carriers to inhibit their transferring from high to low energy state, thus leading to efficiency saturation [117]. Figure 8.16 shows the dependence of the light output power (LOP) on the current density (L-I) for μ-LEDs with different diameters. The

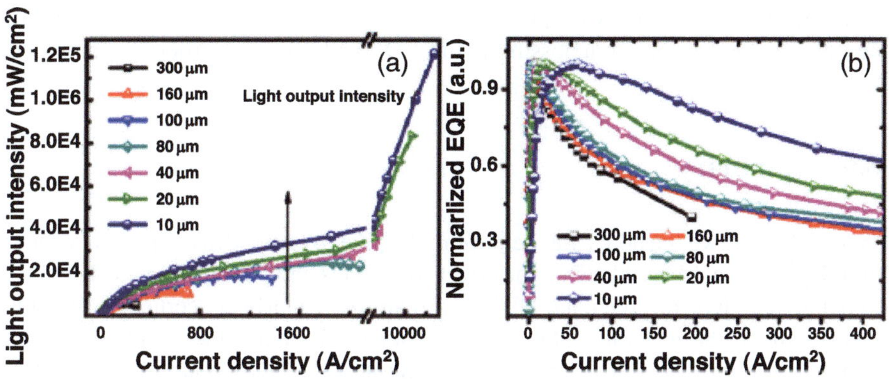

Fig. 8.16 (**a**) The dependence of the light output power in EL spectra on the current density for 10, 20, 40, 80, 160, and 300 μm LEDs. (**b**) The external quantum efficiency (EQE) curves for 10, 20, 40, 80, 160, and 300 μm LEDs with increased current density [7]

10 μm LEDs survive at extremely high current density of 16 kA/cm^2. The droop values of 10 μm and 300 μm LED are 18% and 61%, respectively.

When the size of the LED reduces to nanometer, the droop tends to be zero at the injection level of 640 A/cm^2 [139]. These LEDs operated in the interconnected array. Another work reported that the droop commonly observed in InGaN QW LEDs could also be measured in nanorod devices [140]. They found critical importance of surface recombination on the carrier dynamics and performance of nanorod LEDs. Actually, surface states are very important because of the large ratio of surface area to the volume of nano LEDs. That will be invalid before the current satisfy for emission requirement [141]. It is well known that 3D nano LEDs have the advantages of large emission surface, m-plane growth, and so on [48, 49], which is of potential importance in operation under high injection level. Moreover, we also find the nanocavity effect in these devices [142], which will further improve the droop performance.

At last, SP coupling are also discussed for reduction of the droop. Yang et al. reported that with a current spreading grid pattern on the mesa surface, a smaller grid period leads to more effective carrier transport into the QW regions of Ag deposition for stronger SP–QW coupling such that the droop effect is more significantly reduced, as indicated by the increase in the injection current density of maximum EQE and the decrease in drooping slope [143]. They also get efficiency improvement and low droop of a vertical light-emitting diode through surface plasmon coupling and grating scattering [144]. Based on the reflection measurement from the metal grating structure and numerical simulation result, it is found that the LSP resonance induced around the metal grating crest plays a major role in the SP-QW coupling process. By adding a surface grating structure to the SP-coupled vertical LED on the n-GaN side, the droop effect is further suppressed. By adding a lower-index dielectric interlayer, further reduction of efficiency droop effect is obtained in SP-coupled LEDs [145]. The SP coupling to QWs should reduce the non-radiative recombination possibility in LEDs, and hence the droop is suppressed.

8.6 Summary

In this chapter, we introduce the definition of IQE and some ideas of how to achieve high IQE for GaN-based LEDs. The effects of defects and polarization are discussed in detail. The IQE of green and DUV-LEDs are particularly concerned. In the IQE measurement, the general TDPL method is analyzed, and some significant developments are introduced to obtain a more accurate value. Droop is a very important issue for high-efficiency, low-cost solid-state lighting. The exact origins are still on debates. We tend to attribute the droop to various origins for different situations. At last, we give some remedies to alleviate the droop. Radiative recombination rate enhancement may be more important for low-droop LEDs. LSP and nanocavity may be the potential candidates for novel low-droop devices.

References

1. Martin Strassburg (OSRAM), Perspectives of III-N optoelectronics, in *International Workshop on Nitride Semiconductors (IWN2016), Orlando, USA*, 2016
2. K.P. O'Donnell, M. Auf der Maur, A. Di Carlo, K. Lorenz, Phys. Status Solidi RRL **6**(2), 49–52 (2012)
3. M. Kneissl, T. Kolbe, C. Chua, V. Kueller, N. Lobo, J. Stellmach, A. Knauer, H. Rodriguez, S. Einfeldt, Z. Yang, N.M. Johnson, M. Weyers, Semicond. Sci. Technol. **26**(1), 014036 (2011)
4. Y. Yamada, K. Iwamura, T. Kuronaka, T. N Shinomura, H. Taguchi, H. Kudo, J. Okagawa, Light Vis. Environ. **32**(2), 191–195 (2008)
5. N.F. Gardner, G.O. Müller, Y.C. Shen, G. Chen, S. Watanabe, W. Götz, M.R. Krames, Appl. Phys. Lett. **91**, 243506 (2007)
6. J. Hader, J.V. Moloney, S.W. Koch, Proc. SPIE **8625**, 86251M (2013)
7. Q.Q. Jiao, Z.Z. Chen, J. Ma, S.Y. Wang, Y. Li, S. Jiang, Y.L. Feng, J.Z. Li, Y.F. Chen, T.J. Yu, S.F. Wang, G.Y. Zhang, P.F. Tian, E.Y. Xie, Z. Gong, E.D. Gu, M.D. Dawson, Opt. Exp. **23**(13), 16565 (2015)
8. S.D. Lester, F.A. Ponce, M.G. Craford, D.A. Steigerwald, Appl. Phys. Lett. **66**, 1249 (1995)
9. J.Y. Tsao, IEEE Circuits Devices Mag. **20**(3), 28–37 (2005)
10. M.J. Cich, R.I. Aldaz, A. Chakraborty, A. David, M.J. Grundmann, A. Tyagi, M. Zhang, F.M. Steranka, M.R. Krames, Appl. Phys. Lett. **101**, 223509 (2012)
11. Y. Sun, S. Qi, Z. Chen, X. Kang, G. Zhu, C. Chen, S. Li, Y. Pan, J. Yan, J. Deng, H. Long, M. Hao, T. Yu, G. Zhang, Phys. Stat. Sol. C **6**(S2), S623 (2009)
12. J.S. Speck, M.A. Brewer, G. Beltz, A.E. Romanov, W. Pompe, J. Appl. Phys. **80**, 3808 (1996)
13. D.C. Reynolds, D.C. Look, B. Jogai, J.E. Hoelscher, J. Appl. Phys. **88**, 1460 (2000)
14. L. Gorgens, O. Ambacher, M. Stutzmann, C. Miskys, Appl. Phys. Lett. **76**, 577 (2000)
15. J.-Z. Li, T. Yue-Bin, C. Zhi-Zhong, J. Xian-Zhe, F. Xing-Xing, J. Shuang, J. Qian-Qian, Y. Tong-Jun, Z. Guo-Yi, Chin. Phys. B **23**(1), 016101 (2014)
16. H. Amano, N. Sawaki, I. Akasaki, et al., Appl. Phys. Lett. **48**, 353 (1986)
17. S. Nakamura, Y. Harada, M. Sehno, Appl. Phys. Lett. **58**, 2021 (1991)

18. O.H. Nam, M.D. Bremser, T. Zheleva, R.F. Davis, Appl. Phys. Lett. **71**, 2638 (1997)
19. C.I.H. Ashby, C.C. Mitchell, J. Han, N.A. Missert, P.P. Provencio, D.M. Follstaedt, G.M. Peake, L. Griego, Appl. Phys. Lett. **77**, 3233 (2000)
20. K. Tadatomo, N. Okada, Proc. SPIE **7954**, 795416 (2005)
21. M.T. Wang, K.Y. Liao, Y. Li, IEEE Photon. Technol. Lett. **23**, 962–964 (2011)
22. J.H. Cheng, Y.S. Wu, W.C. Liao, B.W. Lin, Appl. Phys. Lett. **96**, 051109 (2010)
23. X.H. Huang, J.P. Liu, Y.Y. Fan, J.J. Kong, H. Yang, H.B. Wang, IEEE Photon. Technol. Lett. **23**, 944–946 (2011)
24. J.J. Chen, Y.K. Su, C.L. Lin, S.M. Chen, W.L. Li, C.C. Kao, IEEE Photon. Technol. Lett. **20**, 1193–1195 (2008)
25. Y.F. Li, S. You, M.W. Zhu, L. Zhao, W.T. Hou, T. Detchprohm, Y. Taniguchi, N. Tamura, S. Tanaka, C. Wetzel, Appl. Phys. Lett. **98**, 151102 (2011)
26. Y.-K. Ee, X.-H. Li, et al., J. Cryst. Growth **312**, 1311 (2010)
27. J.Z. Li, Z.Z. Chen, Q.Q. Jiao, Y.L. Feng, S. Jiang, Y.F. Chen, T.J. Yu, S.F. Li, G.Y. Zhang, CrystEngComm **17**, 4469 (2015)
28. J.I. Pankove, J. Lumin. **7**, 114 (1973)
29. H. Amano, M. Kito, K. Hiramatsu, et al., Jpn. J. Appl. Phys. **28**(12), L2112–L2114 (1989)
30. S. Nakamura, T. Mukai, M. Sehno, Appl. Phys. Lett. **64**(13), 1687–1689 (1994)
31. J.-W. Ju, E.-S. Kang, H.-S. Kim, et al., J. Appl. Phys. **102**, 053519 (2007)
32. M.G. Cheong, R.J. Choi, E.-K. Suh, H.J. Lee, Appl. Phys. Lett. **82**, 625–626 (2003)
33. Y.B. Tao, Z.Z. Chen, F.F. Zhang, J. Appl. Phys. **107**, 103529 (2010)
34. G. Franssen, T. Suski, P. Perlin, Appl. Phys. Lett. **87**, 041109 (2005)
35. M.-H. Kim, M.F. Schubert, Q. Dai, J.K. Kim, E. Fred Schuberta, J. Piprek, Y. Park, Appl. Phys. Lett. **91**, 183507 (2007)
36. J. Zhang, J. Yang, G. Simin, M. Shatalov, M. Asif Khan, Appl. Phys. Lett. **77**, 2668–2670 (2000)
37. B.A. Haskell, F. Wu, M.D. Craven, S. Matsuda, P.T. Fini, T. Fujii, K. Fujito, S.P. Den Baars, J.S. Speck, S. Nakamura, Appl. Phys. Lett. **84**, 644 (2003)
38. A. Chakraborty, B.A. Haskell, S. KELLER, J.S. Speck, S.P. Denbaars, S. Nakamura, U.K. Mishra, Jpn. J. Appl. Phys. **44**, L173–L175 (2005)
39. T.J. Baker, B.A. Haskell, F. Wu, P.T. Fini, J.S. Speck, S. Nakamura, Jpn. J. Appl. Phys. **44**, L920–L922 (2005)
40. K. Okamoto, A. Kaneta, Y. Kawakami, S. Fujita, J. Appl. Phys. **98**, 064503 (2005)
41. Y.D. Lin, A. Chakraborty, S. Brinkley, H.C. Kuo, T. Melo, K. Fujito, J.S. Speck, S.P. DenBaars, S. Nakamura, Appl. Phys. Lett. **94**, 261108 (2009)
42. H. Sato, A. Tyagi, H. Zhong, N. Fellows, R.B. Chung, M. Saito, K. Fujito, J.S. Speck, S.P. DenBaars, S. Nakamura, Phys. Stat. Sol. (RRL) **1**(4), 162–164 (2007)
43. S.E. Brinkley, Y.-D. Lin, A. Chakraborty, N. Pfaff, D. Cohen, J.S. Speck, S. Nakamura, S.P. DenBaars, Appl. Phys. Lett. **98**, 011110 (2011)
44. H. Sato, R.B. Chung, H. Hirasawa, N. Fellows, H. Masui, F. Wu, M. Saito, K. Fujito, J.S. Speck, S.P. DenBaars, S. Nakamura, Appl. Phys. Lett. **92**, 221110 (2008)
45. F. Brunner, U. Zeimer, F. Edokam, W. John, D. Prasai, O. Krüger, M. Weyers, Phys. Status Solidi B **252**, 1189 (2015)
46. J. Bai, B. Xu, F.G. Guzman, K. Xing, Y. Gong, Y. Hou, T. Wang, Appl. Phys. Lett. **107**, 261103 (2016)
47. H. Li, M. Khoury, B. Bonef, A.I. Alhassan, A.J. Mughal, E. Azimah, M.E.A. Samsudin, P. De Mierry, S. Nakamura, J.S. Speck, S.P. DenBaars, ACS Appl. Mater. Interfaces **9**(41), 36417–36422 (2017)
48. S. Li, A. Waag, J. Appl. Phys. **111**, 071101 (2012)
49. H. Sekiguchi, K. Kishino, A. Kikuchi, Appl. Phys. Lett. **96**, 231104 (2010)
50. S.D. Hersee, X.Y. Sun, et al., J. Appl. Phys. **97**, 124308 (2005)
51. M.L. Kuo, Y.S. Kim, M.L. Hsieh, et al., Nano Lett. **11**(2), 476–481 (2011)
52. S. Neugebauer, S. Metzner, J. Blasing, F. Bertram, A. Dadgar, J. Christen, A. Strittmatter, Phys. Status Solidi B **253**(1), 118 (2016)

53. F. Akyol, D.N. Nath, S. Krishnamoorthy, P.S. Park, S. Rajan, Appl. Phys. Lett. **100**, 111118 (2012)
54. B.-C. Lin, K.-J. Chen, C.-H. Wang, C.-H. Chiu, Y.-P. Lan, C.-C. Lin, P.-T. Lee, M.-H. Shih, Y.-K. Kuo, H.-C. Kuo, Opt. Express **22**(1), 463 (2014)
55. C.H. Chen, S.J. Chang, Y.K. Su, I.E.E.E. Senior Member, G.C. Chi, J.K. Sheu, J.F. Chen, IEEE J. Sel. Topics Quantum Electron. **8**(2), 284 (2002)
56. I.H. Ho, G.B. Stringfellow, Appl. Phys. Lett. **69**(18), 2701 (1996)
57. Y. Arakawa, H. Sakaki, Appl. Phys. Lett. **40**(11), 939–941 (1982)
58. K. Tachibana, T. Someya, Y. Arakawa, R. Werner, A. Forchel, Appl. Phys. Lett. **75**(17), 2605–2607 (1999)
59. K. Tachibana, T. Someya, S. Ishida, Y. Arakawa, Appl. Phys. Lett. **76**(22), 3212–3214 (2000)
60. Y.L. Chang, J.L. Wang, F. Li, Z. Mi, Appl. Phys. Lett. **96**, 013106 (2010)
61. L. Jun-Lin, Z. Jian-Li, W. Guang-Xu, M. Chun-Lan, X. Long-Quan, D. Jie, Q. Zhi-Jue, W. Xiao-Lan, P. Shuan, Z. Chang-Da, W. Xiao-Ming, F. Wen-Qing, J. Feng-Yi, Chin. Phys. B **24**(6), 067804 (2015)
62. X.M. Wu, J.L. Liu, Z.J. Quan, C.B. Xiong, C.D. Zheng, J.L. Zhang, F.Y. Jiang, Appl. Phys. Lett. **104**, 221101 (2014)
63. Z.J. Quan, L. Wang, C.D. Zheng, J.L. Liu, F.Y. Jiang, J. Appl. Phys. **116**, 183107 (2014)
64. J.L. Zhang, C.B. Xiong, J.L. Liu, Z.J. Quan, L. Wang, F.Y. Jiang, Appl. Phys. A Mater. Sci. Process. **114**, 1049 (2014)
65. J.R. Sambles, G.W. Bradbery, F.Z. Yang, Contemp. Phys. **32**(3), 173–183 (1991)
66. K. Okamoto, I. Niki, A. Shvartser, Y. Narukawa, T. Mukai, A. Scherer, Nat. Mater. **3**, 601 (2004)
67. D.M. Yeh, C.F. Huang, C.Y. Chen, Y.C. Lu, C.C. Yang, Nanotechnology **19**, 345201 (2008)
68. S. Jiang, Z. Hu, Z.Z. Chen, X.X. Fu, X.Z. Jiang, Q.Q. Jiao, T.J. Yu, G.Y. Zhang, Opt. Express **21**(10), 12100–12110 (2013)
69. J. Henson, E. Dimakis, J. DiMaria, R. Li, S. Minissale, L.D. Negro, T.D. Moustakas, R. Paiella, Opt. Express **18**(20), 21322–21329 (2010)
70. C.H. Lu, C.C. Lan, Y.L. Lai, Y.L. Li, C.P. Liu, Adv. Funct. Mater. **21**(24), 4719–4723 (2011)
71. S. Jiang, Z. Chen, F. Xingxing, Q. Jiao, Y. Feng, W. Yang, J. Ma, J. Li, S. Jiang, Y. Tongjun, G. Zhang, IEEE Photon. Technol. Lett. **27**(13), 1363 (2015)
72. M. Kneissl, *A Brief Review of III-Nitride UV Emitter Technologies and Their Applications* (Springer, Cham, 2016), pp. 1–25. https://doi.org/10.1007/978-3-319-24100-5_1
73. M. Kneissl, T. Kolbe, C. Chua, et al., Advances in group III-nitride-based deep UV light-emitting diode technology. Semicond. Sci. Technol. **26**, 014036 (2011)
74. K. Ban, J.-i. Yamamoto, K. Takeda, K. Ide, M. Iwaya, T. Takeuchi, S. Kamiyama, I. Akasaki, H. Amano, Appl. Phys. Express **4**, 052101 (2011)
75. M.A. Khan, J.N. Kuzina, D.T. Olson, T. George, W.T. Pike, GaN/AlN digital alloy short-period superlattices by switched atomic layer metalorganic chemical vapor deposition. Appl. Phys. Lett. **63**, 3470–3472 (1993)
76. L.W. Sang, Z.X. Qin, G.Y. Zhang, et al., Appl. Phys. Lett. **93**, 122104 (2008)
77. P. Dong et al., Appl. Phys. Lett. **102**, 241113 (2013)
78. M. Conroy et al., J. Mater. Chem. C **3**, 431–437 (2015)
79. L. Zhang, F. Xu, J. Wang, C. He, W. Guo, M. Wang, B. Sheng, L. Lu, Z. Qin, X. Wang, B. Shen, Sci. Rep. **6**, 35934 (2016)
80. S. Kamiyama et al., J. Cryst. Growth **223**, 83–91 (2001)
81. J. Han et al., Appl. Phys. Lett. **78**, 67–69 (2001)
82. J.P. Zhang et al., Appl. Phys. Lett. **80**, 3542–3544 (2002)
83. M.A. Khan, M. Shatalov, H.P. Maruska, H.M. Wang, E. Kuokstis, Jpn. J. Appl. Phys. **44**, 7191–7206 (2005)
84. G. Zhang, B. Shen, Z. Chen, et al., Chin. Sci. Bull. **59**(12), 1201–1218 (2014)
85. Y. Yang, J. Liu, L. Liu, A. Anders, Appl. Phys. Lett. **106**, 162102 (2015)
86. T.C. Zheng, W. Lin, R. Liu, Sci. Rep. **6**, 21897 (2016)
87. A. Khan, K. Balakrishnan, T. Katona, Nat. Photonics **2**, 77–84 (2008)

88. Y. Lin, B. Liou, J. Chang, Y. Kuo, Proc. SPIE **8619**, 86191V (2013)
89. G. Yang, J. Chang, J. Wang, Q. Zhang, F. Xie, J. Xue, D. Yan, F. Wang, P. Chen, R. Zhang, Y. Zheng, Superlattice. Microst. **83**, 1–8 (2015)
90. Z. Min, L. Yang, C. Shengchang, T. Wu, X. Jintong, L. Xiangyang, W. Zhihao, F. Yanyan, D. Jiangnan, C. Changqing, Superlattice. Microst. **75**, 63–71 (2014)
91. A.Y. Yi, W. Naiyin, F. Guanghan, Z. Yong, Superlattice. Microst. **76**, 149–155 (2014)
92. W. Guo, F. Xu, Y. Sun, L. Lu, Z. Qin, T. Yu, X. Wang, B. Shen, Superlattice. Microst. **100**, 941–946 (2016)
93. T. Akasaka, H. Gotoh, T. Saito, T. Makimoto, Appl. Phys. Lett. **85**, 3089 (2004)
94. C.E. Martinez, N.M. Stanton, A.J. Kent, J. Appl. Phys. **98**, 053509 (2005)
95. Q. Dai, M.F. Schubert, M.H. Kim, J.K. Kim, E.F. Schubert, D.D. Koleske, M.H. Crawford, S.R. Lee, A.J. Fischer, G. Thaler, M.A. Banas, Appl. Phys. Lett. **94**, 111109 (2009)
96. M.F. Schubert, J. Xu, Q. Dai, F.W. Mont, J.K. Kim, E.F. Schubert, Appl. Phys. Lett. **94**, 081114 (2009)
97. C. Xu, T. Yu, J. Yan, Z. Yang, X. Li, Y. Tao, X. Fu, Z. Chen, G. Zhang, Phys. Status Solidi C **9**(3–4), 757–760 (2012)
98. T. Nakano, K. Kawakami, A.A. Yamaguchi, Proc. SPIE **9748**, 97481W (2016)
99. K. Kawakami, T. Nakano, A.A. Yamaguchi, Proc. SPIE **9748**, 97480S (2016)
100. A.A. Yamaguchi, Y. Mochizuki, M. Mizuta, Jpn. J. Appl. Phys. **39**, 2402 (2000)
101. K. Kojima, T. Ohtomo, K.-i. Ikemura, Y. Yamazaki, M. Saito, H. Ikeda, K. Fujito, S.F. Chichibu, J. Appl. Phys. **120**, 015704 (2016)
102. K. Kojima, H. Ikeda, K. Fujito, S.F. Chichibu, Appl. Phys. Lett. **111**, 032111 (2017)
103. G. Verzellesi, D. Saguatti, M. Meneghini, F. Bertazzi, M. Goano, G. Meneghesso, E. Zanoni, J. Appl. Phys. **114**, 071101 (2013)
104. Y.C. Shen, G.O. Mueller, S. Watanabe, N.F. Gardner, A. Munkholm, M.R. Krames, Appl. Phys. Lett. **91**, 141101 (2007)
105. K.T. Delaney, P. Rinke, C.G. Van de Walle, Appl. Phys. Lett. **94**(19), 191109 (2009)
106. E. Kioupakis, P. Rinke, K.T. Delaney, C.G. Van de Walle, Appl. Phys. Lett. **98**, 161107 (2011)
107. A.S. Polkovnikov, G.G. Zegrya, Phys. Rev. B **58**(7), 4039 (1998)
108. J. Iveland, L. Martinelli, J. Peretti, J.S. Speck, C. Weisbuch, Phys. Rev. Lett. **110**, 177406 (2013)
109. K.J. Vampola, M. Iza, S. Keller, S.P. DenBaars, S. Nakamura, Appl. Phys. Lett. **94**(6), 061116 (2009)
110. S. Hammersley, D. Watson-Parris, P. Dawson, M.J. Godfrey, T.J. Badcock, M.J. Kappers, C. McAleese, R.A. Oliver, C.J. Humphreys, J. Appl. Phys. **111**, 083512 (2012)
111. J. Hader, J.V. Moloney, S.W. Koch, Appl. Phys. Lett. **96**(22), 221106 (2010)
112. J. Xie, X. Ni, Q. Fan, R. Shimada, U. Ozgur, H. Morkoc, Appl. Phys. Lett. **93**(12), 121107 (2008)
113. D.sS. Meyaard, G.-B. Lin, Q. Shan, J. Cho, E.F. Schubert, H. Shim, M.-H. Kim, C. Sone, Appl. Phys. Lett. **99**(25), 251115 (2011)
114. X. Ni, X. Li, J. Lee, S. Liu, V. Avrutin, U. Ozgur, H. Morkoc, A. Matulionis, T. Paskova, G. Mulholland, and K. R. Evans, Appl. Phys. Lett. 97(3), 031110 (2010)
115. X. Ni, X. Li, J. Lee, S. Liu, V. Avrutin, U. Ozgur, H. Morkoc, and A. Matulionis, J. Appl. Phys. 108, 033112 (2010)
116. M. Feneberg, S. Osterburg, K. Lange, C. Lidig, B. Garke, R. Goldhahn, E. Richter, C. Netzel, M.D. Neumann, N. Esser, S. Fritze, H. Witte, J. Bläsing, A. Dadgar, A. Krost, Phys. Rev. B **90**(7), 075203 (2014)
117. K. Jarašiūnas, S. Nargelas, R. Aleksiejunas, S. Miasojedovas, M. Vengris, S. Okur, H. Morkoç, Ü. Özgür, C. Giesen, Ö. Tuna, M. Heuken, J. Appl. Phys. **113**(10), 103701 (2013)
118. N.I. Bochkareva, V.V. Voronenkov, R.I. Gorbunov, A.S. Zubrilov, Y.S. Lelikov, P.E. Latyshev, Y.T. Rebane, A.I. Tsyuk, Y.G. Shreter, Appl. Phys. Lett. **96**(13), 133502 (2010)
119. D.-S. Shin, D.-P. Han, J.-Y. Oh, J.-I. Shim, Appl. Phys. Lett. **100**(15), 153506 (2012)
120. A. David, M.J. Grundmann, Appl. Phys. Lett. **96**, 103504 (2010)
121. J. Yun, H.-S. Yeom, J.-I. Shim, J. Korean Phys. Soc. **63**(6), 1218–1221 (2013)

122. X. Shao, H. Lu, D. Chen, Z. Xie, R. Zhang, Y. Zheng, Appl. Phys. Lett. **95**, 163504 (2009)
123. N.L. Ploch, S. Einfeldt, M. Frentrup, J. Rass, T. Wernicke, A. Knauer, V. Kueller, M. Weyers, M. Kneissl, Semicond. Sci. Technol. **28**, 125021 (2013)
124. N.F. Gardner, G.O. Müller, Y.C. Shen, et al., Appl. Phys. Lett. **91**, 243506 (2007)
125. C.S. Xia, Z.M. Simon Li, Z.Q. Li, Y. Sheng, Z.H. Zhang, W. Lu, L.W. Cheng, Appl. Phys. Lett. **100**, 263504 (2012)
126. J.-Y. Chang, F.-M. Chen, Y.-K. Kuo, Y.-H. Shih, J.-K. Sheu, W.-C. Lai, H. Liu, Opt. Lett. **38**(16), 3158 (2013)
127. X. Guo, E.F. Schubert, J. Appl. Phys. **90**(8), 4191 (2001)
128. C.-K. Li, Y.-R. Wu, IEEE Trans. Electron Devices **59**(2), 400 (2012)
129. I.V. Rozhansky, D.A. Zakheim, Phys. Status Solidi A **204**(1), 227–230 (2007)
130. T. Lu, S. Li, C. Liu, K. Zhang, Y. Xu, J. Tong, L. Wu, H. Wang, X. Yangs, Y. Yin, G. Xiao, Y. Zhou, Appl. Phys. Lett. **100**(14), 141106 (2012)
131. S.-C. Ling, T.-C. Lu, S.-P. Chang, J.-R. Chen, H.-C. Kuo, S.-C. Wang, Appl. Phys. Lett. **96**, 231101 (2010)
132. C.-H. Chiu, D.-W. Lin, C.-C. Lin, Z.-Y. Li, W.-T. Chang, H.-W. Hsu, H.-C. Kuo, T.-C. Lu, S.-C. Wang, W.-T. Liao, T. Tanikawa, Y. Honda, M. Yamaguchi, N. Sawaki, Appl. Phys. Express **4**, 012105 (2011)
133. A. Castiglia, D. Simeonov, H.J. Buehlmann, J.-F. Carlin, E. Feltin, J. Dorsaz, R. Butté, N. Grandjean, Appl. Phys. Lett. **90**(3), 033514 (2007)
134. Z. Gong, S. Jin, Y. Chen, J. McKendry, D. Massoubre, I.M. Watson, E. Gu, M.D. Dawson, J. Appl. Phys. **107**(1), 013103 (2010)
135. Y.B. Tao, S.Y. Wang, Z.Z. Chen, Z. Gong, E.Y. Xie, Y.J. Chen, Y.F. Zhang, J. McKendry, D. Massoubre, E.D. Gu, B.R. Rae, R.K. Henderson, G.Y. Zhang, Phys. Status Solidi C **9**(3–4), 616–619 (2012)
136. P. Tian, J.J.D. McKendry, Z. Gong, B. Guilhabert, I.M. Watson, E. Gu, Z. Chen, G. Zhang, M.D. Dawson, Appl. Phys. Lett. **101**(23), 231110 (2012)
137. T.I. Kim, Y.H. Jung, J. Song, D. Kim, Y. Li, H.S. Kim, I.-S. Song, J.J. Wierer, H.A. Pao, Y. Huang, J.A. Rogers, Small **8**(11), 1643–1649 (2012)
138. N. Lobo Ploch, H. Rodriguez, C. Stölmacker, M. Hoppe, M. Lapeyrade, J. Stellmach, F. Mehnke, T. Wernicke, A. Knauer, V. Kueller, M. Weyers, S. Einfeldt, M. Kneissl, IEEE Trans. Electron Devices **60**(2), 782–786 (2013)
139. H.P.T. Nguyen, S. Zhang, K. Cui, X. Han, S. Fathololoumi, M. Couillard, G.A. Botton, Z. Mi, Nano Lett. **11**, 1919 (2011)
140. S. Zhang, Y. Li, S. Fathololoumi, H.P.T. Nguyen, Q. Wang, Z. Mi, Q. Li, G.T. Wang, AIP Adv. **3**, 082103 (2013)
141. Q. Li, K.R. Westlake, M.H. Crawford, S.R. Lee, D.D. Koleske, J.J. Figiel, K.C. Cross, Opt. Express **19**(25), 25528 (2011)
142. Q. Jiao, Z. Chen, Y. Feng, S. Li, S. Jiang, J. Li, Y. Chen, T. Yu, X. Kang, B. Shen, G. Zhang, Nanoscale Res. Lett. **11**, 340 (2016)
143. C.-F. Lu, C.-H. Liao, C.-Y. Chen, C. Hsieh, Y.-W. Kiang, C.C. Yang, Appl. Phys. Lett. **96**, 261104 (2010)
144. C.-H. Lin, C. Hsieh, C.C. Yang, et al., Opt. Express **23**(S2), A842 (2014)
145. C.-H. Lin, C.-Y. Su, Y. Kuo, C.-H. Chen, Y.-F. Yao, P.-Y. Shih, H.-S. Chen, C. Hsieh, Y.-W. Kiang, C.C. Yang, Appl. Phys. Lett. **105**, 101106 (2014)

Chapter 9
On the Light Extraction Efficiency for III-Nitride-Based Light-Emitting Diodes

Zi-Hui Zhang, Yonghui Zhang, Sung-Wen Huang Chen, Wengang Bi, and Hao-Chung Kuo

9.1 Introduction

III-nitride semiconductor has the direct bandgap, the family of which comprises GaN, InN, and AlN, and the ternary/quaternary materials can be made by alloying any of the binary ones. The material alloy for III-nitride semiconductor can tune the energy bandgap from 0.70 to 6.2 eV, which makes III-nitride semiconductor material very promising for optoelectronic devices [1]. One of the important applications for III-nitride semiconductor lies on light-emitting diodes (LEDs), the emission wavelength for which can cover the spectral range between the deep ultraviolet and the infrared region. A very important index for evaluating the excellence of III-nitride-based LEDs is the external quantum efficiency (EQE), which is to measure the ratio of the number of photos emitted into the free space per second and the number of the electrons injected into the device per second. The EQE is the product of the internal quantum efficiency (IQE) and the light extraction efficiency (LEE). The IQE is defined as the ratio of the number of photos emitted from the active region per second and the number of the electrons injected into the device. The LEE is defined as the ratio of the number of the photons emitted into the free space per second and the number of the photons emitted from the active region per second [2, 3]. An improved EQE requires a better IQE and LEE. The IQE is strongly

Z.-H. Zhang · Y. Zhang · W. Bi
Institute of Micro-Nano Photoelectron and Electromagnetic Technology Innovation, School of Electronics and Information Engineering, Hebei University of Technology, Key Laboratory of Electronic Materials and Devices of Tianjin, Tianjin, P. R. China

S.-W. Huang Chen · H.-C. Kuo (✉)
Department of Photonics and Institute of Electro-optical Engineering, National Chiao Tung University, Hsinchu, Taiwan
e-mail: hckuo@faculty.nctu.edu.tw

J. Li, G. Q. Zhang (eds.), *Light-Emitting Diodes*, Solid State Lighting Technology and Application Series 4, https://doi.org/10.1007/978-3-319-99211-2_9

affected by quite a few factors including the dislocation/defect density, the carrier injection efficiency, the current spreading effect, the polarization-induced electric field within the quantum wells, Auger recombination, etc. [3]. There have been several reports that have extensively discussed the IQE for III-nitride-based LEDs [2, 4–7]. Hence this chapter is concentrated on the LEE enhancement.

The large contrast between the refractive indices for the air and the III-nitride film results in a low LEE for III-nitride-based LEDs, and the LEE for the bare nitride film is low and further decreases if the emission wavelength becomes even shorter [8, 9]. For [0001] oriented III-nitride LEDs emitting visible light, the emitted photons are dominated by transverse electric (TE) mode. Numerically simulated results indicate that no more than 30% of the visible photons are able to escape out of the planar LEDs [10]. In the case of deep ultraviolet (DUV) LED, the portion of the TM mode increases as the Al composition increases or the emission wavelength decreases; the LEE can be even lower than 10% if the DUV photonics are not specifically treated [11]. There are a number of approaches aiming to increase the LEE for LED chips, e.g., chip structure engineering, reflection engineering, surface engineering, substrate engineering, photonics crystal, surface plasmon engineering, etc. Details will be addressed subsequently.

9.2 Chip Structure Engineering

The chip structure and the chip package have essentially important impact on the LEE. At the current stage, III-nitride-based LEDs are normally grown on foreign insulating sapphire substrates. Therefore, a conventional lateral LED (CC) structure is adopted (see Fig. 9.1a), which comprises the mesa, the current spreading layer,

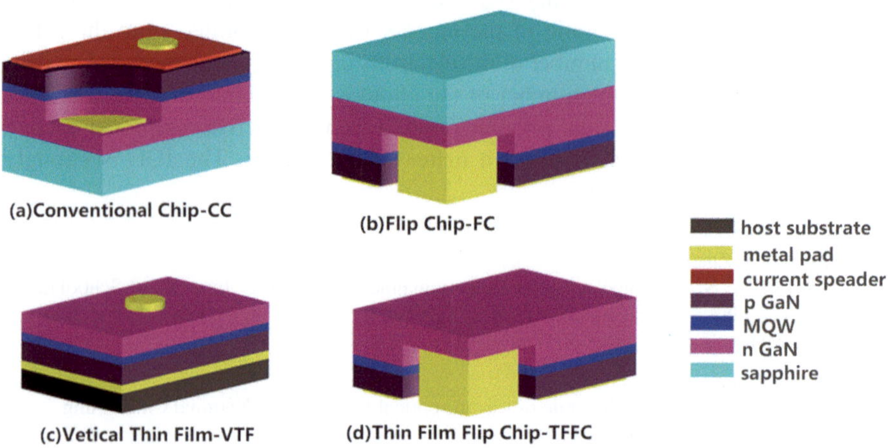

Fig. 9.1 Different chip configurations for sapphire-based InGaN/GaN LEDs: (**a**) conventional lateral chip (CC), (**b**) flip-chip (FC), (**c**) vertical-injection thin-film (VTF), (**d**) thin-film flip-chip (TFFC)

the p-pad, and the n-pad. The conventional lateral LED device is designed to have the photons emitted from the p-type material. This can inevitably cause the optical absorption by the p-type current spreading layer and the p-type bonding metal, leading to the unsatisfied LEE. Fortunately, the previously mentioned concerns can be solved by utilizing a flip-chip (FC) design, the device structure of which is shown in Fig. 9.1b. A reflective mirror is deposited on the p-type layer surface, and by doing so, the light beams can escape from the LED from the sapphire side. The sapphire has the energy bandgap of ~10 eV, leading to a negligible optical absorption. In addition, the distance between the reflective mirror and the quantum wells is in the order of the optical wavelength, and the optical cavity effects can occur that helps to tune the radiation pattern and results in more light into escape cone [12, 13]. Another strategy to improve the LEE can be realized by using the vertical thin-film (VTF) chip, which adopts the substrate removal process and has the N-face GaN layer upward, serving as the out-light plane. More importantly, the random-textured N-face GaN layer can be obtained by the low-cost wet etching technology and can result in the LEE as high as 75%. Besides the high LEE, such design also enables the benefit of almost 100% top-surface emission and a near-Lambertian radiation pattern [13]. However, similar to the CC design, the typical VTF design suffers from the light blocking and the optical absorption effects by the n-type metal electrode as is illustrated in Fig. 9.1c. An alternative design that combines the advantage of the CC chip and the VTF chip is the thin-film flip-chip (TFFC) structure, which enables the 1.3-fold LEE enhancement [13].

9.3 Engineering the Optical Reflections

As has been mentioned in the last section, both the flip-chip and vertical LED structures have to utilize the reflective mirror to tune the light propagation. Therefore, the LEE can be improved by engineering the optical reflections through different types of reflective mirrors. If the effect of the side facet is neglected, ideal reflective mirrors can approximately double the intensity of the light compared with the LED without the reflective mirror. The reflector mainly includes distributed Bragg reflectors (DBRs), metal reflectors, and omnidirectional reflectors (ODRs) as depicted in Fig. 9.2a–c, respectively [14].

A distributed Bragg reflector is a multilayer structure consisting of N pairs of dielectric layers. The periodic dielectric layers possess the high and low refractive indices (n_h and n_l, respectively) which require the respective layer thickness of the "quarter-wave" at a certain wavelength λ, i.e., $n_l d_l = n_n d_n = \frac{1}{4}\lambda$. In an ideal condition, the DBR structure shall feature a large refractive index contrast (Δn), small mismatches for the lattice constant and the thermal expansion coefficient with the underlying epi-layer, a large bandgap, and a high electrical conductivity [15]. The large refractive index contrast can fulfill the better reflectivity; the small mismatches for the lattice constant and the thermal expansion coefficient with the underlying epi-layer can maintain excellent crystal quality for the subsequently

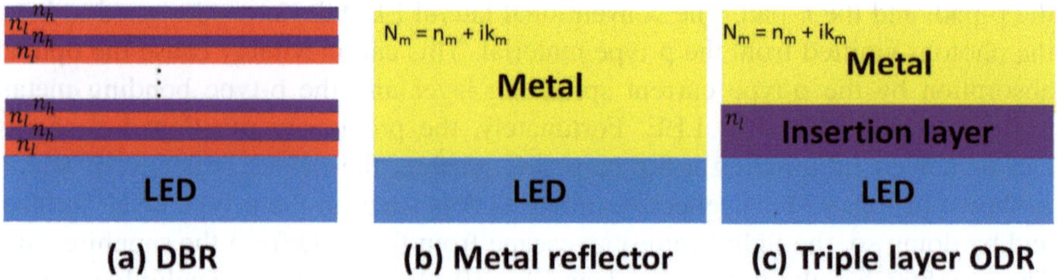

Fig. 9.2 Schematic cross-sectional views for (**a**) distributed Bragg reflector (DBR), the periodic dielectric layers with high and low refractive indices of n_h and n_l, respectively; (**b**) a metal reflector; and (**c**) a triple-layer omnidirectional reflector (ODR). The semiconductor has the refractive index of n_s; the coated material with a low refractive index layer of n_{li}; the metal with a complex refractive index $N_m = n_m + ik_m$, where k_m is the extinction coefficient

grown epi-layer; a reduced optical absorption caused by the DBR can be obtained by using DBR materials with large energy bandgap; the DBR has to be electrically conductive to allow the current flow and reduce the fabrication complexity.

For the purpose of cost saving, it is more advised to in situ grow GaN/AlNDBRs in the metal-organic chemical vapor deposition (MOCVD) system for III-nitride-based LEDs. Nevertheless, the main difficulty in fabricating AlN/GaN DBRs lies on the fact that there is a huge lattice mismatch (2.4%) between the GaN layer and the AlN layer. The very large lattice mismatch may cause cracks for the films [15]. Besides the in situ DBRs, the DBRs can also be obtained by depositing TiO_2/SiO_2 periodic structures by using electron beam evaporator system [16], and the reflectance for the fully optimized TiO_2/SiO_2 DBR can be as high as 99% [17]. Unfortunately, the small bandgap of 3.3 eV for the TiO_2 thin layer limits the application for the TiO_2/SiO_2 DBR in the UV LEDs. Instead, HfO_2/SiO_2 DBRs [18] and Ta_2O_5/SiO_2 DBRs [17] can be excellent candidates for UV LEDs. Moreover, DBRs can be patterned into three-dimensional (3D) photonic crystal reflector to enhance the scattering ability for the incoming light [16, 19]. Huang et al. report to use the self-assembled polystyrene nano-spheres and fabricate the 3D colloidal-photonic-crystal (CPC) bottom reflector for blue GaN-based LEDs [20].

DBRs can serve as excellent reflective mirrors with very high reflectivity if they are fully optimized. Nevertheless, the reflectivity is strongly subject to the incident angle and the wavelength [14]. Therefore, metal reflectors that can be insensitive to the incident angle and the emission wavelength over a wide wavelength range are proposed. However, the disadvantage for the metal reflectors is not competitive when compared with DBRs, e.g., Al and Ag reflective mirrors yield the reflectivity of ~95% for visible spectral range [21]. Their report shows that only Al metal is of high reflectivity even if the wavelength is shorter than 400 nm, and Ag metal can be an excellent reflective mirror for visible wavelength regime. Significant optical absorption can be found for Au metal when the emission falls into the blue spectral range.

The next optical device is called omnidirectional reflector (ODR) which is less affected by the incidence angle, and the ODR can also maintain a very high

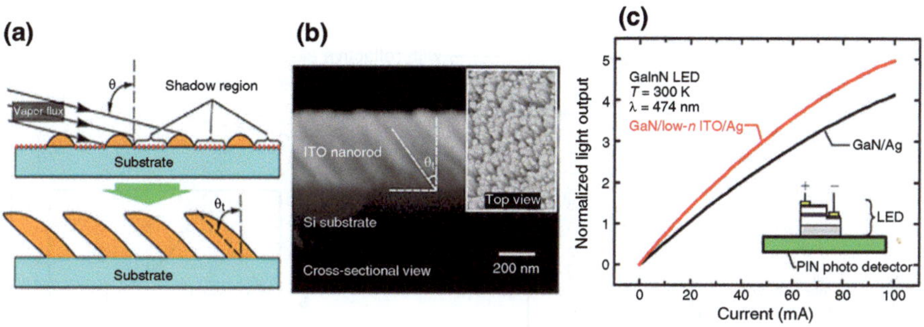

Fig. 9.3 (**a**) Schematic diagram for the oblique-angle deposition, (**b**) SEM images for ITO nanorods deposited by oblique-angle e-beam evaporation on a Si substrate, (**c**) light output power in terms of the injection current for LEDs with Ag reflective mirror and with the proposed ODR structure, respectively. Reproduced from Ref. [24], with the permission of AIP Publishing

reflectivity. The ODR structure is illustrated in Fig. 9.2c that comprises a low refractive indexed insertion layer and the metal. The reflectivity can be further enhanced if the refractive index contrast between the insertion layer and the semiconductor becomes large. To meet that goal, quite a few designs have been demonstrated. It is proved that low refractive indexed nanoporous SiO_2 with the average pore size of ~2 nm can be fabricated by using the sol–gel process [14]. An even lower refractive index of $n = 1.05$ can be obtained by adopting SiO_2 nanorods that are grown by oblique-angle deposition [22]. Alternatively, indium–tin oxide (ITO) nanorods with the refractive index as low as 1.34 at 461 nm via oblique-angle deposition approach (see Fig. 9.3a) also help to form excellent ODR structure as shown in Fig. 9.3b [23, 24]. Compared to the LEDs with merely the Ag reflector, the light output power for the InGaN/GaN-based blue LED with the proposed ODR structure is enhanced by 31.6%, which is well ascribed to the enhanced reflectivity of the ODR and the lower losses of the wave-guided modes [24]. However, during the process for fabrication of the ODR with nanostructures, more care shall be paid to the rough nanostructures that may cause the textured metal mirror and then decrease the reflectivity [25].

Besides III-nitride-based visible LEDs, the reflective mirror is also particularly important for AlGaN-based DUV LEDs. Current DUV LEDs adopt the p-GaN layer to form the good ohmic contact and provide sufficient holes. However, the p-GaN layer can significantly absorb the DUV photons. Therefore, the combination of a transparent contact layer and a highly reflective p-type electrode is desirable to realize a decent LEE. According to Ref. [21], we can get that the reflectance of a conventional Ni/Au p-electrode is low (<30%) and cannot function as a highly reflective mirror for DUV photons. Although the Al mirror can possess a high reflectivity of 92% in DUV spectral range, the ohmic contact is difficult to achieve. As a result, Lobo et al. propose to grow the nanopixel Pd ohmic contacts on the p-AlGaN layer to realize excellent current–voltage characteristics. The Pd nanopixels are then covered by the Al reflective mirror. The better ohmic contact and the Al

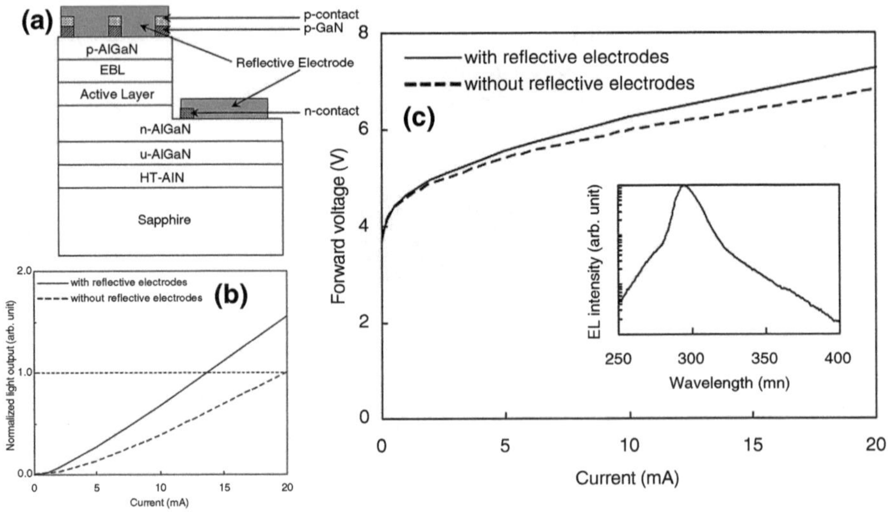

Fig. 9.4 (**a**) Schematic device architecture for the LED structure with the proposed reflective electrode, (**b**) optical power as the function of the injection current, and (**c**) current–voltage characteristics for the two LEDs with and without reflective electrodes, respectively. The inset in (**c**) shows the emission spectra for the two LEDs. Reproduced from Ref. [27], with the permission of IOP Publishing

reflective mirror enhance the light output power by 90% [26]. Inazu et al. partially remove the p-GaN layer so that the Al windows can be directly obtained on the p-AlGaN layer [27]. Ni is deposited on the rest p-GaN layer, and finally the Al/Ni/p-GaN structure is formed (see Fig. 9.4a). By doing so, a 1.55-fold increase in the LEE has been demonstrated which has been depicted in Fig. 9.4b. Nevertheless, these mesh ohmic contacts result in an increase in the forward voltage due to the reduced contact area according to Fig. 9.4c. Takano et al. grow the DUV LED films on the pattern sapphire substrate, which have the Rh mirror electrode, and the chips are finally encapsulated by resin to improve the light extraction. The EQE of 20% at the peak emission wavelength of 275 nm is achieved [28].

Besides the optically absorptive p-GaN layer, another bottleneck limiting the LEE for the Al-rich AlGaN-based DUV LEDs arises from the strong anisotropic light emission from the active region, such that the TM-polarized light propagates in parallel with the growth plane [7]. To address that issue, Wierer et al. report to use the inclined sidewall reflectors which can favor the extraction for the anisotropic light, and especially, the LEE for the TM-polarized light is enhanced up to four times [29]. Zhang et al. also numerically confirm that inclined sidewall reflectors are more beneficial to scatter TM-polarized light than TE-polarized light [30]. Kim et al. utilize the micro-reflectors that are deposited on the sidewalls for the selectively grown n-GaN micro-mesa to improve the LEE. The micro-reflectors are able to

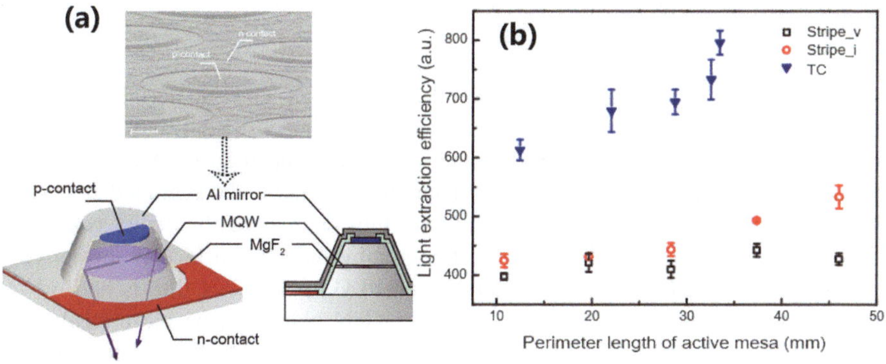

Fig. 9.5 (**a**) Scanning electron microscopic images, 3D and cross-sectional schematic illustrations for truncated cone (TC) AlGaN structure, along which we also schematically draw the MgF_2/Al ODR, (**b**) light extract efficiency for the three device structures with different perimeter length of the active mesa. Reproduced from Ref. [32] with the permission of ACS Publishing

tune the propagation for both the TE- and TM-polarized DUV photons [31]. Kim et al. further demonstrate coating low refraction indexed MgF_2/Al ODR mirror on the inclined sidewalls for the truncated cone (TC)-shaped mesa [32, 33], and the truncated cone-shaped mesa is shown in Fig. 9.5a. They also study another two device architectures, i.e., Strip_v and Strip_i, for comparison. Here, Strip_v means the strip-type DUV LED with vertical sidewalls, and Strip_i means the strip-type DUV LED with inclined sidewalls. The sidewalls for both Strip_v and Strip_i DUV LEDs possess MgF_2/Al ODR mirror. Figure 9.5b reflects that the TC configuration shows the highest LEE.

9.4 Micro-/Nanostructure Engineering

It has been shown that the large refractive index difference at the GaN/air interface results in a small critical angle of 23°. Therefore, a considerable amount of light emitted from the MQWs is trapped by the total internal reflection (TIR) at the GaN/air interface. The trapped light is reabsorbed by MQWs and is eventually converted into heat [34]. This process significantly deteriorates the LEE and reliability for LEDs. Therefore, it is a key step to break the TIR process to enhance the LEE. Various microscale surface roughness approaches have been adopted to enhance LEE of LED due to increased probability of light escaping from LEDs through multiple scattering in the microscale structure [1]. In this section, the progress of surface engineering to enhance photon extraction of GaN-based LEDs is reviewed and investigated.

9.4.1 Surface Texturing

Dry etching is often used to texture the surface of GaN material. However, the preparation for the photomask tends to make the cost high and the process complicated. Moreover, the surface defects that are induced during the dry etching process can sacrifice the electronic characteristics of LEDs. The degraded electronic properties often lead to high leakage current, thus decreasing the device reliability [1]. An alternate strategy of avoiding the etching-induced surface is to wet etch the GaN material. However, the [0001] oriented Ga-polar GaN film is hardly wet etched, whereas the [000-1] oriented N-face GaN plane can be more easily wet etched by using hot KOH solution. This is because the dangling bond density (DBD) for the N-face facet or other planes is larger than that for the Ga-polar facet [35]. Fortunately, the N-face of the GaN layer for the [0001] oriented LED can be obtained by using the laser-liftoff technology.

Nakamura et al. report an n-side-up GaN-based LED with a hexagonal "cone-like" surface [36]. The N-face surface for the GaN layer is exposed by using the laser-liftoff technique. The exposed N-face GaN surface is treated with an anisotropic wet etching process to roughen the surface, which improves the LEE and doubles the light output power. The LEE can be further improved if the periodical corrugated surfaces of the N-face n-GaN surface are fabricated before wet etched [37]. By doing so, integrated surface textures can be generated. Compared to the LED for which the surface is treated by using the conventional wet etching technique, the proposed LED structure increases the light output power by 32%. It is worth noting that the formation of hexagonal pyramid-like structures by the photochemical etching may generate threading dislocations (TDs), which serve as diffusion pathways for metals and act as leakage current pathways. The electrical properties for LEDs will be correspondingly degraded [38]. In addition, the randomly textured patterns on the N-face GaN evolve into the hexagonal pyramid structures with a fixed sidewall angle of about 31.6 ° due to the lower etching rate of the r-surface GaN when compared to the N-face GaN surface [38]. However, the inclination angle is believed to be essentially important in deciding the LEE. Therefore, cone-shaped nanostructures with optimized inclination angle of 23.4° for the sidewalls are further developed by using the SiO_2 nanosphere lithography technique. Experimental and numerically calculated results show that V-LEDs with the proposed design are effective to eliminate TIR and the optical power enhancement of 6% is obtained compared to those using the photochemical etching technique [38]. On the other hand, N-polar wet etching technique can be used to fabricate high-efficiency pyramids LED and bendable LED [39, 40]. Ma et al. report the fabrication for nitride-based hexagonal pyramids array (HPA) vertical-injection light-emitting diodes (V-LEDs) by N-polar wet etching technique [39]. The wet etching process penetrates through the active region, and hence the quantum wells in each pyramid are isolated from the ones in the adjacent pyramids (see Fig. 9.6a–c for the schematic process flow). The deep wet etching eliminates the TDs in hexagonal pyramids. As a result, the performance of HPA V-LEDs devices is

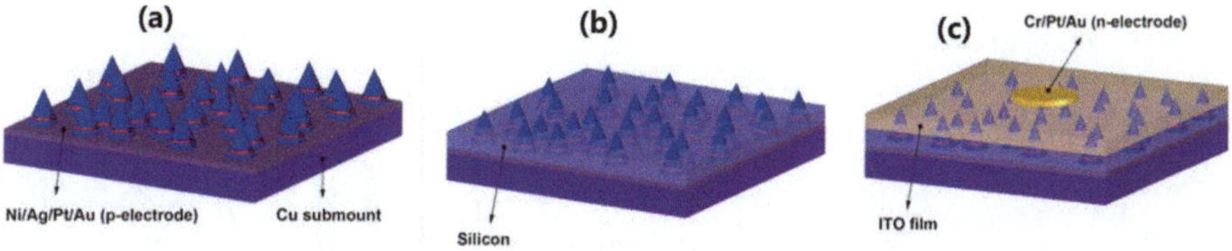

Fig. 9.6 Schematic fabrication processes for obtaining the nitride-based hexagonal pyramids array (HPA) vertical-injection light-emitting diodes: (**a**) pyramids with isolated quantum wells are formed by wet etching the N-face GaN surface, (**b**) silicon gel is spin coated to protect the quantum wells, and (**c**) deposition of ITO film and the Cr/Pt/Au n-electrode. Reproduced from Ref. [39], with the permission of OSA publishing

significantly improved with 30% higher internal quantum efficiency and 20% higher LEE compared with conventional roughened broad area V-LEDs.

Despite the difficulty in wet etching the Ga-polar GaN, the c-plane GaN layer can still be wet etched into pits at defect sites. Ha et al. report that the selectively chemical etching of the p-GaN surface enables the improved LEE and better device reliability [41]. The improved LEE is ascribed to the textured p-GaN surface, and the decrease in the number of defects in the p-GaN interprets the better device reliability. In addition, the p-GaN surface can also be in situ textured by reducing the growth temperature. To in-depth investigate the impact of the growth temperature for the p-GaN layer on the LEE, Liu et al. grew Mg-doped p-GaN epitaxial layers at different temperatures [42]. They find that, compared to the p-GaN layer grown at the temperature of 1000 °C, the p-GaN layer grown at the temperature of 800 °C can most enhance light output power for the LED partially because of the rough p-GaN surface. However, the rough p-GaN surface also aggravates the leakage current and device reliability. Moreover, Si or Mg treatment after high-temperature p-GaN layer also can achieve an in situ textured p-GaN surface [43, 44]. Nevertheless, the formation of the highly resistive thin MgN or SiN layers results in the slightly high forward voltage. A more advanced strategy that can considerably increase the LEE involves the hierarchical structures, which comprise in situ textured p-GaN microdomes and the post-grown SiO_2 nanorods on LED surface [45]. The SEM image for hierarchical structures is depicted in Fig. 9.7b. For better comparison, the SEM image with only in situ textured p-GaN microdomes is shown in Fig. 9.7a. Figure 9.7c shows that the suggested two structures have negligible effects on the current–voltage characteristics. Insightful investigations into Fig. 9.7d–f illustrate that the hierarchical micro-/nanoscale architecture can best enhance the LEE and the corresponding light output for blue LEDs is enhanced by up to 36.8%. The mechanism of enhanced LEE is attributed to the multiple reflections in the microdomes and additional graded refractive index profile provided by SiO_2 nanorods [45].

The surface texturing technology either by wet etching or by in situ epitaxial technique is most economical and convenient. However it is difficult to completely avoid the damages to the epi-layer and the resulting degradation for the electrical

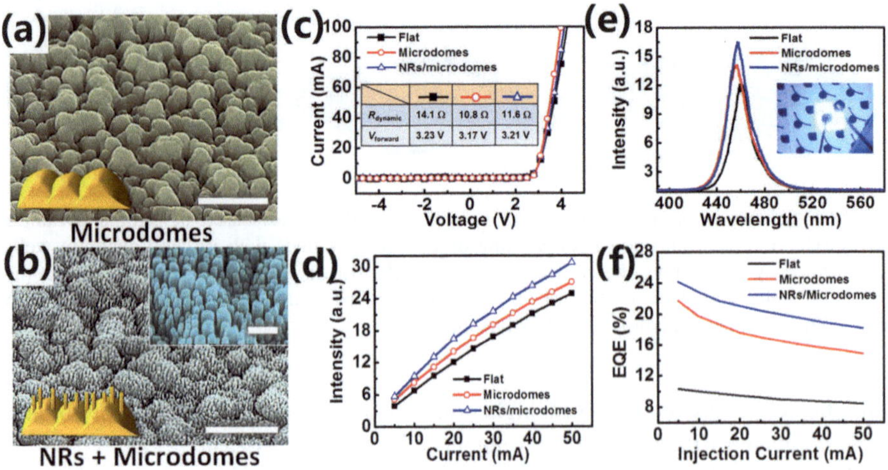

Fig. 9.7 SEM images of the p-GaN surface with (**a**) p-GaN microdomes and (**b**) SiO_2nanorods/p-GaN microdomes, (**c**) current–voltage characteristics, (**d**) measured light output power at different current levels, (**e**) measured electroluminescence spectra at the injection current of 20 mA, and (**f**) EQEs as a function of driving current for the three types of LEDs. Reproduced from Ref. [45], with the permission of AIP publishing

properties. Therefore, it is a good choice to texture the ITO current spreading layer through the wet etching technique. Oh et al. suggest the dual-layer ITO electrode to increase the LEE for InGaN/GaN blue LEDs. The proposed method lies on the fact that the etching selectivity for the as-deposited top ITO layer and the annealed bottom ITO layer varies, and this can produce textured the top ITO layer by rinsing the samples in the BOE solution [46]. Compared to the reference LED, the proposed textured dual-layer ITO configurations can double the optical output power. More importantly, the proposed structure does not degrade the electrical property. Oh et al. further fabricate the ITO nanoballs by using the wet etching technique to increase the LEE of UV LED (see Fig. 9.8a–c) [47]. The ITO nanoballs are finally coated by the Al film to form the nanoball ODR (NODR). The high reflection can be realized by the proposed ODR structure, and in the meantime, the increase of diffuse reflection is enabled by the interfaces between the ITO nanoball layer and Al film. Moreover, the nanovoids in the ITO layer further contribute to the diffuse reflection. Therefore, the enhanced LEE can be obtained for the UV LED, and this simultaneously increases optical power as demonstrated in Fig. 9.8d. Figure 9.8e depicts that the proposed ODR structures have less impact on the electrical properties.

To enhance the LEE for DUV LEDs, the sidewall texture technology is even more important due to the anisotropic light emission from the AlGaN-based active region [48]. Guo et al. roughen the sidewall of the sapphire substrate by using picosecond laser dicing technique [49]. Their results show thatthe LEE does not monotonically

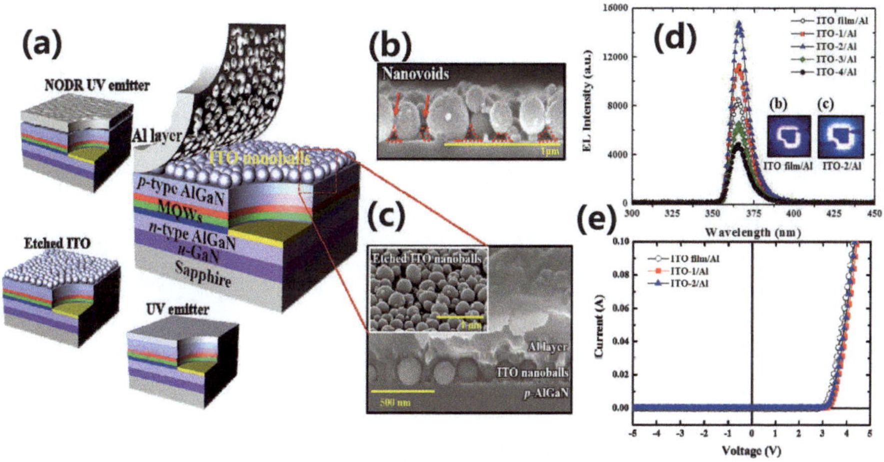

Fig. 9.8 (**a**) Schematic diagrams for the UV LED with the proposed ITO nanoballs/Al film NODR structure; (**b**) cross-sectional SEM image for ITO nanoballs, along with which the nanovoids are also shown; (**c**) ITO nanoballs/Al film NODR structure on a p-type AlGaN layer; (**d**) electroluminescence spectra for UV LEDs with the ITO film/Al ODR and different NODRs; and (**e**) injection current at different applied biases for UV LED with the ITO film/Al ODR and different NODRs. Reproduced from Ref. [47], with the permission of The Royal Society of Chemistry

improve as the number of the diced layer increases as can be indicated from Fig. 9.9. The electrical properties are also not significantly impacted by the dicing process, since the sapphire serves as the insulating layer and provides no current paths. Guo et al. later also suggest fabricating oblique sidewalls for AlGaN-based DUV LEDs by using the picosecond laser multiple-scribing technique [50]. According to their report, the oblique sidewalls help to improve the LEE and increase the power by 13.8% when the inclination angle is set to 60°.

9.4.2 *Photonic Crystal*

Photonic crystal (PhC) is well known as a periodic configuration of the optical medium with a photonic bandgap. The photonic bandgap is able to prevent the light falling into a certain wavelength range from propagating in certain directions [51]. The most popular occurrence of one-dimensional (1D) PhC is DBR, which is a multiple-layer dielectric stack with periodic refractive index and allows the photons to possess very high reflectivity/transmittance with an angular cone around normal incidence. Two-dimensional (2D) PhCs can increase the angular range of the bandgap and limit the propagation of light in plane. In spite that three-dimensional (3D) PhCs are able to support an omnidirectional bandgap, practical implementation for 3D PhCs is more challenged at the current stage. Therefore, significant research effort has been invested to 2D PhCs for III-nitride-based LEDs.

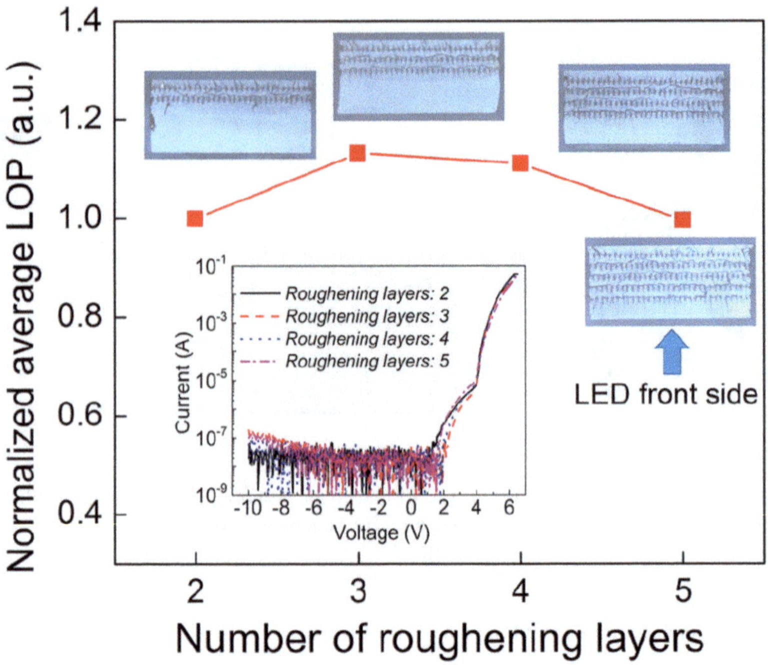

Fig. 9.9 Normalized average light optical power (LOP) for 275 nm DUV LEDs. Inset micrographs show the cross section of the sidewalls for the sapphire substrate sidewalls. Current–voltage characteristics are also presented in the inset. Reproduced from Ref. [49], with the permission from AIP publishing

In the ideal case of a full in-plane bandgap for the 2D PhC, no guided modes exist, and all of the guided light emission is then inhibited. Therefore, the light can only propagate in the out-of-plane modes, and this gives rise to the ideal LEE of 100%. As a matter of fact, the bandgap for the realistic 2D PhC only extends up to a maximum off-plane propagation angle. As a result, it is difficult to design 2D PhC with a bandgap that can cover the whole range of angles to suppress all guided modes especially for very thick LED films [51]. In spite of that, the PhCs that are located very close to the active area can lead to the direct light out-coupling process and the strong light scattering results in a high LEE [52]. However, the 2D PhCs etched through the active region directly are inevitable to reduce the active region area and create new surface defects. Therefore the reduced radiative recombination rate results in the poor internal quantum efficiency. Recently, the bottom-up nanorod LEDs and core-shell LEDs offer good solutions to fabricate 2D PhCs. The active area can be self-assembled embodied into 2D PhCs. The proposed structures are also featured with the advantages of strain relaxation and no dislocations. Djavid et al. demonstrate that the LEE for TM-polarized light can be significantly enhanced if AlGaN-based nanowired DUV LED structures are adopted because the nanorod PhCs inhibit the emission of guided modes and redirect the trapped light into the radiated modes [53]. Core-shell LEDs possess unique features of three-dimensional MQWs and the PhC-effect-caused LEE enhancement, and therefore the photon generation rate can be even more [1].

Although the embedded PhCs can increase the LEE and avoid the dislocations, nevertheless, the LEDs with embedded PhCs demand complicated process, e.g., how to keep excellent electrical properties? Think about above these challenges: 2D PhCs as the diffraction gate is more practical in improving the overall performance for LEDs. For the LEDs with properly designed photonic crystals, the light that is trapped by total internal reflection can be redirected into the radiated modes, which can greatly enhance the LEE [51]. 2D PhCs as the diffraction gate can also be induced onto the surface for the p-GaN layer, the n-GaN layer, and the dielectric material for LEDs. In 2004, Oder et al. fabricate triangular arrays of the PhCs in p-GaN of blue and UV LED using electron-beam lithography and inductively coupled plasma dry etching. The enhancement of optical output power reveals an increase by 63% and 95% for the blue and UV LEDs, respectively. The enhancement for LEDs is attributed to refractive index periodicity creating a cutoff frequency for the guided modes. Modes above the cutoff frequency couple with PhC resulting in leaky mode. Compared with random scattering from a randomly textured surface, the PhCs offer a short escaped length for light, thus avoiding parasitic absorption [54]. The PhC that serves as the diffraction gate only has a high overlap with high-order modes. Unfortunately, low-order modes also carry a large portion of 30% for the rest optical energy [55]. A lower-index $Al_xGa_{1-x}N$ layer can be used to modify low-order modes as proposed by David et al. [55].

The large electrical resistivity for the p-GaN window layer gives rise to the poor current spreading for LEDs. Although the resistance can be reduced with a thicker p-GaN layer, the increased thickness can result in excessive optical absorption. The issue can be partially solved by inserting dielectric material, embedded photonic crystals which can be obtained by secondary epitaxy [56], thin-film fabrication with periodic nanostructures in the less resistive n-GaN layer [57]. Ee et al. report the InGaN/GaN MQW LED structure utilizing a SiO_2/polystyrene lens array on p-GaN layer to significantly enhance the LEE [58]. As a result, the 249% improvement of the light output power is obtained due to the increase in effective photon escape cone and the reduced Fresnel reflection. Liu et al. incorporate air-voided PhCs into the n-GaN layer to enhance the light extraction efficiency by ~32.7% [56]. Wierer et al. fabricate a 700-nm-thick flip-chip thin-film LED with 2D PhCs in the n-GaN layer. The record-breaking LEE is as high as 73% for the unencapsulated device at that time [57]. In addition to the enhanced LEE, PhCs also yield the benefits for increasing the controllability of the optical emission directionality [59] and preserving the excellent light polarization [60], which are strongly desired in the applications of display, projectors, etc.

9.4.3 Patterned Substrate

Patterned sapphire substrates (PSSs) have been widely used in commercial GaN-based LEDs, because the epitaxial layers grown on the PSS possess the decreased TDs and the increased light scattering. At present, the specification of commercial PSSs is the cone shape with 3 μm period, 1.7 μm height, and 0.3 μm space. To further improve the performance for LEDs grown on the PSS, the effective C-plane area that supports the subsequent epi-growth in microscale PSS continues to be reduced in the past few decades. However, more difficulty will be made to grow GaN layer on PSS with too small effective C-plane area. In addition, LEE is substantially affected by the pattern geometries for the PSS [61]. Therefore different pattern geometries for the PSS are continued to be developed by adopting various fabrication processes. Lin et al. transfer the patterns of a cone-shaped patterned sapphire substrate (CPSS) into SiO_2 layer through using nano-imprint lithography (NIL). By doing so, a cone-shaped SiO_2 patterned template is obtained [62]. Compared with the LED grown on CPSS, the LEE of the LED grown on cone-shaped SiO_2 patterned template can be enhanced due to the smaller refractive index for SiO_2 than sapphire. A volcano-shaped patterned sapphire substrate (VPSS in Fig. 9.10a) can be fabricated to improve the LEE of LEDs [63, 64]. The VPSS LED show a top optical flux is 2.8 times higher than that for LEDs grown on planar

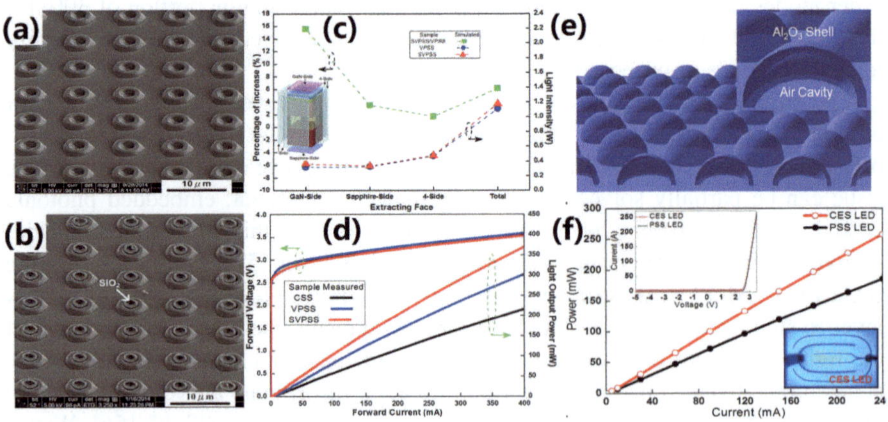

Fig. 9.10 (**a**) SEM images of VPSS and (**b**) SVPSS structures, (**c**) percentage increase in the light intensity of InGaN-based LEDs grown on the VPSS and SVPSS. Inset: schematic representation of different receiver positions considered in the simulation, (**d**) measured light output power and the required voltage at different injection current, (**e**) schematic structures of a hemispherical CES template. Inset: schematically magnifying an inner air cavity covered with a thin alumina shell. (**f**) Light output power in terms of the injection current for CES and PSS LED devices, respectively. Top inset: SEM image magnifying an inner air cavity covered with a thin alumina shell; bottom inset: optical microscopy image for the working LED device. Figures (**a–d**) are reproduced from Ref. [65], with the permission of RSC publishing. Figures (**e**) and (**f**) are reproduced from Ref. [66], with the permission of ACS publishing

substrate and is 1.6 times higher than that for LEDs grown on hemispherical PSS [64]. If a layer of SiO_2 is embodied into the VPSS (i.e., SVPSS in Fig. 9.10b), the LEE and the crystal quality for the LEDs can be further enhanced. Therefore the light output power for LEDs grown on the SVPSS is the strongest as can be seen in Fig. 9.10c, d [65]. Furthermore, Fig. 9.10d also depicts that the electrical properties for different LEDs are not strongly influenced by the substrate. Moon et al. report a novel high-index contrast cavity engineered substrates (CESs) in which hexagonally arranged hemispherical air cavities are covered with a 80-nm-thick crystallized Al_2O_3 shell (see Fig. 9.10e). According to Fig. 9.10f, InGaN/GaN blue LEDs fabricated on the 3 μm pitch CES exhibit ~39% enhancement in the optical power compared to state-of-the-art PSS LEDs [66].

With the development of nanotechnology, the nanoscaled PSS (NPSS) can be fabricated through nano-imprint, e-beam lithography, holographic lithography, metal annealing, and nanosphere lithography. The nanoscaled PSS (NPSS) can provide higher pattern densities than microscaled PSS, leading to the increased possibility of photon scattering and the higher LEE. In the year of 2000, the NPSS is first proposed to tune the strain and to improve the crystal quality for heteroepitaxy films [67]. Later on, Chen et al. report that, due to the efficient optical scattering by NPSS, the output power for LEDs grown on NPSS is 1.11 times higher than that for LEDs grown on micro-patterned sapphire substrates (MPSS) [68]. Kao et al. reveal the further enhancement for the LEE can be achieved by properly increasing the aspect ratio for the NPSS [69]. Note that a too high aspect ratio may result in the decreased LEE and the decrease of the C-plane area, which then also limits the LED epitaxial window [70]. The other advantage for the NPSS is to suppress the quantum-confined Stark effect (QCSE) within the multiple quantum wells [71]. Thus far, various reports show that the NPSS with different designs can increase the LEE. However, a commonly accepted consent implies that a too small pattern size for NPSS cannot promote the LEE [72, 73]. Recently, a hybrid PSS that is obtained by embedding nanopatterns into commercial MPSS has been proposed [74, 75]. Ke et al. fabricate the hybrid PSS with high-density concave nanopattern structures on a PSS (see Fig. 9.11a, b). Comparison between Fig. 9.11c and d depicts that the hybrid PSS can even better support the diffuse reflection. As a result, when comparing conventional PSS LED, the hybrid PSS LED generates a ~22.3% enhancement for the light output power as is shown in Fig. 9.11e, f [75].

Another strategy to further improve the LEE of NPSS is to fill lower refractive index material such as SiO_2 or air voids into the GaN/substrate interface (see Fig. 9.12a). If the material is air void, it can also block or laterally bend TDs. Chiu et al. demonstrate LEDs grown on the SiO_2 nanorod-array patterned sapphire substrate [76]. Compared to conventional LEDs, NPSS LEDs enhance the light output power by 52%. Low cost self-assembled SiO_2 microspheres can be integrated with substrate to relax strain and improve LEE [77, 78]. Ueda et al. demonstrate the high performance of blue LED grown on sapphire substrates coated with a silica-submicron-sphere monolayer film, as shown in Fig. 9.12b [77]. It can be seen from Fig. 9.12a that compared with the LED without silica spheres, the LED output power is increased by 2.5-fold due to the reduced threading dislocation and the enhanced

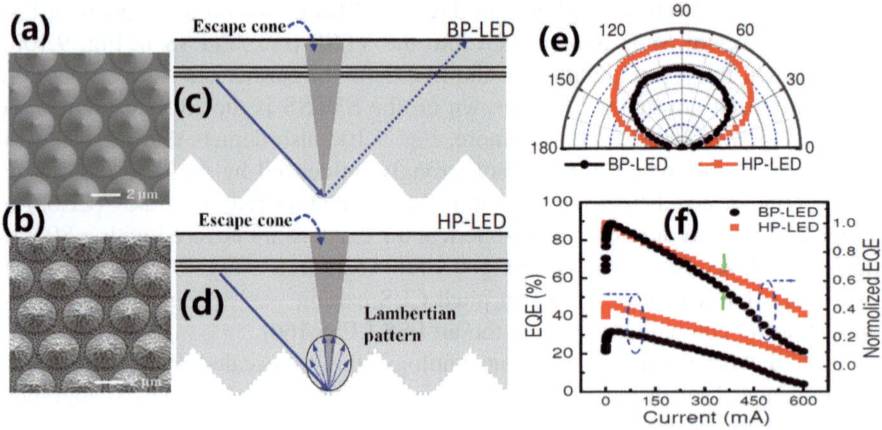

Fig. 9.11 Top-view SEM images of (**a**) bare-PSS (BP) and (**b**) hybrid-PSS (HP), (**c**) light racing schematic diagrams for BP-LED, (**d**) light racing schematic diagrams for HP-LED, (**e**) far-field radiation patterns of the BP-LED and HP-LED, (**f**) EQE and normalized EQE of BP-LED and HP-LED at various injection currents. Reproduced from Ref. [75], with the permission of ACS publishing

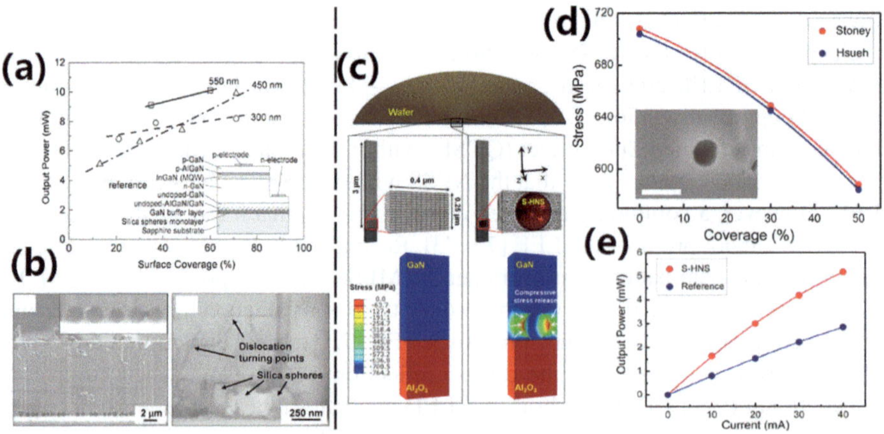

Fig. 9.12 (**a**) LED output power at various silica-sphere surface coverages on substrates. Inset: the GaN-based LED device structure. (**b**) SEM and TEM images of GaN films deposited on silica-coated substrates, (**c**) the stress analysis of GaN without S-HNS (left) and with a S-HNS monolayer with 30% surface coverage (right) obtained by FE simulations. Reduction in compressive stress as well as a strong stress gradient is notable around the S-HNS, (**d**) compressive stress in the GaN epitaxial layer at various S-HNS surface coverages. The stress values for 3 μm thick GaN layers on sapphire substrates were estimated from the measured wafer bowing and radius of curvature by an optical means. Inset: SEM image of ellipsoidal S-HNS taken after GaN epitaxial growth. The deformation of S-HNS from spherical to ellipsoidal shape by compressive stress in GaN is clearly observed (scale bar 200 nm), (**e**) output power of LEDs fabricated with and without S-HNS monolayers at various input currents. Figures (**a**) and (**b**) are reproduced from Ref. [77], with the permission of AIP publishing. Figures (**c**–**e**) are reproduced from Ref. [79], with the permission of Nature Group

light extraction. Pan et al. introduce a monolayer silica hollow nanosphere on the sapphire substrate (S-HNS) (see the insert in Fig. 9.12d). A four times enhancement of output power is attained by well-defined voids embedded at the GaN/sapphire interface (see Fig. 9.12e). The wafer bowing is reduced due to partial alleviation of compressive stress in GaN through the deformation of hollow nanospheres during growth, as depicted in Fig. 9.12c, d [79].

In addition to the enhanced LEE, a thinner coalescence thickness on NPSS can be realized, which reduces the growth time and is especially important for DUV LEDs. Epitaxial lateral overgrowth (ELO) technique on micro-stripe patterned sapphires usually requires a great coalescence thickness close to 10 μm because of the low mobility of Al atom, yielding a long growth time and a high cost. Dong et al. demonstrate 282 nm AlGaN-based DUV LEDs are grown on the novel NPSS fabricated through the nanosphere lithography technique [80]. Full coalescence is achieved at the 3-μm thick AlN layer (see Fig. 9.13a), and dislocation density also

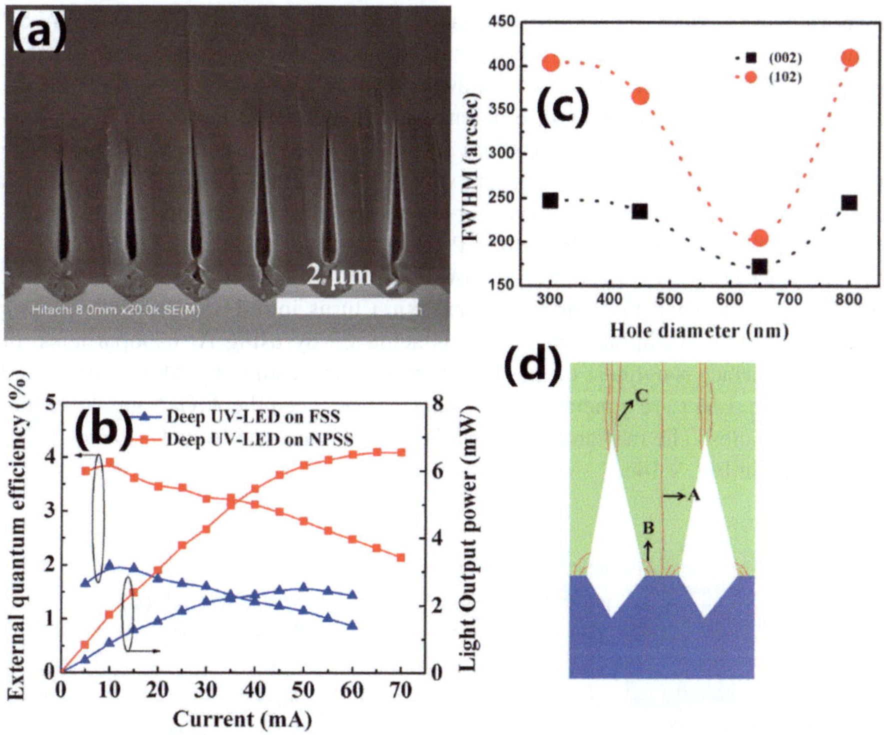

Fig. 9.13 (**a**) A cross-sectional SEM image of AlN grown on NPSS. (**b**) LOP-I-EQE curves of the deep UV LEDs grown on NPSS and flat sapphire substrate (FSS), (**c**) dependence of the XRCFWHM of AlN epi-layers on hole diameter of the NPSS, (**d**) three main competitive processes influencing TDD of AlN epi-layers on NPSS. Figures (**a**) and (**b**) are reproduced from Ref. [80], with the permission of AIP publishing. Figures (**c**) and (**d**) are reproduced from Ref. [81], with the permission of Nature Group

decreases. The EQE is enhanced by 98% (see Fig. 9.13b), which is attributed to the improved IQE and the better LEE [80]. Zhang et al. also declare that, by optimizing NPSS and the growth process, the threading dislocations (TDs) originating from the AlN/sapphire interface are bent because of the image force from the void sidewalls before coalescence. Hence, agreeing well with the findings by Dong et al. in Ref. [80], the dislocation density can be reduced as depicted in Fig. 9.13c, d [81].

9.5 Surface Plasmon-Enhanced LEDs

In the year of 2004, Okamoto et al. first report the possibility of increasing the IQE for blue LEDs by the surface plasmon (SP) [82]. The mechanism of the enhanced IQE by the SP effect can be explained that, when the resonant frequency of the surface plasmon polaritons (SPPs) of rough Ag film overlaps the emission frequency for the quantum well emission, the energy coupled to the mode is significantly increased [82]. Then, an additional channel of energy transfer from the QW-SP coupling can occur, and thus the IQE is enhanced [1, 83]. Since then, surface plasmon (SP)-enhanced LEDs have drawn intensive research efforts. It should be noted that, in order to maintain efficient QW-SP coupling, the distance between QW and metal is no more than 100 nm and the optimum distance is 20 nm due to the exponential decay of SP evanescent waves [84]. Therefore, metal nanoparticles are often embedded into the epi-layer or assembled into the partially etched p-GaN [85, 86]. Besides improving IQE, the SP effect can also modify polarized emission for DUV LEDs (DUV structure is shown in Fig. 9.14a) according to the report by Zhang et al., such that partial TM emission turns into TE emission [86]. The localized surface plasmon (LSP) effect is achieved by using Al nanoparticles, of which the surface roughness of the Al nanoparticles measured by AFM is presented in Fig. 9.14b. LSPs are generated at the interface between the AlGaN cap layer and Al nanoparticles. The in-plane (TE) polarization is improved from 13.7% to 19.7% as depicted in Fig. 9.14e.

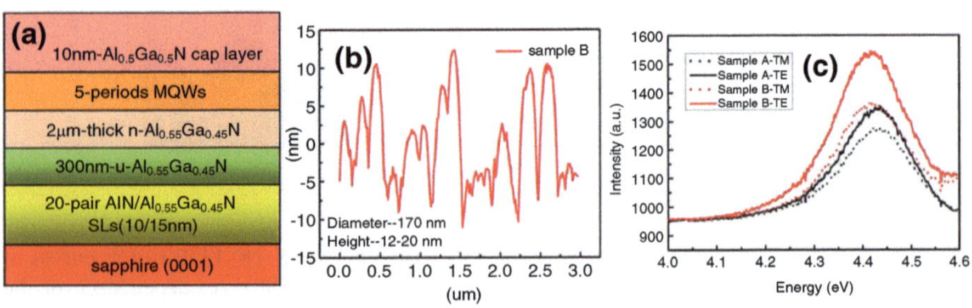

Fig. 9.14 (**a**) Schematic illustration of the structure for sample A, (**b**) AFM depth profile of Al particles on AlGaN surface of sample B, (**c**) wavelength dependent Purcell enhancement ratios of TE mode and TM mode, which are calculated to describe the enhancement ratio of the radiative recombination rate. Reproduced from Ref. [87], with the permission of Nature Group

Recently, it has been found that the SP also can be used to increase LEE by means of out-coupling of the generated photons through both scattering and localized surface plasmon resonance (LSPR) effects [88]. When distance between the metal and QW is more than 100 nm, surface plasmon (SP)-enhanced LEDs can only stem from the improved LEE by the metal SP. Sung et al. fabricate gold NPs on the ITO current spreading layer and lead to an electroluminescence (EL) increase by 1.8-fold [88]. Then, Sung et al. further confirm that the origin for the improved light emission arises from the combined action of the LSPR modes and the light scattering modes [89]. Chuang et al. find that the Ag nanoparticles on the p-GaN layer can be embedded into the subsequent growth ITO layer, and this can result in the three-dimensional distribution of Ag-based nanoparticles (see Fig. 9.15a, b)

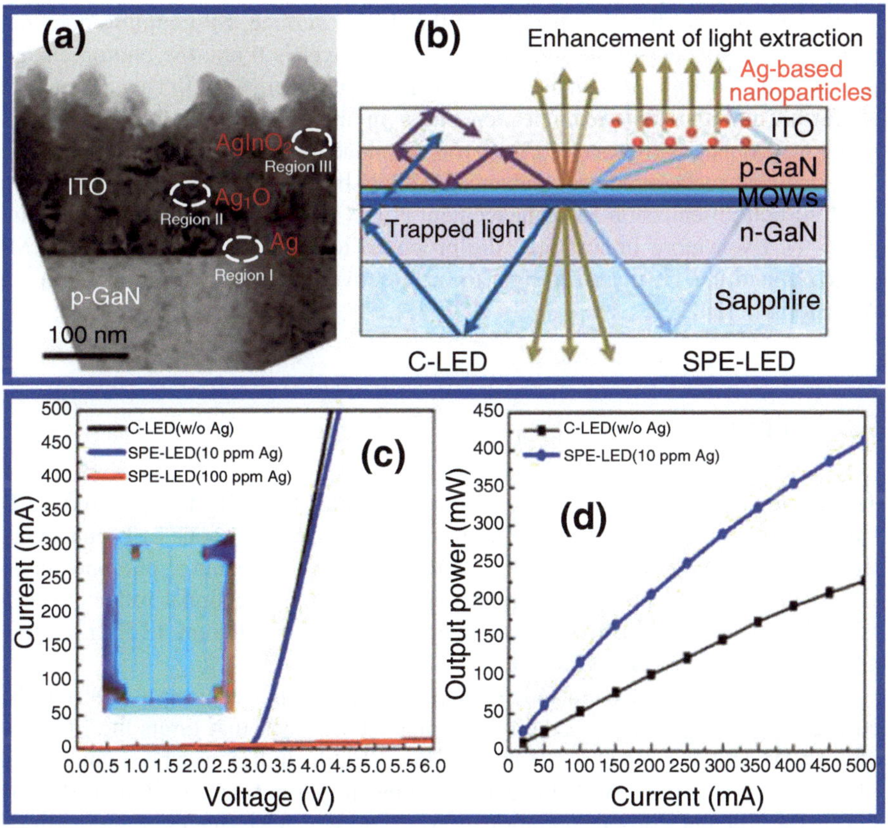

Fig. 9.15 (**a**) Cross-sectional TEM images of ITO film embedded with Ag-based nanoparticles on the p-GaN top layer (using the nano-Ag solution with 10 ppm concentration), (**b**) schematic diagram of light output enhanced mechanisms of surface plasmon-enhanced LED (SPE-LED) in comparison with the conventional LED (C-LED), (**c**) typical current–voltage characteristics of SPE-LED samples fabricated with different concentrations of a nano-Ag solution, (**d**) light output power as a function of injection current for SPE-LED and C-LED. Reproduced from Ref. [90], with the permission of ACS publishing

[90]. Then, an 88.10% enhancement in the light output power is achieved (see Fig. 9.15d). Since the Ag nanoparticles are 200 nm away from the active region, then the enhancement is well attributed to the enhanced LEE. Note, Fig. 9.15c shows that the forward voltage for the LED will become dramatically high if the nano-Ag solution is 100 ppm. The very large forward voltage originates from the resistive $AgInO_2$, and therefore Chuang et al. optimize the nano-Ag solution to 10 ppm in their experiment.

Enhancing the LEE through LSP–TM light coupling becomes increasingly useful in the case of DUV LEDs which mainly have TM-polarized light emission. By means of using the metal Al thin layer for SP coupling, Gao et al. demonstrate the 217% enhancement in the photoluminescence intensity for the 294 nm AlGaN-based DUV LEDs [91]. However, the thin Al layer can only couple the TM-polarized light parallel to the surface of the device. For a TE-polarized light propagating in the direction parallel to the device surface, SP coupling does not occur. Worse still, when the thickness of the Al layer is 6 nm, the photon energy will be reduced to approximately 40% of its initial energy [92]. Then, Huang et al. further use the oblique-angle deposition method (see Fig. 9.16a) to deposit polygonal Al nanoparticles (see Fig. 9.16b–d) on the top surfaces of the DUV LEDs. The localized surface plasmon-enhanced AlGaN-based LEDs showed a maximum 5.1-fold bottom-emission EL enhancement at the wavelength of 287 nm (see Fig. 9.16e, f). Even more importantly, the proposed technique enables both the TM-polarized and the TE-polarized light to be effectively coupled with the LSP in the Al nanoparticles [92].

9.6 Summary

The technologies presented for extracting photons from III-nitride-based LED chips have been reviewed and summarized in this chapter. Various LEE-enhancing techniques, such as anti-reflective coating, surface roughening, photonic crystals, patterned substrate, surface plasmonics, and omnidirectional reflector, have been compared. Insightful physical interpretations for different LEE-enhancing approaches have also been revealed.

The highest LEE of InGaN/GaN-based blue LEDs has been up to 80% at the current stage [13]. Therefore the LEE is not a bottleneck that limits the EQE for blue LEDs, and as a result a 303 lm/W white LED lamp has ever been announced by Cree Company [93]. We believe that the most important issue for blue LED now lies on the efficiency droop and fabrication cost, to which we suggested more efforts have to be devoted.

Techniques that are effective to improve the LEE for InGaN/GaN-based blue LED are not always applicable for AlGaN-based DUV LEDs. One of the essentially critical strategies is to find the alternative material to remove the p-GaN layer, so that

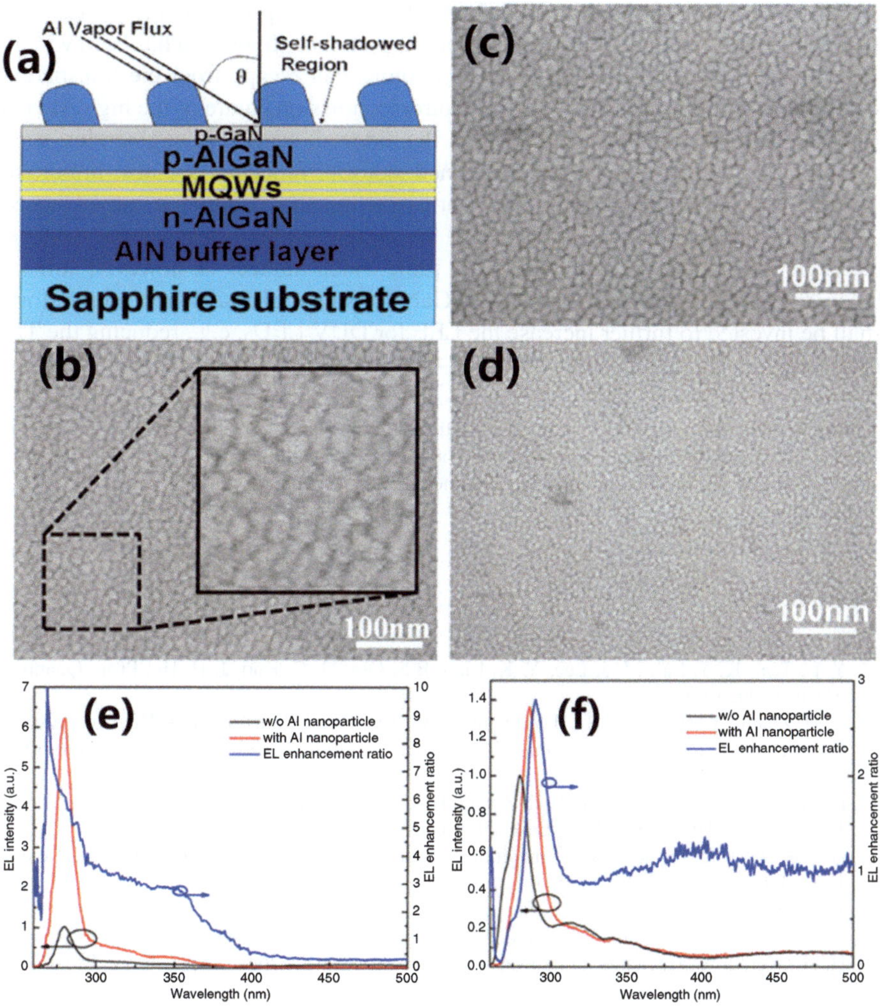

Fig. 9.16 Schematic illustration and top-view SEM images of the oblique-angle deposition of Al NPs. (**a**) Schematic illustration of oblique-angle deposition on complete structural deep UV LEDs; (**b–d**) top-view SEM images of Al NPs deposited on LEDs with deposition angles of 45°, 60°, and 75°, respectively; (**e**) top-emission EL spectra of DUV LED samples without Al NPs (black line) and with Al NPs deposited on the top surface at 60° (red line) as well as the enhancement ratio between them (blue line); (**f**) bottom-emission EL spectra of DUV LED samples without Al NPs (black line) and with Al NPs deposited on the top surface at 60 ° (red line) as well as the enhancement ratio between them (blue line). Reproduced from Ref. [92], with the permission of Nature Group

the optical absorption can be reduced and the excellent p-type ohmic contact can also be obtained simultaneously. At present, flip-chip configuration has been widely employed for DUV LEDs so that DUV photons can escape from the transparent sapphire. In addition, because of the anisotropic emission nature of the high Al-rich AlGaN quantum wells for DUV LEDs, the sidewall scattering engineering such as sapphire sidewall texturing and inclined sidewall reflectors are also important methods for the better TE and TM light extraction. By combining the advanced epitaxial growth technique on PSS with the current photon management designs, the EQE for DUV LED can reach 20% [28], which number is, however, still far below the EQE for the state-of-the-art InGaN/GaN blue LEDs. Hence, more efforts shall be invested to further increase the LEE for DUV LEDs, e.g., reducing the TM portion by managing the stress for the Al-rich AlGaN quantum wells.

Acknowledgments This work is supported by the National Natural Science Foundation of China (Nos. 61604051, 51502074), Natural Science Foundation of Tianjin City (No. 16JCQNJC01000), and Natural Science Foundation of Hebei Province (No. F2017202026).

References

1. Y.-L. Tsai, K.-Y. Lai, M.-J. Lee, Y.-K. Liao, B.S. Ooi, H.-C. Kuo, J.-H. He, Prog. Quantum Electron. **49**, 1 (2016)
2. Z.-H. Zhang, Y. Zhang, W. Bi, H.V. Demir, X.W. Sun, Phys. Status Solidi A **213**, 3078 (2016)
3. E.F. Schubert, *Light-Emitting Diodes*, 2nd edn. (Cambridge University Press, Cambridge, 2006)
4. L. Li, Y. Zhang, S. Xu, W. Bi, Z.-H. Zhang, H.-C. Kuo, Materials **10**, 1221 (2017)
5. J. Cho, E.F. Schubert, J.K. Kim, Laser Photonics Rev. **7**, 408 (2013)
6. G. Verzellesi, D. Saguatti, M. Meneghini, F. Bertazzi, M. Goano, G. Meneghesso, E. Zanoni, J. Appl. Phys. **114**, 071101 (2013)
7. J.-S. Park, J.K. Kim, J. Cho, T.-Y. Seong, ECS J. Solid State Sci. **6**, 42 (2017)
8. A. Khan, K. Balakrishnan, T. Katona, Nat. Photonics **2**, 77–84 (2008)
9. H. Hirayama, N. Maeda, S. Fujikawa, S. Toyoda, N. Kamata, Jpn. J. Appl. Phys. **53**, 100209 (2014)
10. S. Dang, C. Li, T. Li, W. Jia, P. Han, B. Xu, Optik **125**, 3623 (2014)
11. H.-Y. Ryu, I.-G. Choi, H.-S. Choi, J.-I. Shim, Appl. Phys. Express **6**, 062101 (2013)
12. Y.C. Shen, J.J. Wierer, M.R. Krames, M.J. Ludowise, M.S. Misra, F. Ahmed, A.Y. Kim, G.O. Mueller, J.C. Bhat, S.A. Stockman, P.S. Martin, Appl. Phys. Lett. **82**, 2221 (2003)
13. M.R. Krames, O.B. Shchekin, R. Mueller-Mach, G.O. Mueller, L. Zhou, G. Harbers, M.G. Craford, J. Disp. Technol. **3**, 160 (2007)
14. K.P. Streubel, J.K. Kim, H.W. Yao, J.Q. Xi, E.F. Schubert, E.F. Schubert, Proc. SPIE **6134**, 61340D (2006)
15. A.I. Zhmakin, Phys. Rep. **498**, 189 (2011)
16. J.-K. Liou, P.-C. Chou, C.-C. Chen, Y.-C. Chang, W.-C. Hsu, S.-Y. Cheng, J.-H. Tsai, W.-C. Liu, IEEE Trans. Electron Devices **61**, 831 (2014)
17. Y. Ma, B. Fan, Z. Chen, Y. Lao, L. Yan, X. Ma, Y. Zhuo, Y.L. Pei, G. Wang, IEEE Photon. Technol. Lett. **29**, 1564 (2017)
18. Y.K. Song, H. Zhou, M. Diagne, A.V. Nurmikko, R.P. Schneider, C.P. Kuo, M.R. Krames, R.S. Kern, C. Carter-Coman, F.A. Kish, Appl. Phys. Lett. **76**, 1662 (2000)
19. Y.-C. Chang, J.-K. Liou, W.-C. Liu, IEEE Electron Device Lett. **34**, 777 (2013)

20. K.-M. Huang, H.-J. Chang, C.-L. Ho, M.-C. Wu, IEEE Photon. Technol. Lett. **24**, 1298 (2012)
21. M. Kneissl, J. Rass, *III-Nitride Ultraviolet Emitters* (Springer, Berlin, 2016)
22. J.Q. Xi, M.F. Schubert, J.K. Kim, E.F. Schubert, M.F. Chen, S.Y. Lin, W. Liu, J.A. Smart, Nat. Photonics **1**, 176 (2007)
23. J.K. Kim, S. Chhajed, M.F. Schubert, E.F. Schubert, A.J. Fischer, M.H. Crawford, J. Cho, H. Kim, C. Sone, Adv. Mater. **20**, 801 (2008)
24. J.K. Kim, T. Gessmann, E.F. Schubert, J.Q. Xi, H. Luo, J. Cho, C. Sone, Y. Park, Appl. Phys. Lett. **88**, 013501 (2006)
25. A. Zhen, P. Ma, Y. Zhang, E. Guo, Y. Tian, B. Liu, S. Guo, L. Shan, J. Wang, J. Li, Appl. Phys. Lett. **105**, 251103 (2014)
26. N. Lobo, H. Rodriguez, A. Knauer, M. Hoppe, S. Einfeldt, P. Vogt, M. Weyers, M. Kneissl, Appl. Phys. Lett. **96**, 081109 (2010)
27. T. Inazu, S. Fukahori, C. Pernot, M.H. Kim, T. Fujita, Y. Nagasawa, A. Hirano, M. Ippommatsu, M. Iwaya, T. Takeuchi, S. Kamiyama, M. Yamaguchi, Y. Honda, H. Amano, I. Akasaki, Jpn. J. Appl. Phys. **50**, 122101 (2011)
28. T. Takano, T. Mino, J. Sakai, N. Noguchi, K. Tsubaki, H. Hirayama, Appl. Phys. Express **10**, 031002 (2017)
29. J.J. Wierer, A.A. Allerman, I. Montaño, M.W. Moseley, Appl. Phys. Lett. **105**, 061106 (2014)
30. Y. Zhang, R. Meng, Z.-H. Zhang, Q. Shi, L. Li, G. Liu, W. Bi, IEEE Photon. J. **9**, 2736642 (2017)
31. D.Y. Kim, J.H. Park, J.W. Lee, S. Hwang, S.J. Oh, J. Kim, C. Sone, E.F. Schubert, J.K. Kim, Light-Sci. Appl. **4**, e263 (2015)
32. J.W. Lee, J.H. Park, D.Y. Kim, E.F. Schubert, J. Kim, J. Lee, Y. Kim, Y. Park, J.K. Kim, ACS Photon. **3**, 2030 (2016)
33. J.W. Lee, D.Y. Kim, J.H. Park, E.F. Schubert, J. Kim, J. Lee, Y.-I. Kim, Y. Park, J.K. Kim, Sci. Rep. **6**, 22537 (2016)
34. P.-H. Fu, G.-J. Lin, H.-P. Wang, K.-Y. Lai, J.-H. He, Nano Energy **8**, 78 (2014)
35. K.-S. Im, C.-H. Won, S. Vodapally, D.-H. Son, Y.-W. Jo, Y. Park, J.-H. Lee, J. Cryst. Growth **441**, 41 (2016)
36. T. Fujii, Y. Gao, R. Sharma, E.L. Hu, S.P. Den Baars, S. Nakamura, Appl. Phys. Lett. **84**, 855 (2004)
37. H. Kim, K.K. Choi, K.K. Kim, J. Cho, S.N. Lee, Y. Park, J.S. Kwak, T.-Y. Seong, Opt. Lett. **33**, 1273 (2008)
38. J.H. Son, J.U. Kim, Y.H. Song, B.J. Kim, C.J. Ryu, J.L. Lee, Adv. Mater. **24**, 2259 (2012)
39. J. Ma, L. Wang, Z. Liu, G. Yuan, X. Ji, P. Ma, J. Wang, X. Yi, G. Wang, J. Li, Opt. Express **21**, 3547 (2013)
40. L.C. Wang, J. Ma, Z.Q. Liu, X.Y. Yi, H.W. Zhu, G.H. Wang, ACS Photon. **1**, 42 (2014)
41. G.-Y. Ha, T.-Y. Park, J.-Y. Kim, D.-J. Kim, K.-I. Min, S.-J. Park, IEEE Photon. Technol. Lett. **19**, 813 (2007)
42. C.H. Liu, R.W. Chuang, S.J. Chang, Y.K. Su, L.W. Wu, C.C. Lin, Mat. Sci. Eng. B **112**, 10 (2004)
43. Y.P. Hsu, S.J. Chang, Y.K. Su, S.C. Chen, J.M. Tsai, W.C. Lai, C.H. Kuo, C.S. Chang, IEEE Photon. Technol. Lett. **17**, 1620 (2005)
44. C.M. Tsai, J.K. Sheu, W.C. Lai, Y.P. Hsu, P.T. Wang, C.T. Kuo, C.W. Kuo, S.J. Chang, Y.K. Su, IEEE Electron Device Lett. **26**, 464 (2005)
45. C.-H. Ho, Y.-H. Hsiao, D.-H. Lien, M.S. Tsai, D. Chang, K.-Y. Lai, C.-C. Sun, J.-H. He, Appl. Phys. Lett. **103**, 161104 (2013)
46. S. Oh, P.-C. Su, Y.-J. Yoon, S. Cho, J.-H. Oh, T.-Y. Seong, K.-K. Kim, Opt. Express **21**, A970 (2013)
47. S. Oh, K.J. Lee, S.J. Kim, K. Ha, J. Jeong, D. Kim, K.-K. Kimb, S.-J. Park, Nanoscale **9**, 7625 (2017)
48. K.H. Lee, H.J. Park, S.H. Kim, M. Asadirad, Y.T. Moon, J.S. Kwak, J.-H. Ryou, Opt. Express **23**, 20340 (2015)

49. Y. Guo, Y. Zhang, J. Yan, H. Xie, L. Liu, X. Chen, M. Hou, Z. Qin, J. Wang, J. Li, Appl. Phys. Lett. **111**, 011102 (2017)
50. Y. Guo, Y. Zhang, J. Yan, X. Chen, S. Zhang, H. Xie, P. Liu, H. Zhu, J. Wang, J. Li, Appl. Phys. Express **10**, 062101 (2017)
51. A. David, H. Benisty, C. Weisbuch, Rep. Prog. Phys. **75**, 126501 (2012)
52. Y.C. Shin, D.H. Kim, E.H. Kim, J.-M. Park, K.-M. Ho, K. Constant, J.H. Choe, Q.H. Park, H.-Y. Ryu, J.H. Baek, T. Jung, T.G. Kim, IEEE J. Quantum Electron. **46**, 116 (2010)
53. M. Djavid, Z. Mi, Appl. Phys. Lett. **108**, 051102 (2016)
54. T.N. Oder, K.H. Kim, J.Y. Lin, H.X. Jiang, Appl. Phys. Lett. **84**, 466 (2004)
55. A. David, T. Fujii, R. Sharma, K. McGroddy, S. Nakamura, S.P. DenBaars, E.L. Hu, C. Weisbuch, H. Benisty, Appl. Phys. Lett. **88**, 061124 (2006)
56. C.-Y. Liu, C.-Y. Huang, P.-Y. Wu, J.-K. Huang, T.S. Kao, A.-J. Zhou, D.-W. Lin, Y.S. Wu, C.-Y. Chang, IEEE Electron Device Lett. **37**, 452 (2016)
57. J.J. Wierer, A. David, M.M. Megens, Nat. Photonics **3**, 163 (2009)
58. E. Yik-Khoon, P. Kumnorkaew, R.A. Arif, T. Hua, H. Zhao, J.F. Gilchrist, N. Tansu, IEEE J. Sel. Topics Quant. Electron. **15**, 1218 (2009)
59. D. Qing-an, L. Kang, K. Fanmin, C. Xinlian, Z. Jia, J. Disp. Technol. **10**, 909 (2014)
60. Q. Zhang, K.H. Li, H.W. Choi, IEEE Photon. Technol. Lett. **24**, 1642 (2012)
61. C.T. Kuo, L.H. Hsu, B.H. Huang, H.C. Kuo, C.C. Lin, Y.J. Cheng, Appl. Opt. **55**, 7387 (2016)
62. D.-W. Lin, J.-K. Huang, C.-Y. Lee, R.-W. Chang, Y.-P. Lan, C.-C. Lin, K.-Y. Lee, C.-H. Lin, P.-T. Lee, G.-C. Chi, H.-C. Kuo, J. Disp. Technol. **9**, 285 (2013)
63. S.X. Jiang, Z.Z. Chen, X.Z. Jiang, X.X. Fu, S. Jiang, Q.Q. Jiao, T.J. Yu, G.Y. Zhang, CrystEngComm **17**, 3070 (2015)
64. J.B. Kim, S.M. Kim, Y.W. Kim, S.K. Kang, S.R. Jeon, N. Hwang, Y.-J. Choi, C.S. Chung, Jpn. J. Appl. Phys. **49**, 042102 (2010)
65. Y.H. You, F.C. Chu, H.C. Hsieh, W.H. Wu, M.L. Lee, C.H. Kuan, R.-M. Lin, RSC Adv. **5**, 67809 (2015)
66. Y.J. Moon, D. Moon, J. Jang, J.Y. Na, J.H. Song, M.K. Seo, S. Kim, D. Bae, E.H. Park, Y. Park, S.-K. Kim, E. Yoon, Nano Lett. **16**, 3301 (2016)
67. D. Zubia, S.H. Zaidi, S.R.J. Brueck, S.D. Hersee, Appl. Phys. Lett. **76**, 858 (2000)
68. J.J. Chen, Y.K. Su, C.L. Lin, S.M. Chen, W.L. Li, C.C. Kao, IEEE Photon. Technol. Lett. **20**, 1193 (2008)
69. C.-C. Kao, Y.-K. Su, C.-L. Lin, J.-J. Chen, Appl. Phys. Lett. **97**, 023111 (2010)
70. Y. Zhang, T. Wei, J. Wang, D. Lan, Y. Chen, Q. Hu, H. Lu, J. Li, AIP Adv. **4**, 027123 (2014)
71. V.C. Su, P.H. Chen, R.M. Lin, M.L. Lee, Y.H. You, C.I. Ho, Y.-C. Chen, W.-F. Chen, C.-H. Kuan, Opt. Express **21**, 30065 (2013)
72. J.W. Pan, P.J. Tsai, K.D. Chang, Y.Y. Chang, Appl. Opt. **52**, 1358 (2013)
73. Z. Lin, H. Yang, S. Zhou, H. Wang, X. Hong, G. Li, Cryst. Growth Des. **12**, 2836 (2012)
74. Y. Cheng, L. Wang, Y. Zhang, H. Zheng, J. Ma, X. Yi, G. Wang, J. Li, ECS Solid State Lett. **2**, Q93 (2013)
75. W.C. Ke, F.W. Lee, C.Y. Chiang, Z.Y. Liang, W.K. Chen, T.Y. Seong, ACS Appl. Mater. Interfaces **8**, 34520 (2016)
76. C.H. Chiu, H.H. Yen, C.L. Chao, Z.Y. Li, P. Yu, H.C. Kuo, T.C. Lu, S.C. Wang, Appl. Phys. Lett. **93**, 081108 (2008)
77. K. Ueda, Y. Tsuchida, N. Hagura, F. Iskandar, K. Okuyama, Y. Endo, Appl. Phys. Lett. **92**, 101101 (2008)
78. S.J. An, Y.J. Hong, G.C. Yi, Y.J. Kim, D.K. Lee, Adv. Mater. **18**, 2833 (2006)
79. J. Kim, H. Woo, K. Joo, S. Tae, J. Park, D. Moon, et al., Sci. Rep. **3**, 3201 (2013)
80. P. Dong, J. Yan, J. Wang, Y. Zhang, C. Geng, T. Wei, P. Cong, Y. Zhang, J. Zeng, Y. Tian, L. Sun, Q. Yan, J. Li, S. Fan, Z. Qin, Appl. Phys. Lett. **102**, 241113 (2013)
81. L. Zhang, F. Xu, J. Wang, C. He, W. Guo, M. Wang, B. Sheng, L. Lu, Z. Qin, X. Wang, B. Shen, Sci. Rep. **6**, 35934 (2016)
82. K. Okamoto, I. Niki, A. Shvartser, Y. Narukawa, T. Mukai, A. Scherer, Nat. Mater. **3**, 601 (2004)

83. X. Gu, T. Qiu, W. Zhang, P.K. Chu, Nanoscale Res. Lett. **6**, 199 (2011)
84. A. Fujiki, T. Uemura, N. Zettsu, M. Akai-Kasaya, A. Saito, Y. Kuwahara, Appl. Phys. Lett. **96**, 043307 (2010)
85. M.-K. Kwon, J.-Y. Kim, B.-H. Kim, I.-K. Park, C.-Y. Cho, C.C. Byeon, S.-J. Park, Adv. Mater. **20**, 1253 (2008)
86. C.-H. Lu, C.-C. Lan, Y.-L. Lai, Y.-L. Li, C.-P. Liu, Adv. Funct. Mater. **21**, 4719 (2011)
87. C. Zhang, N. Tang, L. Shang, L. Fu, W. Wang, F. Xu, X. Wang, W. Ge, B. Shen, Sci. Rep. **7**, 2358 (2017)
88. J.-H. Sung, B.-S. Kim, C.-H. Choi, M.-W. Lee, S.-G. Lee, S.-G. Park, E.-H. Lee, O. Beom-Hoan, Microelectron. Eng. **86**, 1120 (2009)
89. J.-H. Sung, J.S. Yang, B.-S. Kim, C.-H. Choi, M.-W. Lee, S.-G. Lee, S.-G. Park, E.-H. Lee, B.-H. O, Appl. Phys. Lett. **96**, 261105 (2010)
90. S.H. Chuang, C.S. Tsung, C.H. Chen, S.L. Ou, R.H. Horng, C.Y. Lin, D.-S. Wuu, ACS Appl. Mater. Interfaces **7**, 2546 (2015)
91. N. Gao, K. Huang, J. Li, S. Li, X. Yang, J. Kang, Sci. Rep. **2**, 816 (2012)
92. K. Huang, N. Gao, C. Wang, X. Chen, J. Li, S. Li, X. Yang, J. Kang, Sci. Rep. **4**, 4380 (2014)
93. J. Cho, J.H. Park, J.K. Kim, E.F. Schubert, Laser Photonics Rev. **11**, 1600147 (2017)

Chapter 10
Enhancing Wall-Plug Efficiency for Deep-UV Light-Emitting Diodes: From Crystal Growth to Devices

SM Islam, Vladimir Protasenko, Shyam Bharadwaj, Jai Verma, Kevin Lee, Huili (Grace) Xing, and Debdeep Jena

10.1 Introduction

Sub-280-nm deep ultraviolet (DUV) photonic devices (Light-Emitting Diodes and Lasers) are in strong demand due to their applications in several areas including water disinfection, medical diagnostics, and security. Replacing mercury-based DUV lamps by semiconductor LEDs is environmentally friendly, enables miniaturization, and can ensure higher energy efficiency and portability. The family of high Al-containing $Al_xGa_{1-x}N$ alloy ($x > 0.5$) semiconductors is well suited for such LEDs. DUV emission from LEDs using AlGaN multi-quantum well light-emitting active region has been demonstrated in many reports [1–7]. Though AlGaN UV-C LEDs spanning the 210–280-nm spectral window have been demonstrated, they suffer from low external quantum efficiency (EQE) [7], which drops dramatically for wavelengths shorter than 240 nm.

The reasons for this efficiency drop are manifold. The LEDs are typically grown on nonnative substrates such as sapphire with a large lattice mismatch, leading to dislocation densities $\sim 1\times 10^{10}$ cm^{-2}. Non-radiative recombination in the quantum well active regions causes a significant decrease in the internal quantum efficiency (IQE). Spontaneous and piezoelectric polarization-induced internal electric fields in the polar AlGaN crystal spatially separate the electron and hole wavefunctions, reducing the overlap integral and the oscillator strength for interband radiative transitions [8]. This quantum-confined Stark effect (QCSE) reduces the IQE by lowering the radiative recombination rate.

SM Islam (✉) · V. Protasenko · S. Bharadwaj · K. Lee · H. (Grace) Xing · D. Jena
Cornell University, Ithaca, NY, USA
e-mail: smi29@cornell.edu; vvp7@cornell.edu

J. Verma
University of Notre Dame, Notre Dame, IN, USA

J. Li, G. Q. Zhang (eds.), *Light-Emitting Diodes*, Solid State Lighting Technology and Application Series 4, https://doi.org/10.1007/978-3-319-99211-2_10

Poor doping efficiency especially for p-AlGaN causes ohmic losses at the LED contact regions. Unlike the visible light emitters, the activation energy of acceptor-type dopants for Al-rich AlGaN alloys is $\gg$kT, inhibiting activation of dopants at 300K. Lack of suitable metal for ensuring ohmic contact further reduces the wall-plug efficiency. Low density of holes in the p-region requires higher operating voltages for the UV-LEDs, which further causes overflow of electrons in the p-region and therefore reduction of injection efficiency.

The extraction of generated photons from the LEDs causes further challenge in achieving high efficiency. The light extraction is limited by total internal reflection due to refractive index mismatch. Polarization property of the emitted light from the active region is also an important factor that decides the light extraction efficiency. Use of p-GaN on top of UV-LEDs to reduce the contact resistance to metal further reduces the light extraction by absorption.

In the subsequent sections, an organized approach to achieve efficient deep-UV emission with novel heterostructure design methods is presented. Section 10.2 gives an introduction to various loss mechanisms that limit the overall efficiency of deep-UV LEDs. This motivates the need of new heterostructure design to overcome the losses. Section 10.3 discusses the use of ultrathin GaN/AlN quantum heterostructures to enhance the internal quantum efficiency and the light extraction efficiency components from theoretical perspective. The crystal growth using plasma-assisted molecular beam epitaxy of ultrathin GaN/AlN quantum structures is presented in Sect. 10.4. The experimentally measured IQE enhancement of GaN/AlN compared to commonly used AlGaN-based design is presented. The use of polarization-induced doping to enhance injection and wall-plug efficiency is discussed in Sect. 10.5. The effectiveness of polarization-induced doping over conventional doping is manifested by cryogenic operation of a deep-UV LED. Finally, the enhancement of light extraction efficiency by the use of GaN/AlN quantum structures is discussed in Sect. 10.6.

10.2 UV-LED Efficiency Components

The efficiency of the best reported UV LEDs so far is ~20% for ~280-nm emission [9], which lags far behind their visible counterparts. The efficiency decreases dramatically at shorter wavelengths [10]. A summary of the reasons behind low efficiency of UV LEDs is presented here. The performance metric of UV LEDs is the external quantum efficiency (EQE) which is composed of three components:

$$\eta_{\mathrm{EQE}} = \eta_{\mathrm{IQE}} \times \eta_{\mathrm{INJ}} \times \eta_{\mathrm{EXTR/LEE}} \tag{10.1}$$

where

$$\eta_{\mathrm{EQE}} = \frac{rate\ of\ emitted\ photons}{rate\ of\ carriers\ injected\ into\ LED}$$

$$\eta_{\mathrm{IQE}} = \frac{rate\ of\ generated\ photons}{rate\ of\ carriers\ injected\ into\ active\ region}$$

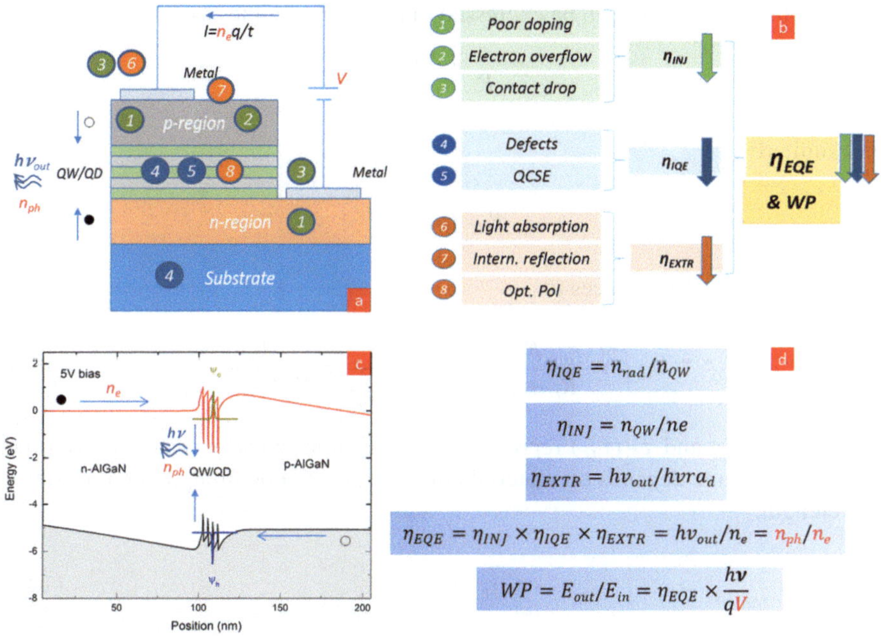

Fig. 10.1 Summary of loss mechanisms in DUV LEDs: (**a**) schematic showing locations where losses occur (V: applied bias, I: current, n_e =number of injected electrons, n_{ph} =number of emitted photons, (**b**) description of losses, (**c**) energy band diagram for a forward-biased LED to visualize the loss mechanisms, (**d**) efficiency equations, n_{rad} =No. of electrons recombining radiatively, n_{QW} =electrons captured in the QW, hν_{out} =released photon energy, hν_{rad} =generated photon energy, WP=Wall-plug efficiency (Output power/Input power)

$$\eta\text{INJ} = \frac{\textit{rate of carriers injected into active region}}{\textit{rate of carriers injected into LED}}$$

$$\eta\text{EXTR/LEE} = \frac{\textit{rate of emitted photons}}{\textit{rate of generated photons}}.$$

Each of these efficiency components is briefly discussed in the following in context to DUV photonic devices (Fig. 10.1).

10.2.1 Internal Quantum Efficiency (IQE)

The internal quantum efficiency (IQE) is defined as the ratio of rates of radiative recombination to total recombination:

$$\eta\text{IQE} = \frac{R_r}{R_r + R_{nr}}. \tag{10.2}$$

The recombination rates can be expressed in terms of lifetimes as:

$$R_r = \frac{1}{\tau_r}\ , \tag{10.3}$$

and similarly,

$$R_{nr} = \frac{1}{\tau_{nr}}\ . \tag{10.4}$$

Therefore, Eq. (10.4) reduces to:

$$\eta_{\mathrm{IQE}} = \frac{R_r}{R_r + R_{nr}} = \frac{\frac{1}{\tau_r}}{\frac{1}{\tau_r} + \frac{1}{\tau_{nr}}} = \frac{1}{1 + \frac{\tau_r}{\tau_{nr}}} \tag{10.5}$$

where R_r (R_{nr}) and τ_r (τ_{nr}) represent radiative (non-radiative) rate and lifetime. Radiative lifetime depends on material constants, oscillator strength, and operating wavelength:

$$\tau_r \propto \frac{\lambda^2}{f_{\mathrm{osc}}}\ . \tag{10.6}$$

Non-radiative recombination through defects is often an Arrhenius function of temperature. This is expressed as:

$$\tau_{nr} = \tau_{nr0} e^{\frac{E_a}{kT}}\ . \tag{10.7}$$

Using this, IQE is expressed as:

$$\eta_{\mathrm{IQE}} = \frac{1}{1 + ae^{-\frac{E_a}{kT}}} \tag{10.8}$$

where a = $\frac{\tau_r}{\tau_{nr0}}$ and E_a is the activation energy for non-radiative recombination centers.

The lifetime represents the average time carriers spend before they recombine either radiatively or non-radiatively. If the radiative lifetime is much shorter, that means an electron radiatively recombines with a hole before it recombines through a defect non-radiatively. In order to have a high $\eta_{\mathrm{IQE}} \sim 1$, the necessary condition is $\tau_r \ll \tau_{nr}$ (Fig. 10.2).

The IQE is also explained with the ABC model [12] by the following equation:

$$\eta_{\mathrm{IQE}} = \frac{Bn}{A + Bn + Cn^2} \tag{10.9}$$

where A is the monomolecular non-radiative recombination coefficient, B is the bimolecular radiative recombination coefficient, C is the Auger recombination coefficient, and n is the electron concentration. SRH recombination process is usually caused by defects or dislocations in the crystal. Auger recombination is

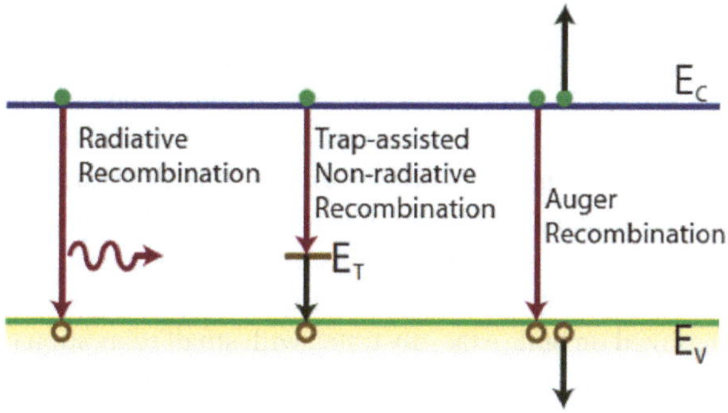

Fig. 10.2 Schematic energy band diagram showing various recombination mechanisms in an LED [11]

only considered at high carrier densities ($>1\times10^{19}$ cm^{-3}). Plenty of studies have been performed on elimination of the efficiency droop in visible LEDs [13–15]. Chichibu et al. have shown that dislocations act as non-radiative centers (NRCs) [16]. However, InGaN-based active region's localization effect reduces trapping of carriers into threading dislocations causing efficient photon emission. Dai et al. have also used photoluminescence to extract the IQE dependence on the dislocation density in the epi-layer [17]. They concluded that the threading dislocations reduce the IQE of the InGaN-based visible light emitters. Similar techniques have been applied to AlGaN-based active region for IQE estimation.

For deep-UV LEDs, the choice of the substrate is limited to AlN due to favorable bandgap. At present, AlN is available both as templates on a foreign substrate or as bulk single crystals. AlN on sapphire (Al_2O_3) is the most commonly used templates for deep-UV LEDs. Due to large lattice mismatch of 13.3% [18] between AlN and sapphire, the templates have threading dislocation density (TDD) of $\sim10^{10}$cm^{-2}. There are several techniques to further reduce the dislocation in the AlN film. For example, Hirayama et al. have used epitaxial lateral overgrowth (ELO) method to reduce the TDD down to 3×10^8cm^{-2} [19, 20]. Jmerik et al. have employed AlGaN/AlN superlattice and insertion of ultrathin GaN layers in the AlN to reduce the TDD to $\sim6\times10^9$cm^{-2} [21]. To achieve low TDD, single crystal bulk AlN is the most desired substrate. Using physical vapor transport (PVT) crystal growth technique, single crystal bulk AlN substrate with a TDD as low as 10^4cm^{-2} has been achieved [22].

10.2.1.1 Threading Dislocations

Internal quantum efficiency (IQE) decreases due to high densities of threading dislocations (TDs, τ_{nr} is shortened) and due to the quantum-confined stark effect (QCSE, τ_r becomes large). In the presence of high ($>10^9$ cm^{-2}) dislocation densities, the IQE decreases rapidly. The effect of TDs on the reduction of IQE is discussed in Speck et al. [23] and Kneissl et al. [10]. Because dislocations act as non-radiative recombination centers, they increase the rate of the non-radiative SRH process in LEDs (increase the A coefficient value).

Increasing IQE with low TDD motivates the use of high-quality bulk AlN substrates for UV applications. Unfortunately, such substrates are extremely expensive compared to AlN templates and currently have significant absorption in the deep-UV range because of carbon-based defects which absorb emission shorter than 265 nm [24]. Continuous improvement of bulk crystal growth techniques is helping to reduce this sub-bandgap absorption, and bulk crystals will eventually replace AlN templates on sapphire for deep-UV applications specifically for lasers. Recently, the growth of thick (~100s of nm) AlGaN with Al composition of 65% has been realized on bulk AlN substrates with small relaxation (<10%) [25]. These promising results motivate the use of bulk AlN for LED structures with low dislocation densities.

10.2.1.2 Quantum-Confined Stark Effect (QCSE)

The other factor leading to low IQE is the reduction of the oscillator strength by QCSE (Fig. 10.3). III-nitride heterostructures are commonly grown on polar c-plane AlN surface. This results in built-in polarization charges at the hetero-interfaces due to both spontaneous and piezoelectric components. The polarization charges result in electric fields in the active region (i.e., quantum wells), causing the spatial separation of the electron and hole wavefunctions and reduction of their wavefunction overlap. This also reduces the oscillator strength which is expressed as:

$$f_{\text{osc}} \propto \lambda \times \mid \int \psi_e \psi_h \mid; \qquad (10.10)$$

The integral in this equation represents wavefunction overlap. Since the IQE is proportional to the oscillator strength, it is reduced because of QCSE.

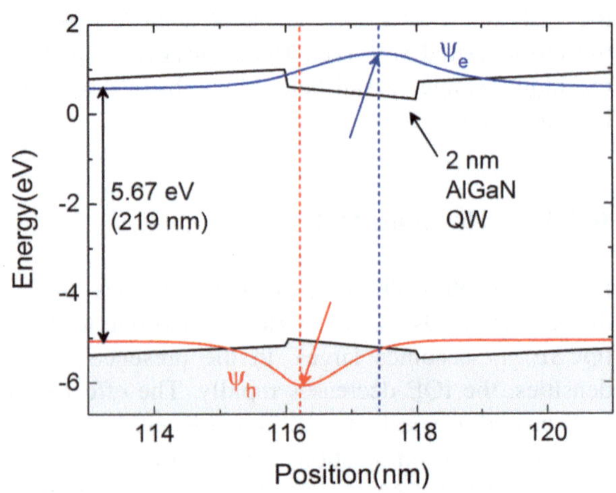

Fig. 10.3 QCSE in thick AlGaN QWs. Separation of electron and hole wavefunction is shown explicitly using energy band diagram

This undesirable reduction in the oscillator strength due to QCSE can be avoided by using nonpolar/semipolar-oriented substrates, where the polarization charges are low, or simply absent. Alternatively, for growth on polar substrates, the use of very thin quantum structures can also be very effective in solving this problem. The ultrathin quantum structures enhance the overlap significantly and therefore can boost the IQE even in heterostructures on polar substrates.

10.2.2 Injection Efficiency (INJ) and Wall-Plug Efficiency (WPE)

Wall-plug efficiency (WPE) is a measure of useful light output power to input electrical power and is limited by ohmic losses in injection regions of the LED. Optimal injection regions are transparent to light emission, requiring an aluminum content greater than 65% for ~250-nm emitters, and leads to high dopant activation energies [4]. Difficulty in incorporating dopants especially acceptor type into these wide-bandgap AlGaN cladding regions compounded with low activation of dopant impurities leads to high sheet resistances and contact resistances and hence large ohmic losses. For high-aluminum composition, n-cladding regions transparent to ~260-nm emission, sheet resistances as low as 0.0075 Ωcm, and contact resistances of $\sim 1 \times 10^{-6}$ Ωcm^2 have been reported [26, 27] as measured by the transmission line method (TLM) and Hall effect. However, p-cladding regions exhibit conductivities that are at best two orders of magnitude lower [28–36], reducing the WPE for 270-nm emitters to ~6% [37].

A strategy widely used to enhance dopant activation and thus WPE is polarization-induced doping. Polarization-induced doping utilizes the inherent spontaneous and piezoelectric polarization properties of the wurtzite crystal structure to ionize dopants more effectively [38]. As shown in Fig. 10.4, polarization-induced doping can enhance conductivity greatly over impurity doping alone. With proper tuning of well and barrier thickness in superlattices, enhanced vertical and lateral transport in p-type layers can be achieved.

Simply enhancing carrier transport is not enough to enhance EQE for LEDs. Equally necessary is capturing the carriers in the active region so that radiative recombination may occur. Injection efficiency (INJ) is a measure of the proportion of carriers in the active region versus total carriers generated. Use of quantum structures such as quantum dots/disks and wells provides carrier confinement, though carriers may still escape beyond the active region into the opposite cladding region. Generally, due to their lower effective mass, electrons are most likely to escape beyond the active region. Use of an abrupt heterojunction electron blocking layer or compositionally graded p-type cladding region can help eliminate carrier escape and enhance the injection efficiency [39].

Another consequence of using Al-rich AlGaN for the p/n regions is the formation of Schottky contact between metal and the doped region. Due to wide bandgap of

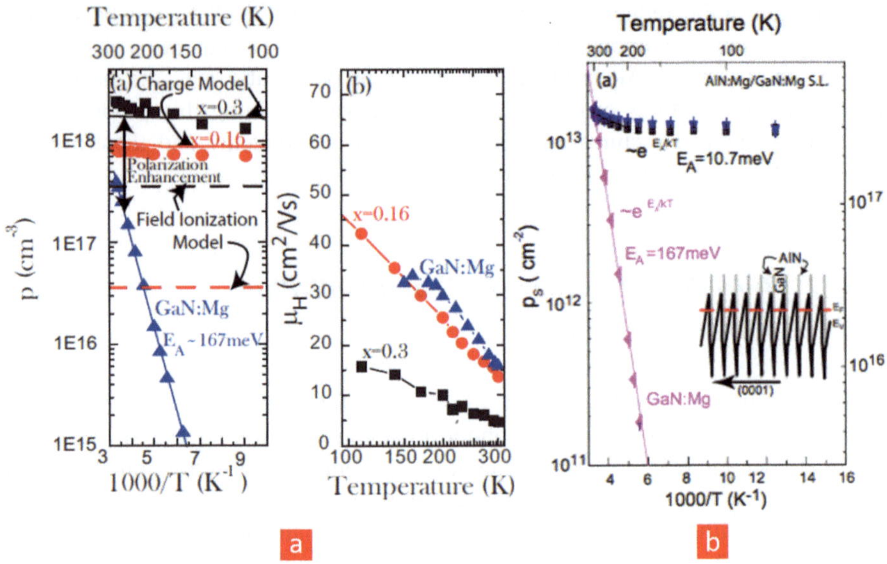

Fig. 10.4 (**a**) Temperature-dependent Hall data for a polarization-doped compositionally graded p-GaN sample on N-polar substrate, (**b**) sheet charge measured by Hall effect for one constant composition GaN sample (pink dots) and two superlattice samples (blue and black dots) showing the lack of freeze-out due to polarization-induced 2DEGs [38]

these AlGaN alloys, metals with desired work functions that can form low-resistance ohmic contacts to either n- or p-type AlGaN are not available. This introduces further increase in undesirable voltage drop at the contacts.

As a solution to these problems, polarization-induced doping is proposed for the n and p regions of deep-UV LEDs. It will be demonstrated next that this technique is very effective in boosting the carrier densities, even at cryogenic temperature. Since the polarization is a very weak function of temperature, the use of polarization-induced doping in deep-UV LED structures has made it possible to observe the operation of the devices at cryogenic temperatures. The heterostructures used for polarization-induced doping also reduce the electron overflow without introducing additional potential barriers for hole injection into the active region. Consequently, the voltage drop and loss outside the active region are reduced and both the INJ and WPE are enhanced.

10.2.3 Light Extraction Efficiency (LEE)

Once light is generated within the active region of an LED, it must escape in order to be collected for any useful application. Light extraction for deep-UV LEDs poses a challenge due to the high refractive index differences between the semiconductor–air interfaces and absorbing contact layers. Such a large difference promotes total internal reflection and reduces the escape cone down to $\sin^{-1}(n_0/n_s) = 23°$ for a

GaN/air interface. The photons emitted from the Al-rich AlGaN alloy are mostly TM polarized. The TE component of emitted light decreases, as the Al composition increases. Above ~70% AlGaN composition, the emission is predominantly TM polarized. This means that the light is emitted perpendicular to the growth direction towards the side of the wafer which poses an added challenge for efficient light extraction. Clever active region design can promote out-of-plane (TE) emission over TM emission [6], though total internal reflection and absorption still limit LEE in conventional devices to ~9% [9]. Light that is total-internally reflected may escape after several round trips, but absorbing defects, metal contact pads, and p-GaN contact layers make such a process unlikely. Light extraction efficiency (LEE) for 260-nm LEDs is a bottleneck for LED EQE, remaining around 10% [9] prior to specific preparations for enhancing light extraction.

Several methods are available for enhancing light extraction. Surface roughening promotes light extraction by changing the angle at which photons reach the semiconductor–air interface. Roughening can be achieved through dry or wet etching or polishing, and can enhance LEE by >40% (demonstrated at 326 nm by Zhang et al. [40]). For flip-chipped LEDs, light extraction can be enhanced through the use of reflective contacts (such as Rh or Al for 275-nm diodes) [9] or dielectric mirror stacks which serve to reflect light that travels up towards the p-layers back to the substrate (or opposite if not flip-chipped). Such a method can be combined with patterning of contact layers to reduce absorbance [4].

Use of p-AlGaN contact layers in place of commonly used p-GaN can reduce absorbance and enhance LEE [41]. Modifying the shape of the LED itself through truncated inverted pyramid (TIP) design has been shown to improve LEE by redirecting light towards the surface more effectively (demonstrated at visible wavelengths by Krames et al. [42]).

In addition to these methods, LED packaging serves to further improve light extraction. Encapsulation with refractive index matched to the semiconductor will allow all emitted light to be coupled into it, and if a spherical shape is chosen, all light will have normal incidence to the surface preventing total internal reflection. The best results to date for wavelengths of 275 nm use a combination of flip-chip packaging with resin, a reflective Rh p-electrode, patterned sapphire substrates, and p-AlGaN contact layers to achieve LEE of ~50%, with device structure shown in Fig. 10.5 [9].

LEE can be improved by using ultrathin GaN quantum structures as the deep-UV emitting active region instead of Al-rich AlGaN. Since GaN has TE polarized emission property, it is expected to be more effective in enhancing the light extraction compared to Al-rich AlGaN alloy active regions.

Based on the discussion on different efficiency components, ultrathin GaN/AlN superlattice is a promising candidate to replace the AlGaN-based active regions to enhance the IQE (by reduction of QCSE) and the LEE (by enhanced TE emission). The following sections of the chapter discuss such efficiency enhancement with ultrathin GaN/AlN quantum structures.

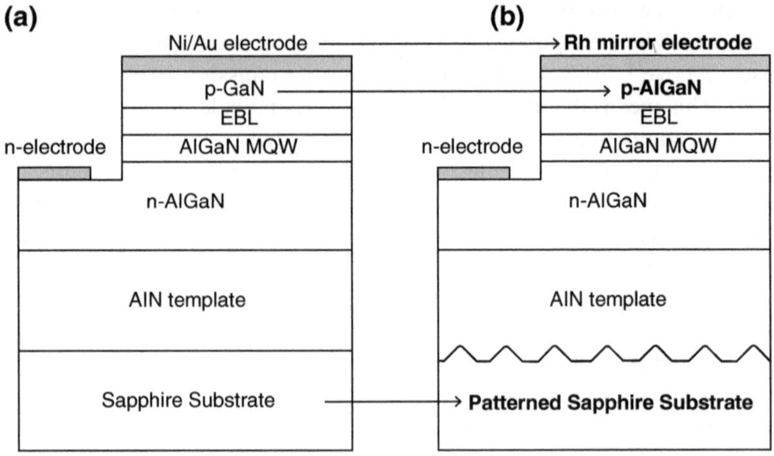

Fig. 10.5 (**a**) Conventional LED structure, (**b**) UV LED structure with patterned sapphire substrate, p-AlGaN contact layer, and Rh mirror electrode using flip-chip geometry and encapsulant resin (not shown) for improved light extraction [9]

10.3 Deep-UV Photon Emission from Extreme Quantum-Confined GaN/AlN Heterostructures

The conventional design of DUV photonic devices (LEDs and LASERs) employs high Al-containing $Al_xGa_{1-x}N$ alloys as the light-emitting active region material. The benefit of using $Al_xGa_{1-x}N$ is the tunability of the emission wavelength over a wide range from 200 nm (x:1.0, 6.2 eV) to 360 nm (x:0.0, 3.4 eV) encompassing the UV-A (400-320 nm), UV-B (320-280 nm), and the UV-C (280-200 nm) spectral regimes. But as the shorter wavelengths (<250 nm) are approached, the AlGaN-active region-based design suffers from several challenges. This section discusses a plausible alternative: use of ultrathin GaN as the active region material to overcome these challenges.

10.3.1 Motivation of Using GaN Instead of AlGaN

To date, sub-280-nm LEDs suffer from very low external quantum efficiency (<10%) [7] due in part to material-based challenges as well as physical challenges. Material-based challenges include high dislocation densities (10^{10} cm^{-2} in AlN and AlGaN layers) [43, 44] which lower EQE by diverting electrons from the active region, and high resistivities in the n- and p-AlGaN cladding regions [11, 40, 45–47] which incur large ohmic losses. Physical challenges for conventional AlGaN QW structures include severe valence band-mixing [48–52] and quantum-confined Stark effect (QCSE) mediated lowering of wavefunction overlap [8, 53, 54].

Specifically, studies have shown that transverse-magnetic (TM)-polarized emission dominates for high Al-containing (> 68%, $\lambda \sim$ 220–230 nm) AlGaN QWs due to a dominant crystal-field split-off (CH) sub-band [48]. TE-polarized spontaneous emission is favored over TM-polarized when Al composition is decreased, as the heavy hole (HH) sub-band gradually moves towards the higher energy light hole (LH) and CH sub-bands, causing conduction band to heavy hole (C-HH) transitions to occur with similar probability as C-LH and C-CH transitions around $\lambda \sim$ 250–300 nm. In this manner, a TE-polarized spontaneous emission rate (TE-R_{sp}) greater than TM-R_{sp} can be obtained, though both are still low at these wavelengths for two reasons: (1) the three valence sub-bands HH, LH, and CH are nearly degenerate leading to low populations in all sub-bands, and (2) QCSE causes a significant spatial separation between electron and HH sub-bands, leading to low wavefunction overlap and low recombination rates. It is therefore apparent that conventional AlGaN QW structures are not sufficient for realizing high-efficiency high-power mid-UV emitters.

Zhang et al. [55] have proposed a novel active region design of AlN-delta-GaN QW to replace the conventional AlGaN QW structures, so as to enhance TE-polarized emission at mid-UV wavelengths. Insertion of a delta-GaN layer in the active region provides two advantages: (1) strong valence band mixing is introduced, restoring the HH sub-band as the highest energy sub-band, leading to high TE-R_{sp} at $\lambda \sim$ 250–300 nm, and (2) QCSE is mitigated due to strong electron and hole wavefunction localization and overlap in the delta-GaN QW.

10.3.2 Evolution of Band Structure of Ultrathin AlN/GaN/AlN Heterostructures

To provide further insight, the band structure of ultrathin GaN/AlN superlattices is analyzed in this section followed by a comparison between GaN/AlN and AlGaN/AlN valence band structure. Basic 8×8 k.p theory fails to provide accurate estimates of the bandgap of GaN QWs/AlN barriers due to non-parabolicity of GaN's band structure and extreme quantum confinement. For this reason, many-body perturbation theory is employed to predict both electronic and optical bandgaps, including excitonic effects. The analysis proceeds as follows: first, the band structure of bulk GaN is calculated, followed by a calculation of the quantum confinement effect through inserting ultrathin layers (1 and 2 monolayers, or MLs) of GaN into an AlN matrix. Figure 10.6 summarizes the three calculations. The band structure of bulk GaN (Fig. 10.6a) follows expectations with a direct gap of 3.37 eV. The 2-ML GaN/AlN heterostructure is calculated to have a gap of 4.65 eV (266 nm), which increases to 5.45 eV for the 1-ML GaN/AlN heterostructure. Both gaps remain direct which is encouraging for use as DUV light emitters [56].

Valence band (VB) structure renormalization in GaN versus in AlGaN is looked at in more detail here. As previously mentioned, low Al-containing (<30% Al)

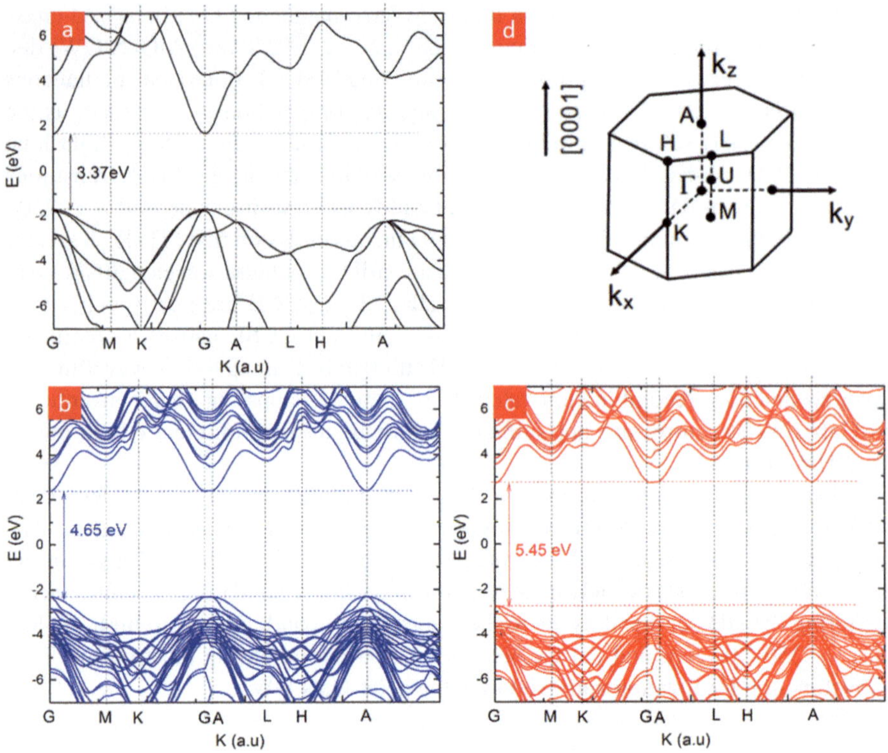

Fig. 10.6 First principle band structure calculation for (**a**) bulk GaN, (**b**) 2-ML GaN/AlN, and (**c**) 1-ML GaN/AlN superlattices. (**d**) The conventional high symmetry points of the band structure. Provided by Dr. Dylan Bayerl and Prof. Emmanouil Kioupakis

AlGaN heterostructures are used for UV-A and UV-B emitters. The VB structure for an AlN/$Al_{0.2}Ga_{0.8}N$ QW for such UV-A and UV-B emitters is shown in Fig. 10.7a. The HH, LH, and CH sub-bands that comprise the VB are indicated. Optical transitions occur between the conduction band (CB) minimum and VB maximum, which in the case of an AlN/$Al_{0.2}Ga_{0.8}N$ QW is the HH1 sub-band. Characteristics of the HH band necessitate TE-polarized emission from this transition. In contrast, Fig. 10.7b describes VB structure for an AlN/$Al_{0.8}Ga_{0.2}N$ QW, in which the CH sub-band is now the VB maximum. The optical transition occurring here is TM-polarized instead of TE-polarized. These high compositions of AlGaN are required for UV-C emission of <280 nm, which at the moment, constitutes the most demanding practical application.

Figure 10.7c shows the VB structure of an AlN/GaN heterostructure to compare to AlN/AlGaN heterostructures. GaN well thickness is held at 0.5 nm (2 monolayers) for this calculation, corresponding to a bandgap of 4.65 eV. In this scenario, the HH1 sub-band is the VB maximum, meaning that TE-polarized emission will be preferred. Such an emission polarization is advantageous for light extraction, and will further be discussed in Sect. 10.6.

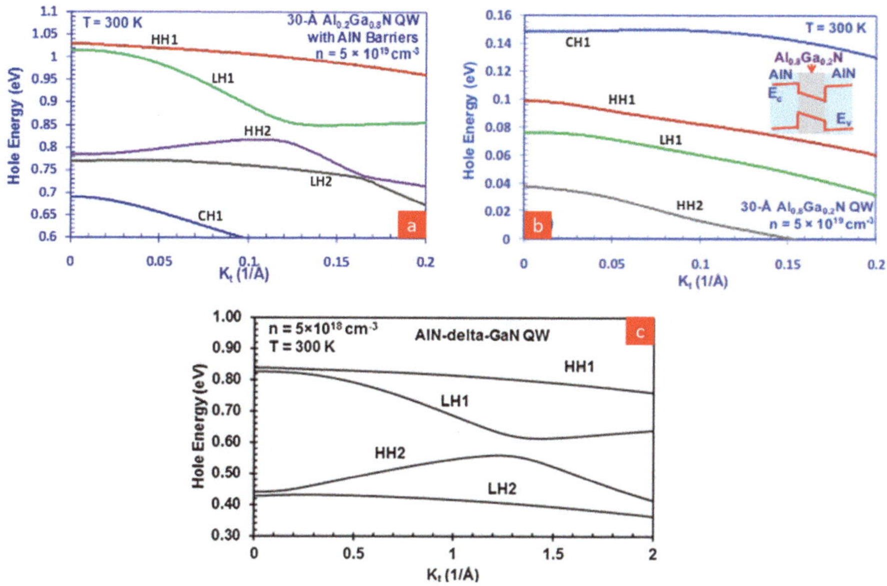

Fig. 10.7 Valence band structure renormalization using first principle band structure calculation for: (**a**) 3 nm $Al_{0.2}Ga_{0.8}N$/AlN QW, (**b**) $Al_{0.8}Ga_{0.2}N$/AlN, and (**c**) 0.5 nm GaN/AlN QW [55]. Reproduced from Appl. Phys. Lett. 98, 171111 (2011), with the permission of AIP Publishing

10.3.3 Achievable Wavelengths and Inhomogeneous Broadening: Theory vs Experiment

A combined approach of theoretical analysis along with experimental work is presented in this section that explores the range of wavelengths GaN/AlN quantum structures are capable of realizing in the DUV spectral region. The effect of change of GaN quantum well thickness and AlN barrier thickness is explored by both theory and experiments. First principle methods based on density functional and many-body perturbation theory were used to calculate the optical and electronic properties [56]; the GW method was used to find electronic gaps for 1- and 2-ML GaN/AlN heterostructures with varying AlN thickness, and the Bethe–Salpeter equation (BSE) method using Quantum ESPRESSO [57], BerkeleyGW [58], and Wannier90 [59] codes were used to find optical gaps. Experimental analysis was carried out through growing a series of samples via plasma-assisted molecular beam epitaxy (MBE). Samples were structurally characterized using high-resolution X-ray diffraction (HR-XRD) along with scanning transmission electron microscopic (STEM) analysis. Photoluminescence (PL) measurements were then performed for optical characterization.

The theoretical analysis is summarized in Fig. 10.8. A structural schematic, energy band diagram, and calculated electronic and optical bandgap is shown for 2 different GaN QW thicknesses (1 and 2 MLs) and AlN barrier thickness ranging

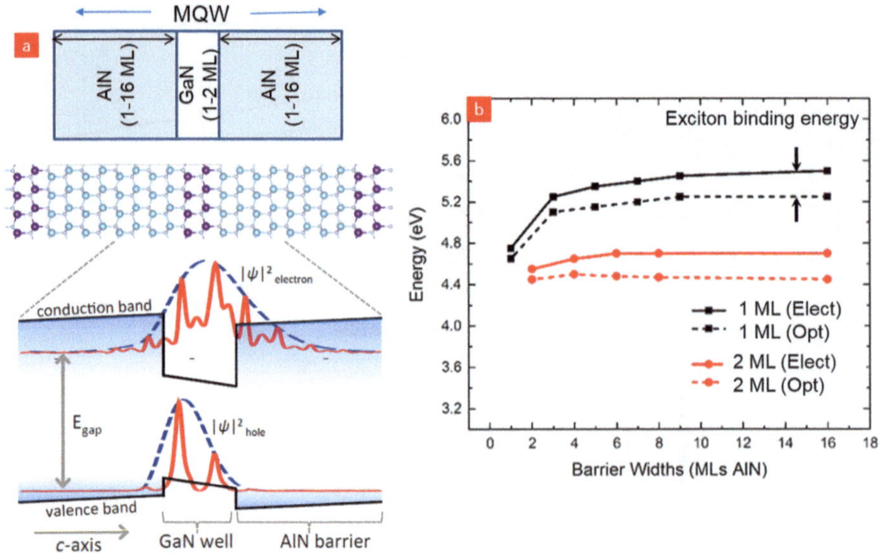

Fig. 10.8 First principle calculation of optical and electronic gap for ultrathin GaN/AlN superlattices: (**a**) simulated structure and the envelope function, (**b**) summary of calculated gap [56]. Reproduced from Appl. Phys. Lett. 109, 241102 (2016), with the permission of AIP Publishing

from 1–9 MLs. Data from 10–16 MLs AlN thickness have been extrapolated. Based on the theoretical prediction, it can be seen that deep-UV emission at ~5.2 eV and ~4.5 eV can be attained with 1-ML and 2-ML GaN QWs, respectively. Beyond 8-ML AlN barrier thickness, the effective bandgap is relatively constant.

Experimental data is summarized in Fig. 10.9 which includes the schematic structure of the MBE-grown samples, structural characterization by HR-XRD and STEM, and optical characterization by PL. The XRD in Fig. 10.9b shows a secondary GaN peak, which is a signature of GaN/AlN superlattice formation. Simulated XRD data agrees with the measured data suggesting that the targeted thicknesses were achieved. STEM analysis in Fig. 10.9c further confirms that ultrathin GaN layers with the desired thicknesses were present within the AlN matrix. From PL data measured at 300K and 77K, the sample with 1-ML GaN QWs had an optical gap around 5.5 eV, while the 2-ML GaN QW sample had a gap around 4.8 eV. Slight blue shift consistent with Varshni's law could be observed at the lower temperature.

Figures 10.8 and 10.10 compare and contrast the experimental data with theoretical simulation. Measured PL results agreed well with theory, especially for 1-ML GaN QWs, with discrepancy below 150 meV (Fig. 10.10). A slightly larger deviation was observed for the 2-ML GaN structures, potentially due to nonuniformity in the well thickness in the plane of the sample.

PL spectra in Fig. 10.11 show significant broadening (0.3–0.4 eV) due to carrier occupation and structural disorder. The occupational structural broadening

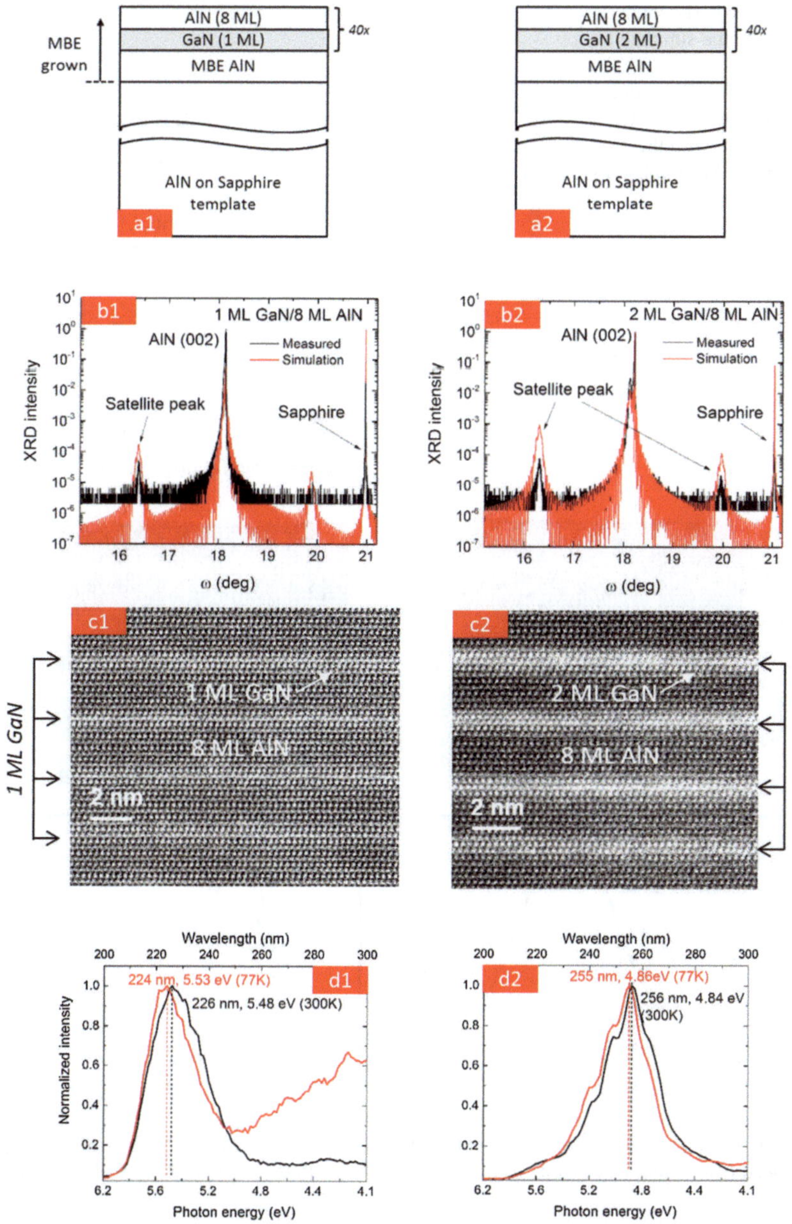

Fig. 10.9 Experimental verification for the simulated data: (**a**) structural layout, (**b**) XRD, (**c**) TEM characterization, and (**d**) PL measurements for some representative samples [56]. Reproduced from Appl. Phys. Lett. 109, 241102 (2016), with the permission of AIP Publishing

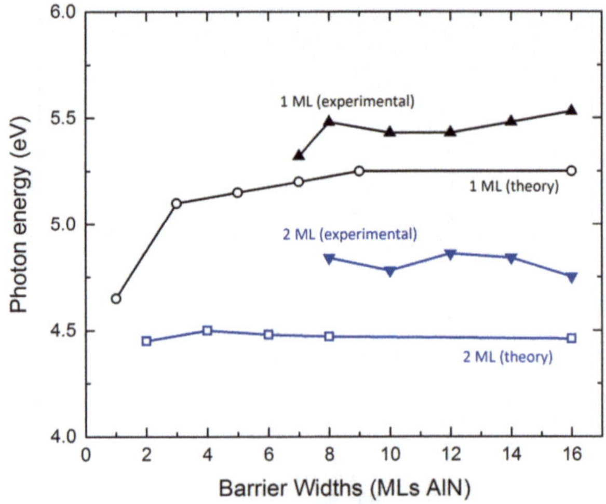

Fig. 10.10 Experimental data added to the simulated data of Fig. 10.8b for comparison [56]. Reproduced from Appl. Phys. Lett. 109, 241102 (2016), with the permission of AIP Publishing

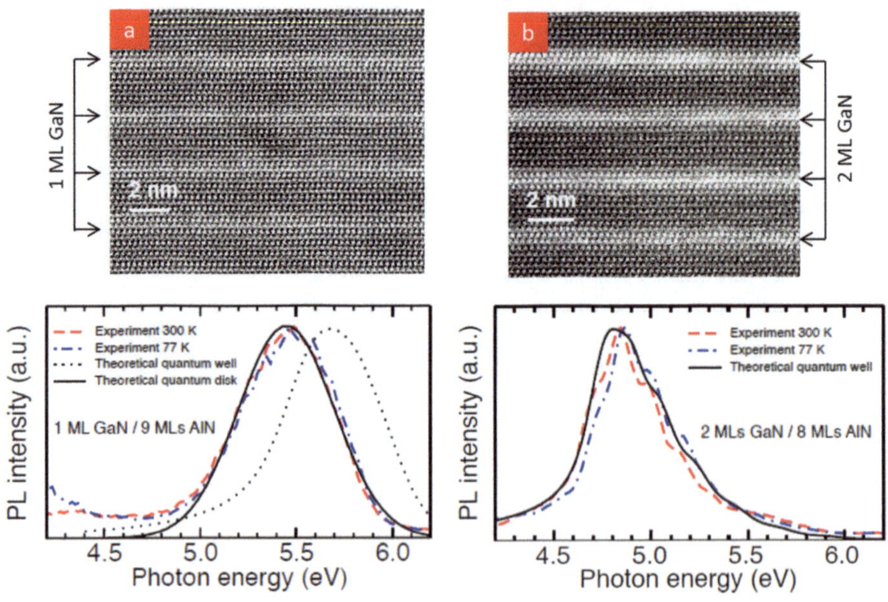

Fig. 10.11 Inhomogeneous broadening of the PL spectra: for (**a**) 2-ML GaN, (**b**) 1-ML GaN [56]. Reproduced from Appl. Phys. Lett. 109, 241102 (2016), with the permission of AIP Publishing

parameter has been estimated for 2-ML GaN QWs to be 0.22–0.32 eV from the slope of the logarithm of the high-energy PL tail, but estimation could not be carried out for 1-ML GaN structures as there was no well-defined linear region in the PL spectra.

Additionally, modeled spectra were compared with the experimental spectra for structures with 8–9-ML AlN barriers. Modeling was carried out with the Shockley van Roosbroeck (SvR) equation. With proper choice of fitting parameters, excellent agreement for the 2-ML GaN structure between the SvR model and experiment was obtained both for lineshapes and peak positions. In contrast, regardless of choice of fitting parameters, the SvR model incorrectly predicted peak position (much higher emission energy than experiment) for 1-ML GaN structures, though the lineshape was accurate. The poor fit suggested that 1-ML GaN quantum structures did not exhibit well-like character, but rather luminesced as quantum disks with homogenously broadened emission at the electronic gap energy [60]. A very accurate fit was obtained by altering the 1 ML GaN structures as disks.

The peak position and lineshape predicted by the quantum disk model are governed by the electronic gap and broadening parameter. Broadening can be attributed to fractional GaN ML formation [61]. Any deviation from ideal monolayer formation leads to fractional ML formation, as observed in InN/GaN heterostructures [62], explaining the different PL characteristics of 1-ML and 2-ML GaN structures. Growth kinetics of ultrathin GaN lead to fractional instead of full coverage for 1-ML GaN, and lead to 1-ML plus 1 fractional ML when 2-ML GaN is intended. Thus, 1-ML GaN layers acted as isolated quantum disks, while 2-ML GaN layers still had well-like behavior as coverage was continuous. Another signature of fractional monolayer formation was broadening of the PL spectrum. Monolayer thickness fluctuations over 5–10-nm lateral distance could cause the observed broadening.

The first principle calculations and experimental data obtained by MBE grown samples suggest that ultrathin GaN/AlN superlattices can achieve DUV emission as short as 224 nm. The following section of the chapter discusses a brief introduction of the MBE system and the growth process of ultrathin GaN/AlN heterostructures using MBE.

10.4 MBE Growth of GaN/AlN Quantum Structures (Wells and Dots) for Enhanced IQE

10.4.1 MBE System in Brief

All samples discussed in the experimental part of this chapter were grown with plasma-assisted MBE system. MBE provides precise control over the thickness and composition of the grown epitaxial layer and therefore is best suited for ultrathin heterostructure epitaxy. The high purity sources (99.99999%) and the high vacuum (1×10^{-11} Torr) condition also ensure high quality of the material. In the following subsections, a short overview of the system itself, in situ RHEED-based characterization, and various MBE growth modes are presented showing how these are relevant to realize the desired GaN/AlN heterostructure.

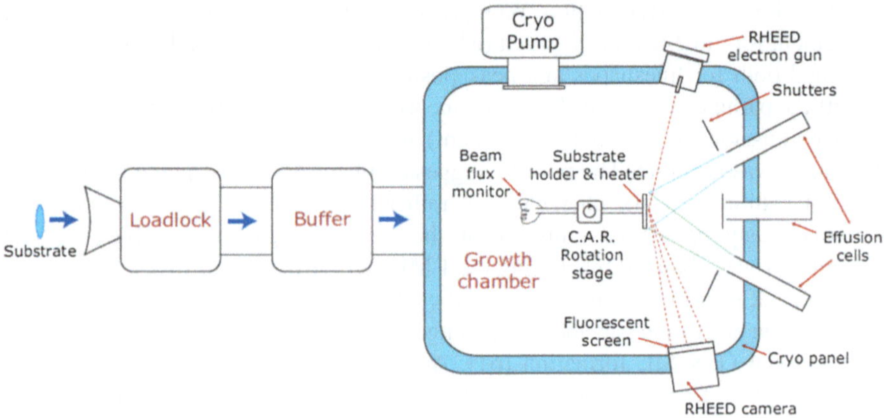

Fig. 10.12 Schematic for a typical MBE system [11]

10.4.1.1 Basic MBE Structure

The main features of an MBE system (Fig. 10.12) for epitaxial growth of crystals are ultrahigh vacuum (UHV) in the growth chamber and the use of ultrahigh purity (UHP) source materials for growth. In the UHV environment, the mean free path for molecules is longer than the physical MBE chamber dimensions. Thus, the molecular beam does not scatter and moves ballistically from the source to the surface of the desired substrate. This ensures a precision in the growth of pure crystals and heterostructures. A combination of cryogenic, turbo-molecular, and ion pumps are used to maintain the high level of vacuum in different sections of the system. The high purity source materials (i.e., group III: B, Al, Ga, and In) are loaded in effusion cells and heated to high temperatures ($\sim$1200°C) using resistive heaters to generate molecular beams. As shown in Fig. 10.12, shutters operated by pneumatic control block/allow the molecular beams from the effusion cells to be deposited on the substrate during growth. The flux of the beam from metal sources is controlled by varying the corresponding cell temperatures with an accuracy of $\sim$1°C. The resulting beam flux is measured with an ion gauge called the beam flux monitor (BFM).

The beam-equivalent pressure (BEP) measured by the BFM helps to control the specific growth conditions, and to reproduce a growth recipe that has been found optimum. During the growth of III-nitride semiconductors using a plasma-assisted MBE, chemically active group V elemental nitrogen is generated using an RF-based plasma source. The stable state of nitrogen gas is the molecular form N_2, which is inert and does not incorporate into the crystal during growth. To make the nitrogen species reactive and available for incorporation in the crystal, it is ionized to form atomic nitrogen using the RF plasma source. Ultrahigh purity nitrogen gas is supplied through a mass flow controller (MFC) into the plasma unit at a controlled flow rate that determines the growth chamber pressure and the growth rate. A typical chamber pressure during growth is $\sim 2\times 10^{-5}$ Torr.

The growth chamber is surrounded by an LN_2 cryo-panel that maintains the wall at a very low temperature. Unwanted molecular species condense on the wall and therefore do not incorporate in the crystal during growth. The substrate holder can be resistively heated by W-Re coils embedded in a PBN disk up to ~1000°C. Typical growth temperatures of III-nitride semiconductor layers are in the 600–900°C range.

The substrate holder and heater are connected to a continuous azimuthal rotation (CAR) stage that is rotated during growth at typically ~20 rpm to ensure uniformity. The CAR also enables the transfer of the samples in and out of the growth chamber. The two other chambers are the buffer chamber and the loadlock chamber. Samples are introduced into the UHV environment of the MBE system from the atmospheric condition through the loadlock chamber, where it is heated at 200°C for a few hours for degassing and primarily the removal of adsorbed surface water. The buffer chamber is placed between the growth and the loadlock chamber and is used for sample preparation prior to growth. This process involves heating the sample to ~450°C for ~1–2 h for surface cleaning and degassing. The standby pressure for buffer chamber is $<5\times 10^{-9}$ Torr. This pressure level can be maintained by an ion pump.

Proper surface cleaning of samples and carrier wafers is essential prior to loading. A typical process involves ultrasonication in organic solvents (acetone, methanol, and iso-propanol). Small (~1×1 cm) substrates are mounted on lapped 3” silicon carrier wafers by melting and resolidifying metallic indium paste. Larger wafers may be mounted directly without using the indium paste. These wafers are then loaded into the UHP N2 vented loadlock chamber and baked at 200°C for at least 5 h till the background chamber pressure reaches $<5\times 10^{-9}$ Torr. This baking expels water molecules from the wafer surface. The samples are then transferred into the buffer chamber where they are baked at 450°C for 1.5 h. The samples from the buffer chamber are transferred into the growth chamber once the pressure difference between these chambers is a minimum. A residual gas analyzer (RGA) is used to detect leaks in the UHV system, or the presence of unwanted elements or gases inside the growth chamber.

10.4.1.2 In Situ Characterization of Growth: Use of RHEED

For in situ characterization during MBE growth, reflection high-energy electron diffraction (RHEED) technique is used. The RHEED electron gun emits an electron beam that is incident on the sample surface at grazing incidence (<5°) at a voltage of ~14 kV (λ ~10pm). The electron beam penetrates only a few nm from the sample surface because of the grazing incidence. The resulting diffraction pattern formed due to Bragg scattering of the incident electrons from the crystal surface is captured on a fluorescent screen. This RHEED diffraction pattern allows real-time

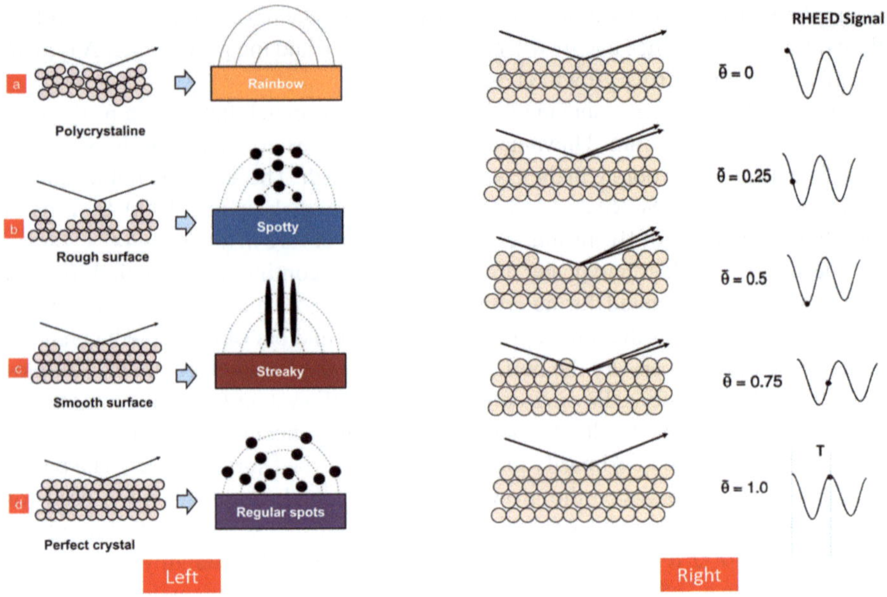

Fig. 10.13 (Left) RHEED pattern from different surfaces (**a**) arc from poly-crystalline, (**b**) spots from rough, (**c**) streaks from smooth, and (**d**) spots on arc from perfect crystals [63]. (Right) Formation of RHEED intensity oscillations with respect to surface layer orientation. T is the time period for one full cycle [63]

tracking of the smoothness and lattice structure of the epitaxial surface. The growth conditions can be adjusted dynamically based on the observed RHEED pattern. As shown in Fig. 10.13 (left), different diffraction patterns are generated based on the surface orientation of the crystal. For example, a polycrystalline surface produces rings/arcs which are Laue circles formed by the merger of specific diffraction spots of randomly oriented crystalline grains. A spotty pattern is produced from a rough surface. The spots overlap and become long steaks in case of a smooth surface. For atomically smooth crystal surfaces, the RHEED pattern consists of bright spots.

Another important application of the RHEED is to measure the growth rate and to track the growth conditions by following the time dynamics of the diffraction pattern, for example using intensity oscillations.

Figure 10.13 (right) shows the time evolution of RHEED intensity with the growth of monolayers of a semiconductor crystal. A bright spot is obtained due to constructive interference of diffracted electrons from a smooth surface, whereas a rough surface causes partial destructive interference, which leads to a lowering of the diffraction beam intensity. The time period for one complete intensity oscillation therefore corresponds to the formation of one monolayer of the crystal. The growth rate is then estimated in nm/hr from this information. For example, if the time period for one RHEED intensity oscillation is 4s during the growth of AlN (1 monolayer of AlN= 0.25 nm), then the corresponding growth rate is 225 nm/hr.

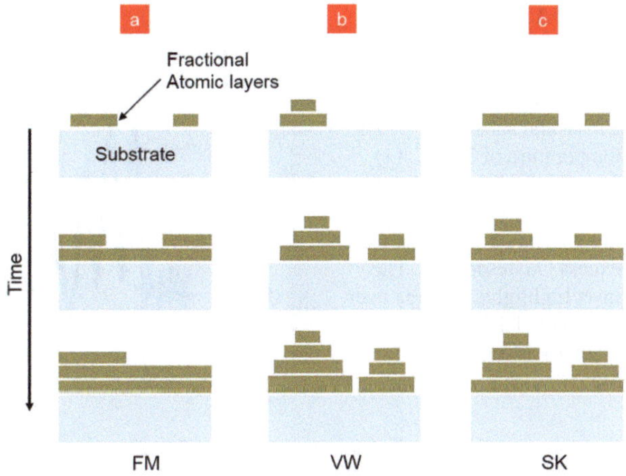

Fig. 10.14 MBE growth modes, (**a**) FM: layer by layer (**b**) VW, and (**c**) SK [63]

10.4.1.3 MBE Growth Modes

The growth mode in MBE is determined by the interaction and forces between the substrate and the adatoms that are deposited on it. There are three growth modes as depicted in Fig. 10.14. These are the Frank–van der Merwe (FM), Volmer–Weber (VW), and Stranski–Krastanov (SK) modes. In the FM mode, the interaction between the crystal surface and the adatom is stronger than between the adatoms themselves, making it energetically favorable for the adatoms to completely "wet" the surface. Thus, a layer-by-layer growth is possible; the second layer does not form till the first layer is completely formed. This mode typically occurs for homoepitaxial growth at high substrate temperatures that provides high kinetic energy for adatoms.

The second growth mode is VW, which occurs at low substrate temperatures when the interaction between adatoms is stronger than the adatom-interface interaction. Consequently, the adatoms do not wet and form 3D islands on the substrate surface. These can coalesce and form films over time.

The third one (SK) is a combination of the FM and VW growth modes. In the SK growth mode, layer-by-layer (2D) growth occurs in the initial phase, and after a few monolayers, the growth mode switches to the formation of 3D islands. The typical driving force for the formation of islands is strain due to lattice mismatch.

10.4.1.4 MBE Growth Condition Markers

To grow GaN/AlN heterostructures with thin (few monolayers) GaN layers, the MBE growth conditions need to be precisely controlled. The growth conditions can be monitored by observing a) RHEED oscillations and b) RHEED side streaks. The growth rate can be precisely determined from RHEED intensity oscillations [46, 64].

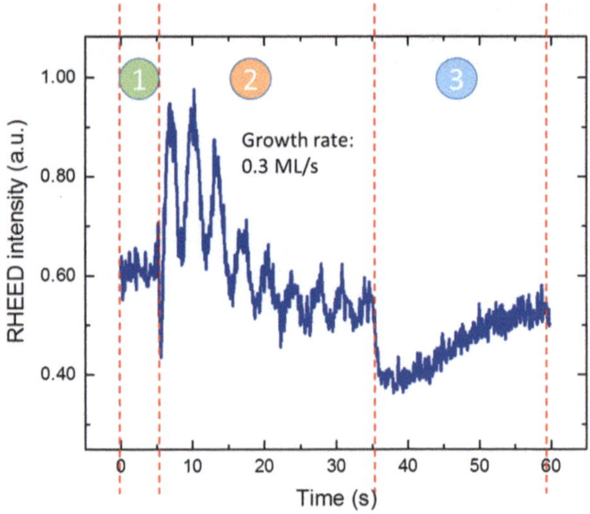

Fig. 10.15 Observation of RHEED intensity oscillations for GaN homo-epitaxial growth with plasma power of 275W and substrate temperature of 730°C. (1) Scan for background intensity, (2) GaN growth in layer-by-layer mode, and (3) excess Ga desorption. The inset highlights the area over which the RHEED beam intensity is tracked [63]

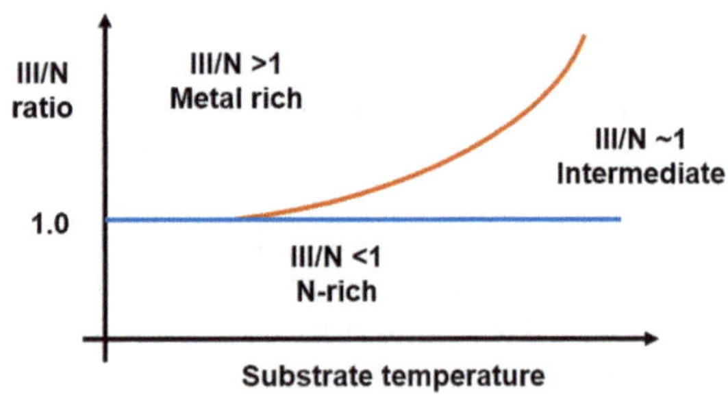

Fig. 10.16 Growth diagram for III-Ns showing three growth regimes based on III/N ratio and the substrate temperature

To calibrate growth rates by RHEED oscillations, epitaxial GaN layers were grown on GaN substrates in a slightly metal-rich condition ensuring smooth surface morphology (2-ML excess Ga) [65]. The substrate temperature was set at 730°C with a plasma RF power of 275W. As shown in Fig. 10.15, both the shutters for Ga and nitrogen were kept open from 5s to 35s within which layer-by-layer growth of GaN takes place, and 9 oscillations were observed. Since 9 monolayers of GaN were deposited over 30s, the growth rate was 0.3 ML/s (equivalent to 4.5 nm/min or 270 nm/hr). As the shutters for Ga and nitrogen were closed at 35s, the RHEED intensity started to increase and approach the initial value of $t = 0$ s. This happened due to the desorption of accumulated Ga from the surface at the growth temperature. Metal-rich growth condition for GaN was the reason for such Ga accumulation. The growth rate is controlled by plasma RF power. A higher growth rate can be achieved with higher plasma power.

For the GaN/AlN superlattice, it is essential to grow the AlN layer with smooth hetero-interfaces. An Al/N ratio of ~1 is maintained to achieve this, which is called the intermediate growth window as depicted in Fig. 10.16. Metal-rich growth provides a smooth growth surface. On the contrary, nitrogen-rich growth makes

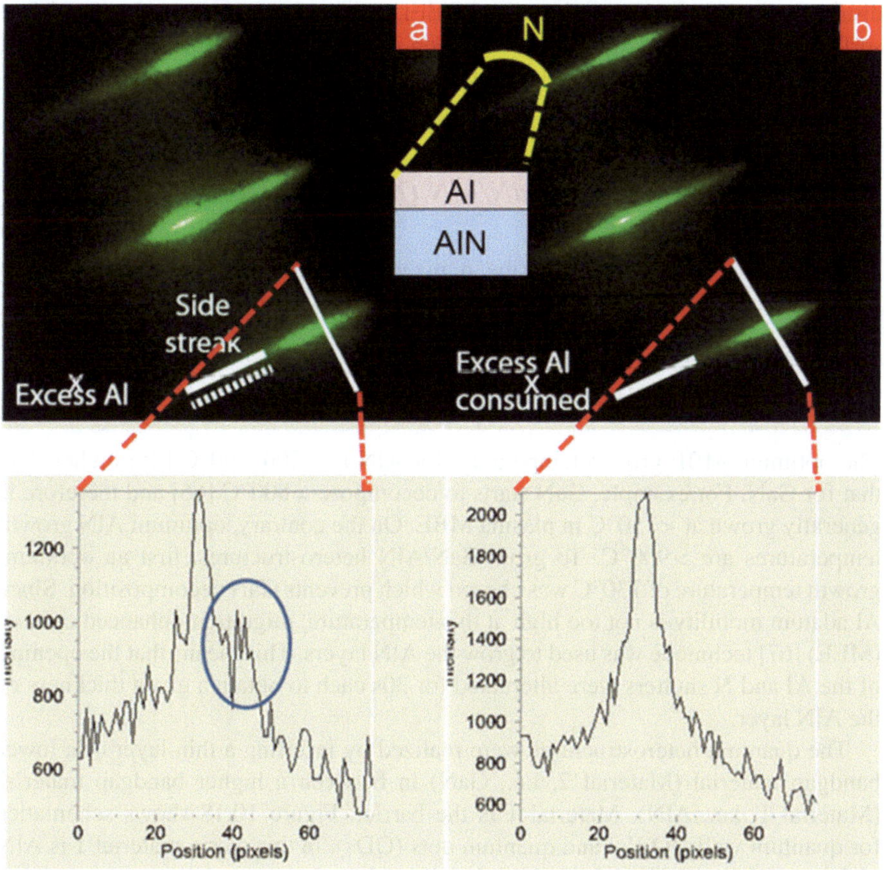

Fig. 10.17 Side streak observation to check AlN growth condition. (**a**) AlN side streak becomes diffused with additional shoulder peak due to accumulation of metal on the surface, (**b**) side streak becomes bright once excess metal is consumed by N and forms AlN. The shoulder peak is no longer present

the surface rough. Excessive supply of metal flux during the MBE growth can cause accumulation on the surface and initiate generation of dislocations and therefore deteriorate crystal quality. This is why growth in the intermediate regime is best-suited to obtain metal-free smooth surfaces. To ensure intermediate growth condition for AlN, RHEED side streak observation was employed. This was done by growing homo-epitaxial AlN. For this growth, the substrate was kept at 730°C with plasma RF power of 275W. First, only the Al shutter is kept open for 30s to deposit Al on the AlN substrate. During this process, RHEED side streak becomes diffused as shown in Fig. 10.17a, and an additional shoulder peak is observed. This peak most likely corresponds to a different lattice parameter of Al metal compared to the underlying AlN layer. To consume this excess metal, Al shutter was closed, and N2-shutter was kept open. Once all the excess metal was consumed and converted to AlN, the side streak became bright and streaky (Fig. 10.17b). The shoulder peak also disappeared. This bright and streaky RHEED pattern is a signature for metal-free surfaces. The intermediate growth condition is achieved when the Al-consumption time is almost as equal as the Al-deposition time.

10.4.2 MBE Growth of GaN/AlN Quantum Heterostructures

The following subsections describe realization of ultrathin GaN/AlN quantum heterostructures with MBE using the RHEED-based growth markers.

10.4.2.1 Quantum Wells/Quantum Dots/Disks

The optimum MBE growth temperature for AlN is ~200–300°C [25] higher than that for GaN. For example, GaN starts to decompose >800°C [66] and therefore is generally grown at <750°C in plasma MBE. On the contrary, optimum AlN growth temperatures are >900°C. To grow GaN/AlN heterostructures, first an optimum growth temperature of 730°C was chosen which prevents GaN decomposition. Since Al adatom mobility is not too high at this temperature, migration-enhanced epitaxy (MEE) [67] technique was used to grow the AlN layers. This means that the opening of the Al and N shutters were alternated for 30s each to obtain a given thickness of the AlN layer.

The quantum heterostructures were realized by inserting a thin layer of a lower bandgap material (Material 2, i.e., GaN) in between a higher bandgap material (Material 1, i.e., AlN). Material 1 is the barrier. Figure 10.18 shows schematics for quantum wells (QWs) and quantum dots (QDs). In this work, material 1 is AlN and material 2 is GaN.

For GaN QWs, the growth condition was set as close as possible to stoichiometry (Ga/N ratio of ~1) to ensure the formation of a smooth 2D layer. Although

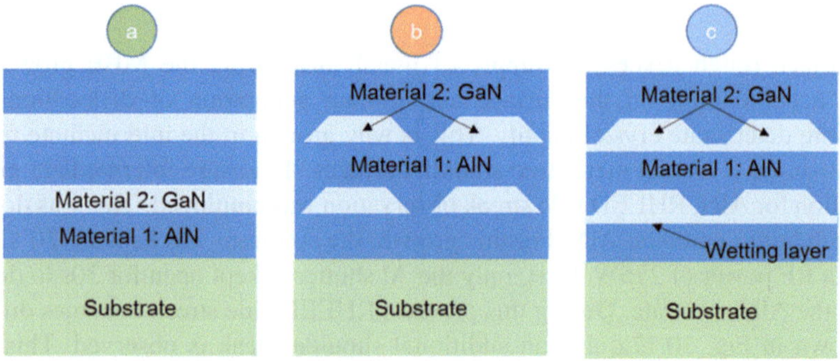

Fig. 10.18 Schematics for (**a**) quantum wells, (**b**) quantum dots/disks, and (**c**) quantum dots/disks with wetting layers

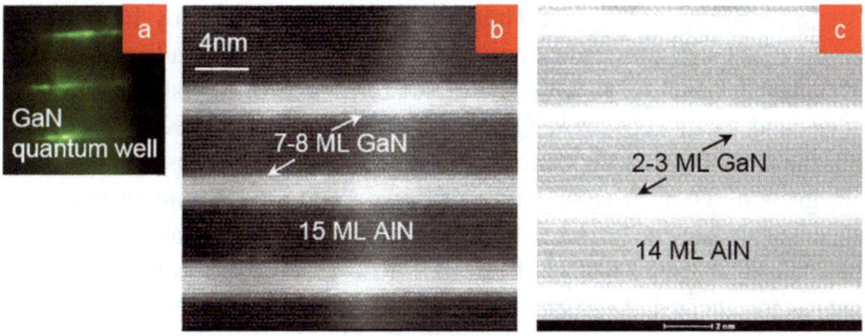

Fig. 10.19 MBE growth of GaN QWs in AlN barriers: (**a**) streaky RHEED during GaN QW growth suggesting a smooth surface, (**b**) STEM for 7–8 ML thick GaN QWs in AlN, and (**c**) 2–3 MLs GaN QWs in AlN

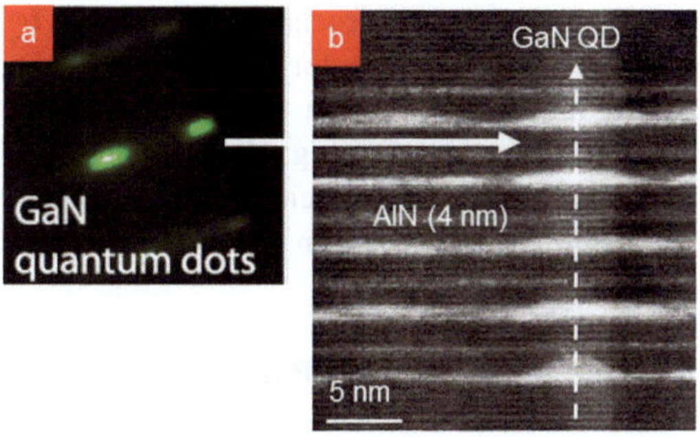

Fig. 10.20 MBE growth of GaN QDs in AlN barriers: (**a**) spotty RHEED during GaN QD growth suggesting surface roughening, and (**b**) STEM image of self-organized GaN QD growth in AlN matrix

smooth layers can be grown with metal-rich growth condition, it poses the risk of accumulation of excess metal and consequently result thicker QWs than expected. Such thickness variations are not desired because they can cause multi-peak emission during luminescence. Figure 10.19a shows the streaky RHEED during growth of a GaN QW. The streaky RHEED confirms formation of a smooth 2-D surface. Some typical thick and thin GaN QW STEM images are shown as examples in Fig. 10.19b–c.

For quantum dots/disks (QDs), material 2 is grown as a 3D structure (Fig. 10.18b). The 3D growth mode can be achieved in either of the two ways:

(a) SK mode
(b) Modified SK mode

To ensure SK (Stranski–Krastanov) growth mode for QDs, the growth condition for GaN needs to be nitrogen rich (Ga/N <1). 3D islands of GaN are formed due to compressive strain from AlN [68, 69]. Self-aligned QD formation occurs due to the existence of underlying thin strained spacer layers of AlN in between GaN QDs (Fig. 10.20b). Besides growing in nitrogen-rich condition, the 3D GaN islands are left in vacuum for a given amount of time (annealed) at the growth temperature before capping with the barrier material (AlN). This process is termed as "ripening" in this chapter. Ripening helps the QDs to rearrange themselves due to the strain from the underlying AlN layer. Thus, QDs can be reshaped by both the ripening process and the Ga/N ratio. The spotty RHEED (Fig. 10.20a) indicates formation of a rough surface, which is also a signature of formation of 3D islands or QDs.

The second method to grow QDs is called the modified SK mode [65]. In this method, the GaN layer is deposited under metal-rich conditions to ensure a smooth 2D film growth. Then, it is ripened for a given amount of time so that the excess Ga can desorb from the surface due to high substrate temperature. If the ripening is done for longer time (>30s), the GaN film can start to decompose and become rough. This roughened 2D layer works as a 3D GaN island or QD. Therefore for this method, the shape of the QDs can be controlled by the duration of ripening time.

For the modified SK mode, the most crucial parameter is the ripening time since it decides whether 3D island formation can take place. On the other hand for SK mode, the nitrogen-rich growth condition decides if dots/disks can form. Therefore, the Ga/N flux ratio is crucial for QD growth by the SK mode.

Schematic structure for an idealistic QD is shown in Fig. 10.18b. But due to SK mode of growth, a thin continuous wetting layer of 1–2-ML thickness is formed below the QDs (Fig. 10.18c). This may lead to some detrimental effects like reduction in quantum efficiency due to extended dislocations in the wetting layer, broadening of the emission spectra, etc.

10.4.2.2 MBE Growth Parameter Optimization

Prior to growing the desired GaN/AlN quantum heterostructures for deep-UV emission, it was essential to investigate the impact of different growth parameters on the emitted light such as the AlN barrier thickness, growth rate, ripening time, etc. The summary of such optimization experiments is presented here.

(a) AlN barrier thickness:

The quantum confinement effect is increased with thicker barriers by reduction of carrier wavefunction penetration into them. The downside of using a thick barrier is the reduction of tunneling probability through the barriers. This has a negative impact on the carrier transport properties needed for LEDs and Lasers. 3 GaN/AlN test structures were grown with different barrier thicknesses to investigate their effect on photoluminescence. 2-ML SK-GaN QD was grown in 2-, 3-, and 4-nm AlN barriers. HR-XRD measurement showed a strong peak from AlN substrate but no satellite peaks were observed due to incoherent reflection from GaN QDs, as shown in Fig. 10.21a. Photoluminescence (PL) measurement was performed at 300K with excitation from a 157-nm excimer laser. The emission energy was found to increase with the thickness of the AlN barrier layers due to higher quantum confinement (2 nm→234 nm, 3 nm→227 nm).

Above 3-nm AlN barrier thickness, the quantum confinement effect was saturated, and the emission energy did not increase further. Based on this observation, a barrier thickness of 2–2.5 nm was used as an optimum value for most of the subsequent experiments to provide quantum confinement with a decent tunneling probability for efficient carrier injection.

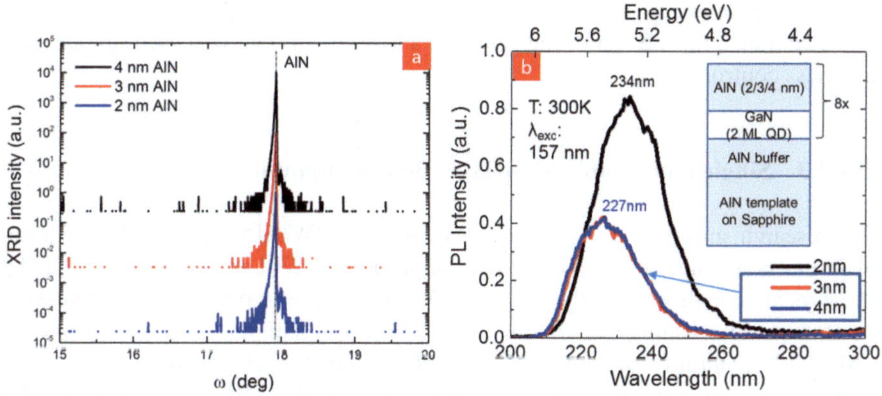

Fig. 10.21 Optimization of AlN barrier thickness: (**a**) (002) ω-2θ HR-XRD showing the absence of secondary peaks due to incoherent reflection from GaN, and (**b**) RTPL spectra for various AlN barriers

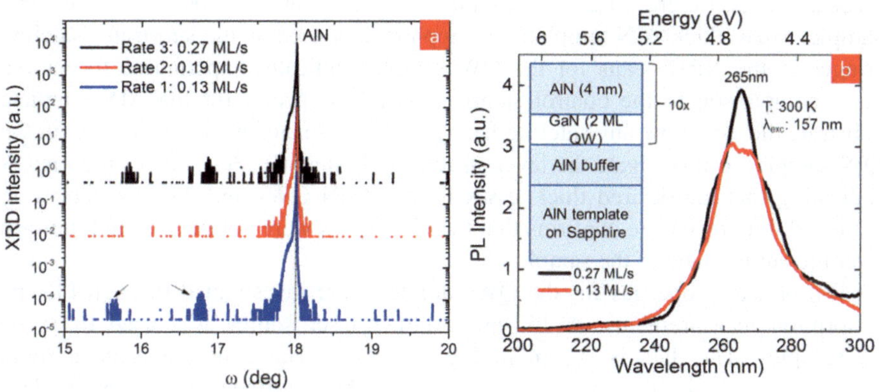

Fig. 10.22 Optimization of growth rate: (**a**) GaN QW reflections are seen from the HRXRD spectra, and (**b**) PL spectra for various growth rates

(b) Growth rate:

For this experiment, test structures containing 2-ML GaN QWs between 4 nm AlN barriers were grown using 3 different growth rates, i.e., 0.13 ML/s, 0.17 ML/s, and 0.27 ML/s. The HR-XRD spectra for the QW samples showed secondary satellite peaks from the GaN layers (Fig. 10.22a). From the PL spectra (Fig. 10.22b), it was found that the sample grown with 0.17 ML/s showed a blue shifted emission compared to the other two (not shown here), most likely because of QD formation instead of QWs.

This claim was further confirmed from the absence of the secondary satellite peaks from the XRD measurement for that sample. The highest PL intensity was achieved from the sample grown with the highest growth rate. Higher growth rates than 0.27 ML/s were not investigated since the ML deposition time reduces below a few seconds, which may cause reduced control over the thickness of a grown layer.

Since the PL intensity dependence was not significant over the investigated range of growth rates, for the rest of the growths, a low growth rate was used for accurate thickness control.

10.4.2.3 Single and Double Monolayer δ-GaN Quantum Wells

Few research groups have realized 1-ML δ-GaN QW/AlN heterostructures, and DUV emission have been reported by Taniyasu et al. [70] and Kamiya et al. [71]. The crystal growth was performed by MOCVD in these reports. 224-nm emission was reported by Kamiya et al. [71] and 237-nm by Taniyasu et al. [70] from the 1-ML GaN quantum structures with a thinner AlN barrier (1.75 nm).

In the current experiment, the structure shown in Fig. 10.23a was grown with MBE. The AlN barrier thickness was 4 nm for both 1- and 2-ML GaN samples. GaN/AlN QW heterostructures were repeated for 10 times.

After the MBE growth, the samples were characterized by XRD triple axis ω-2θ measurements. Figure 10.23b shows the data including the substrate as a control sample. 1-μm-thick AlN template on sapphire was used as the substrate. Satellite fringes in the XRD scans for the QW samples indicates superlattice formation. As a comparison to the control sample, the XRD spectra for the AlN template substrate did not show any satellite fringes. STEM images were taken for both the QW samples to investigate the layer structures. Figure 10.23c–e shows the images indicating that the desired thicknesses for the GaN QWs and AlN barriers were achieved. Figure 10.23e confirms that the GaN QWs were continuous and uniform over the entire plane of the sample.

The optical properties for the QW samples were investigated through PL and absorption measurements. A 157-nm excimer laser source was used to excite all samples. The data is shown in Fig. 10.24a–b. The emission peak at room temperature (RT) is 261 nm for 2-ML QW, which blue shifted to 256 nm at 5K. This is explained due to bandgap increase with temperature by Varshni's law. Similarly, the 1-ML QW showed an emission peak of 224 nm at 300K which also blue shifted to 222 nm at 5K. The absorption data was measured at room temperature and compared with the PL peak emission wavelength to find out the effective bandgap. PL and absorption data showed good agreement.

IQE was estimated from temperature-dependent PL measurement for these samples with the assumption that the non-radiative recombination process ceases out completely at 5K. The IQE was 49% for 2-ML and 17% for the 1-ML GaN QW samples. The high IQE for 2-ML QW is expected because of high overlap integral between the electron and hole wavefunctions, as depicted in Fig. 10.25. The efficiency value is limited by the threading dislocation density of the substrate, which is 1×10^{10} cm^{-2} for this case. A sharp decrease in the measured IQE for the 1-ML QW was observed although the theoretical overlap integral of >80% from Fig. 10.25 predicts high IQE for such thin QWs.

In summary, 1–2-ML GaN QWs can produce deep-UV emission as short as 224 nm at 300K, but lowering of IQE was observed at shorter wavelengths, most likely due to reduced effective band offsets in the δ-QW structures.

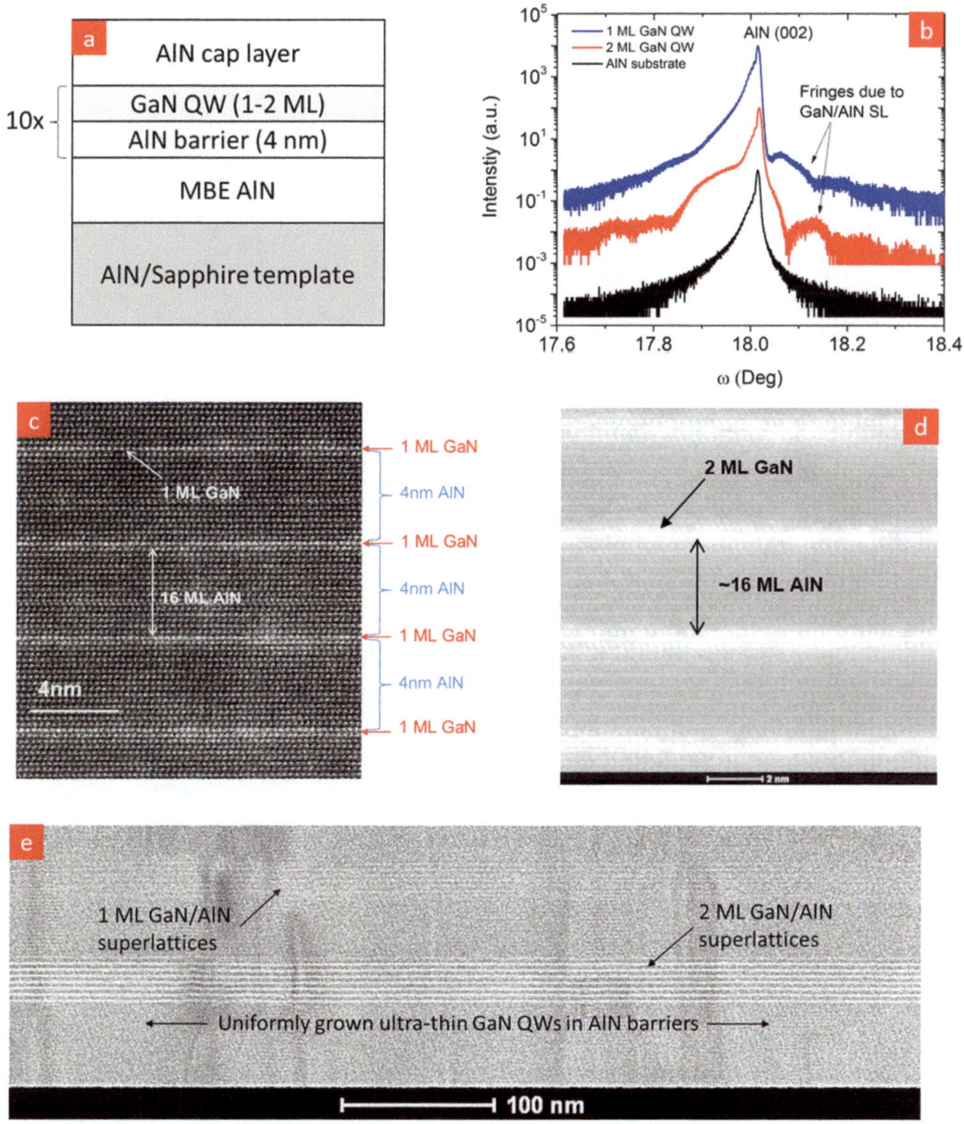

Fig. 10.23 (**a**) Schematic structure for the 1–2-ML GaN/AlN heterostructures, (**b**) measured XRD spectra confirming the presence of GaN/AlN superlattices, STEM images of (**c**) 1-ML GaN and (**d**) 2-ML GaN in AlN matrix, and (**e**) uniformity of the 1- and 2-ML GaN layers over long range [63]

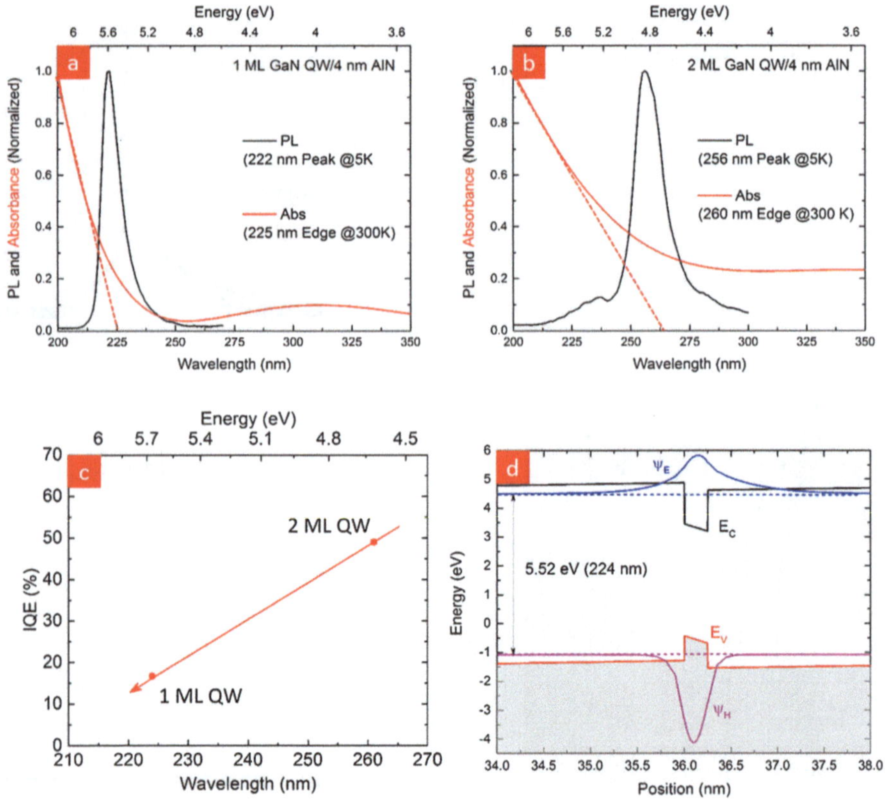

Fig. 10.24 PL and absorption spectra from the (**a**) 1-ML and (**b**) 2-ML GaN/AlN heterostructures, (**c**) IQE measured using temperature-dependent PL, (**d**) simulated energy band diagram to explain the low IQE for 1-ML GaN/AlN heterostructure [63]

10.4.2.4 IQE Improvement with δ-GaN Quantum Disks

Quantum dots/disks (QDs) are confined heterostructures in 3-dimensions (3-D) unlike QWs where the confinement is only along 1 dimension. The additional degree of confinement can help in capturing the carriers in a defect-free region and therefore produce efficient radiative recombination process. The use of QDs to enhance IQE is well known for the InAs/GaAs material system for photonic devices [72]. Also, the state-of-the-art InGaN/GaN heterostructure-based visible photonic devices have reached high EQE [73] owing to the QD-like active regions of InGaN in GaN matrix. In the following sections, an experimental effort to use ultrathin δ-GaN quantum disks to enhance the IQE is presented.

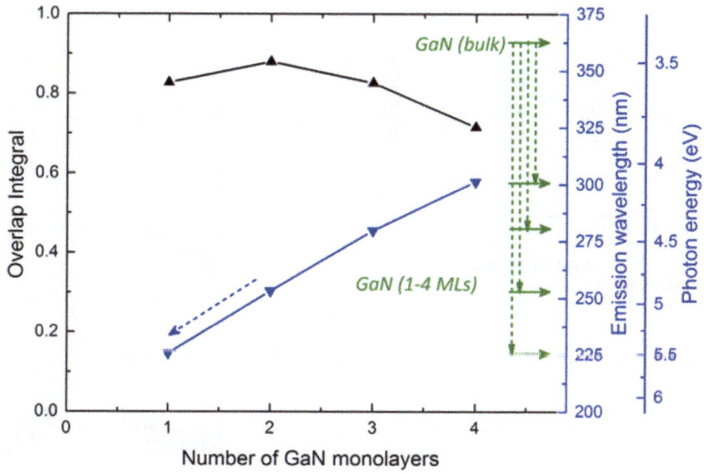

Fig. 10.25 Evolution of simulated overlap integral and emission wavelength as a function of GaN thickness for a GaN/AlN multi-QW heterostructure [63]

10.4.2.4.1 SK Quantum Dots/Disks

For this experiment, the MBE growth and characterization of 1–2-ML GaN QDs in AlN barriers was explored. As it was mentioned earlier (refer to Sect. 10.4.2.1), Ga/N ratio is a very important parameter for growth of QDs in SK mode. Three samples were grown using MBE with different Ga/N ratios for the 2-ML GaN QDs in 4-nm AlN barriers. 10 repeats of the heterostructure were used for enhancing the light emission. All structures were grown on AlN templates on sapphire with threading dislocation densities of 1×10^{10} cm^{-2}.

The schematic of the structure and the growth flow diagram is depicted in Fig. 10.26a–b. The Ga/N ratio was varied over 0.88 (A)-0.75 (B)-0.6 (C) for the three samples. After the MBE growth, structural characterization was done first by HR-XRD measurements. All samples showed a strong AlN peak originating from the substrate. Unlike the XRD spectra of the substrate, the QD samples showed satellite peaks suggesting formation of GaN/AlN superlattices. From the trend of the satellite peak intensities, it was observed that as the Ga/N ratio is reduced, the GaN QD geometry turns more into QW-like producing intense fringes from coherent reflections by the smooth GaN layers [74]. The satellite peak intensities are the strongest for sample C with a Ga/N ratio of 0.6 clearly suggesting QW-like phenomena.

For further structural characterization, TEM imaging was performed. A representative HAADF-STEM image for the sample A is shown in Fig. 10.27. Desired thicknesses of GaN and AlN layers with sharp hetero-interfaces were confirmed from the images. For the GaN layers, a 2-ML QD-like structure was seen with a 1-ML GaN wetting layer. The 1-ML and 2-ML GaN regions are explicitly shown in Fig. 10.27b.

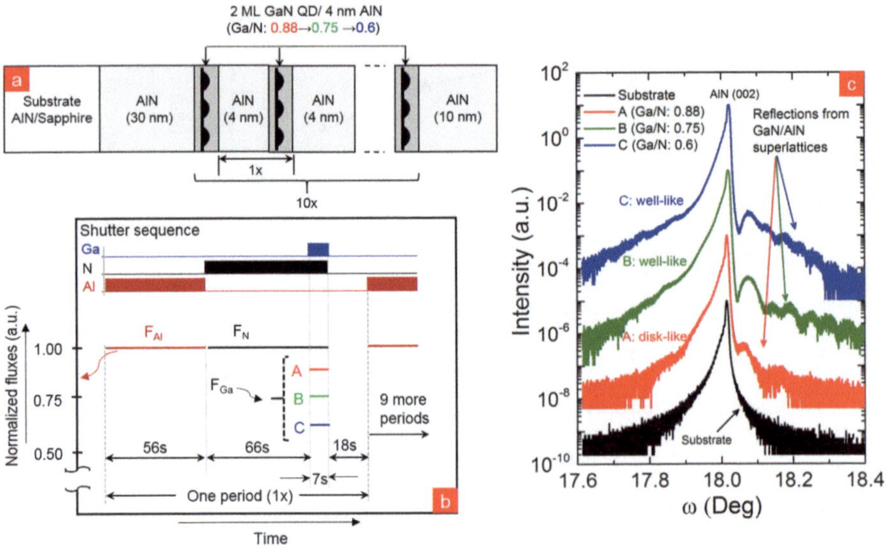

Fig. 10.26 SK-GaN QDs: (**a**) schematic structure, (**b**) MBE growth flow diagram, and (**c**) XRD spectra showing signature of superlattices [47]. Reproduced from Appl. Phys. Lett. 111, 091104 (2017), with the permission of AIP Publishing

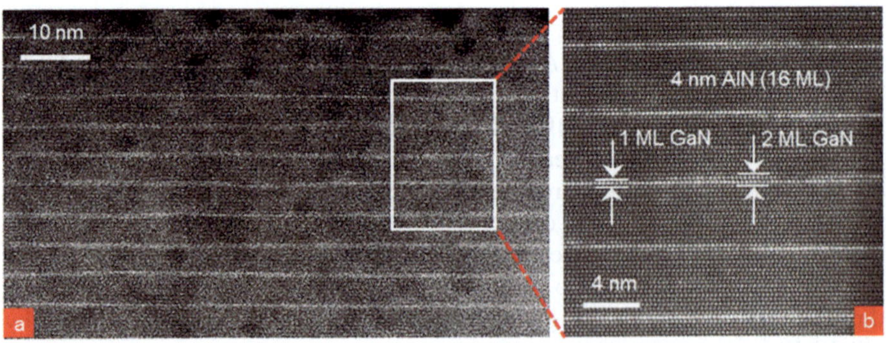

Fig. 10.27 STEM images showing 1–2-ML SK-GaN QDs: (**a**) zoomed-out, and (**b**) zoomed-in [47]. Reproduced from Appl. Phys. Lett. 111, 091104 (2017), with the permission of AIP Publishing

To examine the optical properties, PL measurements were performed for all the samples using a 157-nm excimer laser as a function of temperature. The absorption spectra were measured as well with a standard UV-Vis system at 300K. The measured optical characteristics are shown in Fig. 10.28.

As the Ga/N ratio was reduced from 0.88→0.75→0.6, a monotonic blue shift in the emission wavelengths was observed from 235 nm→222 nm→219 nm. This is the first demonstration of 219-nm emission from GaN-based quantum structures in AlN matrix. The absorption spectra measured at 300K show a similar trend as the PL. But, the absorption edge did not exactly coincide with the PL peak emission

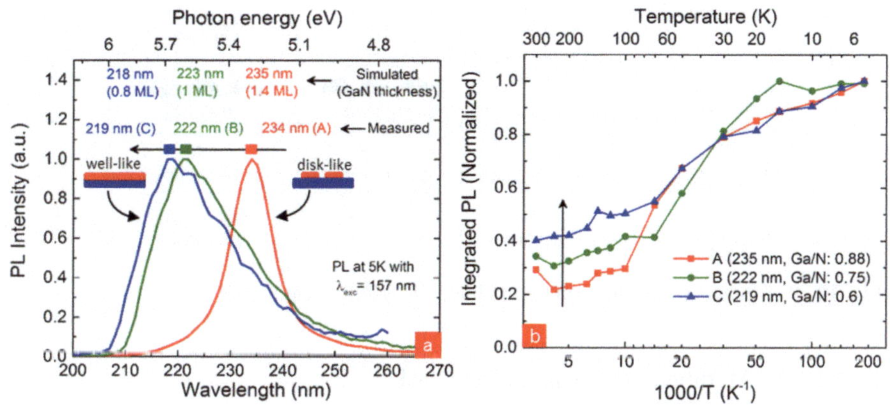

Fig. 10.28 Optical properties of the SK-GaN QDs: (**a**) PL and absorption, and (**b**) IQE [47]. Reproduced from Appl. Phys. Lett. 111, 091104 (2017), with the permission of AIP Publishing

wavelengths. The absorption band edges were also not very sharp most likely because of the small absorption volume of GaN in the whole structure (10 x 2 ML~5 nm of effective GaN in a sample). The measured PL peak positions were fitted with SiLENSe simulations of GaN QWs to extract the effective GaN thicknesses. The extracted GaN thicknesses are shown in Fig. 10.28a. The simulation suggested that the effective GaN thicknesses can be controlled by tuning the Ga/N ratio during the MBE growth, and therefore the emission wavelength could also be changed accordingly.

Internal quantum efficiency (IQE) was measured for these samples using temperature-dependent integrated PL data. Assuming the non-radiative recombination processes ceases at 5K, the IQE can be measured from the ratio of integrated PL taken at 300K and 5K. It was observed that, as the Ga/N ratio was reduced from sample A→B→C, the IQE increased from 29%→34%→40%. The increase of IQE with a reduction in emission wavelength is explained by enhanced quantum confinement effect by the reduction of QD sizes. The size of the QDs was made smaller by lowering the Ga/N ratio. The smaller QDs were also most likely isolated from the non-radiative recombination centers and thus became more efficient. The 40% IQE at 219-nm emission wavelength is the highest reported number for such a short wavelength with any AlGaN composition.

In summary, ultrathin (1–2ML) GaN/AlN SK-QDs were grown by MBE to achieve emission in the DUV regime. With 1-ML QWs, emission as short as 224 nm can be achieved at 300K, but the IQE was only 17%. For SK-QDs, the shortest emission wavelength achieved was 219 nm with an IQE of 40% which is >2× higher than the IQE of the QW emitting at 222 nm at 5K.

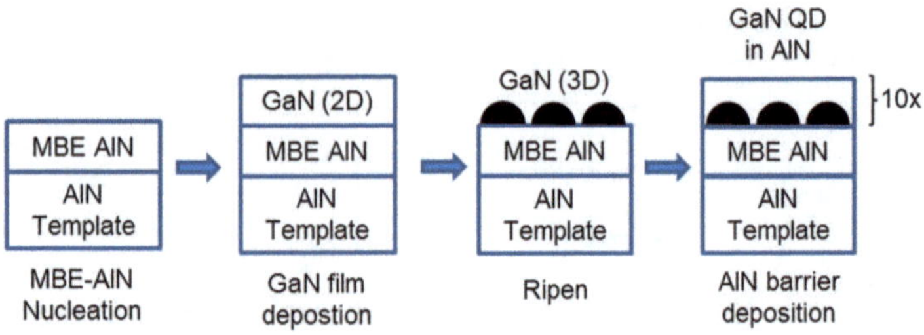

Fig. 10.29 Schematic flow diagram for modified SK growth method [63]

10.4.2.4.2 Modified SK Quantum Dots/Disks

The other method of growing QDs namely the modified SK was examined for deep-UV emission. Previously, Verma et al. [11] demonstrated that QDs grown in modified SK can produce better light emission property with a narrower FWHM for the PL spectra compared to the SK mode.

As explained earlier (refer to Sect. 10.4.2.1), the GaN layers are grown in metal-rich condition first for this modified SK method and then ripened at the growth temperature before capping with AlN. Therefore, the ripening time is a very important parameter to determine the QD formation. To illustrate the process better, the growth flow diagram for modified SK QD is shown in Fig. 10.29.

Based on this modified SK growth scheme, a series of three samples (F/G/H) were grown. 2-ML GaN layers were deposited in 3-nm AlN barriers for each sample, and this GaN/AlN heterostructure was repeated for 10 periods. The ripening time "t" was varied for the three samples: F (30s)→G (45s)→H (60s). As control samples, the two 1-ML and 2-ML GaN QW samples from Sect. 10.4.2.3 were used. Figure 10.30 shows the high-resolution X-ray diffraction (HR-XRD) triple axis (002) ω-2θ scans for the QD samples (F/G/H) and the 2-ML QW control sample. All samples showed a strong AlN substrate peak. The QW sample showed satellite secondary peaks originating from coherent reflection from the smooth GaN layers. QD sample F did not show any such satellite peaks, possibly due to incoherently scattered X-ray reflections from the GaN QDs. QD sample G and H showed satellite peaks like the QW sample suggesting that the GaN layers were more like QWs instead of QDs. This observation is similar to the SK GaN QDs discussed in Sect. 10.4.2.4.1.

To investigate further the structural details, HAADF-STEM images were acquired for the QD samples F, G, and H as shown in Fig. 10.31. Sharp hetero-interfaces were observed for all samples suggesting no intermixing between AlN and GaN. The AlN barrier thicknesses were 3 nm, and GaN QDs were 1–2 MLs for all samples. No clear distinction was observed in the images for the GaN layers grown with different ripening times.

The optical properties were studied through photoluminescence using a 157-nm pulsed excimer laser excitation source. Temperature-dependent PL data was taken down to 5K, as shown in Fig. 10.32. Figure 10.32a is the spectra from the QD

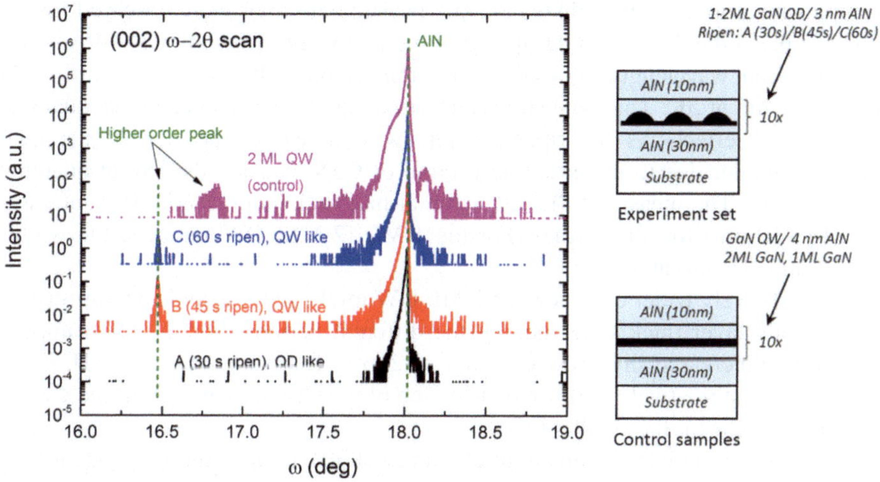

Fig. 10.30 (002) ω-2θ triple axis HR-XRD analysis for QD samples (F/G/H) and the 2-ML QW control sample [63]

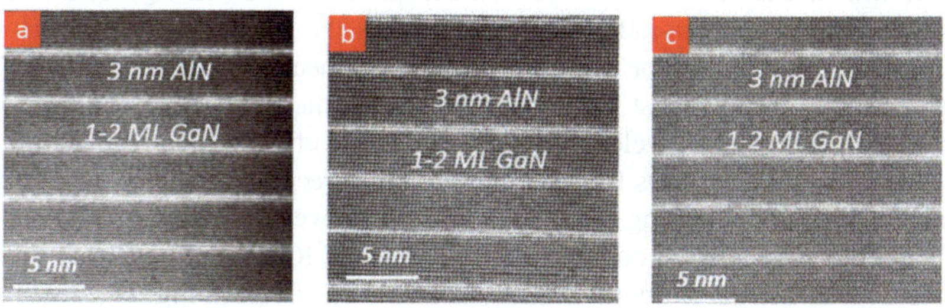

Fig. 10.31 HAADF-STEM images for the samples (**a**) F, (**b**) G, and (**c**) H showing the 1–2-ML GaN layers in 3-nm AlN matrix [63]

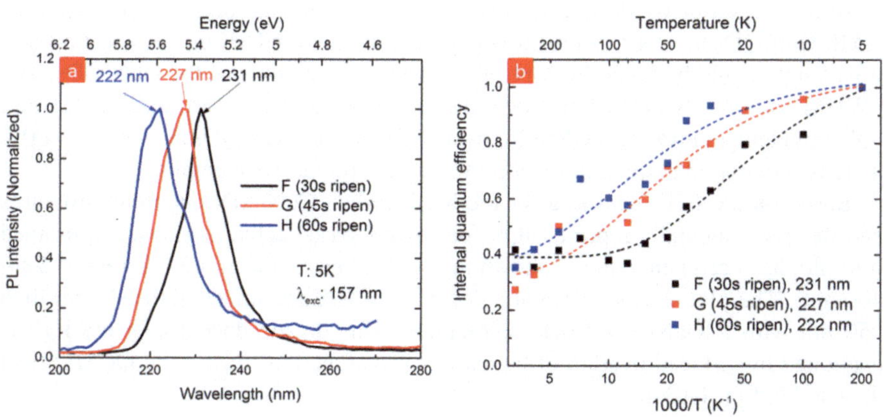

Fig. 10.32 Optical properties of the SK-GaN QDs: (**a**) PL and absorption, and (**b**) IQE [63]

samples measured at 5K. 3 distinct single-peaked emissions were observed for the QD samples F/G/H. As the QD ripening time was increased from 30s→45s→60s, the emission wavelength showed a monotonous blue shift from 231 nm→227 nm→222 nm at 5K. This blue shift is described as a consequence of the quantum confinement effect. As the ripening time was increased, the size of the QDs became smaller due to increased desorption of GaN, causing the lateral quantum confinement. The measured PL full width at half maximum (FWHM) values was small (~10 nm) for all samples. For the 1-ML (2-ML) QW samples, the widths were 9 nm (16 nm) at 5K.

The relatively wider emission for 2-ML QW might suggest some nonuniformity in the GaN layer thicknesses among the different periods. For the QD samples, the FWHM increased slightly from 11nm→12nm→13nm among F→G→H. These widths suggest relatively uniform distribution of the QD sizes and slight increase in the randomness for the smaller QDs.

To estimate the internal quantum efficiency (IQE), temperature-dependent integrated PL measurements were performed. Figure 10.32b shows the trend of the temperature-dependent IQE for the QD samples. From the integrated PL intensity ratio (IPL(300K)/IPL(5K)), the highest IQE was measured for sample F (37%), and the lowest was for sample G (24%). Sample H showed an IQE of 36%. The activation energy for non-radiative processes was the highest for sample H (222 nm) and was the lowest for sample F (231 nm), meaning that the smaller QDs were thermally more robust. The quantum confinement served two purposes: it shortened the emission wavelength by increasing the effective bandgap, and it also most likely isolated the QDs from the non-radiative recombination centers through an energy barrier, increasing the IQE. The IQE between QDs (mod-SK) and the QW control samples were compared. For QWs, the IQE decreased from 49% (2 ML) to 17% (1 ML) at shorter wavelengths as shown earlier in Fig. 10.24c. For the modified SK QD samples, the IQE stayed around 35% for the whole range of 222–231 nm emission, which is ~200% higher than 1-ML GaN QWs at 222 nm. This comparative efficiency boost suggests enhanced 3D quantum confinement in QDs.

In summary, tunable deep-UV emission below 230 nm has been obtained from 1–2-ML GaN QDs in AlN barriers grown by MBE in modified SK mode. The emission wavelength could be tuned by controlling the annealing time of the GaN QDs. The FWHM was narrow (11–13 nm) over the entire range of emission of 222–231 nm. IQE from temperature-dependent PL measurements for GaN QDs showed a ~200% increase compared to quantum wells emitting at similar wavelengths.

Based on the IQE measured for the MBE grown GaN/AlN quantum structures and the experimentally reported IQE for AlGaN-based active regions, a combined IQE plot has been generated as shown in Fig. 10.33 over the entire UV range (200–400 nm). The plot shows a dramatic decrease of IQE for wavelengths shorter than 250 nm with AlGaN-based heterostructures. This drastic decrease in the IQE is explained by reduced overlap of the carrier wavefunctions due to the QCSE effect depicted in Fig. 10.34a.

On the contrary, the GaN-based active regions show markedly enhanced IQE for ~220-nm emission. For example, the IQE of 40% at 219 nm [47], and 36%

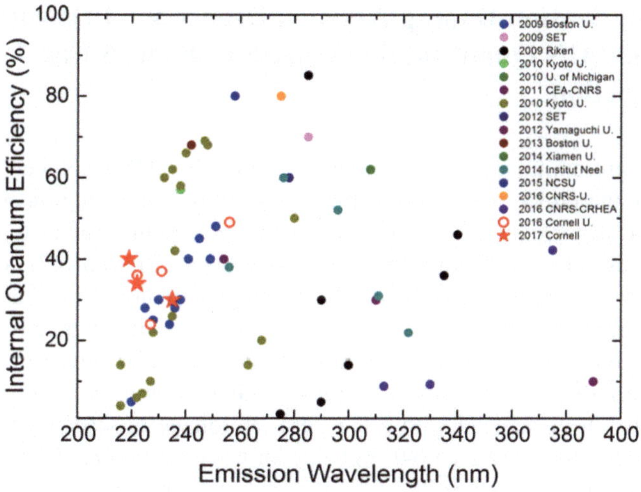

Fig. 10.33 A comparative IQE plot between AlGaN- and GaN-based deep-UV emitters over the entire UV spectral region [7, 19, 57]

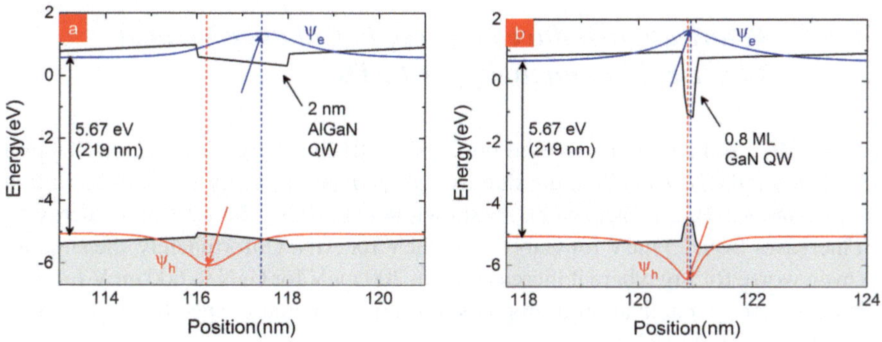

Fig. 10.34 Improvement of wavefunction overlap by using ultrathin GaN quantum structures: (**a**) 2 nm AlGaN QW, and (**b**) 0.25 nm GaN QW [47]. Reproduced from Appl. Phys. Lett. 111, 091104 (2017), with the permission of AIP Publishing

at 222 nm [63] using GaN QDs are almost 8 times higher compared to the 5% at 220 nm using AlGaN QWs (Bhattacharyya et al. 2009). Such enhancement of the IQE is explained by reduction of the QCSE effect by use of very thin "δ" active region. The enhanced overlap of the wavefunctions is illustrated in Fig. 10.34b. This clearly shows the superiority of the GaN-based heterostructures over the AlGaN counterpart to achieve high IQE for sub-250-nm deep-UV emission as was predicted in Sect. 10.3.1.

10.5 Polarization Doping-Assisted Deep-UV LEDs with GaN/AlN Quantum Structures: Enhanced Carrier Injection

Deep-UV light-emitting diodes fabricated using the AlN/GaN heterostructures discussed in the previous sections as the light-emitting active region are presented here. Electrically injected deep-UV LEDs with electroluminescence (EL) emission wavelength as short as 232 nm (5.34 eV) have been achieved [47], which is the shortest EL wavelength with GaN-based active region till now. Polarization-induced doping has been used for the p and n regions for improving carrier injection. The carriers are injected into the light-emitting active region by tunneling mechanism through thin AlN (~2 nm) barriers. Cryogenic operation of a ~260-nm (4.77eV) LED has also been realized with a low turn-on voltage of <6V [75]. Repeatable negative differential resistance (NDR) behavior has been observed for the first time in such light-emitting devices. The following subsections discuss the details of such achievements.

10.5.1 Polarization-Induced Doping to Enhance Vertical Electrical Conductivity for LEDs

Si and Mg are the commonly used dopants for III-nitride semiconductors. Doping in Al-rich AlGaN is challenging due to high activation energy for both Si (n-type donor) and Mg (p-type acceptor). As shown in Fig. 10.35, the activation energy for Si increases from 25 meV for GaN to 250 meV for AlN. Unfortunately, the situation is even worse for Mg where it increases from 200 meV for GaN to 600 meV for AlN. Thus, the thermal activation of dopants at room temperature specifically for Mg is very difficult in Al-rich AlGaN.

This problem can be solved to a great extent by exploiting internal polarization fields of III-nitride semiconductors. As discussed in Simon et al. [76] and Jena et al.

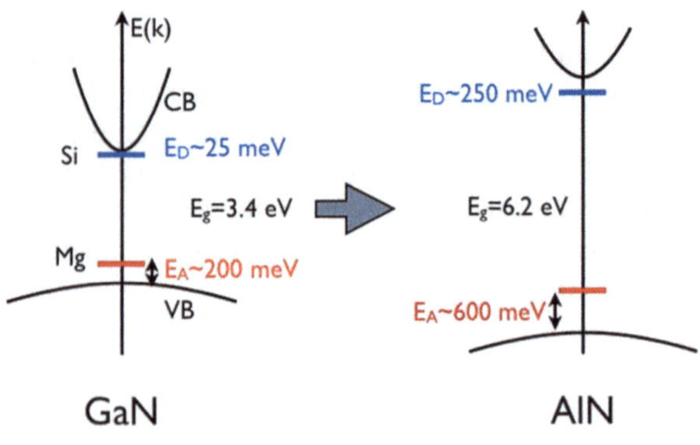

Fig. 10.35 Activation energy of dopants in GaN and AlN [76]

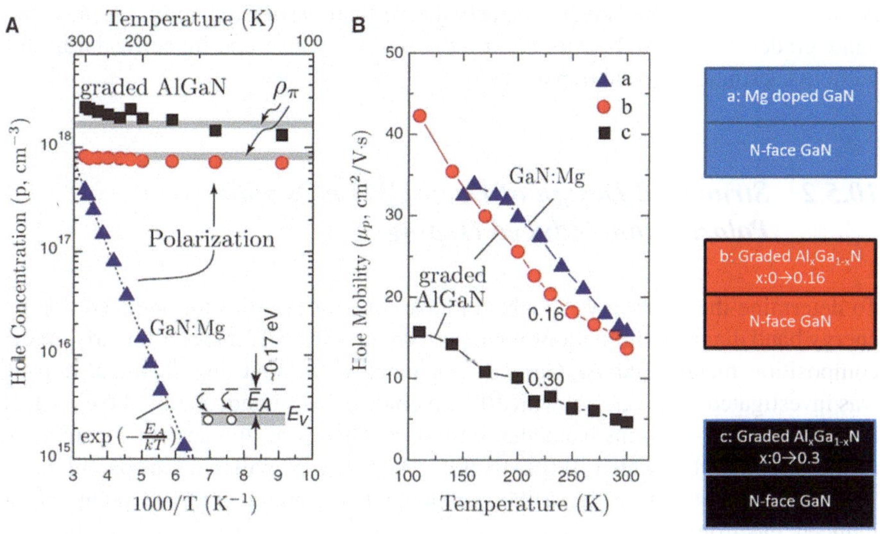

Fig. 10.36 Temperature-dependent hole concentration and mobility comparison between conventional and polarization-induced doping [76]

[77], compositionally graded AlGaN can induce 3-dimensional electron or hole gases (3DEGs or 3DHG). This technique has the potential advantage of reduced ionized impurity scattering, which results in high mobility for carriers even at low temperatures. Dopants are ionized by polarization-induced electric field, and this effect is independent of temperature. Therefore, ionized carriers do not freeze out at low temperatures. This can lead to realization of temperature-independent operation for devices. In Fig. 10.36 (Jena et al. [77]), it is shown that by using a compositionally graded-down $Al_xGa_{1-x}N$ alloy (x: 0.3→0) along the N-polar direction, a hole concentration of $\sim 3\times 10^{18}$ cm^{-3} could be achieved with high mobility. Both the mobility and the carrier concentrations were weakly dependent on temperature. On the contrary, at lower temperature there was complete carrier freeze-out for a conventional Mg-doped p-GaN layer.

This scheme of polarization-induced doping has been used by several research groups [11, 62, 78–81] to improve the carrier injection for deep-UV photonic devices. Besides graded alloys, the other form of using polarization-induced doping is the short period superlattice (SPSL) of $Al_xGa_{1-x}N/Al_yGa_{1-y}N$. The electric field induced by polarization charges at the sharp hetero-interfaces ionizes the dopants. The SPSL method improves the transport of carriers in the plane of the quantum wells, but the vertical transport needs to occur in minibands encountering potential barriers due to the band offsets, leading to a higher vertical resistance than lateral. The SPSL technique has been used widely in device applications [82–85]. The design of SPSLs becomes challenging if one requires these layers to remain optically transparent at shorter wavelengths (<250 nm). To maintain optical transparency, the thickness of AlN layers in the GaN/AlN SPSL needs to be increased, which negatively impacts the vertical carrier transport. On the other hand, graded alloy-based polarization-induced doping has superior vertical transport without sacrificing optical transparency.

10.5.2 Structural Design of Deep-UV LEDs with Polarization-Induced Doping

To determine the necessary AlGaN compositions for polarization-induced doping, energy band diagram simulations were performed using SiLENSe. A linearly graded composition metal polar $Al_xGa_{1-x}N$ p-n junction without any chemical dopants was investigated. $Al_xGa_{1-x}N$ with 50% Al content has a bandgap of ~4.6 eV which corresponds to ~270-nm bandedge emission. Therefore, the structure containing a 100-nm linearly graded 100% to 50% $Al_xGa_{1-x}N$ will be transparent to all wavelength >270 nm. Due to the compositional grading gradient, polarization-induced electrons and holes will be $\sim 2.5\times10^{18}$ cm^{-3} based on the direction of grading (graded-up: electrons; graded-down: holes, with metal polar growth direction). This charge is solely due to the grading gradient. Additional impurity doping (e.g., Si or Mg) can increase the mobile charge densities. In addition to the composition control, the polarization-induced charge density can also be controlled by changing the thickness of the graded layer. For example, if the thickness of the graded region is reduced to 50 nm, the charge density will double to 5×10^{18} cm^{-3}. The polarization-induced (pi-) doped UV LED structures discussed in the later part of this section make use of these design freedoms.

The effectiveness of pi-doping with deep-UV LEDs especially for the p-AlGaN layer has been demonstrated by Verma et al. [46]. As shown in Fig. 10.37, by replacing the nongraded p-AlGaN layer with pi-doped graded p-AlGaN, the light intensity of the deep-UV LED can be enhanced by 23 times. This experiment clearly demonstrates the benefits of pi-doping over the conventional p-AlGaN doping scheme.

Besides pi-doping, a tunnel injection into the active region scheme was used in the device design for the deep-UV LEDs mentioned in this section. The conventional over-the-barrier carrier injection processes into the active region quantum wells lead to the relaxation of hot carriers. The excess energy equal to the band offset minus the ground state energy of the well is dissipated by optical phonon emission. This leads to heat generation, which translates to loss. To overcome this loss, carriers can be cold-injected by tunneling through the thin barriers. In this technique, the heterostructure is designed in a way that the conduction band states of the doped p and n regions are aligned with the electronic ground states of the active region. The barriers are kept thin to enable tunneling.

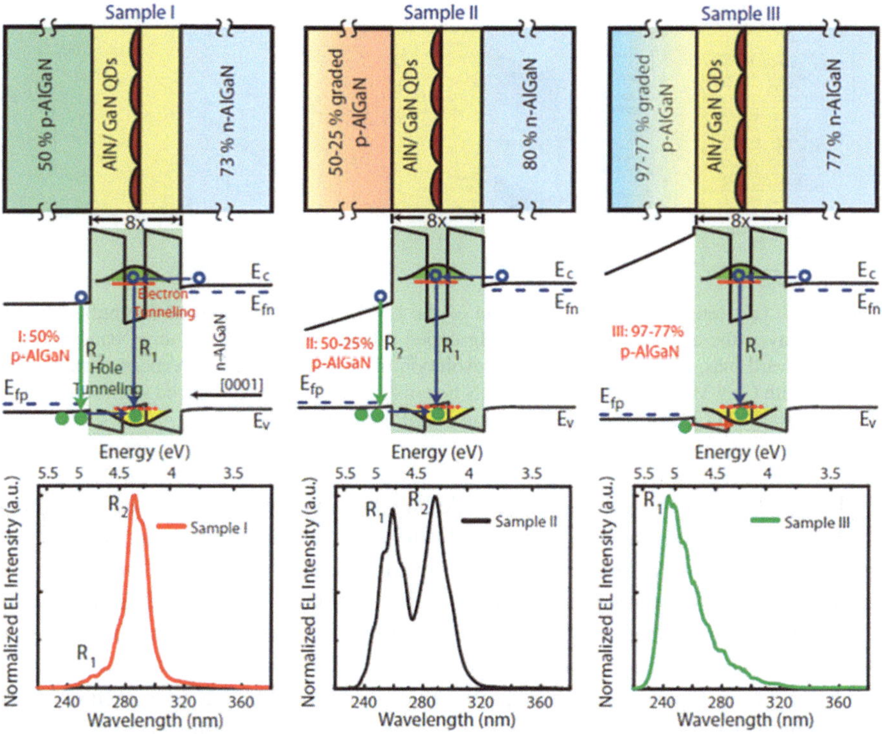

Fig. 10.37 Schematics and corresponding EL data demonstrating enhancement of emission from the active region of a DUV LED due to usage of a polarization-induced graded p-AlGaN cladding layer [11]

10.5.3 Tunable Deep-UV Emission over 232–270 nm

Deep-UV LEDs using ultrathin GaN QDs in the active region are discussed in this section. The emission wavelength has been tuned over 232–270 nm by changing the GaN layer thicknesses with monolayer (ML) precision. The pi-doping scheme is employed for both n and p contact regions to enhance the electrical injection of the carriers into the devices. As outlined in [77], a linearly graded-up metal polar AlGaN layer is used as the pi-doped n region. Similarly, a graded-down layer is used for the p-type doping layer.

The epitaxial structures for the UV-LEDs were grown on commercially available ~1 μm-thick AlN templates on ~430 μm-thick sapphire. These AlN templates have a typical threading dislocation density of 1×10^{10} cm^{-2}. Epilayers were grown by plasma-MBE. A substrate temperature of 700°C (thermocouple reading) was maintained for the crystal growth of the entire LED structure except the Mg-doped region as shown in Fig. 10.38a.

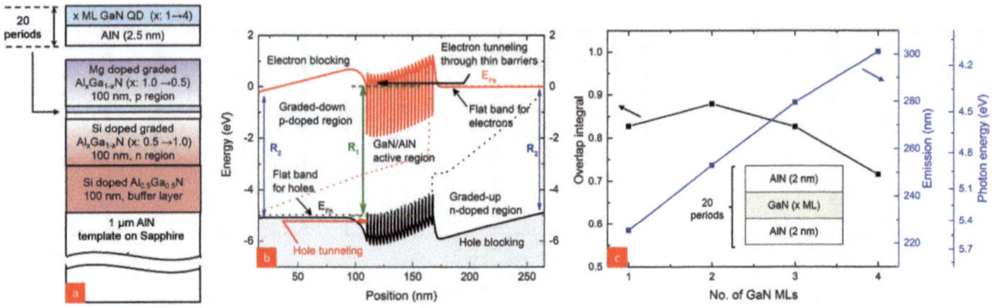

Fig. 10.38 (**a**) Heterostructure layer design of the UV LED devices showing the buffer, pi+Si-doped n layer, the active region, and pi+Mg-doped p layer, (**b**) simulated energy band diagram at 5V forward bias, and (**c**) simulation for evolution of emission wavelength/energy and overlap integral with GaN ML thickness for GaN/AlN heterostructures [47]. Reproduced from Appl. Phys. Lett. 111, 041108 (2017), with the permission of AIP Publishing

First, a 100-nm-thick Si-doped 50% AlGaN layer was grown followed by another 100 nm of linearly graded-up Si-doped AlGaN layer. Si doping concentration of $\sim 1\times 10^{20}$ cm^{-3} was used. This graded-up layer served to produce pi-doping for enhanced n-conductivity. The composition of this layer was varied from 50% to 100% $Al_xGa_{1-x}N$. For the graded layer, both the Al and Ga fluxes were adjusted simultaneously to keep the surface free from excessive Ga accumulation.

After the n-type graded layer, 20 periods of GaN/AlN active regions were grown at 700°C substrate temperature. The growth rate was adjusted to 0.28 ML/s throughout at 400W of nitrogen plasma power. The GaN QDs were grown using SK growth method. To ensure 3D GaN island formation, a Ga/N<1 flux ratio was maintained. The thicknesses of the GaN QDs were adjusted over 1–4 MLs by growth time and Ga/N ratio. Three samples were grown with varying GaN ML thicknesses (D1 (1–2 MLs); D2 (2–3 MLs); D3 (3–4 MLs)). The AlN barrier layer thickness was kept constant at 2 nm to enable tunneling transport through it. The AlN layers were grown using migration-enhanced epitaxy (MEE). Finally, the active region was capped with Mg-doped downgraded AlGaN as a pi-doped p-layer. The composition of the 100-nm p-AlGaN was varied from 100% to 50%. Mg doping concentration of 10^{19} cm^{-3} was used. To incorporate Mg into the crystal, the substrate temperature was reduced to 630°C during the growth of the p-region.

Figure 10.38b shows the simulated energy band diagram at 5V forward bias using SiLENSe. The usefulness of pi-doping is manifested here clearly. Besides ensuring high electron/hole concentrations in the doped regions, there are also built-in electron and hole blocking layers that prevent carriers injected into the active region from spilling over to the p and n regions. The flat bands for holes (p-side) and electrons (n-side) ensure smooth carrier transport into the active region. As only 2-nm-thick AlN barriers were used in the GaN/AlN superlattices, the carriers were injected by tunneling. This mechanism helps reducing the voltage drop across the active region and improves the overall efficiency of the device.

The evolution of the wavelength of emitted photons and the corresponding electron–hole overlap integrals for GaN quantum well (QW)/AlN heterostructures simulated using SiLENSe are shown in Fig. 10.38c. Based on the simulation, as the GaN thickness is reduced from 4ML→3ML→2ML→1ML the emission wavelength evolves as 290 nm→272 nm→250 nm→222 nm.

After growing the epitaxial layers, LED devices were fabricated. Mesa etch was performed to reach the n-type graded AlGaN region. A Cl_2-based reactive-ion etch was used to realize 200-nm-deep mesas. By using e-beam evaporation technique, Ti (20 nm)/Al (100 nm)/Ni (40 nm)/Au (50 nm) contact stacks were deposited as the n-contact metal. For the p-metal contact, first an Ni (5 nm)/Au (5 nm) current spreading layer was used. On top of this layer, a 50-μm diameter circular Ti (20 nm)/Au (100 nm) metal pad was deposited. The typical size of the UV LEDs investigated in this work was 200×200μm. Figure 10.39a shows the schematic of the processed device from the side. The SEM image taken from top of the device is shown in Fig. 10.39b.

Structural analysis using cross-sectional TEM was performed on the processed devices to investigate the layer structures, contacts, mesa, etc. Figure 10.39c shows a typical cross-sectional view of the device D2 (2–3-ML GaN QD). The grown layers matched the targeted thickness values. Grading in the Al composition was observed for both n and p layers by the change in the image contrast. The zoomed-in HAADF-STEM images are shown for devices D1–D3 in Fig. 10.39d–f. The thickness of the AlN barrier was ~2 nm for all the three samples. The GaN layer thicknesses match the target value of 1–2 ML (D1), 2–3 ML (D2), and 3–4 ML (D3).

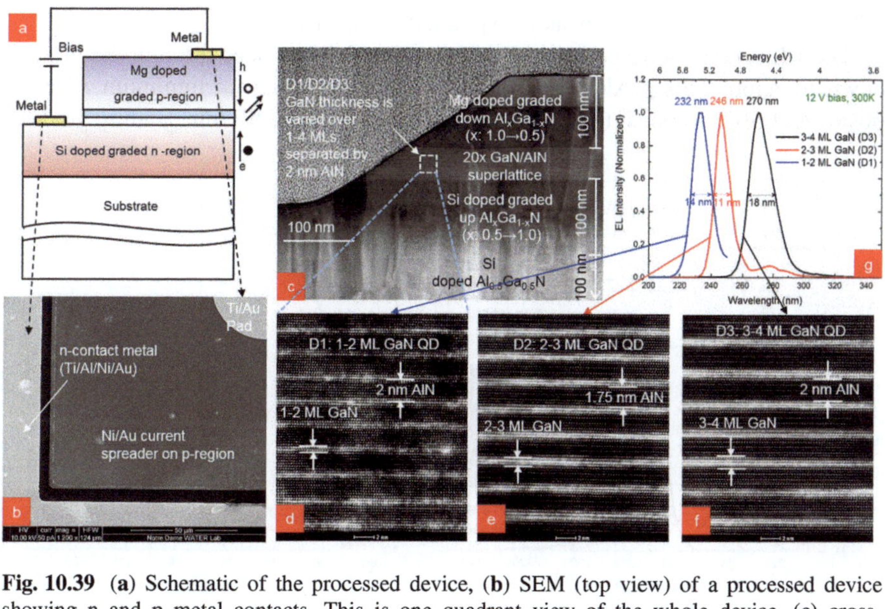

Fig. 10.39 (**a**) Schematic of the processed device, (**b**) SEM (top view) of a processed device showing n and p metal contacts. This is one quadrant view of the whole device, (**c**) cross-section HAADF-STEM image of a processed device showing different layers for a typical sample (D1/D2/D3), zoomed-in active region for (**d**) D1, (**e**) D2, (**f**) D3, and (**g**) EL spectra showing 232–270 nm tunable emission from 1–4-ML GaN active region [47]. Reproduced from Appl. Phys. Lett. 111, 041108 (2017), with the permission of AIP Publishing

For electrical characterization, room temperature electroluminescence (EL) measurements were performed on the samples. To avoid excessive heating, pulsed excitation under 10 kHz, 5% duty cycle was used for the characterization. All LEDs showed strong single peak emission at 300 K (Fig. 10.39g). A blue shift was observed in the EL emission peak wavelength from 270 nm→246 nm→232 nm as the thickness of the GaN layer was reduced. This trend was in good agreement with the simulation shown in Fig. 10.38c. The full width at half maximum (FWHM) was measured to be less than 20 nm (~325 meV) for all devices. Using GaN/AlN active region, the shortest EL wavelength reported before this was 239 nm [70]. This work demonstrates that by changing the thickness of GaN layer with an ML precision, it is possible to achieve tunable deep-UV LEDs emitting as short as 232 nm.

The EL data shown in Fig. 10.39g were measured with a 12-V forward bias. The turn-on voltages for the devices were ~7.5 V, larger than the bandgap and the peak emission wavelength energy. The undesired presence of AlN layers on either side of the active region was observed which can cause an increase in the turn-on voltage because of thicker tunneling barriers for the carriers. This can be avoided by further optimization of the graded layer growth conditions.

Figure 10.40a shows the comparison between the measured and the simulated EL spectra using SiLENSe. The simulation was performed with 1–2–3-ML GaN QW active region. The discrepancy in the peak wavelength between the simulation and measurement was due to the geometry of the GaN region in the experimental devices (QDs instead of uniform QWs). Also in Fig. 10.40b, besides the EL emission peak R1 from the active region, the presence of a secondary EL peak R_2 was seen around ~4.6 eV. The origin of this peak is explained in Fig. 10.38b.

The EL spectra were collected from the top of the devices through the graded-down p-region. This p-AlGaN region was designed to have a minimum Al composition of 50% to boost the polarization doping with a 0.5%/nm grading gradient

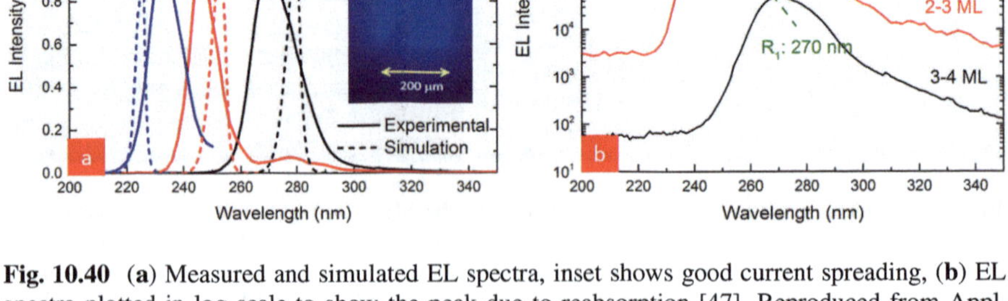

Fig. 10.40 (**a**) Measured and simulated EL spectra, inset shows good current spreading, (**b**) EL spectra plotted in log scale to show the peak due to reabsorption [47]. Reproduced from Appl. Phys. Lett. 111, 041108 (2017), with the permission of AIP Publishing

(graded down from 100% to 50% $Al_xGa_{1-x}N$ over 100 nm). The graded n-AlGaN was designed similarly with a 50% AlGaN as the minimum Al content to boost the polarization-induced doping and also to form a decent contact with the external Ti/Al/Ni/Au metal stack. The 100-nm-thick buffer n-AlGaN therefore had a 50% Al composition as well. $Al_{0.5}Ga_{0.5}N$ has a bandgap of 4.6 eV (~270 nm) and therefore, the 232-nm (246-nm) emission from D1 (D2) active regions was susceptible to absorption in $Al_{0.5}Ga_{0.5}N$. The photons emitted in the active region could get reabsorbed in the 100-nm n-type buffer region and partially in the graded p/n layers. The photons were therefore down converted to generate the secondary R_2 peaks. The secondary emission was relatively stronger for D1 compared to D2, which is explained by a thicker reabsorption volume in the graded layers for D1 due to higher energy R_1 peak (5.34 eV for D1 vs 5.04 eV for D2). On the contrary, D3 showed a single peak (R_1) dominated EL spectra since $Al_{0.5}Ga_{0.5}N$ is transparent to 270 nm. Transparency of the R1 peak can be achieved for D1 and D2 devices by choosing the appropriate minimum composition of p and n-AlGaN, i.e., $Al_{0.7}Ga_{0.3}N$. This will ensure transparency though with a compromise in the boost in polarization doping.

In summary, tunable deep-UV electroluminescence over 232–270 nm has been demonstrated using few monolayers thick GaN quantum structures with AlN barriers on AlN-on-sapphire templates. Emission wavelength as short as 232 nm has been achieved using 1–2-ML GaN quantum structures with a spectral linewidth of 14 nm at 300K. Plasma-assisted molecular beam epitaxy was used to maintain precise control on the thickness of the grown epitaxial layers. Tunneling transport through the barriers and pi-doping was used to enhance electroluminescence in the deep-UV.

10.5.4 Cryogenic Operation of a Deep-UV LED

In this section, a pi-doped and tunneling-injection-assisted deep-UV LED device is discussed that can emit light at cryogenic temperature. The device also showed negative-differential-resistance (NDR) phenomena which is unheard of with regard to a DUV photonic device [75].

The LED structure was grown on a commercially available metal polar 1-μm-thick AlN template on sapphire with a threading dislocation density of $\sim 1\times 10^{10}$ cm^{-2}. For deep-UV light-emitting region, 1–2 monolayers (MLs)-thick GaN layers were inserted in between 10-ML (2.5-nm)-thick AlN barriers. A linearly graded-down buffer $Al_xGa_{1-x}N$ region was introduced between the AlN substrate and the pi-doped upgraded n-region to eliminate any abrupt change in composition and reduce formation of threading dislocations due to lattice mismatch between AlGaN and AlN substrate.

For the MBE growth of the structure, a 5-nm AlN nucleation layer was first grown at 700°C substrate temperature (T_{sub}) using migration-enhanced epitaxy (MEE). Next, as shown in Fig. 10.41a, there were three ~100-nm linearly graded AlGaN layers in the LED structure:

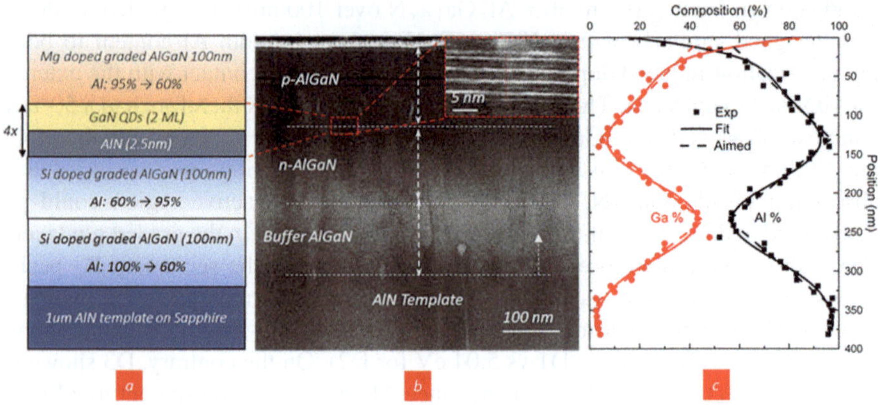

Fig. 10.41 (**a**) Schematic of the deep-UV LED, (**b**) Z-contrast STEM micrograph shows smooth contrast variation for linearly graded AlGaN regions, ML-thick GaN/AlN active region can be observed in the inset, and (**c**) EDX scan across the device showing linear grading of the AlGaN composition with desired compositions [75]

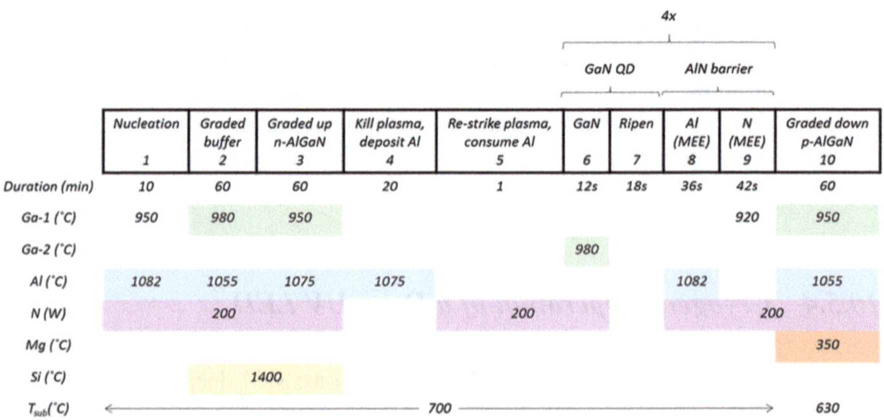

Fig. 10.42 Flux-time diagram showing detailed MBE growth parameters for the LED shown in Fig. 10.41 [63]

(i) Graded-down Si-doped buffer layer (T_{sub} = 700°C),
(ii) Graded-up pi-induced Si-doped n-type layer (T_{sub} = 700°C), and
(iii) Graded-down pi-induced Mg-doped p-type layer (T_{sub} = 630°C).

The substrate temperature was reduced to 630°C for the growth of the p-region to promote the incorporation of Mg into the crystal.

A typical growth flux-time diagram is shown in Fig. 10.42. The duration for growth of each layer, shutter position of effusion cells, cell temperatures, substrate temperature, dopant cell temperature, etc. are mentioned in this diagram. The Al and Ga cell temperatures were controlled by a computer program to obtain a linear variation of fluxes with growth time. After the n-graded-up layer, the plasma was turned-off, and sample was annealed at 700°C for 20 min to desorb any excess Ga from the surface which accumulates due to metal-rich AlGaN layer growth.

The doping concentrations were 5×10^{19} cm^{-3} (Si) and 1×10^{19} cm^{-3} (Mg) as determined by SIMS calibration growths performed separately. For the graded layer growths, both Al and Ga metal fluxes were varied to avoid significant accumulation of metal on the growth surface. This was confirmed by the observation of streaky RHEED pattern throughout the crystal growth. The growth rate for the whole structure was 0.17 ML/sec. This was accurately estimated from RHEED intensity oscillations and TEM analysis. A precise knowledge of growth rate was essential to grow the ultrathin GaN-/AlN-based active region.

Four periods of ultrathin GaN quantum disk (QD)/AlN barrier heterostructure were inserted between the doped regions as the deep-UV light-emitting active region. The 2 MLs GaN QDs were grown in the SK mode. To ensure such a growth mode, Ga/N ratio was kept <1 for the GaN QDs. MEE method was used to grow 10 MLs (2.5 nm) of AlN barriers.

The AlN barriers were kept thin (2.5 nm) to allow sub-barrier tunneling of carriers into GaN QDs. The active region was designed for an emission wavelength around 250 nm. A minimum Al content for the graded p- and n-type AlGaN layers was designed to be 60% to guarantee optical transparency to 250-nm emission because $Al_{0.6}Ga_{0.4}N$ has an absorption edge <250 nm.

STEM micrographs shown in Fig. 10.41b and the EDX profile in Fig. 10.41c show gradual changes in the image contrast associated with Al and Ga compositional grading. The experimentally measured concentration profile matched closely with the values that were targeted. The zoomed-in image for the active region is shown in the inset of Fig. 10.41b. It confirmed 4 layers of GaN QDs inserted in ~2-nm AlN barriers.

After the MBE growth, a mesa etch was performed using Cl_2-based ICP-RIE to reach the graded n-AlGaN region to form the n-type contacts. Ti (20 nm)/Al (100 nm)/Ni (40 nm)/Au (50 nm) metal stacks were deposited as n-type contacts using e-beam evaporation. For the p-contact, first an Ni (5 nm)/Au (5 nm) current spreading layer was deposited followed by deposition of a smaller area Ti (20 nm)/Au (100 nm) metal pads for probing.

The simulated energy band diagram using SiLENSe for the LED device at 5V forward bias is shown in Fig. 10.43a,b. The graded pi-doped regions provide a flat band profile for smooth carrier transport for both holes on the p-side and electrons on the n-side. The tunnel injection mechanism is schematically shown in Fig. 10.43b. At a forward bias voltage of ~5 V, the energy of electrons in conduction band lines up with carrier eigenstates in 2-ML QDs. The AlN barriers in the active region being thin, the carriers can tunnel through the barriers into the GaN QD eigenstates.

As a part of the device characterization, current density vs. voltage (J-V) and electroluminescence (EL) vs. current/voltage measurements were performed. All examined devices had sizes larger than 100 μm $\times$100 μm. The J-V characteristics revealed a very interesting phenomenon not often observed in bipolar optical

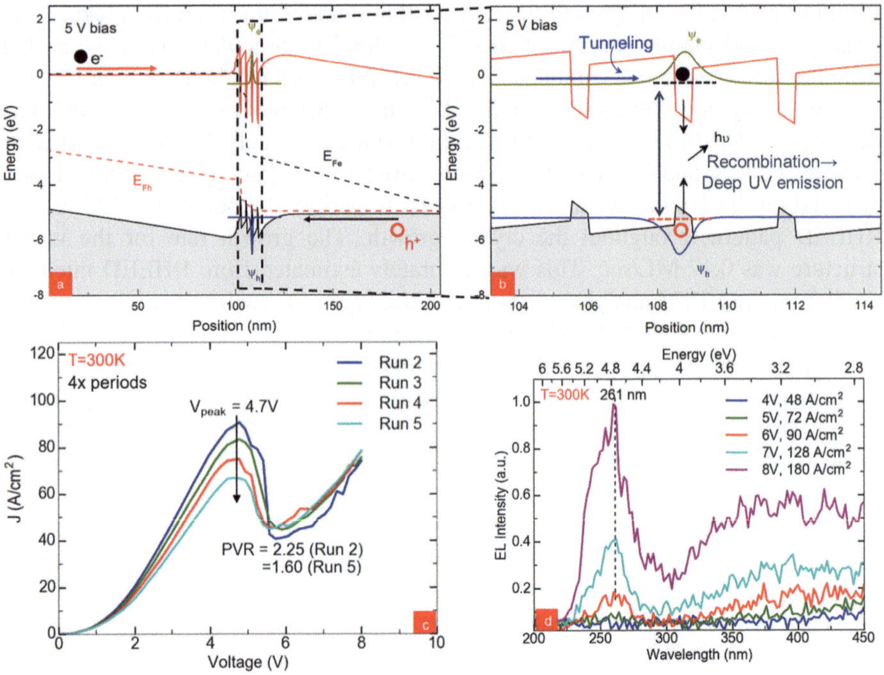

Fig. 10.43 (**a**) Energy band diagram for the active region at 5 V forward bias showing tunnel injection of carriers and (**b**) light emission process due to recombination. (**c**) Tunneling is evidenced by NDR in J-V. (**d**) Light emission turns-ON at NDR voltage [63]

devices, namely negative differential resistance (NDR). Figure 10.43c exhibits J-V sweeps on the UV LED device. Starting from the second sweep, clear NDR is observed at ~4.8 V.

On some devices, the NDR sustained for more than 20 scans, but the peak-to-valley ratio (PVR) degraded over subsequent sweeps. The highest PVR measured was 2.25 for the second scan. The diminishing NDR could be due to charge trapping at defect levels in the AlN barriers. NDR is a signature of tunneling through the barriers into well-defined electronic eigenstates of QDs as often seen in resonant tunnel diodes (RTDs) [86]. For the LEDs, a turn-on voltage around ~5V was expected but the current started to flow at a low bias (<0.5 V). This was probably the leakage current due to the growth on AlN/sapphire templates with high $\sim 1 \times 10^{10}$ cm^{-2} density of threading dislocations (TDs). Each 100 μm × 100 μm LED device therefore should contain $\sim 10^6$ TDs and thus might be leaky. Since the tunneling mechanism and the electroluminescence should proceed through the same electronic levels in the GaN regions, a correlation between the NDR and the EL on-set was expected. Indeed, the EL turn-on took place at the same voltage range where the NDR was detected. This was further confirmed from the L-I characteristics plotted in Fig. 10.44b.

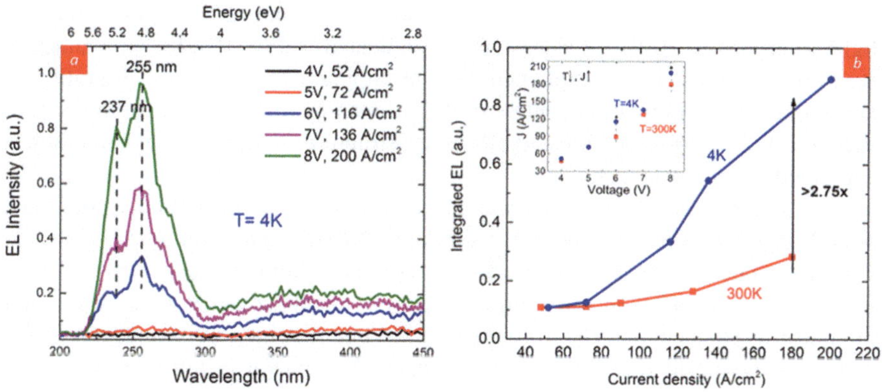

Fig. 10.44 (**a**) Normalized electroluminescence at 4K, and (**b**) comparison of integrated EL between 300K and 4K. Inset shows increase of conductivity of device at low temperature [63]

Figure 10.43d shows the corresponding EL spectra taken at 300K under various injection currents. To avoid excessive heating, pulsed current injection at 10 kHz, 5% duty cycle was used. The 261-nm peak emission with a shoulder around 240 nm appeared first between 5V and 6V biases, in the NDR region. This indicated that the EL and the tunneling mechanisms were closely related, due to tunneling injection of carriers into the GaN active area. A monotonic increase of EL intensity vs. applied bias was observed over 5V→8V range. Above 8V forward bias, the top metal contact started to degrade due to heat dissipation in the p-contact layer. Therefore, for longer measurements the LEDs were limited to biases less than 8V.

Because of the pi-induced doping and the tunneling mechanism for carrier injection both of which are temperature-independent processes, the LEDs examined here were supposed to operate at cryogenic temperatures. To confirm this, the J-V and EL vs. J measurements were performed at temperatures down to 4K. The results are shown in Fig. 10.44a. The low-temperature measurements provided strong evidence that the devices were functional even at 4K. Strong EL was detected and the peak position blue-shifted from 261 nm to 255 nm according to Varshni's law as the temperature was lowered from 300K to 4K. A secondary peak at 237 nm became clearly visible at low temperatures. This is described by the bimodal distribution of GaN QD thicknesses. A prominent 237-nm peak at low temperature indicated suppressed carrier transport between QDs at cryo-temperatures. In contrast, at room temperature the carriers were free to find the lowest energy levels for recombination which were the thicker 2-ML QDs and therefore the ~260-nm emission dominated as seen in Fig. 10.43d. The LED turn-on voltage stayed still between 5–6V at 4K, which coincided with the NDR region. Higher current densities at 4K seen in the inset of Fig. 10.44b indicated that the conductivity of the graded layers increased at lower temperatures. Measurement of EL at 4K with higher contact conductivity than 300K confirmed the effectiveness of polarization doping.

The integrated EL intensity at 4K exhibited more than 275% enhancement of light emission vs. 300K measurements as indicated in Fig. 10.44b. This enhancement of EL emission is explained by the suppression of carrier migration to non-radiative recombination centers such as threading dislocations. A similar ratio was detected for PL of such GaN/AlN heterostructures under 157-nm excitation as explained in Sect. 10.4.2.4.

As a control sample, an LED was fabricated with the conventional nongraded AlGaN layers for contacts but with a similar GaN/AlN light-emitting active region. For such a device, conventional temperature-assisted doping is expected to lead to more resistive contacts at low temperatures due to carrier freeze-out. This should dramatically reduce the electroluminescence intensity at low temperature. The control sample had a 200-nm-thick Si-doped $Al_{0.77}Ga_{0.23}N$ layer as the n-type contact and a 100-nm-thick Mg-doped $Al_{0.5}Ga_{0.5}N$ as the p-type contact. The active region GaN/AlN QD/barrier structure was designed to emit around 300 nm. The EL spectra for the control sample showed strong room temperature EL emission with an emission peak at 298 nm. Temperature-dependent EL measurements showed more than 20 times reduction in the intensity as the device temperature was cooled down to 80K, as shown in Fig. 10.45b.

This reduction is explained by the carrier freeze-out at cryogenic temperatures in the nongraded AlGaN layers. Figure 10.45a on the other hand shows more than 2.75 times enhancement in light emission at 4K. The control sample investigated here had an emission at 298 nm. Carrier injection for <260-nm LEDs using conventional doping becomes even more challenging due to more resistive contacts with higher Al-containing AlGaN layers. Therefore, the carrier freeze-out effect for <260-nm LED with conventional doping can be more severe. Putting this together, it is concluded that the pi-doping enhances the light emission at least 55 times ($>2.75\times20$) at cryogenic temperature compared to the conventional doping scheme.

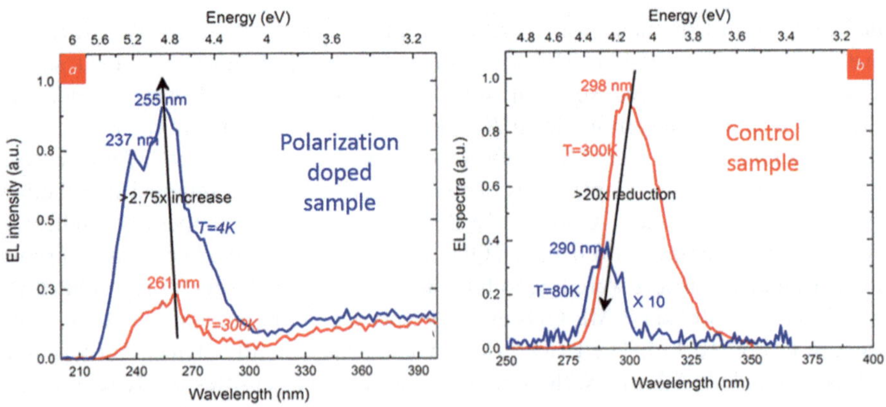

Fig. 10.45 Comparative temperature-dependent EL spectra for (**a**) pi-doped and (**b**) conventionally doped (control sample)

In summary, by exploiting the polarization-induced doping, improved operation of deep-UV LEDs has been achieved at cryogenic temperatures. The LED devices were operated at ~237–260 nm over the temperature range of 4K–300K. 1–2-monolayer-thick GaN quantum disks sandwiched between 10-ML (2.5-nm) AlN barriers were used to generate the deep-UV emission. Carriers were injected into the GaN active layers by tunneling through the AlN barriers. Tunneling was confirmed by the presence of NDR in forward bias J-V characteristics. Improvement in conductivity of the devices due to pi-doping at low temperature was observed. Enhancement in light emission of more than 275% was observed at cryogenic temperature. Also, the effectiveness of pi-doping at lower temperatures was confirmed by comparing the EL with a conventional impurity-doped UV LED device. This demonstration of cryogenic operation of deep-UV LEDs will be attractive in design of light emitters/sensors for low-temperature deep-space applications.

10.6 Enhancement of Light Extraction with Ultrathin GaN/AlN Quantum Heterostructure Deep-UV LEDs

The challenge of light extraction for deep-UV LEDs has been discussed earlier in Sects. 10.2.3 and 10.3.2. For the conventional AlGaN alloy-based active region design, as the Al content is increased to achieve <250-nm deep-UV emission, the valence band structure changes in a way that favors transverse-magnetic (TM) polarized light emission over transverse electric (TE) [51, 52, 70, 87]. For a typical surface-emitting LED, TM polarized light extraction is challenging due to total internal reflection. On the contrary, the valence band structure of GaN allows for isotropic distribution of light emission. The AlN distribution is strongly towards the edge and thus not suitable for a surface-emitting LED.

The ultrathin GaN/AlN quantum heterostructure-based surface-emitting LEDs are therefore promising from the light extraction perspective due to isotropic radiation pattern. Liu et al. [87] performed a careful study of theoretical analysis and experimental measurement of polarization of emitted light for a 298-nm GaN/AlN quantum structure-based UV-LED.

The LED investigated in this study was grown by plasma-assisted MBE on standard metal-polar AlN template on sapphire substrate. First, an MBE AlN nucleation layer is grown followed by Si-doped 200-nm n-AlGaN layer with 77% Al content. The active region of the LED was formed with 3–4-monolayer GaN QDs sandwiched in 2.5-nm AlN tunneling barriers. 8 repeats of the heterostructures were used to enhance the light emission intensity. The AlN barrier layers were formed using migration-enhanced epitaxy technique. The 3–4 monolayers GaN were grown in SK mode aiming for an emission wavelength of 300 nm. The p-AlGaN layer was grown as a 100-nm Mg-doped AlGaN with 50% Al content. The composition of the layers was confirmed based on HR-XRD and HR-TEM analysis. For device processing, standard mesa etch was performed to form n-contact to the

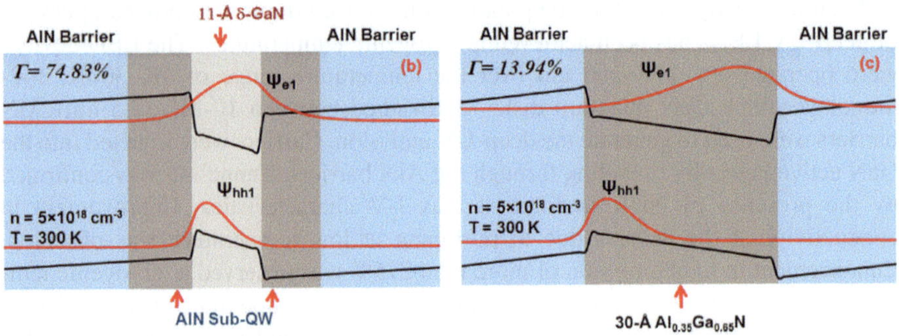

Fig. 10.46 Simulated energy band diagrams and electron–hole wavefunctions comparing ultrathin GaN/AlN and AlGaN/AlN heterostructures for 298-nm emission [87]. Reproduced from Appl. Phys. Lett. 110, 071103 (2017), with the permission of AIP Publishing

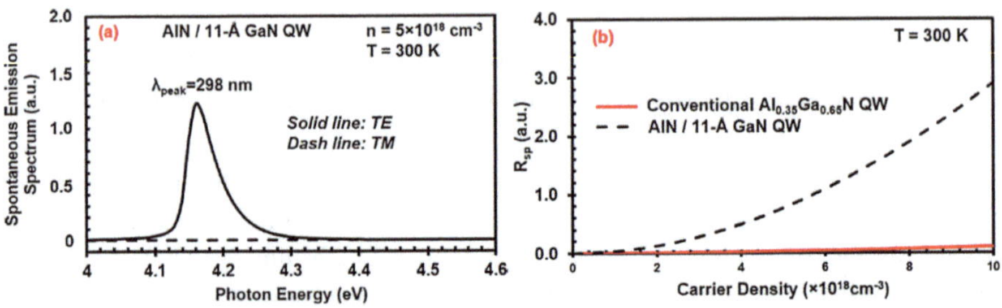

Fig. 10.47 (**a**) Simulated spontaneous emission spectra for the GaN/AlN structure showing strong TE emission with a 298-nm peak wavelength, and (**b**) enhanced spontaneous emission rate is expected for GaN/AlN compared to AlGaN/AlN design [87]. Reproduced from Appl. Phys. Lett. 110, 071103 (2017), with the permission of AIP Publishing

bottom n-AlGaN region. Ti/Al/Ni/Au metal stacks were used for the n-contact. Semitransparent Ni/Au metal stacks were put for contact to the p-AlGaN followed by Ti/Au contact pad. None of the contacts were annealed.

Two different heterostructure designs were simulated for comparison, one containing the GaN/AlN active region and the other having AlGaN/AlN aiming for ∼300-nm emission wavelengths. The simulated energy band diagrams and the corresponding electron–hole wavefunctions are depicted in Fig. 10.46. The significantly higher overlap of 74.83% for the GaN/AlN structure compared to 13.94% for the 3-nm AlGaN/AlN quantum well design is a consequence of QCSE effect as discussed in Sect. 10.2.1.2.

This large overlap integral further predicts higher spontaneous emission rate for the GaN/AlN structure compared to AlGaN/AlN. Figure 10.47a shows the simulated spontaneous emission spectra for the GaN/AlN design. A 298-nm peak emission wavelength is calculated for 1.1-nm GaN QW/AlN barrier design. The light emission pattern is calculated to be strongly TE polarized.

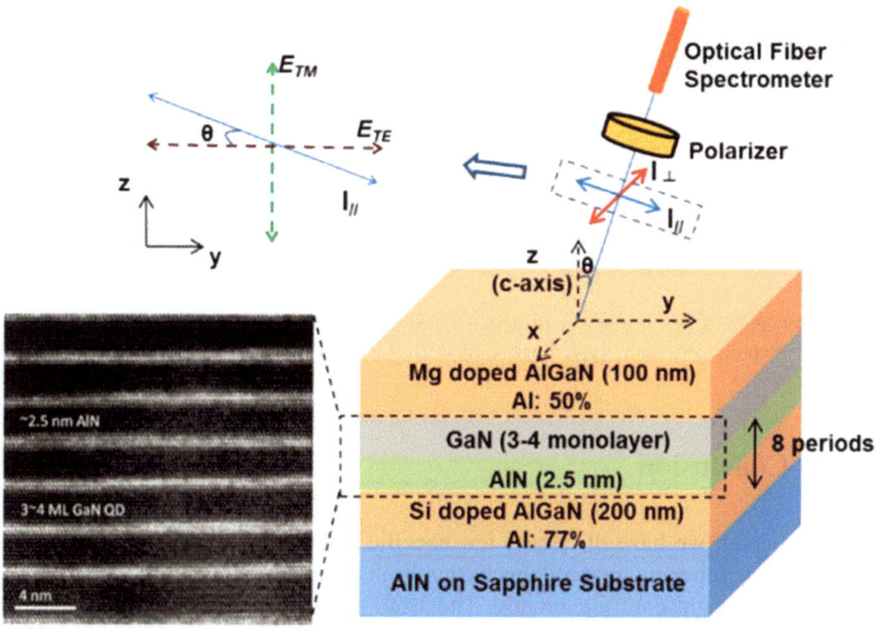

Fig. 10.48 Schematic diagram of the polarization-dependent EL measurement setup and cross-sectional TEM image of the AlN/GaN QW-like structure, the θ stands for the angle between c-axis and collecting fiber. The polarizer is held on a rotational stage to achieve the intensity in y-z plane and the intensity along x direction [87]. Reproduced from Appl. Phys. Lett. 110, 071103 (2017), with the permission of AIP Publishing

Figure 10.47b compares the spontaneous emission rate between GaN/AlN and AlGaN/AlN designs over a wide range of injected carried densities. Figure 10.47a shows the simulated spontaneous emission spectra for the GaN/AlN design. The enhanced spontaneous rate from GaN/AlN confirms what was predicted from the overlap integral simulation from Fig. 10.46.

Angle-resolved electroluminescence measurement was performed on the processed GaN/AlN quantum structure-based LED devices, and polarization of the emitted light was extracted. The schematic setup of such measurement is shown in Fig. 10.48. Strong electroluminescence around 300 nm was measured. The extracted TE/TM ratio at 30° respective to the surface normal to LED was found to be 15.3 as shown in Fig. 10.49. At another collection angle of 45°, the extracted TE/TM ratio of 4.05 still showed TE-dominant emission from the device.

In summary, ultrathin GaN/AlN quantum heterostructure-based LEDs were fabricated as a 298-nm UV-LED, and the polarization property of the emitted light was simulated and measured experimentally. The simulation suggested significantly enhanced TE/TM ratio of emitted light for the GaN/AlN design. The measurement confirmed such prediction over a wide range of angular range. This specific distribution of emitted light pattern will enhance the light extraction efficiency and therefore further boost the overall external quantum efficiency of UV-LEDs.

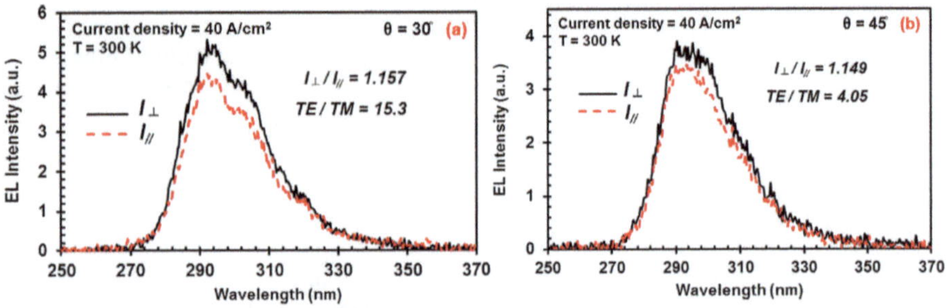

Fig. 10.49 Polarization-dependent EL spectra measured at two different angles, (**a**) 30° and (**b**) 45° [87]. Reproduced from Appl. Phys. Lett. 110, 071103 (2017), with the permission of AIP Publishing

10.7 Conclusion

In this chapter, we have discussed few specific heterostructure design methods to overcome the fundamental bottlenecks that limit external quantum efficiency (EQE) for deep-UV light-emitting diodes. The EQE is composed of internal quantum efficiency (IQE), injection efficiency (INJ), and light extraction efficiency (LEE). Ultrathin GaN/AlN quantum heterostructure can be used to enhance the IQE by improving the electron–hole wavefunction overlap. The use of quantum disks instead of conventional quantum wells can further boost the IQE with additional carrier confinement that prevents non-radiative recombination through defects. Use of compositionally graded AlGaN heterostructures for field-assisted polarization-induced doping can lead to the pathway of enhanced injection efficiency. The doping efficiency for such temperature-independent design is enhanced compared to the conventional thermally assisted dopant activation in AlGaN films, and thus deep-UV LEDs operating at 4K can be realized. Finally, the light emission property of GaN being TE dominated compared to strongly TM-dominated AlGaN in the deep-UV range, the light extraction efficiency can be significantly enhanced with GaN/AlN quantum heterostructures. The enhancement of each of the IQE, INJ, and LEE has been discussed with simulation and experimental data. The experimental analysis consisted of few details of the crystal growth, device processing, and measurement. The combination of all the design rules presented here will help to enhance the current state-of-the-art low external quantum efficiency of deep-UV light-emitting diodes.

References

1. T. Mukai, D. Morita, S. Nakamura, High-power UV InGaN/AlGaN double-heterostructure LEDs. J. Cryst. Growth **189–190**, 778–781 (1998)
2. J.P. Theodore Moustakas, Gallium-Nitride (GaN) II, Vol. 57 (Academic Press, 1998)
3. Y. Taniyasu, M. Kasu, T. Makimoto, An aluminium nitride light-emitting diode with a wavelength of 210 nanometres. Nature **441**(7091), 325–328 (2006)

4. A. Khan, K. Balakrishnan, T. Katona, Ultraviolet light-emitting diodes based on group three nitrides. Nat. Photonics **2**(2), 77–84 (2008)
5. C. Pernot, M. Kim, S. Fukahori, T. Inazu, T. Fujita, Y. Nagasawa, A. Hirano, M. Ippommatsu, M. Iwaya, S. Kamiyama, I. Akasaki, H. Amano, Improved efficiency of 255–280 nm AlGaN-based light-emitting diodes. Appl. Phys. Express **3**(6) (2010)
6. H. Hirayama, Y. Tsukada, T. Maeda, N. Kamata, Marked enhancement in the efficiency of deep-ultraviolet AlGaN light-emitting diodes by using a multiquantum-barrier electron blocking layer. Appl. Phys. Express **3**(3) (2010)
7. M. Shatalov, W. Sun, A. Lunev, X. Hu, A. Dobrinsky, Y. Bilenko, J. Yang, M. Shur, R. Gaska, C. Moe, G. Garrett, M. Wraback, AlGaN deep-ultraviolet light-emitting diodes with external quantum efficiency above 10%. Appl. Phys. Express **5**(8) (2012)
8. F. Bernardini, V. Fiorentini, Nonlinear behavior of spontaneous and piezoelectric polarization in III-V nitride alloys. Phys. Status Solidi (a) **190**(1), 65–73 (2002)
9. T. Takano, T. Mino, J. Sakai, N. Noguchi, K. Tsubaki, H. Hirayama, Deep-ultraviolet light-emitting diodes with external quantum efficiency higher than 20% at 275 nm achieved by improving light-extraction efficiency. Appl. Phys. Express **10**(3) (2017)
10. M. Kneissl, T. Kolbe, C. Chua, V. Kueller, N. Lobo, J. Stellmach, A. Knauer, H. Rodriguez, S. Einfeldt, Z. Yang, N.M. Johnson, M. Weyers, Advances in group III-nitride-based deep UV light-emitting diode technology. Semicond. Sci. Technol. **26**(1) (2011)
11. J.K. Verma, Polarization and band gap engineered III-nitride optoelectronic device structures. Doctoral Dissertation, University of Notre Dame, 2013
12. S. Karpov, ABC-model for interpretation of internal quantum efficiency and its droop in III-nitride LEDs: a review. Opt. Quant. Electron. **47**(6), 1293–1303 (2015)
13. J. Piprek, Efficiency droop in nitride-based light-emitting diodes. Phys. Status Solidi (a) **207**(10), 2217–2225 (2010)
14. M.-H. Kim, M.F. Schubert, Q. Dai, J.K. Kim, E.F. Schubert, J. Piprek, Y. Park, Origin of efficiency droop in GaN-based light-emitting diodes. Appl. Phys. Lett. **91**(18) (2007)
15. G. Verzellesi, D. Saguatti, M. Meneghini, F. Bertazzi, M. Goano, G. Meneghesso, E. Zanoni, Efficiency droop in InGaN/GaN blue light-emitting diodes: Physical mechanisms and remedies. J. Appl. Phys. **114**(7) (2013)
16. S.F. Chichibu, H. Marchand, M.S. Minsky, S. Keller, P.T. Fini, J.P. Ibbetson, S.B. Fleischer, J.S. Speck, J.E. Bowers, E. Hu, U.K. Mishra, S.P. DenBaars, T. Deguchi, T. Sota, S. Nakamura, Emission mechanisms of bulk GaN and InGaN quantum wells prepared by lateral epitaxial overgrowth. Appl. Phys. Lett. **74**(10), 1460–1462 (1999)
17. Q. Dai, M.F. Schubert, M.H. Kim, J.K. Kim, E.F. Schubert, D.D. Koleske, M.H. Crawford, S.R. Lee, A.J. Fischer, G. Thaler, M.A. Banas, Internal quantum efficiency and nonradiative recombination coefficient of GaInN/GaN multiple quantum wells with different dislocation densities. Appl. Phys. Lett. **94**(11) (2009)
18. M. Soltani, R. Soref, T. Palacios, D. Englund, AlGaN/AlN integrated photonics platform for the ultraviolet and visible spectral range. Opt Express **24**(22), 25415–25423 (2016)
19. H. Hirayama, S. Fujikawa, J. Norimatsu, T. Takano, K. Tsubaki, N. Kamata, Fabrication of a low threading dislocation density ELO-AlN template for application to deep-UV LEDs. Phys. Status Solidi (c) **6**(S2), S356–S359 (2009)
20. M. Conroy, V.Z. Zubialevich, H. Li, N. Petkov, J.D. Holmes, P.J. Parbrook, Epitaxial lateral overgrowth of AlN on self-assembled patterned nanorods. J. Mater. Chem. C **3**(2), 431–437 (2015)
21. V.N. Jmerik, E.V. Lutsenko, S.V. Ivanov, Plasma-assisted molecular beam epitaxy of AlGaN heterostructures for deep-ultraviolet optically pumped lasers. Phys. Status Solidi (a) **210**(3), 439–450 (2013)
22. R. Dalmau, Z. Sitar, AlN bulk crystal growth by physical vapor transport, in *Springer Handbook of Crystal Growth*, ed. by G. Dhanaraj, K. Byrappa, V. Prasad, M. Dudley (Springer, Berlin, Heidelberg, 2010), pp. 821–843
23. J.S. Speck, S.J. Rosner, The role of threading dislocations in the physical properties of GaN and its alloys. Phys. B Condens. Matter **273–274**, 24–32 (1999)

24. R. Collazo, J. Xie, B.E. Gaddy, Z. Bryan, R. Kirste, M. Hoffmann, R. Dalmau, B. Moody, Y. Kumagai, T. Nagashima, Y. Kubota, T. Kinoshita, A. Koukitu, D.L. Irving, Z. Sitar, On the origin of the 265 nm absorption band in AlN bulk crystals. Appl. Phys. Lett. **100**(19) (2012)
25. R. Dalmau, B. Moody, R. Schlesser, S. Mita, J. Xie, M. Feneberg, B. Neuschl, K. Thonke, R. Collazo, A. Rice, J. Tweedie, Z. Sitar, Growth and characterization of AlN and AlGaN epitaxial films on AlN single crystal substrates. J. Electrochem. Soc. **158**(5) (2011)
26. K. Zhu, M.L. Nakarmi, K.H. Kim, J.Y. Lin, H.X. Jiang, Silicon doping dependence of highly conductive n-type Al0.7Ga0.3N. Appl. Phys. Lett. **85**(20), 4669–4671 (2004)
27. R. France, T. Xu, P. Chen, R. Chandrasekaran, T.D. Moustakas, Vanadium-based Ohmic contacts to n-AlGaN in the entire alloy composition. Appl. Phys. Lett. **90**(6) (2007)
28. J.P. Zhang, A. Chitnis, V. Adivarahan, S. Wu, V. Mandavilli, R. Pachipulusu, M. Shatalov, G. Simin, J.W. Yang, M.A. Khan, Milliwatt power deep ultraviolet light-emitting diodes over sapphire with emission at 278 nm. Appl. Phys. Lett. **81**(26), 4910–4912 (2002)
29. M.L. Nakarmi, K.H. Kim, M. Khizar, Z.Y. Fan, J.Y. Lin, H.X. Jiang, Electrical and optical properties of Mg-doped Al0.7Ga0.3N alloys. Appl. Phys. Lett. **86**(9) (2005)
30. Y. Bilenko, A. Lunev, X. Hu, J. Deng, T.M. Katona, J. Zhang, R. Gaska, M.S. Shur, W. Sun, V. Adivarahan, M. Shatalov, A. Khan, 10 Milliwatt pulse operation of 265 nm AlGaN light emitting diodes. Jpn. J. Appl. Phys. **44**(3), L98–L100 (2005)
31. H. Yu, E. Ulker, E. Ozbay, MOCVD growth and electrical studies of p-type AlGaN with Al fraction 0.35. J. Cryst. Growth **289**(2), 419–422 (2006)
32. T. Kinoshita, T. Obata, H. Yanagi, S.-i. Inoue, High p-type conduction in high-Al content Mg-doped AlGaN. Appl. Phys. Lett. **102**(1) (2013)
33. T. Mori, K. Nagamatsu, K. Nonaka, K. Takeda, M. Iwaya, S. Kamiyama, H. Amano, I. Akasaki, Crystal growth and p-type conductivity control of AlGaN for high-efficiency nitride-based UV emitters. Phys. Status Solidi (c) **6**(12), 2621–2625 (2009)
34. K.P. Streubel, T. Passow, H. Jeon, R. Gutt, M. Maier, L.-W. Tu, N. Linder, W. Pletschen, M. Kunzer, R. Schmidt, J. Wiegert, D. Luick, S. Liu, K. Köhler, J. Wagner, Ni/Ag as low resistive ohmic contact to p-type AlGaN for UV LEDs, in *Light-Emitting Diodes: Materials, Devices, and Applications for Solid State Lighting* XIV, 2010
35. B.A. Hull, S.E. Mohney, U. Chowdhury, R.D. Dupuis, Ohmic contacts to p-type Al0.45Ga0.55N. J. Appl. Phys. **96**(12), 7325–7331 (2004)
36. H.-K. Kim, T.-Y. Seong, I. Adesida, C.W. Tang, K.M. Lau, Low-resistance Pt/Pd/Au ohmic contacts to p-type AlGaN. Appl. Phys. Lett. **84**(10), 1710–1712 (2004)
37. M. Shatalov, W. Sun, R. Jain, A. Lunev, X. Hu, A. Dobrinsky, Y. Bilenko, J. Yang, G.A. Garrett, L.E. Rodak, M. Wraback, M. Shur, R. Gaska, High power AlGaN ultraviolet light emitters. Semicond. Sci. Technol. **29**(8) (2014)
38. J. Simon, V. Protasenko, C. Lian, H. Xing, D. Jena, Polarization-induced hole doping in wide-band-gap uniaxial semiconductor heterostructures. Science **327**(5961), 60–64 (2010)
39. J. Verma, P.K. Kandaswamy, V. Protasenko, A. Verma, H. Grace Xing, D. Jena, Tunnel-injection GaN quantum dot ultraviolet light-emitting diodes. Appl. Phys. Lett. **102**(4) (2013)
40. Y. Zhang, A.A. Allerman, S. Krishnamoorthy, F. Akyol, M.W. Moseley, A.M. Armstrong, S. Rajan, Enhanced light extraction in tunnel junction-enabled top emitting UV LEDs. Appl. Phys. Express **9**(5) (2016)
41. N. Maeda, H. Hirayama, Realization of high-efficiency deep-UV LEDs using transparent p-AlGaN contact layer. Phys. Status Solidi (c) **10**(11), 1521–1524 (2013)
42. M.R. Krames, M. Ochiai-Holcomb, G.E. Höfler, C. Carter-Coman, E.I. Chen, I.H. Tan, P. Grillot, N.F. Gardner, H.C. Chui, J.W. Huang, S.A. Stockman, F.A. Kish, M.G. Craford, T.S. Tan, C.P. Kocot, M. Hueschen, J. Posselt, B. Loh, G. Sasser, D. Collins, High-power truncated-inverted-pyramid $(Al_xGa_{1-x})_{0.5}In_{0.5}P$/GaP light-emitting diodes exhibiting >50% external quantum efficiency. Appl. Phys. Lett. **75**(16), 2365–2367 (1999)
43. T. Kolbe, F. Mehnke, M. Guttmann, C. Kuhn, J. Rass, T. Wernicke, M. Kneissl, Improved injection efficiency in 290 nm light emitting diodes with Al(Ga)N electron blocking heterostructure. Appl. Phys. Lett. **103**(3) (2013)

44. Z. Ren, Q. Sun, S.Y. Kwon, J. Han, K. Davitt, Y.K. Song, A.V. Nurmikko, H.K. Cho, W. Liu, J.A. Smart, L.J. Schowalter, Heteroepitaxy of AlGaN on bulk AlN substrates for deep ultraviolet light emitting diodes. Appl. Phys. Lett. **91**(5) (2007)
45. M. Jo, N. Maeda, H. Hirayama, Enhanced light extraction in 260 nm light-emitting diode with a highly transparent p-AlGaN layer. Appl. Phys. Express **9**(1) (2016)
46. J. Verma, S.M. Islam, V. Protasenko, P. Kumar Kandaswamy, H. Xing, D. Jena, Tunnel-injection quantum dot deep-ultraviolet light-emitting diodes with polarization-induced doping in III-nitride heterostructures. Appl. Phys. Lett. **104**(2) (2014)
47. S.M. Islam, K. Lee, J. Verma, V. Protasenko, S. Rouvimov, S. Bharadwaj, H. Xing, D. Jena, MBE-grown 232–270 nm deep-UV LEDs using monolayer thin binary GaN/AlN quantum heterostructures. Appl. Phys. Lett. **110**(4) (2017)
48. J. Zhang, H. Zhao, N. Tansu, Effect of crystal-field split-off hole and heavy-hole bands crossover on gain characteristics of high Al-content AlGaN quantum well lasers. Appl. Phys. Lett. **97**(11) (2010)
49. T. Kolbe, A. Knauer, C. Chua, Z. Yang, S. Einfeldt, P. Vogt, N.M. Johnson, M. Weyers, M. Kneissl, Optical polarization characteristics of ultraviolet (In)(Al)GaN multiple quantum well light emitting diodes. Appl. Phys. Lett. **97**(17) (2010)
50. X.-H. Li, T.-T. Kao, M.M. Satter, Y.O. Wei, S. Wang, H. Xie, S.-C. Shen, P.D. Yoder, A.M. Fischer, F.A. Ponce, T. Detchprohm, R.D. Dupuis, Demonstration of transverse-magnetic deep-ultraviolet stimulated emission from AlGaN multiple-quantum-well lasers grown on a sapphire substrate. Appl. Phys. Lett. **106**(4) (2015)
51. C. Reich, M. Guttmann, M. Feneberg, T. Wernicke, F. Mehnke, C. Kuhn, J. Rass, M. Lapeyrade, S. Einfeldt, A. Knauer, V. Kueller, M. Weyers, R. Goldhahn, M. Kneissl, Strongly transverse-electric-polarized emission from deep ultraviolet AlGaN quantum well light emitting diodes. Appl. Phys. Lett. **107**(14) (2015)
52. J.E. Northrup, C.L. Chua, Z. Yang, T. Wunderer, M. Kneissl, N.M. Johnson, T. Kolbe, Effect of strain and barrier composition on the polarization of light emission from AlGaN/AlN quantum wells. Appl. Phys. Lett. **100**(2) (2012)
53. D.A.B. Miller, D.S. Chemla, T.C. Damen, A.C. Gossard, W. Wiegmann, T.H. Wood, C.A. Burrus, Band-edge electroabsorption in quantum well structures: the quantum-confined Stark effect. Phys. Rev. Lett. **53**(22), 2173–2176 (1984)
54. T. Tetsuya, S. Shigetoshi, K. Maki, K. Miho, T. Hideo, A. Hiroshi, A. Isamu, Quantum-confined Stark effect due to piezoelectric fields in GaInN strained quantum wells. Jpn. J. Appl. Phys. **36**(4A), L382 (1997)
55. J. Zhang, H. Zhao, N. Tansu, Large optical gain AlGaN-delta-GaN quantum wells laser active regions in mid- and deep-ultraviolet spectral regimes. Appl. Phys. Lett. **98**(17) (2011)
56. D. Bayerl, S.M. Islam, C.M. Jones, V. Protasenko, D. Jena, E. Kioupakis, Deep ultraviolet emission from ultra-thin GaN/AlN heterostructures. Appl. Phys. Lett. **109**(24) (2016)
57. J. Selles, C. Brimont, G. Cassabois, P. Valvin, T. Guillet, I. Roland, Y. Zeng, X. Checoury, P. Boucaud, M. Mexis, F. Semond, B. Gayral, Deep-UV nitride-on-silicon microdisk lasers. Sci. Rep. **6**(21650) (2016)
58. E. Kioupakis, Q. Yan, C.G. Van de Walle, Interplay of polarization fields and Auger recombination in the efficiency droop of nitride light-emitting diodes. Appl. Phys. Lett. **101**(23) (2012)
59. J.M. Rondinelli, E. Kioupakis, Predicting and designing optical properties of inorganic materials. Annu. Rev. Mater. Res. **45**(1), 491–518 (2015)
60. J.A. Ferrer-Pérez, B. Claflin, D. Jena, M. Sen, R. Vetury, D. Dorsey, Photoluminescence-based electron and lattice temperature measurements in GaN-based HEMTs. J. Electron. Mater. **43**(2), 341–347 (2014)
61. R. Bhattacharya, B. Pal, B. Bansal, On conversion of luminescence into absorption and the van Roosbroeck-Shockley relation. Appl. Phys. Lett. **100**(22) (2012)
62. L.K. Lee, L. Zhang, H. Deng, P.C. Ku, Room-temperature quantum-dot-like luminescence from site-controlled InGaN quantum disks. Appl. Phys. Lett. **99**(26) (2011)

63. S.M. Islam, V. Protasenko, S. Rouvimov, H. Xing, D. Jena, Sub-230 nm deep-UV emission from GaN quantum disks in AlN grown by a modified Stranski–Krastanov mode. Jpn. J. Appl. Phys. **55**(5S) (2016)
64. C.T. Foxon, C.S. Davis, S.V. Novikov, O.H. Hughes, T.S. Cheng, D. Korakakis, N.J. Jeffs, I. Grzegory, S. Porowski, RHEED studies of group III-Nitrides grown by MBE. Phys. Status Solidi (a) **176**(1), 723–726 (1999)
65. N. Gogneau, D. Jalabert, E. Monroy, E. Sarigiannidou, J.L. Rouvière, T. Shibata, M. Tanaka, J.M. Gerard, B. Daudin, Influence of AlN overgrowth on structural properties of GaN quantum wells and quantum dots grown by plasma-assisted molecular beam epitaxy. J. Appl. Phys. **96**(2), 1104–1110 (2004)
66. D. Doppalapudi, E. Iliopoulos, S.N. Basu, T.D. Moustakas, Epitaxial growth of gallium nitride thin films on A-Plane sapphire by molecular beam epitaxy. J. Appl. Phys. **85**(7), 3582–3589 (1999)
67. Y. Horikoshi, Migration-enhanced epitaxy of GaAs and AlGaAs. Semicond. Sci. Technol. **8**(6), 1032 (1993)
68. C. Adelmann, B. Daudin, R.A. Oliver, G.A.D. Briggs, R.E. Rudd, Nucleation and growth of GaN/AlN quantum dots. Phys. Rev. B **70**(12) (2004)
69. B. Daudin, F. Widmann, G. Feuillet, Y. Samson, M. Arlery, J.L. Rouvière, Stranski-Krastanov growth mode during the molecular beam epitaxy of highly strained GaN. Phys. Rev. B **56**(12), R7069–R7072 (1997)
70. Y. Taniyasu, M. Kasu, Polarization property of deep-ultraviolet light emission from C-plane AlN/GaN short-period superlattices. Appl. Phys. Lett. **99**(25) (2011)
71. K. Kamiya, Y. Ebihara, K. Shiraishi, M. Kasu, Structural design of AlN/GaN superlattices for deep-ultraviolet light-emitting diodes with high emission efficiency. Appl. Phys. Lett. **99**(15) (2011)
72. Z. Zheng, H. Ji, P. Yu, Z. Wang, Recent progress towards quantum dot solar cells with enhanced optical absorption. Nanoscale Res. Lett. **11**(1), 266 (2016)
73. Y.-K. Kuo, T.-H. Wang, J.-Y. Chang, Advantages of blue InGaN light-emitting diodes with InGaN-AlGaN-InGaN barriers. Appl. Phys. Lett. **100**(3) (2012)
74. T. Passow, K. Leonardi, A. Stockmann, H. Selke, H. Heinke, D. Hommel, High-resolution x-ray diffraction investigations of highly mismatched II-VI quantum wells. J. Phys. D Appl. Phys. **32**(10A), A42 (1999)
75. S. Islam, V. Protasenko, S. Rouvimov, J. Verma, H. Xing, D. Jena, Deep-UV LEDs using polarization-induced doping: Electroluminescence at cryogenic temperatures, in *2015 73rd Annual Device Research Conference (DRC)*, 21–24 June 2015; 2015; pp 67–68
76. J. Simon, Y. Cao, D. Jena, Short-period AlN/GaN p-type superlattices: hole transport use in p-n junctions. Phys. Status Solidi (c) **7**(10), 2386–2389 (2010)
77. D. Jena, S. Heikman, D. Green, D. Buttari, R. Coffie, H. Xing, S. Keller, S. DenBaars, J.S. Speck, U.K. Mishra, I. Smorchkova, Realization of wide electron slabs by polarization bulk doping in graded III–V nitride semiconductor alloys. Appl. Phys. Lett. **81**(23), 4395–4397 (2002)
78. J. Yan, J. Wang, P. Cong, L. Sun, N. Liu, Z. Liu, C. Zhao, J. Li, Improved performance of UV-LED by p-AlGaN with graded composition. Phys. Status Solidi (c) **8**(2), 461–463 (2011)
79. S. Li, M.E. Ware, V.P. Kunets, M. Hawkridge, P. Minor, J. Wu, G.J. Salamo, Polarization induced doping in graded AlGaN films. Phys. Status Solidi (c) **8**(7–8), 2182–2184 (2011)
80. S.D. Carnevale, T.F. Kent, P.J. Phillips, M.J. Mills, S. Rajan, R.C. Myers, Polarization-induced pn diodes in wide-band-gap nanowires with ultraviolet electroluminescence. Nano Lett. **12**(2), 915–20 (2012)
81. L. Zhang, X.C. Wei, N.X. Liu, H.X. Lu, J.P. Zeng, J.X. Wang, Y.P. Zeng, J.M. Li, Improvement of efficiency of GaN-based polarization-doped light-emitting diodes grown by metalorganic chemical vapor deposition. Appl. Phys. Lett. **98**(24) (2011)
82. A. Yasan, R. McClintock, K. Mayes, S.R. Darvish, P. Kung, M. Razeghi, Top-emission ultraviolet light-emitting diodes with peak emission at 280 nm. Appl. Phys. Lett. **81**(5), 801–802 (2002)

83. A. Yasan, R. McClintock, K. Mayes, D.H. Kim, P. Kung, M. Razeghi, Photoluminescence study of AlGaN-based 280 nm ultraviolet light-emitting diodes. Appl. Phys. Lett. **83**(20), 4083–4085 (2003)
84. A. Chitnis, V. Adivarahan, M. Shatalov, J. Zhang, M. Gaevski, W. Shuai, R. Pachipulusu, J. Sun, K. Simin, G. Simin, J. Yang, M.A. Khan, Submilliwatt operation of AlInGaN based multifinger-design 315 nm Light Emitting Diode (LED) over sapphire substrate. Jpn. J. Appl. Phys. **41**(Part 2, No. 3B), L320–L322 (2002)
85. B. Cheng, S. Choi, J.E. Northrup, Z. Yang, C. Knollenberg, M. Teepe, T. Wunderer, C.L. Chua, N.M. Johnson, Enhanced vertical and lateral hole transport in high aluminum-containing AlGaN for deep ultraviolet light emitters. Appl. Phys. Lett. **102**(23) (2013)
86. F.A.F. Jimy Encomendero, S.M. Islam, V. Protasenko, S. Rouvimov, P. Fay, D. Jena, H.G. Xing, Repeatable room temperature negative differential conductance in GaN/AlN resonant tunneling diodes. arXiv:1606.08100 [cond-mat.mes-hall] (2016)
87. C. Liu, Y.K. Ooi, S.M. Islam, J. Verma, H. Xing, D. Jena, J. Zhang, Physics and polarization characteristics of 298 nm AlN-delta-GaN quantum well ultraviolet light-emitting diodes. Appl. Phys. Lett. **110**(7), 071103 (2017)

Chapter 11
Reliability of Ultraviolet Light-Emitting Diodes

Carlo De Santi, Desiree Monti, Pradip Dalapati, Matteo Meneghini, Gaudenzio Meneghesso, and Enrico Zanoni

11.1 AlGaN-Based Ultraviolet Light-Emitting Diodes

Operating current and temperature are two important parameters that can influence the degradation of UV LEDs. In this last decade, many research groups studied the impact of current and/or temperature on the optical and electrical characteristics during the aging of the devices, trying to identify the physical mechanisms responsible for the reduction in the efficiency and, in some cases, to propose some solutions in order to optimize it.

In 2004, Allerman et al. reported their progress in the development of deep-ultraviolet (DUV) LEDs with emission between 237 and 297 nm [1]. For a 300-μm square device, the observed operating voltage was 6.7 V at 20 mA and 8.2 V at 50 mA, respectively, whereas the wall-plug efficiency and external quantum efficiency were 0.1% and 0.19%, respectively, at 50 mA. An output power of 0.4 mW has been achieved at this current density (∼60 A/cm^2). Stress tests at 20 mA were carried out on the 300 μm × 300 μm devices: analyzing the optical power during the stress for two 295 nm devices, with initial output powers of 92 and 95 μW at 20 mA, a drop of 26–34% was seen.

In 2006, Asif Khan et al. [2, 3] analyzed a group of unpackaged AlGaN-based LEDs with a nominal emission wavelength around 280 nm (at 20 mA), an optical output power of 0.35 mW, and an external quantum efficiency (EQE) of about 0.4%. The use of unpackaged devices had the aim to eliminate the contribution of bonding interfaces and solder from the analysis of the LED degradation process. These samples were mounted on a temperature-controlled chuck in order to carry out stress

C. De Santi (✉) · D. Monti · P. Dalapati · M. Meneghini · G. Meneghesso · E. Zanoni
Department of Information Engineering, University of Padova, Padova, Italy
e-mail: carlo.desanti@dei.unipd.it; matteo.meneghini@dei.unipd.it; gaudenzio.meneghesso@dei.unipd.it

J. Li, G. Q. Zhang (eds.), *Light-Emitting Diodes*, Solid State Lighting Technology and Application Series 4, https://doi.org/10.1007/978-3-319-99211-2_11

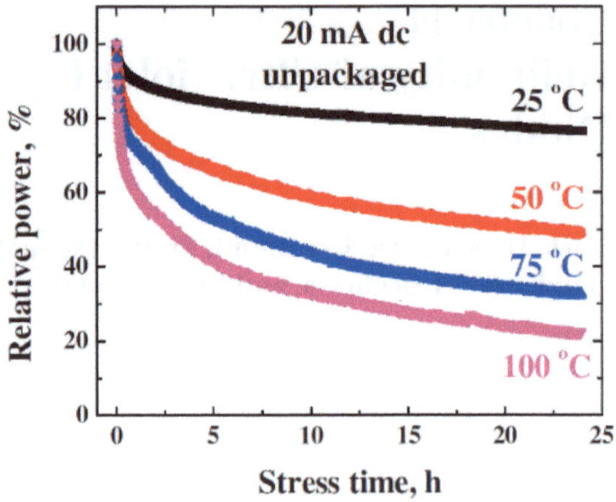

Fig. 11.1 Optical power during the stress at different stress temperatures for 280-nm unpackaged deep UV LED. Reprinted from [2], with the permission of John Wiley & Sons, Inc.

at a fixed current of 20 mA and at different temperatures. Figure 11.1 reports the decrease of the optical power during a 24-h stress at various temperatures, noticing a variation in the degradation rate with the stress temperature. Shatalov et al. [4] analyzed the same samples and observed both catastrophic and gradual output power degradation under DC bias. The catastrophic degradation was found to be caused mostly by the rough surface morphology of the p-GaN contact layer. AFM scans of the 100 μm × 100 μm emission area showed islands morphology, and these islands were attributed to the formation of V-defects in n-AlGaN layers due to the gas phase pre-reaction followed by the overgrowth of these defects with p-AlGaN and p-GaN layers [5]. The open cores of these V-defects work as leakage paths and the high current density will cause a local overheating followed by the alloying of the p-metal, where the EL intensity distribution and the CCD image after catastrophic failure were taken. The propagation of the alloyed region damages the active layer, leading to the shortening of the p–n junction.

Gong et al. continued the work in [6, 7] analyzing the degradation of the optical power also in the case of stress carried out a different pulse current. Current pulses were 1 μs long and of the duty cycle was 5%, in order to provide quasi-steady-state pumping conditions and to minimize device self-heating. Despite self-heating was minimized, the output power degradation was still more prominent for higher pump currents.

The two groups analyzed the stress results and supposed that there are two characteristic time constants discernible in this optical power degradation, associated with the junction temperature and the pump current. The curves were fitted with a double-exponential decay function:

$$P(t) = P_1 \exp(-\beta_1 t) + P_2 \exp(\beta_2 t) \tag{11.1}$$

where P_1 and β_1 are temperature-dependent parameters, whereas P_2 and β_2 are bias-dependent parameters. Parameter β_1 decays exponentially with the junction temperature T_j as:

$$\beta_1 = \beta_0 \exp\left(-E_a/kT_j\right). \tag{11.2}$$

Asif Khan et al. [3] extrapolated, from the Arrhenius plot of the aging test at 20 mA, an activation energy E_a of 0.27 eV, and a decay parameter β_0 of 97.5 h^{-1}, estimating in this way a value of 5.93×10^{-3} h^{-1} for the parameter β_1. Instead, the parameter β_2 does not have a clear dependence from temperature, and they estimated a value of 0.44 h^{-1}.

Their data suggested them that the degradation of the analyzed deep UV LEDs was mostly related to the device temperature.

By using, instead of a device structure with an area of 100 μm × 100 μm, a 10 × 10 micro-pixel UV LEDs, always with a peak emission of 280 nm, the performances improved due to a lower current density, having a larger device area. Additionally, they report a lower operating voltage related to a lower series resistance, and as a consequence also a lower junction temperature, helped also by a lower thermal impedance. Figure 11.2 reports the optical power degradation during the stress for two standard 100 μm × 100 μm and 200 μm × 200 μm square devices and for a 10 × 10 micro-pixel LED. The increase in the junction area increased the reliability of the micro-pixel LED, reducing the optical power degradation during the operation of the device.

In [8], Reed et al. analyzed a group of 280 nm LEDs, studying the impact of self-heating by measuring the output power as function of pulse current for various duty cycles at a fixed pulse width of 1 and 100 μs (Fig. 11.3). In both cases, the output power at 50% approached that of the CW conditions, indicating that heating had a significant impact on the device performance, causing also a small red-shift in wavelength. Stress performed at 100 mA (300 A/cm^2) and room temperature under CW and pulsed conditions of 100 μs and 1% duty cycle have shown that the

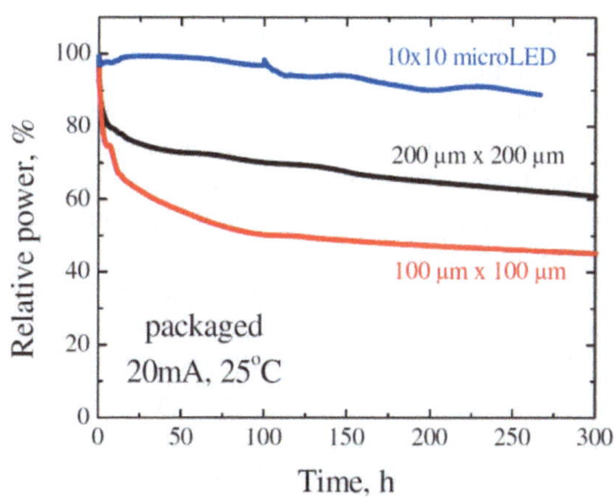

Fig. 11.2 Degradation of the optical power during stress for different type of packaged LEDs stressed at 20 mA and 25 °C. Reprinted from [2], with the permission of John Wiley & Sons, Inc.

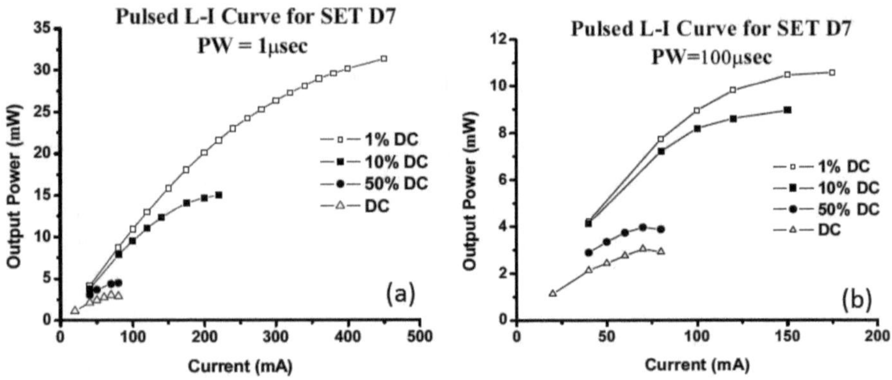

Fig. 11.3 Output power as function of current for (**a**) 1 μs and (**b**) 100 μs pulse at currents from 40 to 450 mA. Reprinted from [8], with the permission of John Wiley & Sons, Inc.

pulsed device presents a half-life of 1400 h, longer than that observed under CW conditions, which produced a half-life of 20 h.

Lifetime measurements on single-chip, packaged 285-nm LEDs were performed by Moe et al. in [9]. In particular, these tests were performed under constant current injection at 20 and 75 mA and were compared to the performance of unbiased LEDs heated in the oven at the equivalent operating junction temperatures. Devices that were thermally stressed showed a lower degradation than those electrically stressed, as reported in Fig. 11.4, and no parasitic peak at longer emission wavelengths (i.e., lower energy) was observed in the EL spectra, indicating that the increased degradation is not due to changes in the active region [10]. Capacitance as function of voltage measurements were carried out for the electrically stressed samples, and allowed to extrapolate the apparent charge profile before and after stress (Fig. 11.5). The results suggested that the depletion region extended further into the p-type layer after electrical degradation, due to the compensation of p-type carriers in the barrier region near the heterojunction.

In another work [11], Asif Khan et al. highlighted the fact that two of the main problems in DUV LEDs are the low efficiency and the long-term reliability. In particular, the low efficiency problem is related to the large amount of threading dislocations that increases with the Al mole fraction in the active layer. The optical power decreases as the emission wavelength decreases from 280 to 250 nm, which means an increase in the Al-mole fraction of the active layer. AlGaN LEDs with a nominal emission wavelength around 280 nm were stressed at two different pump current densities of 100 and 200 A/cm^2. A fast degradation of the output power in the first 1–5 h was detected; this is then followed by a slower degradation. The device lifetime was found to be a strong function of the pump current density and the number of dislocations in the active region.

Since a high threading dislocation density is present in the LED structures due to heteroepitaxial growth on foreign substrates, different groups tried to use different techniques in order to reduce it. Jain et al. report in [12] the growth of low-

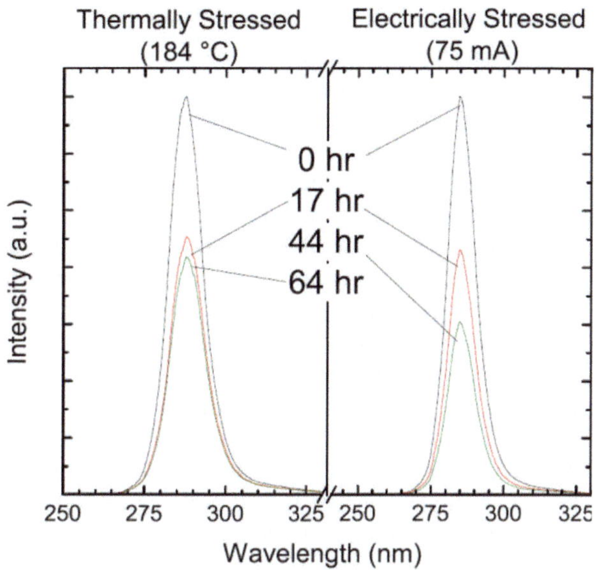

Fig. 11.4 EL spectra of LEDs thermally (left) and electrically (right) stressed over time. Reprinted from "Current-induced degradation of high performance deep ultraviolet light emitting diodes", Craig G. Moe, Meredith L. Reed, Gregory A. Garrett, Anand V. Sampath, Troy Alexander, Hongen Shen, Michael Wraback, Yuriy Bilenko, Maxim Shatalov, Jinwei Yang, Wenhong Sun, Jianyu Deng, and Remis Gaska, Applied Physics Letters 2010 96:21, with the permission of AIP Publishing

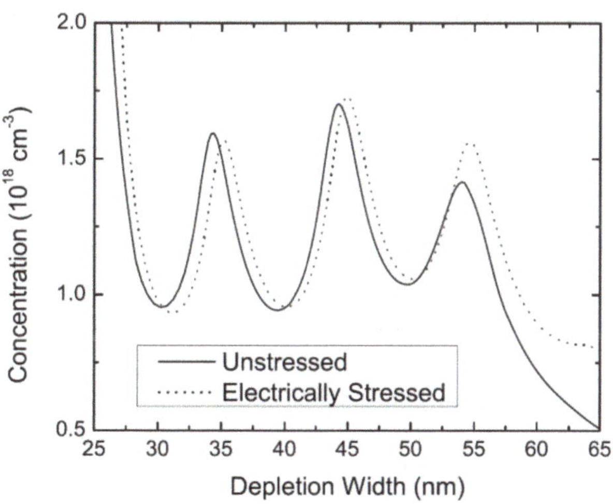

Fig. 11.5 Apparent charge profile as function of depletion width before and after stress. Reprinted from "Current-induced degradation of high performance deep ultraviolet light emitting diodes", C. G. Moe, M. L. Reed, G. A. Garrett, A. V. Sampath, T. Alexander, H. Shen, M. Wraback, Y. Bilenko, M. Shatalov, J. Yang, W. Sun, J. Deng, and R. Gaska, Applied Physics Letters 2010 96:21, with the permission of AIP Publishing

defect thick films of AlN and AlGaN on trenched AlGaN/sapphire templates using migration-enhanced lateral epitaxial overgrowth (MELEO) technique. They were able to realize a DUV LED with a nominal emission wavelength around 310 nm with a CW output power of 1.6 mW at 50 mA. The extrapolated lifetime was estimated to be about 5000 h, with a five times improvement in device reliability, validating the high quality of MELEO grown thick AlN films.

Monitoring common parameters, such as output power, peak wavelength, and current–voltage characteristics, together with additional parameters like the low-frequency current noise at low and high currents, Sawyer et al. [13] characterized a group of AlGaN/AlInGaN LEDs with a nominal emission wavelength of 280 nm.

Devices were aged at 20 mA forward current and showed: (1) no changes in the peak wavelength and full width at half magnitude (FWHM), indicating that the properties of the quantum wells did not change during the stress; (2) a decrease in the power with the square root of time, suggesting the presence of a diffusion process, in this case diffusion of contact metal to the device structure or diffusion of dopant atoms [14]. Low-frequency noise is a sensitive indicator of the semiconductor devices degradation, in particular, at low bias it is sensitive to degradation of the barrier layer, whereas at high bias, it reflects the degradation of the contacts and/or semiconductor layers contributing to the series resistance [15, 16]. Sawyer et al. observed a strong increase in the noise related to the series resistance. Figure 11.6 reports the noise as function of the aging time at two different currents and $f = 10$ Hz. At low current ($I < 10$–50 μA), the noise spectra were close to the $1/f$ noise, and spectral noise density of the short circuit current fluctuations, S_I, was proportional to the first power of the current. At intermediate current (100 μA < I < 1–3 mA), the generation-recombination noise was found in many devices. At higher current, the $1/f$ noise again dominated the spectra and $S_I - I^2$. In general, no correlation between noise, the optical power, and the degradation rate of the optical power was observed.

Pinos et al. investigated the impact of the AlN molar fraction on the AlGaN-based UV LEDs performances [10, 17, 18]. In [10], a 285-nm LED with

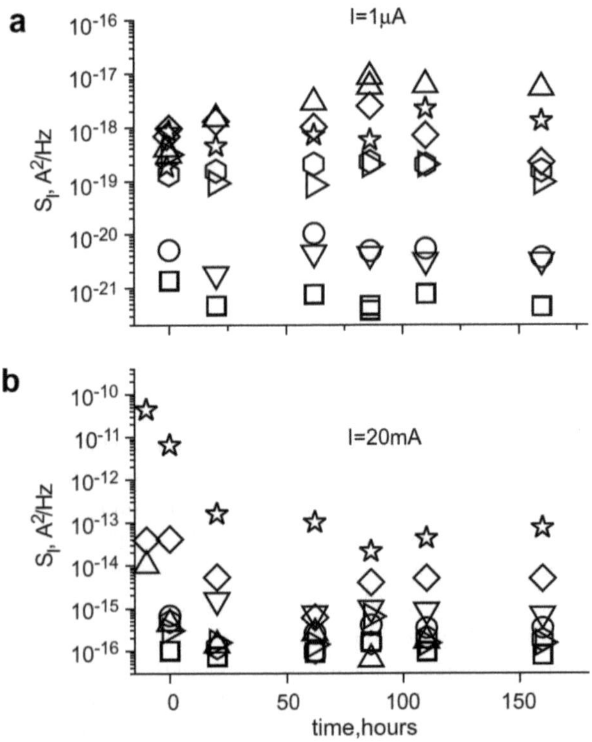

Fig. 11.6 Spectral noise density as function of the stress time measured at (**a**) 1 μA and (**b**) 20 mA. Different symbols correspond to different LEDs, $f = 10$ Hz. Reprinted from Solid-State Electronics, 52, S. Sawyer, S.L. Rumyantsev, M.S. Shur, "Degradation of AlGaN-based ultraviolet light emitting diodes", 968–972, Copyright (2008), with permission from Elsevier

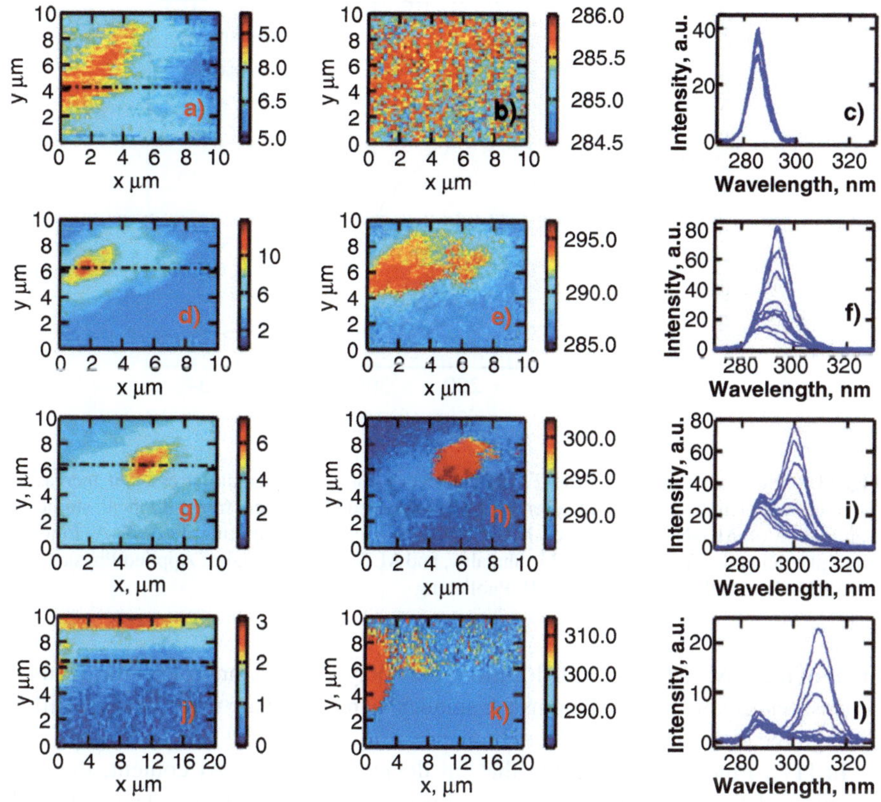

Fig. 11.7 In the first column, the emission intensity of the near-field scan is reported, whereas the second column reports the peak wavelength maps. The third column reports selected spectra taken along the lines shown in the intensity maps. Reprinted from "Aging of AlGaN quantum well light-emitting diode studied by scanning near-field optical spectroscopy", A. Pinos, S. Marcinkevičius, J. Yang, Y. Bilenko, M. Shatalov, R. Gaska, and M. S. Shur, Applied Physics Letters 2010 95:181914, with the permission of AIP Publishing

$Al_{0.35}Ga_{0.65}N/Al_{0.50}Ga_{0.50}N$ multiple QWs in the active region was stressed, and near-field electroluminescence (EL) measurements were performed using a scanning near-field optical microscope (SNOM). From the measurements, they observed that the EL intensity experienced some variation in the scan area, as Fig. 11.7 shows. The emission intensity from the "hot" spots increases, and the wavelength presents a red-shift. This red-shift of the emission wavelength in the domain-like areas was ascribed to an increase in the Ga concentration in the QWs, with a reduction of the AlN molar fraction in the domain due to a vertical diffusion of Al atoms. Since the current crowding occurs only in the "hot" spots, this causes a strong local heating that will cause a change in the atoms diffusion coefficient. This temperature gradient-induced diffusion might overcome diffusion due to Al composition gradient, assuring the net outflow of Al atoms from the domain-like

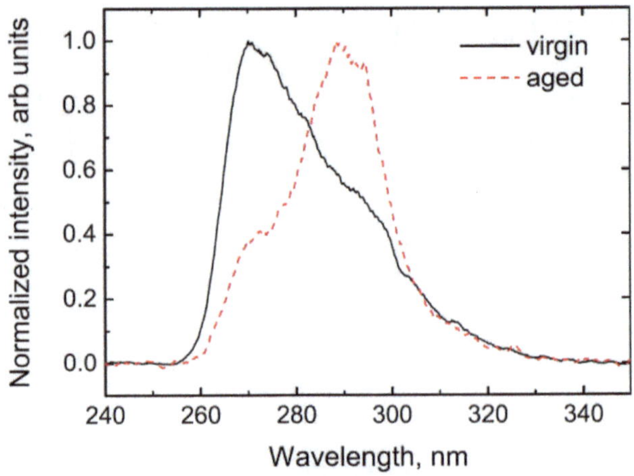

Fig. 11.8 Near-field spectra of the 270-nm device before and after the stress. An increase in the long wavelength component after aging is observed. Reprinted from "Optical studies of degradation of AlGaN quantum well based deep ultraviolet light emitting diodes", A. Pinos, S. Marcinkevičius, J. Yang, R. Gaska, M. Shatalov, and M. S. Shur, Journal of Applied Physics 2010 108:093113, with the permission of AIP Publishing

areas of the QWs. Moreover, dislocations may play an important role in the diffusion process, especially since Al atoms are smaller than Ga and are more likely to diffuse through the dislocations.

In [17], Pinos et al. revealed that, in devices with high Al content, the deterioration of the cladding layers is one important factor in device degradation. In [17], two LEDs with a nominal emission wavelength of 270 and 335 nm were stressed, and, again, near-field optical spectroscopy was performed. The structure of these two LEDs consisted of an AlN buffer, AlN/AlGaN superlattice, a few micrometer-thick n- and p-AlGaN cladding layers and an active region with three to five 2.5–3-nm-wide QWs and 11.5–12 nm barriers. From these measurements, they observed that for 335 nm LED the EL intensity variation did not show any clear tendency with aging and was attributed to nonuniform current injection. Instead, for 270-nm LED permanent changes occurred in the spectra during the aging, as reported in Fig. 11.8: after the stress, the emission peak at longer wavelength becomes dominant. The position of this peak suggested that its origin may be due to either transitions between the conduction band (shallow donor) and Mg acceptor or transitions between ionized nitrogen vacancies and the valence band. The observed increase in the amplitude of this parasitic peak highlighted the fact that the number of states participating in this transition increases. This means that this peak should be ascribed to transition involving the defects in the p-side cladding layer that are most probably nitrogen vacancies.

Other groups analyzed the presence of non-radiative recombination processes carrying out stress tests. Meneghini et al. in [19] analyzed the reliability of a group of DUV LEDs with an AlGaN multiple quantum well (MQW) layer, an n-

AlGaN:Si cladding layer, a p-AlGaN cladding layer, and a p-AlInGaN contact layer [20]. The stress was carried out at the nominal current of 20 mA, by continuously monitoring electrical and optical parameters and carrying out capacitance–voltage measurements in order to extrapolate information on the apparent charge profile. Analyzing the L-I characteristics measured before and after the stress, it was possible to observe that stress induces a decrease in the optical power in the whole current range. Moreover, this degradation took place mostly during the initial hour of stress (first 200 h); instead, the optical power remains quite stable in the remaining stress time. They found that the optical power degradation was more prominent at low measuring current levels, suggesting that the degradation is related to the generation of non-radiative recombination centers.

By analyzing the electrical characteristics, they observed an increase in the reverse leakage current and in the current below the turn-on voltage, indicating that stress induces an increase in the tunneling and generation/recombination processes; effect that was usually attributed to the generation of defects [21, 22]. This hypothesis was better confirmed by the comparison between the optical power degradation and the variation of the forward bias, as reported in [23]. The C–V measurements allowed to analyze the effect of stress on the properties of the active region. In fact, stress induced an increase in the apparent charge profile, probably ascribed to the modification of charge distribution in the active region, which in turn can be related to the worsening in the properties of the active layer that consists in the generation of defective states [24]. The change in the charge took place mostly in the first 250 h of stress, just like the optical power decrease, suggesting a correlation between the optical power degradation and the modification in the charge profile.

In other papers [25, 26], Meneghini et al. analyzed a group of DUV-LEDs with emission wavelength at 310 nm, carrying out several stress at different conditions, in order to evaluate the role of current and temperature in the device degradation. Figure 11.9a reports the electroluminescence spectra carried out during a stress at 17 A/cm^2 and room temperature. It was possible to observe two emission bands: one related to the QW emission, centered around 310 nm, and the second one, a broad green-yellow peak, centered around 520 nm. This parasitic emission is usually attributed to deep-level transitions [27, 28] originating in the AlGaN quantum wells or barrier layers [29]. As the figure shows, stress causes a decrease in the main peak and an increase in the parasitic one. Moreover, the variations in these two peaks are well correlated: Figure 11.9b reports the intensity of the main peak and the intensity of the green-yellow peak during the stress; both of them present very similar kinetics, suggesting that stress induces an increase in the concentration of the defects responsible for the parasitic emission band.

Stress at different current levels and at different junction temperatures allowed to understand if the degradation observed was activated by the temperature or by the flow of carriers through the active region. Figure 11.10 shows the results that they obtained from these measurements. In particular, Fig. 11.10a shows that for these devices, the dependence on the junction temperature is very weak. Instead, by analyzing Fig. 11.10b, it is possible to observe that with increasing stress current, the optical power presents a stronger degradation. With these results, they concluded

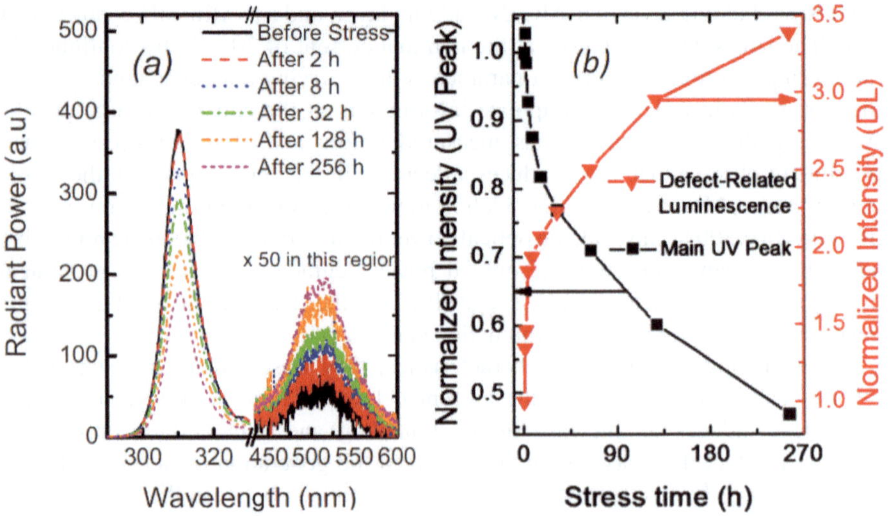

Fig. 11.9 (**a**) EL spectra of one of the analyzed samples during stress. (**b**) Correlation between the intensity of the main peak and of the parasitic one during the stress. Reprinted from "Analysis of the physical processes responsible for the degradation of deep-ultraviolet light emitting diodes", M. Meneghini, D. Barbisan, L. Rodighiero, G. Meneghesso, and E. Zanoni, Applied Physics Letters 2010 97:143506, with the permission of AIP Publishing

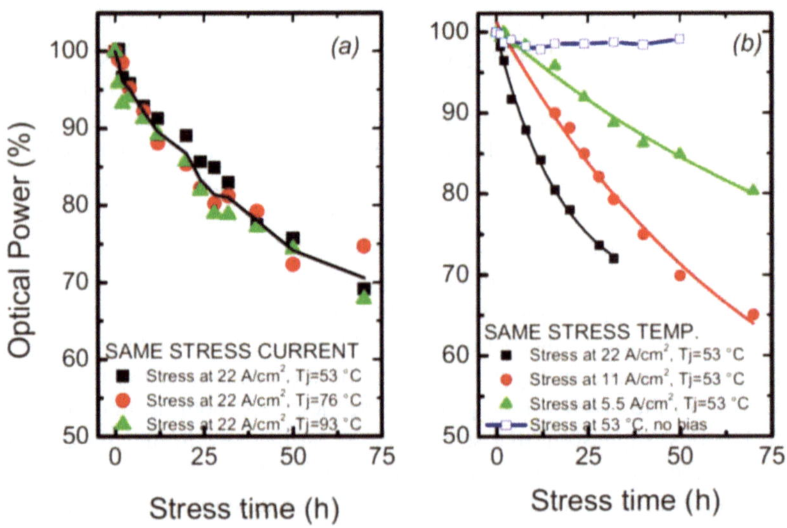

Fig. 11.10 Optical power degradation during stress (**a**) at different junction temperatures and (**b**) at different current levels. Reprinted from "Analysis of the physical processes responsible for the degradation of deep-ultraviolet light emitting diodes", M. Meneghini, D. Barbisan, L. Rodighiero, G. Meneghesso, and E. Zanoni, Applied Physics Letters 2010 97:143506, with the permission of AIP Publishing

in [25] that degradation was not ascribed to a purely thermal effect, but that the flow of the current through the active region may activate this process.

Other performance and reliability measurements were carried out by Grandusky et al. in [30], where they analyzed a group of UV-C LEDs with a 5-period multiple quantum well and with an AlN substrate. Currently, devices are fabricated on sapphire substrate, which cause an increase in dislocation density, lowering the efficiency and lifetime. AlN substrates allow to reduce lattice and thermal mismatch between the substrate and the device layer. Stress at different current levels, 20, 100, and 150 mA, was carried out on these devices. The measurements showed three different mechanisms for power decay. The first mechanism was related to the temperature; this mechanism was reversible; that is, if the device was turned off and allowed to cool, the power would initially recover, followed by a rapid drop until thermal equilibrium was reached. The second mechanism was burn-in: it could be related to impurity diffusion in the material or alloying of the contacts. The third and last mechanism was a slow decay in the optical power. Again, it is possible to observe that the optical power decay is stronger at higher stress current levels. In this case, it was not clear if the degradation was thermally related or current density related. They fitted the data with several functions: the best fit was obtained for a fit with a two-channel exponential decay, even if fitting the first 100 h in order to predict the power after 1000 h, the value was overestimated. In order to predict the correct value, the best function was a log transformation of the x-axis. Grandusky's group expected that these devices had a time to half the initial power greater than 3000 h.

A similar study on UV-C LEDs, with emission wavelength in the range of 250 and 280 nm, grown on AlN substrates by MOCVD, was proposed by Kitamura et al. in [31], where they performed stress tests at various temperatures and currents. For these devices, the degradation in the optical power increased with increasing case temperature above 25 °C, with an acceleration factor of ~0.25% per °C. $I-V$ characteristics measured before and after a stress of 300 mA at room temperature present an increase in the reverse and forward leakage currents after 1000 h of stress. Moreover, in the range between −5 and 5 V after stress, the shape of the $I-V$ curve maintains asymmetry, suggesting that this variation was not caused by metal migration from the electrodes, which may create a connection between electrodes. As explained in [25], the high current stress induces the increase in the leakage current, reducing the radiative recombination. TEM images were able to clarify the correlation between the position of the leakage path and the presence of structural defects. Specifically, they show the cross sectional image at the point of the emission response related to the leakage current: defects in the MQW and n-AlGaN together with a crystal defect in the AlN substrate under the emission point were observed. This suggested that the crystal defect in the substrate may lead to the generation of the other defects in n-AlGaN and MQW during the epitaxial growth. Therefore, stress induced the diffusion of doping/impurities, such as Mg and Si, towards the active region [32, 33] and non-radiative recombination components arranged along these defects.

In 2014, Fujioka et al. realized 1-mm sized DUV LEDs emitting at 255, 280, and 310 nm suitable for high-current operation and tested the lifetime for the 280-nm LEDs, by means of aging tests varying the junction temperature [34]. Tests were performed first under 350 mA DC condition. After an increase in the first hours of stress, the optical power started to decrease with time. The first behavior was related to an enhanced Mg activation in the p-type AlGaN layer within the first hours of stress that causes an improvement in the internal quantum efficiency of the LED. With increasing junction temperature, they observed a faster decay in the optical power: after 1000 h, the optical power dropped to 73, 50, and 16% of its initial value for the junction temperatures of 35, 98, and 152 °C, respectively. Keeping the junction temperature constant at 85 °C, they performed other reliability tests changing the current density: when both the drive and measurement current are equal to 20 or 150 mA, the rate of power decay was comparable to that of the 350 mA current case and maintained 70% of its initial value after 1000 h. In contrast, when measured at 350 mA, the power decay seemed smaller and kept 99 and 72% for a bias current of 20 and 150 mA, respectively. This means that power has a strong decrease for small measurement current levels. The estimated 50% lifetime (L_{50}) was 3000 h at a junction temperature of 30 °C.

In [35], Rass et al. analyzed a group of high-power UV-B LEDs grown by MOCVD on (0001) c-plane sapphire substrate. They discussed the different parameters responsible for achieving a high EQE, i.e., the internal quantum efficiency (IQE), the injection efficiency (η_{inj}), and the extraction efficiency (η_{ext}), trying to improve these parameters in order to improve the device performances. By growing the LEDs on templates with a thick relaxed n-$Al_{0.50}Ga_{0.50}N$ buffer, they reduced the number of threading dislocations (TDD) from about 2×10^{10} cm^{-2} to 9×10^{9} cm^{-2}. Thanks to this improvement, the IQE, that is related to the defect density, achieved a value of 26% at room temperature. Regarding η_{inj}, varying the Al content of the EBL or by employing an electron blocking heterostructure (EBH), that consist in a conventional EBL and a thin high aluminum content interlayer, they obtained an increase in this figure of merit and reduced the parasitic emission from the p-side. An efficient heat extraction and an improvement in the η_{ext} was gained by flip-chip mounting the LED chips on AlN submounts. The final devices were submitted to a stress at high current densities of 140 A/cm^2 (I = 100 mA) at a heatsink of 20 °C. They showed a decrease of about 5–10% of the initial output power in the first 100 h. The degradation kinetic up to 3300 h can be described by a square root time behavior with a low degradation rate. The total output power decrease was about 30% of its initial value and by extrapolating the square root time dependence an L_{50} lifetime (time when the optical power reaches 50% of its initial value) of more than 10,000 h was estimated. Analyzing the electro-optical characteristics after different operation times, they observed that both the leakage current in reverse and at voltages below the band-gap energies (<4 V) have shown an inverse trend compared to the optical power over time; that is, the leakage current at −2.5 V increases, especially in the first 100 h of stress. Moreover, the drive voltage at 100 mA decreased slightly within the first 50–100 h and increased gradually within the rest of the stress. No changes in the position of the main peak emission were observed, indicating that no changes in the material composition in the active region are happening. What Rass's group assumed was that during operation, a temperature-assisted diffusion of point defects into the active region, that leads to an increase of the non-radiative recombination rate, could be the main degradation mechanism in their devices. In particular, Mg-dopant atom may be the involved diffusion species, explaining in this way the increase in the drive voltage.

Glaab et al. in [36, 37] presented further results on a group of (InAlGa)N-based LEDs with a nominal emission wavelength around 308 with the same structure reported in [35], reporting the strong dependence of the degradation on the current density and the presence of non-recombination centers in the active region. They carried out stress at different conditions in order to characterize the effect of current, current density, and temperature. Analyzing the decrease of the optical power of a sample stressed at 100 mA ($J = 140$ mA/cm^2) and 20 °C, they observed that the degradation kinetics over the time presents two different modes. The first degradation mode takes place within the first 25–100 h of operation, where the optical power presents a fast decrease. The second degradation mode takes place for longer operation time (>100 h), and the optical power has a slower decay; moreover, it shows a square root time dependence.

These two different behaviors were observed also for stress carried out at different temperatures (Fig. 11.11). Regarding degradation mode 1, they subtracted from the optical power the background attributed to mode 2, as reported in [36], and observed that $\Delta P_{opt,1}$ increases with the temperature, indicating that this parameter has a strong influence on mode 1. From the $I-V$ characteristics, it is possible to observe an increase in the reverse leakage current and in the current below the turn-on voltage, with a stronger effect in reverse bias. The increase in the current in this last region may be ascribed to point defects and/or threading dislocations in the active region that form parasitic current paths and short the active region [38, 39].

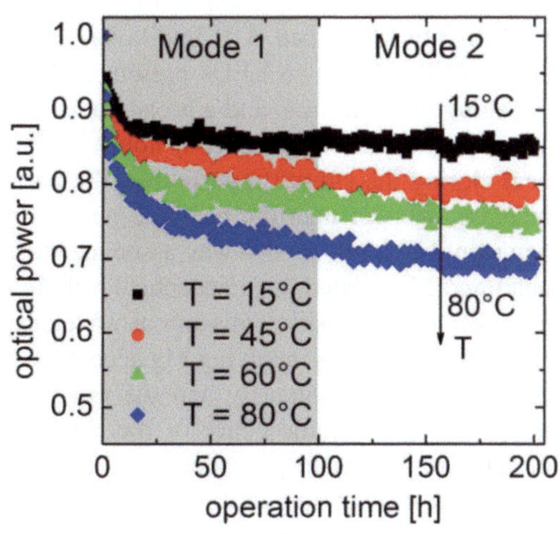

Fig. 11.11 Optical power degradation for different stress carried out at 100 mA (67 A/cm^2) and at different heat sink temperatures between 15 and 80 °C. Reprinted from [37], with the permission of Cambridge University Press

Considering degradation mode 2, they attributed this behavior to the involvement of a diffusion process, due to the square root time dependence of this degradation. The diffusing species are supposed to diffuse towards the active region, increasing the defectiveness within. This species may act as non-radiative recombination centers, causing the reduction of the optical power, and may create parasitic current paths, inducing an increase in the reverse leakage current, as the $I-V$ characteristics reports. In the literature, different diffusing species are proposed: Mg-atoms may diffuse from the p-doped layers [40], hydrogen atoms from the SiN passivation [41], or metal atoms from the contacts [13]. In order to better investigate the dependence of the degradation on the temperature, Glaab's group stressed another group of samples at constant current and different heat sink temperatures, in the range between 15 and 80 °C. The activation energies for the two degradation modes were found to be $E_{a,1} = 0.13$ eV for degradation mode 1, and $E_{a,2} = 0.21$ eV for degradation mode 2, indicating that both the two models have a strong dependence on temperature. However, the degradation rate of mode 2 was found to be more dependent on the current density. In conclusion, degradation mode 2 is probably ascribed to the diffusion of point defects associated with threading dislocations and/or recombination enhanced defect reactions. Since the activation energy of mode 1 cannot be readily attributed to other activation energies in the literature, and since this degradation occurs in a relatively short period of time, this process could be not related to the movement and increase of point defects, but rather could be ascribed to an activation process of fixed non-radiative recombination centers, that saturate in the second phase of the degradation.

The variation of the emission spectra during the degradation was studied in a following work [42], where two InAlGaN-based UV-B LEDs with different emission wavelengths, one at 290 nm and one at 310 nm, were stressed at a nominal current of 100 mA ($J = 140$ A/cm^2). Another difference between the two LED heterostructures can be found in the layers next to the pn-junction. In order to improve the carrier injection in the active region of the 290 nm LEDs, a 4-nm-thick highly Mg-doped AlN interlayer was introduced between the active region and the EBL. The two LEDs present a QW emission peak that decreases during the aging of the device, and a broad parasitic band at wavelength between 310 and 450 nm. This parasitic peak does not show the strong decrease detected for the main peak during the stress. In particular, the parasitic band in 290-nm LED shows an increase stronger than the parasitic peak in the 310-nm LED. The parasitic peak between 310 and 350 nm was associated with transition between the conduction band and Mg-related deep acceptor levels in the p-type superlattice (SL) [43, 44]. Regarding the parasitic emission at wavelength longer than 350 nm, this could be related to transitions involving Mg acceptors and V_N [45]. They report the evolution of the main peak intensity and the ratio of the peak intensities of the parasitic and main peak luminescence of the (a) 290 nm and (b) 310 nm during a stress of 1000 h. The main peak decreases rapidly in the first hours of stress for both LEDs, but stronger and more rapidly for the 290-nm LED. In the following hours, the degradation becomes slower. Again, two degradation modes were proposed as in the previous paper. In particular, QW peak and parasitic peak have opposite trend:

the fast initial reduction of the QW luminescence results from an increased non-radiative recombination within the QWs. The leakage current across the EBL is reduced, but since it is not affected by the non-radiative recombination in the QWs, the parasitic luminescence does not change over time or can slightly increase if the EBL does not block the carriers in an efficient way.

The high III/V molar ratio during growth and the high Mg-concentration in the AlN interlayer present in 290-nm LED could produce additional point defects, such V_N and Mg cluster close to the active region. Moreover, the thickness of the last quantum barrier before the p-region is only 2.5 nm in the 290-nm LED instead of 5.0 nm in the 310-nm LED. This could explain the difference in the parasitic emission band and its change during operation. Mg-dopant atoms may diffuse through the active region due to the temperature and mat act as non-radiative recombination centers or promote optical transition to V_N. For this reason, the lower overall intensity and the higher relative intensity of the parasitic band for wavelength higher than 350 nm in the 290-nm LED compared to the 310-nm LED could be explained by a larger Mg-atom concentration in the active region.

Monti et al. in [46, 47] and De Santi et al. in [48] worked on similar UV-B LEDs, with emission wavelength around 308 nm, grown on a sapphire substrate and with (In)AlGaN multi-quantum wells [35]. In order to study the correlation between the degradation and the generation of defects in the analyzed samples, they carried out a stress at a current of 100 mA and room temperature, performing also deep-level transient spectroscopy (DLTS) analysis and photocurrent (PC) spectroscopy. From the DLTS analysis, the group observed the presence of different peaks associated to different defects. In particular, during the stress, the activation energy of the trap h1 is in the range 0.11–0.24 eV, and one possible origin of this deep level is Mg-related acceptor traps, already proposed as the major gradual degradation mechanism in InAlGaN-based LEDs in [33]. Instead, the PC spectroscopy (Fig. 11.12) allowed to detect the presence of mid-gap states, between 2 and 3.5 eV, that may act as non-radiative recombination centers and increase the trap-assisted tunneling components, detected also by the increase in the current below the turn-on voltage.

An in-depth study on the dependence on temperature of the optical power degradation (thermal droop [49]) was performed in [48], where also an analysis of the origins of the different EL peaks was reported. From the analysis of the QW peak as function of temperature, three different trends were investigated: (1) at low temperature and high currents, where the optical power increases due to an enhanced hole injection from the p side to the n side; (2) at high temperature, where the optical power decreases due to an increase in the SRH recombination rate; and (3) at low temperature and low currents, where, instead, the optical power decay is probably caused by exciton delocalization from potential minima inside the QWs. The emission peak 3 was ascribed to parasitic intra-bandgap radiative transitions beyond the p side and is influenced by the increase in the injection of holes with increasing temperature at higher current levels. The last peak (peak 4) was attributed to a parasitic radiative transition, possibly in the p-AlGaN superlattice, assisted by a mechanism that causes a transfer in the charge.

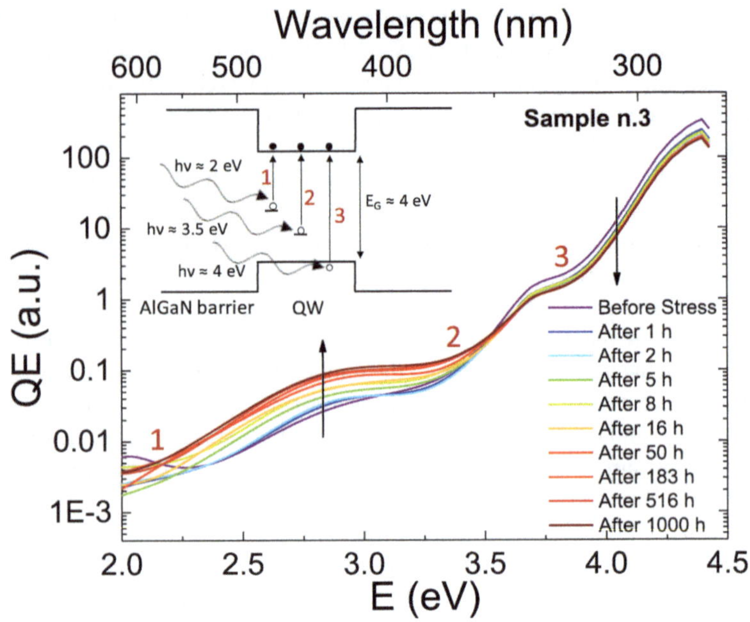

Fig. 11.12 PC spectra of the analyzed samples during the stress. The inset shows the different mechanisms responsible for the changes in the slope of the PC signal. from D. Monti et al., "Defect-Related Degradation of AlGaN-Based UV-B LEDs," in IEEE Transactions on Electron Devices, vol. 64, no. 1, pp. 200–205, Jan. 2017

11.2 InGaN-Based Ultraviolet Light-Emitting Diodes

The UV LEDs in the wavelength range 365–420 nm are commonly manufactured with InGaN/GaN quantum wells. A high degree of effort has been put into the development of high-reliability InGaN-based UV LEDs, but these devices still have open issues, such as the smaller height of the confinement potential barriers (in the range 0.1–0.15 eV or even lower), carrier escape and overflow processes, and high density of the threading dislocations (TDs) during GaN layer growth [50, 51].

In order to increase the long-term reliability, Wang et al. used a patterned sapphire substrate (PSS) to fabricate UV LEDs [50]. During the 1000 h test, they measured the variation of EL intensity of the PSS and conventional UV LEDs under a forward current of 20 mA at room temperature. The EL intensity of PSS and conventional LEDs showed a decay of 18% and 23%, respectively. They suggest that this result indicates a positive effect of the decreased trap density and non-radiative recombination centers (induced by TD) due to the growth on PSS.

Similar long-term reliability test was carried out by Mukai et al. with two different UV LEDs, grown on epitaxial lateral overgrown (ELO) GaN (TDD: $1 \times 10^7/\text{cm}^2$) and standard GaN template (TDD: $1 \times 10^8/\text{cm}^2$) [51]. After 4000 h aging test, the ELO-GaN showed almost no degradation of the output power, whereas a stronger degradation was reported for the standard GaN template. This result further supports the negative effect of the dislocation density on the device reliability.

Liu et al. employed the vertical LED design to fabricate high-power UV LEDs, with wavelengths from 365 to 410 nm [52]. These LEDs were attached to a metal base with silicon submount package and showed very good reliability, thanks to the high-thermal conductivity from the junction to the board and heat sink.

Lin et al. tested the effect on the reliability of an un-doped $Al_{0.23}Ga_{0.77}N$ EBL inserted between the p-type doped $Al_{0.23}Ga_{0.77}N$ EBL and the last barrier [53]. As the un-doped EBL was inserted, the output power (at 19 A/cm^2) of the UV-LED with 365-nm wavelength showed a 400% enhancement. The degradation of EL peak intensity after 1032 h of stress was 17% and 33% for 365-nm UV-LEDs, fabricated with and without inserting an un-doped EBL, respectively. Moreover, the leakage current densities were increased to 4.5×10^{-4} and 4.7×10^{-3} A/cm^2 for 365-nm UV-LEDs, confirming the improvement in the device reliability with the insertion of the un-doped AlGaN EBL. The authors attribute this effect to the lower defect generation rate in the active region caused by the reduced Mg back-diffusion.

Reliability accelerated tests in continuous mode and cycled working mode for two types of commercial UV-LEDs (type A and B) were carried out in [54]. After 1500 h of test at three different temperatures (60, 75, and 90 °C), they found that the degradation is stronger for LEDs working at continuous current mode in all tests. This degradation is related to the worsening of the properties of the chip and of the encapsulation. The cracking of the silicone encapsulation was observed in both types during CW operation, and also during high-temperature test (90 °C) in cycled mode. They relate the faster degradation to a thermos-mechanical stress of silicone lens in contact with the semiconductor chip.

The effect of reverse-bias stress in water vapor on near UV LEDs was investigated in [55]. No reverse-bias luminescence was observed on a device stressed up to 60 min, whereas an increase in reverse-bias EL was detected in devices stressed up to 75 min and more. The result of spatially resolved EL maps revealed that the luminescence was distributed near the chip edge. They attributed the EL to hot electron-induced emission at high-electric field. By means of SEM, EDX, and SIMS analysis on cross-sectioned devices, they detected the down diffusion of Au towards the active region, and attributed to this process a role in the final device degradation.

Monti et al. analyzed extensively the degradation of UV-A LEDs with nominal peak wavelengths from 385 to 420 nm submitted to constant current stress ranging from 1 to 1.3 A at various temperatures (25, 50, and 75 °C) [56]. The time-to-failure (defined as the time to reach a 30% optical power reduction) showed an almost linear dependence on stress current, confirming that the stress current acts as a relevant driving force for the degradation process. The reliability is negatively affected by the temperature, leading to an activation energy of 0.36 eV, confirming that the degradation, ascribed to the generation of defects inside the active region, is a current-driven and thermally activated process.

11.3 Reliability of Packages for Ultraviolet Light-Emitting Diodes

LED chips are usually placed in suitable packages in order to make them compatible with the surface-mounted device (SMD) technology. The empty parts of the package are covered with filler materials, whose goal is to protect the device from air and moisture and to improve the extraction efficiency, lowering the refractive index mismatch with air. Given the high energy of the photons and the high temperature reached during operation, the materials composing the package may be affected by degradation themselves.

A comprehensive study on the properties and reliability of encapsulant materials was presented by Li et al. [57]. They incorporated into the epoxy five different filler materials, namely quartz, fused silica, cristobalite, spherical silica, and spherical silicone and tested their stability under various test conditions.

The first aging was purely thermal at 150 °C, and its results are presented in Fig. 11.13 for the representative cases of epoxy without any filler and with quartz. A decrease of the transmittance is clearly visible in the UV region, and it is more intense in the filled epoxies, leading to a higher yellowing and to a lower optical power. A similar test was carried out under illumination with a 365-nm UV light, and Fig. 11.14 reports the increase in yellowing over time for the various materials. Quartz and cristobalite were the systems with the lowest change in both tests, pure epoxy aside, and the difference was assumed to be related to variations in the reflection, caused by the different refractive index of the various materials.

Yazdan Mehr et al. tested the degradation of thermoplastic Bisphenol A Polycarbonate (BPA-PC) submitted to thermal storage at temperatures comparable with nominal operating conditions [58]. They detected a significant thermally activated increase in absorbance below 400 nm and the presence of a peak at 290 nm, attributed to the formation of phenolic end groups in the polymer. By means of infrared spectra in attenuated total reflection (ATR) mode, they report an increase in the absorption bands of aromatic ketones and cyclic anhydrides in the carbonyl region, resulting from thermal oxidation, and no change in the hydroxyl region, pointing out no major contribution of the Fries rearrangement reaction to the yellowing. Additionally, they detect no band in the ATR spectra caused by carboxylic acids. The proposed reaction pathway for the thermal oxidation causing the yellowing is sketched in Fig. 11.15.

Bae et al. proposed a new oligosiloxane resin as encapsulant, obtained by hydrosilylation of hydrogen-methyl oligosiloxane resin and vinyl-methyl siloxane resin, and compared it to polydimethylsiloxane (PDMS) [59]. The transmittance spectrum of the two materials is similar in a wide wavelength range (see Fig. 11.16a), and the largest benefit is an improved reliability. As can be seen from comparing Fig. 11.16b, c, the degradation resulting from illumination with UVB light at 10 mW/cm^2 is lower for the methyl hybrimer, especially in the UV-A and UV-B spectral range. The comparison between the two materials is reported in Fig. 11.16d, which includes also the result of a thermal stress test at 200 °C, confirming

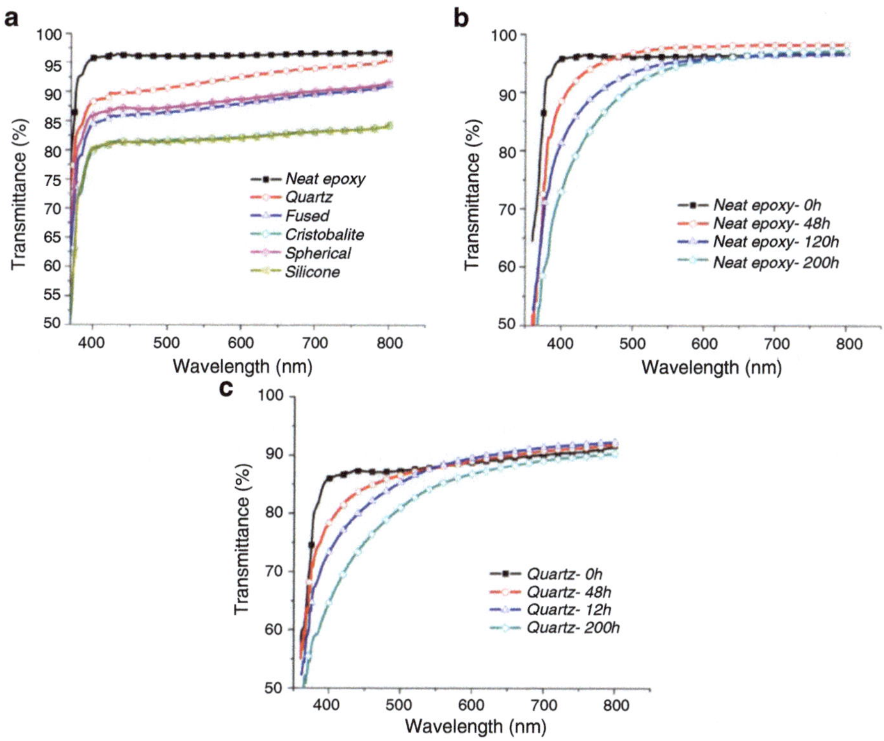

Fig. 11.13 (**a**) Transmittance spectra of various filler materials before stress and corresponding spectra during thermal stress for (**b**) pure epoxy and (**c**) quartz. Reprinted with permission from "High-Performance Light-Emitting Diodes Encapsulated with Silica-Filled Epoxy Materials", T. Li, J. Zhang, H. Wang, Z. Hu, and Y. Yu, ACS Applied Materials & Interfaces 2013 5 (18), 8968–8981. Copyright (2013) American Chemical Society

the stability of the new material. No visible formation of cracks or shrinkage was detected during the stresses, indicating no major light-induced scission of organic groups. The authors relate the higher stability to several factors, such as lower radical sources (such as Pt catalyst and unreacted organic phases), the robust siloxane network, the hyper-branched cross-linking, and the high cross-link density of vinyl and hydrogen groups.

In a following study, the same research group further improved the material creating a fluoro-siloxane hybrid, with superior performance and higher reliability in the UV-C range [60]. Even though the methyl hybrimer has a higher initial transmittance, due to its large free volume and long and homogeneous siloxane backbone, during UV-B aging the transparency decreases due to the formation of methyl radicals and to the radical-activated oxidative degradation of the organic pendants in the siloxane backbone. The fluoro hybrimer showed almost no degradation under the same condition, and the authors suggest that it is related to the high number of vinyl and fluoro groups, low amount of methyl groups, and hydrosilylation-derived short

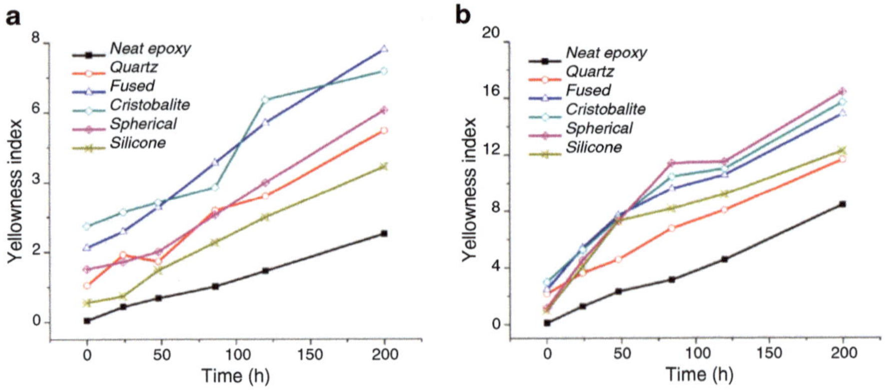

Fig. 11.14 Yellowing of various filler materials as a consequence of (**a**) thermal aging and (**b**) UV aging. Reprinted with permission from "High-Performance Light-Emitting Diodes Encapsulated with Silica-Filled Epoxy Materials", T. Li, J. Zhang, H. Wang, Z. Hu, and Y. Yu, ACS Applied Materials & Interfaces 2013 5 (18), 8968–8981. Copyright (2013) American Chemical Society

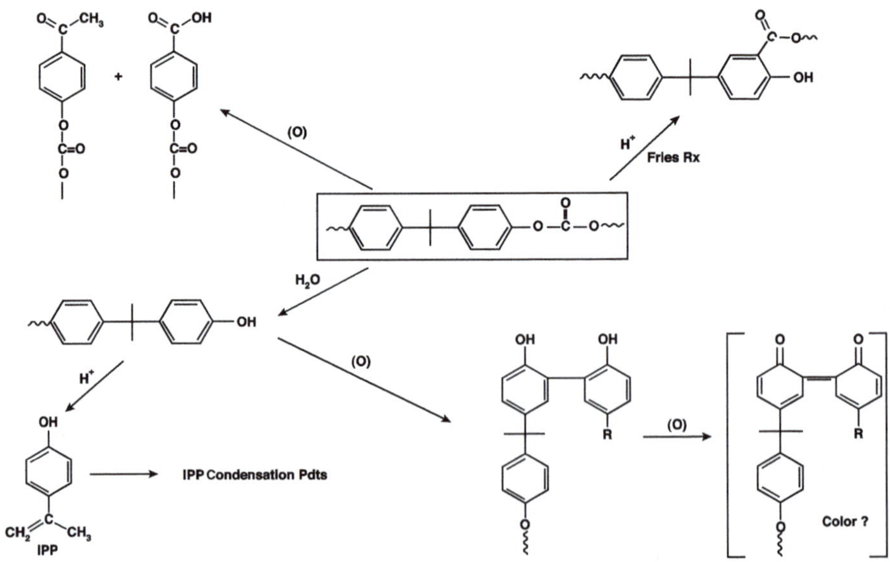

Fig. 11.15 Thermal oxidation reaction causing the yellowing of BPA-PC during thermal aging [58]. Reprinted from Microelectronics Reliability, 54, M. Yazdan Mehr, W.D. van Driel, S. Koh, G.Q. Zhang, "Reliability and optical properties of LED lens plates under high temperature stress", 2440–2447, Copyright (2004), with permission from Elsevier

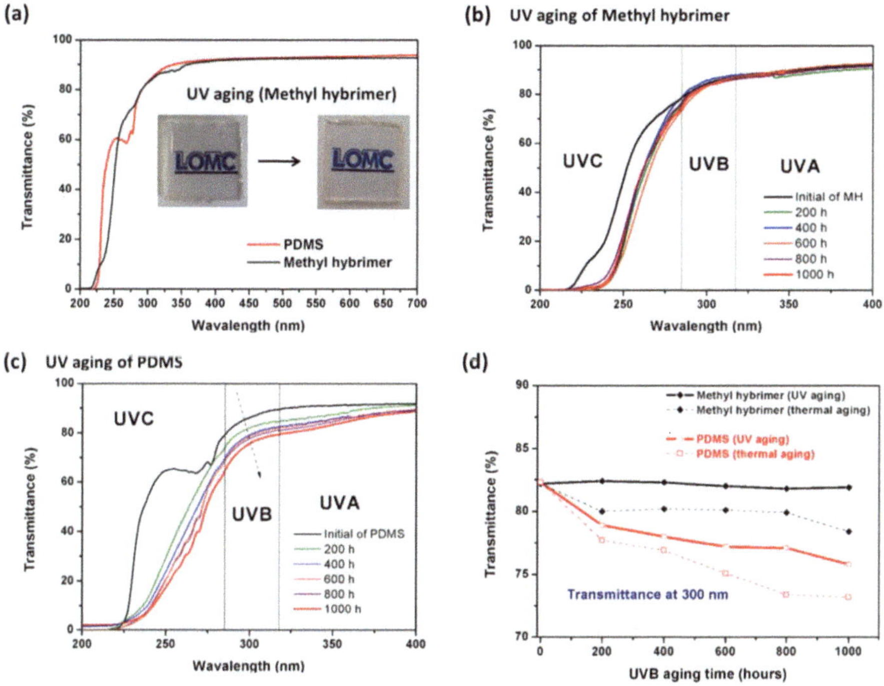

Fig. 11.16 (**a**) Transmittance of common PDMS compared to the proposed methyl hybrimer. (**b**) and (**c**) show its variation during UV-B stress for the two materials, summarized in (**d**) along with the results of a thermal stress at 200 °C [59]. Reprinted with permission from "Ultraviolet Light Stable and Transparent Sol–Gel Methyl Siloxane Hybrid Material for UV Light-Emitting Diode (UV LED) Encapsulant", J., YongHo Kim, H. Kim, Y. Kim, J. Jin, and B. Bae, ACS Applied Materials & Interfaces 2015 7 (2), 1035–1039. Copyright (2015) American Chemical Society

cross-links. The two hybrimers were compared also during UV-C stress, leading to the transmittance changes highlighted in Fig. 11.17a, b for fluoro and methyl hybrimer, respectively. The new material showed an increase in its transmittance as a consequence of irradiation in a wide spectral range (see Fig. 11.17c), attributed to a vinyl radical stabilization induced by the UV light.

The effect of the chain flexibility of the encapsulant was analyzed by Chen et al. [61]. They tested cycloaliphatic epoxy resin (3–4-epoxycy-clohexane) methyl 3–4-epoxycyclohexyl-carboxylate (ECC), both pure and with addition of various quantities of diglycidyl ether (HBADGE). They found that, even though the pure ECC showed the best reliability under thermal and UV stress, up to 30% HBADGE still led to comparable stability (see Fig. 11.18). The 30% HBADGE inclusion proved effective in lowering the water sorption and in maximizing the pass rate during various reliability tests according to the JEDEC J-STD-020 standard, attributed to the lower thermal expansion coefficient and storage modulus at reflow temperature.

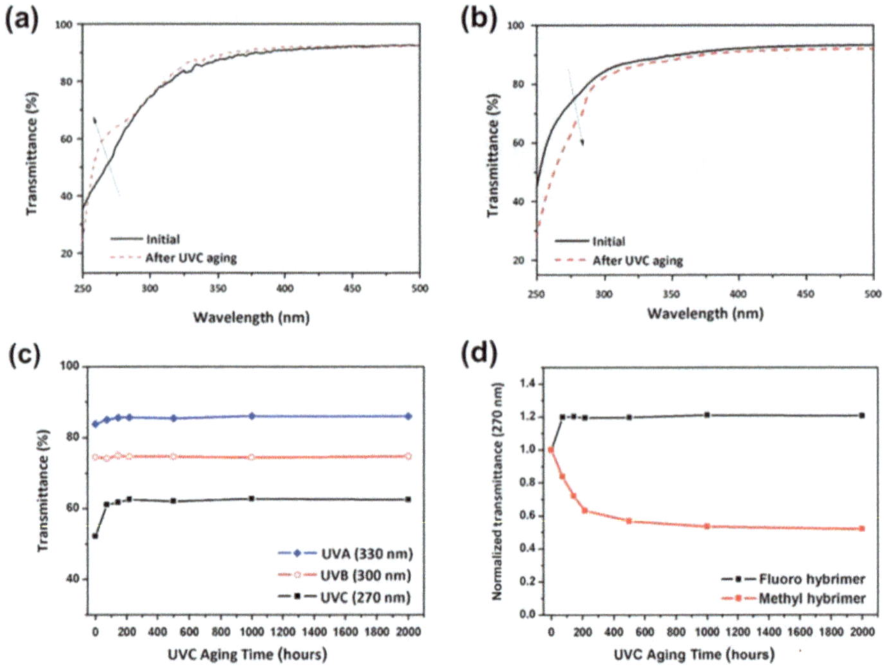

Fig. 11.17 Transmittance of (**a**) fluoro hybrimer and (**b**) methyl hybrimer before and after UV-C aging. The improvement in a wider spectral range is shown in (**c**) as a function of stress time, and the UV-C transmittance of the two materials is compared in (**d**). Reproduced from Ref. [60] with permission of The Royal Society of Chemistry

A similar set of tests was carried out by Wen et al., incorporating silicone instead of HBADGE into ECC [62]. They found better thermal and UV stability with increasing silicone content, due to the higher bond energy of Si–O with respect to C–O and C–C. The added silicone has a beneficial effect on delamination issues, thanks to the lower internal stress of the epoxy matrix, and on water sorption, given the hydrophobic nature of Si–O–Si compounds.

Transparent materials for filling make up most of the reports on package-related issues in the literature. An analysis of the reflectance of the materials for the package substrate was reported by Qiu et al., who tested epoxy molding compounds (EMC) and silicone molding compounds (SMC) during stress at high temperature (150 °C) and under a mercury UV lamp [63]. The SMC reflectance was unaffected by the thermal stress, whereas EMC showed an increase in the 365–375-nm range. SEM analysis of EMC plates highlighted the removal of the epoxy resin from the surface, leaving the filler particles exposed. No change was detected in the surface morphology of SMC plates. During the UV stress, the reflectance of SMC did not change, and the increase in the EMC reflectance was stronger. The surface of the SMC plate became rougher after the stress, and in the EMC case this phenomenon

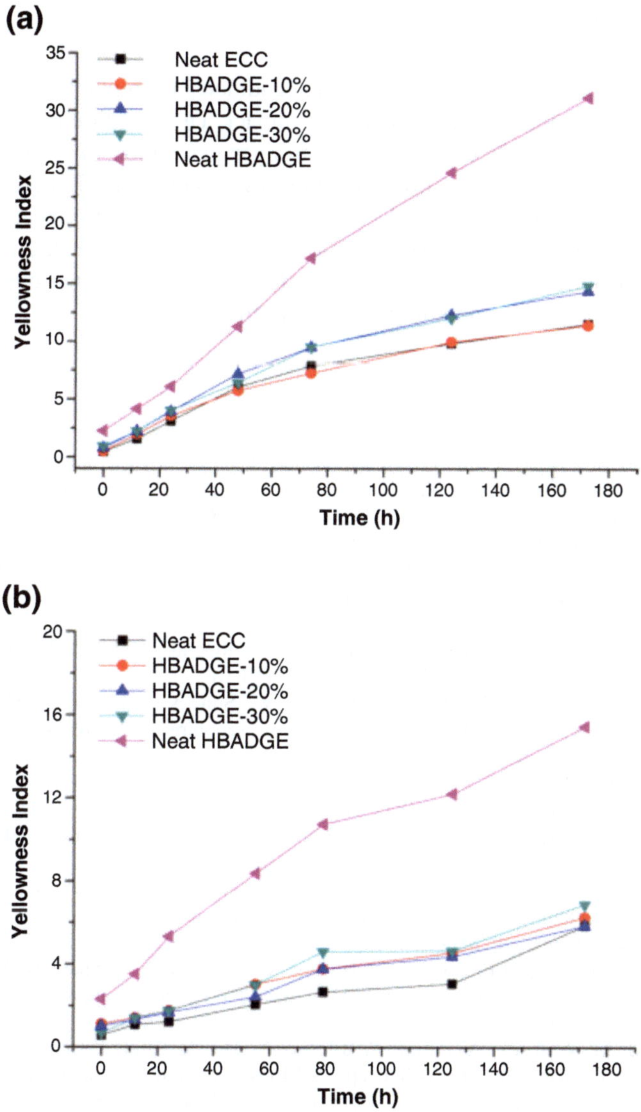

Fig. 11.18 Variation in yellowness index during (**a**) thermal and (**b**) UV aging of various epoxy encapsulants [61]. Reprinted with permission from "Effect of Chain Flexibility of Epoxy Encapsulants on the Performance and Reliability of Light-Emitting Diodes", Z. Chen, Z. Liu, G. Shen, R. Wen, J. Lv, J. Huo, and Y. Yu, Industrial & Engineering Chemistry Research 2016 55 (28), 7635–7645. Copyright (2016) American Chemical Society

was more intense. This effect could be caused by the chalking effect, i.e., the erosion of the top layer of organic molecules caused by the UV light.

11.4 Summary

Ultraviolet GaN-based LEDs are promising candidates for replacing traditional UV light sources, lowering their environmental impact. The adoption is still limited by the nonoptimal reliability, caused by the stronger generation of defects inside the devices containing aluminum alloys, leading to an increase in SRH recombination. The degradation is accelerated by the significant self-heating originated by the high turn-on voltage of the higher bandgap devices. The stability of the package and encapsulant materials against high-energy photons is an additional open issue that must be solved in order to maximize device lifetime.

References

1. A.A. Allerman, M.H. Crawford, A.J. Fischer, K.H.A. Bogart, S.R. Lee, D.M. Follstaedt, P.P. Provencio, D.D. Koleske, Growth and design of deep-UV (240-290 nm) light emitting diodes using AlGaN alloys. J. Cryst. Growth **272**(1–4), 227–241 (2004). https://doi.org/10.1016/j.jcrysgro.2004.08.035
2. M. Asif Khan, AlGaN multiple quantum well based deep UV LEDs and their applications. Phys. Status Solidi Appl. Mater. Sci. **203**(7), 1764–1770 (2006). https://doi.org/10.1002/pssa.200565427
3. M. Asif Khan, Deep ultraviolet light emitting diodes with emission below 300 nm. MRS Proc. **892** (2005). https://doi.org/10.1557/PROC-0892-FF01-01
4. M. Shatalov, Z. Gong, M. Gaevski, S. Wu, W. Sun, V. Adivarahan, M. Asif Khan, Reliability of AlGaN-based deep UV LEDs on sapphire. SPIE **6134**, 61340P (2006). https://doi.org/10.1117/12.647204
5. J.P. Zhang, H.M. Wang, W.H. Sun, V. Adivarahan, S. Wu, A. Chitnis, C.Q. Chen, M. Shatalov, E. Kuokstis, J.W. Yang, M. Asif Khan, High-quality AlGaN layers over pulsed atomic-layer epitaxially grown AlN templates for deep ultraviolet light-emitting diodes. J. Electron. Mater. **32**(5), 364–370 (2003). https://doi.org/10.1007/s11664-003-0159-2
6. Z. Gong, M. Gaevski, V. Adivarahan, W. Sun, M. Shatalov, M. Asif Khan, Optical power degradation mechanisms in AlGaN-based 280 nm deep ultraviolet light-emitting diodes on sapphire. Appl. Phys. Lett. **88**(12), 1–4 (2006). https://doi.org/10.1063/1.2187429
7. Z. Gong, S. Chhajed, M.E. Gaevski, W.H. Sun, V. Adivarahan, M. Shatalov, M. Asif Khan, Reliability and degradation modes of 280 nm deep UV LEDs on sapphire. Mater. Res. Soc. Symp. Proc. **892**, 169–174 (2006)
8. M.L. Reed, M. Wraback, A. Lunev, Y. Bilenko, X. Hu, A. Sattu, J. Deng, M. Shatalov, R. Gaska, Device self-heating effects in deep UV LEDs studied by systematic variation in pulsed current injection. Phys. Status Solidi Curr. Top. Solid State Phys. **5**(6), 2053–2055 (2008). https://doi.org/10.1002/pssc.200778402
9. C.G. Moe, M.L. Reed, G.A. Garrett, A.V. Sampath, T. Alexander, H. Shen, M. Wraback, Y. Bilenko, M. Shatalov, J. Yang, W. Sun, J. Deng, R. Gaska, Current-induced degradation of high performance deep ultraviolet light emitting diodes. Appl. Phys. Lett. **96**(21), 213512 (2010). https://doi.org/10.1063/1.3435485
10. A. Pinos, S. Marcinkevičius, J. Yang, Y. Bilenko, M. Shatalov, R. Gaska, M.S. Shur, Aging of AlGaN quantum well light emitting diode studied by scanning near-field optical spectroscopy. Appl. Phys. Lett. **95**(18), 181914 (2009). https://doi.org/10.1063/1.3262964
11. M. Asif Khan, S. Hwang, J. Lowder, V. Adivarahan, Q. Fareed, Reliability issues in AlGaN based deep ultraviolet light emitting diodes. 2009 IEEE Int. Reliab. Phys. Symp. 89–93. https://doi.org/10.1109/IRPS.2009.5173229

12. R. Jain, W. Sun, J. Yang, M. Shatalov, X. Hu, A. Sattu, A. Lunev, J. Deng, I. Shturm, Y. Bilenko, R. Gaska, M.S. Shur, Migration enhanced lateral epitaxial overgrowth of AlN and AlGaN for high reliability deep ultraviolet light emitting diodes. Appl. Phys. Lett. **93**(5), 110–113 (2008). https://doi.org/10.1063/1.2969402
13. S. Sawyer, S.L. Rumyantsev, M.S. Shur, Degradation of AlGaN-based ultraviolet light emitting diodes. Solid State Electron. **52**(6), 968–972 (2008). https://doi.org/10.1016/j.sse.2008.01.027
14. G. Meneghesso, S. Levada, E. Zanoni, S. Podda, G. Mura, M. Vanzi, A. Cavallini, A. Castaldini, S. Du, I. Eliashevich, Failure modes and mechanisms of DC-aged GaN LEDs. Phys. Status Solidi **194**(2), 389–392 (2002). https://doi.org/10.1002/1521-396X(200212)194:2<389::AID-PSSA389>3.0.CO;2-O
15. S. Bychikhin, D. Pogany, L.K.J. Vandamme, G. Meneghesso, E. Zanoni, Low-frequency noise sources in as-prepared and aged GaN-based light-emitting diodes. J. Appl. Phys. **97**(12), 123714 (2005). https://doi.org/10.1063/1.1942628
16. X. Chen, A. Pedersen, A.D. van Rheenen, Effect of electrical and thermal stress on low frequency noise characteristics of laser diodes. Microelectron. Reliab. **41**(1), 105–110 (2001). https://doi.org/10.1016/S0026-2714(00)00201-8
17. A. Pinos, S. Marcinkevičius, J. Yang, R. Gaska, M. Shatalov, M.S. Shur, Optical studies of degradation of AlGaN quantum well based deep ultraviolet light emitting diodes. J. Appl. Phys. **108**(9), 93113 (2010). https://doi.org/10.1063/1.3506697
18. A. Pinos, S. Marcinkevičius, M.S. Shur, High current-induced degradation of AlGaN ultraviolet light emitting diodes. J. Appl. Phys. **109**(10) (2011). https://doi.org/10.1063/1.3590149
19. M. Meneghini, M. Pavesi, N. Trivellin, R. Gaska, E. Zanoni, G. Meneghesso, Reliability of deep-UV light-emitting diodes. IEEE Trans. Device Mater. Reliab. **8**(2), 248–254 (2008). https://doi.org/10.1109/TDMR.2008.919570
20. J.P. Zhang, X. Hu, Y. Bilenko, J. Deng, A. Lunev, M.S. Shur, R. Gaska, M. Shatalov, J.W. Yang, M. Asif Khan, AlGaN-based 280 nm light-emitting diodes with continuous-wave power exceeding 1 mW at 25 mA. Appl. Phys. Lett. **85**(23), 5532–5534 (2004). https://doi.org/10.1063/1.1831557
21. O. Pursiainen, N. Linder, A. Jaeger, R. Oberschmid, K. Streubel, Identification of aging mechanisms in the optical and electrical characteristics of light-emitting diodes. Appl. Phys. Lett. **79**(18), 2895–2897 (2001). https://doi.org/10.1063/1.1413721
22. T. Yanagisawa, T. Kojima, Degradation of InGaN blue light-emitting diodes under continuous and low-speed pulse operations. Microelectron. Reliab. **43**(6), 977–980 (2003). https://doi.org/10.1016/S0026-2714(03)00093-3
23. M. Meneghini, N. Trivellin, L. Trevisanello, A. Lunev, J. Yang, Y. Bilenko, W. Sun, M. Shatalov, R. Gaska, E. Zanoni, G. Meneghesso, Combined optical and electrical analysis of AlGaN-based deep-UV LEDs reliability. IEEE Int. Reliab. Phys. Symp. Proc. 441–445 (2008). https://doi.org/10.1109/RELPHY.2008.4558925
24. F. Rossi, M. Pavesi, M. Meneghini, G. Salviati, M. Manfredi, G. Meneghesso, A. Castaldini, A. Cavallini, L. Rigutti, U. Strass, U. Zehnder, E. Zanoni, Influence of short-term low current dc aging on the electrical and optical properties of InGaN blue light-emitting diodes. J. Appl. Phys. **99**(5), 10–17 (2006). https://doi.org/10.1063/1.2178856
25. M. Meneghini, D. Barbisan, L. Rodighiero, G. Meneghesso, E. Zanoni, Analysis of the physical processes responsible for the degradation of deep-ultraviolet light emitting diodes. Appl. Phys. Lett. **97**(14), 143506 (2010). https://doi.org/10.1063/1.3497082
26. M. Meneghini, D. Barbisan, Y. Bilenko, M. Shatalov, J. Yang, R. Gaska, G. Meneghesso, E. Zanoni, Defect-related degradation of deep-UV-LEDs. Microelectron. Reliab. **50**(9–11), 1538–1542 (2010). https://doi.org/10.1016/j.microrel.2010.07.089
27. V. Adivarahan, W. Sun, A. Chitnis, M. Shatalov, S. Wu, H. Maruska, M. Asif Khan, 250 nm AlGaN light-emitting diodes. Appl. Phys. Lett. **85**(12), 2175 (2004). https://doi.org/10.1063/1.1796525
28. N. Otsuka, A. Tsujimura, Y. Hasegawa, G. Sugahara, M. Kume, Y. Ban, Room temperature 339 nm emission from Al0.13Ga0.87N/Al0.10Ga0.90N double heterostructure light-emitting diode on sapphire substrate. Jpn. J. Appl. Phys. **39**(Part 2, No. 5B), L445–L448 (2000). https://doi.org/10.1143/JJAP.39.L445

29. J.S. Park, D.W. Fothergill, P. Wellenius, S.M. Bishop, J.F. Muth, R.F. Davis, Origins of parasitic emissions from 353 nm AlGaN-based ultraviolet light emitting diodes over SiC substrates. Jpn J. Appl. Physics, Part 1 Regul. Pap. Short Notes Rev. Pap. **45**(5 A), 4083–4086 (2006). https://doi.org/10.1143/JJAP.45.4083
30. J. Grandusky, Y. Cui, S. Gibb, M. Mendrick, L. Schowalter, Performance and reliability of ultraviolet-C pseudomorphic light emitting diodes on bulk AlN substrates. Phys. Status Solidi Curr. Top. Solid State Phys. **7**(7–8), 2199–2201 (2010). https://doi.org/10.1002/pssc.200983635
31. K. Kitamura, J.R. Grandusky, C.G. Moe, J. Chen, M.C. Mendrick, Y. Li, M. Toita, K. Nagase, T. Morishita, H. Ishii, S. Yamada, L.J. Schowalter, S3-P1: reliability and lifetime of pseudomorphic UVC leds on AlN substrate under various stress condition. Lester Eastman Conference 2014—High Performance Devices, LEC (2014), pp. 2–6. https://doi.org/10.1109/LEC.2014.6951561
32. S. Tomiya, T. Hino, S. Goto, M. Takeya, M. Ikeda, Dislocation related issues in the degradation of GaN-based laser diodes. IEEE J. Sel. Top. Quantum Electron. **10**(6), 1277–1286 (2004). https://doi.org/10.1109/JSTQE.2004.837735
33. S.N. Lee, H.S. Paek, J.K. Son, H. Kim, K.K. Kim, K.H. Ha, O.H. Nam, Y. Park, Effects of Mg dopant on the degradation of InGaN multiple quantum wells in AlInGaN-based light emitting devices. J. Electroceram. **23**(2–4), 406–409 (2009). https://doi.org/10.1007/s10832-008-9478-2
34. A. Fujioka, K. Asada, H. Yamada, T. Ohtsuka, T. Ogawa, T. Kosugi, D. Kishikawa, T. Mukai, High-output-power 255/280/310 nm deep ultraviolet light-emitting diodes and their lifetime characteristics. Semicond. Sci. Technol. **29**(8), 84005 (2014). https://doi.org/10.1088/0268-1242/29/8/084005
35. J. Rass, T. Kolbe, N. Lobo Ploch, T. Wernicke, F. Mehnke, C. Kuhn, J. Enslin, M. Guttmann, C. Reich, A. Mogilatenko, J. Glaab, C. Stoelmacker, M. Lapeyrade, S. Einfeldt, M. Weyers, M. Kneissl, High power UV-B LEDs with long lifetime. Proc. SPIE Gall. Nitride Mater. Devices X **9363**, 93631K (2015). https://doi.org/10.1117/12.2077426
36. J. Glaab, C. Ploch, R. Kelz, C. Stölmacker, M. Lapeyrade, N. Lobo Ploch, J. Rass, T. Kolbe, S. Einfeldt, F. Mehnke, C. Kuhn, T. Wernicke, M. Weyers, M. Kneissl, Degradation of (InAlGa)N-based UV-B light emitting diodes stressed by current and temperature. J. Appl. Phys. **118**(9) (2015). https://doi.org/10.1063/1.4929656
37. J. Glaab, C. Ploch, R. Kelz, C. Stölmacker, M. Lapeyrade, N. Lobo Ploch, J. Rass, T. Kolbe, S. Einfeldt, F. Mehnke, C. Kuhn, T. Wernicke, M. Weyers, M. Kneissl, Temperature induced degradation of InAlGaN multiple-quantum well UV-B LEDs. MRS Proc. **1792** (2015). https://doi.org/10.1557/opl.2015.446
38. Q. Shan, D.S. Meyaard, Q. Dai, J. Cho, E. Fred Schubert, J. Kon Son, C. Sone, Transport-mechanism analysis of the reverse leakage current in GaInN light-emitting diodes. Appl. Phys. Lett. **99**(25), 253506 (2011). https://doi.org/10.1063/1.3668104
39. M.W. Moseley, A.A. Allerman, M.H. Crawford, J.J. Wierer, M.L. Smith, A.M. Armstrong, Detection and modeling of leakage current in AlGaN-based deep ultraviolet light-emitting diodes. J. Appl. Phys. **117**(9) (2015). https://doi.org/10.1063/1.4908543
40. K. Orita, S. Takigawa, M. Yuri, T. Tanaka, M. Meneghini, N. Trivellin, L.-R. Trevisanello, E. Zanoni, G. Meneghesso, Analysis of diffusion involved in degradation of InGaN-based laser diodes. 2009 IEEE International Reliability Physics Symposium (2009), pp. 736–740. https://doi.org/10.1109/IRPS.2009.5173340
41. M. Meneghini, L.-R. Trevisanello, S. Levada, G. Meneghesso, G. Tamiazzo, E. Zanoni, T. Zahner, U. Zehnder, V. Harle, U. Strauss, Failure mechanisms of gallium nitride leds related with passivation. IEEE International Electron Devices Meeting. IEDM Technical Digest (2005), pp. 1009–1012. https://doi.org/10.1109/IEDM.2005.1609534
42. J. Glaab, N. Lobo Ploch, J. Rass, T. Kolbe, T. Wernicke, F. Mehnke, C. Kuhn, J. Enslin, C. Stölmacker, V. Kueller, A. Knauer, S. Einfeldt, M. Weyers, M. Kneissl, Influence of the LED heterostructure on the degradation behavior of (InAlGa)N-based UV-B LEDs. SPIE Opto. **9748**, 97481O (2016). https://doi.org/10.1117/12.2208906

43. J.C. Zhang, Y.H. Zhu, T. Egawa, S. Sumiya, M. Miyoshi, M. Tanaka, Suppression of the subband parasitic peak by 1 nm i-AlN interlayer in AlGaN deep ultraviolet light-emitting diodes. Appl. Phys. Lett. **93**(13) (2008). https://doi.org/10.1063/1.2996580
44. T. Kolbe, J. Stellmach, F. Mehnke, M.A. Rothe, V. Kueller, A. Knauer, S. Einfeldt, T. Wernicke, M. Weyers, M. Kneissl, Efficient carrier-injection and electron-confinement in UV-B light-emitting diodes. Phys. Status Solidi Appl. Mater. Sci. **213**(1), 210–214 (2016). https://doi.org/10.1002/pssa.201532479
45. M.L. Nakarmi, N. Nepal, J.Y. Lin, H.X. Jiang, Photoluminescence studies of impurity transitions in Mg-doped AlGaN alloys. Appl. Phys. Lett. **94**(9), 91903 (2009). https://doi.org/10.1063/1.3094754
46. D. Monti, M. Meneghini, C. De Santi, G. Meneghesso, E. Zanoni, J. Glaab, J. Rass, S. Einfeldt, F. Mehnke, J. Enslin, T. Wernicke, M. Kneissl, Defect-related degradation of AlGaN-Based UV-V LEDs. IEEE Trans. Electron Devices **64**(1), 200–205 (2017). https://doi.org/10.1109/TED.2016.2631720
47. D. Monti, M. Meneghini, C. De Santi, G. Meneghesso, E. Zanoni, J. Glaab, J. Rass, S. Einfeldt, F. Mehnke, T. Wernicke, M. Kneissl, L. Institut, U. Berlin, I. Festkörperphysik, Defect generation in deep-UV AlGaN-based LEDs investigated by electrical and spectroscopic characterisation. Proc. SPIE **10124**, 1–9 (2017). https://doi.org/10.1117/12.2253843
48. C. De Santi, M. Meneghini, D. Monti, J. Glaab, M. Guttmann, J. Rass, S. Einfeldt, F. Mehnke, J. Enslin, T. Wernicke, M. Kneissl, G. Meneghesso, E. Zanoni, Recombination mechanisms and thermal droop in AlGaN-based UV-B LEDs. Photonics Res. **5**(2), 44–51 (2017). https://doi.org/10.1364/PRJ.5.000A44
49. C. De Santi, M. Meneghini, M. La Grassa, B. Galler, R. Zeisel, M. Goano, S. Dominici, M. Mandurrino, F. Bertazzi, D. Robidas, G. Meneghesso, E. Zanoni, Role of defects in the thermal droop of InGaN-based light emitting diodes. J. Appl. Phys. **119**(9), 94501 (2016). https://doi.org/10.1063/1.4942438
50. W.K. Wang, D.S. Wuu, S.H. Lin, P. Han, R.H. Horng, T.C. Hsu, D.T.C. Huo, M.J. Jou, Y.H. Yu, A. Lin, Efficiency improvement of near-ultraviolet InGaN LEDs using patterned sapphire substrates. IEEE J. Quantum Electron. **41**(11), 1403–1409 (2005). https://doi.org/10.1109/JQE.2005.857057
51. T. Mukai, D. Morita, M. Yamamoto, K. Akaishi, K. Matoba, K. Yasutomo, Y. Kasai, M. Sano, S.I. Nagahama, Investigation of optical-output-power degradation in 365-nm UV-LEDs. Phys. Status Solidi Curr. Top. Solid State Phys. **3**(6), 2211–2214 (2006). https://doi.org/10.1002/pssc.200565354
52. W.H. Liu, C.F. Chu, C.C. Cheng, K.H. Hsu, Y.T. Chung, Y.K. Wang, C.C. Li, J.Y. Chu, F.H. Fan, H.C. Cheng, Y.W. Chen, Y.H. Chang, L.W. Shan, T. Doan, C. Tran, Development of high-power UV LEDs for epoxy curing applications. Proc. SPIE **7602**, 76021K (2010). https://doi.org/10.1117/12.845472
53. W. Lin, T. Wang, S. Ou, J. Liang, D. Wuu, Improved performance of 365-nm LEDs by inserting an un-doped electron-blocking layer. IEEE Electron Device Lett. **35**(4), 467–469 (2014). https://doi.org/10.1109/LED.2014.2306711
54. F.J. Arques-Orobon, N. Nuñez, M. Vazquez, V. González-Posadas, UV LEDs reliability tests for fluoro-sensing sensor application. Microelectron. Reliab. **54**(9–10), 2154–2158 (2014). https://doi.org/10.1016/j.microrel.2014.07.059
55. H. Chen, H.Y. Shen, S.C. Shei, N.C. Kang, H.C. Lai, Y.C. Chu, H.W. Chang, Exploring failure mechanisms of near ultraviolet AlGaN/GaN light-emitting diodes by reverse-bias stress in water vapour. Int. J. Nanotechnol. **12**(1/2), 38 (2015). https://doi.org/10.1504/IJNT.2015.066192
56. D. Monti, M. Meneghini, C. De Santi, G. Meneghesso, E. Zanoni, Degradation of UV-A LEDs: physical origin and dependence on stress conditions. IEEE Trans. Device Mater. Reliab. **16**(2), 213–219 (2016). https://doi.org/10.1109/TDMR.2016.2558473
57. T. Li, J. Zhang, H. Wang, Z. Hu, Y. Yu, High-performance light-emitting diodes encapsulated with silica-filled epoxy materials. ACS Appl. Mater. Interfaces **5**(18), 8968–8981 (2013). https://doi.org/10.1021/am402035r

58. M. Yazdan Mehr, W.D. Van Driel, S. Koh, G.Q. Zhang, Reliability and optical properties of LED lens plates under high temperature stress. Microelectron. Reliab. **54**(11), 2440–2447 (2014). https://doi.org/10.1016/j.microrel.2014.05.003
59. J.Y. Bae, Y.H. Kim, H.Y. Kim, Y.B. Kim, J. Jin, B.S. Bae, Ultraviolet light stable and transparent sol-gel methyl siloxane hybrid material for UV light-emitting diode (UV LED) encapsulant. ACS Appl. Mater. Interfaces **7**(2), 1035–1039 (2015). https://doi.org/10.1021/am507132a
60. J.-Y. Bae, H.-Y. Kim, Y.-W. Lim, Y.-H. Kim, B.-S. Bae, Optically recoverable, deep ultraviolet (UV) stable and transparent sol–gel fluoro siloxane hybrid material for a UV LED encapsulant. RSC Adv. **6**(32), 26826–26834 (2016). https://doi.org/10.1039/C6RA01346E
61. Z. Chen, Z. Liu, G. Shen, R. Wen, J. Lv, J. Huo, Y. Yu, Effect of chain flexibility of epoxy encapsulants on the performance and reliability of light-emitting diodes. Ind. Eng. Chem. Res. **55**(28), 7635–7645 (2016). https://doi.org/10.1021/acs.iecr.6b01159
62. R. Wen, J. Huo, J. Lv, Z. Liu, Y. Yu, Effect of silicone resin modification on the performance of epoxy materials for LED encapsulation. J. Mater. Sci. Mater. Electron. **28**(14522), 1–14 (2017). https://doi.org/10.1007/s10854-017-7316-5
63. X. Qiu, J.C.C. Lo, A.W. Shang, S.W.R. Lee, Investigation of reliability of EMC and SMC on reflectance for UV LED applications. 17th International Conference on Thermal Mechanical Multi-Physics Simulation and Expriments in Microelectronics and Microsystems, EuroSimE 2016, vol. 2 (2016), pp. 1–7. https://doi.org/10.1109/EuroSimE.2016.7463398

Chapter 12
Nitride Nanowires for Light Emitting Diodes

Nan Guan, Xing Dai, François H. Julien, Joël Eymery, Christophe Durant, and Maria Tchernycheva

12.1 Introduction

Nitride-based light emitting diodes (LEDs) are of tremendous importance for future high-efficiency lighting technologies. More than 20 years have passed since the demonstration of the first InGaN/GaN LEDs [1]. Nitride LEDs represent today a huge market reaching $30.5 Billion in 2016. Their fabrication technology continuously progresses targeting higher efficiencies and lower production costs.

A standard nitride LED consists of functional thin films typically grown on a Sapphire substrate along the c-direction of the wurtzite crystal. It contains the n-doped GaN layer for electron injection, the InGaN/GaN quantum well active region followed by an electron blocking layer, and finally the p-doped GaN layer for hole injection. The whole structure is contacted and appropriately packaged. Despite the relative maturity of this technology, certain issues for nitride LEDs remain unsolved. The most important one is the efficiency droop at high injection. This is a major problem in high power LEDs, which operate beyond the optimal efficiency point. The principal physical phenomenon responsible for the efficiency droop is believed to be Auger recombinations [2]. These are three-body processes (mainly, interactions implying two electrons and one hole, but also processes mediated by

N. Guan · X. Dai · F. H. Julien · M. Tchernycheva (✉)
Centre de Nanosciences et de Nanotechnologies, UMR9001 CNRS, University Paris Sud, University Paris Saclay, Palaiseau, France
e-mail: maria.tchernycheva@u-psud.fr

J. Eymery
Univ. Grenoble Alpes, CEA, INAC, MEM, Grenoble, France

C. Durant
Univ. Grenoble Alpes, CEA, INAC, PHELIQS, "Nanophysique et Semiconducteurs", Grenoble, France

J. Li, G. Q. Zhang (eds.), *Light-Emitting Diodes*, Solid State Lighting Technology and Application Series 4, https://doi.org/10.1007/978-3-319-99211-2_12

interactions with phonons [3]). The losses due to Auger recombinations can only be reduced by reducing the carrier density. The second important problem is the so-called green gap, which transduces the decrease of efficiency in high indium content LEDs with emission wavelengths from green to red. This is due to the increased defect density in In-rich InGaN/GaN quantum wells, but also to the enhanced internal field spatially separating electrons and holes and increasing the radiative recombination time (quantum confined Stark effect).

The above-mentioned open issues of thin film LEDs have motivated a strong research effort on alternative LED architectures. Nanostructured LEDs have been proposed as a possible solution. The main-stream approach consists in nanostructuring of the substrate (for example, LED growth on a nanopatterned sapphire is today actively explored [4, 5]). It allows to reduce the dislocation density and to improve the light extraction while keeping a flat geometry for the active region and contact layers [5]. This approach is attractive for LED manufacturers since it requires only moderate modifications in the existing LED fabrication technology. Alternatively, one can further exploit the nanostructuring and fully change the device architecture from two-dimensional films to three-dimensional nanowires (NWs).

NWs can be obtained by either a top-down etching or a bottom-up growth. The top-down approach uses lithography and dry etching of a thin film LED to form either needles [6] or NWs [7] at well-controlled positions. It yields a NW LED with an axial architecture, i.e., an LED with the InGaN/GaN heterostructure along the wire axis sandwiched between the p-doped and n-doped GaN extremities as illustrated in Fig. 12.1 left. This approach benefits from the high degree of control over the layer thickness and doping profile achieved in thin film growth. Etching also allows to relax the strain and thus reduce the internal field in axial InGaN/GaN heterostructures, which is particularly important for green LEDs. Moreover, the light extraction is increased due to scattering in the array. As a drawback, for the axial geometry the active region intersects the NW surface, which increases the surface non-radiative recombination. In addition, dry etching induces surface damage, which, however, may be healed by wet etching and surface passivation [8].

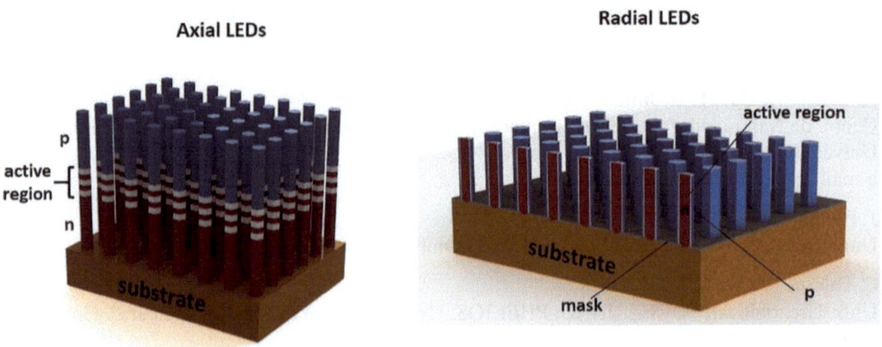

Fig. 12.1 Schematics of axial (left) and radial (right) LED architectures

Compared to the top-down etching, the bottom-up paradigm gives a larger design freedom since both axial and radial architectures can be implemented. Radial (also called core/shell) LEDs fully exploit the three-dimensionality of the NWs by forming the active region in the radial direction (Fig. 12.1 right). In this case, the InGaN/GaN cylindrical quantum wells are formed around an n-doped GaN core and are radially covered with a p-GaN shell. We note that a hybrid approach combining a top-down etching of the NWs with a subsequent radial overgrowth to form a core/shell active region has also been investigated [9–11].

Different types of functional nanostructures can be formed by a bottom-up approach such as pyramids, walls, or NWs. A common property of these nanostructures is the drastic reduction of the dislocation density compared to the planar case [12]. Micro- and nano-pyramids have been largely studied for LED application [13–16]. They have a hexagonal pyramidal shape defined by semipolar facets and can be obtained by selective area epitaxy. InGaN quantum wells deposited on their facets exhibit a lower internal electric field compared to c-plane and can incorporate more In, which is favorable for long wavelength emission [17]. The In content is reported to vary over the pyramid facets, which may yield a broad emission and can be used to realize phosphor-free white LEDs [18]. Other types of nanostructures such as vertical wall-shaped nanostructures [19] also called fins [20] or triangular nano-stripes [21] have been demonstrated and are quickly developing, however, light emitting devices based on these nanostructures remain today at an early stage [22].

NWs present a large number of advantages for three-dimensional LEDs. Thanks to their small footprint, NWs can accommodate the strain [23] arising from the lattice mismatch or from the thermal expansion mismatch and thus the defect density is reduced. The strain relaxation can take place not only at the NW/substrate interface, but also inside the active region (which is especially beneficial for axial In-rich heterostructures). This allows to avoid the dislocation formation in thin NWs with diameters of several tens of nanometers. For thicker NWs, dislocations can be formed at the NW/substrate interface, however, they bend to join the free lateral surface and can only be found in the immediate vicinity of the NW base [24]. The reduction of dislocations in NWs is predicted to lead to an enhancement of the quantum efficiency of NW devices compared to thin films.

Additional advantages for LEDs are provided by the core-shell geometry. This geometry can significantly increase the active area keeping the same device footprint as for a thin film LED [25]. This leads to a smaller carrier density in the quantum wells for a given injected current and thus to a droop reduction via lowering Auger effects. In addition, analyses of Auger coefficients in core/shell m-plane quantum wells revealed values two orders of magnitude lower than in c-plane thin film quantum wells [26]. These properties have motivated a strong research on core/shell NWs for high power high-efficiency LEDs. In addition, for core/shell LEDs the active region is usually grown on non-polar m-plane sidewalls, which allows to suppress the quantum confined Stark effect in InGaN/GaN quantum wells.

Finally, improved light extraction can be achieved in NW LEDs. Indeed, the three-dimensional geometry provides more escape passes for the light generated inside the NWs than for the light generated in a thin film. However, care should be taken about the possible light reabsorption. In particular, in core/shell NWs having a several hundreds of nanometers diameter, the NW can efficiently guide light and it can be reabsorbed by the quantum wells. Depending on the quantum efficiency, the absorbed light can be either reemitted or lost via non-radiative mechanisms. For axial NW LEDs, reabsorption is less important and light extraction is reported to be improved with respect to thin films [27, 28].

Accounting for these advantages, NW arrays are today investigated as potential competitors to thin film LEDs for general lighting market. Many works focus on the white light emission either using phosphor downconversion [29] or by mixing red, green, and blue (RGB) colors [30]. Optimization of spectrally narrow nitride NW LEDs emitting blue, green, and red light [31] is also on going for the development of all-nitride RGB displays.

In addition to lighting and display applications, where NW LEDs compete with the standard thin film technology, bottom-up NWs also open the way for devices with new functionalities. For example, using selective area epitaxy one can exploit the dependence of the indium incorporation on the NW diameter to tune the emission color of nanowire LEDs grown in the same run from blue to red [32, 33]. Such tuning favorable for monolithic RGB integration is not possible in thin films devices. Thanks to their small footprint and mechanical flexibility, NWs can be reported on plastic or metal substrates in order to fabricate mechanically flexible LEDs [34–36]. Single NW LEDs also present a unique possibility to achieve bright sub-micrometer emitters that can be integrated into a functional photonic circuit [37, 38]. Vertical single NW LEDs can be used for optical stimulation in biomedical applications [39] or can be functionalized as a matrix of micro-pixels [40].

Bottom-up NWs can be formed using catalytic growth [41], self-assembled growth [42], or selective area epitaxy [43, 44]. Dominating growth techniques for their elaboration are MOVPE and MBE, however HVPE has also been used in early works [45]. MOVPE is better suited for core/shell LED geometry whereas MBE yields axial heterostructures. The NW LED performance critically depends on the wire-to-wire dispersion and intra-wire compositional variations. The complexity of the three-dimensional active region of NW LEDs makes the growth challenging. Selective area epitaxy allows to achieve a good control over the NW size distribution which in turn reduces the wire-to-wire fluctuations. However, the homogeneous indium incorporation within the NW remains an important issue even in organized NW arrays. Multicolor emission shifting with injection current has been observed not only in self-assembled NW LEDs [46] but also in organized NW LEDs [47–49]. In particular, in NWs with intrinsic compositional inhomogeneities the electroluminescence depends on the contact layout [50].

LEDs with a nanostructured active region bring new characterization challenges. Indeed, standard macroscopic characterization techniques provide only average values of important parameters, which is not sufficient for in-depth understanding of the device operation. First, electroluminescence of representative individual wires can be analyzed. This is performed using either single NWs detached from their substrate and electrically contacted in planar geometry [41, 46, 49] or vertical

NWs electrically contacted using nanoprobes in SEM [51] or with a nanoscale contact [50]. In addition, optical and electrical properties of individual NWs can be analyzed directly within an operational macroscopic LED using high-resolution techniques like cathodoluminescence (CL) and electron beam induced current (EBIC) microscopy [52].

In the following, different points highlighted in the introduction will be detailed. The main focus of this chapter will be on bottom-up NW LEDs. In the first part, we will describe the material growth. Next, the LED fabrication will be outlined and the early works on the NW LED fabrication will be enumerated. We will then review the present status of NW LEDs giving representative examples. Open issues of NW LEDs will be described. Finally, mechanically flexible LEDs based on nitride NWs will be presented.

12.2 NW Growth for LEDs

From the yearly reports of GaN NW synthesis [53, 54], intense efforts have been accomplished to control the NW growth in terms of length, size, density, doping, and heterostructure design. MBE and MOVPE are the two key growth techniques for the fabrication of nitride NW LEDs. These two approaches will be described separately to highlight the differences in NW morphologies and growth mechanisms.

12.2.1 Spontaneous Growth of GaN NWs by MBE

12.2.1.1 Growth Conditions

Compared to reactive MBE using NH_3 for N-precursor, the plasma-assisted MBE based on active N from RF plasma N_2 source is the most efficient method to grow GaN NWs. The MBE growth of self-assembled GaN NWs requires N-rich condition (i.e., V/III ratio > 1) and elevated growth temperature (730–850 °C). Figure 12.2 actually shows the domain (red area in panel c), where the NW growth (Fig. 12.2b) is obtained instead of a compact GaN layer (Fig. 12.2a) [55]. These conditions limit the nucleation site density and avoid the coalescence due to the reduction of the Ga adatom diffusion length [56].

The growth is usually performed on Si(111) substrates, even if the NW growth has also been reported on other substrates such as Si(001) [57], sapphire [54], or diamond [58]. The addition of an AlN layer on Si(111) that reduces the lattice mismatch is also commonly used to grow vertical GaN NWs. The thickness, crystalline granularity, and flatness of this AlN intermediate layer play a key role to control the orientation, diameter, density, and the polarity of GaN NWs [59–61]. The typical diameter of GaN NWs spontaneously grown by MBE is between 10 and 100 nm for lengths in the range of 0.5–3 μm depending on the growth

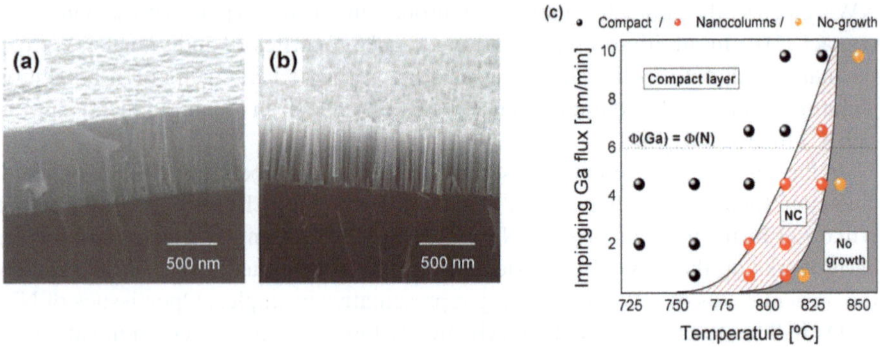

Fig. 12.2 Typical growth of GaN on Si(111) by MBE. (**a**, **b**) Bird-view SEM images of compact GaN layer and self-assembled GaN NWs. (**c**) Growth diagram showing different growth modes (layer, nanocolumns (NC), no growth) as a function of the growth temperature and V/III ratio (i.e., impinging Ga flux for a given impinging N flux of 6 nm/min). Adapted with permission from [55]

duration. Spontaneous growth by MBE produces high density of NWs (typically 10^9 to 10^{10} NWs/cm^2). For the direct growth on Si(111) substrate, the MBE GaN NWs polarity is mainly N-polar (i.e., with vertical chemical bonds pointing from N to Ga atoms along the $\bar{c}$ growth direction of the hexagonal lattice) [62, 63]. The growth conditions and morphology of an intermediate AlN layer have a direct impact on the wire polarity [64–66], but again by controlling its structural defect density, N-polar wires are preferentially grown [62, 67]. With the bending of misfit and threading dislocations to the sidewall surfaces, the GaN wire crystal is dislocation-free along most of its length. It results in excellent optical properties with a typical near band edge (NBE) emission broadening in the range of a meV or less. Moreover, no defect band emission is generally measured in these materials such as the yellow band (at ~550 nm) that can occur in planar materials [68, 69].

Recently, the use of alternative substrates such as glass [70], 2D materials as graphene [71], or metals [72] was reported for the GaN NW MBE growth. These emerging approaches are potentially promising for future flexible NW devices.

12.2.1.2 Growth Mechanisms Involved in MBE Growth

In general, growth mechanism highly depends on the surface properties: composition, amorphous or crystalline state, and morphology [73]. Actually, seed nucleation is strongly favored by the structural defects or morphological irregularities. On the contrary, if the nucleation surface is smooth with a low defect density, the nucleation is mainly related to specific growth mechanisms. It is, for example, the case for the growth performed on Si(111) substrate where an intermediate Si_xN_y layer is spontaneously formed on top of the Si substrate due to the reaction of the nitrogen plasma with the substrate surface [74].

More generally, the GaN NW growth on defect-free AlN or SiN_x surface can be decomposed into three subsequent phases: incubation, transition, and elongation. The incubation corresponds to the first moments of the nucleation where the nuclei are not completely stabilized [75]. The duration of this phase strongly depends on the growth conditions. At the end of the incubation period, the nucleation of stable "spherical" islands is observed both on AlN and SiN_x surfaces [75]. Then, a transition period occurs corresponding to a size increase of the nuclei associated with a shape modification that will lead to the NW formation. Two situations illustrated in Fig. 12.3 can be distinguished depending on the nature of the initial surfaces:

- For AlN nucleation surface, the elastic stress increases when the nucleus site rises due to the AlN/GaN lattice mismatch. Several modifications of the nucleus shape are observed and can be related to partial elastic relaxation: the spherical islands are converted into concatenated pyramids and then in full pyramidal islands. For nuclei having a size higher than the critical radius estimated to be 13 nm, a plastic relaxation happens with the formation of interfacial misfit dislocations [76]. The dislocation moves toward the nucleus center to relieve the residual strain, resulting in the last shape transition that corresponds to a NW-shaped nucleus. This mechanism shows that plastic relaxation plays a key role in the NW formation by the MBE method [60, 76, 77].

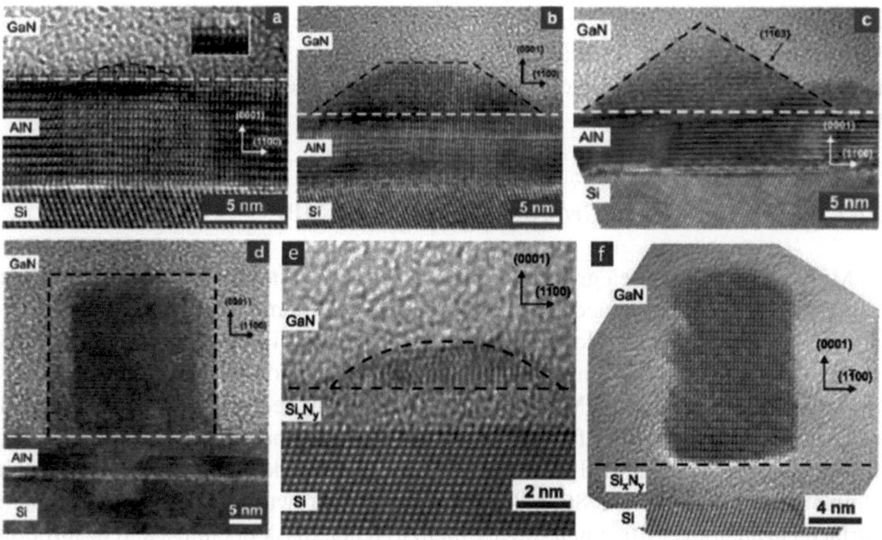

Fig. 12.3 Evolution of the GaN nucleus from spherical to NW shape during the transition phase based on a series of TEM images for two scenarios: GaN NW growth on Si(111) substrate with AlN buffer layer (**a–d**) (adapted with permission from [76]) and on Si(111) substrate with spontaneously formed SiN_x interlayer (**e, f**) (adapted with permission from [75])

- For SiN_x nucleation surface, epitaxial strain effects are not expected due to the amorphous nature of the SiN_x interlayer. Contrary to the AlN surface, the spherical islands keep the same shape when the size is increasing, until a critical value estimated at 5 nm is reached. Then, an abrupt shape transition (mainly driven by surface energy) is observed to reach directly the NW shape. An epitaxial relationship between GaN NWs and the Si substrate may be observed and depends on the conditions of SiN_x interlayer formation. The presence of crystalline domains in SiN_x interlayer could be a medium to drive substrate/NW epitaxial relationships [64, 74, 78].

The final NW shape of the nuclei on both AlN and SiN_x may be explained by the surface energy anisotropy, due to the lowest surface energy of m-plane surface (estimated to be $\sim$118 meV/Å^2) compared to the c-plane surface ($\sim$123 meV/Å^2) [79, 80]. In addition, the edge energy can also be introduced to explain the final NW shape since it forms an energy barrier that increases with the facet angle of nucleus.

The last phase of the growth corresponds to the elongation of the NW nuclei. Direct adatom impinging and adatom diffusion on the substrate and along vertical NW sidewalls contribute to the NW elongation. Due to the high density of NWs grown by MBE, the elongation is also dependent on shadowing and collective effects [81]. The NW elongation is mainly governed by the Ga species flux coming from NW vertical sidewalls by surface diffusion. As a consequence, the NW length is rapidly increasing when the NW size is decreasing [82]. Moreover, the incorporation of Ga adatoms being likely more efficient on the c-plane surfaces than on the m-plane facets, the axial growth is significantly increased [83] and as the diffusion length of Ga adatoms is limited to about 50 nm on the NW sidewall [82, 84, 85], only the Ga adatoms impinging close to the NW top will participate to the NW elongation.

12.2.1.3 Selective Area Growth of GaN NWs in MBE

In MOVPE, selective area growth (SAG) is usually used to organize nanostructures using SiO_2 or SiN_x patterned mask deposited on GaN template. This technique has also been successfully developed in MBE growth using Ti patterned mask deposited on GaN templates by Kishino's group [86] and then followed by other groups [87, 88]. Realization of SAG by MBE is highly demanding because an excellent control of growth parameters and of the surface flatness of the patterned masks is needed. In contrast to self-assembled GaN NW growth, SAG mode requires a higher growth temperature (900–950 °C) [86, 89]. The increase of growth temperature favors both surface Ga adatoms diffusion allowing them to reach mask openings and desorption helping to limit the GaN overgrowth on the mask. Higher the temperature is, higher the Ga flux is required. Figure 12.4 shows that the reduction of N flux limits the formation of organized GaN NWs. A balance between the growth temperature and the N flux acting in opposite ways needs to be well controlled in order to obtain an effective selectivity [90].

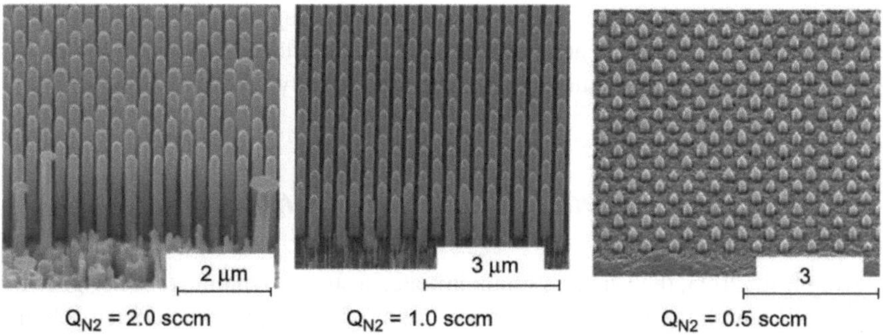

Fig. 12.4 SAG of GaN NWs grown by MBE using a Ti patterned mask on GaN/sapphire substrates. An increase of N flux promotes the NW growth and the mask selectivity at 900 °C. Adapted with permission from [86]

Other types of masks, for example, SiO_x [91, 92] and SiNx [59, 93], have also been employed for GaN NW SAG; however, the Ti mask remains the most frequent choice for NW LEDs. The SAG is usually performed on MOVPE-grown GaN template that imposes the Ga-polarity to GaN NWs. The SAG approach is an efficient way to control the LED emission color via the control of NW diameter and density as will be discussed later [32, 94].

12.2.1.4 Axial Growth of Heterostructures

As MBE is a physical deposition technique with sources positioned in front of the sample holder, the growth of heterostructures mainly results in axial geometry. Additionally, the high NW density prevents the radial growth by shadowing effect, even if a residual lateral growth can be present in some cases. Complex heterostructures with Al(Ga)N and In(Ga)N alloys can be axially grown by MBE for device fabrication. For Al(Ga)N/GaN heterostructures, sharp interfaces with a precise control of insertion thickness can be achieved. The realization of In(Ga)N/GaN heterostructures is more demanding in terms of interface and axial/radial geometry controls, nevertheless heterostructures with a high In composition can be achieved due to the strain relaxation in the NW geometry. Depending on the NW polarity, relatively flat InGaN insertions grown on the c-plane for N-polar NWs [95] or insertions of a complex shape following the semipolar planes in Ga-polar NW [94] were developed for LED fabrication. To enhance the quality of In-rich structures, growth of nano-umbrellas has been reported in view of fabricating red InGaN LEDs with improved efficiency [96, 97]. Note that the control of In(Ga)N/GaN heterostructures can be also improved by the SAG approach by mastering diffusion paths [32].

Regarding the heterostructures with different doping, we have to mention that the deposition of p-GaN is accompanied with a significant lateral growth that can be exploited to favor the coalescence of the high density NW layer.

12.2.2 *MOVPE Growth of Catalyst-free Nitride NWs*

Starting from 2005, the Lieber's group initiated a strong dynamic to study growth of nitride NWs with the first demonstrations of single NW photonic devices obtained by MOVPE growth assisted by nickel catalyst particles [98]. The catalyzed growth consists in using metallic nanoparticles to promote axial growth following the vapor-liquid-solid (VLS) process discovered in 1964 by Wagner and Ellis [99]. In the case of MOVPE-grown GaN NWs, catalyzed growth leads to triangular shaped nanoobjects with preferential growth directions along either a- or m-axes. Quantum wells deposited on such NWs (illustrated in Fig. 12.5) exhibit different properties depending on the facet orientation, which is undesirable for photonic applications. A better control over the NW shape can be achieved with a catalyst-free approach. The first demonstration of catalyst-free GaN NWs by MOVPE was achieved in 2003 based on organo-metallic azide precursors [100] and different methods have been developed later on.

12.2.2.1 Methods and Growth Conditions

Contrary to MBE where spontaneous growth of GaN NWs is easily achievable under N-rich condition, the conditions to spontaneously form NWs in MOVPE took a certain time to be determined and several ways have been opened. In general, a low V/III ratio (using trimethyl-gallium TMGa and ammonia) and high growth

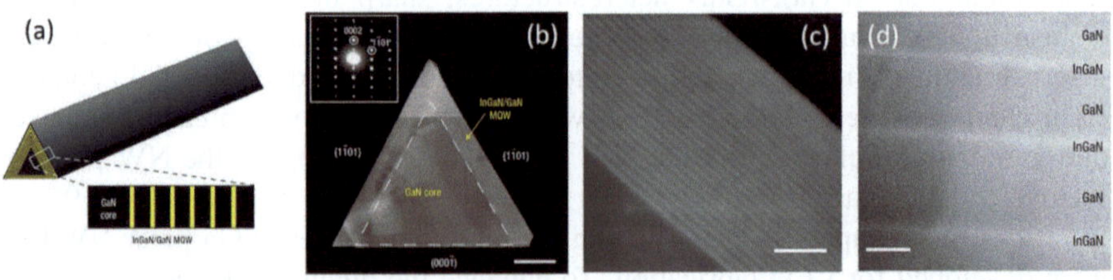

Fig. 12.5 Example of triangular Ni-catalyzed GaN NWs with core-shell InGaN/GaN multiple quantum wells (MQWs). (**a**) Schematic diagram of a GaN NW with a cross-sectional view showing InGaN/GaN core-shell MQWs (InGaN wells in yellow). (**b**) Dark-field cross-section STEM image along the [11–20] zone axis indicating the triangular GaN core with a shell composed of 26 MQWs (the scale bar is 100 nm). (**c**, **d**) Enlarged STEM images at the {1-101} facet corresponding to a detailed view of MQWs (the scale bar is equal to 20 and 2 nm, respectively). Adapted with permission from [41]

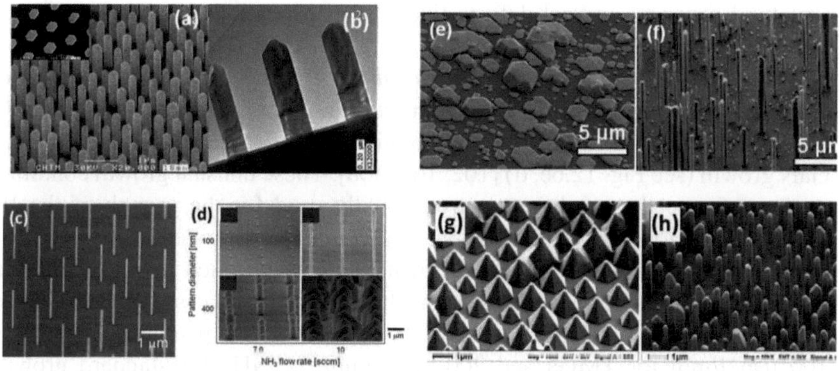

Fig. 12.6 Main MOVPE methods to grow catalyst-free GaN NWs. (**a, b**) SEM and TEM images of ordered GaN wire arrays grown by pulsed precursor mode (Adapted with permission from [101]). (**c, d**) SEM images of ordered GaN NW array grown by ultra-low precursor flux method as a function of the NH_3 flux and the pattern pitch (Adapted with permission from [102]). (**e, f**) SEM images of self-assembled GaN wires grown without (**e**) and with (**f**) silane flux addition. (**g, h**) SEM images of ordered GaN wire arrays grown without (**g**) and with (**h**) hydrogen dilution (2:1) of N_2 carrier gas (Adapted with permission from [103])

temperatures (>950 °C) are used to favor the vertical growth, but each method requires specific growth conditions including usually a preliminary seed nucleation stage followed by wire growth steps. The main MOVPE growth methods illustrated in Fig. 12.6 are detailed hereafter.

- *Growth based on pulsed precursor flows*: The Hersee's group was the first in 2006 to achieve the vertical growth of GaN wires along the c-axis without catalyst by introducing pulsed growth (see Fig. 12.6a, b) [101]. Beyond the absence of a catalyst, this pioneering work also demonstrated for the first time the use of SAG, opening the possibility to control the diameter, length, and density of the wire arrays. The growth is carried out in a pulsed manner, where the precursors III and V are injected alternately into the growth chamber. The principle is based on the discovery in the 2000s of the possibility to enhance vertical growth by injecting in a pulsed way the V precursor (NH_3) that also allows tuning the nanostructure shape [104]. The length of the wires is simply controlled by the number of cycles, leading to moderate growth rates in the order of 1–2 μm/h, corresponding to a growth of 3–6 nm of GaN by cycle. Such pulsed growth is performed on GaN templates on c-sapphire substrates leading to GaN NWs with Ga-polar crystal orientation imposed by the template polarity. The growth is usually performed at high temperature (>950 °C), low pressure (100–200 mbar), high V/III ratio (1500), and generally under H_2 carrier gas. The key point of this approach is the optimization of injection and waiting durations for elements III and V, which will depend on the reactor type, precursor fluxes, and growth temperature [30, 105, 106]. The pros and cons of this approach is the high growth control of complex NW arrays and the low growth rate, respectively. In addition, the optical and structural properties reveal defects (including dislocations) mainly localized at the base of NWs [101].

- *Growth based on ultra-low precursor flux*: An effective approach to grow catalyst-free NWs consists in using an extremely low precursor flux and V/III ratio within a range of two orders of magnitude less compared to standard planar GaN growth (see Fig. 12.6c, d) [102, 107, 108]. These unusual growth conditions are opposite to the conditions conventionally used for the growth of the III–V semiconductors where the element V precursor is saturated in the gas phase to avoid the desorption of the V-elements, which are much more volatile than the III-elements. For the NW growth, the flux of TMGa is typically reduced to only 0.1–1.0 μmol/min and the NH_3 flux to 10–100 μmol/min (instead of 100 μmol/min for TMGa and 200 mmol/min for NH_3 for standard growth). Such a low precursor flux results in very low V/III ratio between 1 and 50, in comparison to the typical V/III ratio used for planar growth in the range of 1500–4000. The growth is usually performed at high temperature (950–1100 °C), low pressure (100 mbar), and generally under N_2 carrier gas with hydrogen dilution. As the ultra-low flux mode leads to Ga-polar NWs, this approach is achievable on surface favorable to Ga-polar growth, such as GaN templates grown on sapphire substrates [108]. The NW diameter in the range of 100 nm can be achieved with a low growth rate (typically 1–4 μm/h depending on the NW density) due to low precursor flux. This type of growth has been studied extensively by Arakawa's group, which has developed the selective growth of ordered NWs with a diameter of only 50 nm [102]. Excellent optical properties related to a high structural quality are reported, the near band edge emission width being similar to those of the wires obtained by MBE [108].
- *Growth based on silane flux addition*: In the case of a planar growth, it is known that Si doping using a high flux of silane favors vertical growth, whereas Mg doping favors lateral growth [109]. This idea has been used by the Eymery's group to grow NWs using a silane flux more than 10 times higher than the flux used for typical n doping (i.e., 10 nmol/min for standard n-doping at 10^{18} at./cm^2). Combining low V/III ratio (<50), high growth temperature (1050 °C), high pressure (800 mbar), and rich N_2 carrier gas, this approach is highly efficient with growth rate larger than 100 μm/h for silane flux of 200 nmol/min (see Fig. 12.6e, f) [110]. This approach is well suited to enhance the vertical growth of N-polar NWs. Such N-polar wires require N-polar GaN substrates or specific surface preparation of c-sapphire substrates to get, for example, AlN and SiN_x ultra-thin layers to promote N-polar growth. After a given wire length (>5 μm), it is possible to stop the silane flux keeping the NW morphology, but the growth rate is tremendously decreased as a function of length. The use of silane imposes n-type doping with a high doping level (>5×10^{19} at./cm^3) [110, 111]. The silane approach can also be employed on Ga-polar GaN NWs to enhance the vertical growth [20, 112]. The high doping level induces a broad emission peak of GaN accompanied by a blue shift due to band filling effect. These near band edge optical properties can be significantly

improved when a part of the NW is grown without silane, but we generally observe in parallel an increase of the yellow band related to deep defects [110].

- *Growth based on the mixing of H_2/N_2 carrier gas*: Another key approach that favors the occurrence of vertical m-plane facets for the formation of GaN NWs by MOVPE is the H_2/N_2 ratio of the carrier gas [103]. As shown by the Waag's group, selective growth on c-sapphire under N_2 results in a pyramid growth regime (Fig. 12.6g), whereas a progressive dilution of the carrier gas with hydrogen (H_2/N_2 ratio between 1:2 and 2:1) leads to the formation of m-facets associated with a strong limitation of lateral growth (Fig. 12.6h). For given growth conditions (V/III ratio <50 and growth temperature near 1050 °C), the control of the H_2/N_2 ratio carrier gas favors vertical planes and supports the formation of GaN NWs. This effect is very sensitive for Ga-polar NWs.

12.2.2.2 Growth Mechanisms Involved in MOVPE NW Formation

Three key mechanisms have been proposed to explain the GaN wire catalyst-free growth involved in the different MOVPE methods described before.

- *Preferential growth along the c-axis direction*: The enhanced elongation rate by pulsed growth can be understood by a larger adhesion coefficient of Ga adatoms to the c-plane than to the m-plane facets. Consequently, for a high enough temperature, the desorption rate of Ga adatoms from m-plane surfaces is much higher than from the c-plane surface. Therefore, when the Ga precursor is injected, the Ga adatoms first cover all the surfaces, but then preferentially desorb from the m-plane facets or diffuse toward the top c-plane surface during the waiting step, which causes an accumulation of adatoms on the upper c-plane facet. Then, the exposure to the ammonia flow forms GaN preferentially on the c-plane facet favoring the vertical growth [113, 114]. The same mechanism can also be proposed for the ultra-low precursor flux method, if the diffusion of Ga adatoms is considered faster than the GaN formation due to the low kinetic induced by the low precursor flux.
- *Self-passivation of NW sidewalls by SiN_x mask*: In the case of growth under a high silane flux, an accumulation of Si on wire sidewalls is observed related to the formation of an SiN_x passivation layer [22, 115, 116]. Once the SiN_x layer is formed spontaneously on the vertical facets, the lateral growth is blocked favoring the axial growth [115]. The formation of SiN_x is expected because the typical silane flux used for the in situ deposition of planar SiN_x is about 200 nmol/min, which is exactly the flux used in NW growths [117]. This mechanism is based on the fact that the SiN_x layer is formed preferentially on the m-plane surfaces compared to the c-plane surface. Consequently, Ga adatoms diffuse along the m-plane surfaces covered by an SiN_x shell until reaching the upper surface to contribute to the vertical growth. Since this surface passivation layer is formed continuously during the wire growth, this method is well suited to grow ultra-long GaN wires (>100 μm) just by increasing the growth time [118].

– *Surface hydrogen passivation*: It has been shown by the Waag's group that a flow of hydrogen has a passivation effect for the r-plane $(0\,1\,\overline{1}\,\overline{1})$semipolar surfaces in the case of Ga-polar crystals that significantly reduce the growth rate, leading to a pyramidal shape. On the contrary, the H-passivation mainly occurs on the c-plane surface compared to r-plane surfaces for the N-polarity revealing wire-shaped nanostructures [119]. In addition, hydrogen tends to preferentially etch the r-plane surfaces at the benefit of m-plan $(0\,1\,\overline{1}\,0)$ surfaces corresponding to non-polar vertical facets to spontaneously form wire nanostructures. This preferential etching most easily occurs with N-polar crystal. It is possible to form vertical m-plane facets on Ga-polar crystals if the r-plane H-passivated surfaces are exposed to a high Ga flux. In this case, the r-plane surfaces become unstable and promote the m-plane surface formation, resulting in wire-shaped nanostructure with a Ga-polarity. This phenomenon under high Ga flux can also explain the wire formation in the case of pulsed growths or low precursor flux methods that both allow growing Ga-polar wires.

The above-described three mechanisms are highly dependent on the polarity of the GaN crystal and can simultaneously play their role in GaN NW formation by MOVPE. Therefore, an accurate control of crystal polarity together with growth conditions is required to promote NW growth.

12.2.2.3 NW Growth by Selective Area Growth in MOVPE

SAG through regular hole array in a dielectric SiO_2 or Si_3N_4 mask is an effective way to control density, size, and height of the wires. This approach is largely used in MOVPE to manage the GaN wire growth with the benefit of high wire-to-wire uniformity. A preliminary nucleation step of GaN is usually required to get a perfect filling of all mask apertures [101, 120, 121]. Then, the growth conditions are switched to promote the vertical growth regime in order to achieve perfect ordered wire arrays. The selective growth is traditionally performed on a GaN surface, but can be also managed on c-sapphire or AlN/Si(111) substrates, however, with a lower selectivity. In the case of SAG, the Ga adatoms diffuse firstly on the mask and secondly along the wire sidewall to contribute to the wire growth (the diffusion is typically about 1–2 μm for the mask and 3–5 μm on sidewall) [43]. The control of the wire growth remains a key issue, because the SAG is highly dependent on the growth conditions (V/III ratio, flux of precursors, temperature, pressure, etc.) and also on the pattern design, as shown in Fig. 12.6d.

12.2.2.4 Radial Growth of Heterostructures

The principle of MOVPE growth is based on pyrolysis chemical reaction of precursors coming from the gas phase. Consequently, the growth occurs on all surfaces in contact with the gaseous species, which leads to a conformal coverage of nanostructures. In the case of NWs, MOVPE growth of heterostructures results in a core-shell geometry. The radial growth is favored by reducing the growth temperature to limit the diffusion length of adatoms [122]. The radial growth is observed for InGaN/GaN multiple quantum wells under a low growth temperature for efficient In incorporation (700–800 °C) [110]. In the same way, the growth of GaN layer in core-shell structures is performed at a lower growth temperature (<1000 °C). The growth rate and the In incorporation of radial heterostructures are highly dependent on the facet crystalline orientation and also on the crystal polarity of GaN NWs. For pencil shaped Ga-polar NWs, the InGaN/GaN multiple quantum wells grow at a much higher rate with higher In incorporation on the horizontal c-plane facets localized on the NW summits compared to vertical m-plane facets corresponding to the NW sidewalls, with almost no growth on inclined r-plane facets of the NW upper part [48]. In the case of long NWs (>5 μm), a decrease of thickness and of In content can be observed along the NW from its top to its bottom leading to an emission wavelength shift, which has a negative effect on the LED performance as described later on in Sect. 12.9. The heterostructure growth conditions need to be carefully optimized in order to get an excellent uniformity of optical and electrical properties of NW ensembles.

12.2.3 Comparison of Catalyst-free Growth Method of Nitride NWs

A summary of the main methods to grow GaN NWs along the c-axis direction is proposed in Table 12.1 with a description of the related NW features. We stress that the morphology and properties of the produced NWs are different depending on the growth approach. On one hand, MBE allows getting high optical quality NWs with a diameter of few tens of nanometers. On the other hand, MOVPE produces larger NWs with the possibility to extend the NW length to several hundreds of micrometers by using silane flux.

Even if MBE is an efficient method to grow high quality GaN NWs with pioneer demonstrations of LED devices, MOVPE is a key growth technique used in the industry. The attempts of industrial development of NW devices are actually mostly based on MOVPE methods.

12.3 NW LED Fabrication

The architecture of NW array LEDs containing a large number of parallel-connected NWs is schematically shown in Fig. 12.7a. This schematic was elaborated following a typical morphology of organized core/shell NW arrays illustrated in the SEM image of Fig. 12.7b [49]. The growth substrate is normally used as a bottom

Table 12.1 Comparative table of GaN NW features as a function of the growth method

GaN NW growth methods	MBE	MOVPE Pulsed precursor mode	 Ultra-low precursor flux	 Silane flux addition	 Hydrogen dilution of carrier gas
NW diameter	10–100 nm	200–500 nm	30–200 nm	0.5–2 μm	0.2–1.5 μm
NW length	0.1–3 μm	0.3–2 μm	0.5–3 μm	5–500 μm	1–3 μm
NW density	$\sim 10^{10}$ cm^{-2}	$\sim 10^{8}$ cm^{-2} (SAG)	$\sim 10^{8}$ cm^{-2} (SAG)	10^{7}–10^{8} cm^{-2}	$\sim 10^{8}$ cm^{-2} (SAG)
Standard substrate and polarity	SiN/Si(111): ↓	GaN/c-sapphire: ↑	GaN/c-sapphire: ↑	c-sapphire: ↓ (↑)	c-sapphire: ↓ (↑)
↑: Ga-polarity	AIN/Si(111): ↑ ↓			AIN/Si(111): ↓ (↑)	AIN/Si(111): ↑
↓: N-polarity				GaN/c sapphire: ↑	GaN/c sapphire: ↑
Optical properties	Narrow NBE ~1–10 meV	Large NBE ~100 meV	Narrow NBE ~1–25 meV	Large NBE ~120 meV + band filling shift	Data not available
NBE: near band edge					
YB: yellow band	No YB	Presence of YB	No YB	Narrow YB	

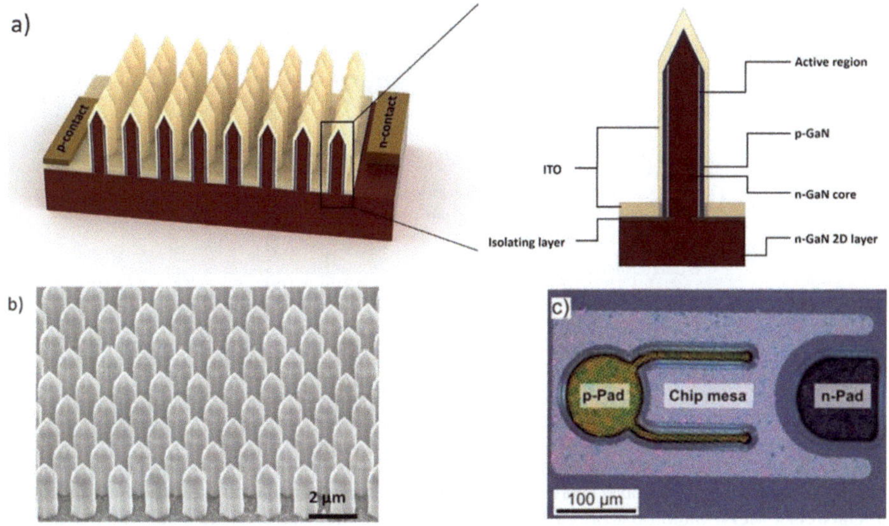

Fig. 12.7 (**a**) Schematic of a core/shell LED. (**b**) SEM image of as-grown organized core/shell NWs grown by GLO-AB. Adapted from [49]. (**c**) Blue NW LEDs fabricated by OSRAM. Reproduced with permission from [29]

contact (Fig. 12.7a). For axial LEDs, to ensure electrical insulation, the NWs are encapsulated into a polymer matrix (e.g., hydrogen silsesquioxane, BCB, polyimide, etc.) and the NW tips are exposed for contacting by etching the excess of the encapsulating material. For radial LEDs fabricated by a selective area growth using a dielectric mask (as exemplified in Fig. 12.7b), the encapsulation step can be omitted since the mask provides electrical insulation. In this case, a transparent contact can be deposited all around the NWs as shown in Fig. 12.7a. Lithography is used to define devices with a desired size, which may range from tens of microns to millimeters. If back contacting is not possible (e.g., when an insulating sapphire substrate is used), etching is used to access to the bottom contact layer as shown in the device schematic of Fig. 12.7a. A transparent conductive layer (typically indium tin oxide (ITO)) is deposed onto the LED front surface to form the upper contact suitable for light extraction. To reduce the series resistance, a metal current-spreading grid can be added. Fig. 12.7c shows a top-view optical microscopy image of a fully processed LED based on core/shell NWs, where the ITO layer and the top and bottom metal contacts are clearly visible [29].

12.4 Early Demonstrations of Nitride NW LEDs

First demonstrations of nitride NW LEDs have appeared at the beginning of years 2000. They were using self-assembled NWs. Kim et al. [123] reported

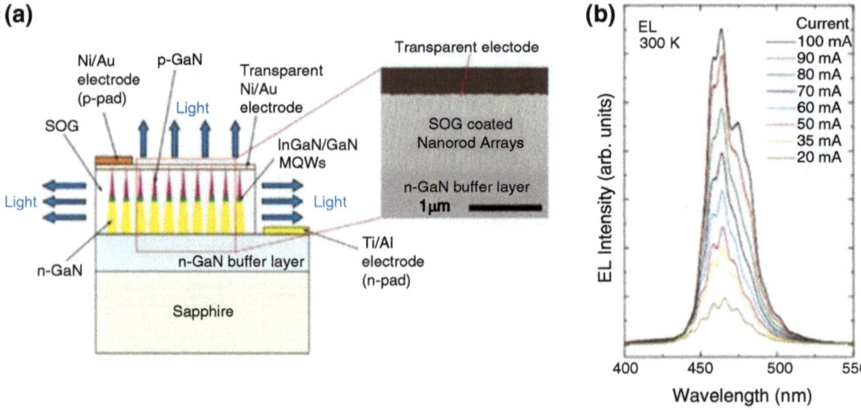

Fig. 12.8 (**a**) Cross-sectional schematic diagram (left) and SEM image (right) of InGaN/GaN multiple quantum well nanorod array LEDs. (**b**) Room temperature electroluminescence spectra of InGaN/GaN MQW nanorod array LEDs under different applied currents. Adapted with permission from [124]

the metalorganic-hydride vapor phase epitaxy (MO-HVPE) growth of axial GaN p–n junctions. Individual NWs were contacted using focused ion beam (FIB) lithography. They produced electroluminescence around 390 nm attributed to DAP recombination in the p-GaN segments. The same team reported one year later a demonstration of NW array LEDs containing quantum wells [124]. Axial InGaN/GaN nanoneedles with 6 $In_{0.25}Ga_{0.75}N$/GaN embedded axial quantum wells were grown by MO-HVPE. The structure was synthesized on a 2D n-GaN buffer layer on sapphire substrate used as a bottom contact. The empty space between the nanoneedles was filled with a spin-on glass and the p-GaN tips were contacted with a semi-transparent Ni/Au (20 nm/40 nm) electrode as illustrated in Fig. 12.8a. The device presented a broad blue electroluminescence peaked around 460 nm. The peak wavelength (Fig. 12.8b) was stable with injected current.

The same year Kikuchi et al. [125] have demonstrated an LED based on MBE-grown NWs [32, 126]. The structure illustrated in Fig. 12.9a contained 8 axial InGaN/GaN quantum wells grown on an n-doped GaN NW base. The growth of the p-doped GaN cap was performed at lower temperature under Mg flux, which led to the wire broadening and coalescence as seen in the SEM image of Fig. 12.9b. The coalesced top surface allowed to directly deposit a semi-transparent Ni/Au (2 nm/3 nm) contact without a planarization step. The emission wavelength was adjusted from green (530 nm) to red (645 nm) by changing the In content in the wells. Figure 12.9c shows optical microscopy images of the NW LEDs under operation. Due to wire-to-wire fluctuations the electroluminescence has a spotty nature with intensity and color variations within the array.

The group of C. Lieber has demonstrated in 2003 LEDs based on crossed p-GaN and n-GaN NWs synthesized by metal-catalyzed chemical vapor deposition

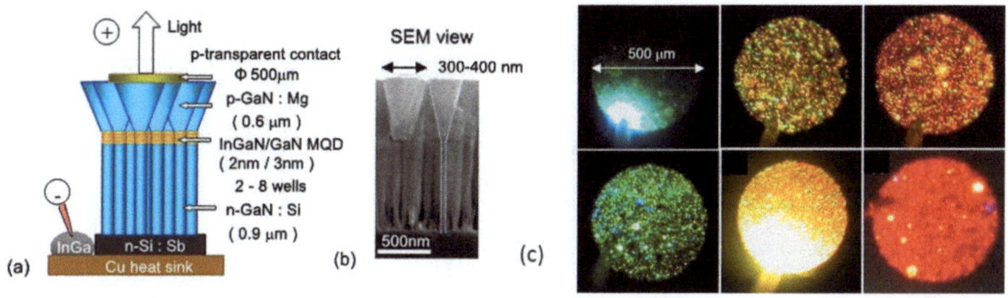

Fig. 12.9 (**a**) Schematic diagram of the MBE-grown InGaN/GaN NW LEDs, (**b**) their cross-sectional SEM image, and (**c**) their electroluminescence optical microscope images. Adapted with permission from [126]

[127]. During the following year, core/shell NW LEDs were realized by this team [41, 98]. The wires were synthesized by MOCVD using Ni particles as a catalyst. As mentioned in Sect. 12.2, this method yields NWs with the long axis aligned with a⟨11-20⟩direction and a triangular cross-section with one polar c-facet and two {1-10-1} semipolar facets. The GaN core was n-doped, then a radial single InGaN quantum well was deposited and capped with a p-doped GaN shell. Due to their different crystallographic orientation, the facets were non-equivalent and the InGaN layer was thicker on semipolar facets than on the (0001) facet. As illustrated in Fig. 12.10a, single NWs were contacted by the p-GaN shell and by the core, which was revealed by FIB etching of the shell on one of the NW extremities. Diodic current–voltage characteristics with a diode opening voltage around 3 V were achieved. The electroluminescence of single wires was imaged (Fig. 12.10b) and spectrally characterized (Fig. 12.10c). The emission wavelength was tuned by adjusting the In content in the quantum well (function of the growth temperature) from UV to yellow–green spectral range (namely, from 367 to 577 nm). The external quantum efficiency of single NW LED was estimated to be ~5.8% at 440 nm and ~3.9% at 540 nm.

12.5 NW LEDs With Emission Color in the Visible Spectral Range

Since the early demonstrations of NW LEDs, these nanostructures are believed to become serious competitors of thin films. A large number of research groups, but also start-ups and big industrial groups launched research activities on this topic exploring different designs and different growth techniques. In the following we present a non-extensive summary of the main achievements of the past ten years with a focus on the current progress.

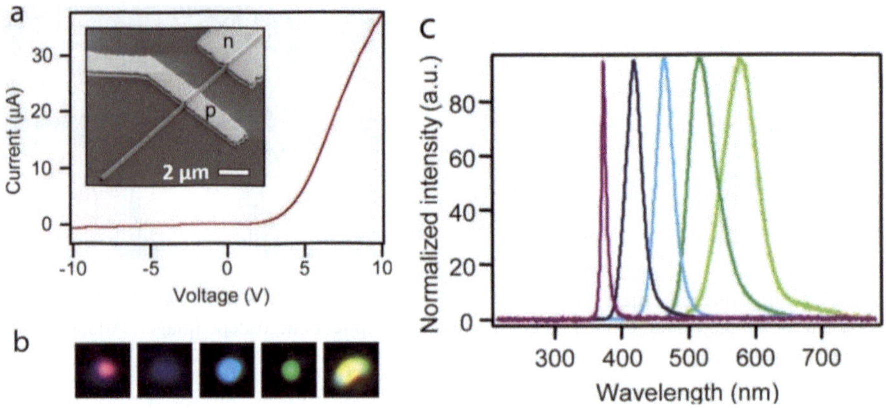

Fig. 12.10 (**a**) I–V curve of a core/multishell NW LED. Inset: SEM image of the device. (**b**) Optical microscopy images of electroluminescence emissions under forward bias, showing purple, blue, greenish-blue, green, and yellow emission, respectively. (**c**) Normalized electroluminescence spectra of five representative multicolor LEDs. Adapted with permission from [41]

12.5.1 MOVPE-grown NW LEDs

The potential advantages of the core/shell NW LEDs over the conventional planar technology were clearly highlighted by Waag et al. [128], the main one being the tremendous increase of the active surface. By taking a circular cylinder approximation for NWs and assuming that the radial quantum wells are close to the NW surface (i.e., neglecting the p-GaN shell), the increase of the active area in a NW LED with respect to the active area in a thin film LED of the same chip surface can be calculated as:

$$\frac{S_{\mathrm{NW}}}{S_{\mathrm{film}}} = \frac{2\pi \mathrm{RH}}{\pi R^2/F} = \frac{2H}{R}F,$$

where S_{NW} is the active area of a NW LED, S_{film} is the active area of a thin film LED with an equivalent chip surface, R is the NW radius, H is the NW height, and F is the fill factor, i.e., the ratio of the NW cumulated footprint to the total projected device surface. Thanks to the larger active area, the carrier concentration in the quantum wells of a core/shell NW LED is lower with respect to a thin film device. Therefore, NW emitters will show efficiency droop at a higher current [25].

Early works on MOVPE-grown core/shell NW LEDs mainly employed self-assembled catalyzed [129] or non-catalyzed [110] growth; however, organized growth on sapphire [128] and on GaN templates [48] was also reported.

Non-catalyzed GaN NWs with core/shell InGaN/GaN quantum wells (Fig. 12.11a) were used to fabricate single NW LEDs [46, 110]. In these devices, five radial InGaN/GaN quantum wells were grown around the upper part of the GaN NW stems. It was shown that quantum wells are formed not only on the lateral m-planes,

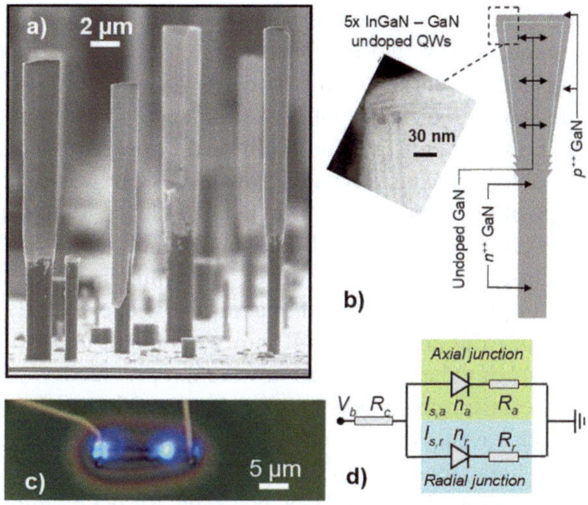

Fig. 12.11 (**a**) SEM image of MOVPE-grown core/shell NWs. (**b**) Schematic of the wire internal structure and a TEM image of the top region of a wire, evidencing the presence of radial and axial quantum wells. (**c**) Optical microscopic image of a single NW LED in operation. (**d**) Equivalent electrical circuit of the LED representing the axial and radial junctions. Adapted with permission from [46]

but also on the top-c plane as can be seen in the schematic and in the TEM image of Fig. 12.11b. Line defects propagating perpendicular to quantum well planes with an enriched In composition were evidenced in these type of wires [130]. The NWs were dispersed on SiO_2 on Si substrates and contacted using e-beam lithography as illustrated in Fig. 12.11c. The In content and the well thickness were different on the non-polar (radial) and on the polar (axial) planes. This structural dissimilarity, but also the difference in the internal field in the axial and radial quantum wells, resulted in different luminescence wavelengths as evidenced by cathodoluminescence [46]. The competition of the carrier injection in the axial and radial quantum wells caused the change of the electroluminescence color as a function of the injected current and as a function of temperature [46]. An electrical model describing the axial and radial junctions as parallel-connected diodes with different parameters (Fig. 12.11d) allowed to explain the observed electroluminescence color change.

Using self-assembled NWs of a similar geometry, Koester et al. [131] demonstrated LEDs based on NW arrays with an emission wavelength around 410 nm. In particular, the authors have characterized the device bandwidth as described later in Sect. 12.7.

The group at Chonbuk National University employed Au + Ga catalyst assisted MOCVD to demonstrate blue NW LEDs with an axial InGaN/GaN active region [129]. We note that this is one of the rare reports on purely axial NW LEDs grown by MOCVD. Dense mostly vertical 2 μm long NWs were formed. At the end of the growth, Au catalyst was no longer observed at the top of the NWs, which allowed to deposit quantum wells with well-defined interfaces. Pulsed-flow precursor method was used to grow axial multiple quantum wells at a reduced growth temperature. The quantum well emission wavelength was varied from 415 to 520 nm by adjusting the In content. A p-doped GaN cap segment was deposited on top of the quantum wells. To fabricate LED devices, polyimide encapsulation of the NWs and an ITO top contact were used. Devices showed diodic behavior with an opening voltage around 2.75 V and exhibited blue electroluminescence varying linearly with the injected current up to an emitted power of 88 μW for 35 mA. Several months later the same group has applied a similar growth technique using Au catalyst to form LEDs with radial quantum wells on the non-polar and on the semipolar planes [132]. To obtain coaxial NWs the growth pressure was reduced during the quantum well deposition. A coalesced p-doped GaN layer was deposited under an increased Ga precursor flow. A p-GaN layer covered mainly the NW tops and the upper part of the m-plane sidewalls; however, the dominant electroluminescence originated from the m-planes. An external quantum efficiency (EQE) value of 27.9% was reported for an LED emitting at 440 nm wavelength under 50 mA injection; however, the details of the EQE estimation were not provided.

Self-assembled NWs suffer from geometrical and compositional fluctuations which have a dramatic impact on the electroluminescence. To improve the wire-to-wire homogeneity, SAG was applied. Hong et al. [48] demonstrated color-tunable LEDs using SAG of GaN NWs by MOVPE with a nanopatterned SiO_2 layer on GaN-on-sapphire templates. After the growth of the NW cores, InGaN/GaN quantum wells were deposited on the NW template yielding a complex axial and radial heterostructure with thick In-rich (In content $\approx$0.6) quantum wells on the top polar plane, thinner quantum wells with a lower In content ($\approx$0.15) on the m-planes, and in addition a curved thick InGaN layer beneath the axial quantum wells. The NW structure was buried into a continuous p-doped GaN layer forming a nanorod-embedded thin film as seen in the SEM image of Fig. 12.12d. The LEDs were fabricated using a semi-transparent Ni/Au layer applied to a coalesced p-GaN surface for the top contact (Fig. 12.12c). The LED color changed with applied bias from a single broad peak around 690 nm under 3 V to a blue-shifted double peak (at 500 nm and 440 nm) under 10 V as shown in Fig. 12.12a, b. The authors explained the electroluminescence color change with a field distribution model. The equipotential surface in the highly resistive p-GaN layer was initially flat corresponding to the injection in the NW tops and then changed to a curved shape along the NWs promoting the injection in the m-plane quantum wells. The internal quantum efficiency (IQE) of the quantum well was estimated by temperature-dependent photoluminescence to be around 32%.

The variation of the emission color with bias [48] may be regarded as an advantage and used for LED color tuning; however, for standard narrow-band LEDs this phenomenon is undesirable. Chen et al. [133] demonstrated a NW LED with a stable emission color. Organized NWs were selectively grown by pulsed MOCVD on GaN on sapphire templates (Fig. 12.13a). Similar to the work of Hong et al. [48], quantum wells were deposited on the m-plane facets and on the c-plane in the tapered NW top part; however, the structural quality of the top wells was quite poor. Instead of a coalesced p-GaN layer, a conformal p-GaN shell was grown around the NWs. An original approach for the p-contacting was developed based on a deposition of a transparent conductive GaZnO by molecular beam epitaxy, which first formed a conformal shell but then coalesced at the NW tops as illustrated in the schematic of Fig. 12.13b. The LED turn-on voltage was around 6 V with

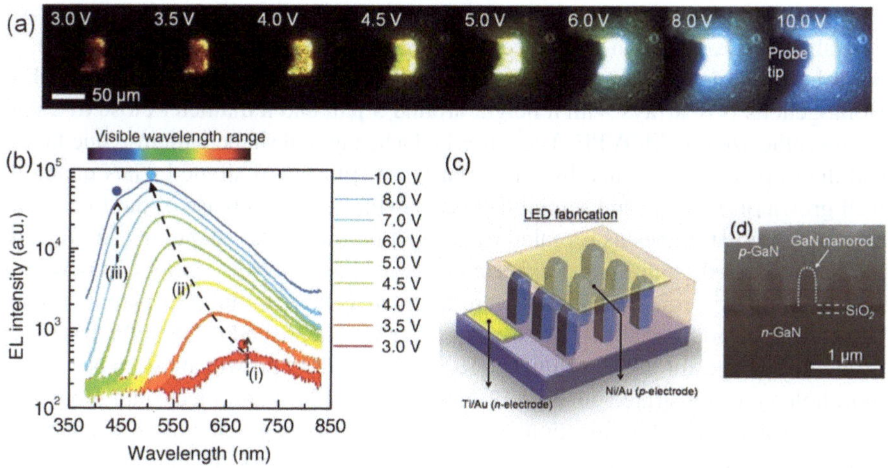

Fig. 12.12 (**a**) Optical microscopy images of LEDs operating under different applied biases. (**b**) Electroluminescence spectra for biases from 3 to 10 V. (**c**) Schematic of the LED showing core/shell NWs, the planarizing p-GaN layer, and the contacts. (**d**) Cross-sectional SEM image of a p-doped GaN layer burying GaN NWs. Adapted with permission from [48]

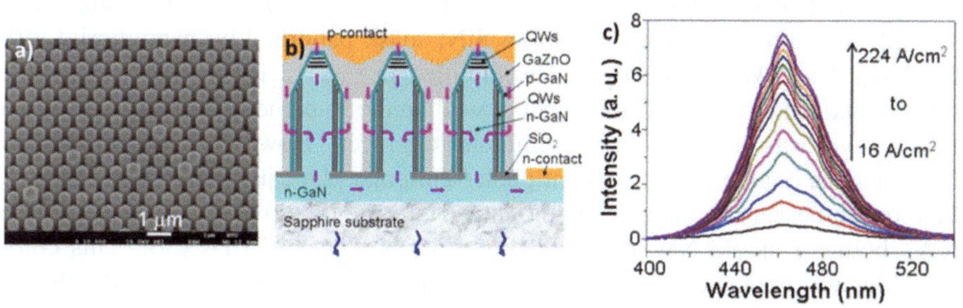

Fig. 12.13 (**a**) Tilted SEM image of a NW array. (**b**) Schematic of the NW structure and contact geometry. Pink arrows indicate the possible current flow directions. (**c**) Electroluminescence spectra when the injected current density increases from 16 to 224 A/cm^2. The insert shows the photograph of the LED in operation. Adapted with permission from [133]

a broad electroluminescence spectrum centered at 463 nm. Contrary to [48], the emission wavelength was remarkably stable with increasing injection (Fig. 12.13c), no redistribution of the injected current in the quantum wells along the wire axis was taking place.

Wang et al. [43] also developed NW SAG on GaN on sapphire templates covered with SiO_x mask patterned by nanoimprint lithography. Fixed 400 nm mask openings were chosen to define the NW diameter and growth positions. The statistical evaluation of the geometry of the GaN wire arrays after the growth showed that the standard deviation of the NW height is in the range of 4–8% and for the diameter it is close to 3% depending on the NW aspect ratio (higher standard deviations are observed for higher aspect ratio SAG wires).

The OSRAM group [29] extensively described the fabrication and properties of core/shell LEDs based on organized NWs (SAG is described in detail in [122]). A core/shell NW design was adopted to demonstrate both blue and white NW LEDs. Homogenous NW arrays with a height around 3 μm and a diameter close to 2 μm were synthesized by MOVPE. The wires had a hexagonal shape with m-plane facets and their tips were delimited by semipolar {10-12} facets. A single InGaN quantum well grown on both m- and semipolar-facets exhibited an increase of the In content on the semipolar planes as revealed by cathodoluminescence mapping. The IQE of the wires derived from optical experiments was close to 10%. Array LEDs with a 150 × 250 μm^2 projected surface were fabricated. Prior to the transparent contact deposition, an additional SiO_x insulating layer was deposited on the NW base part to avoid leakage through eventual defects of the SiO_x growth mask. The device morphology is illustrated in Fig. 12.13b in Sect. 12.3. The electroluminescence spectra evolved with the injection current showing first a green emission attributed to In-rich semipolar quantum wells and then turning quickly to the blue emission from the m-plan quantum wells. A further blue shift of the electroluminescence was observed, attributed to an In gradient in the m-plane quantum wells. The LEDs demonstrated an EQE of 3% for 400 mA/mm^2 and a maximum wall-plug efficiency of 1.5% at 300 mA/mm^2.

Core/shell NW LEDs were also developed by Glo-AB company using SAG on GaN-on-sapphire templates with an SiN_x mask patterned by nanoimprint [50, 52, 134]. The NW morphology is exemplified in Fig. 12.13b of Sect. 12.3. The NWs contained a single InGaN quantum well on the m-planes. Contrary to [29], no InGaN layer was observed on the semipolar top facets. However, an In-rich region located in the quantum well at the m-plane/semipolar plane junction was evidenced. Electroluminescence color switching shown in Fig. 12.14a was reported in these NWs both for single NW LEDs [49] and for NW array LEDs [50]. This color switching was explained within a simple electrical model [49]. A core/shell NW was represented by a network of parallel-connected diodes and the current distribution in this network was calculated for different applied biases and different contact layouts. To account for the compositional inhomogeneities, a different saturation current was ascribed to the diodes, corresponding to the top-most In-rich part of the core/shell quantum well. The saturation current depends on the quantum well bandgap energy following the equation [49]:

$$J_{\mathrm{sat}} = \int_{\mathrm{QW}} \mathrm{d}z \left(B \times {n_i}^2\right) = wBN_cN_v \exp\left(\frac{E_g}{k_BT}\right)$$

and therefore increases in the In-rich regions, where the bandgap is smaller. For an electrical circuit shown in Fig. 12.14c, the current distribution in the individual diodes (Fig. 12.14d) showed that at low bias first the In-rich segment was injected. For higher biases, parallel and radial shell resistances started to play an important role and the current concentrated in the m-plane quantum well under the injecting p-contact, which resulted in the electroluminescence color switching. Similar effect was predicted for an extended contact covering the entire m-plane [49].

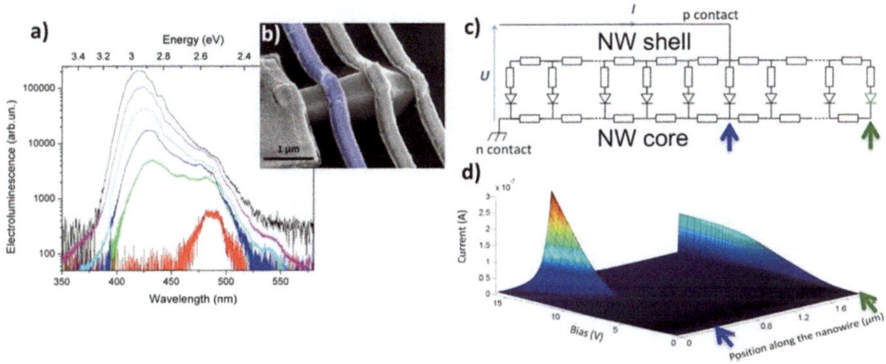

Fig. 12.14 (**a**) Electroluminescence spectra of a single NW LED with a metal contact on the m-plane. (**b**) SEM image of a single wire LED with the injecting contact highlighted in violet. (**c**) Equivalent electrical circuit of a core/shell NW LED. (**d**) Current distribution in a diode network as a function of the applied bias and diode position in the wire. Adapted with permission from [49]

Zhang et al. [50] showed that the emission color of NW LEDs exhibiting a compositional inhomogeneity could be controlled by modifying the contact layout. Core/shell NW LEDs with an In-rich region in the top part of the m-plane quantum well were examined. By removing the ITO contact from the top part of the NWs and by inhibiting the conductivity of the exposed p-GaN shell by CHF_3 plasma treatment, the undesired long wavelength emission could be suppressed. Figure 12.15 shows the electroluminescence spectra of a nominally blue NW LEDs for three different processing schemes: conformal ITO contact, ITO contact etched away from the NW top part, and the NW top part removed by ICP etching. Electroluminescence spectra for the same injected electrical power were compared. Two distinct emission peaks were observed for the conformal contact. The long wavelength peak was strongly suppressed after partial ITO removal and CHF_3 plasma exposure while the intensity of the blue emission was increased. ICP etching of the NW top part completely removed the undesired long wavelength peak, which proved that it originated from the top region.

12.5.2 *MBE-grown NW LEDs*

Thanks to its low growth temperature, MBE is a technique well suited to fabricate In-rich alloys. Therefore, a strong effort was committed to demonstrate and optimize MBE-grown NW LEDs emitting in the long wavelength spectral range (from the green to the near infrared (IR)).

The pioneer of the MBE NW LEDs is the group of Kishino. Various electroluminescence emission colors from green to red were achieved with axial InGaN/GaN multiple quantum discs grown in self-assembled GaN NWs on Si

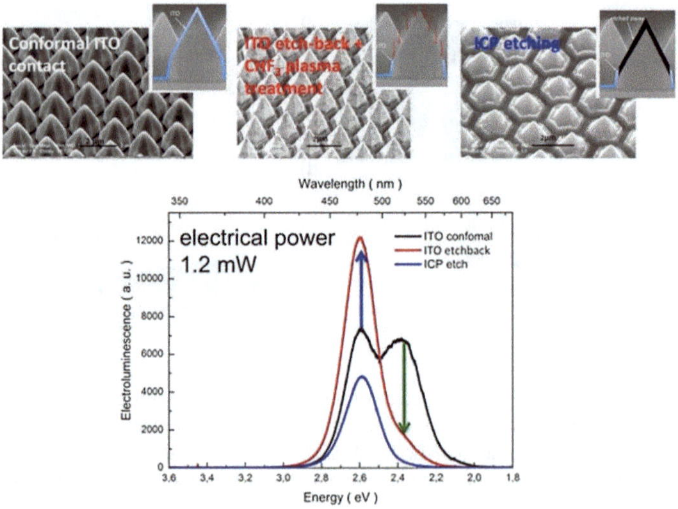

Fig. 12.15 Comparison of the electroluminescence spectra under the same electrical injected power of core/shell NW array LEDs processed with a conformal ITO contact, with a partially etched ITO contact and with the NW top part removed by ICP etching. The top row shows SEM images of the corresponding LEDs with a schematical drawing of the contact

substrates illustrated in Fig. 12.16a, b [125]. Later on, their research interests moved to SAG, which brings new functionalities for monolithic integration owing to the more precise control of the NW morphology and of the In composition [32]. We note that the SAG NWs have a different morphology with respect to self-assembled NWs. As illustrated in Fig. 12.16c, d, SAG NWs due to their Ga-polarity exhibit a pyramidal top. The multiple quantum wells grown on SAG wires are deposited on the sixfold side faces of the pyramidal top, which are semipolar planes with a low internal field. For self-assembled NWs, axial multiple quantum wells grow on the polar c-plane as shown in Fig. 12.16a, b.

By changing the NW diameter by SAG, InGaN quantum wells with different In composition were demonstrated within the same growth run [32]. With NW diameters ranging from 143 to 270 nm, these monolithically integrated NWs emitted light from blue to red. In 2013, a monolithic integration of 544, 583, and 597 nm emitting InGaN-based NW LEDs was demonstrated on the same Ti-masked GaN substrate in a single growth [136]. In 2015, a monolithic integration of four-color InGaN-based NW LEDs was achieved [33]. Figure 12.17 illustrates this spectacular realization. The Ti-mask was processed with different diameters of nano-holes in each region. The NWs were encapsulated into a spin-on glass and contacted with a transparent ITO layer on top as shown in Fig. 12.17a. Four electroluminescence colors (465, 489, 510, and 579 nm) were emitted from the NWs of different diameters as shown in Fig. 12.17b. A red LED with 670 nm emission wavelength was also fabricated using a larger pitch. It is noteworthy that this color control method can also be performed using SAG MOVPE growth, although the underlying

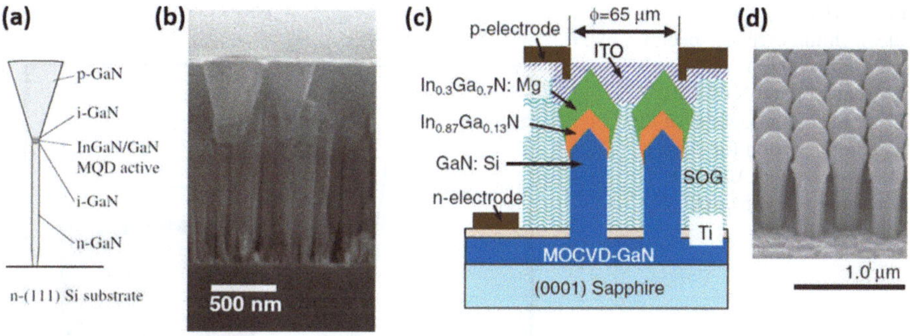

Fig. 12.16 (**a**) Schematic and (**b**) SEM image of self-assembled InGaN/GaN NW LEDs grown on Si substrate. Adapted with permission from [125]. (**c**) Schematic and (**d**) SEM image of SAG InGaN/GaN NW LEDs grown on Ti masked GaN substrate. Adapted with permission from [135]

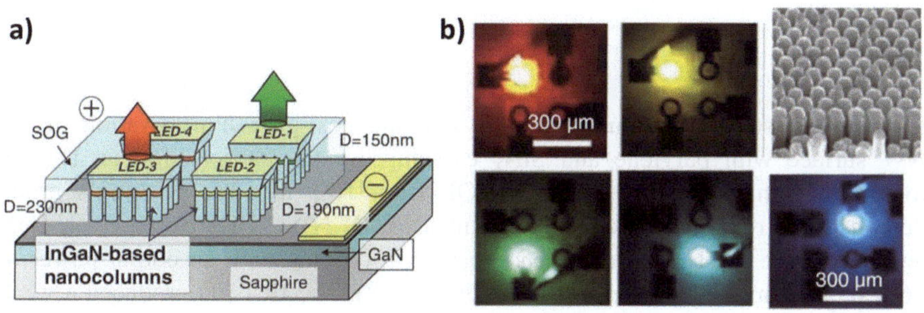

Fig. 12.17 (**a**) Schematic of the NW LED with SAG NW of different emission color. (**b**) Photographs of the NW LEDs in operation and an SEM image of the green LED. Reproduced by permission of the Institution of Engineering and Technology from [33]

mechanisms are different and the wavelength tunability range (440–520 nm) is smaller [137].

Angular profiles of the emission from the MBE SAG NW LEDs with NWs arranged into a photonic crystal were studied on the example of a yellow device [138]. The LED exhibited a high directionality with a radiation angle of ±20° due to the light diffraction at the photonic band edge. Flip-chip structure NW array LEDs were demonstrated [139]. This configuration showed an improvement of the thermal management of NW devices.

Many other groups have investigated green–yellow NW LEDs using MBE growth. Limbach et al. [140] demonstrated a self-assembled InGaN/GaN NW LED grown on an Si substrate. The electroluminescence was peaked at 540 nm. Bavencove et al. [141] demonstrated an MBE axial InGaN/GaN NW LED showing a spotty polychromatic emission from 566 nm to 585 nm under different injection currents. The EQE for a current density of the order of 50 kA cm^{-2} was 2.7%.

There are also several reports on red and near-IR MEB-grown NW LEDs. Jahangir et al. [142] demonstrated a red NW LED with InGaN/GaN discs on

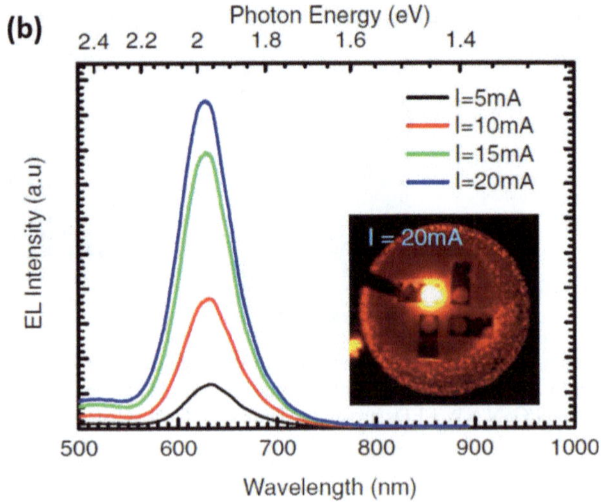

Fig. 12.18 Electroluminescence spectra under different injection currents of a red NW SAG LED. Inset: photograph of the LED in operation for 20 mA injected current. Adapted with permission from [136]

a silicon substrate. The authors found that the post-growth passivation of the NW surface with parylene and SiN_x can improve the carrier lifetime, leading to a significant improvement of the IQE from 10% to 52%. Kishino's group demonstrated red [136] and IR [135] InGaN/GaN NW LEDs by SAG on GaN templates. The red LED showed an emission at 633 nm with an output power of 16 μW at 20 mA injected current (see Fig. 12.18). The IR LED produced a weak electroluminescence at 1.46 μm at room temperature. The In concentration in the active region was 86%. Recently, following previous studies of Eu doped quantum dots [143], an alternative approach for the red-emitting NW LEDs was reported using Eu doped GaN NWs grown on Si substrate by MBE [144]. A GaN:Eu segment was grown axially between p-GaN and n-GaN extremities. This LED showed two main peaks at 620.3 and 633.8 nm, originating from the Eu ions.

A strong effort for wavelength tuning was performed by the group of Mi, who proposed a number of original approaches. The growth of quaternary AlInGaN insertions in GaN NWs was investigated [145] showing that a spontaneous Al-rich shell is formed during the growth, which protects the carriers from recombination on surface states. These core/shell structures were embedded into axial p–n GaN NWs to demonstrate LEDs with a color tunable from blue to red. Lateral integration of RGB LEDs was demonstrated using a three-step approach in which a protective SiO_2 mask was deposited allowing to form first the blue-emitting NWs, then another area was opened to deposit green-emitting NWs and finally red-emitting wires were grown during the third growth run [146]. Nguyen et al. [147] demonstrated that a high energy AlGaN barrier shell around the active InGaN/AlGaN quantum discs can eliminate the non-radiative surface recombinations and electron overflow. Recently, Philip et al. [148] also showed the enhancement of the carrier lifetime and photoluminescence intensity of a white NW LED by using this core-shell structure.

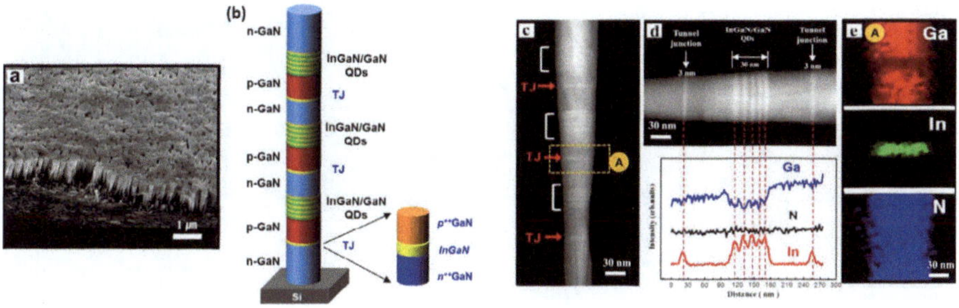

Fig. 12.19 AC multi-active-region tunnel junction InGaN/GaN NW LED (a) SEM image, (b) schematic, (c) HR-TEM of tunnel junction dot-in-a-wire LED structure, (d) top: high angle annular dark field (HAADF) image, bottom: EDXS line profile, and (e) EDXS elemental mapping image of tunnel junction region showing the In, Ga, and N variations. Adapted with permission from [150]

The Shockley–Read–Hall non-radiative recombinations in NW LEDs can be also reduced by octadecylthiol surface passivation [149].

An alternating current (AC) powered multi-active-region InGaN/GaN NW LED [150] was reported. Figure 12.19 shows the NW morphology, internal structure, and electron microscopy images revealing the internal structure. The NWs contained three emitting stages connected by tunnel junction. The tunnel junction design followed the strategy developed for thin films. In nitride thin film devices, a tunnel junction is typically based on polarization induced electric field of an AlN [151] or InGaN [152] insertion, which enhances tunneling in a GaN p–n junction. One advantage of the tunnel junction for LEDs is the possibility to take an ohmic contact for hole injection on n-GaN instead of a high-resistive p-GaN. In addition, it enables stacking of multiple LEDs [153] in series to produce a higher emission power for a given current. Sadaf et al. [150] have transferred this approach to NW LEDs. This realization is further described in Sect. 12.5.3.2.

MBE growth of nitride NWs was also demonstrated on metal substrates. Nitride visible NW LEDs were grown on Ti [154] and Mo [155, 156] films, showing a comparable opto-electronic performance to those grown on Si substrates. These approaches bring a new pathway not only for large-scale fabrication of NW devices with improved thermal management, but also for mechanically flexible devices. Further description of flexible LEDs grown on metal foils [157] is given in Sect. 12.10.

12.5.3 White NW LEDs

One of the envisioned applications of NW LEDs is general lighting. Two main routes for white NW LEDs are today explored: a conventional approach using phosphor down conversion and a direct mixing of red, green, and blue emissions from NWs of different composition.

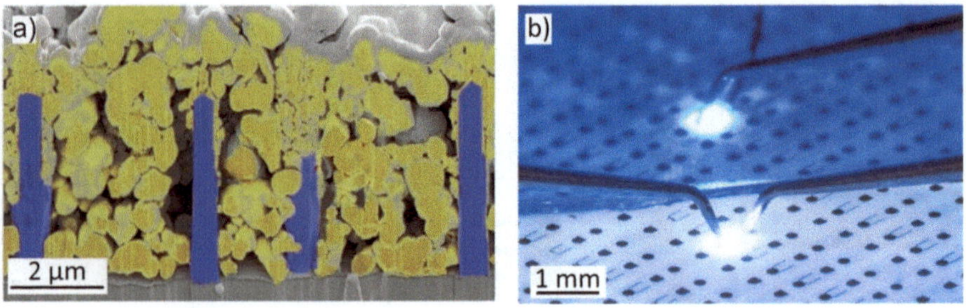

Fig. 12.20 (**a**) SEM image in artificial colors of phosphor particles filling the space between NWs. (**b**) Photograph of white NW LEDs in operation. Adapted with permission from [29]

12.5.3.1 Phosphor Converted White LEDs

The first approach consists in mimicking the well-developed technology of thin film LEDs, where an efficient blue LED is covered with a phosphor, which absorbs a part of the blue light end reemits a yellow light. By adjusting the relative intensities of the remaining blue emission and of the yellow luminescence of the phosphor one can obtain white light from the two-color mixture [158]. A white light of better quality can be obtained by using two different phosphors of two colors, e.g., yellow and red [159]. The OSRAM team has transferred this technology to NW LEDs [29]. In conventional thin film white LEDs, the phosphor is typically embedded in a silicone matrix with a bad thermal conductivity, which leads to phosphor heating. For NW white LEDs, the phosphor particles were integrated in contact with the active NWs. A special procedure was developed to form a phosphor material with a submicron grain size, which can be fitted into the space between the NWs. As illustrated in Fig. 12.20a, the phosphor particles fill the space between the NWs, which guaranties an efficient light conversion and also a more efficient phosphor cooling. By using red nitrido-silicate and yellow YAG:Ce phosphors, white emission (Fig. 12.20b) close to the Planckian locus with a coordinated color temperature of 6600 K and a color rendering index of 73 was achieved. Similar approach employing microscale phosphors for downconversion has been applied to demonstrate mechanically flexible white NW LEDs as will be discussed in Sect. 12.10.

12.5.3.2 Phosphor-free White LEDs by RGB Color Mixing

The energy down conversion process has a fundamental drawback: the difference in energy between the pump and the phosphor emission is lost as heat. This reduces the overall efficiency of phosphor-converted devices. Additional losses appear from the light backscattering in the phosphor layer and from the sub-100% efficiency of the phosphor converters. These drawbacks motivate a continuous search for converter-free white emitters based on mixing red, green, and blue colors. In principle, thanks to the bandgap tuning over the whole visible range, the InGaN alloy with different compositions is well suited for fabricating monolithic white LEDs. A large number of investigations aiming on white all-InGaN LEDs by RGB color mixing were performed using thin film devices with both quantum wells and quantum dots in the active region [160–163]. However, high indium content InGaN layers suffer from a poor crystal quality due to a strong lattice mismatch. Therefore, the radiative efficiency of the green and red emitters is much weaker compared to the blue ones. Theoretical analyses of a monolithic InGaN RGB white source show that to be competitive with the standard phosphor-converted LEDs, a significant improvement of the efficiency of the green and red emitters beyond the state of the art is required [164]. Recently, the topic of monolithic RGB integration has gained a renewed interest with development of NW LEDs. Thanks to the efficient strain relaxation leading not only to a lower defect density but also to a lower internal field in NWs, these nanostructures are expected to improve the quantum efficiency for red and green InGaN LEDs. Many efforts are focused toward this objective.

The growth technique well suited for In-rich alloy formation is MBE, therefore most of the investigations of monolithic white LEDs make use of this growth method. Lin et al. [30] used spontaneous NW growth by MBE on an Si(111) substrate to form axial InGaN/GaN quantum discs embedded in a GaN p–n junction (Fig. 12.21a). The In content in the discs was varied within the same NW by changing the growth temperature to achieve blue, yellow, and red emission. As seen in Fig. 12.21b, the spatially averaged emission had a white color with a correlated color temperature close to 6000 K. However, the spontaneous NW growth resulted in compositional variations from wire to wire, therefore a significant fluctuation of the emission wavelength was observed at the microscale (Fig. 12.21c).

Guo et al. [165] followed the approach of Lin et al. [30], while transferring the growth to Si(001) substrate. Three groups of axial quantum discs with different In content were grown along the wire axis within an axial p–n junction. The structure

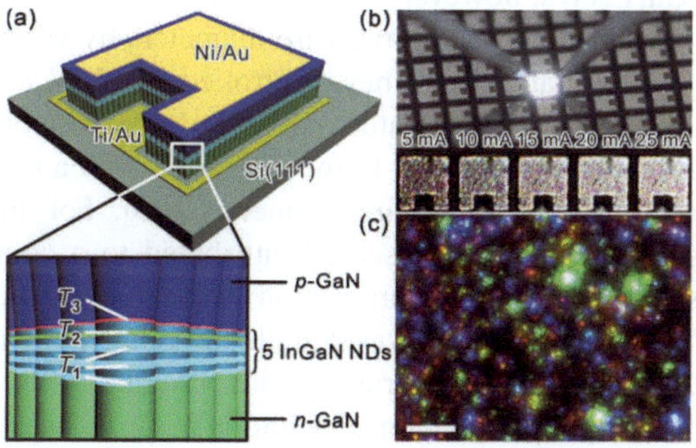

Fig. 12.21 (**a**) Schematic of NW white LED with axial InGaN/GaN quantum discs grown at different temperatures. (**b**) Photograph of a NW array LED emitting white light at 20 mA injection current. (**c**) Micro-electroluminescence image at 20 mA revealing wire-to-wire color variations. The scale bar is 10 μm. Adapted with permission from [30]

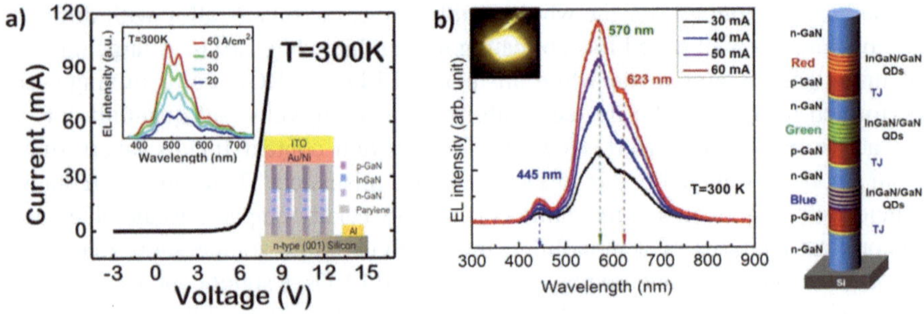

Fig. 12.22 (**a**) I–V characteristics of a white NW LED. The insets show the electroluminescence spectra for increasing injection current and the LED schematics. Adapted with permission from [165]. (**b**) Electroluminescence spectra of a white NW LED composed of axially stacked RGB stages and a schematic of the device showing the RGB stages connected with tunnel junctions. Inset shows the photograph of operating LED. Adapted with permission from [150]

was planarized using parylene, which is reported also to provide surface passivation [165], and contacted with a Ni/Au/ITO semi-transparent electrode (see inset of Fig. 12.22a). The LED showed a diodic behavior with a rather large opening voltage around 6 V (Fig. 12.22a) and a broad visible emission with a correlated color temperature ranging from 5500 to 6500 K. The electroluminescence spectra for increasing injection current shown in inset of Fig. 12.22a demonstrated a stable spectral shape with no spectral shift when varying the injection. The EQE analyses under pulsed injection with 1% duty cycle showed that the efficiency droop appeared at higher current densities (around 400 A/cm^2) compared to thin film LEDs; however, no absolute values of the EQE were reported.

Armitage and Tsubaki [95] demonstrated white NW LEDs by mixing emissions from axially and radially grown InGaN insertions in MBE-grown NWs. Similar to Sekiguchi et al. [32], the authors analyzed the In incorporation in InGaN/GaN quantum wells in GaN NWs; however, their analyses were based on self-assembled NWs. The authors controlled the average wire diameters by varying the Ga/N flux ratio and substrate temperature and they controlled the NW density by varying the GaN buffer layer growth temperature. Similar to [32], spontaneously grown NWs also showed diameter-dependent In incorporation with a trend toward shorter emission wavelengths for smaller wire diameters [95]. For low density NWs, two distinct emission peaks were observed, attributed to a combination of self-shadowing effects and different In incorporation efficiencies on the top and sidewall planes of the NWs. By adjusting the wavelength and relative intensity of these two peaks, Armitage and Tsubaki demonstrated white NW LEDs.

A strong effort toward optimization of white NW LEDs was performed in the group of Mi. Following thin film LED architecture, an AlGaN electron blocking layer was introduced in NW LEDs to prevent electron overflow into the p-GaN segment [166]. Surface recombination was identified as the main source of losses and the growth of a protective AlGaN shell around the active InGaN/GaN region was optimized [167, 168]. The replacement of InGaN/GaN active region by InGaN/AlGaN quantum discs allows to form this AlGaN shell in a spontaneous way, and it can also act as a distributed electron blocking layer reducing the carrier overflow into the p-GaN top segment [147]. Finally, a complex design using tunnel junctions to connect red, green, and blue emitters was proposed [150]. The LED architecture is illustrated in Fig. 12.22b. The tunnel junction design follows the strategy developed for thin films [152]. The authors used a tunnel junction with an InGaN interlayer, which is placed in the bottom part of the LED to connect n-GaN to p-GaN segments and also between individual blue, green, and red LED sections (Fig. 12.22b). The multi-junction LED showed a light-up voltage approx. equal to three times value for a single junction LED (around 16 V). Electroluminescence spectra for different injection currents and a device photograph are shown in Fig. 12.22b. The LED produced a nearly white-light with a correlated color temperature ~3000 K.

12.6 Ultraviolet NW LEDs

For applications like water purification, phototherapy, or medical diagnostics, nitride UV LEDs are an appealing alternative for replacing mercury lamps. However, the efficiency of deep-UV nitride emitters remains low due to a high density of dislocations lowering the IQE to poor electrical injection issues because of the difficulty to dope AlGaN and fabricate ohmic contacts. In addition, the light extraction is impeded by the absorption in the contact and the predominant TM light polarization [169]. AlGaN NWs can potentially improve the IQE of UV emitters by improving the crystalline quality. Another advantage of NWs is the predicted improvement of light extraction efficiency [170, 171].

The light extraction efficiency (LEE) of AlGaN NW UV LED was investigated by finite-difference time-domain simulations [170, 171]. These numerical studies show that the LEE strongly depends on the geometrical parameters of the NWs, e.g., diameter, height. With an optimized structure, the theoretical LEE can be higher than 40%, both for transverse-magnetic and for transverse-electric modes.

Regarding experimental realizations, Sekiguchi et al. [172] demonstrated a GaN/AlGaN NW near-UV LED on Si substrate with a continuous planar p-GaN top layer by MBE. GaN/AlGaN multiple quantum discs were grown along the axial direction, emitting UVA light at 354 nm under an electrical injection of 50 mA. To explore UVB and UVC bands, growth of ternary AlGaN NWs on top of a GaN base segment was analyzed by different authors (e.g., [173]). Carrier localization in AlGaN composition inhomogeneities was observed and an optical (room temperature over low temperature) IQE of 50% was reported for emission from deep localized states in AlGaN alloys with a peak energy of 4.3 eV [173]. We note, however, that the IQE of the near band edge peak of these structures was much lower [173]. Himwas et al. [174] compared optical IQE for long AlGaN NW segments and for AlGaN/AlN quantum discs in NWs: 5% optical IQE was reported for near band edge emission of AlGaN segments with 50% Al content and 30% IQE was achieved for quantum discs, thanks to 3D carrier confinement [174]. These undoped NWs were developed in view of deep-UV emission using electron beam pumping to avoid the problems of p-conductivity and contacting of AlGaN. Core/shell GaN/AlGaN [175] and GaN/InAlN [176] NWs were also investigated for UV emission.

The group of Mi reported a series of spectacular results on the NW deep UV emitters, which however have not yet been reproduced by other groups. Highly conductive Mg-doped AlN NWs grown by MBE on GaN NW templates were reported and used for realization of electrically injected near-vacuum-UV NW LEDs [177]. Electroluminescence at ~205 nm with a turn-on voltage of ~5.5 V was reported [177]. Electrically injected UV-B band NW lasers based on Anderson localization were reported [178, 179]. Electrically injected room temperature lasing was observed at 289 nm, with a small threshold of 300 A/cm^2 [178]. Recently, the same group also reported a NW UV-C LED with a tunnel junction between n-GaN and p-AlGaN sections inserting a metallic Al layer, which allowed to avoid p-AlGaN contacting [180]. This LED showed an optical IQE of ~80% and a wall plug efficiency of ~0.3%. Finally, Laleyan et al. [181] reported Mg dopant-free Al(Ga)N/h-BN NW UV-C LED exploiting strong p-type conductivity of h-BN in the presence of B vacancies, which allowed to avoid the AlGaN p-doping. These impressive results remain to be confirmed by other research teams.

12.7 Operation Speed of NW LEDs

Widespread application of LEDs to general lighting creates new functionalities, one of them being data transfer with visible light. High speed visible light communications (VLC) such as light fidelity (Li-Fi) are today a topic of intense research. VLC allows a simultaneous data transfer and lighting and it shows a good data safety. One main obstacle is that the theoretical maximal bandwidth for an LED transmitter is limited by the carrier lifetime in the active region, which is slowed down in c-plane LEDs due to the internal field. NW core/shell LEDs are considered as an alternative to thin film LEDs enabling ultra-fast operation. Indeed, in core/shell structures the active quantum wells are grown on the non-polar m-planes avoiding the internal electrical fields from the spontaneous and piezoelectric polarization.

In 2015, Koester et al. [131] demonstrated a high-speed m-planar core-shell InGaN/GaN NW LEDs grown on Si substrate by catalyst-free MOVPE. The NW morphology is illustrated in Fig. 12.23a. Carrier dynamics in core/shell InGaN/GaN quantum wells grown around the top part of the n-GaN wires was probed by time-resolved photoluminescence. Low-temperature carrier lifetime of 110 ps and room temperature value of 20 ps were reported. The devices were processed by partially encapsulating NWs into a spin-on-glass and depositing a Ni/Au contact on one side of the NWs (using a tilted deposition method (Fig. 12.23b, c). The devices showed an electroluminescence around 410 nm. For time-resolved electroluminescence

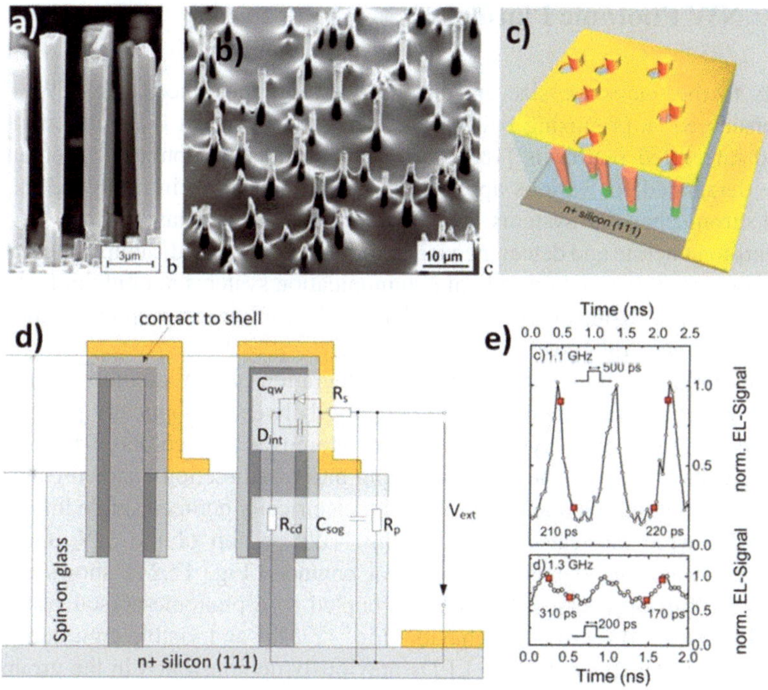

Fig. 12.23 SEM images of (**a**) as-grown core-shell InGaN/GaN NWs and (**b**) of NW encapsulated into spin-on-glass. (**c**) LEDs schematic. (**b**) Electroluminescence spectrum at 300 K of the nanowire array LED. (**d**) Device electrical circuit. (**e**) Time-resolved electroluminescence signal recorded for $V_{DC} = 8$ V and $V_{Pulse} = 6$ V square pulses for a frequency of 1.1 GHz (top) and for a frequency of 1.3 GHz (bottom). Adapted with permission from [131]

measurements the device was subjected to a DC bias superimposed with an RF input from a pulse pattern generator. An on/off modulation was achieved at room temperature at 1.1 GHz with an on/off ratio of ~5. Ten ninety per cent rise- and fall-times close to 200 ps were recorded. For a modulation frequency increased to 1.3 GHz, the modulation depth strongly decreased down to about 1.6 (Fig. 12.23e). The electrical scheme of the NW LED was proposed (Fig. 12.23d). The authors showed that the RC low-pass behavior limiting the operation frequency originated not from the capacitance of the top/bottom contacts, but from the series resistance and the junction capacitance of the core/shell NW p–n junctions.

Different approaches can be implemented to increase the modulation bandwidth of NW LEDs. Zhu et al. [182] used top-down NWs etched from a c-plane LED (i.e., presenting an internal field in the quantum wells) to evidence the effect of carrier coupling with surface plasmons as a way to enhance the bandwidth. Deposition of Ag nanoparticles was shown to increase the -3dB cut-off frequency by a factor of ~2 (going up to 30 MHz, which still remains rather low).

12.8 NW Photonic Platforms

Thanks to the nanoscale size and bright electroluminescence, single NW LEDs are considered as promising building blocks of the compact integrated platforms. Hybrid integrated platforms [183] with photonic and non-photonic elements are highly demanded for various applications in bio-detection, in optogenetics, or in optoelectromechanical systems. For example, NWs can play the role of active elements to generate and detect the visible light on a microfluidic chip. Alternatively, NW emitters can be used for optical communication systems not only in free space, but also on chip. III-nitride NWs represent an ideal solution for optical components in the visible spectral range.

An important technological step is the integration of NW emitters with waveguides. Park et al. [184] demonstrated coupling between an optically pumped CdS NW and a 2D photonic-crystal waveguide. Moreover, two colors of light emitted from the CdS/CdSe core/shell NW were split into two directions, showing the possibilities for on-chip light manipulation. Next, electrically pumped single InGaN/GaN NW LEDs were fabricated by dispersing NWs on top of the SiN_x membrane photonic crystal and depositing electrical contacts. Fig. 12.24a shows an SEM image of a fully processed NW LEDs coupled to a photonic-crystal waveguide. NW LEDs showed rectifying behavior (Fig. 12.24b) and visible emission peaked at 480 nm similar to single wire LEDs previously demonstrated in the group [41]. Electroluminescence from the NW was efficiently guided by the photonic-crystal waveguide over about 20 μm as shown in Fig. 12.24c, which superimposes an optical and an SEM image.

Brubaker et al. [37] realized an on-chip optical coupling between an LED and a photoconductive detector both based on GaN NWs. Axial p–n junction NWs were grown by MBE. The NWs were dispersed and closely lying pairs of wires were contacted to fabricate an LED (contacts on the p- and n-doped segments) and a photoconductive detector (both contacts on the n-doped segment). The communication took place in free space as schematized in Fig. 12.25a, b. GaN p–n junction NW (without quantum wells) presented a near sub-bandgap emission under electrical injection for biases above 20 V as illustrated in Fig. 12.25c. The NW detector showed a high photoconductive gain with the sensitivity down to $\sim 10^{-7}$ W/cm^2 but suffered from a slow switching speed (several tens of seconds). Figure 12.25d shows the detector current traces when the LED is consecutively switched on and off with a different injected current. The correlation between the photodetector signal and the LED switching was demonstrated; however, a continuously increasing baseline was observed due to a slow return to equilibrium in NW photoconductive detectors.

In [37] the LED-detector communication was taking place in free space. A waveguide for light channeling is expected to improve the LED-detector coupling efficiency. In addition, a fast photodetector with no persistent photoconductivity effects is also desirable to achieve high speed communication. In 2014, an integrated photonic platform consisting of an InGaN/GaN NW LED and a p–n NW photodiode

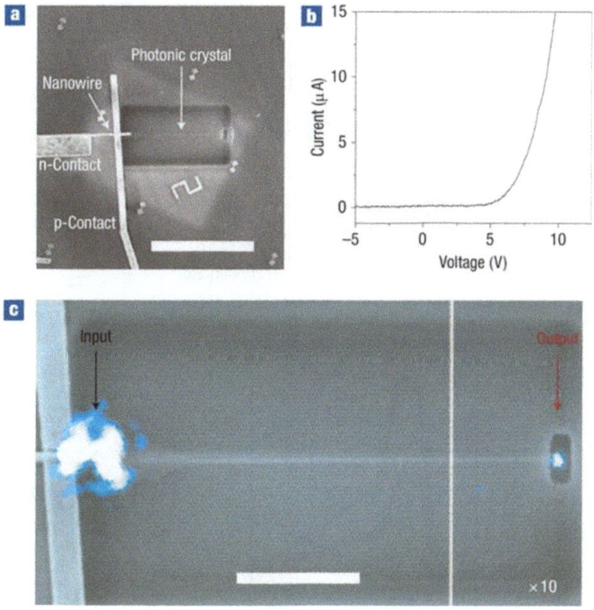

Fig. 12.24 (**a**) SEM image of single NW LED coupled to an SiN photonic-crystal waveguide (scale bar 20 μm). (**b**) I–V characteristic of the LED. (**c**) Electroluminescence image superimposed to the SEM image for photonic platform in operation (scale bar 5 μm). Adapted with permission from [184]

interconnected with an SiN_x waveguide was demonstrated [38]. MOVPE-grown core/shell InGaN/GaN NWs were used to fabricate the LED and the photodiode. As illustrated in Fig. 12.26a, the LED wires contained InGaN/GaN quantum wells whereas the photodiode wires were composed of a thick InGaN layer with an increased In composition in order to achieve a good spectral matching between the LED and the detector. The emitter and detector NWs were dispersed on an SiO_2 on Si templates, electrically contacted and connected with a multimode SiN_x waveguide, which was designed as a function of the wire positions, deposited by PECVD, and then structured using e-beam lithography. Figure 12.26a displays an SEM image of the resulting photonic platform. Observation of the platform in operation (Fig. 12.26b) demonstrated that the emitted light from the NW LED was coupled into an SiN_x multimode waveguide as evidenced by the light diffraction at the opposite end of the waveguide. At the end of the SiN_x waveguide, a NW photodetector realized light collection and optical-electrical signal conversion. Calculations of the LED to photodiode coupling efficiency estimated that only ~8.7% of the light intensity is coupled into the detecting NW. Despite this low coupling efficiency, the current trace of the NW photodiode under zero bias shown in Fig. 12.26c demonstrated a clear correlation with the LED on/off switching. Contrary to [37], no baseline due to slow photoconductive effects was observed in the current trace.

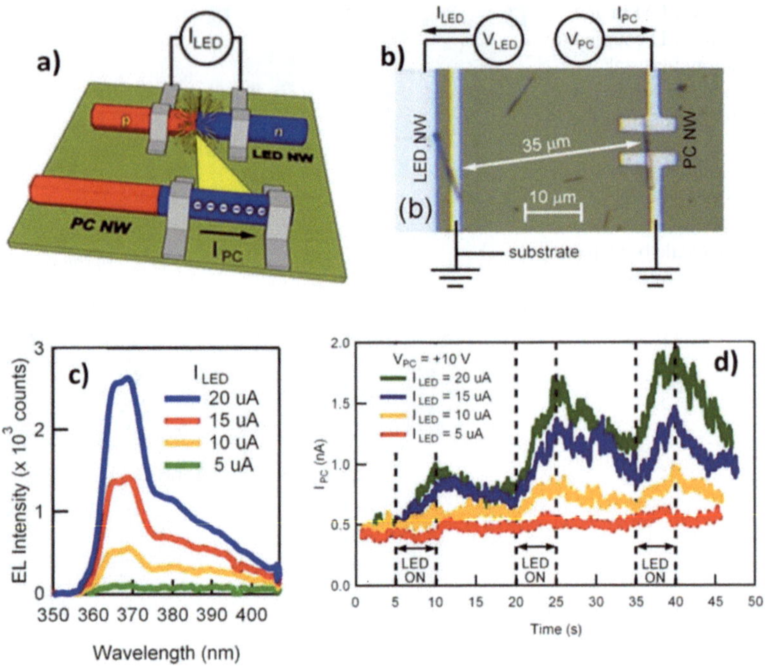

Fig. 12.25 (**a**) Schematic illustration of a free space coupled NW LED/photodetector platform. (**b**) Optical microscopy image of the LED-detector couple illustrating the electrical connections. (**c**) Electroluminescence spectra of the NW LED for increasing injection current. (**d**) Photodetector current trace for LED on/off switching for different LED currents. Adapted with permission from [37]

12.9 Open Issues of NW LEDs

Despite more than 10 years passed since the demonstration of the first NW LEDs and the constant efforts toward optimization of their performance, NW LEDs have not yet been commercialized and their performance remains behind the thin film values. Due to the complexity of the 3D architecture, nanoscale characterization of the NWs is necessary to point out the origin of existing problems of this new technology and to propose solutions. In this section, we consider the present open issues of NW LEDs together with the characterization techniques enabling their analyses.

12.9.1 Low IQE Values

One important problem of NW LEDs is the relatively low value of the IQE reported for core/shell InGaN/GaN quantum wells. Based on the temperature and

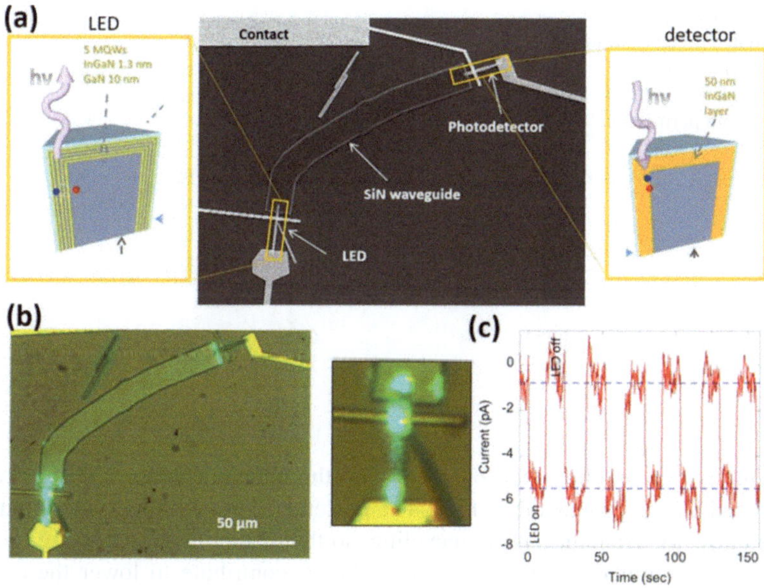

Fig. 12.26 (**a**) SEM images of the integrated photonic platform consisting of a waveguide-coupled NW LED and photodetector with the schematics of the internal structures of the constituting elements. (**b**) Photograph of the full system in operation and a close-up on the NW LED. (**c**) Current temporal trace of the detector when the LED is repeatedly switched on and off. Adapted with permission from [38]

power dependent photoluminescence with a resonant excitation of the quantum wells, Li et al. reported an IQE of 28% [119]. Hong et al. [48] reported an IQE of 32%; Schimpke et al. [29] reported 10%; Mounir et al. [185] used confocal photoluminescence to probe the IQE at different locations and reported local IQEs between 15 ± 1% near the NW tip and 44 ± 5% near the NW base. For comparison, IQE of m-plane quantum wells on a bulk m-GaN substrate can reach 80% [186] and for c-plane blue-emitting quantum wells it exceeds 90% [187]. It is not fully clear, why the IQE of core/shell NW quantum wells is below that of thin films. Mounir et al. [185] concluded that the radiative coefficient B is the main limiting factor for the efficiency and suggested that growing thinner multiple quantum wells instead of a single thick quantum well might improve the efficiency.

12.9.2 Reabsorption in Core/Shell LEDs

It is generally believed in the NW community that the NW geometry is good for light extraction properties. However, low IQE is a problem for light extraction from core/shell LEDs. For wire diameters D larger than 0.35λ, where λ is the emitted wavelength, GaN NWs sustain guided modes. Part of the emitted light is guided

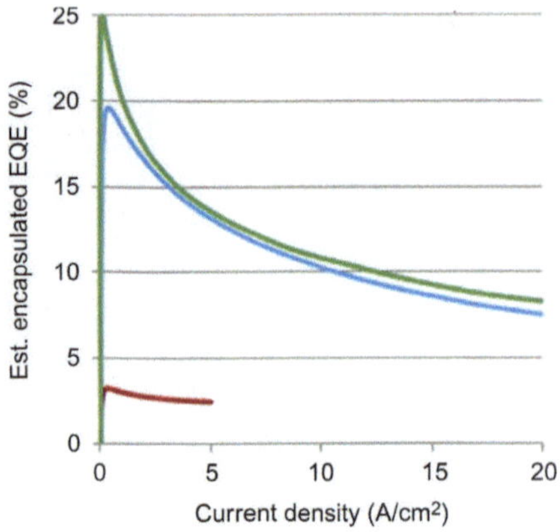

Fig. 12.27 EQE for encapsulated green (520 nm), blue (465 nm), and red (600 nm) nanowire LEDs. Adapted with permission from [31]

along the wire axis, the mode overlaps with the quantum wells and light can be reabsorbed during propagation [188]. The presently reported IQE values are rather low to allow for efficient photon recycling, so that a significant part of reabsorbed light is not reemitted. These reabsorption losses contribute to lower the external quantum efficiency.

12.9.3 Low EQE Values

The reported EQE values remain also lower than for thin film devices. An EQE of 3% was measured for the emission wavelength of 450 nm at a current density of 400 mA/mm^2 for highly homogeneous arrays of core/shell NWs [29]. Monemar et al. [31] reported the EQE of nitride NW core/shell LEDs as a function of current density (Fig. 12.27). The efficiency of the blue LED was quite poor, whereas the green LED showed an improved performance with a maximal EQE of 25%. However, a rather strong efficiency droop was observed: at 20 A/cm^2 the efficiency already dropped by a factor of 3. This fast decrease could have been related to thermal management in NW LEDs, since heat dissipation may be more difficult than in thin films. However, the authors suggested a different origin of the droop, which is attributed to an inhomogeneous incorporation of indium along the length of the wire resulting in inhomogeneous electron-hole injection.

12.9.4 Wavelength Control

Compositional homogeneity is of great importance for NW LEDs. It is generally accepted that SAG is the best way to reduce the wire-to-wire size and compositional fluctuations and therefore to control the NW LEDs wavelength. Indeed, in

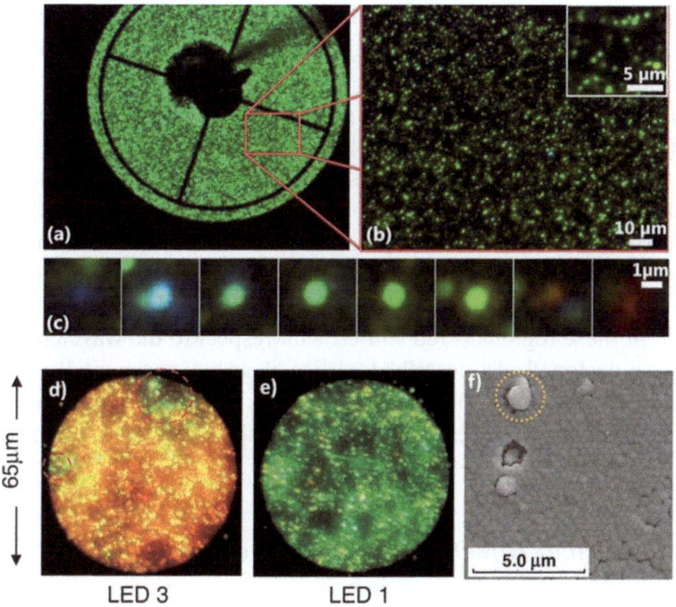

Fig. 12.28 Electroluminescence maps of an axial NW LED under 10 V forward bias imaged through a microscope at (**a**) 200× and (**b**) 500× magnification. The inset in (**b**) shows an enlarged part of the image. Panel (**c**) shows close-ups of individual luminescence spots taken from the image with 500× magnification. Adapted with permission from [140]. (**d**, **e**) Optical microscopy images of SAG MBE NW orange and green LEDs. (**f**) Top-view SEM image. Adapted with permission from [94]

spontaneously grown NWs, optical microscopy images evidence wire-to-wire color fluctuations [140] while a more homogeneous color is observed in SAG NW LED [29, 33]. Figure 12.28 compares MBE grown In-rich NW LEDs composed of self-assembled (Fig. 12.28a–c) and organized (Fig. 12.28d–f) NWs. Strong wire-to-wire fluctuations as well as a large percentage of dark NWs were observed in the self-assembled sample [140]. In addition, for MBE-grown LEDs with multiple axial quantum discs the disc shape and composition (and thus the emission wavelength) change within the stack, which increases the electroluminescence broadening [189]. Growth imperfections can still be present in SAG NWs leading to local color changes of the electroluminescence as illustrated in Fig. 12.28d–f [94]; however, a reduction of the spectral broadening and a higher yield of NWs producing electroluminescence were observed for SAG NWs.

Even for a perfectly homogenous array of core/shell NWs with no apparent growth imperfections obtained either by SAG or by top-down etching and regrowth, intra-wire gradients are still present in the quantum wells [10, 29, 49]. The presence of compositional inhomogeneity was evidenced by different groups by structural analyses (such as energy dispersive X-ray spectroscopy in STEM [52], atomic probe tomography [10], nanofocus X-ray diffraction [190], and by optical techniques (confocal photoluminescence [185, 191], cathodoluminescence [10, 52, 190], time-resolved cathodoluminescence [192]).

In-rich regions locally change the parameters of the p–n junction and favor current injection in these locations [49]. Because of the current redistribution, the emission wavelength shifts with increasing injection [29, 49]. Therefore, the usual assumption of a homogeneous injection made in core/shell NW LED modelling [188, 193] is not fulfilled in the experiments. An inhomogeneous current density with a high carrier concentration in the In-rich regions results in an early onset of Auger recombinations and thus in efficiency droop. In addition, the emission wavelength of these regions is red shifted with respect to the wavelength for which the LED was designed. As described earlier, it was experimentally demonstrated that by inhibiting the injection in In-rich regions responsible for long wavelength emission it is possible to increase the intensity of the blue electroluminescence [50].

12.9.5 *Electrical Injection Inhomogeneities*

The problem of injection homogeneity (both intra-wire and wire-to-wire) is considered as a key challenge of NW LEDs.

12.9.5.1 Intra-wire Injection Inhomogeneities

Regarding intra-wire homogeneity, the top NW region where m-plane facets meet semipolar facets or the top c-plane may present injection singularities [50]. In addition to In-content variations in the quantum wells discussed in the previous section, electrical homogeneity of the core and the shell is important. It is a challenging task to analyze doping at the nanoscale. Micro-Raman [194] analyses were applied to access doping in the n-doped GaN cores, which can reach high values (10^{20} cm^{-3}) without degrading material quality. For these high doping values, the core resistance should not be a limiting factor for homogeneous electrical injection even in tens of micron long NWs. However, electrical activity at the interface between the axially grown n-GaN core and a radially grown n-GaN underlayer was reported [52], which may create a parasitic barrier for electron injection from the core to the active region. A good control of thickness and p-doping of the GaN shell along the wire axis is also important for homogeneous injection. It is more difficult to assess the doping of the shell compared to the core. Optical methods like cathodoluminescence [195] can be applied, however, it is difficult to achieve quantitative results. SIMS analyses are very challenging for NWs [110]. EBIC mapping of the NW cross-section was used to probe the p-GaN shell conductivity (Fig. 12.29a, b, d, e) [52, 196]. As illustrated in Fig. 12.29c, doping

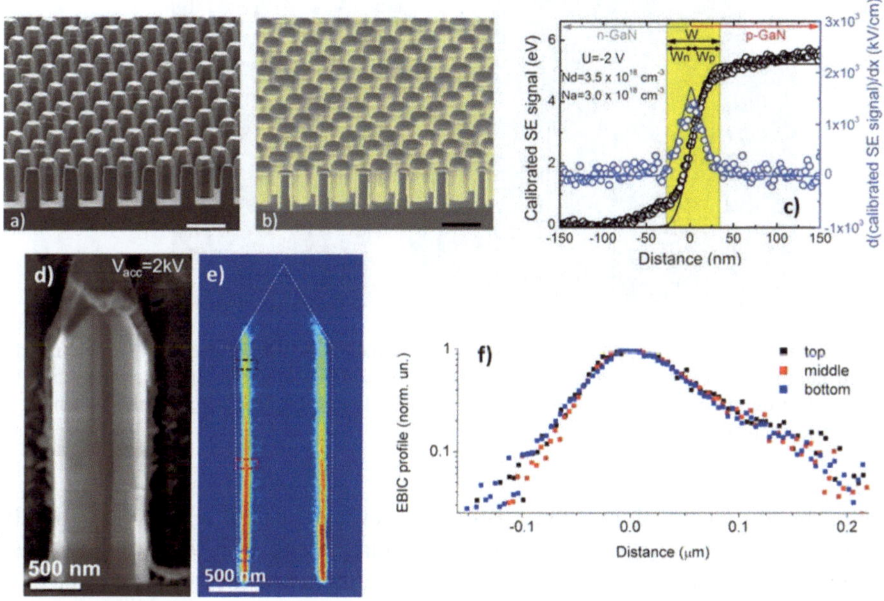

Fig. 12.29 (**a**) SEM image and (**b**) EBIC map at 10 keV of a cleaved processed array of core/shell n-p GaN NWs. (**c**) Secondary electron contrast profile and its derivative under reverse bias. Good agreement with an abrupt p–n junction theory is obtained for Na = 3.0×10^{18} cm^{-3} and Nd = 3.5×10^{18} cm^{-3}. Reproduced with permission from [196]. (**d**) SEM image and (**e**) EBIC map at 2 keV of an industrial-grade LED grown by GLO-AB based on InGaN/GaN core-shell NWs. (**f**) EBIC profiles at the top, middle and bottom part of the NWs. Adapted with permission from [52]

values in the 10^{18} cm^{-3} range were reported for core/shell p–n junction GaN wires using EBIC and secondary electron contrast profiles [196]. Figure 12.29f compares the EBIC profiles at different positions along the axis of a core/shell LED wire. It was concluded that in these LEDs the p-shell doping was homogeneous over the 2 μm NW length [52].

12.9.5.2 Wire-to-wire Injection Inhomogeneities

Fluctuations of the In composition and of the electrical parameters from wire to wire lead to a non-uniform current distribution in the NW array.

Limbach et al. [140] analyzed axial NW LEDs based on self-assembled MBE grown NWs by EBIC, CL, and electroluminescence techniques. The authors found that only 1% of NWs produced electroluminescence whereas a much larger amount of NWs yielded a CL signal. Higher CL homogeneity compared to electroluminescence shows that the limiting factor is not the fluctuation of IQE of individual

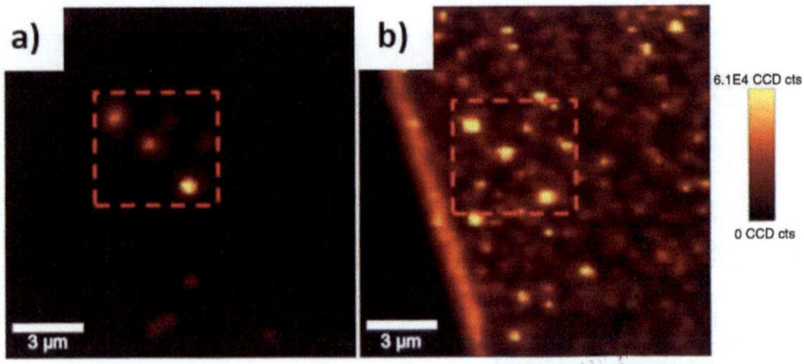

Fig. 12.30 Nanoscale mapping of MBE-grown axial NW LEDs. (**a**) Top-view micro-electroluminescence confocal mapping ($I = 50$ mA, $V = 9.3$ V) without laser excitation and (**b**) micro-electroluminescence–PL confocal mapping under laser excitation (excitation wavelength = 407 nm). Adapted with permission from [141]

wires, but the electrical injection. EBIC maps revealed that all NWs were correctly contacted. Therefore, the absence of electroluminescence in 99% of NWs was attributed to the differences in the individual currents in the NWs connected in parallel, which are due to fluctuations of individual series resistances and turn-on voltages. Similar conclusion was reached by Bavencove et al. [141] by analyzing micro-photoluminescence and micro-electroluminescence confocal maps. Figure 12.30a represents the micro-electroluminescence mapping, carried out solely under electrical injection compared with a map collected under a simultaneous electrical and laser excitations (Fig. 12.30b). The number of NWs producing intense photo-luminescence is much larger than that producing electroluminescence spots. This result clearly evidences that the electrical injection issue is the main factor of losses in these LEDs: despite a good IQE most of the wires remain inactive under electrical excitation.

The wire-to-wire injection homogeneity improves in SAG NW arrays, however, electroluminescence intensity fluctuations can still be observed (e.g., from [140] and [29]).

12.10 Flexible NW LEDs

Despite a certain number of challenges described in the previous section, NW LEDs are very attractive in particular, thanks to new functionalities that they can offer. In this section, we highlight one of them, namely fabrication of mechanically flexible light sources.

12.10.1 Motivation

Mechanically flexible LEDs are today a topic of intense research since they can provide novel functionalities and have the potential to open up a new branch of industry. Their applications include rollable displays, wearable intelligent electronics, deformable light sources, biomedical devices, e.g., lighting systems that can be ultimately integrated on a soft surface for biological and optogenetic purposes, etc. "Photonics Multiannual Strategic Roadmap 2014–2020" mentions flexible electronics, flexible light sources, displays, and solar cells as key emerging technologies with high expected growth of the market share. It is noteworthy that in the above-mentioned roadmap the notion of a "flexible device" is inseparable from an "organic device." Indeed, flexible devices mainly use organic materials integrated on lightweight and flexible plastic substrates. Thanks to the flexibility, relative ease of processing, compatibility with various flexible substrates, low cost, and their potential for large-scale manufacture, organic LEDs (OLEDs) are today the key technology for flexible displays [197]. However, organic devices are facing instabilities caused by oxidation, recrystallization, and temperature variations, which degrade the electrical conductivity of organic layers and interfaces in the active regions [198]. Moreover, organic LEDs present limitations in the short visible wavelength range [199]. They have a much lower luminance and a shorter lifetime in comparison to the LEDs based on inorganic semiconductors.

It is therefore advantageous to use inorganic semiconductors as the active material in flexible LEDs. However, the inorganic semiconductor devices are mechanically rigid. The fabrication of flexible devices from conventional thin film structures is challenging and requires additional processing steps to micro-structure the active layer. Moreover, the emission wavelengths of transferred thin film devices are shifting with the bending radii due to the strain induced shifts in the bandgap [200].

To avoid the micro-structuring step, it is advantageous to shrink the active element dimensions and to use bottom-up nanostructures, such as NWs, instead of two-dimensional films [201]. Polymer-embedded NWs offer an elegant solution to create flexible opto-electronic devices, which combine the high efficiency and the long lifetime of inorganic semiconductor materials with the high flexibility of polymers. NW arrays embedded in a flexible film and lifted-off from their native substrate can sustain large deformations, thanks to the high flexibility of individual NWs and to their footprint much smaller than the typical bending radius of the device [202–205]. Moreover, the lift-off and transfer procedure enables the assembly of free-standing layers of NW materials with different bandgaps without any constraint related to lattice-matching or growth conditions compatibility. This novel concept therefore allows for a large design freedom and modularity since it enables combination of materials with very different physical and chemical properties, which cannot be achieved by monolithic growth.

12.10.2 Fabrication Approaches

12.10.2.1 Direct Growth

A straightforward method for flexible LED fabrication would be a direct growth of NWs on flexible substrates. This method can readily be applied to some semiconductor materials such as ZnO. ZnO NW can be grown in solution on plastic substrates covered with a conductive layer at low temperature [206]. ZnO NW arrays grown directly on a flexible substrate have been used to demonstrate LEDs with a broad visible emission band. In 2008, Nadarajah et al. [207] reported single-crystalline ZnO NWs grown on flexible polymer-based indium-tin-oxide (ITO) substrates.

The direct growth of NWs on flexible materials eases the process of integration; however, it is limited to few kinds of semiconductors. As far as nitride NWs are concerned, the main limitation arises from the high growth temperature, which excludes the use of any plastic substrates. However, nitride NWs can be directly grown on metal foils. For example, MBE-grown GaN and AlGaN NWs were demonstrated on refractory metal foils like Ti and Ta foils [72, 157] and InN NWs were grown on brass [208]. The first LED from NWs grown on metal was demonstrated by May et al. in 2016 [157]. Self-assembled AlGaN NWs were grown on Ta and Ti foils by MBE. Within one metal grain, the NWs stand straight along the surface normal and have a good radius and height uniformity, as shown in Fig. 12.31. With an electroluminescence peak emission at around 350 nm, this LED showed a comparable opto-electronic performance to the NW LEDs grown on monocrystalline silicon substrate.

12.10.2.2 In-plane Transfer

As an alternative to a direct growth, one can transfer NWs to an exotic substrate to form an in-plane device.

Furthermore, the advances in the alignment of dispersed NWs with the assistance of electric field [209], magnetic field, fluid flow, capillary force, and so on offer new opportunities for the bottom-up approach to nanoelectronics. Lieber et al. [210] demonstrated a crossed-NW ultraviolet LED assembled from n-type GaN NWs and p-type Si NWs, which are sequentially dispersed on a plastic substrate through the orthogonal fluid-directed assembly. This NW LED maintained its emissive property upon multiple bending/unbending cycles [210]. However, the complexity of positioning single NWs and integrating with nonconventional substrates restricts their widespread applications.

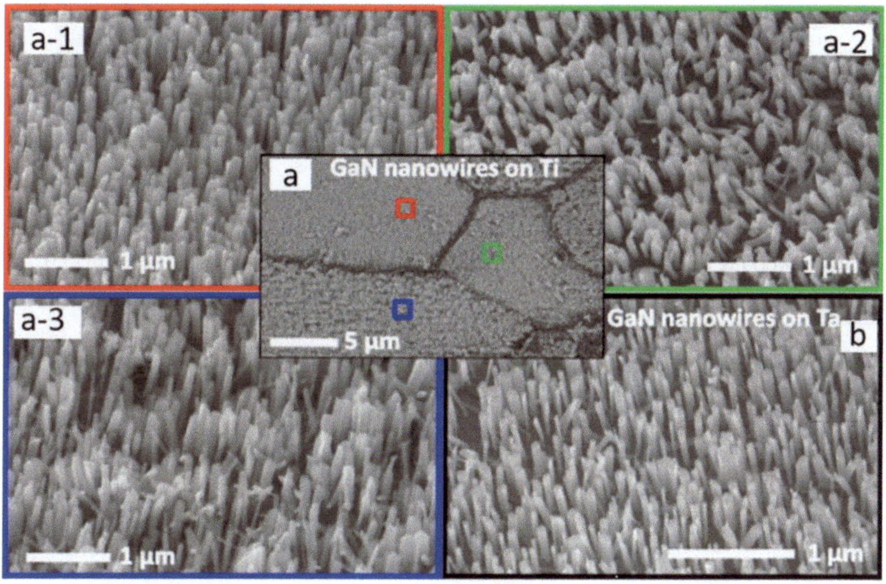

Fig. 12.31 SEM images of GaN NWs on flexible (**a**) Ti foil within different grains of the Ti foil and on (**b**) Ta foil. Adapted with permission from [157]

12.10.2.3 Vertical Transfer

NW arrays can also be transferred to flexible substrates preserving their vertical orientation. There have been several reports on light emitters based on polymer-embedded NW arrays, where active NW components are first embedded in a flexible film and then lifted-off from their native substrate [34–36, 201, 211]. First realizations used GaN micro-rods [35] and GaN/ZnO coaxial nanorod heterostructures [34] synthesized on graphene films. As shown in Fig. 12.32, active nanorods grown on graphene-covered Si/SiO_2 substrate were encapsulated in an insulating polymer layer. Then a lift-off process of the nanorods embedded in a polymer layer was realized by wet etching of the SiO_2 layer. Finally, the whole layer was transferred onto a metal coated flexible substrate [34]. In these realizations, a Ni/Au semi-transparent conductive layer was used as a top electrode; however, this semi-transparent contact can induce losses due to its non-ideal transmittance.

The transparency of the top contact can be improved by replacing Ni/Au layer by a conductive material presenting lower absorption such as Ag nanowire network. Blue and green flexible LEDs with Ag nanowire contact were elaborated using the following process flow [36]. First, the NW array was embedded into a polymer layer (e.g., PDMS). Plasma etching was used to expose and clean the top of the NWs. Then, the whole layer was peeled off to form a composite membrane with NWs maintaining their orientation and positions as illustrated in Fig. 12.33a. Two types of devices can be fabricated following this protocol: semi-transparent

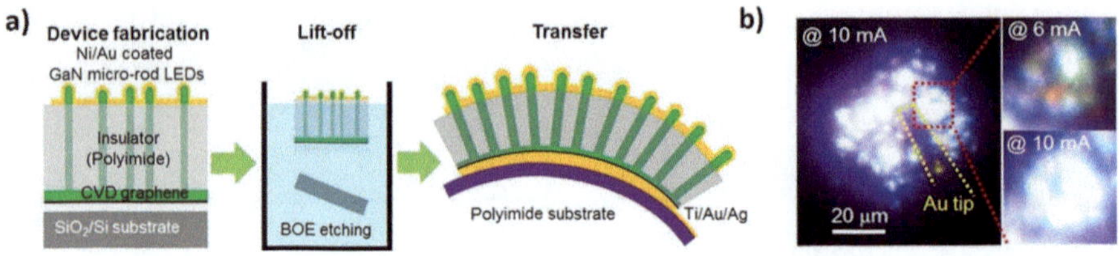

Fig. 12.32 (**a**) Fabrication process of flexible micro-rod LEDs grown on graphene substrate. (**b**) Photos of working LED under different injected currents. Adapted with permission from [35]

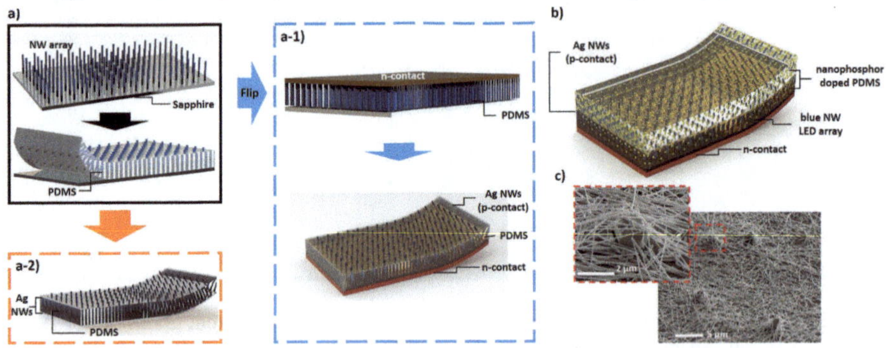

Fig. 12.33 (**a**) Process flows of flexible (**a-1**) semi-transparent NW LED and (**a-2**) fully transparent NW LED. (**b**) Schematic of the flexible NW white LED. (**c**) SEM images of Ag nanowire mesh spin-coated on PDMS layer with embedded NW LEDs. Tip of a core/shell NW protruding from the PDMS is shown by the red dashed rectangle. Adapted with permission from [36]

LEDs (Fig. 12.33a-1) and fully transparent LEDs (Fig. 12.33a-2). In the case of a semi-transparent device, the membrane is flipped onto an arbitrary receiver for the metallization of the NW bottom part. After that, a flexible substrate (e.g., PET or metal foil) is brought in contact with the backside metal layer which is mounted using silver epoxy. Next, Ag nanowires are spin-coated on the top side of the LED. In the case of a fully transparent device, considering the opacity of the backside metal layer, it is replaced by the Ag nanowires. Figure 12.33a-2 shows a schematic drawing of a fully transparent flexible NW LED with Ag nanowires as both the top and bottom transparent electrodes. The morphology of the transparent contact is illustrated in Fig. 12.33c, which presents an SEM image of the LED surface. Large objects correspond to protruding InGaN/GaN NWs, which are contacted by a large number of Ag nanowires.

MOVPE-grown core-shell InGaN/GaN NWs with different In concentration in the quantum wells were used as the active materials to generate light in the blue or green spectral range. As shown in Fig. 12.34a, these LEDs showed a good flexibility with a bending curvature less than 1 cm without showing any observable degradation of current and electroluminescence. The electroluminescence spectra of the blue and green LEDs are shown in Fig. 12.34b peaked at 425 nm and 530 nm, respectively.

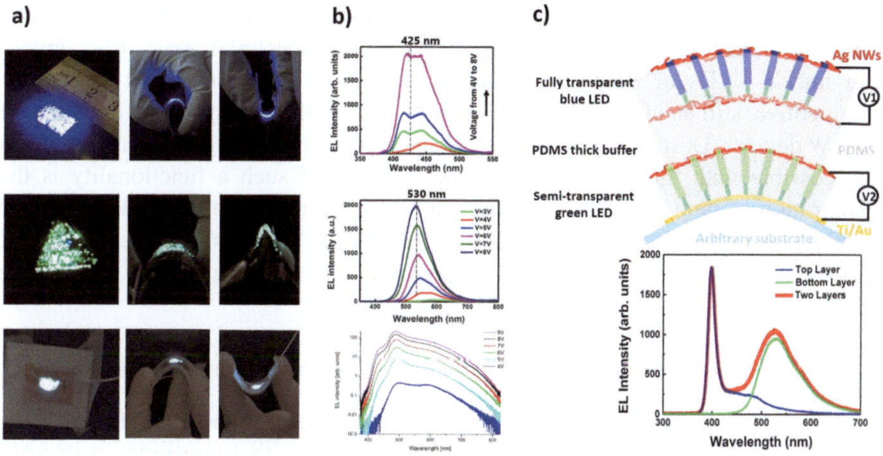

Fig. 12.34 (**a**) Photos of electroluminescence emissions and (**b**) electroluminescence spectra under different biases of blue, green, and white flexible NW LEDs. (**c**) Schematic (top) and electroluminescence spectra (bottom) of flexible bi-color LED. Adapted with permission from [36]

By assembling NW membranes with different In-content in the core/shell quantum wells, a blue–green two-color flexible LED was realized by mounting a fully transparent blue LED on top of a semi-transparent green LED. The electroluminescence spectra of the two-color LED are shown in Fig. 12.34c, corresponding to a superposition of the blue and green emissions of individual layers.

Similar method was applied to demonstrate a flexible phosphor converted white LED [211]. The nano-phosphor particles were mixed in the PDMS to fill the gaps between the blue LED NWs. An additional phosphor-doped PDMS cap layer was deposited on the top surface of the LED to increase the downconversion. The emission of the white LED presented in Fig. 12.34b covered a broad spectral range from 400 nm to 800 nm (i.e., almost the whole visible spectrum). The correlated color temperature of 6306 K and a color rendering index (CRI) of 54 were reported.

12.11 Summary

In summary, NW LEDs represent a promising approach for lighting and display applications. The addition of a third dimension to the device allows for complex architectures, namely core/shell, axial, or mixed geometries. Nitride NW LEDs covering UV to near IR spectral range as well as white emitters were demonstrated. Selective area epitaxy was shown to improve the NW LED performance by reducing wire-to-wire dispersion. However, it is still difficult for NW devices to compete with the main-stream thin film LED technology. Today, only prototypes of NW LEDs exist showing a lower efficiency compared to mature thin film emitters. For general lighting applications, a strong boost of efficiency is necessary to become competitive with thin films. For niche applications in challenging spectral domains (e.g., green-red nitride LEDs or deep-UV LEDs) it should be easier for NWs to become competitive, still an efficiency improvement is desirable for large-scale deployment of NW devices. Despite the existing challenges, NW LEDs bring new functionalities inaccessible with thin film devices. One example of such a functionality is the elaboration of blue to red NW emitters in a single growth run for RGB all-nitride displays. Another attractive opportunity offered by NW LEDs is the realization of flexible inorganic light sources with a high luminance and long lifetime for lighting of displays.

Acknowledgements The authors acknowledge fruitful discussions with V. Consonni and A. V. Maslov.

This work has been partially financially supported by ANR "Investissement d'Avenir" programmes "GaNeX" (ANR-11-LABX-2014) and "NanoSaclay" (ANR-10-LABX-0035), by ANR-14-CE26-0020-01 project "PLATOFIL,", by EU ERC project "NanoHarvest" (grant no. 639052), and FP7 ITN Marie Curie project "INDEED" (grant no. 722176).

References

1. S. Nakamura, T. Mukai, M. Senoh, Candela- class high- brightness InGaN/AlGaN double-heterostructure blue- light- emitting diodes. Appl. Phys. Lett. **64**, 1687 (1994)
2. J. Iveland, L. Martinelli, J. Peretti, J.S. Speck, C. Weisbuch, Direct measurement of Auger electrons emitted from a semiconductor light-emitting diode under electrical injection: identification of the dominant mechanism for efficiency droop. Phys. Rev. Lett. **110**, 177406 (2013)
3. E. Kioupakis, P. Rinke, K.T. Delaney, C.G. Van de Walle, Indirect Auger recombination as a cause of efficiency droop in nitride light-emitting diodes. Appl. Phys. Lett. **98**, 161107 (2011)
4. C.I. Ashby, C.C. Mitchell, J. Han, N.A. Missert, P.P. Provencio, D.M. Follstaedt, G.M. Peake, L. Griego, Low-dislocation-density GaN from a single growth on a textured substrate. Appl. Phys. Lett. **77**, 3233 (2000)
5. Z. Li-Xia, Y. Zhi-Guo, S. Bo, Z. Shi-Chao, A. Ping-Bo, Y. Chao, L. Lei, W. Jun-Xi, L. Jin-Min, Progress and prospects of GaN-based LEDs using nanostructures. Chin. Phys. B **24**, 068506 (2015)
6. L.-W. Jang et al., Enhanced light output of InGaN/GaN blue light emitting diodes with Ag nanoparticles embedded in nano-needle layer. Opt. Express **20**, 6036 (2012)
7. Q. Li et al., Optical performance of top-down fabricated InGaN/GaN nanorod light emitting diode arrays. Opt. Express **19**, 25528 (2011)
8. M. Latzel, P. Büttner, G. Sarau, K. Höflich, M. Heilmann, W. Chen, X. Wen, G. Conibeer, S. Christiansen, Significant performance enhancement of InGaN/GaN nanorod LEDs with multilayer graphene transparent electrodes by alumina surface passivation. Nanotechnology **28**, 055201 (2016)
9. J.-H. Park, U. Chatterjee, S. Kang, K. Lee, J.-S. Kim, C.-R. Lee, Synthesis of hybrid nanowires comprising uniaxial and coaxial InGaN/GaN MQWs with a nano-cap. J. Mater. Chem. C **4**, 10005 (2016)
10. J.R. Riley, S. Padalkar, Q. Li, P. Lu, D.D. Koleske, J.J. Wierer, G.T. Wang, L.J. Lauhon, Three-dimensional mapping of quantum wells in a GaN/InGaN core–shell nanowire light-emitting diode array. Nano Lett. **13**, 4317 (2013)

11. Y.-J. Li, J.-R. Chang, S.-P. Chang, B.-W. Lin, Y.-H. Yeh, H.-C. Kuo, Y.-J. Cheng, C.-Y. Chang, Multifacet microrod light-emitting diode with full visible spectrum emission. J. Disp. Technol. **12**, 951 (2016)
12. S.D. Hersee, A.K. Rishinaramangalam, M.N. Fairchild, L. Zhang, P. Varangis, Threading defect elimination in GaN nanowires. J. Mater. Res. **26**, 2293 (2011)
13. Y.H. Ko, J.H. Kim, L.H. Jin, S.M. Ko, B.J. Kwon, J. Kim, T. Kim, Y.H. Cho, Electrically driven quantum dot/wire/well hybrid light- emitting diodes. Adv. Mater. **23**, 5364 (2011)
14. J. Kang, Z. Li, H. Li, Z. Liu, X. Li, X. Yi, P. Ma, H. Zhu, G. Wang, Pyramid array InGaN/GaN core–shell light emitting diodes with homogeneous multilayer graphene electrodes. Appl. Phys. Express **6**, 072102 (2013)
15. W. Chen et al., High-performance, single-pyramid micro light-emitting diode with leakage current confinement layer. Appl. Phys. Express **8**, 032102 (2015)
16. T. Kim, J. Kim, M.-S. Yang, S. Lee, Y. Park, U.-I. Chung, Y. Cho, Highly efficient yellow photoluminescence from {11–22} InGaN multiquantum-well grown on nanoscale pyramid structure. Appl. Phys. Lett. **97**, 241111 (2010)
17. T. Wernicke et al., Indium incorporation and emission wavelength of polar, nonpolar and semipolar InGaN quantum wells. Semicond. Sci. Technol. **27**, 024014 (2012)
18. K. Wu et al., Fabrication and optical characteristics of phosphor-free InGaN nanopyramid white light emitting diodes by nanospherical-lens photolithography. J. Appl. Phys. **115**, 123101 (2014)
19. K. Kishino, A. Kikuchi. III nitride structure and method for manufacturing III nitride semiconductor fine columnar crystal, (Google Patents, 2014).
20. J. Hartmann et al., High aspect ratio GaN fin microstructures with nonpolar sidewalls by continuous mode metalorganic vapor phase epitaxy. Cryst. Growth Des. **16**, 1458 (2016)
21. A.K. Rishinaramangalam, M. Nami, M.N. Fairchild, D.M. Shima, G. Balakrishnan, S. Brueck, D.F. Feezell, Semipolar InGaN/GaN nanostructure light-emitting diodes on c-plane sapphire. Appl. Phys. Express **9**, 032101 (2016)
22. J. Hartmann et al., Study of 3D-growth conditions for selective area MOVPE of high aspect ratio GaN fins with non-polar vertical sidewalls. J. Cryst. Growth **476**, 90 (2017)
23. F. Glas, Critical dimensions for the plastic relaxation of strained axial heterostructures in freestanding nanowires. Phys. Rev. B **74**, 121302 (2006)
24. P. Coulon, M. Mexis, M. Teisseire, M. Jublot, P. Vennéguès, M. Leroux, J. Zuniga-Perez, Dual-polarity GaN micropillars grown by metalorganic vapour phase epitaxy: Cross-correlation between structural and optical properties. J. Appl. Phys. **115**, 153504 (2014)
25. S. Li, A. Waag, GaN based nanorods for solid state lighting. J. Appl. Phys. **111**, 5 (2012)
26. S. Boubanga-Tombet, J.B. Wright, P. Lu, M.R. Williams, C. Li, G.T. Wang, R.P. Prasankumar, Ultrafast carrier capture and Auger recombination in single GaN/InGaN multiple quantum well nanowires. ACS Photon. **3**, 2237 (2016)
27. M. Djavid, Z. Mi, Enhancing the light extraction efficiency of AlGaN deep ultraviolet light emitting diodes by using nanowire structures. Appl. Phys. Lett. **108**, 051102 (2016)
28. Y.-H. Hsiao, M.-L. Tsai, J.-H. He, GaN-based multiple-quantum-well light-emitting diodes employing nanotechnology for photon management. IEEE Trans. Ind. Appl. **51**, 1277 (2015)
29. T. Schimpke et al., Phosphor- converted white light from blue- emitting InGaN microrod LEDs. Phys. Status Solidi A **213**, 1577 (2016)
30. H.-W. Lin et al., InGaN/GaN nanorod array white light-emitting diode. Appl. Phys. Lett. **97**, 073101 (2010)
31. B. Monemar, B.J. Ohlsson, N.F. Gardner, L. Samuelson, Chapter seven-nanowire-based visible light emitters, present status and outlook. Semicond. Semimetals **94**, 227 (2016)
32. H. Sekiguchi, K. Kishino, A. Kikuchi, Emission color control from blue to red with nanocolumn diameter of InGaN/GaN nanocolumn arrays grown on same substrate. Appl. Phys. Lett. **96**, 231104 (2010)
33. K. Kishino, A. Yanagihara, K. Ikeda, K. Yamano, Monolithic integration of four-colour InGaN-based nanocolumn LEDs. Electron. Lett. **51**, 852 (2015)

34. C.H. Lee, Y.J. Kim, Y.J. Hong, S.R. Jeon, S. Bae, B.H. Hong, G.C. Yi, Flexible inorganic nanostructure light- emitting diodes fabricated on graphene films. Adv. Mater. **23**, 4614 (2011)
35. K. Chung, H. Beak, Y. Tchoe, H. Oh, H. Yoo, M. Kim, G.-C. Yi, Growth and characterizations of GaN micro-rods on graphene films for flexible light emitting diodes. Appl. Mater. **2**, 092512 (2014)
36. X. Dai, A. Messanvi, H. Zhang, C. Durand, J. Eymery, C. Bougerol, F.H. Julien, M. Tchernycheva, Flexible light-emitting diodes based on vertical nitride nanowires. Nano Lett. **15**, 6958 (2015)
37. M.D. Brubaker et al., On-chip optical interconnects made with gallium nitride nanowires. Nano Lett. **13**, 374 (2013)
38. M. Tchernycheva et al., Integrated photonic platform based on InGaN/GaN nanowire emitters and detectors. Nano Lett. **14**, 3515 (2014)
39. C. Goßler et al., GaN-based micro-LED arrays on flexible substrates for optical cochlear implants. J. Phys. D. Appl. Phys. **47**, 205401 (2014)
40. D. Massoubre, E. Xie, B. Guilhabert, J. Herrnsdorf, E. Gu, I.M. Watson, M.D. Dawson, Micro-structured light emission from planar InGaN light-emitting diodes. Semicond. Sci. Technol. **29**, 015005 (2013)
41. F. Qian, S. Gradecak, Y. Li, C.-Y. Wen, C.M. Lieber, Core/multishell nanowire heterostructures as multicolor, high-efficiency light-emitting diodes. Nano Lett. **5**, 2287 (2005)
42. R. Köster, J.-S. Hwang, C. Durand, D.L.S. Dang, J. Eymery, Self-assembled growth of catalyst-free GaN wires by metal–organic vapour phaseepitaxy. Nanotechnology **21**, 015602 (2009)
43. X. Wang et al., Continuous-flow MOVPE of Ga-polar GaN column arrays and core–shell LED structures. Cryst. Growth Des. **13**, 3475 (2013)
44. Y.T. Lin, T.W. Yeh, Y. Nakajima, P.D. Dapkus, Catalyst- free GaN nanorods synthesized by selective area growth. Adv. Funct. Mater. **24**, 3162 (2014)
45. H.-M. Kim, D.S. Kim, Y.S. Park, D.Y. Kim, T.W. Kang, K.S. Chung, Growth of GaN nanorods by a hydride vapor phase epitaxy method. Adv. Mater. **14**, 991 (2002)
46. G. Jacopin et al., Single-wire light-emitting diodes based on GaN wires containing both polar and nonpolar InGaN/GaN quantum wells. Appl. Phys. Express **5**, 014101 (2011)
47. J.-R. Chang, S.-P. Chang, Y.-J. Li, Y.-J. Cheng, K.-P. Sou, J.-K. Huang, H.-C. Kuo, C.-Y. Chang, Fabrication and luminescent properties of core-shell InGaN/GaN multiple quantum wells on GaN nanopillars. Appl. Phys. Lett. **100**, 261103 (2012)
48. Y.J. Hong, C.H. Lee, A. Yoon, M. Kim, H.K. Seong, H.J. Chung, C. Sone, Y.J. Park, G.C. Yi, Inorganic optoelectronics: Visible- color- tunable light- emitting diodes (Adv. Mater. 29/2011). Adv. Mater. **23**, 3224 (2011)
49. M. Tchernycheva et al., InGaN/GaN core–shell single nanowire light emitting diodes with graphene-based p-contact. Nano Lett. **14**, 2456 (2014)
50. H. Zhang, G. Jacopin, V. Neplokh, L. Largeau, F.H. Julien, O. Kryliouk, M. Tchernycheva, Color control of nanowire InGaN/GaN light emitting diodes by post-growth treatment. Nanotechnology **26**, 465203 (2015)
51. J. Ledig, X. Wang, S. Fündling, H. Schuhmann, M. Seibt, U. Jahn, H.H. Wehmann, A. Waag, Characterization of the internal properties of InGaN/GaN core–shell LEDs. Phys. Status Solidi A **213**, 11 (2016)
52. M. Tchernycheva et al., Core–shell InGaN/GaN nanowire light emitting diodes analyzed by electron beam induced current microscopy and cathodoluminescence mapping. Nanoscale **7**, 11692 (2015)
53. M. Sanchez-Garcia, E. Calleja, E. Monroy, F. Sanchez, F. Calle, E. Munoz, R. Beresford, The effect of the III/V ratio and substrate temperature on the morphology and properties of GaNand AlN-layers grown by molecular beam epitaxy on Si (1 1 1). J. Cryst. Growth **183**, 23 (1998)

54. M. Yoshizawa, A. Kikuchi, M. Mori, N. Fujita, K. Kishino, Growth of self-organized GaN nanostructures on Al2O3 (0001) by RF-radical source molecular beam epitaxy. Jpn. J. Appl. Phys. **36**, L459 (1997)
55. S. Fernández-Garrido, J. Grandal, E. Calleja, M. Sánchez-García, D. López-Romero, A growth diagram for plasma-assisted molecular beam epitaxy of GaN nanocolumns on Si (111). J. Appl. Phys. **106**, 126102 (2009)
56. J. Ristić, E. Calleja, S. Fernández-Garrido, L. Cerutti, A. Trampert, U. Jahn, K.H. Ploog, On the mechanisms of spontaneous growth of III-nitride nanocolumns by plasma-assisted molecular beam epitaxy. J. Cryst. Growth **310**, 4035 (2008)
57. L. Cerutti, J. Ristić, S. Fernández-Garrido, E. Calleja, A. Trampert, K. Ploog, S. Lazic, J. Calleja, Wurtzite GaN nanocolumns grown on Si (001) by molecular beam epitaxy. Appl. Phys. Lett. **88**, 213114 (2006)
58. M. Hetzl, F. Schuster, A. Winnerl, S. Weiszer, M. Stutzmann, GaN nanowires on diamond. Mater. Sci. Semicond. Process. **48**, 65 (2016)
59. K.A. Bertness, A.W. Sanders, D.M. Rourke, T.E. Harvey, A. Roshko, J.B. Schlager, N.A. Sanford, Controlled nucleation of GaN nanowires grown with molecular beam epitaxy. Adv. Funct. Mater. **20**, 2911 (2010)
60. O. Landré, C. Bougerol, H. Renevier, B. Daudin, Nucleation mechanism of GaN nanowires grown on (111) Si by molecular beam epitaxy. Nanotechnology **20**, 415602 (2009)
61. R. Songmuang, O. Landré, B. Daudin, From nucleation to growth of catalyst-free GaN nanowires on thin AlN buffer layer. Appl. Phys. Lett. **91**, 251902 (2007)
62. J. Zúñiga-Pérez et al., Polarity in GaN and ZnO: Theory, measurement, growth, and devices. Appl. Phys. Rev. **3**, 041303 (2016)
63. K. Hestroffer, C. Leclere, C. Bougerol, H. Renevier, B. Daudin, Polarity of GaN nanowires grown by plasma-assisted molecular beam epitaxy on Si (111). Phys. Rev. B **84**, 245302 (2011)
64. K. Hestroffer, C. Leclere, V. Cantelli, C. Bougerol, H. Renevier, B. Daudin, In situ study of self-assembled GaN nanowires nucleation on Si (111) by plasma-assisted molecular beam epitaxy. Appl. Phys. Lett. **100**, 212107 (2012)
65. L. Largeau, E. Galopin, N. Gogneau, L. Travers, F. Glas, J.-C. Harmand, N-polar GaN nanowires seeded by Al droplets on Si (111). Cryst. Growth Des. **12**, 2724 (2012)
66. T. Auzelle, B. Haas, A. Minj, C. Bougerol, J.-L. Rouvière, A. Cros, J. Colchero, B. Daudin, The influence of AlN buffer over the polarity and the nucleation of self-organized GaN nanowires. J. Appl. Phys. **117**, 245303 (2015)
67. S. Fernández-Garrido, X. Kong, T. Gotschke, R. Calarco, L. Geelhaar, A. Trampert, O. Brandt, Spontaneous nucleation and growth of GaN nanowires: The fundamental role of crystal polarity. Nano Lett. **12**, 6119 (2012)
68. C. Chèze et al., Direct comparison of catalyst-free and catalyst-induced GaN nanowires. Nano Res. **3**, 528 (2010)
69. O. Brandt, C. Pfüller, C. Chèze, L. Geelhaar, H. Riechert, Sub-meV linewidth of excitonic luminescence in single GaN nanowires: Direct evidence for surface excitons. Phys. Rev. B **81**, 045302 (2010)
70. V. Kumaresan, L. Largeau, F. Oehler, H. Zhang, O. Mauguin, F. Glas, N. Gogneau, M. Tchernycheva, J.-C. Harmand, Self-induced growth of vertical GaN nanowires on silica. Nanotechnology **27**(13), 135602 (2016)
71. V. Kumaresan, L. Largeau, A. Madouri, F. Glas, H. Zhang, F. Oehler, A. Cavanna, A. Babichev, L. Travers, N. Gogneau, M. Tchernycheva, J.-C. Harmand, Epitaxy of GaN nanowires on graphene. Nano Lett. **16**(8), 4895 (2016)
72. G. Calabrese, P. Corfdir, G. Gao, C. Pfüller, A. Trampert, O. Brandt, L. Geelhaar, S. Fernández-Garrido, Molecular beam epitaxy of single crystalline GaN nanowires on a flexible Ti foil. Appl. Phys. Lett. **108**, 202101 (2016)
73. V. Consonni, A. Trampert, L. Geelhaar, H. Riechert, Physical origin of the incubation time of self-induced GaN nanowires. Appl. Phys. Lett. **99**, 033102 (2011)

74. C. Chèze, L. Geelhaar, A. Trampert, H. Riechert, In situ investigation of self-induced GaN nanowire nucleation on Si. Appl. Phys. Lett. **97**, 043101 (2010)
75. V. Consonni, M. Hanke, M. Knelangen, L. Geelhaar, A. Trampert, H. Riechert, Nucleation mechanisms of self-induced GaN nanowires grown on an amorphous interlayer. Phys. Rev. B **83**, 035310 (2011)
76. V. Consonni, M. Knelangen, L. Geelhaar, A. Trampert, H. Riechert, Nucleation mechanisms of epitaxial GaN nanowires: Origin of their self-induced formation and initial radius. Phys. Rev. B **81**, 085310 (2010)
77. M. Knelangen, V. Consonni, A. Trampert, H. Riechert, In situ analysis of strain relaxation during catalyst-free nucleation and growth of GaN nanowires. Nanotechnology **21**, 245705 (2010)
78. A. Wierzbicka et al., Influence of substrate nitridation temperature on epitaxial alignment of GaN nanowires to Si (111) substrate. Nanotechnology **24**, 035703 (2013)
79. J.E. Northrup, J. Neugebauer, Theory of GaN (10 1¯ 0) and (11 2¯ 0) surfaces. Phys. Rev. B **53**, R10477 (1996)
80. V.G. Dubrovskii, V. Consonni, A. Trampert, L. Geelhaar, H. Riechert, Scaling thermodynamic model for the self-induced nucleation of GaN nanowires. Phys. Rev. B **85**, 165317 (2012)
81. N.V. Sibirev, M. Tchernycheva, M.A. Timofeeva, J.-C. Harmand, G.E. Cirlin, V.G. Dubrovskii, Influence of shadow effect on the growth and shape of InAs nanowires. J. Appl. Phys. **111**, 104317 (2012)
82. R. Debnath, R. Meijers, T. Richter, T. Stoica, R. Calarco, H. Lüth, Mechanism of molecular beam epitaxy growth of GaN nanowires on Si (111). Appl. Phys. Lett. **90**, 123117 (2007)
83. L. Lymperakis, J. Neugebauer, Large anisotropic adatom kinetics on nonpolar GaN surfaces: Consequences for surface morphologies and nanowire growth. Phys. Rev. B **79**, 241308 (2009)
84. E. Galopin, L. Largeau, G. Patriarche, L. Travers, F. Glas, J. Harmand, Morphology of self-catalyzed GaN nanowires and chronology of their formation by molecular beam epitaxy. Nanotechnology **22**, 245606 (2011)
85. V. Consonni, V. Dubrovskii, A. Trampert, L. Geelhaar, H. Riechert, Quantitative description for the growth rate of self-induced GaN nanowires. Phys. Rev. B **85**, 155313 (2012)
86. K. Kishino, H. Sekiguchi, A. Kikuchi, Improved Ti-mask selective-area growth (SAG) by rf-plasma-assisted molecular beam epitaxy demonstrating extremely uniform GaN nanocolumn arrays. J. Cryst. Growth **311**, 2063 (2009)
87. Z.A. Gačević, D. Gomez Sanchez, E. Calleja, Formation mechanisms of GaN nanowires grown by selective area growth homoepitaxy. Nano Lett. **15**, 1117 (2015)
88. F. Schuster, M. Hetzl, S. Weiszer, J.A. Garrido, M. de la Mata, C. Magen, J. Arbiol, M. Stutzmann, Position-controlled growth of GaN nanowires and nanotubes on diamond by molecular beam epitaxy. Nano Lett. **15**, 1773 (2015)
89. H. Sekiguchi, K. Kishino, A. Kikuchi, Ti-mask selective-area growth of GaN by RFplasma-assisted molecular-beam epitaxy for fabricating regularly arranged InGaN/GaN nanocolumns. Appl. Phys. Express **1**, 124002 (2008)
90. A. Bengoechea-Encabo et al., Understanding the selective area growth of GaN nanocolumns by MBE using Ti nanomasks. J. Cryst. Growth **325**, 89 (2011)
91. T. Schumann, T. Gotschke, F. Limbach, T. Stoica, R. Calarco, Selective-area catalyst-free MBE growth of GaN nanowires using a patterned oxide layer. Nanotechnology **22**, 095603 (2011)
92. J. Kruse et al., Selective-area growth of GaN nanowires on SiO2-masked Si (111) substrates by molecular beam epitaxy. J. Appl. Phys. **119**, 224305 (2016)
93. M.D. Brubaker, S.M. Duff, T.E. Harvey, P.T. Blanchard, A. Roshko, A.W. Sanders, N.A. Sanford, K.A. Bertness, Polarity-controlled GaN/AlN nucleation layers for selective-area growth of GaN nanowire arrays on Si (111) substrates by molecular beam epitaxy. Cryst. Growth Des. **16**, 596 (2015)
94. K. Kishino, K. Nagashima, K. Yamano, Monolithic integration of InGaN-based nanocolumn light-emitting diodes with different emission colors. Appl. Phys. Express **6**, 012101 (2012)

95. R. Armitage, K. Tsubaki, Multicolour luminescence from InGaN quantum wells grown over GaN nanowire arrays by molecular-beam epitaxy. Nanotechnology **21**, 195202 (2010)
96. T. Kouno, K. Kishino, Well-arranged novel InGaN hexagonal nanoplates at the tops of nitrogen-polarity GaN nanocolumn arrays. AIP Adv. **2**, 012140 (2012)
97. X. Zhang, B. Haas, J.-L. Rouvière, E. Robin, B. Daudin, Growth mechanism of InGaN nano-umbrellas. Nanotechnology **27**, 455603 (2016)
98. F. Qian, Y. Li, S. Gradecak, D. Wang, C.J. Barrelet, C.M. Lieber, Gallium nitride-based nanowire radial heterostructures for nanophotonics. Nano Lett. **4**, 1975 (2004)
99. R. Wagner, W. Ellis, Vapor- liquid- solid mechanism of single crystal growth. Appl. Phys. Lett. **4**, 89 (1964)
100. J. Khanderi, A. Wohlfart, H. Parala, A. Devi, J. Hambrock, A. Birkner, R.A. Fischer, MOCVD of gallium nitride nanostructures using (N 3) 2 Ga {(CH 2) 3 NR 2}, R= Me, Et, as a single molecule precursor: morphology control and materials characterization. J. Mater. Chem. **13**, 1438 (2003)
101. S.D. Hersee, X. Sun, X. Wang, The controlled growth of GaN nanowires. Nano Lett. **6**, 1808 (2006)
102. K. Choi, M. Arita, Y. Arakawa, Selective-area growth of thin GaN nanowires by MOCVD. J. Cryst. Growth **357**, 58 (2012)
103. W. Bergbauer et al., Continuous-flux MOVPE growth of position-controlled N-face GaN nanorods andembedded InGaN quantum wells. Nanotechnology **21**, 305201 (2010)
104. R. Qhalid Fareed, J. Yang, J. Zhang, V. Adivarahan, V. Chaturvedi, M. Asif Khan, Vertically faceted lateral overgrowth of GaN on SiC with conducting buffer layers using pulsed metalorganic chemical vapor deposition. Appl. Phys. Lett. **77**, 2343 (2000)
105. Y.-S. Chen et al., Threading dislocation evolution in patterned GaN nanocolumn growth and coalescence overgrowth. J. Appl. Phys. **106**, 023521 (2009)
106. S.-Y. Bae, J.-Y. Lee, J.-H. Min, D.-S. Lee, Morphology evolution of pulsed-flux Ga- polar GaN nanorod growth by metal organic vapor phase epitaxy and its nucleation dependence. Appl. Phys. Express **6**, 075501 (2013)
107. W. Seifert, D. Asoli, Z. Bi, Nitride nanowires and method of producing such, (Google Patents, 2010).
108. X. Chen, B. Gayral, D. Sam-Giao, C. Bougerol, C. Durand, J. Eymery, Catalyst-free growth of high-optical quality GaN nanowires by metal-organic vapour phase epitaxy. Appl. Phys. Lett. **99**, 251910 (2011)
109. S. Haffouz, B. Beaumont, P. Gibart, Effect of magnesium and silicon on the lateral overgrowth of GaN patterned substrates by metal organic vapor phase epitaxy. MRS Internet J. Nitride Semicond. Res. **3**, e8 (1998)
110. R. Koester et al., M-plane core–shell InGaN/GaN multiple-quantum-wells on GaN wires for electroluminescent devices. Nano Lett. **11**, 4839 (2011)
111. P. Tchoulfian, F. Donatini, F. Levy, B. Amstatt, A. Amstatt, P. Dussaigne, E.B. Ferret, J. Pernot, Thermoelectric and micro-Raman measurements of carrier density and mobility in heavily Si-doped GaN wires. Appl. Phys. Lett. **103**, 202101 (2013)
112. B. Foltynski, N. Garro, M. Vallo, M. Finken, C. Giesen, H. Kalisch, A. Vescan, A. Cantarero, M. Heuken, The controlled growth of GaN microrods on Si (111) substrates by MOCVD. J. Cryst. Growth **414**, 200 (2015)
113. Y.-T. Lin, T.-W. Yeh, P.D. Dapkus, Mechanism of selective area growth of GaN nanorods by pulsed mode metalorganic chemical vapor deposition. Nanotechnology **23**, 465601 (2012)
114. B.O. Jung, S.-Y. Bae, Y. Kato, M. Imura, D.-S. Lee, Y. Honda, H. Amano, Morphology development of GaN nanowires using a pulsed-mode MOCVD growth technique. Cryst. Eng. Commun. **16**, 2273 (2014)
115. J. Eymery, D. Salomon, X. Chen, C. Durand, Method of selective growth without catalyst on a semiconducting structure, (Google Patents, 2015)
116. C. Tessarek, M. Heilmann, E. Butzen, A. Haab, H. Hardtdegen, C. Dieker, E. Spiecker, S. Christiansen, The role of Si during the growth of GaN micro-and nanorods. Cryst. Growth Des. **14**, 1486 (2014)

117. S. Haffouz, H. Lahreche, P. Vennégués, P. De Mierry, B. Beaumont, F. Omnes, P. Gibart, The effect of the Si/N treatment of a nitridated sapphire surface on the growth mode of GaN in low-pressure metalorganic vapor phase epitaxy. Appl. Phys. Lett. **73**, 1278 (1998)
118. S. Salomon, J. Eymery, E. Pauliac-Vaujour, GaN wire-based Langmuir–Blodgett films for self-powered flexible strain sensors. Nanotechnology **25**, 375502 (2014)
119. S. Li et al., Nitrogen-polar core-shell GaN light-emitting diodes grown by selective area metalorganic vapor phase epitaxy. Appl. Phys. Lett. **101**, 032103 (2012)
120. X. Wang, J. Hartmann, M. Mandl, M. Sadat Mohajerani, H.-H. Wehmann, M. Strassburg, A. Waag, Growth kinetics and mass transport mechanisms of GaN columns by selective area metal organic vapor phase epitaxy. J. Appl. Phys. **115**, 163104 (2014)
121. X.J. Chen, J.-S. Hwang, G. Perillat-Merceroz, S. Landis, B. Martin, D.L.S. Dang, J. Eymery, C. Durand, Wafer-scale selective area growth of GaN hexagonal prismatic nanostructures on c-sapphire substrate. J. Cryst. Growth **322**, 15 (2011)
122. T. Schimpke, A. Avramescu, A. Koller, A. Fernando-Saavedra, J. Hartmann, J. Ledig, A. Waag, M. Strassburg, H.-J. Lugauer, The influence of MOVPE growth conditions on the shell of core-shell GaN microrod structures. J. Cryst. Growth **465**, 34 (2017)
123. H.M. Kim, T.W. Kang, K.S. Chung, Nanoscale ultraviolet- light- emitting diodes using wide-bandgap gallium nitride nanorods. Adv. Mater. **15**, 567 (2003)
124. H.-M. Kim, Y.-H. Cho, H. Lee, S.I. Kim, S.R. Ryu, D.Y. Kim, T.W. Kang, K.S. Chung, High-brightness light emitting diodes using dislocation-free indium gallium nitride/gallium nitride multiquantum-well nanorod arrays. Nano Lett. **4**, 1059 (2004)
125. A. Kikuchi, M. Kawai, M. Tada, K. Kishino, InGaN/GaN multiple quantum disk nanocolumn light-emitting diodes grown on (111) Si substrate. Jpn. J. Appl. Phys. **43**, L1524 (2004)
126. K. Kishino, A. Kikuchi, H. Sekiguchi, S. Ishizawa, InGaN/GaN nanocolumn LEDs emitting from blue to red. Proc. SPIE **2017**, 64730T (2007)
127. Z. Zhong, F. Qian, D. Wang, C.M. Lieber, Synthesis of p-type gallium nitride nanowires for electronic and photonic nanodevices. Nano Lett. **3**, 343 (2003)
128. A. Waag et al., The nanorod approach: GaN NanoLEDs for solid state lighting. Phys. Status Solidi C **8**, 2296 (2011)
129. Y.-H. Ra, R. Navamathavan, J.-H. Park, C.-R. Lee, High-quality uniaxial in x Ga1–x N/GaN multiple quantum well (MQW) nanowires (NWs) on Si (111) grown by metal-organic chemical vapor deposition (MOCVD) and light-emitting diode (LED) fabrication. ACS Appl. Mater. Interfaces **5**, 2111 (2013)
130. L. Rigutti et al., Correlation of microphotoluminescence spectroscopy, scanning transmission electron microscopy, and atom probe tomography on a single nano-object containing an InGaN/GaN multiquantum well system. Nano Lett. **14**, 107 (2013)
131. R. Koester et al., High-speed GaN/GaInN nanowire array light-emitting diode on silicon (111). Nano Lett. **15**, 2318 (2015)
132. Y.-H. Ra, R. Navamathavan, J.-H. Park, C.-R. Lee, Coaxial In x Ga1–x N/GaN multiple quantum well nanowire arrays on Si (111) substrate for high-performance light-emitting diodes. Nano Lett. **13**, 3506 (2013)
133. H.-S. Chen, Y.-F. Yao, C.-H. Liao, C.-G. Tu, C.-Y. Su, W.-M. Chang, Y.-W. Kiang, C. Yang, Light-emitting device with regularly patterned growth of an InGaN/GaN quantum-well nanorod light-emitting diode array. Opt. Lett. **38**, 3370 (2013)
134. N. Anttu, Modifying the emission of light from a semiconductor nanowire array. J. Appl. Phys. **120**, 043108 (2016)
135. K. Kishino, J. Kamimura, K. Kamiyama, Near-infrared InGaN nanocolumn lightemitting diodes operated at 1.46 μm. Appl. Phys. Express **5**, 031001 (2012)
136. R. Vadivelu, Y. Igawa, K. Kishino, 633 nm red emissions from InGaN nanocolumn lightemitting diode by radio frequency plasma assisted molecular beam epitaxy. Jpn. J. Appl. Phys. **52**, 08JE18 (2013)
137. M. Nami, R.F. Eller, S. Okur, A.K. Rishinaramangalam, S. Liu, I. Brener, D.F. Feezell, Tailoring the morphology and luminescence of GaN/InGaN core–shell nanowires using bottomup selective-area epitaxy. Nanotechnology **28**, 025202 (2016)

138. A. Yanagihara, S. Ishizawa, K. Kishino, Directional radiation beam from yellow-emitting InGaN-based nanocolumn LEDs with ordered bottom-up nanocolumn array. Appl. Phys. Express **7**, 112102 (2014)
139. H. Hayashi, D. Fukushima, T. Noma, D. Tomimatsu, Y. Konno, M. Mizuno, K. Kishino, Thermally engineered flip-chip InGaN/GaN well-ordered nanocolumn array LEDs. IEEE Photon. Technol. Lett. **27**, 2343 (2015)
140. F. Limbach et al., Current path in light emitting diodes based on nanowire ensembles. Nanotechnology **23**, 465301 (2012)
141. A. Bavencove et al., Submicrometre resolved optical characterization of green nanowirebased light emitting diodes. Nanotechnology **22**, 345705 (2011)
142. S. Jahangir, M. Mandl, M. Strassburg, P. Bhattacharya, Molecular beam epitaxial growth and optical properties of red-emitting (λ= 650 nm) InGaN/GaN disks-in-nanowires on silicon. Appl. Phys. Lett. **102**, 071101 (2013)
143. T. Andreev, E. Monroy, B. Gayral, B. Daudin, N.Q. Liem, Y. Hori, M. Tanaka, O. Oda, L. Si Dang, Eu locations in Eu-doped In Ga N/ Ga N quantum dots. Appl. Phys. Lett. **87**(2), 021906 (2005)
144. H. Sekiguchi, T. Imanishi, R. Matsuzaki, K. Ozaki, K. Yamane, H. Okada, K. Kishino, A. Wakahara, Stable-wavelength operation of europium-doped GaN nanocolumn light-emitting diodes grown by rf-plasma-assisted molecular beam epitaxy. Electron. Lett. **53**, 666 (2017)
145. R. Wang, X. Liu, I. Shih, Z. Mi, High efficiency, full-color AlInGaN quaternary nanowire light emitting diodes with spontaneous core-shell structures on Si. Appl. Phys. Lett. **106**, 261104 (2015)
146. R. Wang, H.P. Nguyen, A.T. Connie, J. Lee, I. Shih, Z. Mi, Color-tunable, phosphorfree InGaN nanowire light-emitting diode arrays monolithically integrated on silicon. Opt. Express **22**, A1768 (2014)
147. H.P.T. Nguyen et al., Engineering the carrier dynamics of InGaN nanowire white lightemitting diodes by distributed p-AlGaN electron blocking layers. Sci. Rep. **5**, 14745 (2015)
148. M.R. Philip, D.D. Choudhary, M. Djavid, M.N. Bhuyian, J. Piao, T.T. Pham, D. Misra, H.P. Nguyen, Controlling color emission of InGaN/AlGaN nanowire light-emitting diodes grown by molecular beam epitaxy. J. Vac. Sci. Technol. B: Nanotechnol. Microelectron. Mater. Process. Meas. Phenom. **35**, 02B108 (2017)
149. C. Zhao et al., An enhanced surface passivation effect in InGaN/GaN disk-in-nanowire light emitting diodes for mitigating Shockley–Read–Hall recombination. Nanoscale **7**, 16658 (2015)
150. S. Sadaf, Y.-H. Ra, H. Nguyen, M. Djavid, Z. Mi, Alternating-current InGaN/GaN tunnel junction nanowire white-light emitting diodes. Nano Lett. **15**, 6696 (2015)
151. J. Simon, Z. Zhang, K. Goodman, H. Xing, T. Kosel, P. Fay, D. Jena, Polarizationinduced Zener tunnel junctions in wide-band-gap heterostructures. Phys. Rev. Lett. **103**, 026801 (2009)
152. S. Krishnamoorthy, D.N. Nath, F. Akyol, P.S. Park, M. Esposto, S. Rajan, Polarizationengineered GaN/InGaN/GaN tunnel diodes. Appl. Phys. Lett. **97**, 203502 (2010)
153. M.J. Grundmann, U.K. Mishra, Multi- color light emitting diode using polarizationinduced tunnel junctions. Phys. Status Solidi C **4**, 2830 (2007)
154. A. Sarwar, S.D. Carnevale, F. Yang, T.F. Kent, J.J. Jamison, D.W. McComb, R.C. Myers, Semiconductor nanowire light- emitting diodes grown on metal: A direction toward large-scale fabrication of nanowire devices. Small **11**, 5402 (2015)
155. C. Zhao et al., Droop-free, reliable, and high-power InGaN/GaN nanowire light-emitting diodes for monolithic metal-optoelectronics. Nano Lett. **16**, 4616 (2016)
156. C. Zhao, T.K. Ng, N. Wei, A. Prabaswara, M.S. Alias, B. Janjua, C. Shen, B.S. Ooi, Facile formation of high-quality InGaN/GaN quantum-disks-in-nanowires on bulk-metal substrates for high-power light-emitters. Nano Lett. **16**, 1056 (2016)
157. B.J. May, A.G. Sarwar, R.C. Myers, Nanowire LEDs grown directly on flexible metal foil. Appl. Phys. Lett. **108**, 141103 (2016)

158. K. Bando, K. Sakano, Y. Noguchi, Y. Shimizu, Development of high-bright and purewhite LED lamps. J. Light. Vis. Environ. **22**, 1_2 (1998)
159. Y. Hu, W. Zhuang, H. Ye, D. Wang, S. Zhang, X. Huang, A novel red phosphor for white light emitting diodes. J. Alloys Compd. **390**, 226 (2005)
160. M. Yamada, Y. Narukawa, T. Mukai, Phosphor free high-luminous-efficiency white light-emitting diodes composed of InGaN multi-quantum well. Jpn. J. Appl. Phys. **41**, L246 (2002)
161. B. Damilano, N. Grandjean, C. Pernot, J. Massies, Monolithic white light emitting diodes based on InGaN/GaN multiple-quantum wells. Jpn. J. Appl. Phys. **40**, L918 (2001)
162. D. Yang, L. Wang, W.-B. Lv, Z.-B. Hao, Y. Luo, Growth and characterization of phosphor-free white light-emitting diodes based on InGaN blue quantum wells and green–yellow quantum dots. Superlattice. Microst. **82**, 26 (2015)
163. B. Damilano, B. Gil, Yellow–red emission from (Ga, In) N heterostructures. J. Phys. D. Appl. Phys. **48**, 403001 (2015)
164. D. Schiavon, M. Binder, A. Loeffler, M. Peter, Optically pumped GaInN/GaN multiple quantum wells for the realization of efficient green light-emitting devices. Appl. Phys. Lett. **102**, 113509 (2013)
165. W. Guo, A. Banerjee, P. Bhattacharya, B.S. Ooi, InGaN/GaN disk-in-nanowire white light emitting diodes on (001) silicon. Appl. Phys. Lett. **98**, 193102 (2011)
166. H.P.T. Nguyen, K. Cui, S. Zhang, M. Djavid, A. Korinek, G.A. Botton, Z. Mi, Controlling electron overflow in phosphor-free InGaN/GaN nanowire white light-emitting diodes. Nano Lett. **12**, 1317 (2012)
167. H.P.T. Nguyen, S. Zhang, A.T. Connie, M.G. Kibria, Q. Wang, I. Shih, Z. Mi, Breaking the carrier injection bottleneck of phosphor-free nanowire white light-emitting diodes. Nano Lett. **13**, 5437 (2013)
168. Z. Mi, H. Nguyen, M. Djavid, S. Zhang, A. Connie, S. Sadaf, Q. Wang, S. Zhao, I. Shih, High power phosphor-free InGaN/GaN/AlGaN core-shell nanowire white light emitting diodes on Si substrates. ECS Trans. **61**, 9 (2014)
169. S.-I. Inoue, T. Naoki, T. Kinoshita, T. Obata, H. Yanagi, Light extraction enhancement of 265 nm deep-ultraviolet light-emitting diodes with over 90 mW output power via an AlN hybrid nanostructure. Appl. Phys. Lett. **106**, 131104 (2015)
170. Y.K. Ooi, C. Liu, J. Zhang, Analysis on polarization-dependent light extraction and effect of passivation layer for 230 nm AlGaN nanowire light-emitting diodes. IEEE Photon. J. **9**, 1–12 (2017)
171. H.-Y. Ryu, Large enhancement of light extraction efficiency in AlGaN-based nanorod ultraviolet light-emitting diode structures. Nanoscale Res. Lett. **9**, 58 (2014)
172. H. Sekiguchi, K. Kato, J. Tanaka, A. Kikuchi, K. Kishino, Ultraviolet GaN- based nanocolumn light- emitting diodes grown on n- (111) Si substrates by rf- plasma- assisted molecular beam epitaxy. Phys. Status Solidi A **205**, 1067 (2008)
173. A. Pierret, C. Bougerol, S. Murcia-Mascaros, A. Cros, H. Renevier, B. Gayral, B. Daudin, Growth, structural and optical properties of AlGaN nanowires in the whole composition range. Nanotechnology **24**, 115704 (2013)
174. C. Himwas, M. Den Hertog, L.S. Dang, E. Monroy, R. Songmuang, Alloy inhomogeneity and carrier localization in AlGaN sections and AlGaN/AlN nanodisks in nanowires with 240–350 nm emission. Appl. Phys. Lett. **105**, 241908 (2014)
175. G. Jacopin et al., Photoluminescence polarization in strained GaN/AlGaN core/shell nanowires. Nanotechnology **23**, 325701 (2012)
176. C. Durand et al., M-Plane GaN/InAlN multiple quantum wells in core–shell wire structure for UV emission. ACS Photon. **1**, 38 (2013)
177. S. Zhao, M. Djavid, Z. Mi, Surface emitting, high efficiency near-vacuum ultraviolet light source with aluminum nitride nanowires monolithically grown on silicon. Nano Lett. **15**, 7006 (2015)
178. S. Zhao, S. Woo, M. Bugnet, X. Liu, J. Kang, G. Botton, Z. Mi, Three-dimensional quantum confinement of charge carriers in self-organized AlGaN nanowires: A viable route to electrically injected deep ultraviolet lasers. Nano Lett. **15**, 7801 (2015)

179. Z. Mi et al., Molecular beam epitaxial growth and characterization of Al (Ga) N nanowire deep ultraviolet light emitting diodes and lasers. J. Phys. D. Appl. Phys. **49**, 364006 (2016)
180. S. Sadaf, S. Zhao, Y. Wu, Y.-H. Ra, X. Liu, S. Vanka, Z. Mi, An AlGaN core–shell tunnel junction nanowire light-emitting diode operating in the ultraviolet-C band. Nano Lett. **17**, 1212 (2017)
181. D.A. Laleyan, S. Zhao, S.Y.-M. Woo, H.N. Tran, H.B. Le, T. Szkopek, H. Guo, G.A. Botton, Z. Mi, AlN/h-BN heterostructures for Mg dopant-free deep ultraviolet photonics. Nano Lett. **17**, 3738–3743 (2017)
182. S.-C. Zhu et al., Enhancement of the modulation bandwidth for GaN-based light-emitting diode by surface plasmons. Opt. Express **23**(11), 13752 (2015)
183. W. Pfaff, A. Vos, R. Hanson, Top-down fabrication of plasmonic nanostructures for deterministic coupling to single quantum emitters. J. Appl. Phys. **113**, 024310 (2013)
184. F. Qian, Y. Li, S. Gradečak, H.-G. Park, Y. Dong, Y. Ding, Z.L. Wang, C.M. Lieber, Multi-quantum-well nanowire heterostructures for wavelength-controlled lasers. Nat. Mater. **7**, 701 (2008)
185. C. Mounir, T. Schimpke, G. Rossbach, A. Avramescu, M. Strassburg, U.T. Schwarz, Optical properties and internal quantum efficiency of InGaN/GaN core-shell microrods for solid state lighting. J. Appl. Phys. **120**, 155702 (2016)
186. J. Lee, X. Li, X. Ni, Ü. Özgür, H. Morkoç, T. Paskova, G. Mulholland, K. Evans, On carrier spillover in c-and m-plane InGaN light emitting diodes. Appl. Phys. Lett. **95**, 201113 (2009)
187. T. Sano, T. Doi, S.A. Inada, T. Sugiyama, Y. Honda, H. Amano, T. Yoshino, High internal quantum efficiency blue-green light-emitting diode with small efficiency droop fabricated on low dislocation density GaN substrate. Jpn. J. Appl. Phys. **52**, 08JK09 (2013)
188. C. Kölper, M. Sabathil, F. Römer, M. Mandl, M. Strassburg, B. Witzigmann, Core–shell InGaN nanorod light emitting diodes: Electronic and optical device properties. Phys. Status Solidi A **209**, 2304 (2012)
189. G. Tourbot et al., Growth mechanism and properties of InGaN insertions in GaN nanowires. Nanotechnology **23**, 135703 (2012)
190. T. Krause et al., Nanofocus x-ray diffraction and cathodoluminescence investigations into individual core–shell (In, Ga) N/GaN rod light-emitting diodes. Nanotechnology **27**, 325707 (2016)
191. C. Mounir, T. Schimpke, G. Rossbach, A. Avramescu, M. Strassburg, U.T. Schwarz, Polarization-resolved micro-photoluminescence investigation of InGaN/GaN core-shell microrods. J. Appl. Phys. **121**, 025701 (2017)
192. M. Shahmohammadi, J.-D. Ganière, H. Zhang, R. Ciechonski, G. Vescovi, O. Kryliouk, M. Tchernycheva, G. Jacopin, Excitonic diffusion in InGaN/GaN core–shell nanowires. Nano Lett. **16**, 243 (2015)
193. C.-K. Li, H.-C. Yang, T.-C. Hsu, Y.-J. Shen, A.-S. Liu, Y.-R. Wu, Three dimensional numerical study on the efficiency of a core-shell InGaN/GaN multiple quantum well nanowire light-emitting diodes. J. Appl. Phys. **113**, 183104 (2013)
194. M. Mohajerani et al., Evaluation of local free carrier concentrations in individual heavily-doped GaN: Si micro-rods by micro-Raman spectroscopy. Appl. Phys. Lett. **108**, 091112 (2016)
195. V. Hortelano, O. Martínez, R. Cuscó, L. Artús, J. Jiménez, Cathodoluminescence study of Mg activation in non-polar and semi-polar faces of undoped/Mg-doped GaN core–shell nanorods. Nanotechnology **27**, 095706 (2016)
196. P. Tchoulfian, F. Donatini, F. Levy, A. Dussaigne, P. Ferret, J. Pernot, Direct imaging of p–n junction in core–shell GaN wires. Nano Lett. **14**, 3491 (2014)
197. A. Sugimoto, H. Ochi, S. Fujimura, A. Yoshida, T. Miyadera, M. Tsuchida, Flexible OLED displays using plastic substrates. IEEE J. Sel. Top. Quant. Electron. **10**, 107 (2004)
198. D. Kondakov, W. Lenhart, W. Nichols, Operational degradation of organic light-emitting diodes: Mechanism and identification of chemical products. J. Appl. Phys. **101**, 024512 (2007)

199. A. Fallahi, F.A. Taromi, A. Mohebbi, J.D. Yuen, M. Shahinpoor, A novel ambipolar polymer: from organic thin-film transistors to enhanced air-stable blue light emitting diodes. J. Mater. Chem. C **2**, 6491 (2014)
200. S.I. Park, A.P. Le, J. Wu, Y. Huang, X. Li, J.A. Rogers, Light emission characteristics and mechanics of foldable inorganic light- emitting diodes. Adv. Mater. **22**, 3062 (2010)
201. N. Guan, X. Dai, A. Babichev, F.H. Julien, M. Tchernycheva, Flexible inorganic light emitting diodes based on semiconductor nanowires. Chem. Sci. **8**, 7904–7911 (2017)
202. Z. Fan et al., Three-dimensional nanopillar-array photovoltaics on low-cost and flexible substrates. Nat. Mater. **8**, 648 (2009)
203. J.M. Spurgeon, S.W. Boettcher, M.D. Kelzenberg, B.S. Brunschwig, H.A. Atwater, N.S. Lewis, Flexible, polymer- supported, Si wire array photoelectrodes. Adv. Mater. **22**, 3277 (2010)
204. K.E. Plass, M.A. Filler, J.M. Spurgeon, B.M. Kayes, S. Maldonado, B.S. Brunschwig, H.A. Atwater, N.S. Lewis, Flexible polymer- embedded Si wire arrays. Adv. Mater. **21**, 325 (2009)
205. M.E. Reimer, G. Bulgarini, N. Akopian, M. Hocevar, M.B. Bavinck, M.A. Verheijen, E.P. Bakkers, L.P. Kouwenhoven, V. Zwiller, Bright single-photon sources in bottom-up tailored nanowires. Nat. Commun. **3**, 737 (2012)
206. M. Willander, O. Nur, Q.X. Zhao, L.L. Yang, M. Lorenz, B.Q. Cao, J.Z. Pérez, C. Czekalla, G. Zimmermann, M. Grundmann, A. Bakin, A. Behrends, M. Al-Suleiman, A. El-Shaer, A.C. Mofor, B. Postels, A. Waag, N. Boukos, A. Travlos, H.S. Kwack, J. Guinard, D.L.S. Dang, Zinc oxide nanorod based photonic devices: recent progress in growth, light emitting diodes and lasers. Nanotechnology **20**, 332001 (2009)
207. A. Nadarajah, R.C. Word, J. Meiss, R. Könenkamp, Flexible inorganic nanowire lightemitting diode. Nano Lett. **8**, 534 (2008)
208. H. Li, G. Zhao, L. Wang, Z. Chen, S. Yang, Morphology controlled fabrication of InN nanowires on brass substrates. Nanomaterials **6**, 195 (2016)
209. H.K. Park, S.W. Yoon, Y.J. Eo, W.W. Chung, G.Y. Yoo, J.H. Oh, K.N. Lee, W. Kim, Y.R. Do, Horizontally assembled green InGaN nanorod LEDs: scalable polarized surface emitting LEDs using electric-field assisted assembly. Sci. Rep. **6**, 28312 (2016)
210. M.C. McAlpine, R.S. Friedman, S. Jin, K.-H. Lin, W.U. Wang, C.M. Lieber, Highperformance nanowire electronics and photonics on glass and plastic substrates. Nano Lett. **3**, 1531 (2003)
211. N. Guan et al., Flexible white light emitting diodes based on nitride nanowires and nanophosphors. ACS Photon. **3**, 597 (2016)

Chapter 13
Light-Emitting Diodes for Healthcare and Well-being

Ying Gu, Haixia Qiu, Ying Wang, Naiyan Huang, and Timon Cheng-Yi Liu

13.1 Basic Theories and Mechanisms of LED Phototherapy

13.1.1 Basic Theories and Mechanisms of Low-Level Light Therapy

13.1.1.1 Development History of LED for Low-Level Light Therapy

In 1903, the Danish scientist Finsen pioneered in treating skin tuberculosis with ultraviolet light, which turned over a new chapter for modern phototherapy. With the advancement of science and technology, the light source of phototherapy has experienced the development from sunlight, wide spectrum artificial light source to single spectral line laser light source, which makes the treatment more effective. At present, due to its high light intensity and better monochromaticity and coherence, laser has become the best source of monochromatic light, which in recent years has made remarkable breakthroughs in the fields of treatment. Nevertheless, the available laser wavelengths for clinical use are still limited at present.

In recent years, with the rapid progress of LED production, especially in the aspects of lighting intensity, wavelength, and price, LED now has more advantages than laser. The LED lighting efficiency has been improved markedly. For instance, the lighting efficiency of an orange red LED of a wavelength of 620 mn reaches 100 lm/W with a luminous flux amounting to that of 4mW helium–neon gas laser, which can meet the ordinary clinical demands.In respects of wavelength,

Y. Gu (✉) · H. Qiu · Y. Wang · N. Huang
Chinese PLA General Hospital, Beijing, China

T. C.-Y. Liu
South China Normal University, Guangzhou, China

J. Li, G. Q. Zhang (eds.), *Light-Emitting Diodes*, Solid State Lighting Technology and Application Series 4, https://doi.org/10.1007/978-3-319-99211-2_13

practicability, and price, the LED far outperforms laser. Laser can only emit the light of a few different wavelengths, but LED spans nearly the whole visible light region. With regard to practicability, due to its simple structure, small power, lower consumption, long lifetime, reliability, and easy promotion, the price of LED is much less than that of any kind of laser in the market. For these reasons, LED has promising prospects and the higher practicability in the fields of medical treatment. With the further development of LED chip technology, the increasing light intensity of LED units, extension of the LED spectrum range from ultraviolet to infrared, the appearance of more LED wavelengths, and LED phototherapy will certainly play a more and more important role in the area of low-intensity phototherapy.

The application of LED in the field of medical treatment can be traced back to the treatment of neonatal jaundice. In 1968, Lucy made use of light to cure the jaundice of premature infants for the first time, but the limitation of light source resulted in a low treatment efficiency. Since 2000, blue LED has been utilized in the treatment of neonatal jaundice, and gradually replaced the traditional method of soft laser therapy. At present, blue LED has been recommended by the American Academy of Pediatrics as the best light source for the treatment of neonatal unconjugated hyperbilirubinemia. In 2001, in the process of curing leukemia pediatric patients by Whelan H. T. and some people, prior to the bone marrow implantation matched with their cell antigen, the patients had to accept radiotherapy and chemotherapy for the maximum dose to kill the neoplastic bone marrow in the bodies. However, the drugs used in the chemotherapy and radiotherapy will indiscriminately kill rapidly dividing cells, such as oral cavity membrane cells and gastrointestinal tract cells, which will cause serious negative gastrointestinal effects. As a result, the patients given the chemotherapy usually suffered acute oral ulcer and food dysphagia. On the other hand, once these patients were given the last chemotherapy and 688mn LED 4 J/cm^2 per day, 47% of them were relieved of oral pain, and the healing rate of oral ulcers was better than expected. In 2012, it was reported that some Japanese scholars discovered that LED light can assuage bone metabolism disorder in children with cerebral palsy. Subsequently, in 2014, it was reported by the Korean University in South Korea that 635 nm LED could reduce the resorption induced by osteoclasts, and mediate the homeostasis of bones. In 2013, Park KY used 830 nm LED light source to treat acute herpes zoster ophthalmicus, confirming that LED phototherapy can promote wound healing and reduce pain and discovering that LED phototherapy can reduce pigmentation and scars in the wake of inflammation, and LED light source also has regulatory effect on the inflammatory immune system. In 2016, Sohn, H. and M. and others found that 635 nm red light LED can inhibit osteoclast activity and bone resorption and delay the progress of osteoporosis.

13.1.1.2 Interactivity Between Light and Human Tissues

When shining on the tissues of a human body, most of the light will be absorbed by the tissues except some being reflected or penetrating the tissues. Following the absorption of the photons in human tissues, the light energy will be transformed into other forms of energy, such as thermal energy, mechanical energy, and chemical energy. A series of biological effects will be induced, such as thermal effect, mechanical effect, and photochemical reaction. The absorption of photons in human tissues can also cause photon re-emission. The photons haven't undergone energy exchange with human tissues remain constant regarding frequency, and the spatial distribution of these photons is also closely related to the structure and function of the tissues.

13.1.1.3 Light Dosiology for LED Used in Low-Level Light Therapy

The main parameters of LED mediated low-level light therapy are wavelength, average output, power density, irradiance or treatment time, energy density, accumulated energy, output mode, etc. In terms of the wavelengths, nanometer (nm) is the main unit. The available wavelengths for LED cover the visible lights and parts of the invisible lights. The power of LED is usually between 1 and 1000 mW. For LLLT, a few or dozens of mW is usually used. The main unit for power density is mW/cm^2, which can also be called irradiance. The output modes include continuous wave and pulsed wave. If the light beam is pulsed wave, the reported power should be average power. The calculation is as follows: peak power (W) $\times$ pulse width (s) $\times$ pulse frequency (Hz) $=$ average power (mW). Talking about dose–effect relationship, there is a two-way stoichiometric reaction between irradiation time and energy density. If the light does not have enough energy density or the irradiation time is too short, no effect will appear. On the other hand, should the energy density be too high or the irradiation time too long, the response may be suppressed. Nevertheless, there is always a best combination of energy density and irradiation time in certain place, which will produce the best effect. This two track reaction can be traced back into an article published in 1887 by Hugo Schultz, after which Rudolf introduced his own theory that lower dose of light irradiation will bring in stimulating effect and higher dose inhibiting effect [1]. The division of high and low doses hinges on the change of biological structure and function.

13.1.1.4 Mechanisms of Low-Level Light Therapy

According to current researches, it is known that low-level light therapy mainly leads to functional changes at cellular, subcellular, and molecular levels. Its role is complex due to the diversity of parameters and observational indicators and the lack of consistency in the results of studies on the dose–effect relationship. In terms of mechanism researches, since the cellular signals and pathways involved in low-level light therapy have not yet been fully explored, non-invasive detection techniques that can accurately detect functional changes at cellular and molecular levels are very limited, which restricts the exploration of the mechanisms of low-level light therapy. Currently, despite the existence of some hypotheses, the exact mechanisms have not yet been fully elucidated.

13.1.1.4.1 Photobiomodulation

Low-level light therapy is also called photobiomodulation therapy. The mechanism of photobiomodulation (PBM) lies in the fact that non-ionizing optical radiation in the visible and near-infrared spectral range can be absorbed by endogenous chromophores to elicit photophysical and photochemical events at various biological scales.

Firstly, there is a set of parameters (wavelength, irradiance, energy density, pulsed vs continuous wave, exposure time, duty cycle–on time/dark period, frequency of exposure, time interval between exposures, etc.) that determine the types of the cellular effects as well as the relative intensity and duration of these effects.

Secondly, PBM modulates the activities of cells by stimulating/inhibiting or activating/deactivating natural cellular processes. There are multiple mechanisms of action for photobiomodulation of living cells. Mitochondrial mechanism is the most recognized mechanism, which will be introduced in the following part.

Thirdly, different cell types may respond differently to the same photomodulation parameters. Homeostasis is one of the most remarkable and typical properties of a highly complex open biosystem [2]. It is a negative-feedback response of a biosystem to maintain constant conditions inside the biosystem. It has been extended to be function-specific homeostasis (FSH) by Liu et al. [3–5]. An FSH is a negative-feedback response of a biosystem to maintain the function-specific conditions inside the biosystem so that the function is perfectly performed. The functions are finely categorized into two types: the normal and dysfunctional. The transformation of a function from a dysfunctional one to a normal one is called the normalization of the dysfunctional function. It has been found that PBM can promote the normalization of a dysfunctional function [6]. The FSH has been quantitatively represented with quantitative difference (QD) by Liu et al. [7–9].

13.1.1.4.2 Mitochondrial Mechanism

There were many theories on the precise molecular mechanism of PBM. Among them, the cytochrome c oxidase (COX) theory was very popular [1, 10–13]. Cytochrome c and COX represent the terminal step of the electron transport chain in mitochondria, the proposed rate-limiting reaction in mammals. Cytochrome c and COX show unique regulatory features including allosteric regulation, isoform expression, and regulation through cell signaling pathways [14]. The COX theory [11] suggested that COX in mitochondria was the primary photoacceptor upon LLLT exposure of cells, and PBM was mediated by LLLT increased COX activities. The LLLT increased COX activity may be mediated by low-level mitochondrial reactive oxygen species (ROS). It was demonstrated by Wu et al. [15] that the initial reaction after photon absorption of moderate intensity monochromatic light was the photosensitization of COX, which would inhibit enzymatic activity of COX in situ and cause respiratory chain ROS burst.

On the other hand, low-level mitochondrial ROS has been found with the ability to keep cells alive under stressful conditions [16, 17]. High level mitochondrial ROS may induce apoptosis through the intrinsic pathway to protect from severely damaged cells as being observed by Wu et al. [15]. However, Yee et al. [17] demonstrated that sensing of low-level mitochondrial ROS with apoptotic pathway would, independently of apoptosis, elicit the protective mechanism that keeps the nematode alive under stressful conditions and extend the nematode longevity. De Haes et al. [16] showed that metformin extended life span through the process of mitohormesis and proposed a signaling cascade in which metformin-induced production of ROS increased overall life expectancy of nematodes. They also revealed that this beneficial signal of the mitohormetic pathway was propagated by the peroxiredoxin PRDX-2. Because of its evolutionary conservation, peroxiredoxin signaling might underlie a general principle of prolongevity signaling.

In summary, COX-induced low-level ROS production in mitochondria under LLLT may promote the normalization of a dysfunctional function.

13.1.1.4.3 Phototransduction

The COX theory on PBM does not universally hold. There are strong evidences suggesting that LLLT acts on mitochondria will promote adenosine triphosphate (ATP) production. It was shown by Drew et al. [18] that an age-associated decline in ATP content and rate of ATP production was tissue specific as it occurred in skeletal muscles but not heart, and mitochondrial ATP production was not altered by caloric restriction in both tissues. The observation that nitric oxide (NO) was released from cells during LLLT irradiation has led to a speculation that COX and NO release are linked. Montoro et al. [19] found lipopolysaccharide (LPS)-induced stress resulted in significant increase in NO production by human dental pulp cells without causing damage to cell respiratory metabolism and NO production was significantly reduced when LPS-stressed cells were irradiated with infrared light-emitting diode irradiation irrespective of energy dose delivered. Moreover, the COX theory cannot explain the PBM on erythrocytes [20]. In view of this background, the phototransduction mechanism of PBM was introduced.

Phototransduction is the process by which a photon of light captured by a molecule of visual pigment generates an electrical response in a photoreceptor cell [21]. In this way, the photoreceptor-specific G protein mediates between the membrane visual pigment, rhodopsin, and the effector enzyme, cGMP phosphodiesterase. The possibility of extraocular phototransduction (EPT) was put forward by Campbell and Murphy [22] in their report, according to which 3 h of bright light exposure to the area behind the knee caused phase shifts of the circadian rhythms of both body temperature and saliva melatonin in humans. However, as early as in 1996, Liu et al. [23] have put forward the quasi-hormone model of laser biostimulation from the viewpoint of EPT, which was developed as BIMP [24]. The first EPT phenomenon was observed in low-intensity He–Ne laser irradiation induced respiratory burst of polymorphonuclear neutrophils in our laboratory in 2001 [25]. Since then, many LLLT induced EPT phenomena have been observed [26]. Moreover, BIMP has held for the EPT phenomena. It may be further understood with FSH [5]. A cellular normal function in its FSH can resist not only external perturbations under its threshold but also the activation of other signal transduction pathways so that there should be normal function-specific signal transduction pathways (NSPs). For a cellular function, there are many redundant genes and redundant NSPs. The full expression of one redundant gene may fully activate its redundant NSP to maintain its normal cellular function. The full activation of one redundant NSP maintains the first-order normal function. The synergistic activation of n redundant NSPs maintains the *n*th-order normal function. The more the fully activated redundant NSPs, the higher the fitness of the normal function. A cellular PBM can promote the activation of a partially activated NSP, and can be classified into direct and indirect PBM (dPBM and iPBM). For a cellular dysfunctional function, dense signal transduction pathways are partially activated, but there is no fully activated redundant NPS. A cellular dPBM can promote the activation of one partially activated redundant NSP until it is fully activated and then the dysfunctional function becomes a normal one. For an *n*th-order normal function, n redundant NSPs have been fully and synergistically activated. A cellular iPBM may promote the activation of the partially activated $(n + 1)$th redundant NSP until it is fully activated and then the *n*th-order normal function is transformed into its $(n + 1)$th-order one.

13.1.2 Basic Theories and Mechanisms of LED Mediated Photodynamic Therapy

13.1.2.1 Basic Theories of LED-Mediated Photodynamic Therapy

Photodynamic therapy (PDT) is a promising targeting treatment for both malignant and benign diseases. Its function is based on the reactive oxygen species (ROS) generated through a series of photochemical reaction, which is mediated through the interaction of photosensitizing agents, light and oxygen. Currently, PDT is performed on body surface or via endoscopic and interstitial methods in clinic. The clinical application and basic research mainly focus on anti-tumor PDT, vascular targeted PDT, and anti-microbial PDT.

13.1.2.1.1 Light Sources for Photodynamic Therapy

So far, laser, filtered xenon arc, metal halogen lamp, fluorescent lamp and LED, and even sun light have been used as light sources for PDT. The question which light source is the best for PDT has always been a topic of debate. Laser is the most commonly used PDT light source as it may greatly increase the penetration depth in tissues because of its high intensity and collimation. At the same time, the monochromaticity of laser will enable a concentration of energy during the absorption peak of photosensitizers. Laser may also be transferred to digestive tracts and respiratory tracts via fibers. For these reasons, laser is undoubtedly the most suitable light source for endoscopic and interstitial PDT. In recent years, LED has been used in the PDT for superficial skin mucosal lesion. For example, the combination of LED and tropical applied 5-ALA or MAL (Metvix; Galderma) to treat actinic keratosis (AKs) is effective, which is much less painful than laser guided PDT [27].

LED is more cost-effective than laser because of its low manufacture cost and long service life. Moreover, a LED array usually has a big shiny surface. Its curvature, size, and shape can be designed according to the geometry characteristics of targeted body surface. Lots of portable and battery-equipped wearable LEDs have been applied in clinical PDT or undergoing clinical trials.

13.1.2.1.2 Dosimetry of Photodynamic Therapy

The dosimetry of PDT is more complicated than that of photothermal treatment. Not only light but also photosensitizer, oxygen, and target lesion play important roles in the process of PDT. Single factor (for example, light and photosensitizer) doesn't show linear relationship with the therapeutic effect. Studies have indicated that only when all the above factors match well can PDT achieve the best therapeutic effect. At present, only light dose and photosensitizer dose are adjustable in clinical practice. The dose of a photosensitizer includes the type of the photosensitizer, dosage (concentration) applied, time interval between drug administration and light delivery. The dose of light involves wavelength, power density or irradiance, irradiation time or treatment time, and continuous wave or pulsed mode. As light sources for PDT, the main difference between LED and laser is the penetration depth [28]. It is mainly caused by the different collimation of LED and laser. The collimation of laser enables it to illuminate the lesion surface vertically and generate a large number of forward scattering in the tissue, indicating a deeper penetration. To the contrary, light generated from LED is not collimated. The photons of LED irradiate the lesion surface from various directions, and more backscatter is induced within the tissue. As a result, most of the photons deposit only in the shallow part of the tissue.

The absorption peak of photosensitizer and the depth of target lesion are the main factors guiding the choice of light wavelength. From UV to infrared band, the penetration depth of light in tissue grows with the increase of wavelength. Lights with longer wavelength such as red and infrared are more suitable for tumor targeted PDT. For vascular target PDT, green light and blue light are widely used in clinic and red light is selected sometimes. The selection of light wavelength should be mainly based on the depth of the targeted vessel. The effect of the hemoglobin absorption on light penetration depth and the excitation effect on photosensitizer should also be taken into consideration. In general, the effect of light dose on efficacy/side effect is more significant for PDT with photosensitizer intravenously injected than for the topical PDT. As to topical PDT, the main factor affecting the therapeutic effect is the diffusion depth and concentration gradient of photosensitizer in target tissues.

The comparison between pulsed laser and continuous laser guided PDT in therapeutic efficiency has been controversial. Current clinical experience and experimental results in solution suggest that under the same fluence rate, the PDT effect of pulsed laser of nanosecond pulse is lower than that of continuous laser. For the pulsed laser of microseconds pulse, no obvious difference is observed [29].

In conclusion, the dose–response relationship of PDT is complex. Real-time monitoring is an important mean to improve the effectiveness and accuracy of PDT. Many studies have been carried out with a view to establish non-invasive monitoring platform for the detection of fluence rate, photosensitizer fluorescence intensity, oxygen content, and the signals of single oxygen yield during PDT.

13.1.2.2 Mechanisms of LED-Mediated Photodynamic Therapy

Theoretically, both LED and laser are electromagnetic waves although they are different in collimation and monochromaticity. The mechanisms of PDT reactions are consistent. The following is a brief introduction on the mechanisms of PDT for different biological effects.

13.1.2.2.1 Anti-Tumor Photodynamic Therapy

At present, the mechanism of anti-tumor effect mainly includes: direct killing of tumor cells, tumor vascular damage, inflammatory response, and subsequent anti-tumor immune response. The latter three functions play important roles for a sustainable tumor destruction and long-term control of tumor.

13.1.2.2.1.1 Direct Killing of Tumor Cells

PDT-induced direct killing of tumor cells is mainly divided into three types: necrosis, apoptosis, and autophagy related cell death. The direct damage is confined to the locations where photosensitizers are localized because of the limited oxygen lifetime (approximately 10–320 ns) and the limited diffusion distance (10–55 nm in cells) [30]. For most photosensitizers, cell membrane, mitochondria, and lysosome are the most sensitive subcellular structure of PDT, and other organelles, such as endoplasmic reticulum, microtubules, ribosomes, and nucleuses might also be damaged. In general, the photosensitizers located in mitochondria and endoplasmic reticulum will induce apoptosis, and the photosensitizers located in the cell membrane and lysosomes may induce necrosis. Nevertheless, it is not absolute as mitochondrial damage or high doses of PDT can also lead to necrosis.

13.1.2.2.1.2 Vascular Closure Effect

In the 1960s, scientists discovered the vascular effect of PDT. It was observed that PDT can damage tumor cells as well as cause slowing-down of microvascular blood flow, aggregation of platelet, and shutdown of lumen [31]. Further studies have shown that PDT can damage endothelial cells and shut down microvascular.

13.1.2.2.1.3 Inflammatory Response and Anti-Tumor Immunity

PDT-induced inflammation is initiated from the increase of tumor vascular permeability, which leads to the gathering of circulating neutrophil and a large release of inflammatory mediators, affects inflammatory cells (neutrophils, mast cells, macrophages, etc.) and expands the inflammatory response [32]. Despite the fact that the photosensitizers are mainly located in tumor cells, PDT still have the same effect. Its inflammatory response is mainly activated by the photooxidation damage around the blood vessels, and it also reaches the vascular endothelium through the gradient of chemotaxis. Local tissue edema after PDT is an expression of inflammatory response. Inflammatory cells will quickly gather in large quantities in tumor tissues when the damaged tumor cells are removed. The role of neutrophils is particularly important. Neutrophils can release a large number of oxygen free radicals, peroxidase, and lysosomal enzyme in tumor tissues, which are directly involved in the destruction of tumor tissues. It can also release cell chemotactic agent to raise more inflammatory cells and immune cells to enhance anti-tumor immune inflammatory response and induce the follow-up reactions. The anti-tumor immune response induced by PDT is based on the inflammatory response. There are two types of immunity: innate immunity and acquired immunity.

13.1.2.2.2 Vascular Targeted Photodynamic Therapy

The mechanism of the selective microvascular effect of PDT is based on the difference of the concentration of photosensitizers inside and outside the vessel. The difference will appear in a period of time after intravenous injection of photosensitizers. When light irradiation is initiated, photosensitizers in normal tissues surrounding the vessels will be rapidly photobleached while intravascular photosensitizers being constantly supplied with the blood flow. Then, a large amount of reactive oxygen will be generated only in the vessel. Vascular targeted PDT is applied in the treatment of Port Wine Stains [33] age-related macular degeneration [34] and vascular lesions in gastrointestinal mucous [35].

13.1.2.2.3 Anti-Microbial Photodynamic Therapy

At present, most of the researches on anti-microbial PDT are to explore the anti-microbial effect of various novel photosensitizers. Few studies focus on the mechanisms of anti-microbial effect. It is generally believed that the anti-microbial effect of PDT is mainly induced by the oxidative damage of biological macromolecular (such as protein and lecithin) or functional unit within the microbial cells. The photophysical and photochemical properties of photosensitizers and their targeting towards certain microbial cells are key factors for excellent anti-microbial effect.

13.2 Clinical Application of LED Light

13.2.1 LED Light in Low Level Light Therapy

Low level light therapy (LLLT) makes use of a relative low power density of light to treat diseases with the biological stimulation of light, which will not produce any irreversible damage. Being non-invasive, safe and painless, and, among other things, having no adverse reactions, LLLT has a wide range of clinical applications due to its anti-inflammatory, pain-relieving and detumescent functions and its role in promoting wound healing.

The most commonly used therapeutic light source for LLLT is laser light source, such as semiconductor lasers and He–Ne lasers. Therefore, LLLT is also called low level laser therapy. In recent years, some non-laser light sources like LED have also gradually been used clinically as LLLT light sources. LLLT is a physical therapy highly dependent on irradiation parameters (wavelength, power, power density, energy density, time, etc.) and more clinical evidences are needed to prove whether the efficacy of LED-based LLLT is comparable to that of laser-based LLLT. At present, LED-based LLLT is often used in the areas given below.

13.2.1.1 Neonatal Jaundice

Neonatal jaundice is attributed to abnormal metabolism of bilirubin that improves the neonatal blood bilirubin level, leading to typical emergence of yellow staining of skin, mucous membrane, and sclera. Blue LED is most effective in degrading bilirubin (28%), which is followed by cyan LED (18%), white LED (14%), and green LED (11%) in a decreasing order.

Kumar et al. evaluated whether LED light therapy is superior to conventional light therapy in reducing serum total bilirubin level of postoperative and preterm neonates (wavelength 430–490 nm, power density $\geq$30mW/cm^2) [36]. The result indicated that LED therapy had better therapeutic effect. Mohammadizadeh M compared LEDs with fluorescent phototherapy in the treatment of preterm infants with neonatal jaundice. The duration of phototherapy was the same. The results showed no significant difference was found in the rate of fall of bilirubin between these two methods. However, in terms of side effects, energy saving, service life, and cost, LED phototherapy was better and more promising than conventional phototherapy [37].

13.2.1.2 Emotion Cognitive Impairment

Recent studies highlight that human circadian rhythm systems are LED-sensitive [38]. The efficacy of LED phototherapy to treat seasonal affective disorder (SAD) was tested in a randomized, double-blind, and placebo-controlled multicenter trial. Subjects were 18–65 years old and randomly isolated for a week, The patients were placed 20 in. apart from the LED (1350 Lux visible white with a spectral peak of 464 nm and 564 nm). They were wakened up before 8 am every morning for 30 min' treatment. The observation was conducted 1, 2, 3, and 4 weeks, respectively, after the treatment. The results revealed that the subjects' symptoms were remarkably relieved.

There are evidences proving that three major types of brain diseases, namely traumatic events (stroke, traumatic brain injury, and global ischemia), degenerative diseases (dementia, Alzheimer's disease, and Parkinson's disease), and mental disorders (depression, anxiety, post-traumatic stress disorder) may benefit from the photobiomodulation (PBM) produced by skull illumination. PBM can be used even for cognitive enhancement in healthy people [10]. Schiffer et al. applied LED to patients with major depressive disorder and anxiety by having LED light on their forefronts (wavelength 810 nm, output power 1 W, power density 250 mW/cm^2, 4 min, energy density 60 J/cm^2) [39]. It was found that their Hamilton depression rating scale (HAM-D) and Hamilton anxiety rating scale (HAM-A) improved 2 weeks after a single treatment. According to a study report, the transcranial posterior LED therapy for chronic and mild traumatic brain injury (mTBI) would lead to a significant improvement of cognitive performance [40].

The details on the application of blue narrow-band LED phototherapy to treat SAD by Gena Glickman et al. were as follows: blue LED light, 468 nm, 607 mW/cm^2; red LED, 654 nm, 34 mW/cm^2; 45 min' treatment each morning for 3 week [41]. SIGH-SAD assessment and scoring was performed weekly. The result of the mixed-effect analysis showed that the SIGH-SAD score of the blue short-wavelength light treatment was remarkably higher than that of the red light treatment, and the depressive symptoms of the patients would be reversed when the light intensity was 607 mW/cm^2.The efficacy of the LED treatment may be related to increased cerebral blood flow [42, 43].

13.2.1.3 Wound Healing

Currently, the application of LED in wound healing is gradually increasing, including diabetic foot, refractory ulcer, adjuvant therapy after laser treatment, and surgery. Dixit et al. successfully cured a refractory ulcer about 3.6 cm × 3.8 cm of a 69-year-old male with Klippel-Trenaunay syndrome by using He–Ne laser in combination with LED [44]. The treatment parameters were as follows: LED 660 nm, power density 50 mW/cm^2, 6 min, energy density 22.8 J/cm^2; He–Ne laser 632.8 nm, 15 min, energy density 1.58 J/cm^2; twice a week; shift application of LED and He–Ne laser at an interval of 10 min. After about 7 months' treatment, the ulcers completely healed and no recurrence was found, which was followed up for 8 months. Sutterfield reported that Charcot, a 66-year-old diabetic patient, was treated with LED (4.9 J/cm^2, 1.44 min) and his plantar ulcers was cured completely within 8 weeks after the treatment [45].

LED can effectively promote post-laser and post-operative wound healing. Trelles MA performed LED (633 nm, 20 min, 96 J/cm^2) adjuvant treatment for 5 metatarsophalangeal patients given Er: YAG laser therapy immediately and 2, 6, and 10 days after the laser treatment, respectively. On the sixth day after the operation, the wounds were significantly reduced and granulation tissues formed [46].

13.2.1.4 Acne

Acne is a common sebaceous gland disease with a prevalence of up to 80% in adolescents. In recent years, LED has been widely used in the treatment of acne. The combination of 415 nm blue and 633 nm red is the most commonly used LED group. Lee et al. treated 24 patients with mild to moderate and severe facial acne with 415 nm blue light (40 mW/cm^2, 48 J/cm^2) and 633 nm red light (80 mW/cm^2, 96 J/cm^2), which was given twice a week for 4 weeks, 20 min each time with one of the LED lights in shifts [47]. The final mean percentages of improvement in non-inflammatory and inflammatory lesions were 34.28% and 77.93%, respectively. In another study [48], subjects were given 8 alternating treatments with LED 415 nm blue light (20 min/time, 48 J/cm^2) and 633 nm red light (20 min/time, 96 J/cm^2), 3 times a week at an interval of 3 days. At week 4 and 12 after the treatment, the average number of lesions was reduced by 46% and 81%, respectively, and no apparent side effect was found. Sadick evaluated the efficacy of a combination of 415 nm blue and 830 nm near-infrared LEDs in the treatment of moderate acne [49]. The blue light (40 mW/cm^2) was applied for 20 min firstly and near-infrared light (55 mW/cm^2) was used for 20 min after an interval of 2–3 days, which lasted for 4 weeks. The results showed that the IR/blue LED combination did not improve the inflammatory lesions better than the red–blue LED combination, but had better efficacy on non-inflammatory lesions.

Compared with drugs and other treatments, LED has unique advantages in acne treatment. de Arruda et al. evaluated the safety and efficacy of blue light therapy and topical 5% benzoyl peroxide on grade II and grade III acne patients in a prospective, randomized, open, and comparative study [50]. The treatment parameters of the therapy were: 407–420 nm, 40 mW/cm^2, twice a week for a total of 8 treatments. The results showed that both the treatments were effective for grade II and III acne, but the adverse effects of light therapy were smaller. Another randomized trial of patients with acne vulgaris in China compared the efficacy of three light treatments for moderate to severe acne vulgaris [51]. Totally 150 patients were randomly divided into three groups and given PDT, IPL, or red–blue LED phototherapy on their right half faces, respectively. The patients were examined 1 and 3 months after the treatment. It was found that LEDs did not behave better than the other two phototherapy methods in terms of lesion clearance, treating course required, and relapse rate. The PDT group had the highest rate of lesion clearance or the highest improvement of moderate acne vulgaris and the shortest average course of treatment a month after the treatment. In the PDT group, only 4 cases developed papules and pustules and no nodular pustular recurred 3 months after the treatment.

13.2.1.5 Facial Rejuvenation

Weiss et al. observed the effect of LEDs on photoaging of skin in a prospective study. A total of 90 patients were included in the study [52]. The parameters were set at: LED wavelength 590 nm, energy density 0.1 J/cm^2, continuous treatment of 8 times for 4 weeks in total. The follow-up was performed 4, 8, 12, and 18 weeks and 6 and 12 months after the treatment, respectively. The digital imaging data showed that 90% of the subjects had reduced signs of photoaging, periorbital wrinkles, erythema, and pigmentation as well as smoother skin texture. The results of optical profilometry revealed a 10% increase in surface topography measurements. Histological data also showed a significant increase in collagen in the papillary dermis of all treated skin specimens.

In a spilt-face single-blind study, Barolet et al. evaluated the efficacy of 660 nm LEDs on 40 patients with skin photoaging [53]. The treatment was given 3 times a week, 4 weeks in total. The results showed that the wrinkle depth and surface roughness decreased in more than 90% of the patients, and the results of blinded assessment also revealed a decrease in severity Fitzpatrick wrinkle score in 87% of the patients.

In a prospective, randomized, placebo-controlled, double-blind clinical study, Lee et al. divided 76 patients with facial wrinkles randomly into four groups, of which the patients were given respectively 830 nm, 633 nm, 830 nm plus 633 nm light treatment and pseudo light treatment twice a week for 4 weeks [54]. The follow-up period was 3 months, during which the efficacy was evaluated via continuous photography, contour determination, and skin elasticity and objective measurement of melanin. The results showed a significant reduction of wrinkles (up to 36%) and increase of skin elasticity (up to 19%) in the three treatment groups based on a comparison to the baselines.

13.2.1.6 Scars

In Barolet's study, prophylactic low-level LED at the wavelength of 805 nm was used to treat keloids and hypertrophic scars or such scars left after carbon dioxide laser ablation. After the initial treatment, each scar of each patient was treated with LED for 15 min a day for 30 days (805 nm, 30 mW/cm^2). The therapeutic effect was determined with VSS score, measurement of scar height, and quantitative assessment of skin morphology photographs, and the result was satisfactory [55]. Park et al. used 830/590 nm LED-LLLT to treat 35 patients with scars attributing to thyroidectomy [56]. From the first day after the surgery, the patients were given 60 J/cm^2 (11 min) treatment per day for 1 week and then 3 times a week for 3 weeks. Colorimetric assessments were performed with tristimulus colorimeter and an assessment on scar formation was performed 1 and 3 months after the thyroidectomy, respectively. The results indicated that LED-LLLT was a safe method to inhibit scar formation after thyroidectomy.

13.2.1.7 Motor Functions

Leal-Junior et al. performed a systematic review with meta-analysis to investigate the effects of low-level laser therapy (LLLT) or narrow-band light-emitting diode therapy (LEDT) in modulating the activity-induced skeletal muscle fatigue or subsequently protecting against muscle injury [57]. Totally 13 randomized controlled trials (RCTs) (4 trials had the LED as light source) were included. Based on a comparison with the placebo, the time until exhaustion increased significantly by 4.12 s (95% CI 1.21–7.02, $p < 0.005$) and number of repetitions by 5.47 (95% CI 2.35–8.59, $p < 0.0006$) after the phototherapy. The most significant and consistent results were found with the application of red or infrared phototherapy before exercises with a power output of 50–200 mW and dose of 5–6 J per point (spot). Foley et al. assessed the effect of 830 nm light-emitting diode (LED) phototherapy on injured university athletes in a pilot study [58]. A total of 395 injuries were treated with 1669 sessions of 830 nm LED phototherapy (mean of 4.3 treatments per injury, range 2–6). The result showed that the average LED-mediated return to play (RTP) in the 65 subjects was 9.6 days, significantly shorter than the mean anticipated RTP of 19.23 days. Eighty-eight patients (78.5%) were either very satisfied or satisfied, and only 8 patients (7.2%) were dissatisfied. One limitation of the present study was the subjective nature of the assessments and the lack of control groups. Further controlled studies are expected to enable the confirmation and generalization of the desirable results from the present studies.

13.2.1.8 Pains

Leal-Junior et al. investigated the effects of the phototherapy combining the light sources of laser (905 nm) and LEDs (875 nm and 640 nm) in treating nonspecific knee pain [57]. Eighty-six patients rated 30 or greater on the pain VAS were recruited and included in the study. The patients in the LLLT group were given 12 treatments of active phototherapy (with 905 nm super-pulsed laser and 875 and 640 nm LEDs) and conventional treatment (physical therapy or chiropractic care), and those in the placebo group were treated in the same way but with placebo phototherapy devices. Pain assessments (VAS) were performed respectively at baseline and the fourth, seventh, and tenth treatments, after the completion of the treatments and at 1-month's follow-up. The results demonstrated that the combination of super-pulsed laser, red, and infrared LEDs was effective to decrease the knee pain of the patients and improve their quality of life.

13.2.2 Application of LED Light Sources in Photodynamic Therapy

The light source to be used for PDT has to meet these demands: (1) the wavelength of the light source should correspond well with the photosensitizer selected to achieve maximum excitation efficiency; (2) it has certain output power to meet the needs of PDT; and (3) it has accurate and convenient method of measurement. Besides a wide range of wavelength and large output powers, LED has such advantages as low cost and flexible structure. It can be predicted that LED will be more and more widely used in PDT.

13.2.2.1 Actinic Keratosis

Actinic keratosis (AK) is a kind of pre-cancerous damage caused by long-term sun exposure and damages to skin. Boasting excellent cosmetic effects, PDT is a first-line method currently recommended for treating actinic keratosis. PDT treatment of AK is an early use of LED as a source of light for PDT.

Pariser et al. evaluated the efficacy of the MAL (methyl-aminolevulinic acid) mediated PDT using red LED light [59]. Totally 49 patients with 363 AK lesions were given the illumination (630 nm, light dose 37 J/cm^2) at a distance of 5–8 cm from the skin for an average period of 8 min after MAL was applied at least 3 h. Repeated treatment was conducted 1 week later. At 3 months after the last treatment, the complete response of MAL-PDT of the lesions and patient were 86.2% and 59.2%, respectively.

In another study, the combined application of 633 nm LED and ALA was used to treat head and facial multiple non-hypertrophic AK. The lesions were exposed to the LED light of an irradiance of 1.3 mW/cm^2 for 45 min with a total fluence of 3.5 J/cm^2. All the 15 patients were improved after the first PDT session and only 2 patients required a second session to obtain satisfactory remission. None of the patients relapsed during the 3 months' follow-up period [60].

13.2.2.2 Acne

PDT is mainly used to treat severe acne at grade III–IV. The combination of 633 nm LED (50–70 J/cm^2, 66 mW/cm^2) and 10% ALA cream is the most commonly used PDT regimen for acne. Twenty-two percent of the patients had excellent effect on their first treatment, 34% on the second treatment, and some patients required more treatments. The treatment had trivial side effects, neither ulcer scar nor other obvious side effects were observed, the symptoms of recurrent patients were mild and the routine local treatment had a good response [61, 62].

In addition, the combination of LED (633 $\pm$ 3 nm, 126 J/cm^2, 66 mw/cm^2) and low concentration of ALA (3.6%) was also quite effective in treating acne. ALA was locally applied for 1.5 h, and the irradiation was given for 20 min at an interval of 2 weeks, 3 times in total. The results showed no difference between the acne of different degrees in therapeutic effect, and the therapy was applicable to moderate to severe acne with mild side effects [63, 64].

The combination of LED with chlorophyll for acne has also been reported. After topical application of chlorophyll for 30 min, blue light (430 ± 10 nm, 600 mW/cm^2, 1170 J/cm^2) combined with red light (660 ± 10 nm, 650 mW/cm^2, 1080 J/cm^2) was used to Illuminate the face for 30 min twice a week for 8 times. The results showed that the acne lesion count, severity grade, and sebum level were significantly lower in the PDT group than in the LED group and the side effects were tolerable [65].

13.2.2.3 Basal Cell Carcinoma

Basal cell carcinoma is a type of low-grade basal cell malignancy. Boasting good curative and cosmetic effect, PDT is the recommended first-line therapy for low-risk multi-focal superficial basal cell carcinoma.

The LED parameters used for the routine treatment include: 630 nm, 37 J/cm^2; illumination for 7–9 min; 20% ALA or MAL as the photosensitizer and partially encapsulated for 3 h; twice of treatment at an interval of 1 week [66]. There is a study on the comparison between the LED-based PDT (630 nm, 30 mW/cm^2, 150 J/cm^2) and 585 nm PDL-based PDT (spot diameter 7 mm, 10 J/cm^2, 10% overlap, 3 scans) in the treating effect on basal cell carcinoma. The parameters adopted in the study include: topical application of 20% 5-ALA; PDL-PDT treatment of 3–5 min; LED-PDT treatment of 65 min. The treating effect of the two methods is similar. Basal cell carcinoma completely disappeared 3 or 12 months after the treatment. Despite a slightly higher pain index and longer treating duration, LED-PDT had a lower recurrence rate than PDL-PDT [67].

For nodular basal cell carcinoma, certain physical methods are needed as supplementary treatment. There were studies on combined application of LED-PDT and skin grinding to treat nodular basal cell carcinoma of nose. Firstly, tumor lesions were removed via skin grafting, and then 20% aminolevulinic acid was locally applied for 3–4 h. Finally, the site of the treatment was irradiation with LED (630 ± 10 nm, 100 J/cm^2) for 20 min. The treatment was given once a week for 3 weeks. One year later, the tumor recurrence and scar formation were evaluated. The results showed that the wound healed quickly, the postoperative recurrence rate was low, and patient satisfaction was high.

13.2.2.4 Oral Antibacterial Therapy

In a study, a combination of LED and curcumin as a photosensitizer was adopted for oral antibacterial treatment. The subjects were divided into three groups: LED plus curcumin group, LED group, and curcumin group. The LED parameters were:

455 ± 30 nm, 600 mW/cm^2, 200 J/cm^2; irradiation area: 0.6 cm^2; irradiation time: 5 min; curcumin concentration: 30 mg/l. The patients in the PDT group were asked to gargle with curcumin firstly and given LED irradiation 5 min later. The colony forming units in saliva samples were counted respectively before the treatment, immediately after the treatment, and 1 and 2 h later. Compared with that before the treatment, the count of the colony forming units of the PDT group immediately and 1 and 2 h after the treatment was significantly smaller. No effect was observed in the LED group. As to the curcumin group, the unit count of the colony formation increased rather than decreased 1 h after the treatment and dropped to the baseline level after 2 h. The results showed that the combined application of LED and curcumin was effective for oral disinfection [68].

There is a study on a combined usage of LED and curcumin for the antibacterial treatment of children given orthodontic tooth treatment. The subjects were divided into three groups: 2% chlorhexidine group, placebo group, and LED plus curcumin (1.5 mg/ml) group (LED parameters: 450 nm, 165 mW/cm^2, 96 J/cm^2). The patients were given the treatment weekly for 4 weeks. A follow-up was performed 1 and 3 months after the treatment, respectively, the plaque accumulation was evaluated with plaque index (PI), and gingival conditions were evaluated with gingival bleeding index (GBI). The results showed that there was no significant difference in dental plaque index between the LED plus curcumin group and the other two groups at month 1 follow-up, but the plaque index was significantly higher than the other two groups at month 3 follow-up. The result of gingival bleeding index evaluation showed that the gingival bleeding level of the LED plus curcumin group and the chlorhexidine group at month 1 follow-up was significantly lower than the basic level of bleeding gums, but no significant difference existed at month 3 follow-up. In conclusion, with a main function of reducing gum bleeding, the anti-microbial treatment of LED in combination with curcumin has short-term effect for children given orthodontic treatment.

13.2.2.5 Bactericidal Treatment for Deep Caries

Deep dental caries sterilization is rather difficult. There is a study on the combined application of LED and toluidine blue to kill bacteria caused by dental caries. In the experiment, the subjects were divided into two groups: LED plus toluidine blue group and LED plus saline group as the control group. After the infiltration of 10 μl toluidine blue (100 μg/ml) or 10 μl saline in the cavities for 5 min, irradiation of LED (parameter: 630 nm, output power 150 mW, 94 J/cm^2) was given 2 mm away from the cavities. The amount of live bacteria, Streptococcus mutans, and lactic acid bacteria colonies were counted immediately before and after the treatment. The results showed that, compared with the control group, the amount of live bacteria and counts of colony forming units of Streptococcus mutans and Streptococcus lactis significantly decreased in the LED plus toluidine blue group. It was proven via the research that the combination of LED and toluidine blue was effective in killing caries-caused bacteria [69]. However, a long-term study is needed to find out whether the method may prevent the further development of dental caries and even avoid the removal of dental caries.

13.2.2.6 Periodontitis

Periodontitis often leads to gingival recession and bleeding. Treating periodontitis with PDT is a clinical exploration. There is a study on treating periodontitis jointly with LED and methylaniline blue. In the experiment, the subjects were divided into two groups: LED plus toluidine blue group and LED group. The LED parameters used in the treatment were: 628 nm; 2000 mW/cm^2; 10 s per exposure; 40 s in total. The therapeutic effect was evaluated with the detection results of the periodontal bag depth and the bleeding and microorganism under the gums before the treatment and 1 week after the treatment. The results showed that the treatment of both the two groups improved significantly the depth of pocket and probe bleeding 1 week after the treatment, but the therapeutic effect of the PDT group was better. While both the methods being able to reduce the amount of a large variety of microorganisms, the effect of the PDT and red LED group was more obvious [70].

13.2.2.7 Denture Stomatitis

Denture use often induces denture stomatitis and is difficult to cure. In some studies, PDT was adopted to treat denture stomatitis, which had achieved good results via local fungal killing. One of the researches was carried out with LED and porphyrin derivatives. In the experiment, the subjects were divided into two groups: LED plus Photogem (500 mg/l) group and mycostatin group. The LED parameters were 455 nm and 102 mW/cm^2. The energy density of the denture irradiation was 37.5 J/cm^2 and that of the maxillary irradiation 122 J/cm^2. The treatment was given 3 times a week for 2 weeks. Nystatin (100,000 IU) was topically applied, 1 min at a time, 4 times per day for 2 weeks. To evaluate the therapeutic effect, caries and maxillary fungal culture and standard photographic treatment of the upper jaw were performed before the treatment and 15, 30, 60, and 90 days after the treatment, respectively. For both groups, the number of fungal colonies at the end of treatment was reduced significantly and the reduction was most pronounced 30 days after the treatment, and the comparison via pictures showed that the success rates of the two groups were similar. It was proven that PDT was as effective as topical application of nystatin in treating denture stomatitis, but the operating process of PDT was relatively cumbersome [71].

13.3 LED Medical Equipment

13.3.1 Application of LED Equipment in Medical Field

LED is currently the most efficient and energy-saving artificial light source, its light efficiency can reach 254 lm/W, the light spectrum of LED is relatively pure, and the half-wave width is only 10 nm or even more less. LED is also characterized by good direction, monochrome performance, fast response, little flicker, easy adjustment of color temperature, high reliability, long life, and so on, making it suitable for medical application as a light source. After decades of development, the medical application of LED focuses mainly in the following areas.

13.3.1.1 Lighting

Lighting includes indoor and outdoor lighting in hospital buildings, lesion observation lighting, and surgical procedure lighting. Lighting is the most common function of LED at present, but, compared with general lighting, medical lighting is somewhat particular in terms of market demand and access. For example, when a lesion is illuminated, the light source used has to be healthy and easy to adjust and control for the sake of preventing the damage from ultraviolet and infrared light. For operation process lighting, the requirements include good illumination, no shadow, high uniformity, easy adjustment of color temperature, and light intensity and low heat.

13.3.1.2 Disinfection and Sterilization

The reason to use LED for disinfection and sterilization lies mainly in the fact that the ultraviolet light passing through microbial cell membrane and nucleus may damage their DNA or RNA molecular chains, making them lose their ability to copy and other functions.

13.3.1.3 Phototherapy

As a new type of light source, LED, with its biological effects gradually recognized by people, is playing an active role in medical treatment. Phototherapy achieves therapeutic effects via a series of chemical reactions caused by the thermal and photochemical effects that appear as a result of the absorption of the energy of lights of different wavelengths by biological tissues.

13.3.2 LED Phototherapy Device

In recent years, with the rapid development of the third-generation semiconductor materials and technologies, new types of LED are introduced continuously. While the luminous intensity of LED being increasing, the peak wavelength is more and more stable, the half-wave bandwidth more and more narrow, and the monochromatic performance and direction function better and better.

LED also has a wide coverage of wavelength from UV to infrared, covering basically the full spectrum of visible light. At the same time, two-color light, three-color light, and white LED with excellent performance have also been developed. According to the different needs of phototherapy, single wavelength light source such as blue light, green light, red light and infrared light, or the combination of multi-wavelength light source are designed to treat the target diseases. It provides a solid hardware foundation for the wide application of LED device in phototherapy.

At the same time, LED can be used to form complicated geometric shapes, including but not limited to point, line, and plate, and can even be made into flexible soft light sources so as to change shape according to the shape of the treatment site, making the light irradiation more evenly. There are now three common LED-based phototherapy modes: top projection, LED bed, and LED blanket based on patched LED [72].

At present, LED phototherapy equipment mainly includes low-lever light therapy devices and photodynamic therapy devices. The main difference between them is that photodynamic therapy devices have to be used jointly with photosensitizers. Moreover, they need higher power, and their wavelength should match with the photosensitizer selected.

13.3.2.1 Low-Level Light Therapy (LLLT) Devices

13.3.2.1.1 Single Wavelength LED LLLT Devices

13.3.2.1.1.1 Blue LED Phototherapy Devices

Treating neonatal jaundice is a typical application of blue light device. Currently, there are many companies producing LED phototherapy systems, which are designed in diversified types such as blanket, projection, and bed type. The LED phototherapy systems designed in blanket type include the Bilisoft LED phototherapy system manufactured by the US GE Corporation and the neoBLUE blanket by the US Natus Company. The Bilisoft LED phototherapy system contains two sizes of blanket LED cold light source, their power densities are 35 and 50 $\mu W/cm^2/nm$, and their wavelengths range from 440 to 460 nm. Thanks to the soft carpet design and fiber conduction mode, phototherapy can be performed from a position close to children's skin (http://www3.gehealthcare.co.uk/en-gb/products/categories/perinatal_care/phototherapy/bilisoft_phototherapy_system). The neoBLUE blanket has a wavelength ranging from 450 to 475 nm and a power density of 30–35 μW/cm^2/nm at the factory, which can be adjusted to more than 50 μW/cm^2/nm (http://www.natus.com/index.cfm?page=products_1&crid=583&contentid=657).

Projection design is currently used in most Blu-ray phototherapy devices, such as the Lullaby LED Phototherapy System of GE, which has a wavelength of 450–465 nm and a power density of 22–45 μW/cm^2/nm. The irradiation is in the range of 50 × 30 cm when the projection distance is 35 cm (http://www.natus.com/index.cfm?page=products_1&crid=189&contentid=90). Other projection phototherapeutic devices include Bilitron (wavelength: 400–550 nm) produced by Fanem, Select BiliBee (wavelength: 430–490 nm) by Medical Company, BiliLED (wavelength: 453–487 nm) by Controles Company, and so on.

Bilitron™ Bed 4006 produced by Fanem Company uses 17 Super LEDs with a wavelength of 450 nm and a power density of 47.5 ± 5 μW/cm^2/nm (http://www.ssemmthembu.co.za/index.php/product-ranges/phototherapy-led/fanem-bilitron-bed-4006). Bilitron 3006, another device designed by the company, is an ultra-compact portable phototherapy device using 5 Super LEDs with a wavelength of 460 nm. Bilitron 3006 can be used alone or be fixed on an adjustable bracket, which makes it highly adaptable (http://www3.gehealthcare.co.uk/en-GB/Products/Categories/Perinatal_Care/Phototherapy/Lullaby_LED_Phototherapy_System#tabs/tabD7FF1A1F7C8A425DB477CD6C8AC4FCAE). The neoBLUE® cozy LED phototherapy system produced by Natus is a typical LED bed device, which has a wavelength of 450–470 nm and a power density greater than 30 μW/cm^2/nm (http://www.ssemmthembu.co.za/index.php/product-ranges/phototherapy-led/bilitron-3006-the-super-led-phototherapy).

In addition to the treatment of neonatal jaundice, blue LEDs are also used for patients with psoriasis such as the LED based on rechargeable battery array developed by Philips company in the Netherlands, which consists of 40 high-intensity array LED light source of a wavelength of 453 ± 5 nm and a power density up to 50 mW/cm^2. The device was inserted into the hole in the middle of the textile fixing belt and bound through the fixing belt on the surface of the lesion skin to be irradiated [73].

13.3.2.1.1.2 Red LED Phototherapy Devices

Red light has a wavelength ranging from 600 nm to 760 nm, and the LEDs of this wavelength are used to treat various diseases. For example, the FDA has approved the use of Omnilux Revive™ at 633 nm for skin rejuvenation, which has an output power density of 105 mW/cm^2. Having been actively committed to the development of non-invasive LED phototherapy equipment since the late nineteenth century, the laser medicine department of the PLA General Hospital developed a wearable red LED phototherapy device in 2000, and a series of red band LED wearable products on this basis, such as the flexible and wearable hair-growing and sleeping-assisting head phototherapy system (Patent Application number 201610697781.3), the flexible and wearable phototherapy device (Patent Application No. 201610694853.9) for lumbar muscle injury, the flexible and wearable phototherapy device (Patent Application No. 201610694879.3) for osteoarthrotic diseases, and the flexible and wearable phototherapy device (Patent Application No. 201610695946.3) for elbow joint injuries.

13.3.2.1.1.3 Infrared LED Phototherapy Devices

At present, there are few independent infrared LED phototherapy devices, which are usually used jointly with lights of other wavelengths. The 980 nm near-infrared MIRE (monochromatic infrared energy) treatment system (Anodyne® Therapy/MIRE™) produced by Anodyne® Therapy was approved for clinical use by the FDA in 1994. Being the first near-infrared therapeutic apparatus approved by FDA, it is widely used in the USA. Anodyne® Therapy/MIRE™ contains multiple flexible touch pads. The device is mainly used to improve circulation and relieve pain, stiffness, and muscle spasm.

Moreover, Omnilux Plus™ (830 nm, power 55 mW/cm^2), one of the near-infrared phototherapy equipment, is used for rejuvenation and promoted wound healing. The Omnilux Rehabilitation System (Photo Therapeutics, Limited, Altrincham, UK) with a wavelength of 830 nm $\pm$ 8 nm is used in clinical studies on various sports injuries, including tendons, muscle tears and injuries, back pain, and joint pain.

13.3.2.1.2 Multi-Wavelength LED Phototherapy Devices

With the progress of science and technology, multiple LED chips with different wavelengths can integrate into one LED device, which can meet the needs of different combinations of phototherapy.

For example, Lutronic's Healite II uses 1800 LEDs having the wavelengths of 830/590 nm, 633 nm, and 415 nm, which is supported with the patented optical lens array technology (OLAT) to collimate the light output from the LED and improve the optical transmission efficiency. In addition, the photo-sequencing technology (PST) of the wavelength of 830/590 nm is also used. The working principle of the PST is to sequentially treat the cells with 590 nm light and continuously irradiate them with 830 nm after 1960s, which can be used for wound healing, pain relief, arthritis, hair regeneration, and other treatment (http://www.international.lutronic.com/healiteii) [74]. Dermalux's Dermalux® LED Phototherapy comprises 8 LED arrays, each having multiple 415 nm, 633 nm, and 830 nm LEDs. The main feature of the device is that the three wavelengths can be used independently or jointly depending on clinical purposes (http://www.dermaluxled.com/led-phototherapy.php).

The Omnilux (TM) System, a product of the USA, comprises 2000 LED arrays that provide 415 nm blue light, 633 nm red light, and 830 nm near-infrared light. It can be used to treat severe acnes based on a combined application of 415 nm blue light and 633 nm red light, and skin photoaging by combining 830 nm near-infrared and 633 nm red light. Delphia Del SolTM manufactured by Edge Systems LLC of the USA is also a triple therapy system providing 430 nm blue light, 600–700 nm red light, and 700–1000 nm near-infrared light in a continuous manner, which can be used to treat skin photoaging. In addition, LIGHTWAVE Professional of the USA, consisting of the LED of 417 nm blue light, 630 nm red light, and 880 nm near-infrared light and providing the light in a pulsed manner, is also used for treating skin photoaging. Furthermore, devices of more wavelength composites have been developed such as Max7 (Flip4, Inc., Quebec, Canada), which is based on the visible light of 7 wavelengths ranging from 420 nm to 700 nm and also used for facial photoaging.

The Litebook® LED Phototherapy Instrument (The Litebook Company Ltd., Alberta, Canada) consists of 60 LEDs, having a wavelength ranging from 400 nm to 700 nm, a peak wavelength at 464 nm, and a sub-peak wavelength at 564 nm, is mainly used for the treatment of seasonal affective disorder (SAD) [38].

13.3.2.2 LED Photodynamic Therapeutic Devices

13.3.2.2.1 LED Photodynamic Therapeutic Devices of Single Wavelength

At present, the most common light source of PDT is laser due to desirable monochromaticity and the excitation wavelength of the absorption peak of suitable photosensitizers that may guarantee a high efficiency of the photosensitizers. In addition, for endoscopic and inter-organizational PDT, laser is undoubtedly the most suitable light source because it can be transferred to digestive tracts, respiratory tracts, and so on. For body surface irradiation, in addition to laser, LED is more and more used in the PDT for superficial skin-mucosal lesions.

13.3.2.2.1.1 Blue LED Photodynamic Therapeutic Devices

The Omnilux blue™ (Photo Therapeutics, Limited, Altrincham, UK) has a wavelength of 415 nm with the output intensity of 40 mW/cm^2 and dose range from 1 to 150 J/cm^2, which has been approved by the FDA for the treatment of superficial actinic keratosis based on 5-ALA-PDT [75, 76].

13.3.2.2.2 Red LED Photodynamic Therapeutic Devices

Red LED is mainly used in the photodynamic therapy of cancer. For example, in the USA, Omnilux PDT™ (633 nm) is to be approved for the treatment of nonmelanoma skin cancer (NMSC). Recently, Ambicare Health Ltd. has developed a portable adhesive PDT device for nonmelanoma skin cancer [77].

13.3.2.2.3 Hybrid Multi-Wavelength LED Photodynamic Therapeutic Devices

The BS-LED3E LED treatment instrument developed by Sincery International Limited may provide 940 nm infrared light, 630 nm red light, 590 nm yellow light, and 460 nm blue light with a maximum power density of 90 J/cm^2. The output light energy is divided into 10 levels and can be adjusted. The instrument has two independent treatment arms that allow simultaneous treatment of two different regions of the same patient or two patients, each having LEDs of 4 independent wavelengths (http://www.sincery.com/productdetails/beauty/44.html). The BS-LED2, a product of single-arm dual-wavelength design, provides both 630 nm and 470 nm light with an energy density of 64 J/cm^2 and 48 J/cm^2, respectively, and works for a set period of 1–90 min (http://www.sincery.com/productdetails/beauty/30.html). In addition, the LED-IA (Yage Optic and Electronic Technique, Wuhan, China) adopts a half-arc projection mode and uses lever-type electric control to adjust its height. It may provide the lights of 633 nm, 590 nm, and 417 nm with a power of 20–100 mW/cm^2, 10–40 mW/cm^2, and 25–120 mW/cm^{2,} respectively (http://www.yagelaser.com/content/?91.html).

References

1. H. Chung et al., The nuts and bolts of low-level laser (light) therapy. Ann. Biomed. Eng. **40**(2), 516–533 (2012)
2. W. Cannon, *The Wisdom of the Body* (W. W. Norton, New York, 1932)
3. T.C.Y. Liu et al., Homeostatic photobiomodulation. Front. Optoelectron. China **2**(1), 1–8 (2009)
4. T.C.Y. Liu et al., Photobiomodulation on stress. Int. J. Photoenergy **2012**, 628649 (2012). https://doi.org/10.1155/2012/628649
5. T.C.Y. Liu et al., Microenvironment dependent photobiomodulation on function-specific signal transduction pathways. Int. J. Photoenergy **2014**, 904304 (2014). https://doi.org/10.1155/2014/904304
6. T.C.Y. Liu, Y.Y. Kang, Functional photobiomodulation. Photomed. Laser Surg. **32**(9), 479–480 (2014)
7. T.C.Y. Liu et al., The mitochondrial $Na^+/Ca2^+$ exchanger is necessary but not sufficient for $Ca2^+$ homeostasis and viability. Adv. Exp. Med. Biol. **1072**, 281–285 (2018)
8. T.C.Y. Liu et al., Quantitative biology of exercise-induced signal transduction pathways. Adv. Exp. Med. Biol. **977**, 419–424 (2017)
9. C.Y. Liu et al., Self-similarity constant and quantitative difference and their applications in sports science. J. Phys. Educ. **24**(6), 72–78 (2017). in Chinese
10. M.R. Hamblin, Shining light on the head: photobiomodulation for brain disorders. BBA Clin. **6**, 113–124 (2016)
11. T. Karu, *The Science of Low-Power Laser Therapy* (Gordon and Breach Science, Amsterdam, 1998)
12. N. Lane, Cell biology - power games. Nature **443**(7114), 901–903 (2006)
13. X.L. Wang et al., Interplay between up-regulation of cytochrome-c-oxidase and hemoglobin oxygenation induced by near-infrared laser. Sci. Rep. **6**, 30540 (2016)

14. M. Huttemann et al., Phosphorylation of mammalian cytochrome c and cytochrome c oxidase in the regulation of cell destiny: respiration, apoptosis, and human disease, in *Mitochondrial Oxidative Phosphorylation: Nuclear-Encoded Genes, Enzyme Regulation, and Pathophysiology*, vol. 748, (Springer, New York, 2012), pp. 237–264
15. S.A. Wu et al., Cancer phototherapy via selective photoinactivation of respiratory chain oxidase to trigger a fatal superoxide anion burst. Antioxid. Redox Signal. **20**(5), 733–746 (2014)
16. W. De Haes et al., Metformin promotes lifespan through mitohormesis via the peroxiredoxin PRDX-2. Proc. Natl. Acad. Sci. U. S. A. **111**(24), E2501–E2509 (2014)
17. C. Yee, W. Yang, S. Hekimi, The intrinsic apoptosis pathway mediates the pro-longevity response to mitochondrial ROS in C. elegans. Cell **157**(4), 897–909 (2014)
18. B. Drew et al., Effects of aging and caloric restriction on mitochondrial energy production in gastrocnemius muscle and heart. Am. J. Physiol. Regul. Integr. Comp. Physiol. **284**(2), R474–R480 (2003)
19. L.A. Montoro et al., Infrared LED irradiation photobiomodulation of oxidative stress in human dental pulp cells. Int. Endod. J. **47**(8), 747–755 (2014)
20. G.Y. Luo et al., The effects of low-intensity He-Ne laser irradiation on erythrocyte metabolism. Lasers Med. Sci. **30**(9), 2313–2318 (2015)
21. V.Y. Arshavsky, T.D. Lamb, E.N. Pugh, G proteins and phototransduction. Annu. Rev. Physiol. **64**, 153–187 (2002)
22. S.S. Campbell, P.J. Murphy, Extraocular circadian phototransduction in humans. Science **279**(5349), 396–399 (1998)
23. T.C.Y. Liu, Y.Q. Gao, S.H. Liu, Light-cell interaction: quasi-hormone model and time theory. Proc. SPIE **2887** (1996). https://doi.org/10.1117/12.251944
24. Liu, T. C. Y., et al., Membrane mechanism of low intensity laser biostimulation on a cell. Simunovic Z. Lasers in Medicine, Surgery and Dentistry. Zagreb: European Medical Laser Association. 2003: p. 83–105
25. R. Duan et al., Signal transduction pathways involved in low intensity He-Ne laser-induced respiratory burst in bovine neutrophils: a potential mechanism of low intensity laser biostimulation. Lasers Surg. Med. **29**(2), 174–178 (2001)
26. S.N. Wu, D. Xing, Intracellular signaling cascades following light irradiation. Laser Photonics Rev. **8**(1), 115–130 (2014)
27. L.R. Braathen et al., Guidelines on the use of photodynamic therapy for nonmelanoma skin cancer: an international consensus. International Society for Photodynamic Therapy in Dermatology, 2005. J. Am. Acad. Dermatol. **56**(1), 125–143 (2007)
28. M.H. Abdel-Kader, *Photodynamic Therapy. From Theory to Application* (Springer, New York, 2014), pp. 102–103
29. D.F. Chen et al., Effects of pulse width and repetition rate of pulsed laser on kinetics and production of singlet oxygen luminescence. J. Innov. Opt. Health Sci. **9**(6), 1650019 (2016)
30. J.S. Dysart, M.S. Patterson, Characterization of Photofrin photobleaching for singlet oxygen dose estimation during photodynamic therapy of MLL cells in vitro. Phys. Med. Biol. **50**(11), 2597–2616 (2005)
31. V.H. Fingar et al., The role of microvascular damage in photodynamic therapy: the effect of treatment on vessel constriction, permeability, and leukocyte adhesion. Cancer Res. **52**(18), 4914–4921 (1992)
32. M. Firczuk, D. Nowis, J. Golab, PDT-induced inflammatory and host responses. Photochem. Photobiol. Sci. **10**(5), 653–663 (2011)
33. H.X. Qiu et al., Twenty years of clinical experience with a new modality of vascular-targeted photodynamic therapy for port wine stains. Dermatol. Surg. **37**(11), 1603–1610 (2011)
34. U. Schmidt-Erfurth, T. Hasan, Mechanisms of action of photodynamic therapy with verteporfin for the treatment of age-related macular degeneration. Surv. Ophthalmol. **45**(3), 195–214 (2000)
35. H.X. Qiu et al., Vascular targeted photodynamic therapy for bleeding gastrointestinal mucosal vascular lesions: a preliminary study. Photodiagn. Photodyn. Ther. **9**(2), 109–117 (2012)

36. P. Kumar, D. Chawla, A. Deorari, Light-emitting diode phototherapy for unconjugated hyperbilirubinaemia in neonates. Cochrane Database Syst. Rev. (12), CD007969 (2011). https://doi.org/10.1002/14651858.CD007969.pub2
37. M. Mohammadizadeh, F.K. Eliadarani, Z. Badiei, Is the light-emitting diode a better light source than fluorescent tube for phototherapy of neonatal jaundice in preterm infants? Adv. Biomed. Res. **1**(1), 51 (2012)
38. P.H. Desan et al., A controlled trial of the litebook light-emitting diode (LED) light therapy device for treatment of seasonal affective disorder (SAD). BMC Psychiatry **7**(1), 38 (2007)
39. F. Schiffer et al., Psychological benefits 2 and 4 weeks after a single treatment with near infrared light to the forehead: a pilot study of 10 patients with major depression and anxiety. Behav. Brain Funct. **5**(1), 46 (2009)
40. M.A. Naeser et al., Significant improvements in cognitive performance post-transcranial, red/near-infrared light-emitting diode treatments in chronic, mild traumatic brain injury: open-protocol study. J. Neurotrauma **31**(11), 1008–1017 (2014)
41. G. Glickman, B. Byrne, C. Pineda, et al., Light therapy for seasonal affective disorder with blue narrow-band light-emitting diodes (LEDs). Biol. Psychiatry **59**(6), 502–507 (2006)
42. A.S. Salgado et al., The effects of transcranial LED therapy (TCLT) on cerebral blood flow in the elderly women. Lasers Med. Sci. **30**(1), 339–346 (2015)
43. H. Nawashiro, K. Wada, K. Nakai, et al., Focal increase in cerebral blood flow after treatment with near-infrared light to the forehead in a patient in a persistent vegetative state. Photomed. Laser Surg. **30**(4), 231–233 (2012)
44. S. Dixit et al., Closure of non-healing chronic ulcer in Klippel–Trenaunay syndrome using low-level laser therapy. BMJ Case Rep. **2012**, bcr2012006226 (2012). https://doi.org/10.1136/bcr-2012-006226
45. R. Sutterfield, Light therapy and advanced wound care for a neuropathic plantar ulcer on a Charcot foot. J. Wound Ostomy Continence Nurs. **35**(1), 113–115 (2008)
46. M.A. Trelles, I. Allones, E. Mayo, Er:YAG laser ablation of plantar verrucae with red LED therapy-assisted healing. Photomed. Laser Surg. **24**(4), 494–498 (2006)
47. S.Y. Lee, C.E. You, M.Y. Park, Blue and red light combination LED phototherapy for acne vulgaris in patients with skin phototype IV. Lasers Surg. Med. **39**(2), 180–188 (2007)
48. D.J. Goldberg, B.A. Russell, Combination blue (415 nm) and red (633 nm) led phototherapy in the treatment of mild to severe acne vulgaris. J. Cosmet. Laser Ther. **8**(2), 71–75 (2006)
49. N. Sadick, A study to determine the effect of combination blue (415 nm) and near-infrared (830 nm) light-emitting diode (led) therapy for moderate acne vulgaris. J. Cosmet. Laser Ther. **11**(2), 125–128 (2009)
50. L.H. de Arruda et al., A prospective, randomized, open and comparative study to evaluate the safety and efficacy of blue light treatment versus a topical benzoyl peroxide 5% formulation in patients with acne grade II and III. An. Bras. Dermatol. **84**(5), 463–468 (2009)
51. L.H. Liu et al., Randomized trial of three phototherapy methods for the treatment of acne vulgaris in Chinese patients. Photodermatol. Photoimmunol. Photomed. **30**(5), 246–353 (2014)
52. R.A. Weiss et al., Clinical trial of a novel non-thermal led array for reversal of photoaging: clinical, histologic, and surface profilometric results. Lasers Surg. Med. **36**(2), 85–91 (2005)
53. D. Barolet et al., Regulation of skin collagen metabolism in vitro using a pulsed 660 nm led light source: clinical correlation with a single-blinded study. J. Invest. Dermatol. **129**(12), 2751–2759 (2009)
54. S.Y. Lee et al., A prospective, randomized, placebo-controlled, double-blinded, and split-face clinical study on led phototherapy for skin rejuvenation: clinical, profilometric, histologic, ultrastructural, and biochemical evaluations and comparison of three different treatment settings. J. Photochem. Photobiol. B **88**(1), 51–67 (2007)
55. D. Barolet, A. Boucher, Prophylactic low-level light therapy for the treatment of hypertrophic scars and keloids: a case series. Lasers Surg. Med. **42**(6), 597–601 (2010)
56. Y.J. Park et al., Prevention of thyroidectomy scars in Asian adults with low-level light therapy. Dermatol. Surg. **42**(4), 526–534 (2016)

57. E.C. Leal-Junior et al., Adjunctive use of combination of super-pulsed laser and light-emitting diodes phototherapy on nonspecific knee pain: double-blinded randomized placebo-controlled trial. Lasers Med. Sci. **29**(6), 1839–1847 (2014)
58. J. Foley et al., 830 nm light-emitting diode (led) phototherapy significantly reduced return-to-play in injured university athletes: a pilot study. Laser Ther. **25**(1), 35 (2016)
59. D. Pariser et al., Topical methyl-aminolevulinate photodynamic therapy using red light-emitting diode light for treatment of multiple actinic keratoses: a randomized, double-blind, placebo-controlled study. J. Am. Acad. Dermatol. **59**(4), 569–576 (2008)
60. C.D. Enk, A. Levi, Low-irradiance red LED traffic lamps as light source in PDT for actinic keratoses. Photodermatol. Photoimmunol. Photomed. **28**(6), 332–334 (2012)
61. X.L. Wang et al., Topical ALA PDT for the treatment of severe acne vulgaris. Photodiagn. Photodyn. Ther. **7**(1), 33–38 (2010)
62. L. Ma et al., Low-dose topical 5-aminolevulinic acid photodynamic therapy in the treatment of different severity of acne vulgaris. Photodiagn. Photodyn. Ther. **10**(4), 583 (2013)
63. S.Q. Tao et al., Low-dose topical 5-aminolevulinic acid photodynamic therapy in the treatment of different severity of acne vulgaris. Cell Biochem. Biophys. **73**(3), 701–706 (2015)
64. S.Q. Tao et al., Efficacy of 3.6% topical ALA-PDT for the treatment of severe acne vulgaris. Eur. Rev. Med. Pharmacol. Sci **20**(2), 225–231 (2016)
65. B.H. Song et al., Photodynamic therapy using chlorophyll-a in the treatment of acne vulgaris: a randomized, single-blind, split-face study. J. Am. Acad. Dermatol. **71**(4), 764–771 (2014)
66. E. Christensen et al., Guidelines for practical use of MAL-PDT in non-melanoma skin cancer. J. Eur. Acad. Dermatol. Venereol. **24**(5), 505–512 (2010)
67. A. Čarija et al., Single treatment of low-risk basal cell carcinomas with pulsed dye laser-mediated photodynamic therapy (PDL-PDT) compared with photodynamic therapy (PDT): a controlled, investigator-blinded, intra-individual prospective study. Photodiagn. Photodyn. Ther. **16**, 60–65 (2016)
68. D.P. Leite et al., Effects of photodynamic therapy with blue light and curcumin as mouth rinse for oral disinfection: a randomized controlled trial. Photomed. Laser Surg. **32**(11), 627–632 (2014)
69. M.A. Melo et al., Photodynamic antimicrobial chemotherapy and ultraconservative caries removal linked for management of deep caries lesions. Photodiagn. Photodyn. Ther. **12**(4), 581–586 (2015)
70. C. Mongardini, G.L. Di Tanna, A. Pilloni, Light-activated disinfection using a light-emitting diode lamp in the red spectrum: clinical and microbiological short-term findings on periodontitis patients in maintenance. A randomized controlled split-mouth clinical trial. Lasers Med. Sci. **29**(1), 1–8 (2014)
71. E.G. Mima et al., Comparison of photodynamic therapy versus conventional antifungal therapy for the treatment of denture stomatitis: a randomized clinical trial. Clin. Microbiol. Infect. **18**(10), E380–E388 (2012)
72. H.J. Vreman, R.J. Wong, D.K. Stevenson, Phototherapy: current methods and future directions. Semin. Perinatol. **28**(5), 326–333 (2004)
73. S. Pfaff et al., Prospective randomized long-term study on the efficacy and safety of UV-free blue light for treating mild psoriasis vulgaris. Dermatology **231**(1), 24–34 (2015)
74. W.S. Kim, R.G. Calderhead, Is light-emitting diode phototherapy (LED-LLLT) really effective? Laser Ther. **20**(3), 205–215 (2011)
75. H.S. Xiang et al., Led is applied to the development of light therapy. Beijing Biomed. Eng. **24**(4), 311–315 (2005)
76. T. Yan, LED light source is in the application of skin medicine. China Illuminating Society light biology, photochemical application research forum and ecological lighting forum (2013)
77. S.K. Attili, An open pilot study of ambulatory photodynamic therapy using a wearable low-irradiance organic light-emitting diode light source in the treatment of nonmelanoma skin cancer. Br. J. Dermatol. **161**(1), 170–173 (2009)

Chapter 14
Light-Emitting Diodes for Horticulture

Dongxian He, Toyoki Kozai, Genhua Niu, and Xin Zhang

14.1 Fundamentals and Challenges of LED Lighting Technology for Horticulture

Applications of light-emitting diodes (LEDs) in agriculture, horticulture, and fishery have been expanding since around 2010. Especially, its applications in greenhouse horticulture as supplemental light source to solar light and in plant factories with artificial lighting (PFAL) or indoor vertical farms as sole light source have been becoming more and more popular since around 2015.

In this section, fundamental concepts, methodology, challenges, and opportunities of LED applications in the PFAL are discussed. Useful information on LED lighting in the greenhouse horticulture can be obtained from a comprehensive and informative review paper consisting of 87 pages by Mitchel et al. [1] and from a book by Kozai et al. [2]. This section is a kind of concept article and a summary of recently published books by Kozai et al. [3] and Kozai [4].

D. He (✉)
Key Laboratory of Agricultural Engineering in Structure and Environment of Ministry of Agriculture and Rural Affairs, China Agricultural University, Haidian, Beijing, China
e-mail: hedx@cau.edu.cn

T. Kozai
Japan Plant Factory Association, c/o FC Center of Chiba University, Kashiwa, Chiba, Japan
e-mail: kozai@faculty.chiba-u.jp

G. Niu
Texas A&M AgriLife Research at El Paso, Texas A&M University System, El Paso, TX, USA
e-mail: gniu@ag.tamu.edu

X. Zhang
Beijing Research Center for Information Technology in Agriculture, Haidian, Beijing, China
e-mail: zhangx@nercita.org.cn

J. Li, G. Q. Zhang (eds.), *Light-Emitting Diodes*, Solid State Lighting Technology and Application Series 4, https://doi.org/10.1007/978-3-319-99211-2_14

14.1.1 Role of Light and Major Light Environmental Factors

In the greenhouse horticulture and the PFAL, light is provided to plants for controlling the growth and development, for protecting plants from pest insects and harmful small animals, and for providing comfortable working environments for humans. Role of light for plants can be categorized into 2 groups: energy source and signal source. Light as energy source is primarily provided for photosynthesis of plants where the light energy is converted to chemical energy of carbohydrates in plants. While, light as a signal source is provided to stimulate photoreceptors for photomorphogenesis, photoperiodism, phototropism, secondary metabolite production, and stomatal functioning, etc. Light signals captured by the photoreceptors initiate various kinds of biochemical metabolisms.

In the energy conversion phase, less than 10% of light energy captured by chlorophylls and carotenoids is converted into chemical energy in leaves. In other words, more than 90% of light energy absorbed by plants is converted to sensible or latent heat. In addition, a significant portion of light energy incident to the plant canopy is not absorbed by leaves, but absorbed by its surroundings. Thus, in the PFAL, improving the ratio of light energy received by leaves to the light energy emitted by the light source is of primary importance as well as the ratio of light energy fixed by leaves to the light energy absorbed by leaves. The former ratio is influenced by the plant canopy architecture and the design of cultivation space. The latter ratio is influenced by the environments and genetic traits of plants.

Major light environmental factors include:

1. Photosynthetic photon flux density (PPFD, μmol m^{-2} s^{-1}) with waveband from 400 nm to 700 nm;
2. Spectral distribution in the range between 300 nm and 800 nm;
3. Lighting cycle (photoperiod/dark periods);
4. Lighting direction (downward, sideward, and upward lighting).

In the PFAL, these light environmental factors at the plant canopy surface can be controlled fairly well. However, PPFD and its spectral distribution in the plant canopy are significantly influenced by the plant canopy structure and its optical characteristics [2].

14.1.2 Dry Mass Increase and Value-Addition of Plant by Lighting

Amount of light energy required in the energy conversion phase for photosynthesis to increase the dry mass or amount of carbohydrates of plants is roughly 100 times greater than that required as light signals for morphogenesis, etc. After receiving the light signals, plants switch on or off a biochemical pathway and affect the economic value of the produce.

In the plant production process, the energy conversion phase and signal transmission phase often occur concurrently. A basic strategy for LED lighting in plant production is to maximize the economic value of total produce, which is expressed by a product of unit economic value of produce and volume of salable (marketable) produce. Unit economic value of produce is influenced by, for example, the contents of particular secondary metabolites in the produce. The secondary metabolite content often increases under stressful environment for improving quality formation. While, the content of secondary metabolite such as hypericin in St John's wort (*Hypericum perforatum*), a medicinal plant, increases with the increasing temperature stress [5].

Only 30% of electric energy is reported to be converted by current LEDs for horticultural lighting to photosynthetic radiation (from 400 nm to 700 nm) in practice, although the percentage will be improved to around 40% in the near future. Furthermore, only a portion of photosynthetic radiation is received and absorbed by leaves of plants in the cultivation space [2]. As a result, the electricity cost in the PFAL accounts for about 25% of the total production cost [3]. Thus, there is much space to improve the electric and light energy use efficiencies in the PFAL and the greenhouse.

14.1.3 Basic Properties of LEDs Necessary for Design and Operation of PFAL

Basic properties of LED lamp and LED lighting system (consisting of LED lamps and luminaire) for the PFAL to be released by LED manufacturing companies as product label to the customers are listed in Table 14.1. The data on these properties are minimum requirements for lighting system design of the PFAL. A similar product label but with a wider view has been proposed for horticultural lamps including high pressure sodium, fluorescent, incandescent, and LED lamps by Both et al. [6].

The data on LEDs necessary for plant lighting but unnecessary for office/home lighting include: photosynthetic photon flux (μmol s^{-1}), photosynthetic radiation flux (W), photosynthetic photon number efficacy (μmol J^{-1}), and photosynthetic radiation energy efficiency (J J^{-1}). The photon-related data are necessary because photosynthesis, photomorphogenesis, etc. respond to photons, while energy-related data are necessary for energy management of the PFAL.

Table 14.1 Characteristics of LED lamp or LED lighting system (consisting of lamps and luminaire), to be released by LED manufacturing companies as product label, which are required for lighting design of PFAL (revised after Goto [7]). The figure in the right-hand column is strictly an example, and does not imply a representative or standard value

Item	Unit	Example
Temperature of ambient air	°C	25
Power characteristics		
Voltage	V (voltage)	200
Current	A (ampere)	0.16
Effective power consumption	W (watt)	32.0
Light characteristics		
Spectral distribution (300–800 nm) as a dataset	μmol m^{-2} s^{-1} nm^{-1}	In excel file
Percentages of UV (300–399 nm), blue (400–499), green (500–599), red (600–699), far-red (700–799), and near-infrared (800–1500)	–	0.25, 20, 40, 35, 4.5, 0.25
Angular distribution curves perpendicular to and parallel to LED tube[a], as a dataset	mol s^{-1} rad^{-1} (see note 1 for rad.)	In excel file
Photosynthetic photon flux (PPF)	μmol s^{-1}	48.0
Photosynthetic radiation energy (radiant) flux	W (=J s^{-1})	8.0
Luminous flux[b]	lm	450
Correlated color temperature (CCT)[c]	K	3000
Color rendering index (CRI or Ra)[c]	–	87.0
Efficiency/Efficacy		
Photosynthetic radiation energy efficiency (also called PAR energy efficiency)	J J^{-1}	0.35
Photosynthetic photon number efficacy	μmol J^{-1}	3.2
Luminous efficacy	lm W^{-1}	200
Maintainability: Product age (life time) (h) of 10% decrease in PPF, product age of malfunction (h), and waterproof and dust-proof characteristics		
Thermal characteristics: Lamp or package surface temperature—PPF curve		
Luminaire: The entire electrical light fitting, including all the components needed to mount, operate, and use		
Size, shape, and weight: Drawing and weight of package, and drawing of LED lamp		

[a]Radian: SI unit of plane angle; 1 rad (1 radian) = 57.3° = 180/π = 180/3.14
[b]lm: Lumen
[c]CCT and CRI are omissible for monochromatic LEDs

Table 14.2 Factors affecting the economic value of vegetables

	Factor	Example
1	Weight, size	Fresh/dry weight of salable and unsalable parts of plants
2	Shape, color, and texture	Plant height, leaf shape, stem/internode length, leaf area index, leaf thickness, pigment formation, color, and appearance
3	Functional components	Vitamin C (ascorbic acid), antioxidant substances such as polyphenols, carotenoids, polysaccharides, and minerals such as Fe and Mn
4	Physiological disorder	Edema (intumescence), white/black spots due to micronutrient deficiencies
5	Taste, flavor	Sugar (glucose, fructose, and sucrose), polysaccharides, aromas, amino acids, proteins, lipids, NO_3^-, K^+, and bitterness
6	Mouth feeling	Percent water (turgor pressure), cell wall thickness, and smoothness
7	Shelf life	Colony formation unit per gram

14.1.4 Complexity of Light Environmental Control

14.1.4.1 Purpose of Environmental Control

Purpose of environmental control of the PFAL is to obtain a maximum yield of produce sustainably with highest quality using minimum resources, resulting in minimum emission of wastes and minimum costs. In other words, its purpose is to achieve a maximum cost performance (CP) over a long time period. The CP is expressed by the equation:

$$\mathrm{CP} = \text{total sales/total costs} = \Sigma\,(\mathrm{UV}\,(i) \times \mathrm{AP}\,(i))\,/\,\Sigma\,(\mathrm{UC}\,(j) \times \mathrm{ARS}\,(j))$$

where UV (i) and AP (i) are, respectively, unit economic value of produce (i) and amount of salable produce (i) ($i = 1, 2, \ldots$), and UC (j) and ARS (j) are, respectively, unit cost of resource (j) and amount of resource (j) supplied to the PFAL ($j = 1, 2, 3, \ldots$). Thus, environmental control for maximum yield alone and/or highest quality alone do not bring about the highest CP. The unit economic value of produce and unit costs of resources affect the CP considerably, both of which change with time. Unit economic values of vegetables are affected positively or negatively by, for example, factors shown in Table 14.2.

14.1.4.2 Optimal PPFD as Affected by Other Environmental Factors

Net photosynthetic rate generally increases with increasing PPFD but is gradually saturated at a high PPFD, so that 80–90% of saturated PPFD is often considered to be optimal to obtain the highest CP. This optimal PPFD changes with other environmental factors such as air temperature, vapor pressure deficit (relative humidity), CO_2 concentration, air current speed, and nutrient solution composition.

The optimal PPFD changes also with other light environmental factors (light quality, photoperiod, and lighting direction). Furthermore, the optimal PPFD changes significantly with growth stage, plant canopy architecture (leaf area, leaf angle, planting density, etc.), physiological status, and genetic characteristics (cultivar) of plants, etc.

As discussed above, the relationships among the environment, plant growth, quality of produce, and the CP are complicated, so that it is necessary to conduct a series of experiments with minimum labor hours to find an optimal combination of environmental factors (the number of factors is as many as 16 or more). In order to conduct a factorial experiment for 16 factors each with 3 levels, we need an experiment with $3^{\hat{}16}$ (=43 millions) treatments. On the other hand, in the current LED lighting system, the electricity cost is significantly high, and light energy use efficiency is considerably low. Thus, developing a smart LED lighting system for the next-generation PFAL is crucial.

14.1.5 Challenges for Smart LED Lighting Systems

14.1.5.1 Starting Points

Some starting points to develop smart LED lighting systems are given as examples below:

1. The traditional idea of providing the light downward from the top to the plants is discarded. We may find unexpected plant responses useful for humans under controlled environments which have never occurred in nature. For examples, upward lighting, upward air current from underneath, monochromatic light lighting, dark period temperature much higher than photoperiod temperature, fluctuating CO_2 concentrations between 500 μmol mol^{-1} and 2000 μmol mol^{-1}, etc.
2. A large-scale experiment to reveal the environmental effects on plant traits is conducted during commercial production in a PFAL. This is possible by using an automated phenotyping system described later to measure the plant traits under unintentional or unavoidable fluctuations of environments and/or faint genomic variations.
3. During the above experiment, light quality, lighting cycle, lighting direction, and other environmental factors in a limited cultivation space are dynamically changed, and their time-series data on spatial fluctuations are recorded, and the plant responses are measured.
4. For the big data analysis, artificial intelligence (AI), multivariate statistical and mechanistic models are all used.
5. The database is connected with Internet, and the big data are shared with other groups.

14.1.5.2 Basic Ideas

As the first step, we assume that our objective of plant production in the PFAL is to maximize the dry weight of produce multiplied by its unit economic value per unit production cost. Then, basic ideas for the LED lighting system would be as follows:

1. Light energy be absorbed by plant leaves only;
2. Light energy be absorbed by all parts of all leaves equally even at high planting density with high LAI;
3. Environmental factors except light environmental factors be controlled to maximize the light energy use efficiency;
4. Time-dependent set points of light environmental factors (PPFD, light quality, lighting cycle, and lighting direction) be optimized to meet the objective of PFAL operation mentioned above.

When LEDs are installed at the ceiling of cultivation space, and the light is provided downward to the plant canopy with high LAI, most light energy is absorbed by upper leaves. In this case, total net photosynthetic rate or dry mass increase of plant canopy should be significantly lower compared to that of the plant canopy which receives the light equally by all parts of all leaves as mentioned in the above paragraph.

14.1.5.3 Simple Examples

A simple example of LED lighting system based on the basic ideas described above is a soft and movable plastic string with many LEDs at 1–2 cm intervals used for upward lighting in the PFAL. This LED lamp placed on the culture panel for upward lighting from underneath increased the dry and fresh weights of marketable portion of cos lettuce (*Lactuca sativa* L.) plants by retarding the senescence of lower leaves [8]. A similar but rigid bar-type LED lamp has been used for sideward (interplant) supplemental lighting in the greenhouse [9].

A small LED lamp with a small battery or fixed at one end of thin string can be placed at the central part of the crisp head lettuce or cabbage seedlings or young plants. Then, the LED illuminates the plant from the bottom center of head to all. With this inside-out lighting, the color of whole head will become green due to the production of chlorophylls in inner leaves. A similar effect can be expected by illuminating the head downward using green LED which penetrates the head (transmittance of a green leaf for green light is 30–40%) (Lunatic Asylumz et al., unpublished).

14.1.5.4 Smart LED Lighting System

Recent advancement of information and communication technology (ICT) are remarkable, which includes AI, big data mining, Internet of Things (IoT), virtual reality, camera/laser image capturing, and processing for intact measurement of plant architecture and physiological status. Advancement of omics (metabolomics, proteomics, transcriptomics, and genomics) is also remarkable. By introducing the abovementioned ICT and omics, we can challenge for developing a smart LED lighting system to search the time-dependent set points of environmental factors to maximize the CP under given constraints.

The smart LED lighting system for the PFAL is considered to be advantageous over the current LED lighting system, because the environmental factors, resource (electricity, light energy, CO_2, nutrient solution, and labor) supply rates and their use efficiencies, and plant responses can be measured continuously. These data are relatively accurate and can be used for big data mining using AI, multivariate analysis, and mechanistic models. Recent advancement of phenotyping technology is making it possible to get the time-series data of plant traits.

Plant phenotyping is defined as application of methodologies and protocols to measure a plant-specific trait noninvasively, ranging from the cellular level to canopy level, related to plant structure and function [10]. Plant traits include contents of water, nitrogen, chlorophyll, photosynthetic efficiency (PSII), leaf area, canopy structure, etc. These traits are often measured using image sensors including visible camera, spectroradiometry, light detection and ranging (LiDAR) device, visible to far-red hyperspectral camera, thermal camera, fluorescence, and 3D laser scanning, trichromatic (RGB) sensors.

The data on phenome (whole sets of expressions of the genome) obtained by phenotyping are related to the environmental data, and the related results can be used for improving the set points of environmental factors. The analyzed results can also be related with genome data to find a specific gene or a group of genes which are expressed under the specific environment. Then, we can use the gene or a gene group as a gene marker or gene markers for breeding and selection of a plant having a special trait.

Spatial variations of plant growth and other traits will become measurable by plant phenotyping as described above. Spatial variations of environmental factors in the PFAL are measurable already using relatively inexpensive sensors. On the other hand, it is costly to make the spatial variations of environments and resulting plant traits being negligibly small. Under the above conditions, we may discriminate the variations of plant traits due to environmental variations from those due to genetic variations. Then, the former information can be used to find the environmental effects on plant traits, which can be used for improving the set points of environmental factors. The latter information can be used for selecting plants with elite traits for breeding and genomic analysis.

14.2 Case Study: LED Lighting for Lettuce Seedlings in PFAL

In PFAL, seedling stage is often produced in an area separate from production area due to the differences in planting intensity and environmental conditions [11–14]. High-quality seedlings should be compact, with short internodes and firm stems, dark green leaves, and large white roots. Lighting environment (PPFD, quality, and photoperiod) influences seedling growth and quality, which affects subsequent growth and yield after transplanting. Effect of light quality on plant morphology and physiology is more complex compared with that of PPFD and photoperiod [15]. Lettuces have different morphological and physiological responses to specific spectrum, and they are more sensitive to red and blue lights than the other wavelengths [16]. However, monochromatic red or blue lights could not satisfy the requirement of normal plant growth. Leaf length, leaf width of the maximum leaf blade, stem length, and hypocotyl length were decreased with an increase of blue photon flux in lettuce seedlings, and the pattern is likely attributed to blue photon flux which simulated cryptochrome photoreceptor [17, 18]. Cucumber seedlings under red light had unexpected greater plant height, hypocotyl, and epicotyl length than seedlings under the combination of red and blue light [13]. Red light at seedling stage also decreased leaf thickness, palisade tissue, and inhibited chloroplast development and delayed flowering time of female flower of cucumber plant [19]. The combination of red and blue lights had been investigated by many researchers in lettuce [16, 20, 21]; as a general trend, the biomass accumulation of lettuce was decreased with the increase of blue photon flux. Nevertheless, many researchers focus on the light environment at vegetative stage of lettuce, and different PPFDs, photoperiods, or light qualities were used at seedling stage [16, 21–23]. However, little is known about different PPFDs and photoperiods at seedling stage of hydroponic lettuce with its subsequent growth and nutritional values under LEDs with different light spectra.

In this and next sections, we are presenting two case studies, which are parts of a master graduate research project [24] and a doctoral dissertation [25]. The objectives were to investigate the effects of PPFD, photoperiod, and light quality on seedling quality and growth and the residual effects after transplanting on growth and nutritional values of hydroponic lettuce.

14.2.1 Plant Materials and Experiment Design

14.2.1.1 Plant Materials and Growth Conditions

Seeds of lettuce (*Lactuca sativa* L. cv. Frill ice) were sown in sponge cube (23 mm × 23 mm × 23 mm) filled with water in plastic containers (52 cm × 36 cm × 9 cm). After sowing, lettuce seedlings were randomly divided into nine treatments (Table 14.3 as Expt. 1) or 12 treatments (Table 14.4 as Expt. 2)

Table 14.3 Details of treatments: nine combinations of three photosynthetic photon flux density (P) and three photoperiods (H) for hydroponic lettuce seedlings grown under LED with R:B ratio of 1.2

Photosynthetic photon flux density $\mu mol\ m^{-2}\ s^{-1}$	Photoperiod h day^{-1}	DLI mol m^{-2} day^{-1}	Treatment symbol
150	12	6.48	P150-H12
200	12	8.64	P200-H12
250	12	10.80	P250-H12
150	14	7.56	P150-H14
200	14	10.08	P200-H14
250	14	12.60	P250-H14
150	16	8.64	P150-H16
200	16	11.52	P200-H16
250	16	14.40	P250-H16

Note: DLI (daily light integral, mol m^{-2} day^{-1}) = Photosynthetic photon flux density ($\mu mol\ m^{-2}\ s^{-1}$) × photoperiod (h day^{-1}) × 3600 (s h^{-1}) × 10^{-6}

Table 14.4 Details of treatments: photosynthetic photon flux density (PPFD), photoperiod (H), and light quality (light source with R:B ratio) for hydroponic lettuce at seedling stage

Lighting source	R:B ratio	PPFD $\mu mol\ m^{-2}\ s^{-1}$	Photoperiod h day^{-1}	DLI mol m^{-2} day^{-1}	Treatment symbol
Fluorescent lamp	1.8	200	14	10.08	F1.8-P200-H14
		200	16	11.52	F1.8-P200-H16
		250	14	12.60	F1.8-P250-H14
		250	16	14.40	F1.8-P250-H16
LED	1.2	200	14	10.08	L1.2-P200-H14
		200	16	11.52	L1.2-P200-H16
		250	14	12.60	L1.2-P250-H14
		250	16	14.40	L1.2-P250-H16
LED	2.2	200	14	10.08	L2.2-P200-H14
		200	16	11.52	L2.2-P200-H16
		250	14	12.60	L2.2-P250-H14
		250	16	14.40	L2.2-P250-H16

for 20 days in a walk-in growth chamber (China Agricultural University, Beijing, China). One hundred lettuce seedlings were grown in each treatment. When the cotyledons were fully expanded, lettuce seedlings were transplanted to 128-cell trays, and they were placed in plastic containers (52 cm × 36 cm × 9 cm) containing nutrient solution. After 20 days of treatments, 35 uniform seedlings were transplanted from 128-cell trays to hydroponic cultivation beds with a planting density of 32 plants m^{-2}, and grown under PPFD of 250 $\mu mol\ m^{-2}\ s^{-1}$ with photoperiod of 16 h day^{-1} provided by LEDs with R:B ratio of 1.2 (WR-LED-16W, Beijing Lighting Valley Technology Co., Ltd., China) for 20 days.

The air temperature and relative humidity at photoperiod and dark period were maintained at 22 ± 1 °C/18 ± 1 °C and 70 ± 5%/65 ± 5%, respectively, and CO_2 concentration was controlled at 800 ± 50 μmol mol^{-1} from seeding to harvest. A Yamasaki lettuce nutrient solution in pH of 6.0–6.5 and EC of 1.0–1.2 mS cm^{-1} was utilized as a standard solution. Pure water was used for the first 2 days after seeding. Then, a quarter strength of standard nutrient solution was used for the cotyledon stage, and one third strength was used for the 1–2 true leaves stage. When lettuce had 2–3 true leaves, half strength nutrient solution was used. After transplanting, a full strength nutrient solution was applied, and the nutrient solution was replaced every 10 days.

14.2.1.2 Treatment and Experiment Design

In Expt. 1, lettuce seedlings were grown under nine combination of three levels of PPFD at 150, 200, or 250 μmol m^{-2} s^{-1} and three photoperiods of 12, 14, or 16 h day^{-1} provided by LED lamps with R:B ratio of 1.2 (WR-LED-16W, Beijing Lighting Valley Technology Co., Ltd., China) for 20 days, a total of nine treatments (Table 14.3). In Expt. 2, lettuce seedlings were grown under combination of two levels of PPFD at 200 and 250 μmol m^{-2} s^{-1} and two photoperiods of 14 and 16 h day^{-1} under three kinds of artificial light with different R:B ratios for 20 days, a total of 12 treatments (Table 14.4). The three light sources are fluorescent lamp (T5-28W, Shanghai Flower and Biology Lighting Co., Ltd., China) with R:B ratio of 1.8 and two LED lamps with R:B ratios of 1.2 and 2.2 (WR-16W, Beijing Lighting Valley Technology Co., Ltd., China), respectively. Figure 14.1 shows the spectral distribution of different lamps determined for light wavebands ranging from 300 nm to 800 nm at 15 cm below the lamps with a PPFD of 250 μmol m^{-2} s^{-1} by using fiber spectrometer (AvaField-2, Avantes Inc., Apeldoorn, the Netherlands). The PPFDs were measured at 15 cm above the bench top which is the same to 15 cm below the lamps using a quantum sensor (LI-190SA, LI-COR Inc., Lincoln, NE, USA) connected to a data-logger (LI-1400, LI-COR Inc., Lincoln, NE, USA). Daily light integral (DLI) was calculated as follows:

$$\text{DLI}\left(\text{mol m}^{-2}\,\text{day}^{-1}\right) = \text{PPFD}\left(\mu\text{mol m}^{-2}\,\text{s}^{-1}\right) \times \text{photoperiod}\left(\text{h day}^{-1}\right) \times 3600\left(\text{s h}^{-1}\right)/10^6.$$

14.2.1.3 Measurements for Lettuce Growth and Quality

Leaf length (LL) and leaf width (LW) of maximum leaf blade, leaf dry weight, and root dry weight were measured for 6 plants selected from each treatment at 20 days after sowing (DAS). Leaf fresh and dry weights, root fresh and dry weights, and contents of nitrate N, vitamin C, and soluble protein were determined for 6 lettuce

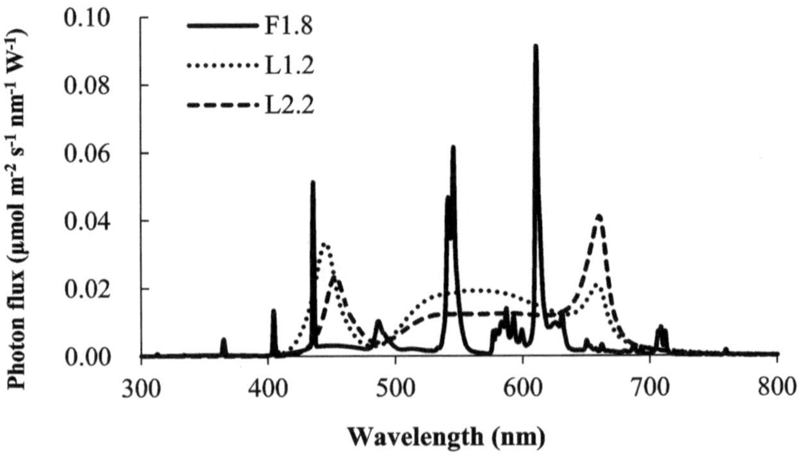

Fig. 14.1 Spectral distribution of fluorescent lamp with R:B ratio of 1.8 (F1.8), and two LED lamps with R:B ratios of 1.2 and 2.2, respectively (L1.2 and L2.2)

plants selected from each treatment at 40 DAS in Expt. 1. Lettuce quality indicators such as nitrate N, vitamin C, soluble sugar, soluble protein, and anthocyanin contents were determined for all treatments, six replicates per treatment, sampled on the seventeenth day after transplanting in Expt. 2. Nitrate N content was determined at wavelength of 410 nm by using a spectrophotometer (UV3150, Shimadzu Productions Co., Japan) based on coloration method of sulfosalicylic acid [26]. Vitamin C content was determined by the 2,6-dichlorophenol indophenol titration method [26]. Soluble sugar content was determined at wavelength of 630 nm by using the same spectrophotometer based on anthrone-sulfuric acid colorimetry method [26]. Coomassie Brilliant Blue G-250 dye method [26] was used to measure soluble protein content at wavelength of 595 nm by using the same spectrophotometer.

14.2.2 *Effect of Lighting Environment at Seedling Stage on Leaf Morphology and Growth of Lettuce Seedlings*

In Expt. 1, PPFD and photoperiod significantly affected morphology and growth of lettuce seedlings (Fig. 14.2). Leaf length was longest and ratio of leaf length to leaf width was greatest in the treatment of P150-H12, and they decreased with the increase of PPFD or photoperiod. LL and LL/LW decreased with the increase of DLI in a logarithmic fashion. LL and LL/LW at initial DLI of 6.48 mol m^{-2} day^{-1} were 46.5% and 70.2% higher than those under DLI of 14.4 mol m^{-2} day^{-1}, which was in treatment of P250-H16, the highest DLI among all treatments. Leaf and root dry weights at 20 DAS in Expt. 1 were highest in treatment of P250-H16 and lowest

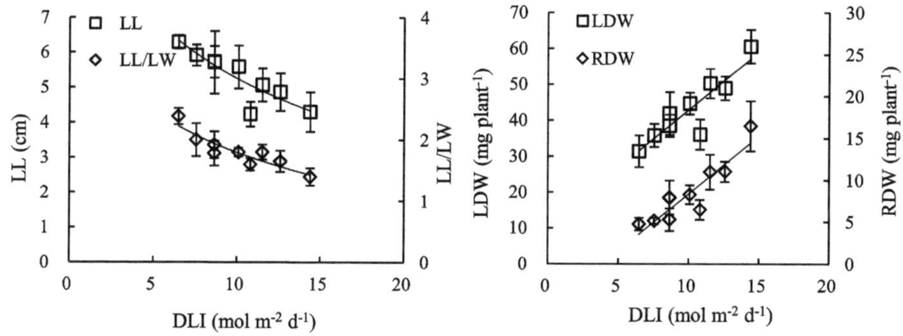

Fig. 14.2 Relationships between daily light integral (DLI) and leaf length (LL), ratio of leaf length to leaf width (LL/LW), and leaf dry weight (LDW) and root dry weight (RDW) of hydroponic lettuce transplants at 20 days after sowing. Regression equations are: LL = −2.53 * ln(DLI) + 11.07 (R^2 = 0.79), and LL/LW = −ln(DLI) + 4.09 (R^2 = 0.84), LDW = 3.22 × DLI + 10.65 (R^2 = 0.81), and RDW = 1.38 × DLI-5.41 (R^2 = 0.83)

in treatment of P150-H12. Both leaf and root dry weights increased linearly with DLI. Leaf and root dry weights at DLI of 14.4 mol m^{-2} day^{-1} were 93% and 244% higher than those grown at DLI of 6.48 mol m^{-2} day^{-1}, respectively.

In Expt. 2, light quality significantly influenced LL and LW but not LL/LW (Table 14.5). Generally, lettuce seedlings grown under FL were bigger regarding leaf size and accumulated more dry weight in both leaves and roots compared to those grown under LEDs. Higher PPFD led to a lower LL with the same photoperiod under FL and LEDs with R:B ratio of 2.2, and no significant difference was found in LL under LEDs with R:B ratio of 1.2. Under LED with R:B ratio of 1.2, higher PPFD led to lower LL/LW with the same photoperiod. Under FL and LED at R:B ratio of 1.2, no significant differences in LW were found among treatments. At LEDs with R:B ratio of 2.2, LW was smallest in the treatment of L2.2-P200-H16, followed by L2.2-P250-H16. Generally, higher PPFD resulted in lower LL/LW, regardless of light source. Leaf dry weight was higher in the treatment of F1.8-P200-H16, followed by F1.8-P250-H14 and F1.8-P250-H16. Root dry weight was highest in the treatment of F1.8-P200-H16 among all treatments. Higher PPFD led to higher root dry weight under LEDs with R:B ratio of 1.2.

Seedlings with high quality should have a higher dry mass and lower LL/LW. Our results showed that leaf and root dry weight increased linearly with the increase of DLI, and LL and LL/LW decreased with DLI in a logarithmic fashion, which agreed with the study reported by Kitaya et al. [27]. PPFD plays an important role for seedling morphology and growth. Low PPFD inhibits plant growth and productivity, whereas excess PPFD has detrimental effects on the photosynthetic apparatus, and plants have developed sophisticated biochemical and physiological mechanisms in order to adapt to the light environment [28–30]. Health index, dry weight, and fresh weight of young tomato plants increased with the increase of PPFD, and they were superior in plants grown under 300, 450, and 500 μmol m^{-2} s^{-1} than those under

Table 14.5 Effects of photosynthetic photon flux density (PPFD), photoperiod (H), and light quality (LQ) during seedling stage on leaf morphology and dry weight of hydroponic lettuce transplants at 20 days after sowing

Treatment	Leaf length	Leaf width	LL/LW	Leaf dry weight	Root dry weight
	cm	cm		mg plant^{-1}	mg plant^{-1}
F1.8-P200-H14	7.75 ± 0.63 a	6.12 ± 0.67 a	1.28 ± 0.14 ab	242.0 ± 44.3 c	31.7 ± 7.4 c
F1.8-P200-H16	7.55 ± 0.33 ab	6.50 ± 0.42 a	1.16 ± 0.11 bc	317.7 ± 34.3 a	42.5 ± 6.9 a
F1.8-P250-H14	7.07 ± 0.41 b	6.42 ± 0.47 a	1.13 ± 0.10 cd	274.5 ± 41.1 b	40.5 ± 10.0 a
F1.8-P250-H16	6.73 ± 0.69 c	6.44 ± 0.74 a	1.02 ± 0.05 d	267.2 ± 50.3 bc	34.1 ± 10.3 b
L1.2-P200-H14	6.78 ± 0.25 c	5.30 ± 0.22 d	1.27 ± 0.12 ab	177.2 ± 25.7 e	23.1 ± 5.4 d
L1.2-P200-H16	6.70 ± 0.28 cd	5.00 ± 0.52 d	1.34 ± 0.08 a	222.1 ± 34.4 cd	27.4 ± 3.5 cd
L1.2-P250-H14	6.40 ± 0.29 d	5.33 ± 0.25 d	1.14 ± 0.05 c	192.9 ± 21.7 de	31.9 ± 4.0 c
L1.2-P250-H16	6.38 ± 0.41 d	5.36 ± 0.24 cd	1.16 ± 0.04 bc	255.4 ± 32.6 bc	34.8 ± 5.1 ab
L2.2-P200-H14	7.73 ± 0.28 a	6.00 ± 0.43 ab	1.29 ± 0.08 ab	197.1 ± 13.4 de	23.1 ± 5.1 d
L2.2-P200-H16	6.82 ± 1.02 c	5.33 ± 0.73 d	1.29 ± 0.21 ab	224.3 ± 24.7 cd	30.9 ± 5.1 c
L2.2-P250-H14	7.03 ± 0.50 b	5.94 ± 0.26 b	1.17 ± 0.07 b	233.8 ± 34.6 cd	31.6 ± 6.5 c
L2.2-P250-H16	5.98 ± 0.25 e	5.50 ± 0.37 c	1.09 ± 0.09 cd	208.7 ± 14.7 d	30.8 ± 11.2 c
ANOVA					
PPFD	*	NS	*	NS	*
H	*	NS	NS	*	NS
LQ	*	*	NS	*	*
PPFD × H	NS	NS	NS	*	*
PPFD × LQ	NS	NS	NS	NS	NS
H × LQ	*	*	NS	NS	NS
PPFD×H × LQ	NS	NS	NS	NS	NS

Note: Means with different letters in the same columns were significantly different tested by LSD's multiple comparison at $P \leq 0.05$. NS, * represent no significant and significant difference at $P \leq 0.05$, respectively

50, 150, 200 μmol m^{-2} s^{-1} [31]. Our study showed that lettuce seedlings grown at PPFD of 150 μmol m^{-2} s^{-1} with photoperiod of 12 h day^{-1} resulted in longest LL and greatest LL/LW, an undesirable trait for seedlings. These adaptations maximized the capture of available light to meet the demand for photosynthesis [32]. Kitaya et al. [27] found that dry mass, percent dry mass, and leaf number increased linearly as DLI increased at 21 DAS. In addition to lettuce, linear relationships between shoot dry weight and DLI were also found in petunia seedlings and pansy seedlings [33].

14.2.3 Effects of PPFD and Photoperiod at Seedling Stage on Mature Lettuce Growth and Quality

Leaf and root fresh weights at 40 DAS were significantly affected by PPFD and photoperiod during seedling stage (Table 14.6). Leaf and root fresh weights were greatest in treatment of P200-H16. Leaf fresh weight reached to 96 g $plant^{-1}$ in treatment of P200-H16, which was 26.8% and 17.3% higher compared with those grown in treatments of P150-H16 and P250-H16, respectively. The PPFD levels during seedling stage did not affect the leaf and root dry weights of mature lettuce plants at 40 DAS. However, photoperiod of 16 h day^{-1} resulted in higher leaf and root dry weights compared to that at photoperiod of 12 and 14 h day^{-1}.

PPFD and photoperiod during seedling stage affected vitamin C content significantly but not nitrate content. Lettuce seedlings grown under a photoperiod of 12 h day^{-1} had lower vitamin C content compared to those under a photoperiod of 16 h day^{-1}. No differences in vitamin C content were found among those under a photoperiod of 16 h day^{-1}. For soluble protein content, generally lower PPFD led to lower content, and photoperiod had no significant effect.

14.2.4 Effects of PPFD, Photoperiod, and Light Quality at Seedling Stage on Mature Lettuce Growth and Quality

PPFD during seedling stage affected leaf and root weights of mature lettuce at 40 DAS (Table 14.7). Among the three light sources, FL generally led to higher leaf fresh weight, followed by LEDs with R:B ratio of 2.2. Leaf fresh weights were biggest in treatments of L2.2-P200-H16, L2.2-P250-H14, and F1.8-P200-H16 and smallest in treatments of L1.2-P200-H14, L1.2-P250-H16, and L2.2-P250-H16. Lettuce seedlings grown at PPFD of 200 μmol m^{-2} s^{-1} with photoperiod of 16 h day^{-1} had higher leaf fresh weight than those grown at 250 μmol m^{-2} s^{-1} with photoperiod of 16 h day^{-1} at 40 DAS regardless of light quality. Root fresh weight grown under FL and LEDs with R:B ratio of 1.2 was higher than those under

Table 14.6 Effects of photosynthetic photon flux density (PPFD) and photoperiod (H) during seedling stage on subsequent growth and nutritional qualities of hydroponic lettuce at 40 days after sowing

Treatment	Leaf fresh weight	Root fresh weight	Leaf dry weight	Root dry weight	Nitrate content	Vitamin C content	Soluble protein content
	g plant^{-1}	g plant^{-1}	g plant^{-1}	g plant^{-1}	mg kg^{-1}	mg 100 g^{-1}	mg g^{-1}
P150-H12	59.59 ± 7.88 d	5.52 ± 0.52 d	1.74 ± 0.48 c	0.19 ± 0.04 bc	3066 ± 514 a	23.75 ± 1.64 d	4.48 ± 0.70 c
P200-H12	65.97 ± 3.81 cd	6.54 ± 0.70 d	1.95 ± 0.35 bc	0.20 ± 0.03 b	2594 ± 295 ab	26.86 ± 3.64 c	5.13 ± 1.04 b
P250-H12	65.98 ± 3.46 cd	7.60 ± 0.82 c	1.73 ± 0.18 cd	0.21 ± 0.03 b	2398 ± 398 b	28.77 ± 2.80 bc	5.35 ± 0.34 ab
P150-H14	56.51 ± 1.70 d	5.43 ± 0.38 d	1.86 ± 0.15 c	0.16 ± 0.02 c	2225 ± 353 b	24.61 ± 1.32 d	4.88 ± 0.56 bc
P200-H14	74.09 ± 9.73 bc	6.67 ± 0.69 cd	2.03 ± 0.59 bc	0.22 ± 0.04 b	2553 ± 372 b	30.63 ± 3.75 ab	5.23 ± 0.78 ab
P250-H14	68.88 ± 7.00 c	6.26 ± 0.48 d	1.36 ± 0.06 d	0.23 ± 0.03 b	2443 ± 322 b	32.95 ± 1.61 a	5.22 ± 0.70 ab
P150-H16	75.69 ± 8.42 bc	7.90 ± 0.89 bc	2.04 ± 0.44 b	0.34 ± 0.04 a	2517 ± 433 b	29.27 ± 2.32 b	4.92 ± 0.66 bc
P200-H16	95.99 ± 13.31 a	9.93 ± 0.62 a	2.63 ± 0.64 a	0.36 ± 0.03 a	2679 ± 792 ab	32.03 ± 5.13 a	5.69 ± 0.90 ab
P250-H16	81.86 ± 14.00 b	8.28 ± 1.84 b	2.56 ± 0.62 a	0.38 ± 0.10 a	2539 ± 243 b	29.46 ± 1.17 ab	6.21 ± 1.40 a
ANOVA							
PPFD	*	*	NS	NS	NS	*	*
H	*	*	*	*	NS	*	NS
PPFD × H	NS	*	NS	NS	NS	NS	NS

Note: Means with different letters in the same columns were significantly different tested by LSD's multiple comparison at $P \leq 0.05$. NS, * represent no significant and significant difference at $P \leq 0.05$, respectively

Table 14.7 Effects of photosynthetic photon flux density (PPFD), photoperiod (H), and light quality (LQ) during seedling stage on subsequent growth and nutritional qualities of hydroponic lettuce at 40 days after sowing

Treatment	Leaf fresh weight	Root fresh weight	Leaf dry weight	Root dry weight	Nitrate content	Vitamin C content	Soluble protein content
	g $plant^{-1}$	g $plant^{-1}$	g $plant^{-1}$	g $plant^{-1}$	mg kg^{-1}	mg 100 g^{-1}	mg g^{-1}
F1.8-P200-H14	79.85 ± 8.52 bc	12.82 ± 3.00 a	3.39 ± 0.45 b	0.33 ± 0.04 de	4030 ± 270 ab	20.32 ± 2.91 cd	5.48 ± 0.75 c
F1.8-P200-H16	92.85 ± 13.02 a	12.91 ± 2.49 a	3.52 ± 0.36 ab	0.40 ± 0.04 c	4265 ± 186 a	30.52 ± 2.59 a	7.45 ± 0.70 a
F1.8-P250-H14	78.16 ± 5.51 bc	10.01 ± 0.52 bc	3.23 ± 0.28 c	0.39 ± 0.03 cd	3507 ± 239 ab	29.19 ± 4.59 ab	6.97 ± 1.17 ab
F1.8-P250-H16	80.52 ± 9.25 bc	11.89 ± 1.70 ab	3.37 ± 0.12 bc	0.53 ± 0.08 b	3414 ± 359 ab	31.25 ± 6.87 a	7.30 ± 1.02 ab
L1.2-P200-H14	70.09 ± 2.39 d	11.13 ± 2.18 ab	2.99 ± 0.33 d	0.35 ± 0.01 d	3290 ± 447 b	18.54 ± 3.40 cd	6.42 ± 0.63 b
L1.2-P200-H16	82.79 ± 10.76 b	12.98 ± 2.40 a	3.90 ± 0.23 a	0.51 ± 0.07 b	3618 ± 399 ab	21.37 ± 6.49 cd	6.35 ± 0.55 b
L1.2-P250-H14	75.34 ± 10.56 c	10.63 ± 1.26 b	3.16 ± 0.35 cd	0.35 ± 0.04 d	3375 ± 337 b	18.32 ± 3.73 d	6.15 ± 0.92 bc
L1.2-P250-H16	72.47 ± 12.28 cd	11.29 ± 1.05 ab	3.24 ± 0.22 bc	0.63 ± 0.07 a	3003 ± 244 b	18.30 ± 1.93 d	5.57 ± 1.08 bc
L2.2-P200-H14	76.25 ± 4.71 c	7.77 ± 1.20 d	3.05 ± 0.27 cd	0.31 ± 0.05 e	4008 ± 806 a	22.93 ± 1.52 c	5.36 ± 1.02 c
L2.2-P200-H16	93.20 ± 13.15 a	9.19 ± 0.75 c	3.65 ± 0.44 ab	0.40 ± 0.02 c	3547 ± 239 ab	27.19 ± 6.04 ab	5.48 ± 0.76 c
L2.2-P250-H14	84.36 ± 7.96 b	7.97 ± 1.03 d	3.33 ± 0.43 bc	0.30 ± 0.04 e	3822 ± 1067 ab	23.98 ± 3.48 b	5.88 ± 1.33 bc
L2.2-P250-H16	70.29 ± 7.44 cd	8.65 ± 1.26 cd	3.00 ± 0.28 cd	0.41 ± 0.04 c	3450 ± 531 ab	26.46 ± 3.61 ab	7.69 ± 1.12 a
ANOVA							
PPFD	*	*	*	*	*	NS	*
H	NS	*	*	*	NS	*	*
LQ	*	*	NS	*	NS	*	*
PPFD × H	*	NS	*	*	NS	NS	NS
PPFD × LQ	NS	NS	NS	*	*	NS	*
H × LQ	NS	NS	NS	*	NS	NS	*
PPFD×H × LQ	NS	NS	*	NS	NS	NS	*

Note: Means with different letters in the same columns were significantly different tested by LSD's multiple comparison at $P \leq 0.05$. NS, * represent no significant and significant difference at $P \leq 0.05$, respectively

LEDs with R:B ratio of 2.2, and no significant difference was found between FL and LEDs with R:B ratio of 1.2 at the same DLI. Lettuce grown under PPFD at 200 μmol m^{-2} s^{-1} with a photoperiod of 16 h day^{-1} had higher leaf dry weight than other treatments regardless of light quality, and no significant difference was found in leaf dry weight between two kinds of LEDs at the same DLI; however, lettuce grown under LEDs with R:B ratio of 1.2 had higher root dry weight than those under LEDs with R:B ratio of 2.2 at the same DLI.

Under each light source, there was no significant difference in nitrate content among treatment, that is, no significant difference between the two PPFDs and two photoperiods. Photoperiod and light quality affected vitamin C and soluble protein contents significantly. Similar to Expt. 1, no significant differences were found as the photoperiod extended from 14 h day^{-1} to 16 h day^{-1} in vitamin C and soluble protein contents under LEDs with R:B ratio of 1.2. Lettuce grown under FL or LEDs with R:B ratio of 2.2 had higher vitamin C content than those under LEDs with R:B ratio of 1.2. Generally, FL led to higher vitamin C and soluble protein content than those under LEDs.

Light quality shows much more complex effects on plant morphology and physiology compared with PPFD and photoperiod [16]. Continuous red light can elicit photomorphogenic seedling development, and led to robust cotyledon expansion in wild-type *Arabidopsis* seedlings [34]. Blue light is a very effective inhibitor of hypocotyl elongation and inducer of cotyledon expansion because blue light stimulates the phytochromes and the blue light photoreceptors [35, 36]. In the present study, lettuce seedlings had shorter LL under LEDs with R:B ratio of 1.2 compared with those under LEDs with R:B ratio of 2.2 with a fixed photoperiod of 14 h day^{-1}, which indicated that elongation of lettuce leaf was inhibited by LEDs containing more blue photon flux under shorter photoperiod (Table 14.7).

The optimal R:B ratio differed in the plants species, growth periods, and the targeted qualities. Shoot fresh weight and dry weight decreased with the increased blue photon flux under the combination of red and blue photon flux in cucumber seedlings [13]. Shoot and root dry weights of pepper seedlings grown under R:B ratio of 7.0 were higher compared with those under R:B ratio of 1.0, which indicated that a high proportion of red light positively affected the morphology and biomass accumulation of pepper seedlings [36]. Dry weight of lettuce seedlings was higher under the R:B ratio of 4.0 than those under R:B ratio of 9.0, 2.3, and 1.5, suggesting that lettuce seedlings was in favor of red light compared with blue light, and dry weight was inhibited by excessively high ratio of red to blue photon flux [37].

Leaf vegetables accumulate nitrate excessively, and nitrate can be converted to nitrite in human body, which is considered as a potential threat to human health [38]. Nitrate levels in leafy vegetables can be controlled by light conditions during seedling stage. Chang et al. investigated the effects of multiple wavelength-band LED lighting during different growth stages on nitrate content of lettuce, and their results indicated that a reduction of nitrate content was attributed to UV-A lighting during seedling stage [39]. In our study, no significant difference was found in nitrate content of mature lettuce in both experiments, suggesting that lettuce plants might have adjusted by the environment after transplanting. Johkan et al. reported that significant difference was found in total phenols and anthocyanin during seedling stage, while the differences diminished at the final cultivation [11].

Soluble proteins provide essential proteins for human body and are considered as an important nutrient in vegetables [40]. An increase in soluble protein content is important for improving the nutritional quality of vegetables for human consumption. PPFD at 250 μmol m^{-2} s^{-1} led to higher soluble protein content in lettuce than those grown under PPFD at 150 μmol m^{-2} s^{-1} (Table 14.6). The soluble protein content was higher under LEDs with R:B ratio of 1.2 than those under LEDs with R:B ratio of 2.2 with a fixed PPFD at 200 μmol m^{-2} s^{-1} (Table 14.7). Zhang et al. found that the soluble protein content of pea seedling grown under blue light was the highest compared with red light, white light, or mixture of red and blue lights [41].

Vitamin C must be supplemented by food because it cannot be synthesized and accumulated in human body [42]. Therefore, vitamin C content is an important index for evaluating vegetable nutritional quality [43]. In the present study, higher vitamin C content was obtained in lettuce grown under a higher PPFD or longer photoperiod. No significant differences were found in vitamin C content between those grown under FL and LEDs with R:B ratio of 2.2 with photoperiod of 16 h day^{-1}. Qian et al. reported that no significant difference was observed in vitamin C content of lettuce grown under FL and LEDs with R:B ratio of 0.9 [44]. Vitamin C content of mature lettuce grown under LEDs with R:B ratio of 2.2 was higher than those under LEDs with R:B ratio of 1.2 (Table 14.7), which indicated that a high proportion of red light positively affected the vitamin C accumulation. This conclusion agreed with the findings of study reported by Qian et al. [44] that red light enhanced the vitamin C content compared with blue light in Chinese kale sprout. However, Pinho et al. found that there was no significant difference in vitamin C content in lettuce grown under white-red LEDs with W:R ratio of 1.0 or 4.0 [43]. Ohashi et al. found that vitamin C contents were higher in lettuce grown under blue FL and the combination of red and blue FL with R:B ratio of 2.0 than those under red FL under PPFD at 300 μmol m^{-2} s^{-1} with a photoperiod of 12 h day^{-1} [45]. These studies indicated that nutritional response to light quality is dependent on the species or experimental conditions.

Nutritional values of lettuce can be enhanced by employing different spectra strategically, and the administration of a combination of LEDs at seedling stage led to favorable plant growth [22, 39, 46]. Based on seedling quality and its subsequent growth and nutritional values, PPFD at 200 μmol m^{-2} s^{-1} with photoperiod of 16 h day^{-1} provided by FLs or LEDs with R:B ratio of 2.2 was recommended for seedling production of hydroponic lettuce under controlled environment. Our results can be used as basic information for future studies using LEDs with different R:B ratios at seedling stage. Furthermore, studies on effects of supplemental light quality, including ultraviolet A or far-red under LEDs as the main light source at seedling stage on growth and its late nutritional qualities of hydroponic lettuce may be required.

14.3 Case Study: LED Lighting for Hydroponic Lettuce in PFAL

In this section, which is part of a doctoral dissertation [25], effects of light intensity, photoperiod, and light quality on growth and quality of hydroponic lettuce were investigated to find out the optimal LED lighting condition for commercial lettuce production in PFAL. In addition, the photosynthetic property was examined continuously for 2 days under a combination of light intensity and photoperiod.

14.3.1 Plant Materials and Experiment Design

14.3.1.1 Plant Materials and Growth Conditions

Seeds of purple leaf lettuce (*Lactuca sativa* L. cv. "Ziwei") were sown in sponge cube (23 mm × 23 mm × 23 mm) in a hydroponic system. After germination, seedlings were grown for 18 days in a closed plant factory system (China Agricultural University, Beijing, China) under PPFD of 200 μmol m^{-2} s^{-1} provided by LEDs with R:B ratio of 1.2 (WR-LED-16W, Beijing Lighting Valley Technology Co., Ltd. China) with photoperiod of 16 h day^{-1}. When the seedlings had 5 true leaves, they were transplanted to the hydroponic cultivation bed under different lighting environment for 17 days. The air temperature and relative humidity at photoperiod and dark period were maintained at 22 ± 1 °C/18 ± 1 °C and 70 ± 5%/65 ± 5%, respectively, and CO_2 concentration was controlled at 800 ± 50 μmol mol^{-1} from seeding to harvest. The Yamasaki lettuce nutrient solution in pH of 6.0–6.5 and EC of 1.0–1.2 mS cm^{-1} was utilized to cultivate hydroponic lettuce as shown above in Sect. 14.2.1.

14.3.1.2 Treatment and Experiment Design

Twenty-four treatments were created by a combination of four levels of PPFD at 150, 200, 250, and 300 μmol m^{-2} s^{-1} and two photoperiods of 12 and 16 h day^{-1} (factorial design) under three kinds of artificial light with different R:B ratios as shown in Table 14.8. Fluorescent lamps (FLs) with an R:B ratio of 1.8 (T5-28W, Shanghai Flower and Biology Lighting Co., Ltd., China) and LED lamps with R:B ratios of 1.2 and 2.2 (WR-LED-16W, Beijing Lighting Valley Technology Co., Ltd. China) were placed at 35 cm above the cultivation bed. The spectral distribution of different lamps is shown above in Fig. 14.1.

Table 14.8 A list of treatments created by combinations of different photosynthetic photon flux density (PPFD), photoperiod (H), and three different lamps with different R:B ratios

Lighting source	R:B ratio	PPFD (μmol m^{-2} s^{-1})	Photoperiod (h day^{-1})	DLI (mol m^{-2} day^{-1})	Treatment symbol
Fluorescent lamp	1.8	150	12	6.48	F1.8-P150-H12
		200	12	8.64	F1.8-P200-H12
		250	12	10.80	F1.8-P250-H12
		300	12	12.96	F1.8-P300-H12
		150	16	8.64	F1.8-P150-H16
		200	16	11.52	F1.8-P200-H16
		250	16	14.40	F1.8-P250-H16
		300	16	17.28	F1.8-P300-H16
LED	1.2	150	12	6.48	L1.2-P150-H12
		200	12	8.64	L1.2-P200-H12
		250	12	10.80	L1.2-P250-H12
		300	12	12.96	L1.2-P300-H12
		150	16	8.64	L1.2-P150-H16
		200	16	11.52	L1.2-P200-H16
		250	16	14.40	L1.2-P250-H16
		300	16	17.28	L1.2-P300-H16
LED	2.2	150	12	6.48	L2.2-P150-H12
		200	12	8.64	L2.2-P200-H12
		250	12	10.80	L2.2-P250-H12
		300	12	12.96	L2.2-P300-H12
		150	16	8.64	L2.2-P150-H16
		200	16	11.52	L2.2-P200-H16
		250	16	14.40	L2.2-P250-H16
		300	16	17.28	L2.2-P300-H16

The three lamps are fluorescent lamp with R:B ratio of 1.8 (F1.8) and two LEDs with R:B ratio of 1.2 and 2.2 (L1.2 and L2.2), respectively. Daily light integral (DLI) is a product of PPFD and photoperiod

14.3.1.3 Growth Quality and Photosynthetic Measurements

In each treatment, one layer of cultivation shelves with a total of 18 plants was employed in experiments. At the end of the treatment, six plants were selected randomly in each treatment for evaluating growth and quality properties. The fresh weights (FW) of leaves and roots were recorded by an electronic analytical balance with accuracy of 0.1 mg (FA1204B, Shanghai Jingke Electr. Inc., Shanghai, China). The leaves and roots were dried in an oven at 105 °C for 3 h and subsequently set to 80 °C and dried for 72 h, and dry weight (DW) of leaves and roots was recorded. The third vigorously growing leaf from the top of the lettuce was used to determine the quality indicators.

Lettuce quality indicators such as nitrate N, vitamin C, soluble sugar, soluble protein, and anthocyanin contents were determined for all treatments, six replicates per treatment, sampled on the seventeenth day after transplanting. The measurement methods of nitrate N content, vitamin C, soluble sugar, and soluble protein were described above in Sect. 14.2.1. The anthocyanin content was measured at 530 nm and 600 nm by using the spectrometric method [46].

Net photosynthetic rates of lettuce leaves were measured continuously for 2 days in treatments of F1.8-P250-H12, F1.8-P250-H16, F1.8-P300-H12, and F1.8-P300-H16, one leaf per treatment, during the last 2 days before harvest by using two open path dynamic photosynthetic systems with four leaf chambers (MD-100A, Yizongqi Technology Co., Beijing, China). In this continuous photosynthesis measurement system, the differential CO_2 concentrations of the inlet and outlet were measured automatically by an infrared gas analyzer (LI-7000, LI-COR Inc., Lincoln, NE, USA) for calculating and recording the net photosynthetic rate every 8 min [47]. The light environment, CO_2 concentration, and temperature in the leaf cuvette were the same as those in the above same cultivation environment. The net photosynthetic rate and light response curve were measured by using two portable photosynthesis systems (LI-6400XT, LI-COR Inc., Lincoln, NE, USA).

Smart electronic metering (TP-9004, Shenzhen Northmeter Co., Ltd., Shenzhen, China) was used to monitor the total power consumption for each treatment. The recorded power consumption was divided by the value of 100 g fresh weight or 1 g dry weight of lettuce leaves. Light energy use efficiency (LEU) was determined as follows:

$$\text{LEU} = (k \times D)\,/\text{PAR}$$

where k is the conversion coefficient from dry weight to chemical energy (about 20 MJ kg^{-1}), D is the increase of dry weight in specific growth time per unit area (kg m^{-2}), and PAR is photosynthetically active radiation (W m^{-2}).

The electric energy use efficiency (EUE) was defined as:

$$\text{EUE} = h \times \text{LEU}$$

where h is the conversion factor from electric energy to PAR energy, which is around 0.195 for FLs (R:B ratio of 1.8), 0.382 for LEDs with R:B ratio of 1.2, and 0.326 for LED lamps with R:B ratio of 2.2 [48].

14.3.2 LED Lighting Affects Growth of Hydroponic Lettuce

The growth of hydroponic lettuce was significantly influenced by PPFD, photoperiod, and light quality (Table 14.9). Leaf fresh weights were highest in the treatments of F1.8-P250-H16 and F1.8-P300-H16. However, there were no differences in leaf fresh weight between the two LED lamps, regardless of PPFD and photoperiod.

Table 14.9 Effects of different photosynthetic photon flux density (PPFD), photoperiod (H), and light quality (LQ) on fresh weight (FW) and dry weight (DW) of leaves and roots of hydroponic lettuce at day 18 after transplanting ($n = 6$)

Treatment	Leaf FW ($g \cdot plant^{-1}$)	Root FW	Leaf DW	Root DW
F1.8-P150-H12	28.1 ± 1.1 de	3.06 ± 0.30 f	0.91 ± 0.06 e	0.14 ± 0.01 e
F1.8-P200-H12	34.5 ± 1.0 cd	3.90 ± 0.25 e	1.34 ± 0.04 d	0.17 ± 0.01 de
F1.8-P250-H12	40.3 ± 2.8 b	4.61 ± 0.45 cd	1.66 ± 0.13 c	0.19 ± 0.02 d
F1.8-P300-H12	36.2 ± 1.1 c	5.18 ± 0.18 c	1.48 ± 0.07 cd	0.23 ± 0.01 c
F1.8-P150-H16	37.7 ± 2.3 bc	3.82 ± 0.13 e	1.27 ± 0.08 d	0.16 ± 0.01 e
F1.8-P200-H16	47.3 ± 0.1 ab	5.32 ± 0.25 c	1.69 ± 0.20 bc	0.19 ± 0.01 d
F1.8-P250-H16	50.1 ± 2.9 a	6.01 ± 0.34 b	1.85 ± 0.09 bc	0.28 ± 0.02 b
F1.8-P300-H16	49.3 ± 6.5 a	6.41 ± 0.40 b	2.00 ± 0.08 ab	0.30 ± 0.02 b
L1.2-P150-H12	23.2 ± 1.9 e	2.94 ± 0.28 f	1.04 ± 0.09 e	0.11 ± 0.03 f
L1.2-P200-H12	28.4 ± 2.5 de	3.61 ± 0.35 ef	1.35 ± 0.01 d	0.15 ± 0.02 e
L1.2-P250-H12	31.1 ± 3.0 d	3.73 ± 0.23 e	1.44 ± 0.05 d	0.17 ± 0.01 de
L1.2-P300-H12	30.6 ± 3.6 d	5.14 ± 0.52 c	1.54 ± 0.16 cd	0.23 ± 0.02 c
L1.2-P150-H16	28.3 ± 2.9 de	4.23 ± 0.30 d	1.37 ± 0.13 d	0.21 ± 0.02 cd
L1.2-P200-H16	32.8 ± 2.8 cd	4.80 ± 0.40 cd	1.64 ± 0.12 cd	0.26 ± 0.04 bc
L1.2-P250-H16	39.9 ± 2.1 b	5.95 ± 0.41 b	1.94 ± 0.18 b	0.28 ± 0.02 b
L1.2-P300-H16	38.2 ± 1.8 bc	5.50 ± 0.49 bc	2.18 ± 0.17 a	0.27 ± 0.04 b
L2.2-P150-H12	25.7 ± 2.3 e	3.51 ± 0.20 ef	1.26 ± 0.09 d	0.14 ± 0.01 e
L2.2-P200-H12	34.9 ± 2.4 cd	4.64 ± 0.11 cd	1.60 ± 0.13 cd	0.22 ± 0.02 cd
L2.2-P250-H12	37.3 ± 2.2 bc	6.01 ± 0.51 b	1.89 ± 0.10 b	0.25 ± 0.01 bc
L2.2-P300-H12	31.4 ± 3.3 d	6.35 ± 0.67 b	1.80 ± 0.10 bc	0.25 ± 0.02 bc
L2.2-P150-H16	30.7 ± 2.3 d	5.02 ± 0.64 c	1.38 ± 0.13 d	0.16 ± 0.02 e
L2.2-P200-H16	37.0 ± 3.7 bc	5.93 ± 0.60 b	1.88 ± 0.11 b	0.18 ± 0.01 de
L2.2-P250-H16	42.7 ± 2.2 b	7.34 ± 0.62 a	2.03 ± 0.31 ab	0.29 ± 0.04 b
L2.2-P300-H16	40.0 ± 5.0 b	6.60 ± 0.94 b	2.28 ± 0.26 a	0.35 ± 0.04 a
ANOVA				
PPFD	*	*	*	*
H	*	*	*	*
LQ	*	*	*	*
LQ × PPFD	NS	*	NS	*
LQ × H	*	NS	*	*
PPFD × H	NS	*	*	*
PPFD × H × LQ	NS	*	NS	*

Note: Different letters in the same columns were significantly different tested by LSD's multiple comparison at $P \leq 0.05$. NS, * represent no significant and significant difference at $P \leq 0.05$, respectively

With photoperiod of 12 h day^{-1}, leaf fresh weights were higher under FL than LED at PPFD of 250 and 300 μmol m^{-2} s^{-1}. Root fresh weight was highest in treatment of L2.2-P250-H16, while there were no differences between FL and LED with R:B ratio of 1.2. No substantial differences were found in leaf fresh weight between FL and LED with R:B ratio of 1.2, while some treatments in LED with R:B ratio of 2.2 had higher leaf dry weight compared to the other two lamps. Similar tendencies were observed for root dry weight among treatments.

Increasing light intensity generally increases photosynthesis and hence plant growth [16, 18], as shown in this experiment. The biomass of lettuce was higher under longer photoperiod at the same PPFD. Longer photoperiod led to more carbohydrates used for physiological metabolism and growth. Longer photoperiod could stimulate the photosensitive pigments to induce gene expression in lettuce growth, quality, and nutrition uptake [49, 50]. Based on the growth parameters, plants performed better in treatments of P250-H16 and P300-H16 than other treatments regardless of light quality. Our results were slightly different from other studies. For example, Jeong et al. reported no differences in lettuce growth among PPFDs of 200, 230, and 260 μmol m^{-2} s^{-1}, but PPFD of 290 μmol m^{-2} s^{-1} resulted in better growth and development of lettuce [51]. No differences in fresh weight of romaine lettuce were found between PPFD of 400 and 200 μmol m^{-2} s^{-1} [35]. These differences may be due to the different types or cultivars of lettuce used in these studies. Based on our results of biomass, light intensity between 250 and 300 μmol m^{-2} s^{-1} was reasonable for lettuce production in PFAL.

In treatments with both photoperiods, fresh and dry weights were higher at PPFD of 200 μmol m^{-2} s^{-1} than that at 150 μmol m^{-2} s^{-1}. With photoperiod of 16 h day^{-1}, PPFD at 250 μmol m^{-2} s^{-1} resulted in similar or even higher fresh and dry weights compared to that at 300 μmol m^{-2} s^{-1}, except for leaf dry weight in L1.2 treatments and root dry weight in L2.2 treatments. Based on the relationships between DLI and leaf fresh and dry weights, it was obvious that as DLI increased, leaf fresh and dry weights increased linearly (Fig. 14.3). Significant interactions were observed between PPFD and photoperiod on root fresh weight, leaf dry weight, and root dry weights; PPFD and light quality on root fresh and dry weights; light quality and photoperiod on leaf fresh and dry weights and root dry weight; and PPFD, photoperiod, and light quality on root fresh and dry weights.

Dry mass accumulation was proportional to DLI which is the product of light intensity and photoperiod. The linear relationships between DLI and leaf fresh and dry weights were obtained in this study, which were consistent with other report in lettuce greenhouse production [52]. DLI has been used to manage light environment in greenhouse production for many years. The growth and quality of lettuce are closely related to DLI. Longer photoperiod could compensate for a low PPFD at the same DLI [28]. In commercial lettuce production, Both suggested that a daily light integral of 17 mol m^{-2} day^{-1} should be maintained in the greenhouse [53]. Our experiments showed a lower DLI (14.4 mol m^{-2} day^{-1}) that was sufficient. The DLI of 14.4 mol m^{-2} day^{-1} (P250-H16) had similar growth and quality compared with P300-H16, DLI of 17.28 mol m^{-2} day^{-1}.

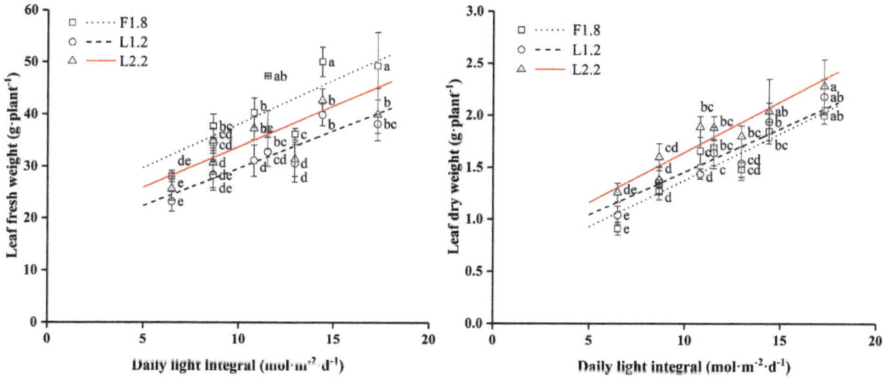

Fig. 14.3 Relationships between daily light integral and leaf fresh and dry weights of hydroponic lettuce. The vertical bars represent the standard deviation. Data points with different letters above the bar were significantly different tested by LSD's multiple comparison at $P \leq 0.05$

14.3.3 LED Lighting Affects Hydroponic Lettuce Quality

PPFD, photoperiod, and light quality had significant effects on the contents of nitrate N, vitamin C, soluble sugar, soluble protein, and anthocyanin (Table 14.10). Nitrate N content was lower at higher PPFD and longer photoperiod, while vitamin C, soluble sugar, soluble protein, and anthocyanin contents were higher. The nitrate N content was lowest in plants grown under higher PPFD at 250 and 300 μmol m^{-2} s^{-1} and longer photoperiod. No differences in the contents of nitrate N, vitamin C, and soluble sugar were found between treatments of P250-H16 and P300-H16, regardless of light quality. Vitamin C, soluble protein, and soluble sugar contents were lower in F1.8 treatments than those in L2.2 treatment. There were no differences in anthocyanin content between PPFD at 250 and 300 μmol m^{-2} s^{-1} with a photoperiod of 16 h day^{-1} except for the treatment of F1.8-P300-H16 which had the highest anthocyanin content. Generally, the differences in anthocyanin among different light quality were small compared to vitamin C, soluble sugar, and protein contents. PPFD and photoperiod interactively affected nitrate N, vitamin C, and anthocyanin contents. PPFD, photoperiod, and light quality interactively affected soluble protein content but not for other quality parameters.

Higher PPFD and longer photoperiod could achieve higher vitamin C, soluble sugar, and soluble protein contents while reducing nitrate N content, which agreed with other reports [54, 55]. The activity of nitrate reductase was higher under high PPFD, which could decrease the nitrate accumulation. High PPFD would increase the photosynthetic production (sucrose concentration), which promoted assimilation of nitrate ion [56]. Genetics, temperature, and light environment affected the anthocyanin content [57]. Our results on anthocyanin was similar to other researches which indicated that PPFD, photoperiod, and light quality significantly affected anthocyanin content [58, 59]. Plants in treatments of P300-H16 had the best quality,

Table 14.10 Effects of different photosynthetic photon flux density (PPFD), photoperiod (H), and light quality (LQ) on contents of nitrate N, vitamin C, soluble sugar, soluble protein, and anthocyanin of hydroponic lettuce at 18 days after transplanting ($n = 6$)

	Nitrate N content	Vitamin C content	Soluble sugar content	Soluble protein content	Anthocyanin content
Treatments	mg kg^{-1}	mg 100 g^{-1}	%	mg g^{-1}	ΔOD·g^{-1} FW
F1.8-P150-H12	783.1 ± 45.0 a	21.0 ± 2.2 d	1.01 ± 0.06 e	9.9 ± 0.7 de	0.62 ± 0.14 d
F1.8-P200-H12	697.5 ± 182.9 bc	22.4 ± 2.2 d	1.11 ± 0.36 de	10.4 ± 1.5 de	0.87 ± 0.11 cd
F1.8-P250-H12	565.6 ± 89.6 cd	22.1 ± 3.8 d	1.48 ± 0.13 d	9.3 ± 1.3 e	0.95 ± 0.18 c
F1.8-P300-H12	470.0 ± 75.5 de	23.5 ± 2.0 d	1.47 ± 0.16 d	11.8 ± 1.1 cd	1.31 ± 0.28 bc
F1.8-P150-H16	756.2 ± 104.0 bc	21.6 ± 2.3 d	0.90 ± 0.47 e	10.3 ± 0.9 de	0.96 ± 0.26 c
F1.8-P200-H16	563.2 ± 79.1 cd	23.4 ± 4.7 d	1.50 ± 0.40 cd	11.6 ± 0.7 d	1.09 ± 0.30 bc
F1.8-P250-H16	292.5 ± 43.1 e	30.3 ± 2.0 cd	2.11 ± 0.55 c	13.7 ± 2.0 bc	1.35 ± 0.09 b
F1.8-P300-H16	339.5 ± 72.0 e	34.7 ± 3.9 bc	2.16 ± 0.66 c	11.6 ± 1.1 d	2.30 ± 0.61 a
L1.2-P150-H12	667.5 ± 121.7 ab	24.4 ± 3.7 cd	1.04 ± 0.44 e	9.5 ± 1.8 e	0.33 ± 0.05 e
L1.2-P200-H12	626.6 ± 99.0 bc	26.9 ± 3.0 cd	0.94 ± 0.15 e	11.9 ± 0.9 cd	0.59 ± 0.15 de
L1.2-P250-H12	589.7 ± 121.7 bc	29.3 ± 7.4 cd	1.28 ± 0.45 de	12.4 ± 1.6 cd	0.61 ± 0.25 d
L1.2-P300-H12	546.3 ± 73.7 cd	29.5 ± 4.3 cd	1.88 ± 0.64 cd	12.7 ± 1.4 c	0.78 ± 0.18 cd
L1.2-P150-H16	654.8 ± 27.4 b	25.7 ± 4.4 cd	1.08 ± 0.33 de	12.8 ± 1.8 c	0.57 ± 0.17 de
L1.2-P200-H16	539.2 ± 65.3 cd	32.0 ± 3.8 c	1.55 ± 0.28 cd	11.9 ± 1.4 cd	0.73 ± 0.13 cd
L1.2-P250-H16	528.4 ± 20.1 cd	41.7 ± 8.8 ab	1.92 ± 0.17 cd	14.3 ± 2.1 b	1.12 ± 0.18 bc
L1.2-P300-H16	506.7 ± 38.2 d	43.3 ± 3.0 a	2.02 ± 0.24 c	13.3 ± 1.3 bc	1.50 ± 0.58 b
L2.2-P150-H12	810.4 ± 78.4 a	33.5 ± 3.2 bc	1.55 ± 0.31 cd	10.0 ± 1.8 de	0.33 ± 0.07 e

(continued)

Table 14.10 (continued)

Treatments	Nitrate N content	Vitamin C content	Soluble sugar content	Soluble protein content	Anthocyanin content
	mg kg^{-1}	mg 100 g^{-1}	%	mg g^{-1}	ΔOD·g^{-1} FW
L2.2-P200-H12	797.8 ± 119.7 a	34.1 ± 6.9 bc	2.98 ± 0.29 b	12.9 ± 1.2 c	0.51 ± 0.12 de
L2.2-P250-H12	779.3 ± 39.6 ab	35.4 ± 4.0 b	3.09 ± 0.78 b	13.4 ± 1.1 bc	0.83 ± 0.09 cd
L2.2-P300-H12	520.8 ± 90.1 cd	35.8 ± 4.9 bc	3.96 ± 0.72 a	14.3 ± 1.4 b	0.82 ± 0.07 cd
L2.2-P150-H16	624.7 ± 70.0 bc	31.2 ± 2.7 c	2.56 ± 0.59 bc	13.3 ± 0.8 bc	0.60 ± 0.08 de
L2.2-P200-H16	590.7 ± 110.0 c	35.6 ± 6.2 b	4.01 ± 1.24 a	14.1 ± 2.1 bc	0.86 ± 0.18 cd
L2.2-P250-H16	479.2 ± 43.3 de	38.4 ± 3.7 ab	4.27 ± 0.60 a	15.6 ± 1.2 b	1.47 ± 0.30 b
L2.2-P300-H16	456.8 ± 59.5 de	40.1 ± 4.9 ab	4.26 ± 0.44 a	18.9 ± 1.7 a	1.34 ± 0.26 b
PPFD	*y	*	*	*	*
H	*	*	*	*	*
LQ	*	*	*	*	*
PPFD×H	*	*	NS	NS	*
PPFD×LQ	*	NS	*	*	*
H × LQ	NS	*	NS	NS	NS
PPFD×H × LQ	NS	NS	NS	*	NS

Note: Means with different letters in the same columns were significantly different tested by LSD's multiple comparison at $P \leq 0.05$. NS, * represent no significant and significant difference at $P \leq 0.05$, respectively

and those in P250-H16 had similar contents of soluble sugar, vitamin C, and nitrate N. Considering the higher power consumption, in treatments of P300-H16 and P250-H16, LED lighting at PPFD of 250 μmol m^{-2} s^{-1} with photoperiod of 16 h day^{-1} was recommended for commercial lettuce production in PFAL.

Effect of light quality on lettuce growth, development, and quality has been investigated using monochromatic LED lights, a combination of red and blue LED lights, and other light sources such as white LED and fluorescent lamp [22, 60]. Many reports have concluded that combination of red and blue lights could effectively promote the lettuce growth and development than monochromatic light. For example, Lin et al. reported that combined red, blue, and white LEDs were better than red and blue LEDs and FL based on growth, nutrition, appearance, and the edible quality of lettuce plants grown in a growth chamber at PPFD of 210 μmol m^{-2} s^{-1} [22]. Continuous light exposure at pre-harvest could effectively reduce nitrate accumulation and increase phytochemical concentrations in lettuce plants [61]. In this experiment, LEDs with R:B ratios of 1.2 and 2.2 using white and red chips had higher quality and lower power consumption than fluorescent lamps.

14.3.4 Continuous Photosynthesis and Light Responses of Hydroponic Lettuce

The two-day continuous measurement of photosynthesis indicated that highest net photosynthetic rates of lettuce leaves were observed in treatments of F1.8-P250-H16 and F1.8-P300-H16 (Fig. 14.4). The differences between the two PPFD with a shorter photoperiod (F1.8-P250-H12 and F1.8-P300-H12) were smaller compared to those with a longer photoperiod. Although the treatment of F1.8-P250-H16 had the greatest net photosynthetic rate, no statistical differences in either leaf fresh or

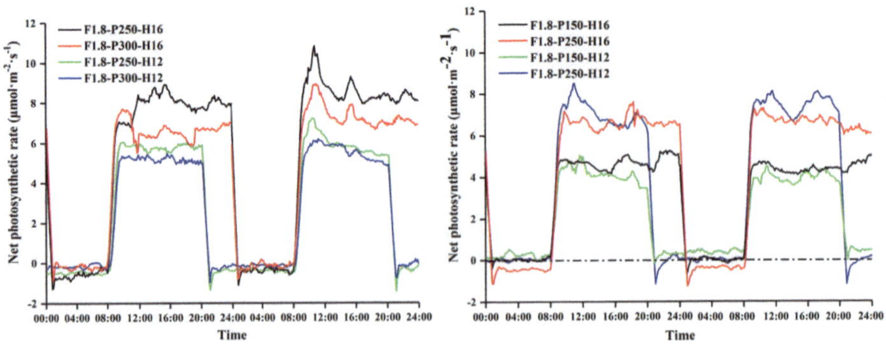

Fig. 14.4 Continuous measurement of net photosynthetic rate of hydroponic lettuces for 2 days before harvest under fluorescent lamps with R:B ratio of 1.8 at PPFD of 150–300 μmol m^{-2} s^{-1} with photoperiod of 12 and 16 h day^{-1}

Table 14.11 Effect of lighting environment at PPFD from 150 μmol m^{-2} s^{-1} to 300 μmol m^{-2} s^{-1} with a photoperiod of 12 and 16 h day^{-1} using fluorescent lamps with R:B ratio of 1.8 on daily CO_2 assimilation of the hydroponic lettuces

Treatments	Daily CO_2 assimilation at 16 days after transplanting	Daily CO_2 assimilation at 17 days after transplanting
	mmol m^{-2} day^{-1}	
F1.8-P250-H12	**219.1 (100%)**	**229.6 (100%)**
F1.8-P300-H12	205.5 (−6%)	221.5 (−4%)
F1.8-P250-H16	429.3 (+96%)	472.2 (+106%)
F1.8-P300-H16	374.0 (+71%)	402.4 (+75%)
F1.8-P150-H12	**183.3 (100%)**	**180.4 (100%)**
F1.8-P250-H12	295.1 (+61%)	310.6 (+72%)
F1.8-P150-H16	269.6 (+47%)	248.6 (+38%)
F1.8-P250-H16	357.9 (+95%)	355.0 (+97%)

Note: Bold identification in the table are references for comparison for treatments of same light quality. Percentage of positive and negative showed above was compared with the treatment of F1.8-P250-H12 or F1.8-P150-H12 at the same day

dry weights were observed at PPDF from 250 μmol m^{-2} s^{-1} to 300 μmol m^{-2} s^{-1}. The photosynthetic data in last two days do not necessarily reflect the whole production period; however, the daily CO_2 assimilation in treatments of F1.8-P250-H16 and F1.8-P300-H16 calculated based on the two-day continuous measurement of photosynthesis was greater in 71–106% increasing than treatment of F1.8-P250-H12 at 16 days and 17 days after transplanting (Table 14.11). The daily CO_2 assimilation at 16 days and 17 days after transplanting at PPFD of 250 μmol m^{-2} s^{-1} with photoperiod of 16 h day^{-1} were higher 60–97% compared with that at 150 μmol m^{-2} s^{-1} with photoperiod of 12 h day^{-1}, respectively, but long photoperiod in 16 h day^{-1} has improved by 30–40% increasing only at the same PPFD.

Figure 14.5 shows the light response curves of hydroponic lettuce at 17 days after transplanting at temperature of 22 °C and CO_2 concentration of 800 μmol mol^{-1} in treatments of L1.2-P250-H16 and L1.2-P300-H16. The fitting curves of rectangular hyperbola model and exponential model are diffusive progressive line, and the difference between the fitting value and measured value increases gradually when the light intensity is larger than 1200 μmol m^{-2} s^{-1}. The coefficient of determination of the non-rectangular hyperbola model is highest in all model simulations, but it is difficult to fit the photo-inhibition phenomenon. The photo-inhibition and photo-saturation phenomena were well explained by the modified rectangular hyperbola model by Ye et al. [62].

The modified rectangular hyperbola model was used to fit the net photosynthetic rate in treatment of L1.2-P250-H16 at 8 days and 17 days after transplanting (Fig. 14.6). The results showed that the maximum net photosynthetic rate, initial quantum efficiency, and saturated light intensity of the hydroponic lettuce were in decreasing trends over time, but the dark respiration rate and light compensation

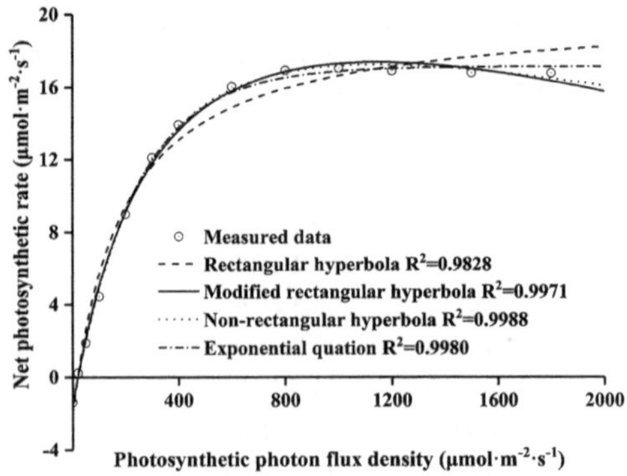

Fig. 14.5 Net photosynthetic rate of hydroponic lettuce response to photosynthetic photon flux density and fitted by rectangular hyperbola, modified rectangular hyperbola, non-rectangular hyperbola, and exponential models compared with the measured data

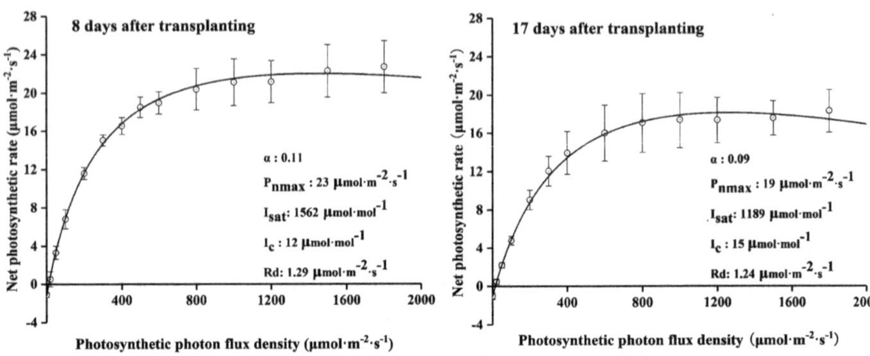

Fig. 14.6 Light response curves of hydroponic lettuce at 8 days and 17 days after transplanting under PPFD of 250 μmol m^{-2} s^{-1} with photoperiod of 16 h day^{-1}, using LED lamps with R:B ratio of 1.2. α means initial slope in the light response curve, P_{nmax} means maximal net photosynthetic rate, I_{sat} means saturation of light intensity, and Rd means rate of dark respiration

point were not changed which agreed with a practical measurement in single leaf. The net photosynthetic rates of lettuce leaves were rapidly increased with increasing photosynthetic photon flux density from 0 to 400 μmol m^{-2} s^{-1} because of the effective light qualitum efficiency. Therefore, high lettuce growth rate could be obtained with less light energy consumption when light intensity was less than 400 μmol m^{-2} s^{-1} under controlled environment.

14.3.5 Energy Use Efficiency of Artificial Light for Lettuce Production

Power consumption per 100 g leaf fresh weight and 1 g leaf dry weight of the lettuce were highest under fluorescent lamps (Table 14.12). LEDs had the advantage of higher LEU and EUE than fluorescent lamps. The EUE of LEDs was lower at higher PPFDs and longer photoperiod. The LEU for LEDs did not change substantially in

Table 14.12 Energy utilization efficiency of the fluorescent lamps and LEDs of the hydroponic lettuces produced in the closed plant factory system

Treatment	Power consumption per 100 g leaf fresh mass (kW h 100 g^{-1} FW)	Power consumption per 1 g leaf dry mass (kW h g^{-1} DW)	Light energy use efficiency (LEU)	Electric energy use efficiency (EUE)
F1.8-P150-H12	4.35	1.32	0.019	0.0038
F1.8-P200-H12	4.60	1.18	0.021	0.0041
F1.8-P250-H12	4.82	1.18	0.021	0.0041
F1.8-P300-H12	6.28	1.53	0.016	0.0031
F1.8-P150-H16	4.30	1.27	0.020	0.0039
F1.8-P200-H16	4.43	1.25	0.020	0.0039
F1.8-P250-H16	5.11	1.36	0.018	0.0036
F1.8-P300-H16	6.17	1.50	0.016	0.0032
L1.2-P150-H12	2.46	0.56	0.025	0.0096
L1.2-P200-H12	2.78	0.59	0.025	0.0091
L1.2-P250-H12	2.90	0.63	0.021	0.0086
L1.2-P300-H12	3.54	0.72	0.020	0.0076
L1.2-P150-H16	2.60	0.54	0.026	0.0101
L1.2-P200-H16	3.13	0.62	0.024	0.0087
L1.2-P250-H16	2.91	0.60	0.022	0.0090
L1.2-P300-H16	3.83	0.68	0.021	0.0079
L2.2-P150-H12	2.47	0.52	0.031	0.0099
L2.2-P200-H12	2.40	0.52	0.030	0.0097
L2.2-P250-H12	2.74	0.56	0.029	0.0092
L2.2-P300-H12	3.48	0.64	0.023	0.0080
L2.2-P150-H16	2.67	0.62	0.026	0.0082
L2.2-P200-H16	2.94	0.61	0.026	0.0083
L2.2-P250-H16	3.15	0.68	0.024	0.0075
L2.2-P300-H16	3.74	0.66	0.022	0.0078

Note: FW means fresh weight of leaves and DM means dry weight of leaves

the PPFD range from 150 μmol m^{-2} s^{-1} to 300 μmol m^{-2} s^{-1} with a photoperiod from 12 h day^{-1} to 16 h day^{-1}. For commercial lettuce production in a closed plant factory system, light intensity, photoperiod, and light quality are controlled collectively in consideration of plant growth or yield, energy consumption, and initial and operational costs. Although many researchers have studied the effect of single factor (PPFD or photoperiod) or two factors (PPFD and photoperiod) on growth and quality of lettuce, limited attention has been paid on the effects of multiple factors as shown in this experiment.

Acknowledgments The research was supported by NEDO (New Energy and Industrial Technology Development Organization) project of Preliminary Study on Plant Phonemics and its Application by Artificial Intelligence, Japan (2017–2019), National High Technology Research and Development Program of China (2013AA103005), and National Key Research and Development Program of China (2017YFB0403901), respectively.

References

1. C.A. Mitchell, M.P. Dzakovich, C. Gomez, R. Lopez, J.F. Burr, R. Hernandez, C. Kubota, C.J. Currey, Q. Meng, E.S. Runkle, C.M. Bourget, R.C. Morrow, A.J. Both, Light-emitting diodes in horticulture. Hortic. Rev. **42**, 1–87 (2015)
2. T. Kozai, K. Fujiwara, E. S. Runkle (eds.), *LED Lighting for Urban Agriculture* (Springer, Singapore, 2016), pp. 3–448
3. T. Kozai, G. H. Niu, M. Takagaki (eds.), *Plant Factory: An Indoor Vertical Farming System for Efficient Quality Food Production* (Academic, Amsterdam, 2015), pp. 3–399
4. T. Kozai (ed.), *Smart Plant Factory: Next-Generation Indoor Vertical Farms* (Springer, Singapore, 2018)
5. S.M.A. Zobayed, F. Afreen, T. Kozai, Temperature stress can alter the photosynthetic efficiency and secondary metabolites concentrations in St. John's wort. Plant Physiol. Biochem. **43**, 977–984 (2005)
6. A.J. Both, B. Bugbee, C. Kubota, R.G. Lopez, C. Mitchell, E.S. Runkle, C. Wallace, Proposed product label for electric lumps used in the plant sciences. HortTechnology **27**(4), 544–549 (2017)
7. Goto E (2016) Measurement of photonmetric and radiometric characteristics of LEDs for plant cultivation. In: Kozai T, Fujiwara K, Runkle ES. LED Lighting for Urban Agriculture. Springer, Singapore, pp 395-402
8. G. Zhang, S. Shen, M. Takagaki, T. Kozai, W. Yamamoto, Supplemental upward lighting from underneath to obtain higher marketable lettuce (*Lactuca sativa*) leaf fresh weight by retarding senescence of outer leaves. Front. Plant Sci. **6**, 1–9 (2015)
9. Lu N, Mitchell CA (2016) Supplemental lighting for greenhouse grown fruiting vegetables. In: Kozai T, Fujiwara K, Runkle ES. LED Lighting for Urban Agriculture. Springer, Singapore, pp 219-232
10. M.E. Ghanem, H. Marrou, T.R. Sinclair, Physiological phenotyping of plants for crop improvement. Trends Plant Sci. **20**(3), 139–144 (2015)
11. M. Johkan, K. Shoji, F. Goto, Blue light-emitting diode light irradiation of seedlings improves seedling quality and growth after transplanting in red leaf lettuce. HortScience **45**(12), 414–415 (2010)
12. X.Y. Liu, T.T. Chang, S.R. Guo, Z.G. Xu, J. Li, Effect of different light quality of LED on growth and photosynthetic character in cherry tomato seedling. Acta Hortic. (907), 325–330 (2011). https://doi.org/10.17660/ActaHortic.2011.907.53
13. R. Hernández, C. Kubota, Physiological responses of cucumber seedlings under different blue and red photon flux ratios using LEDs. Environ. Exp. Bot. **121**(1), 66–74 (2016)
14. J. Song, Q. Meng, W. Du, D. He, Effects of light quality on growth and development of cucumber seedlings in controlled environment. Int. J. Agric. Biol. Eng. **10**(3), 312–318 (2017)
15. X.L. Chen, Q.C. Yang, W.P. Song, Growth and nutritional properties of lettuce affected by different alternating intervals of red and blue LED irradiation. Sci. Hortic. **223**, 44–52 (2017)
16. J. Wang, W. Lu, Y.X. Tong, Q.C. Yang, Leaf morphology, photosynthetic performance, chlorophyll fluorescence, stomatal development of lettuce (*Lactuca sativa* L.) exposed to different ratios of red light to blue light. Front. Plant Sci. **7**, 250 (2016)

17. K. Okamoto, T. Yanagi, S. Kondo, Growth and morphogenesis of lettuce seedlings raised under different combinations of red and blue light. Acta Hortic. (435), 149–158 (1997). https://doi.org/10.17660/actahortic.1997.435.14
18. M.E. Hoenecke, R.J. Bula, T.W. Tibbitts, Importance of 'blue' photon levels for lettuce seedlings grown under red-light-emitting diodes. HortScience **127**(5), 427 (1992)
19. Y.X. Miao, Physiological response mechanism of cucumber seedlings to red and blue light, Ph.D. Thesis of China Agriculture University, 2015, pp. 17–25
20. K.H. Son, M.M. Oh, Leaf shape, growth, and antioxidant phenolic compounds of two lettuce cultivars grown under various combinations of blue and red light-emitting diodes. HortScience **48**(8), 988–995 (2013)
21. W.H. Kang, J.S. Park, K.S. Park, Leaf photosynthetic rate, growth, and morphology of lettuce under different fractions of red, blue, and green light from light-emitting diodes (LEDs). Hortic. Environ. Biotechnol. **57**(6), 573–579 (2016)
22. K.H. Lin, M.Y. Huang, W.D. Huang, The effects of red, blue, and white light-emitting diodes on the growth, development, and edible quality of hydroponically grown lettuce (*Lactuca sativa* L. var. capitata). Sci. Hortic. **150**(2), 86–91 (2013)
23. M. Johkan, K. Shoji, F. Goto, Effect of green light wavelength and intensity on photomorphogenesis and photosynthesis in *Lactuca sativa*. Environ. Exp. Bot. **75**, 128–133 (2012)
24. H. Zhai, Effect of lighting environment at seedling stage on growth of hydroponic lettuce transplant and its late harvest, Master Thesis of China Agricultural University, 2017, pp. 10–43
25. X. Zhang, Technical foundation of environmental feedback control of hydroponic lettuce based on photosynthesis simulation, Ph.D Thesis of China Agricultural University, 2017, pp. 12–28, 58–82
26. H.S. Li, *Experimental Principle and Technology of Plant Physiology and Biochemistry* (High Education Press, Beijing, 2000), pp. 122–123, 190–192, 202–204, 267–268
27. Y. Kitaya, G.H. Niu, T. Kozai, M. Ohashi, Photosynthetic photon flux, photoperiod, and CO_2 concentration affect growth and morphology of lettuce plug transplants. HortScience **33**(6), 988–991 (1998)
28. L. Gaudreau, J. Charbonneau, L.P. Vézina, Photoperiod and photosynthetic photon flux influence growth and quality of greenhouse-grown lettuce. HortScience **29**(11), 1285–1289 (1994)
29. J.A. Zavala, D.A. Ravetta, Allocation of photoassimilates to biomass, resin and carbohydrates in *Grindelia chiloensis* as affected by light intensity. Field Crop Res. **69**(2), 143–149 (2001)
30. H.K. Lichtenthaler, A. Alexander, M.V. Marek, Differences in pigment composition, photosynthetic rates and chlorophyll fluorescence images of sun and shade leaves of four tree species. Plant Physiol. Biochem. **45**(8), 577–588 (2007)
31. X.X. Fan, Z.G. Xu, X.Y. Liu, Effects of light intensity on the growth and leaf development of young tomato plants grown under a combination of red and blue light. Sci. Hortic. **153**, 50–55 (2013)
32. T. Steinger, B.A. Roy, M.L. Stanton, Evolution in stressful environments II: adaptive value and costs of plasticity in response to low light in *Sinapis arvensis*. J. Evol. Biol. **16**(2), 313–323 (2003)
33. W. Oh, E.S. Runkle, R.M. Warner, Timing and duration of supplemental lighting during the seedling stage influence quality and flowering in petunia and pansy. HortScience **45**(9), 1332–1337 (2010)
34. T.W. McNellis, X.W. Deng, Light control of seedling morphogenetic pattern. Plant Cell **7**(11), 1749 (1995)
35. W. Fu, P.P. Li, Y. Wu, J. Tang, Effects of different light intensities on anti-oxidative enzyme activity, quality and biomass in lettuce. Hortic. Sci. **39**, 129–134 (2012)
36. Z. Yang, W. He, S. Mou, X. Wang, D. Chen, X. Hu, L. Chen, J. Bai, Plant growth and development of pepper seedlings under different photoperiods and photon flux ratios of red and blue LEDs. Trans. Chin. Soc. Agric. Eng. **33**(17), 173–180 (2017)

37. R. Wojciechowska, O. Długosz-Grochowska, A. Kołton, M. Zupnik, Effects of LED supplemental lighting on field and some quality parameters of lamb's lettuce grown in two winter cycles. Sci. Hortic. **187**, 80–86 (2015)
38. C.T. Della Valle, C.R. Daniel, B. Aschebrook-Kilfoy, A.R. Hollenbeck, A.J. Cross, R. Sinha, M.H. Ward, Dietary intake of nitrate and nitrite and risk of renal cell carcinoma in the NIH-AARP Diet and Health Study. Br. J. Cancer **108**, 205–212 (2013)
39. C.L. Chang, K.P. Chang, The growth response of leaf lettuce at different stages to multiple wavelength-band light-emitting diode lighting. Sci. Hortic. **179**, 78–84 (2014)
40. R. Breessani, J.E. Braham, L.G. Elias, All vegetable protein mixtures for human feeding. Can. J. Biochem. **42**(4), 631 (1964)
41. L.W. Zhang, S.Q. Liu, Z.K. Zhang, Dynamic effects of different light qualities on pea sprouts quality. North. Hortic. **8**, 4–7 (2010)
42. S. Nobile, J. M. Woodhill (eds.), *Vitamin C in the Human Body* (Springer, Dordrecht, 1981), pp. 57–75
43. P. Pinho, K. Jokinen, L. Halonen, The influence of the LED light spectrum on the growth and nutrient uptake of hydroponically grown lettuce. Light. Res. Technol. **49**(7), 866–881 (2017)
44. H.M. Qian, T.Y. Liu, M.D. Deng, Effects of light quality on main health-promoting compounds and antioxidant capacity of Chinese kale sprouts. Food Chem. **196**, 1232–1238 (2015)
45. K. Ohashi, M. Takase, N. Kon, Effect of light quality on growth and vegetable quality in leaf lettuce, spinach and komatsuna. Environ. Control. Biol. **45**(3), 189–198 (2007)
46. J. K. Cao, W. B. Jiang, Y. M. Zhao (eds.), *Physiological and Biochemical Experimental Guidance of Fruits and Vegetables* (China Light Industry Press, Beijing, 2007), pp. 49–50
47. Z.J. Zhang, D.X. He, G.H. Niu, R.F. Gao, Concomitant CAM and C3 photosynthetic pathways in *Dendrobium officinale* plants. J. Am. Soc. Hortic. Sci. **139**(3), 290–298 (2014)
48. T. Kozai, Resource use efficiency of closed plant production system with artificial light: concept, estimation and application to plant factory. Proc. Jpn. Acad. Ser. B Phys. Biol. Sci. **89**(10), 447–461 (2013)
49. N.N. Cometti, M.Q. Martins, C.A. Bremen Kamp, J.A. Nunes, Nitrate concentration in lettuce leaves depending on photosynthetic photon flux and nitrate concentration in the nutrient solution. Hortic. Bras. **29**, 548–553 (2011)
50. J.Z. Mao, Q. Qiu, F. Zhang, N. Li, Y.G. Hu, X.Z. Xue, Impact of different photoperiods on the morphological index, quality and absorptive amount to ions of lettuce in fluorescent light source. North. Hortic. **15**, 24–28 (2013). in Chinese
51. H.K. Jeong, K.K. Sugumaran, L.S.A. Sarah, B.R. Jeong, J.H. Seung, Light intensity and photoperiod Influence the growth and development of hydroponically grown leaf lettuce in a closed-type plant factory system. Hortic. Environ. Biotechnol. **54**(6), 501–509 (2013)
52. L.D. Albright, A.J. Both, A. Chiu, Controlling greenhouse light to a constant daily integral. Trans. ASAE **43**(2), 421–431 (2000)
53. A.J. Both, Ten years of hydroponic lettuce research. Knowledgecenter, illumitex.com, 2001, pp. 1–14
54. A. Scaife, S. Schloemer, The diurnal pattern of nitrate uptake and reduction by spinach (Spinacia oleracea L.). Ann. Bot. **73**(3), 337–343 (1994)
55. N. Gruda, Impact of environmental variables on product quality of greenhouse vegetables for fresh consumption. Crit. Rev. Plant Sci. **24**, 227–247 (2005)
56. D. McCall, J. Willumsen, Effects of nitrogen availability and supplementary light on the nitrate content of soil-grown lettuce. J. Hortic. Sci. Biotechnol. **174**(4), 458–463 (1999)
57. G. Aparna, M.D. Kleinhenz, J.C. Scheerens, P.P. Ling, Anthocyanin levels in nine lettuce (*Lactuca sativa*) cultivars: influence of planting data and relations among analytic, instrumented, and visual assessments of color. HortScience **42**(2), 232–238 (2007)

58. J. Zhong, T. Seki, S. Kinoshita, T. Yoshida, Effect of light irradiation on anthocyanin production by suspended culture of *Perilla frutecens*. Biotechnol. Bioeng. **38**, 653–658 (1991)
59. J.E. Park, Y.G. Park, B.R. Jeong, S.J. Hwang, Growth and anthocyanin content of lettuce as affected by artificial light source and photoperiod in a closed-type plant production system. Korean J. Hortic. Sci. Technol. **30**(6), 673–679 (2012)
60. X.L. Chen, W.Z. Guo, X.Z. Xue, L.C. Wang, Growth and quality response of 'Green Oak Leaf' lettuce as affected by monochromic or mixed radiation provided by fluorescent lamp (FL) and light-emitting diode (LED). Sci. Hortic. **172**, 168–175 (2014)
61. W. Zhou, W. Liu, Q. Yang, Reducing nitrate concentration in lettuce by elongated lighting delivered by red and blue LEDs before harvest. J. Plant Nutr. **36**, 481–490 (2013)
62. Z.P. Ye, J.D. Suggett, P. Robakowski, A mechanistic model for the photosynthesis-light response based on the photosynthetic electron transport of PS II in C3 and C4 species. New Phytol. **152**, 1251–1262 (2013)

Chapter 15 The Effect and Mechanism of Light on the Growth, Food Intake, and Gonad Development of Atlantic Salmon (*Salmo salar*) Reared in RAS

Ying Liu, Xiaolong Gao, and Liang Chi

15.1 Photoperiod Regulate Gonad Development via Kisspeptin/Kissr in Hypothalamus and Saccus Vasculosus of *Salmo salar*

15.1.1 Introduction

Kisspeptin and its receptor GPR54 (kissr) have been identified as key factors in controlling the reproductive cycle by stimulating gonadotropin-releasing hormone (GnRH) secretion in mammals. Loss of functional mutations in mouse and human GPR54 shows that GPR54 is absolutely required for reproduction in mammals [1–3]. In teleost, the study of kisspeptin and its receptor is still in its infancy. Parhar and co-workers were the first to report the importance of kissr and GnRH in fish reproduction and found that kissr was involved in the sexual development of a cichlid fish [4]. Since then, kisspeptin and kissr have become the subject of active research in fish reproduction. It is believed that kisspeptin and kissr perform similar roles in fish to kisspeptin/GPR54 reported in mammals. Kisspeptin has been reportedly associated with the onset of puberty in some fish species [5–7] and has GnRH regulatory functions in several others [8]. In addition, Martinez-Chavez et al. found that a long photoperiod could delay the onset of puberty and suppress GPR54 expression in Nile tilapia by reducing the expression of kisspeptin. This was the first study of teleost to suggest a possible connection between photoperiod

Y. Liu (✉) · X. Gao
Dalian Ocean University, Dalian, China
e-mail: yingliu@dlou.edu.cn

L. Chi
Qingdao Agricultural University, Qingdao, China

J. Li, G. Q. Zhang (eds.), *Light-Emitting Diodes*, Solid State Lighting Technology and Application Series 4, https://doi.org/10.1007/978-3-319-99211-2_15

and kisspeptin. Similar results have also been found in other fish species [9–11]. Furthermore, most research into photoperiods and GnRH have focused on the stage of puberty. Furthermore, there is very limited reference to the effects of kisspeptin/kissr on other stages of gonad development [12, 13].

Atlantic salmon (*Salmo salar* L.), are native to the North Atlantic and its surrounding rivers, and were introduced into Chinese using Recirculating Aquaculture Systems (RAS). Atlantic salmon are short-day seasonal breeders and are very sensitive to the photoperiod [14]. Furthermore, salmonids are characterized by their direct sensitivity to daylight and lack of endogenous rhythms found in other fish [15, 16]. In a previous study, we found that photoperiod can significantly affect the gonadal development of Atlantic salmon reared in RAS. In order to enrich our knowledge of the functions of the SV in teleost fish, we set out to clarify the relationship between kissr and GnRH in the hypothalamus and SV of Atlantic salmon under the different photoperiod regimes.

15.1.2 Materials and Methods

Atlantic salmon with an average body mass of 1071.70 ± 155.54 g were purchased from the Shandong Oriental Ocean Sci-Tech Co., Ltd., Shandong province, China. The fish were then allowed to acclimate for 4 weeks in an RAS under 24L:0D photoperiod, after which they were distributed between the experimental RAS tanks (130 cm height ×200 cm diameter). Each experimental group contained 60 fish. During the experimental period, the water temperature was maintained at 16.27 ± 0.54 °C, pH 7.2 ± 7.5, total ammonia-nitrogen <0.25 mg/L, and salinity 24–26 over the course of the experiment.

The fish were divided into six photoperiod treatment groups. Four of the photoperiod treatments remained constant throughout the experiment 24L:0D, 18L:6D, 12L:12D, and 16L:8D. The remaining two photoperiod groups had varying photoperiods during the experiment. In the first treatment, the photoperiod changed from 24L:0D to 8L:16D (the LL-SL treatment), and in the second the photoperiod changed from 8L:16D to 24L:0D (the SL-LL treatment), the photoperiod being changed by 5 min per day in both cases. There were three replicate tanks (initially $n = 60$ fish/tank). The experiment was performed over a 7-month period from September to the following March, a period spanning the first reproductive period. Fish were sampled every month. Nine female fish were anesthetized until death in seawater with 0.05% MS-222 (3-aminobenzoic acid ethyl ester methanesulfonate). Body weight, length, and gonad weight were recorded, and the brains were immediately weighed, frozen in liquid nitrogen, and stored at −80 °C. The gonads and brains were placed in Bouin's solution for 24 h and then stored in 70% ethanol for later histological examination. Blood was centrifuged, and plasma was stored at −80 °C.

Furthermore, the gonadal development in this paper was determined based on GSI, external morphological and histology observation as follows: stage II (beginning of vitellogenesis with primary yolk vesicles, GSI = 0.15% ± 0.05%), stage III (secondary yolk stage, GSI = 0.31% ±0.18%), stage IV (accumulation of yolk happened, GSI = 11.87% ± 8.85%), and stage V (cytoplasm of oocyte was filled with yolk granules, GSI = 17.57% ± 3.01%). All of the procedures described in this study were reviewed and approved by the ethical committee of the Institute of Oceanology, Chinese Academy of Sciences.

Antisense digoxigenin (DIG) probes and antisense fluorescein isothiocyanate (FITC) were transcribed for salmon sGnRH gene and skissr gene by using a DIG and FITC-Labeling Kit (Roche, USA). Sense probe for salmon sGnRH and skissr was transcribed as negative control. The brains from Atlantic salmon reared in the different photoperiod treatments were fixed in 4% paraformaldehyde in 0.1 M PBS (phosphate buffered saline, pH 7.4) at 4 °C overnight. The samples were then dehydrated using a graded methanol series. After that, the samples were mixed with warm paraffin to embed. Sections of paraffin-embedded brains were prepared on 5-μM glass slides coated with 0.1% poly-L-lysine solution. The partial CDS of sGnRH3 and skissr were cloned into pGEM-T vectors for preparing sense and antisense RNA probes from a T7 or SP6 promoter using a FITC or digoxigenin (DIG) RNA-Labeling Kit (Roche). The sections were hybridized with the sense or antisense probes at 66 °C for 18 h. After hybridization, the samples were incubated overnight at 4 °C with horseradish peroxidase (POD)-conjugated anti-FITC antibody (Roche) at a 1:2000 dilution in the blocking solution to detect the FITC signal. After three washes in PBST, the samples were incubated for 1 h in TSA-Fluorescein at a 1:150 dilution in TSA Amplification Buffer. The samples were then subjected to detect the DIG signal. They were incubated overnight at 4 °C with POD-conjugated anti-DIG antibody (Roche) at a 1:2000 dilution in blocking buffer with 1% H_2O_2. Following three PBST washes, the samples were incubated in TSA-plus tetramethylrhodamine for 1 h. Double-color fluorescence in situ hybridization was performed using tyramide signal amplification (TSA™) plus fluorescein and tetramethylrhodamine (TMR), according to the manufacturer's instructions (NEL756, PerkinElmer). The nuclei were stained using 4′-6-diamidino-2-phenylindole (DAPI) and embedded in ProLong1 Gold Antifade reagent (Invitrogen). The slides were then mounted and photographed using a Nikon Eclipse 50i fluorescence microscope. In this procedure, two antisense RNA probes were co-incubated in a single sample during the hybridization step, and developed red and green fluorescence.

15.1.3 Results

15.1.3.1 The Location of Skissr and sGnRH3 in the Brain of Atlantic Salmon

First, the mRNA expressions of skissr and sGnRH3 were detected in the different regions of the Atlantic salmon brain using qPCR (quantitative real-time PCR) with

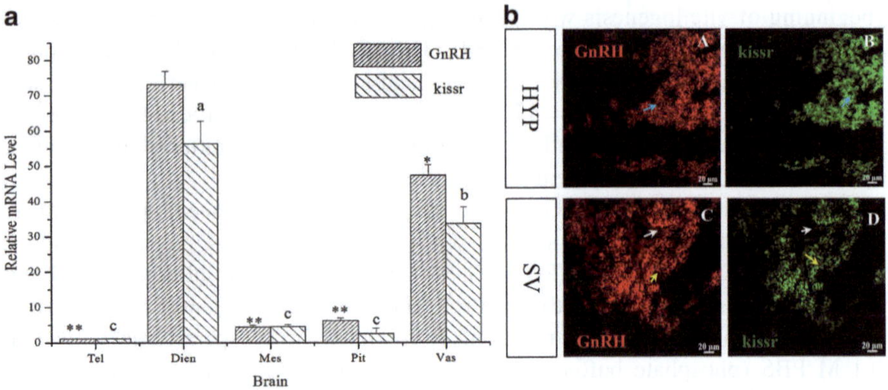

Fig. 15.1 (**a**) The distribution of sGnRH3 and skissr in the Atlantic salmon brain. *Tel* telencephalon, *Dien* diencephalon, *Mes* mesencephalon, *Pit* pituitary gland, *SV* saccus vasculosus. (**b**) The expression of sGnRH3 and skissr in the hypothalamus (A, B) and SV (C, D). The cells which express sGnRH3 mRNA are indicated by blue arrows; the white arrows indicate the supporting cells of the SV; and the yellow arrows indicate the coronet cells of the SV. The results show that skissr is mainly expressed in the cerebrospinal fluid-contacting (CSF-c) cells of the SV. Hyp: hypothalamus; SV: saccus vasculosus; Results are presented as mean ± SD. Significant differences were found at $p \leqq 0.5^*$ and $p \leqq 0.1^{**}$; and different letters indicate statistical significance at $p < 0.05$

β-actin mRNA as a reference gene. The results showed that both sGnRH3 and skissr transcripts were primarily expressed in the diencephalon. The transcription levels of sGnRH3 and skissr were higher in the SV than in other parts of the brain except the diencephalon (Fig. 15.1a). Second, the diencephalon and SV of Atlantic salmon were isolated and used to perform in situ hybridization in order to confirm the precise location of sGnRH3 and skissr transcripts. The results showed that sGnRH3 and skissr transcripts were mainly expressed in the Hyp of the diencephalon. In the SV, the sGnRH3 and skissr also showed the same expression pattern. Both sGnRH3 and skissr were found in the cells close to the ventricles (Fig. 15.1b).

15.1.3.2 Changes in Skissr in the Hyp and SV During Gonad Development

The experiment ran throughout virtually every stage of development of the Atlantic salmon, from stage II to maturity. The expression levels of kissr were taken as the mean value in each photoperiod treatment. In the early and late stages of gonad development, when the ovaries were at stage II and stage V, the expression levels of kissr transcripts were significantly higher than at stage III and stage IV (Fig. 15.2a and c). A similar phenomenon was also observed in the SV (Fig. 15.2b and c).

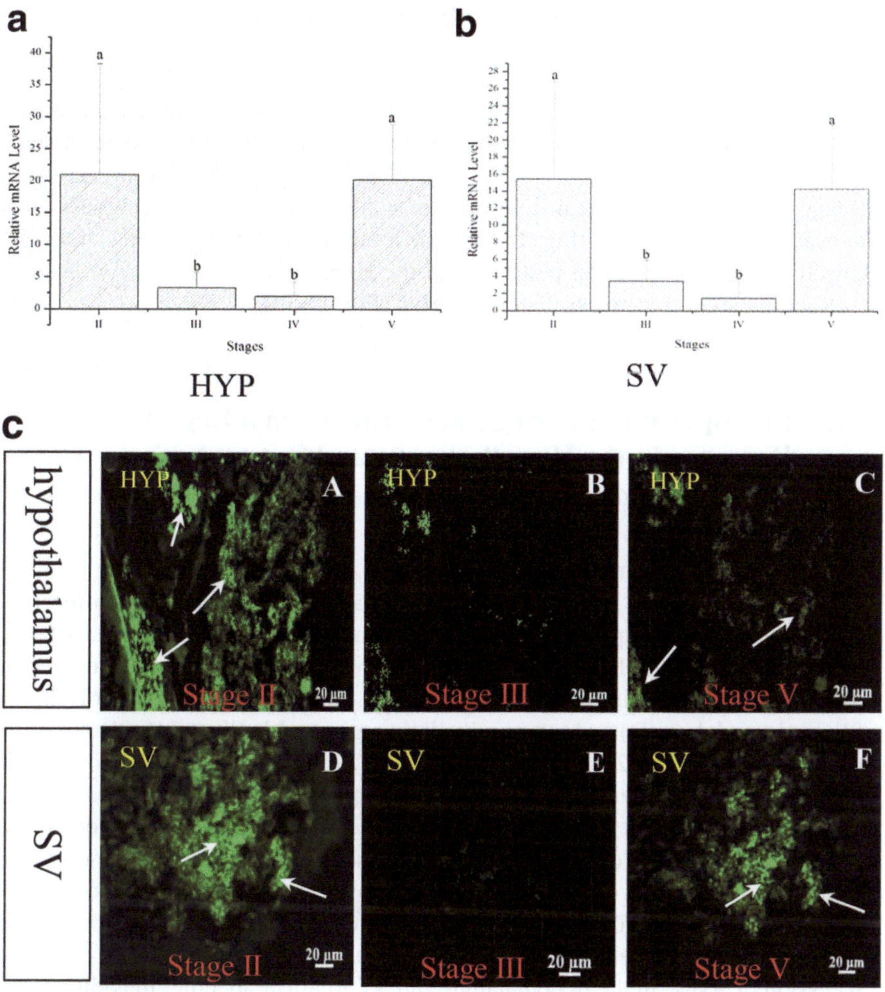

Fig. 15.2 The expression pattern of skissr in the hypothalamus (**a**) and SV (**b**) during the development process of Atlantic salmon. II: the gonad at stage II; III: the gonad at stage III; IV: the gonad at stage IV; and V: the gonad at stage V. (**c**) The expression of kissr in the Atlantic salmon hypothalamus (A, B, and C) and SV (D, E, and F) when fish with different stages of gonad development are assayed using in situ hybridization. *SV* saccus vasculosus. The kissr was mainly expressed in the early and late stages of gonad development in Atlantic salmon both in the hypothalamus and the SV. Data are shown as mean ± SD. One-way ANOVA is performed to determine the significant differences between means. Columns sharing different letters show significant difference ($p < 0.05$)

15.1.3.3 Changes in sGnRH3 in the Hyp and SV During Gonadal Development

The expression of sGnRH3 in the Hyp and SV increased as the gonads developed. The expression of sGnRH3 transcripts was lowest when the ovaries were at stage II. The expression of sGnRH3 mRNA was relatively stable when the ovaries were at stage III and stage IV. The mRNA level of sGnRH3 increased to its highest when the ovaries were at stage V (Fig. 15.3a). The level of sGnRH3 transcripts in the SV showed the same expression pattern as in the Hyp, while the expression level of sGnRH3 in the SV was lower than in the Hyp (Fig. 15.3b).

15.2 Photoperiod May Regulate Growth via a Leptin Receptor in the Hypothalamus and Saccus Vasculosus of *Salmo salar*

Photoperiod is believed to regulate growth in fish, although the mechanism involved is still unclear. In this paper, we report a relationship between leptin-receptor (AsLR), melatonin-receptor (AsMR), and photoperiod in Atlantic salmon. Atlantic salmon (mean weight 1071.70 ± 155.54 g) were reared under six photoperiod regimes, four constant light regimes 24L:0D, 18L:6D, 12L:12D, and 8L:16D, hours of light (L) and dark (D) and two varying light regimes, LL-SL = 24L:0D–8L:16D, and SL-LL = 8L:16D–24L:0D over a period of 7 months. The results showed that AsLR transcripts were mainly existed in the hypothalamus and saccus vasculosus (SV), while AsMR was mainly expressed in the hypothalamus. Long photoperiod inhibited the expression of AsLR and AsMR transcripts in the Atlantic salmon brain. The expression pattern of AsLR was similar to the expression pattern of AsMR in the hypothalamus. Food intake was higher in fish with lower AsLR transcript levels. This demonstrated that photoperiod influenced somatic growth by changing expression of AsLR in the hypothalamus and SV to affect appetite. In addition, the SV appears to act as a seasonal sensor regulating reproduction in a similar way to the hypothalamus.

15.2.1 Introduction

Leptin is secreted by adipose tissue and has an important role in regulating appetite, adiposity, food intake, and energy expenditure in mammals [17]. Leptin interacts with several neuropeptides to regulate food intake in the hypothalamus [18]. The physiological functions of leptin are mediated the leptin receptor (LR) in mammals [19]. In fish, leptin or LRs have been identified in many species including zebrafish, medaka, Arctic charr, rainbow trout, and Atlantic salmon [20, 21]. The functions

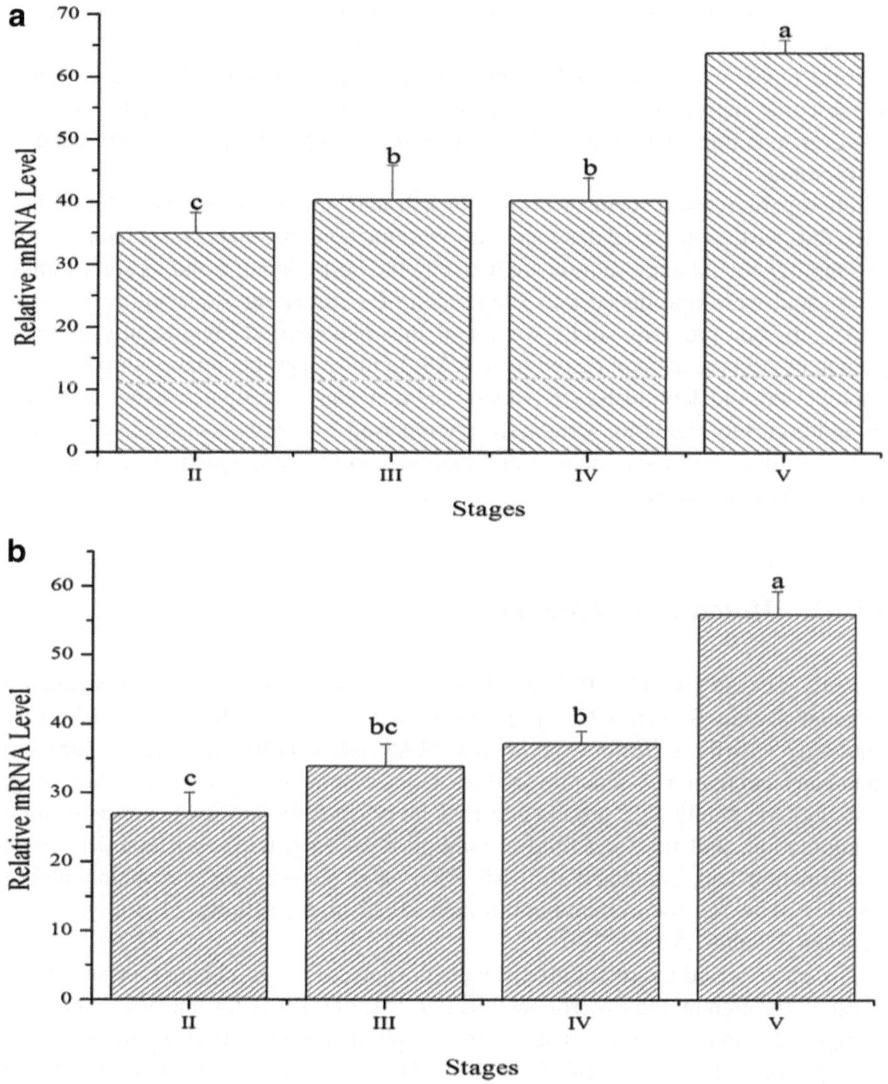

Fig. 15.3 The expression pattern of sGnRH3 in the hypothalamus (**a**) and saccus vasculosus (**b**) during the development of Atlantic salmon. II: the gonad at stage II; III: the gonad at stage III; IV: the gonad at stage IV; and V: the gonad at stage V

of leptin and LRs in fish are similar to those in mammals. The action of leptin is mediated through LRs expressed on appetite-related neurons and circuits in the hypothalamus [22]. Studies in mammals have found that the expression of leptin is rhythmic, which is related to the pineal melatonin axis in ruminants [23]. In fish, the rhythm of leptin/LRs is mainly focused on the feeding regime. Daily changes in leptin mRNA were first studied in Atlantic salmon, in which changes were seen in white muscle, belly flap, visceral adipose tissue, and liver, when fish are exposed to short-term feeding restrictions [24]. Meanwhile, in goldfish, hepatic leptin expression peaks appear at 9 h post-feeding [25]. These results suggest that leptin/LRs could be affected by environmental factors. Up to now, however, our understanding of the relationship of leptin/LRs and photoperiod is still limited.

Photoperiod could regulate growth in fish and growth is related to leptin and LRs, thus we hypothesized that there may be some relationship between photoperiod and leptin/LRs. Atlantic salmon are native to the north Atlantic and its surrounding rivers, and were introduced into China using Recirculating Aquaculture Systems (RAS). In our previous study, we found that photoperiod significantly affected growth of Atlantic salmon reared in an RAS. In this paper, the relationship between photoperiod and leptin/LRs of Atlantic salmon was investigated using an RAS. In addition, the regulation center for photoperiod is mainly located in the brain, and therefore, this study focused on Atlantic salmon LRs, which are also mainly expressed in the brain.

15.2.2 Materials and Methods

Atlantic salmon (weight: 1071.70 ± 155.54 g) were collected from Shandong Oriental Ocean Sci-Tech Co. Ltd., Shandong province, China. The fish were randomly distributed into experimental RAS tanks (130 cm high × 200 cm diameter) and reared by satiation feeding with a commercial salmon diet (Skretting, Norway) containing 48% protein and 18% fat twice daily during the period of light manipulation, and the total food consumption of each tank was recorded. Each experimental tank contained 60 fish. The water temperature was maintained at 16.27 ± 0.54 °C, total ammonia-nitrogen <0.25 mg/L, salinity 24–26, and a pH between 7.2 and 7.5.

Six photoperiod treatment groups were designed. Four of the photoperiods were constant throughout the experiment 24L:0D, 18L:6D, 12L:12D, and 16L:8D [hours of light (L) and dark (D)]. The other two photoperiods varied. In the first, the photoperiod changed from 24L:0D to 8L:16D (LL-SL treatment), and in the second changed from 8L:16D to 24L:0D (SL-LL treatment), with the lighting period changing 5 min per day in both cases. Each group contained three replicate tanks (60 fish/tank). The experiment was performed from September to the following March, a period almost spanning the first reproductive period. First, all fish in each tank were anesthetized using 0.05% MS-222 and measured for body weight and body length every month individually. Then, fish were sampled, and three fish in each tank were anesthetized to death in seawater using 0.05% MS-222. Body mass and length were recorded, and the brain were frozen in liquid nitrogen, and immediately stored at −80 °C. The gonads were stored in Bouin's fixative for 24 h and then in 70% ethanol for histological examination to confirm the specific stage of the experimental fish. All of the procedures described in this study were reviewed and approved by the ethical committee of the Institute of Oceanology, Chinese Academy of Sciences.

The feeding ratio (FR) was calculated as FR (%) = 100 × F/[0.5 × (BW2 + BW1) × (T2 − T1)], where BW1 and BW2 were the average individual weights at days T1 and T2. F was total food consumption.

The brains of Atlantic salmon were fixed in 4% paraformaldehyde in 0.1 M PBS (phosphate buffered saline, pH 7.4) overnight at 4 °C. The samples were dehydrated in a graded series of methanol. Sections of paraffin-embedded brains were prepared on 5-μM glass slides coated with 0.1% 164 poly-L-lysine solution. The partial CDS of AsMR and AsLR were cloned into pGEM-T vectors for preparing sense and antisense RNA probes from a T7 or SP6 promoter by using FITC or digoxigenin (DIG) RNA-Labeling Kit (Roche), respectively. The sections were hybridized with the sense or antisense probes at 66 °C for 18 h. After hybridization, the samples were incubated overnight at 4 °C with horseradish peroxidase (HRP)-conjugated anti-FITC antibody (Roche) at a 1:2000 dilution in blocking solution to detect the FITC signal. After three washes in PBST, the samples were incubated 1 h in tyramide signal amplification (TSA)-fluorescein at a 1:150 dilution in TSA buffer. The DIG signal was detected in samples. They were incubated overnight at 4 °C with HRP-conjugated anti-DIG antibody (Roche) at a 1:2000 dilution in blocking buffer with 1% H_2O_2. Following three PBST washes, the samples were incubated in TSA-plus tetramethylrhodamine (TMR) for 1 h. Double-color fluorescence in situ hybridization was performed using TSA plus fluorescein and TMR according to the manufacturer's instruction (NEL756, PerkinElmer). The nuclei were stained using 4′-6-diamidino-2-phenylindole (DAPI) and embedded in ProLong Gold Antifade reagent (Invitrogen, Carlsbad, CA, USA). The slides were then mounted and photographed by Nikon Eclipse 50i fluorescence microscope (Tokyo, Japan). In this procedure, two antisense RNA probes were co-incubated in a single sample during the hybridization step, to develop red and green fluorescence.

The Atlantic salmon brain was fixed in Bouin's fixative for 24 h and preserved in 70% ethanol. The samples were stained using hematoxylin 186 and eosin (H&E), and sections were observed by a light microscope (NikonYS-100, Japan). Photographs were taken with a digital camera (Nikon coolpix-4500, Japan).

15.2.3 Results

15.2.3.1 Location of AsMR and AsLR in the Brain of Atlantic Salmon

The location of AsMR and AsLR were examined by quantitative real-time PCR with β-actin mRNA as a loading control. The results showed that AsLR was primarily expressed in the diencephalon, pituitary gland, and SV, and AsMR were mainly

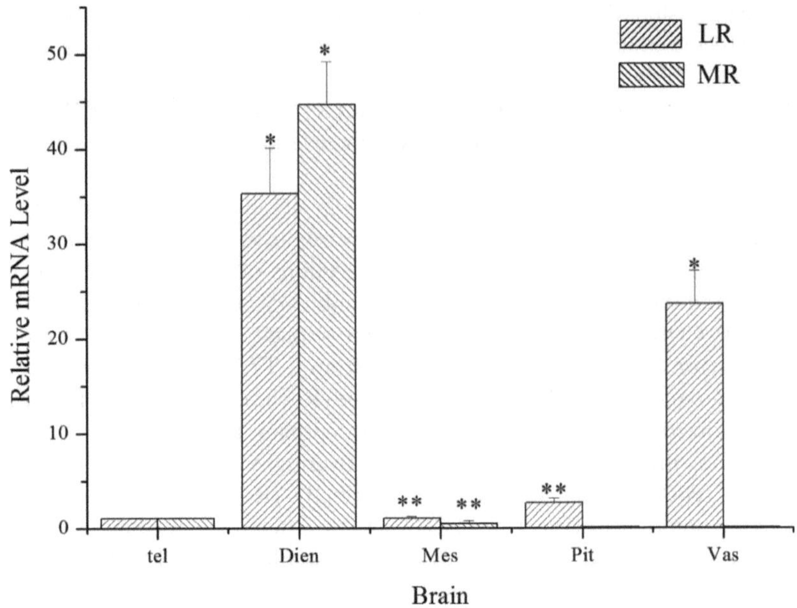

Fig. 15.4 Distribution of AsLR and AsMR in the Atlantic salmon brain. *Tel* telencephalon, *Dien* diencephalon, *Mes* mesencephalon, *Pit* pituitary gland, *Vas* saccus vasculosus

expressed in the diencephalon in the Atlantic salmon brain (Fig. 15.4). To confirm the precise location of AsLR, the diencephalon and SV were isolated to perform in situ hybridization. The results showed that both AsLR and AsMR transcripts were mainly expressed in the hypothalamus of the diencephalon (Fig. 15.5g–i). In the SV, the LR transcripts mainly appeared in the cerebrospinal fluid-contacting (CSF-c) cells (Fig. 15.5d–f). Besides the brain, the LR transcripts were mainly expressed in the gills, gut muscle, and ovaries (Fig. 15.6).

15.2.3.2 Expression Pattern of AsMR in the Different Photoperiod in the Hypothalamus of Atlantic Salmon

Expression of AsMR in the hypothalamus could be influenced by photoperiod. At the beginning of the experiment, the AsMR transcripts levels were lowest in the 24L:0D group, followed by the LL-SL group. The expression level of AsMR transcripts was highest in the 8L:16D group followed by the SL-LL group (Fig. 15.7a). At the end of the experiment, the lowest level of AsMR transcripts appeared in the 24L:0D and SL-LL groups and the highest level in the 8L:16D and LL-SL groups (Fig. 15.7b).

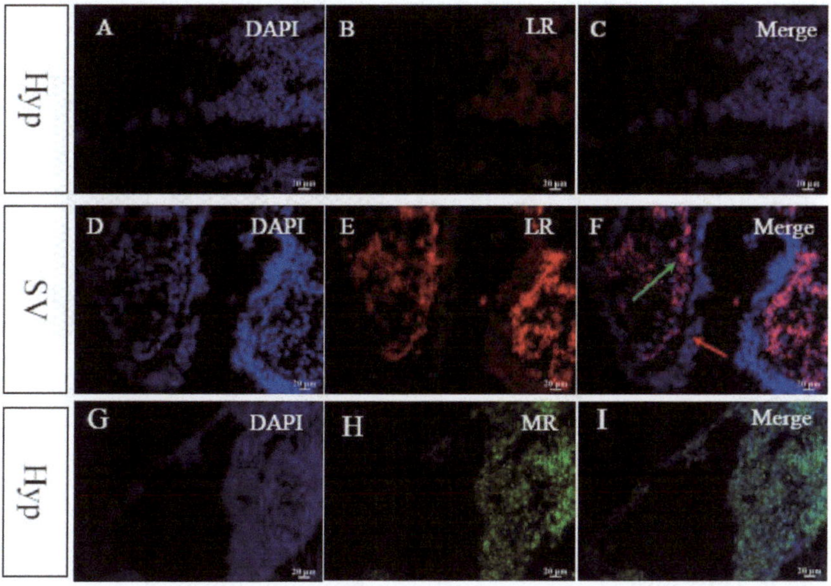

Fig. 15.5 Expression of AsLR and AsMR in the hypothalamus and saccus vasculosus. (**a–c**) Expression pattern of AsLR in the hypothalamus of Atlantic salmon brain; (**d–e**) expression pattern of AsLR in the saccus vasculosus of Atlantic salmon brain; and (**g–i**) location of AsMR mRNA in the hypothalamus using in situ hybridization. The green arrow indicates the cerebrospinal fluid-contacting cells; and the red arrow indicates the coronet cells. *SV* saccus vasculosus, *Hyp* hypothalamus

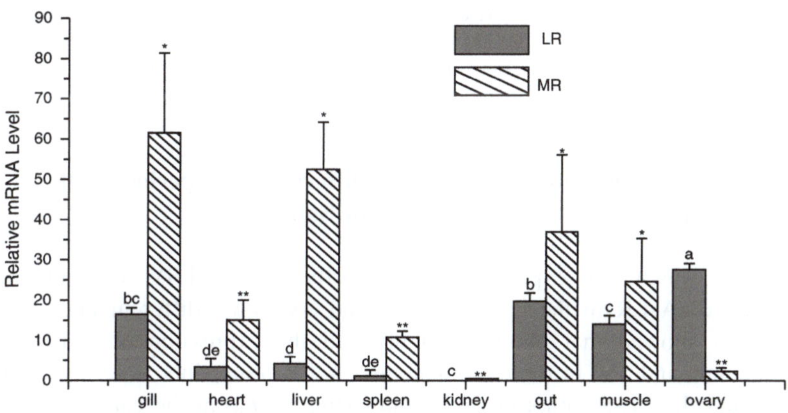

Fig. 15.6 Distribution of AsLR and AsMR transcripts of adult Atlantic salmon. Results are presented as mean ± SD. Significant differences were found at $p < 0.5$* and $p < 0.1$**; and different letters indicate statistical significance at $p < 0.05$

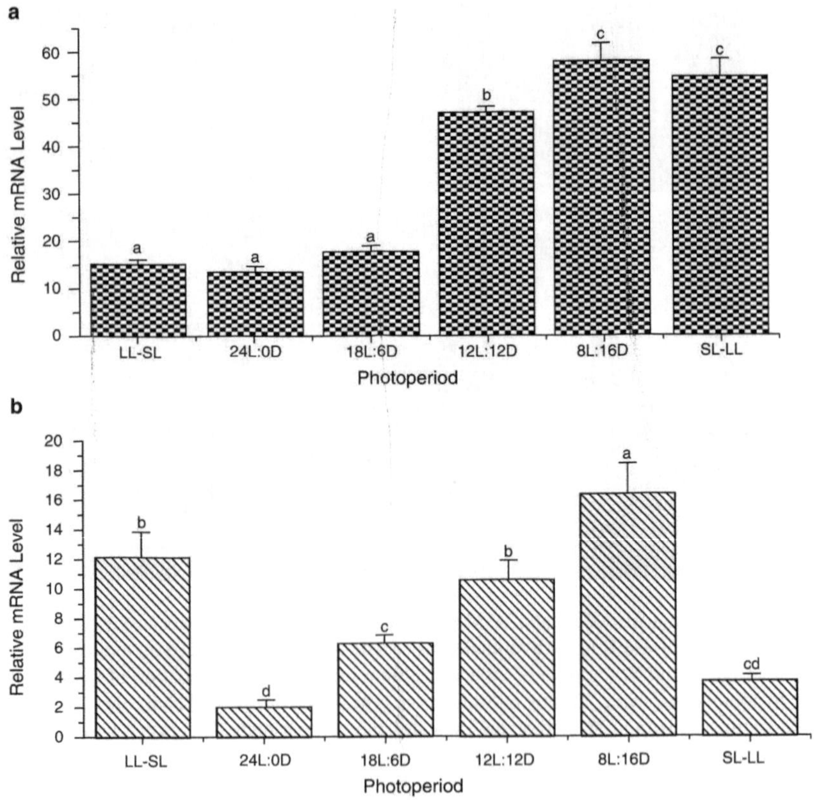

Fig. 15.7 Expression pattern of AsMR transcripts in different photoperiods. (**a**) Expression pattern of AsMR transcripts in different photoperiods at the early stage of the experiment; (**b**) expression pattern of AsMR transcripts in different photoperiods at the end of the experiment. Different letters indicate statistical significance at $p < 0.05$

15.2.3.3 Expression Pattern of AsLR in Different Photoperiods in the Hypothalamus and SV of Atlantic Salmon

Since AsLR was mainly expressed in the hypothalamus and SV, we examined the expression of AsLR in the hypothalamus and SV under different photoperiods. The results showed that photoperiod affected the expression of AsLR in the hypothalamus and SV. In the hypothalamus, AsLR transcripts were lowest in the 24L:0D and the LL-SL photoperiod groups (Fig. 15.8a) at the early stage of the experiment. At this time, these two treatments had the longest photoperiod. At the end time of the experiment, the lowest AsLR transcripts levels were in the 24L:0D group followed by the SL-LL group (Fig. 15.8b). The expression pattern of AsLR in the SV is similar to the hypothalamus. However, the expression level of AsLR in the SV is lower than that in the hypothalamus (Fig. 15.9a and b).

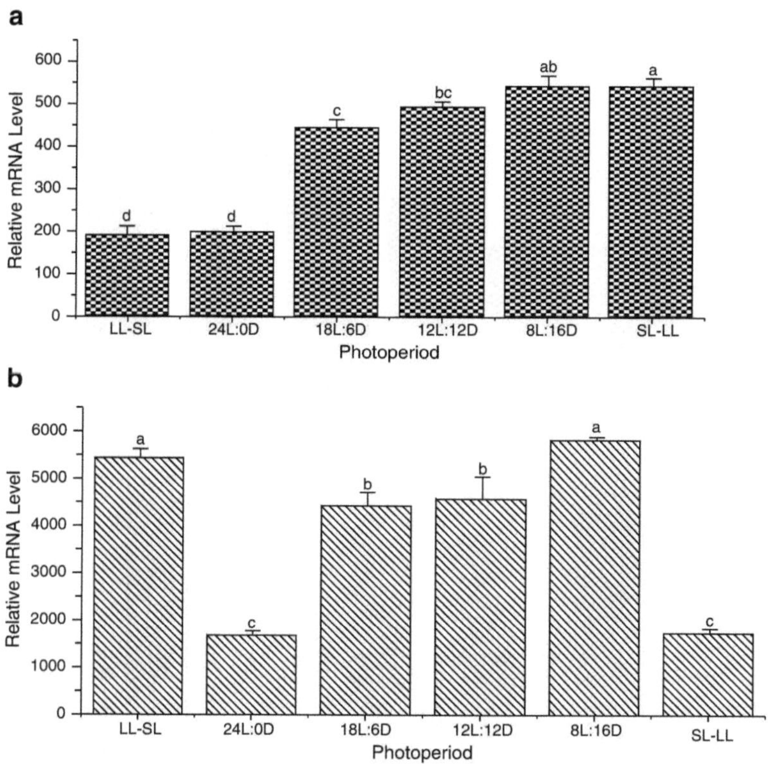

Fig. 15.8 Expression pattern of AsLR transcripts in the Atlantic salmon hypothalamus in different photoperiods. (**a**) Expression pattern of AsLR transcripts in different photoperiods at the early stage of the experiment; and (**b**) expression pattern of AsLR transcripts in different photoperiods at the end of the experiment. Different letters indicate statistical significance at $p < 0.05$

15.3 The Effect and Mechanism of Light on the Growth, Food Intake, and Energy Budget of Abalone (*Haliotis discus hannai Ino*)

Light is one of the key environmental factors influencing the growth, development, and survival of aquatic organisms, which have evolved to be able to change their physiology and behavior to adapt to changes in light conditions. In this study, a bioenergetics approach was used to examine the effects of different LED light qualities (red, orange, blue, green light, natural light, and a dark setting) on the growth and survival of the abalone *Haliotis discus discus*, and its physiological response mechanism under different light qualities. The results suggest that under blue or green light, the survival rate, specific growth rate, food intake, and food conversion efficiency of *H. d. discus* were significantly lower than in those groups under red or orange light ($P < 0.05$). Under red or orange light, the protein content,

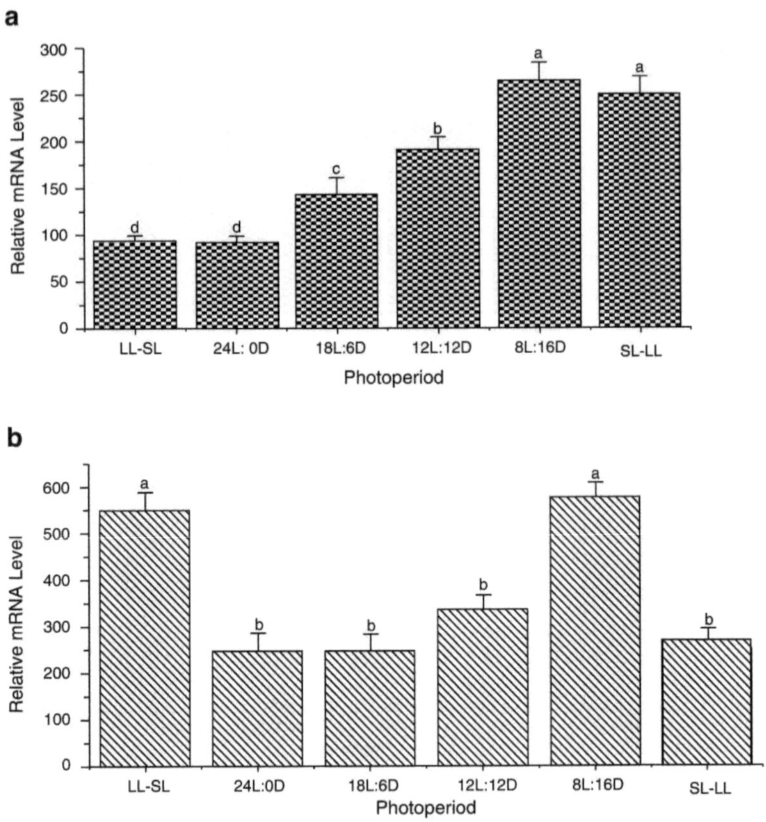

Fig. 15.9 Expression pattern of AsLR transcripts in the Atlantic salmon SV in different photoperiods. (**a**) Expression pattern of AsLR transcripts in different photoperiods at the early stage of the experiment; and (**b**) expression pattern of AsLR transcripts in different photoperiods at the end of the experiment. Different letters indicate statistical significance at $p < 0.05$

ash, and tissue:shell ratio of *H. d. discus* were significantly higher than in those groups under other light qualities ($P < 0.05$). Under red or orange light, pepsin, amylase, and cellulose activity was significantly higher than those in any other light quality group ($P < 0.05$), whereas lipase activity exhibited no significant difference among the light quality groups ($P > 0.05$). Under blue or green light, hexokinase and pyruvate kinase activity was higher than in any other group ($P < 0.05$). Under red or orange light, or in dark setting, succinate dehydrogenase activity was significantly higher than in any other group ($P < 0.05$), indicating the enhanced aerobic metabolism that was conductive to the continuous crawling of *H. d. discus*. Under blue or green light, lactate dehydrogenase activity and lactic acid content were higher ($P < 0.05$), suggesting enhanced anaerobic metabolism. Under blue or green light, *H. d. discus* lost more energy via excretion, feces, and respiration than was acquired from its food; as a consequence, there was no energy available for its growth. Under red or orange light, *H. d. discus* acquired more energy from its food and lost less energy via excretion and feces; as a result, its assimilation efficiency (K1) and net growth efficiency (K2) were significantly higher than those of any other group ($P < 0.05$). Therefore, we suggest that red or orange light should be used as a light source for the aquaculture of *H. d. discus* to reduce costs, increase the yield per water body, and improve the welfare culture of this and other species of abalone.

15.3.1 Introduction

The disc abalone *Haliotis discus discus* is an economically important shellfish in China, with demand growing for this high-protein and low-fat seafood. In 2012, 90,694 tons of *H. d. discus* were harvested from aquacultural sources, but supply cannot currently meet consumer demand. In recent years, the natural habitat of *H. d. discus* has declined as a result of overfishing, marine reclamation, and water pollution, among others, leading to a sharp decrease in the numbers of *H. d. discus* in the wild [26]. Thus, the development of an aquacultural system for farming *H. d. discus*, as well as for its protection in the wild, is the focus of current research. As a result, there is a need to identify the tolerance limits of *H. d. discus* seedlings to different environmental factors and how it responds physiologically to such stress factors. Currently, most studies on the growth and survival of abalone have focused on the effects of temperature [27], salinity [28], flow velocity [29], ammoniacal nitrogen [30], and dissolved oxygen [31], whereas only a few studies [32] have considered the effects of light on the growth and physiology of *H. d. discus*.

In this study, light-emitting diode (LED) lights were used as a light source instead of traditional fluorescent lamps; LEDs have become a mainstay in many research areas and are known to be safe to use, with few, if any, detrimental environmental impacts. LEDs produce less heat, have longer life, more precise wavelength emission, and more precise control of light level [33] making them more efficient sources of light. In addition, LED lights have been shown to not only increase the growth rate of aquatic organisms but also produce high-quality pollution-free food. Therefore, we adopted a bioenergetic approach to examine the effects of different light qualities on the growth and survival of *H. d. discus*, and its physiological response mechanism under different light qualities. Our results will provide a reference on which to base the development of the rapid and biologically friendly aquaculture of *H. d. discus*.

15.3.2 Materials and Methods

LED lights were purchased from the Hangzhou YD Illumination Co. Ltd., China. Four treatment groups under different light qualities and one fully dark setting were established. Light qualities included blue (450 nm), green (565 nm), red (620 nm), and orange light (600 nm). In addition, abalones were reared in natural light ("NL") as a control group, following the natural photoperiod of 14:10 light:dark. Light cycle was controlled by fully masking the sunlight using the blackout cloth in the control group. The LED light was hung above the aquarium, and the light intensity in the different light-quality treatment groups was kept as constant as possible by adjusting the distance from the LED light to the water surface. A spectroradiometer (JETI LiVal Spectroradiometer Model Specbos 1xx1, Jena, Germany) was used to measure the wave peak of light quality and light intensity, and five different points inside the aquarium were selected and then averaged to ensure that the wave peak of each light quality was within the required range. During this experiment, the light cycle was 14:10 light:dark, and lights were turned on at 06.00 h and off at 20.00 h via a clock controller.

Four repetitions of each treatment group per light quality were set up, with a total of 24 39-L plastic aquariums (45 cm × 25 cm × 35 cm). Ten individuals were placed in each aquarium, and a blackout cloth was used to separate the treatment groups of different light qualities, creating six separate square small chambers (1.5 m × 0.8 m × 0.5 m) to avoid light pollution. The approximate extent of light intensity was 1.82 ± 0.34 μmol/m^2/s in the different chambers. In the chamber under a dark setting, the abalones were exposed to no light except briefly during water exchange, feeding, or feces collection. The water temperature was kept at 17 °C with the same salinity and pH as described above. Aeration was almost continuously provided in each aquarium to keep the concentration of dissolved oxygen >6 mg/L; it was stopped for 30 min before collecting the food residues and feces, during which time the ammonia and nitrogen concentrations were < 0.1 mg/L.

All abalones were starved for 24 h before starting the experiment; 240 juvenile abalones (shell length: 42.51 ± 3.66 mm, body weight: 14.72 ± 2.39 g) were randomly divided into treatment groups under different light qualities. In total, 20 abalone samples were randomly selected for pre-treatment analysis to determine their initial weight, energy, and crude protein content before the experiment. As mentioned above, the abalones were fed daily at 17.00 h; at 07.00 h, food residue and feces were siphoned out, dried in the oven at 70 °C until a constant weight was achieved, and then stored in the refrigerator at −20 °C for the determination of energy and nitrogen values following the study. Simultaneously, *Lonicera japonica* (quantity equivalent to 6% of the wet body weight of abalone) were placed in another three aquariums without abalone to determine any possible effects that this food source may have on the experimental setup (e.g., nutrient damage, changes in chemical composition, etc.). The following morning, the food residue were collected, weighed, and dried to determine the value of food intake. During this experiment, two-thirds of the water volume was exchanged daily. After the experiment had finished, four abalones were collected from each aquarium and dried in the oven at 75 °C until a constant weight was achieved, to determine the energy content of the abalones. In addition, three abalones were collected from each aquarium to analyze and determine the enzyme activity and body composition.

Moisture on the surface of the abalones was removed using absorbent paper, and the abalones were then rapidly dissected on an ice tray. After shelling, the hepatopancreas was put into a 1.5-mL centrifuge tube, quickly placed in liquid nitrogen, and later moved to the refrigerator at −80 °C to be tested. Then, 0.2–0.4 g tissues were suspended in 1.8 mL 0.86% normal saline and fully ground in an ice-water bath. The tissue suspension was centrifuged at 3500 r/min for 10 min, until a 10% tissue homogenate were produced for further investigations. The activity of hexokinase (HK), pyruvate kinase (PK), succinic dehydrogenase (SDH), and lactic dehydrogenase (LDH), and the protein content of each sample were determined via kits supplied by Nanjing Jiancheng Bioengineering Institute (Nanjing, PR China), as per the manufacturer's instructions. PK activity was defined as: tissue protein per g converting 1 μmol phosphoenolpyruvate to pyruvic acid at 37 °C, pH 7.6 every 1 min (i.e., 1 activity unit). HK activity was defined as: at 37 °C, pH 7.6, the amount of issuc protein per g generating 1 mmol NADPH every 1 min (i.e., 1 enzyme activity unit). SDH activity was defined as: the amount of protein per mg reducing the absorbency of the reaction medium (comprising 0.1 mol/L sodium succinate, 0.01% methylene blue, 0.04 mol/L disodium malonate, and pH 7.0 phosphate buffer) at 37 °C every 1 min by 0.01 (i.e., 1 specific activity unit). LDH activity was defined as: the amount of tissue protein per g reacting with the substrate (4 mol/L NaOH, 2,4-dinitrophenylhydrazine, 2 mmol/L sodium pyruvate, and coenzyme I) at 37 °C for 15 min to produce 1 μmol pyruvic acid (i.e., 1 activity unit). The SDH activity unit was expressed in U/mg prot, whereas those for PK, HK, and LDH were in U/g prot. Meanwhile, the activity of pepsin (PES), lipase (LPS), cellulase (CL), and amylase (AMS) and the protein content of the sample were also determined via kits supplied by Nanjing Jiancheng Bioengineering Institute (Nanjing, PR China). For the determination of PES, CL, and AMS, 0.2–0.4 g tissue was fully ground with 1.8 mL 0.86% normal saline in an ice-water bath. The suspended tissues were then centrifuged at 2500 r/min, 8000 r/min, and 2500 r/min at 4 °C for 15 min, until a 10% tissue homogenate was produced for further investigations. PES activity was defined as: the amount of tissue protein per mg decomposed into 1 μg amino acid at 37 °C in 1 min (i.e., 1 enzyme activity unit). CL activity was defined as: the amount of tissue per g catalyzed to 1 μg glucose in 1 min (i.e., 1 enzyme activity unit). AMS activity was defined as: the amount of protein per mg in the tissues reacted with the substrate at 37 °C for 30 min to hydrolyze 10 mg amylon (i.e., 1 AMS activity unit). To determine LPS activity, 0.2–0.4 g tissue was fully ground with 1.8 mL 0.86% normal saline in an ice-water bath. The tissues were then mechanically homogenized at 2500 r/min, centrifuged for 10 min, and the supernatant then used to determine LPS activity. This was defined as the amount of tissue protein per g reacted with the substrate in this response system at 37 °C for 1 min to consume 1 μmol substrate (i.e., 1 enzyme activity unit). AMS and PES activity units were expressed in U/mg prot, whereas those of CL and LPS were expressed in U/g prot.

Abalones and *L. japonica* were dried at 105 °C to a constant weight to determine their moisture content. An elemental analyzer (Vario EL III, Germany) was used to determine the nitrogen content. The Soxhlet method was used to determine the fat content. Food and abalone were put in a muffle furnace (SX2-5-12, Shanghai, China) at 550 °C for 12 h to determine the resulting ash content [34]. An oxygen bomb calorimeter (Parr 1281, USA) was used to determine the energy value of food, feces, and abalone body. Each sample was analyzed at least three times. The content of lactic acid was determined via kits supplied by Nanjing Jiancheng Bioengineering Institute (Nanjing, PR China), when NAD was used as hydrogen acceptor, LDH catalyzed the dehydration of lactic acid to pyruvic acid and converted NAD^+ into NADH. PMS transferred hydrogen to reduce nitrotetrazolium blue chloride (NBT) to a purple chromogenic substrate. When the absorbency of the chromogenic substrate equaled 530 nm, it correlated linearly with the content of lactic acid, thus enabling us to determine the lactic acid content of the sample.

Energy budgets were based on the model proposed by Carfoot [35]: $C = F + U + R + G$ was converted to $G = C - F - U - R$, where C was the energy contained in the food intake, G was the energy accumulated for growth, F was the energy lost in feces, U was the excretion energy, and R was the energy lost through respiration. The energy values of C, G, and F were determined using an oxygen bomb calorimeter (Parr 1281, USA) after tissue samples had been dried at 70 °C until they had reached a constant weight.

The energy of excretion was calculated as: $U = U_N \times 24{,}830 = (C_N - G_N - F_N) \times 24{,}830$ [36], where U_N was the excreted nitrogen, C_N was the nitrogen contained in the food intake, G_N was the nitrogen stored in the abalone, and F_N was the nitrogen excreted in feces; the factor 24,830 represents the energy value contained in ammonia nitrogen per g (J g^{-1}). Given that most shellfish are ammonotelic, $NH_4{}^+$–N is the major form of excreted nitrogen, with only a low amount carbamide; thus, the latter was deemed negligible in this study. An elemental analyzer (Vario EL III, Germany) was used to determine the nitrogen content, and the energy of respiration was expressed as: $R = C - G - F - U$.

The assimilation efficiency (K_1) and net growth efficiency (K_2) were calculated using the following equations: K_1 (%) = 100 × $(G + R)/(G + R + U)$ and K_2 (%) = 100 × $G/(G + R)$.

During the study, the survival rate (S%), shell length/body weight-specific growth rate (SGR, %day^{-1}), food intake (FI, % body weight day^{-1}), and food conversion efficiency (FCE, %) of abalone were calculated every other month, as follows:

$$S\ (\%) = [(N_1 - N_2)\,/N_1] \times 100$$

$$\text{SGR}\ \left(\%\text{day}^{-1}\right) = (\ln W_2 - \ln W_1)\,/T \times 100 \text{ and}$$

$$\text{FI}\ \left(\%\text{body weight day}^{-1}\right) = F/\,[T \times (W_2 + W_1)\,/2] \times 100$$

$$\text{FCE}\ (\%) = (W_2 - W_1)\,/\text{F} \times 100,$$

where N_1 and N_2 were the number of abalones at the beginning and end of the experiment, respectively, W_1 and W_2 were the wet weight (g) of abalones at the beginning and end of the experiment, respectively, L_1 and L_2 were the shell length at the beginning and end of the experiment, respectively, T was the length of the experiment in days, and F was the weight of food intake during the experimental process (dry weight, g).

15.3.3 Results

15.3.3.1 Specific Growth Rate

Different light qualities had significant effects on the body weight and shell length of SGR of *H. d. discus*. Under orange or red light, the body-weight SGR of *H. d. discus* was significantly higher than that of the treatment group under any other light quality (Fig. 15.10, $F = 124.27$, $P < 0.001$). Under natural light or in a dark setting, the body-weight SGR was >0.2 mg day^{-1}. Under blue or green light, *H. d. discus* lost weight, resulting in body-weight SGRs that were significantly lower than the treatment group under any other light quality ($F = 163.29$, $P < 0.001$). Under orange light, the shell-length SGR was also significantly longer compared with that of the treatment group under any other light quality ($F = 85.33$, $P < 0.001$). Under natural or red light, or in dark setting, the shell-length SGR showed no significant difference ($F = 3.07$, $P = 0.114$), whereas under green or blue light, it was the lowest recording at <0.05 μm day^{-1}.

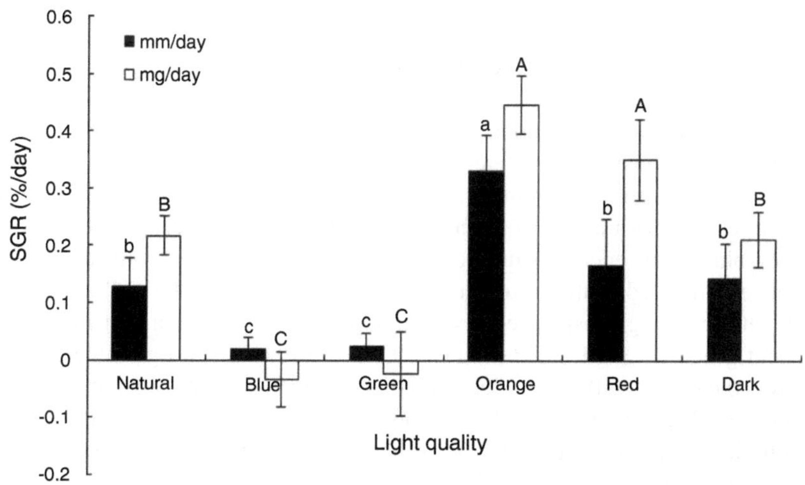

Fig. 15.10 Effect of LED light quality on the specific growth rate of juvenile abalone (*H. d. discus*). Capital letters indicate the difference among the mean body-weight-specific growth rate, whereas lowercase letters indicate the difference among the mean shell length specific growth rate. There was no significant difference between bars with the same letter, whereas significant differences occurred between bars with different letters

15.3.3.2 Biochemical Composition

Under natural, green, or blue light, or in dark setting, the moisture content of *H. d. discus* was significantly higher than that of the treatment group under red light ($F = 15.38$, $P = 0.013$), but no significant difference in moisture content occurred in *H. d. discus* under natural, green, or blue light (Table 15.1, $F = 2.71$, $P = 0.121$). Under orange light, the moisture content of *H. d. discus* was also low (mean 81.87%). The protein content of *H. d. discus* under natural or orange light was significantly higher than that of any other treatment group ($F = 32.58$, $P < 0.001$). The protein content under blue or green light was the lowest and significant differently compared with that under red light or in dark setting ($F = 28.57$, $P = 0.001$). The amount of ash of *H. d. discus* under orange light or in dark setting was significant greater than under any other light quality ($F = 11.82$, $P = 0.017$), but there was no difference in the ash content from treatment groups under any other light quality ($F = 2.02$, $P = 0.145$). Under orange or red light or in dark setting, the fat content of *H. d. discus* was significantly higher compared with that of treatment groups under any other light quality ($F = 18.95$, $P = 0.003$). Under natural, green, or blue light, the amount of lactic acid accumulated in *H. d. discus* was significantly higher than under any other light quality ($F = 49.26$, $P < 0.001$). Under orange or red light or in dark setting, the lactic acid content was lower, but with no significant difference ($F = 1.64$, $P = 0.173$) compared with the other groups. Under green or blue light, the tissue:shell ratio of abalone was significantly less than that of any other treatment group (Table 15.1, $F = 30.17$, $P < 0.001$).

15.3.3.3 Determination of Energy Parameters

As shown in Table 15.2, under different light qualities, there were significant differences in the energy parameters of *H. d. discus*. Under orange or red light or in a dark setting, energy acquired from food was significantly higher than that of the treatment groups under any other light quality ($F = 169.22$, $P < 0.001$). Under green or blue light, energy acquired from food was the lowest with a significant difference compared with any other treatment group ($F = 94.37$, $P < 0.001$). In a dark setting, energy lost in feces was significantly higher than that in any other treatment group ($F = 61.29$, $P < 0.001$). Under blue or green light, energy lost in feces was the lowest with a significant difference compared with any other treatment group ($F = 45.12$, $P < 0.001$). Under orange light, energy accumulated for growth was significantly higher than that in any other treatment group, followed by red light, a dark setting, and natural light ($F = 39.28$, $P < 0.001$). Under blue or green light, abalones exhibited negative growth. In a dark setting, energy lost via excretion was the highest ($F = 19.36$, $P = 0.003$). However, under orange light, energy lost via excretion was significantly less than that of the treatment group under any other light quality ($F = 40.16$, $P < 0.001$). Under orange light, energy lost via respiration was significantly higher than in any other treatment group ($F = 28.08$, $P = 0.001$). Under natural or red light, energy lost via respiration was higher than under any other

Table 15.1 Mean biochemical content (± standard error) and the tissue:shell ratio of juvenile *H. d. discus*, reared under six different light qualities[a]

Variable	Light quality					
	Natural	Blue	Green	Orange	Red	Dark
Tissue moisture (%)	84.63 ± 1.51[ab]	88.12 ± 2.65[a]	88.71 ± 0.82[a]	81.87 ± 1.47[b]	75.42 ± 2.24[c]	84.06 ± 1.71[ab]
Protein (%DTW)	28.31 ± 2.84[a]	15.35 ± 1.57[c]	17.72 ± 1.55[c]	32.92 ± 0.97[a]	24.76 ± 1.12[b]	21.31 ± 2.44[b]
Ash	23.82 ± 2.24[b]	24.62 ± 1.38[b]	22.75 ± 1.62[b]	27.70 ± 2.14[a]	22.83 ± 1.06[b]	29.91 ± 1.15[a]
Lipid	2.01 ± 0.10[b]	1.77 ± 0.14[bc]	1.42 ± 0.13[c]	2.91 ± 0.32[a]	2.79 ± 0.21[a]	3.09 ± 0.26[a]
Lactic acid content (mmol/g prot)	0.38 ± 0.02[a]	0.47 ± 0.03[a]	0.49 ± 0.06[a]	0.19 ± 0.02[b]	0.22 ± 0.03[b]	0.17 ± 0.06[b]
Tissue:shell ratio	0.44 ± 0.01[a]	0.31 ± 0.04[b]	0.28 ± 0.02[b]	0.44 ± 0.04[a]	0.46 ± 0.07[a]	0.42 ± 0.02[a]

[a]The same superscript means that the difference between biochemical content and tissue:shell ratio was not significant, whereas different superscripts means that the differences were significant

Table 15.2 Energy parameters in juvenile abalone, *H. d. discus*, under different light qualities[a]

Light quality	Energy parameters				
	C (J g^{-1} day^{-1})	F (J g^{-1} day^{-1})	G (J g^{-1} day^{-1})	U (J g^{-1} day^{-1})	R (J g^{-1} day^{-1})
Natural	827.61 ± 92.74^{b}	411.36 ± 65.42^{c}	87.52 ± 64.34^{c}	72.21 ± 15.64^{bc}	256.64 ± 72.10^{bc}
Blue	389.38 ± 85.74^{c}	206.42 ± 94.54^{d}	-65.74 ± 28.62^{e}	85.93 ± 10.24^{b}	162.72 ± 64.47^{d}
Green	293.57 ± 102.82^{c}	195.97 ± 61.16^{d}	-104.87 ± 69.50^{d}	57.93 ± 14.23^{c}	144.55 ± 83.27^{d}
Orange	1247.62 ± 139.51^{a}	606.53 ± 122.31^{b}	271.67 ± 72.92^{a}	38.77 ± 9.82^{d}	327.82 ± 89.19^{a}
Red	1032.12 ± 127.37^{ab}	471.27 ± 96.55^{c}	204.57 ± 92.15^{b}	69.16 ± 8.75^{bc}	289.37 ± 80.51^{b}
Dark	1451.74 ± 158.62^{a}	1002.50 ± 106.72^{a}	107.65 ± 47.21^{c}	117.59 ± 12.64^{a}	224.13 ± 54.90^{c}

[a]Values were expressed as mean $\pm$ SE; $n = 4$; C feeding energy, F energy loss in feces discharge, G energy deposit for growth, U energy loss in excretion, R energy loss in respiration. The same superscript means that the differences were not significant, whereas different superscripts mean that the differences were significant

treatment except for orange light, and was the lowest among blue or green light, although there was no significant difference among groups ($F = 1.39$, $P = 0.187$).

15.4 Effects of Light Quality and Photoperiod on the Growth and Energy Metabolism of *H. discus hannai*

The goal of this experiment was to examine the effects of different light qualities (red, white, and blue) and cycles (12L:12D, 8L:16D, 4L:20D, 0L:24D, and 16L:8D) on the survival, growth, metabolism, and antioxidant defense system of *Haliotis discus hannai Ino*. Under red and white light at 4L:20D, the body-weight-specific growth rate of abalones was significantly higher than that of the 0L:24D group ($P < 0.05$). At 16L:8D, succinate dehydrogenase activity decreased and lactate dehydrogenase and alanine aminotransferase activity and the lactic acid content significantly rose ($P < 0.05$) in the red and white light groups. Under blue light, reactive oxygen species and malondialdehyde content for each light cycle were significantly higher than those of the red and white groups ($P < 0.05$). Under blue, white, and red light, total antioxidant capacity (T-AOC), superoxide dismutase (SOD) activity, glutathione peroxidase (GPX) activity, and reduced glutathione (GSH) content gradually rose with increased duration of light exposure. Under red light, significant differences in these parameters were detected between the 4L:20D and 16L:8D groups ($P < 0.05$). Under blue and white light at 16L:8D, values of T-AOC, SOD, GPX activity, and GSH content were significantly lower than those of the 12L:12D group ($P < 0.05$). Overall, these results indicate that red light at light cycles of 4L:20D and 8L:16D is optimal for the culture and production of *H. discus hannai*, as these conditions promoted food intake, growth, and health maintenance as well as normal physiological metabolism of the organism.

15.4.1 Introduction

Most of the natural rhythmicity (day and night or seasonal) observed in aquatic organisms is closely related to the light cycle. For example, abalone adjust their physiological activities by hiding in cracks in the rocks during the day and feeding at night [37]. García-Esquivel et al. [38] reported that the growth rate of *Haliotis fulgens* was highest under the 00:24 Light:Dark (L:D) light cycle. Gorrostieta-Hurtado et al. [39] found that the survival and food intake rates of *Haliotis corrugata* larvae were significantly higher in the dark setting than under constant lighting. In addition, the oxygen consumption rate and ammonia excretion rate of *Haliotis discus discus* were found to be significantly higher under light than in a dark setting [32]. Day et al. [40] reported that the growth rate of *Haliotis rubra* was 15.65% faster in a lightproof water tank than that in the light. Gao et al. [41] found that

Haliotis discus hannai exhibited significant tropism to red and orange light and a significant escape behavior under blue, green, and purple light. Under blue light, the energy appropriated for the growth of *Litopenaeus vannamei* was low and that for excretion was high, thus the specific growth rate (SGR) under blue light was significantly lower than that of groups under any other light quality [42]. The growth of *Salmo salar* larvae can be promoted if the duration of light application increases after food intake begins [43]. Thorpe et al. [44] found that *S. salar* had the fastest growth rate in summer, which is characterized by a long period of light each day. Other researchers have noted that a gradual increase in the light cycle is more effective at influencing the development of the fish gonad compared to a gradual decrease in the light cycle [45].

To date, little is known about the effects of the interaction between light quality and cycle on the food intake, growth, metabolism, and other physiological activities of *H. discus hannai*. *H. discus hannai* is an economically important shellfish species in China, but culture of this species has been experiencing numerous problems as the aquaculture industry develops and expands (e.g., incomplete seed development, worsening summer mortality, and rise of feeding costs). In traditional Chinese aquaculture, producers often use sunshade nets in aquaculture hatcheries to provide a dark for abalone. However, in the abalone's natural environment, the light quality and cycle are subject to the seasonal rhythm and variations in concentration of suspended particles, zooplankton, and phytoplankton. Abalones are able to survive, grow, and reproduce in the ever-changing natural environment, and light may have important regulatory effects on growth and development in this setting.

In this study, the food intake, growth, metabolism, and antioxidant defense system of *H. discus hannai* under different light qualities and cycles were examined. The results should provide a theoretical basis for the optimization of aquaculture conditions to produce healthy and productive *H. discus hannai* in a sustainable aquaculture setting.

15.4.2 Materials and Methods

This experiment was carried out using LED lights supplied by the Institute of Semiconductors CAS (Beijing, China). The lights were suspended above each aquarium. The following three light quality treatments were used: blue, at 450 nm, light intensity 1.81 ± 0.02 μmol m^{-2} s^{-1}; red, at 620 nm, intensity 1.80 ± 0.03 μmol m^{-2} s^{-1}; and white, main peak at 460 nm, intensity 1.78 ± 0.03 μmol m^{-2} s^{-1}. White light was set using the full spectrum to simulate natural light. The light intensity in different light quality groups was kept as constant as possible by adjusting the distance between the light and the water surface. A spectroradiometer (JETI LiVal Spectroradiometer Model Specbos 1xx1, Jena, Germany) was used to measure the wave peak of light quality and light intensity, and five different points inside the aquarium were selected and then averaged to ensure that the wave peak of each light quality was within the required range. Five different light cycle groups (12L:12D, 8L:16D, 4L:20D, 0L:24D, and 16L:8D) were set up, and the lights were controlled using a timer. In total, 60 aquariums (45 cm × 25 cm × 35 cm, water volume: 39 L) were used in this experiment, with four replicates for each of the light quality–cycle combinations.

Twelve abalones were placed in each aquarium. Blackout cloths were used to separate the different light quality and cycle groups to avoid interference. Water exchange, feeding, and collection of residual food were performed during the dark period. During this experiment, the room temperature was kept at 18 °C, and the salinity and pH were the same as those used during acclimation. The water was kept clean by sand filtration, ultraviolet radiation disinfection, a protein-foam separator, and biofiltration using the circulating aquaculture system. One-third of the water volume was exchanged each day, and each aquarium was scrubbed every 7 days. In addition, each aquarium was provided with continuous aeration to maintain a dissolved oxygen concentration >6 mg/L. However, aeration was stopped 20 min before the collection of residual food. The ammonia concentration was always <0.1 mg/L in this experiment. The experiment lasted for 100 days.

All experimental abalones were sourced from the same batch after artificial hatching. They were starved for 24 h before the start of the experiment. Thirty abalones not used in the experiment were randomly selected for the analysis of crude protein content, and 720 abalones (shell length: 3.62 ± 0.13 cm, body weight: 6.85 ± 0.19 g) were randomly divided into the different treatments (12 individuals in each of 60 tanks). During the experiment, these abalones were fed with *L. japonica* at 5% of their wet body weight each day at 17:00H. The food residue was collected and weighed the following morning at 8:00. Samples of the fresh *L. japonica* used to feed these abalones were weighed and then placed in another three aquariums to determine the quantity that might be lost due to disintegration and dissolution rather than consumption. The next morning, the samples were collected and weighed, and the quantity lost was calculated and used to correct the food intake values. After the completion of the experiment, three abalones were randomly collected from each aquarium for the analyses described below.

The water on the body surface of each abalone was removed using absorbent paper, and the specimen then was rapidly dissected on ice. After shelling, the hepatopancreas was removed and placed in a 1.5-mL centrifuge tube, which was quickly placed in liquid nitrogen and later moved to a −80 °C freezer. The hepatopancreas was used for the enzyme activity and biochemical composition analyses.

For enzyme activity analysis, 0.2–0.4 g of hepatopancreas was mixed with 1.8 mL 0.86% normal saline and the sample was fully ground in an ice-water bath. The tissues were centrifuged at 3500 r/min for 10 min until a 10% tissue homogenate was produced. The activities of succinate dehydrogenase (SDH), lactic dehydrogenase (LDH), superoxide dismutase (SOD), and lactic acid content in the homogenate were determined following the method described by Gao et al. [46]. The protein content in the homogenate was determined using Coomassie Blue Staining as described by Bradford [47], and bovine serum albumin was used as the protein marker.

Alanine aminotransferase (GPT) activity, reactive oxygen species (ROS) content, total antioxidant capacity (T-AOC), reduced glutathione (GSH) content, glutathione peroxidase (GP_X) activity, and malondialdehyde (MDA) content were determined using kits purchased from Nanjing Jiancheng Bioengineering Institute (Nanjing, Jiangsu, China). Contents and activities were determined as follows:

- GPT activity: At 37 °C and pH = 7.4, GPT reacts with a substance composed of alanine and α-ketoglutaric acid in the homogenate, generating pyruvic acid and glutamic acid. After 30 min, 2,4-dinitrophenylhydrazinehydrochloric acid is used to terminate the reaction. 2,4-Dinitrophenylhydrazine reacts with the carbonyl group of ketonic acid and generates pyruvic acid phenylhydrazone. Phenylhydrazone appears red-brown in color under the alkaline condition, and the absorbance is tested at 505 nm using a spectrophotometer (721G-100, INESA.CC, Shanghai, China). Absorbance data then are used to calculate the enzyme activity, according to the standard curve shown in the kit instructions. The activity of GPT (U/g prot) in tissue = the activity of GPT in the homogenate, which was obtained from the standard curve divided by the concentration of protein (g prot/L) in the homogenate.
- ROS content: The tissue homogenate was centrifuged at 1000 r/min for 10 min, and then 1 mmol/L DCFH-DA (2′,7′-dichlorofluorescin diacetate) was added to the supernatant. The mixture was incubated at 37 °C for 30 min after being fully and evenly mixed. A fluorescence spectrophotometer (960MC, INESA.CC) was used to determine the fluorescence intensity at the optimum excitation wavelength of 500 nm and at the optimum emission wavelength of 525 nm, and the results were expressed as fluorescence arbitrary units (A.U.)
- T-AOC: T-AOC was determined using the ferric reducing ability of plasma method. The antioxidant substances in an organism can reduce Fe^{3+} to Fe^{2+}, which can form a stable complex with phenanthroline-like substances. The level of T-AOC of a sample was determined using the colorimetric method. At 37 °C, a unit (U/mg) of total antioxidant capacity (T-AOC) was defined as the increase of 0.01 of optical density of all chemical reaction substances by 1 mg of tissue protein per minute.
- GSH content: Following the method described by Anderson [48], one yellow compound can be generated when DTNB (5,5′-dithiobis-(2-nitrobenzoic acid)) reacts with sulfhydryl compounds. The colorimetric method then was used to measure the absorbance of the yellow compound.
- GP_X activity: The DTNB method described by Li et al. [49] was used to measure GPx activity. GPx was assayed spectrophotometrically using glutathione (GSS) as the substrate and measuring the decrease of GSH at 412 nm. One unit (U) of GPx activity was defined as the amount needed to reduce 1 μmol/L of GSH in the reaction system (minus any nonenzymatic reaction) by 1 mg of tissue protein/min.
- MDA content: MDA was determined by measuring the amount of red product produced by MDA and TBA, which has a maximum absorption peak at 532 nm. The absorbance was measured in nmol/mg protein using the spectrophotometer (721G-100, INESA.CC, Shanghai, China).

In addition, the activities of pepsin (PES), cellulase (CL), and amylase (AMS) and the protein content of each sample were determined following the methods described by Gao et al. [46]. In brief, 0.2–0.4 g of tissue was added to 1.8 mL 0.86% normal saline and fully ground in an ice-water bath. The tissues then were centrifuged at 2500 r/min, 8000 r/min, and 2500 r/min at 4 °C for 15 min until 10% tissue homogenates were produced. These homogenates then were used for the assays.

15.4.3 Results

15.4.3.1 Food Conversion Efficiency

The food conversion efficiency (FCE) of abalones in the blue group was significantly lower than that in the white and red groups for all light cycles except 0L:24D (Fig. 15.11, $F = 67.43$, $P < 0.001$), and no significant difference among light cycles was identified for the blue light group ($F = 1.59$, $P = 0.141$). Under red and white light, the FCE at 4L:20D was significantly higher than that of any other light cycle ($F = 12.71$, $P = 0.005$). The FCE in the red and white groups differed significantly under each light cycle except 8L:16D and 0L:24D ($F = 16.04$, $P = 0.002$).

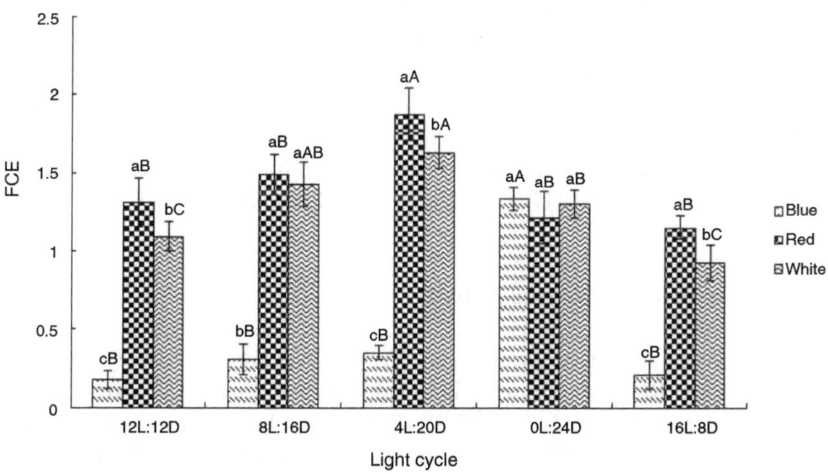

Fig. 15.11 Effects of light qualities and light cycles on the food conversion efficiency of juvenile abalone (*H. discus hannai*). Values are expressed as mean ± s.e. ($n = 4$). The same small letters indicate similar food conversion efficiency, and different small letters indicate significantly different ($P < 0.05$) food conversion efficiency in relation to light quality. The same capital letters indicate similar food conversion efficiency, and different capital letters indicate significantly different ($P < 0.05$) food conversion efficiency in relation to light cycle

15.4.3.2 Digestive Enzyme Activity

Under blue light, the PES activity of abalones in each light cycle group decreased with increased duration of light exposure, but no significant differences were identified (Table 15.3, $F = 1.34$, $P = 0.157$). Under red light, the PES activity was significantly higher at 4L:20D and 8L:16D than at 0L:24D ($F = 23.25$, $P = 0.001$). Under white light, the activity of PES was highest at 4L:20D, and it differed significantly from the activities at 8L:16D and 16L:8D ($F = 15.15$, $P = 0.004$). The interaction between light quality and light cycle also had a significant effect on the PES activity (light quality × light cycle, $F = 17.39$, $P = 0.002$). Under red and white light at 4L:20D, 16L:8D, and12L:12D, the PES activity of abalones was significantly higher than that under blue light ($F = 28.87$, $P = 0.001$).

Under red light, the CL activity of abalones was significantly higher at 8L:16D than at any other light cycle ($F = 11.97$, $P = 0.008$). Under white light, the CL activity was significantly higher at 4L:20D than at any light cycle ($F = 7.81$, $P = 0.015$). Under blue light, no significant differences in CL activity were detected among any of the light cycles ($F = 1.28$, $P = 0.163$) except 0L:24D. At 4L:20D, 16L:8D, and 12L:12D, the CL activity in the red and white groups was significantly higher than that in the blue group ($F = 50.45$, $P < 0.001$). At 8L:16D, the CL activity in the red group was significantly higher than that in the white group ($F = 17.94$, $P = 0.002$).

Light quality and light cycle also had a significant effect on the AMS activity of abalones. Under blue light, the AMS activity at 4L:20D was significantly higher than at 16L:8D, 12L:12D, and 8L:16D (Table 15.3, $F = 15.48$, $P = 0.004$). Under red and white light, the AMS activity at 4L:20D was also significantly higher than at 16L:8D and12L:12D ($F = 24.71$, $P = 0.001$); however, no significant difference between 16L:8D and 12L:12D under red light was detected ($F = 2.04$, $P = 0.117$). The AMS activity under red and white light was significantly higher than that under blue light for all light cycles ($F = 31.64$, $P < 0.001$) except 0L:24D.

15.4.3.3 Antioxidant Enzyme Activity

The ROS content increased as duration of light exposure increased. Under blue and red light, the ROS content was significantly higher at 16L:8D than that at any other light cycle (Table 15.4, $F = 85.27$, $P < 0.001$), but no significant difference between the 8L:16D and 12L:12D groups was detected ($F = 2.94$, $P = 0.092$). Under white light, the ROS content did not differ significantly between the 16L:8D and12L:12D groups ($F = 2.11$, $P = 0.108$), but the contents in these two groups were significantly higher than that at any other light cycle ($F = 71.08$, $P < 0.001$). The ROS content in the blue light group was significantly higher than that in red and white light groups for all light cycles except 0L:24D ($F = 97.39$, $P < 0.001$).

Light quality and light cycle also had a significant effect on the T-AOC of abalones. Under blue and white light, the T-AOC was significantly higher at 12L:12D than at any other light cycle ($F = 41.28$, $P < 0.001$); however, the T-AOC

Table 15.3 Effects of light qualities and light cycles on the activity of digestive enzyme of juvenile abalone (*H. discus hannai*)

Light quality	Light cycle	Variable		
		Pepsase (PES) (U mg^{-1} prot)	Cellulase (CL) (U g^{-1} prot)	Amylase (AMS) (U mg^{-1} prot)
Blue	16L:8D	2.11 ± 0.33^{bB}	2.33 ± 0.19^{bB}	1.85 ± 0.22^{cC}
	12L:12D	1.97 ± 0.15^{bB}	3.01 ± 0.41^{bB}	2.46 ± 0.46^{cB}
	8L:16D	2.46 ± 0.28^{bB}	4.51 ± 0.29^{bC}	2.05 ± 0.16^{cB}
	4L:20D	3.15 ± 0.24^{bB}	3.64 ± 0.57^{bB}	5.91 ± 0.81^{bB}
	0L:24D	5.52 ± 0.39^{aA}	10.29 ± 1.31^{aA}	14.17 ± 1.28^{aA}
Red	16L:8D	5.98 ± 0.66^{abA}	9.45 ± 1.35^{bA}	9.31 ± 1.65^{bA}
	12L:12D	6.22 ± 0.75^{abA}	8.79 ± 1.62^{bA}	9.04 ± 1.44^{bA}
	8L:16D	7.41 ± 0.49^{aA}	12.37 ± 1.25^{aA}	12.41 ± 1.69^{abA}
	4L:20D	7.09 ± 0.28^{aA}	9.02 ± 1.08^{bA}	14.22 ± 1.50^{aA}
	0L:24D	4.59 ± 0.42^{bA}	10.55 ± 1.94^{bA}	12.58 ± 1.11^{abA}
White	16L:8D	4.25 ± 0.87^{bA}	7.86 ± 0.92^{bA}	4.33 ± 0.83^{cB}
	12L:12D	6.85 ± 0.93^{aA}	7.59 ± 0.84^{bA}	8.62 ± 1.34^{bA}
	8L:16D	4.11 ± 0.34^{bB}	8.92 ± 1.03^{bB}	11.74 ± 0.79^{abA}
	4L:20D	6.47 ± 0.51^{aA}	11.79 ± 1.65^{aA}	16.29 ± 1.67^{aA}
	0L:24D	5.36 ± 0.27^{abA}	11.28 ± 1.77^{aA}	12.95 ± 1.41^{abA}

Values are expressed as mean ± s.e. ($n = 4$). Different small letters indicate significant differences in activity of digestive enzyme for the same light cycle at different light qualities, $P < 0.05$. Different capital letters indicate significant differences in activity of digestive enzyme for the same light quality with different light cycles, $P < 0.05$

Table 15.4 Effects of light qualities and light cycles on the T-ACO, contents of ROS, GSH and MDA, and activities of SOD and GPX in juvenile abalones (H. discus hannai)

		Variable					
Light quality	Light cycle	Reactive oxygen species (ROS) (A.U.)	Total antioxidant capacity (T-AOC) (U/mg prot)	Superoxide dismutase (SOD) (U/mg prot)	Glutathione peroxidase (GPx) (U/mg prot)	Glutathione (GSH) (mg/g prot)	Malondialdehyde (MDA) (nmol/mg prot)
Blue	16L:8D	562.16 ± 32.19^{aA}	0.69 ± 0.05^{bA}	9.38 ± 1.17^{bA}	24.15 ± 4.11^{bB}	3.73 ± 0.25^{bA}	14.75 ± 2.02^{aA}
	12L:12D	344.81 ± 25.88^{bA}	0.93 ± 0.08^{aA}	17.29 ± 1.26^{aA}	55.64 ± 2.09^{aA}	6.54 ± 0.81^{aA}	10.01 ± 1.51^{aA}
	8L:16D	261.42 ± 15.97^{bA}	0.51 ± 0.05^{bA}	15.76 ± 1.52^{aA}	49.51 ± 6.14^{aA}	5.38 ± 0.55^{aA}	6.22 ± 0.47^{abA}
	4L:20D	159.03 ± 14.21^{cA}	0.45 ± 0.04^{bA}	8.43 ± 1.13^{bA}	27.38 ± 3.34^{bA}	3.95 ± 0.24^{bA}	3.40 ± 0.65^{bA}
	0L:24D	50.77 ± 8.03^{dA}	0.21 ± 0.01^{cA}	6.27 ± 0.94^{bA}	14.22 ± 1.37^{cA}	2.17 ± 0.32^{bA}	2.96 ± 0.14^{bA}
	16L:8D	109.72 ± 14.76^{aB}	0.56 ± 0.03^{aB}	9.22 ± 1.20^{aA}	30.56 ± 3.72^{aA}	4.02 ± 0.69^{aA}	5.26 ± 1.44^{aB}
	12L:12D	84.13 ± 7.33^{bB}	0.42 ± 0.03^{abC}	6.04 ± 0.85^{bC}	24.77 ± 2.08^{abB}	3.79 ± 0.48^{aB}	4.13 ± 0.23^{abB}
Red	8L:16D	75.94 ± 11.25^{bB}	0.34 ± 0.01^{bB}	6.35 ± 0.61^{bB}	26.48 ± 1.55^{aB}	2.88 ± 0.36^{bB}	3.72 ± 0.19^{abB}
	4L:20D	50.09 ± 5.79^{cB}	0.20 ± 0.02^{cB}	5.92 ± 0.72^{bB}	19.05 ± 2.67^{bB}	2.31 ± 0.22^{bB}	2.61 ± 0.33^{bB}
	0L:24D	54.26 ± 10.08^{cA}	0.25 ± 0.01^{cA}	6.14 ± 0.88^{bA}	16.17 ± 1.83^{bA}	2.05 ± 0.16^{bA}	2.48 ± 0.18^{bA}
	16L:8D	131.27 ± 10.75^{aB}	0.53 ± 0.06^{bB}	9.97 ± 1.09^{aA}	32.95 ± 3.95^{bA}	3.19 ± 0.40^{bB}	7.62 ± 0.85^{aB}
	12L:12D	125.39 ± 16.08^{aB}	0.75 ± 0.02^{aB}	10.69 ± 1.38^{aB}	47.28 ± 5.94^{aA}	4.84 ± 0.47^{aB}	5.74 ± 0.77^{abB}
White	8L:16D	60.88 ± 5.21^{bB}	0.47 ± 0.07^{bA}	7.49 ± 1.02^{bB}	28.39 ± 4.26^{bB}	2.47 ± 0.34^{bB}	4.25 ± 0.26^{abB}
	4L:20D	48.17 ± 4.30^{bB}	0.29 ± 0.02^{cB}	6.52 ± 0.84^{bB}	20.55 ± 3.18^{cB}	2.59 ± 0.19^{bB}	3.58 ± 0.34^{bA}
	0L:24D	42.38 ± 8.16^{bA}	0.25 ± 0.04^{cA}	6.83 ± 1.21^{bA}	13.98 ± 2.67^{cA}	2.33 ± 0.20^{bA}	2.21 ± 0.56^{bA}

value in the red and white groups was significantly lower at 4L:20D than at any other light cycle ($F = 29.61$, $P < 0.001$). Under red light, the T-AOC was significantly higher at 16L:8D and 12L:12D than at any other light cycle, and no significant difference between these two light cycles was detected ($F = 1.15$, $P = 0.195$). The T-AOC under blue light was significantly higher than that in the red and white light groups at 4L:20D and 16L:8D ($F = 24.46$, $P = 0.001$). No significant difference in the T-AOC was detected between the blue and white groups at 8L:16D, but the values were significantly higher than that in red light group ($F = 18.47$, $P = 0.002$).

Under blue light, the SOD activity of abalones was significantly higher at 12L:12D and 8L:16D than at any other light cycle ($F = 23.27$, $P = 0.001$), and it decreased significantly at 16L:8D ($F = 20.35$, $P = 0.002$). Under red light, the SOD activity was significantly higher at 16L:8D than at any other light cycle ($F = 15.49$, $P = 0.004$). Under white light, the SOD activity was significantly higher at 12L:12D and 16L:8D than at any other light cycle ($F = 14.68$, $P = 0.005$). At 8L:16D and 4L:20D, the SOD activity in the blue group was significantly higher than that in the red and white light groups, but no significant difference between the red and white groups was detected ($F = 1.37$, $P = 0.155$). At 16L:8D, the SOD activity in the red and white light groups was significantly higher than at the other light cycles, and no significant difference among these two light qualities was detected ($F = 2.79$, $P = 0.096$).

Under white light, the GP_X activity and GSH content were significantly higher at 12L:12D than that at any other light cycle ($F = 17.36$, $P = 0.002$); however, as duration of light exposure increased to 16 h, the GP_X activity and GSH content decreased significantly ($F = 8.27$, $P = 0.013$). Under red light at 16L:8D, the GP_X activity and GSH content significantly increased ($F = 16.43$, $P = 0.002$), but the values did not differ significantly from those of the 12L:12D group. The GP_X activity and GSH content in the blue group were significantly higher than those in the red and white groups at 4L:20D and 8L:16D ($F = 30.57$, $P < 0.001$). At 12L:12D, the GP_X activity in the blue and white groups was significantly higher than that in the red group ($F = 26.85$, $P = 0.001$).

For each light quality, the MDA content was higher at 16L:8D and 12L:12D than at the other light cycles, and no significant difference between these two groups was detected ($F = 1.90$, $P = 0.128$). The MDA content for each light quality was lower at 0L:24D and 4L:20D than at the other light cycles, and no significant difference between these two groups was detected ($F = 1.52$, $P = 0.147$). The interaction between light quality and light cycle also had a significant effect on the MDA content (light quality $\times$ light cycle, $F = 19.35$, $P = 0.002$). At 4L:20D, the MDA content in the blue and white groups was significantly higher than that in the red group ($F = 12.73$, $P = 0.006$). However, the MDA content in the blue group was higher than that in the red and white groups at 8L:16D, 12L:12D, and16L:8D ($F = 39.46$, $P < 0.001$).

References

1. S.B. Seminara, S. Messager, E.E. Chatzidaki, R.R. Thresher, J.S. Acierno Jr., J.K. Shagoury, et al., The GPR54 gene as a regulator of puberty. N. Engl. J. Med. **349**(17), 1614–1627 (2003)
2. S. Messager, E.E. Chatzidaki, D. Ma, A.G. Hendrick, D. Zahn, J. Dixon, et al., Kisspeptin directly stimulates gonadotropin-releasing hormone release via G protein-coupled receptor 54. Proc. Natl. Acad. Sci. U. S. A. **102**(5), 1761–1766 (2005)
3. S. Funes, J.A. Hedrick, G. Vassileva, L. Markowitz, S. Abbondanzo, A. Golovko, et al., The KiSS-1 receptor GPR54 is essential for the development of the murine reproductive system. Biochem. Biophys. Res. Commun. **312**(4), 1357–1363 (2003)
4. I.S. Parhar, S. Ogawa, Y. Sakuma, Laser-captured single digoxigenin-labeled neurons of gonadotropinreleasing hormone types reveal a novel G protein-coupled receptor (Gpr54) during maturation in cichlid fish. Endocrinology **145**(8), 3613–3618 (2004)
5. C.C. Martinez-Chavez, M. Minghetti, H. Migaud, GPR54 and rGnRH I gene expression during the onset of puberty in Nile tilapia. Gen. Comp. Endocrinol. **156**(2), 224–233 (2008)
6. B.H. Beck, S.A. Fuller, E. Peatman, M.E. McEntire, A. Darwish, D.W. Freeman, Chronic exogenous kisspeptin administration accelerates gonadal development in basses of the genus Morone. Comp. Biochem. Physiol. A Mol. Integr. Physiol. **162**(3), 265–273 (2012)
7. N. Zmora, J. Stubblefield, Z. Zulperi, J. Biran, B. Levavi-Sivan, J.-A. Muñoz-Cueto, et al., Differential and gonad stage-dependent roles of kisspeptin1 and kisspeptin2 in reproduction in the modern teleosts, morone species. Biol. Reprod. **86**, 177 (2012)
8. A. Elizur, The KiSS1/GPR54 system in fish. Peptides **30**(1), 164–170 (2009)
9. S. Kanda, T. Karigo, Y. Oka, Steroid sensitive kiss2 neurones in the goldfish: evolutionary insights into the duplicate kisspeptin gene-expressing neurones. J. Neuroendocrinol. **24**(6), 897–906 (2012)
10. A. Servili, Y. Le Page, J. Leprince, A. Caraty, S. Escobar, I.S. Parhar, et al., Organization of two independent kisspeptin systems derived from evolutionary-ancient kiss genes in the brain of zebrafish. Endocrinology **152**(4), 1527–1540 (2011)
11. M. Tena-Sempere, A. Felip, A. GoÂmez, S. Zanuy, M. Carrillo, Comparative insights of the kisspeptin/kisspeptin receptor system: lessons from non-mammalian vertebrates. Gen. Comp. Endocrinol. **175**(2), 234–243 (2012)
12. A.S. Mechaly, J. Viñas, F. Piferrer, Identification of two isoforms of the kisspeptin-1 receptor (kiss1r) generated by alternative splicing in a modern teleost, the Senegalese sole (Solea senegalensis). Biol. Reprod. **80**(1), 60–69 (2009)
13. S. Selvaraj, H. Kitano, Y. Fujinaga, H. Ohga, M. Yoneda, A. Yamaguchi, et al., Molecular characterization, tissue distribution, and mRNA expression profiles of two kiss genes in the adult male and female chub mackerel (Scomber japonicus) during different gonadal stages. Gen. Comp. Endocrinol. **169**(1), 28–38 (2010)
14. G.L. Taranger, C. Haux, S.O. Stefansson, B.T. BjoÈrnsson, B.T. Walther, T. Hansen, Abrupt changes in photoperiod affect age at maturity, timing of ovulation and plasma testosterone and oestradiol-17β profiles in Atlantic salmon, Salmo salar. Aquaculture **162**(1), 85–98 (1998)
15. M. Iigo, T. Abe, S. Kambayashi, K. Oikawa, T. Masuda, K. Mizusawa, et al., Lack of circadian regulation of in vitro melatonin release from the pineal organ of salmonid teleosts. Gen. Comp. Endocrinol. **154**(1-3), 91–97 (2007a)
16. M. Iigo, T. Azuma, M. Iwata, Lack of circadian regulation of melatonin rhythms in the sockeye salmon (Oncorhynchus nerka) in vivo and in vitro. Zool. Sci. **24**(1), 67–70 (2007b)
17. E.N. Fuentes, D. Safian, I.E. Einarsdottir, J.A. Valdes, A.A. Elorza, A. Molina, B.T. Bjornsson, Nutritional status modulates plasma leptin, AMPK and TOR activation, and mitochondrial biogenesis: implications for cell metabolism and growth in skeletal muscle of the fine flounder. Gen. Comp. Endocrinol. **186**, 172–180 (2013)
18. Y. Minokoshi, B.B. Kahn, Role of AMP-activated protein kinase in leptin-induced fatty acid oxidation in muscle. Biochem. Soc. Trans. **31**, 196–201 (2003)

19. S.H. Bates, R.N. Kulkarni, M. Seifert, M.G. Myers, Roles for leptin receptor/STAT3-dependent and -independent signals in the regulation of glucose homeostasis. Cell Metab. **1**, 169–178 (2005)
20. M. Gorissen, N.J. Bernier, S.B. Nabuurs, G. Flik, M.O. Huising, Two divergent leptin paralogues in zebrafish (Danio rerio) that originate early in teleostean evolution. J. Endocrinol. **201**, 329–339 (2009)
21. E. Froiland, K. Murashita, E.H. Jorgensen, T. Kurokawa, Leptin and ghrelin in anadromous Arctic charr: cloning and change in expressions during a seasonal feeding cycle. Gen. Comp. Endocrinol. **165**, 136–143 (2010)
22. Q. Liu, Y. Chen, D. Copeland, H. Ball, R.J. Duff, B. Rockich, R.L. Londraville, Expression of leptin receptor gene in developing and adult zebrafish. Gen. Comp. Endocrinol. **166**, 346–355 (2010)
23. D.A. Zieba, M. Szczesna, R. Klocek-Gorka, E. Molik, T. Misztal, G.L. Williams, K. Romanowicz, E. Stepien, D.H. Keisler, M. Murawski, Seasonal effects of central leptin infusion on secretion of melatonin and prolactin and on SOCS-3 gene expression in ewes. J. Endocrinol. **198**, 147–155 (2008)
24. A.G.G. Moen, R.N. Finn, Short-term, but not long-term feed restriction causes differential expression of leptins in Atlantic salmon. Gen. Comp. Endocrinol. **183**, 83–88 (2013)
25. A.B. Tinoco, L.G. Nisembaum, E. Isorna, M.J. Delgado, N. de Pedro, Leptins and leptin receptor expression in the goldfish (Carassius auratus). Regulation by food intake and fasting/overfeeding conditions. Peptides **34**, 329–335 (2012)
26. J.K. Wang, Z.L. Yan, Current status and development trend on abalone culture. Fish. Sci. **31**(12), 749–753 (2012). in Chinese
27. Z.M. Romo, A.D. Re, F. Díaz, A. Mena, Physiological responses of pink abalone Haliotis corrugata (Gray, 1828) exposed to different combinations of temperature and salinity. Aquac. Res. **41**, 953–960 (2010)
28. W. Cheng, S.P. Yeh, C.S. Wang, J.C. Chen, Osmotic and ionic changes in Taiwan abalone Haliotis diversicolor supertexta at different salinity levels. Aquaculture **203**, 349–357 (2002)
29. M. Wassnig, R.D. Roberts, A. Krsinich, R.W. Day, Effects of water flow rate on growth rate, mortality biomass return of abalone in slab tanks. Aquac. Res. **41**, 839–846 (2010)
30. M.A. Naylor, H. Kaiser, C.L.W. Jones, The effect of free ammonia nitrogen, pH and supplementation with oxygen on the growth of south African abalone, Haliotis midae L. in an abalone serial-use raceway with three passes. Aquac. Res. **45**, 213–224 (2014)
31. J.O. Harris, G.B. Maguire, S.J. Edwards, D.R. Johns, Low dissolved oxygen reduces growth rate and oxygen consumption rate of juvenile greenlip abalone, Haliotis laevigata Donovan. Aquaculture **174**, 265–278 (1999)
32. F. Ahmed, S. Segawa, M. Yokota, S. Watanabe, Effect of light on oxygen consumption and ammonia excretion in Haliotis discus discus, H. gigantea, H. Madaka and their hybrids. Aquaculture **279**, 160–165 (2008)
33. J.K. Kim, Y.X. Mao, G. Kraemer, C. Yarish, Growth and pigment content of Gracilaria tikvahiae McLachlan under fluorescent and LED lighting. Aquaculture **436**, 52–57 (2015)
34. Y.W. Dong, S.L. Dong, X.L. Tian, F. Wang, M.Z. Zhang, Effects of diel temperature fluctuations on growth, oxygen consumption and proximate body composition in the sea cucumber Apostichopus japonicus Selenka. Aquaculture **255**, 514–521 (2006)
35. T.H. Carfoot, *Animal energetics* (Academic Press, New York, 1987), pp. 407–515
36. D. Lemos, V.N. Phan, Energy partitioning into growth, respiration, excretion and exuvia during larval development of the shrimp Farfantepenaeus paulensis. Aquaculture **199**, 131–143 (2001)
37. R. Jan, K. Shao, K. Chang, A study of diurnal periodicity in oxygen consumption of the small abalone Haliotis diversicolor supertexta Lischke. Bull. Inst. Zool. Acad. Sin. **20**, 1–8 (1981)
38. Z. García-Esquivel, S. Montes-Magallon, M.A. Gonzalez-Gomez, Effect of temperature and photoperiod on the growth, feed consumption, and biochemical content of juvenile green abalone, Haliotis fulgens, fed on a balanced diet. Aquaculture **262**, 129–141 (2007)

39. E. Gorrostieta-Hurtado, R. Searcy-Bernal, C. Anguiano-Beltrán, Z. García-Esquivel, E. Valenzuela-Espinoza, Effect of darkness on the early postlarval development of Haliotis corrugata abalone fed different diatom densities. Cienc. Mar. **35**, 113–122 (2009)
40. R. Day, P. Gilmour, S. Huchette, Effects of density and food supply on postlarval abalone: behaviour, growth and mortality. J. Shellfish Res. **23**, 1009–1018 (2004)
41. X.L. Gao, M. Zhang, J.M. Zheng, X. Li, L. Chi, C.B. Song, Y. Liu, Effect of LED light quality on the phototaxis and locomotion behaviour of Haliotis discus hannai. Aquac. Res. **47**, 3376–3389 (2016)
42. B. Guo, F. Wang, S.L. Dong, Q.F. Gao, The effect of rhythmic light color fluctuation on the molting and growth of Litopenaeus vannamei. Aquaculture **314**, 210–214 (2011)
43. A. Berg, T. Hansen, S. Stefansson, First feeding of Atlantic salmon (Salmo salar L.) under different photoperiods. J. Appl. Ichthyol. **8**, 251–256 (1992)
44. J.E. Thorpe, C.E. Adams, M.S. Miles, D.S. Keay, Some influences of photoperiod and temperature on opportunity for growth in juvenile Atlantic salmon, Salmo salar L. Aquaculture **82**, 119–126 (1989)
45. M. Carrillo, S. Zanuy, A. Felip, M. Bayarri, G. Molés, A. Gómez, Hormonal and environmental control of puberty in perciform fish. Ann. N. Y. Acad. Sci. **1163**, 49–59 (2009)
46. X.L. Gao, M. Zhang, X. Li, C. Shi, C.B. Song, Y. Liu, Effects of LED light quality on the growth, metabolism, and energy budgets of Haliotis discus discus. Aquaculture **453**, 31–39 (2016)
47. M. Bradford, A rapid and sensitive method for the quantitation of microgram quantities of protein utilizing the principle of protein-dye binding. Anal. Biochem. **72**, 248–254 (1976)
48. M.E. Anderson, Determination of glutathione and glutathione disulfide in biological samples. Methods Enzymol. **113**, 548–555 (1985)
49. L.J. Li, F. Zhang, X.M. Liu, Y.P. Guo, E.B. Ma, Oxidative stress related enzymes in response to chromium (VI) toxicity in Oxya chinensis (Orthoptera: Acridoidae). J. Environ. Sci. (China) **17**, 823–826 (2005)

Index

J. Li, G. Q. Zhang (eds.), *Light-Emitting Diodes*, Solid State Lighting Technology and Application Series 4, https://doi.org/10.1007/978-3-319-99211-2

H

I

T

U

V

MIX
Papier aus verantwortungsvollen Quellen
Paper from responsible sources
FSC® C105338

If you have any concerns about our products,
you can contact us on
ProductSafety@springernature.com

In case Publisher is established outside the EU,
the EU authorized representative is:
Springer Nature Customer Service Center GmbH
Europaplatz 3, 69115 Heidelberg, Germany

Printed by Libri Plureos GmbH
in Hamburg, Germany